Lecture Notes in Physics

Edited by H. Araki, Kyoto, J. Ehlers, München, K. Hepp, Zürich
R. Kippenhahn, München, D. Ruelle, Bures-sur-Yvette
H. A. Weidenmüller, Heidelberg, J. Wess, Karlsruhe and J. Zittartz, Köln
Managing Editor: W. Beiglböck

323

D. L. Dwoyer M. Y. Hussaini
R. G. Voigt (Eds.)

11th International Conference on Numerical Methods in Fluid Dynamics

Springer-Verlag
Berlin Heidelberg New York Tokyo

Editors

D. L. Dwoyer
M. Y. Hussaini
R. G. Voigt
ICASE, NASA Langley Research Center
Hampton, VA 23665-5335, USA

ISBN 3-540-51048-6 Springer-Verlag Berlin Heidelberg New York
ISBN 0-387-51048-6 Springer-Verlag New York Berlin Heidelberg

Printed in Germany

Printing: Druckhaus Beltz, Hemsbach/Bergstr.
Binding: J. Schäffer GmbH & Co. KG., Grünstadt
2158/3140-543210 – Printed on acid-free paper

DEDICATION

This symposium is dedicated to Professor Maurice Holt on the occasion of his seventieth birthday in tribute to his four decades of research in applied mathematics and fluid mechanics. Since he is still very active in research and considers the last four decades as but the first phase of his research career, a final evaluation of his contributions is premature. However, it is only proper that his biography be briefly sketched with a mention of honors and recognitions he has already won.

Professor Maurice Holt was born on May 16, 1918 in Wildboarclough, Cheshire, England. He was educated at the Manchester Grammar School and the University of Manchester in England. He did his Masters thesis with T. G. Cowling in 1944, and his doctorate with Sydney Goldstein in 1948. Immediately after graduation, he joined the University of Liverpool as a lecturer in Mathematics, and a year later moved to the University of Sheffield in England. In 1952, he joined the Ministry of Supply where he served as the Principal Scientific Officer in charge of the Theoretical Aerodynamics Section of the Applied Mathematics Division, Armament Research and Development Establishment, Fort Halstead, Kent. In 1955, he was a visiting lecturer in the Mathematics Department at Harvard University at the invitation of Garrett Birkhoff. In 1956, he entered the United States, and joined the faculty of the Division of Applied Mathematics at Brown University. Since 1960, he has been Professor of Aeronautical Sciences at the University of California at Berkeley.

Professor Maurice Holt's scientific work is too extensive to be discussed in detail here. Suffice it to say, then, that his research in fluid mechanics has encompassed a diverse range of physical problems. Supersonic and transonic aerodynamics, blast waves, underwater explosions and supersonic separated flows are some of the subjects on which his work has received international recognition. He is one of the pioneers in the field of computational fluid dynamics. Not only did he develop some numerical techniques that were widely used during the 1960's for supersonic blunt body problems, conical flows and separated flows, but he also is a leading figure in the establishment of East-West cooperation in computational fluid dynamics. In 1969, he and Academician Belotserkovskii together started the highly successful conference series, the International Conference on Numerical Methods in Fluid Dynamics.

As an excellent applied mathematician, Professor Holt's keen insight into the physics of fluids has become a trademark of his work. His work has been fundamental and applied, in the best scientific tradition. He has been a valuable consultant to both industry and government at various times. Lockheed Aircraft Corporation, Northrop Corporate Laboratories, The Aerospace Corporation, and Lawrence Berkeley Laboratory are some of the organizations which have profited from his expertise. He has been a visiting professor at a number of international institutions such as Université Paris, and Université Pierre et Marie Curie. As an excellent teacher, he has taught and trained nearly three dozen students who have themselves become leaders in research. He has served as editor of several journals, including the ASME Journal of Applied Mechanics. In recognition of his pioneering research contributions to fluid mechanics and his training of two generations of students, he has been elected a fellow of the American Society of Mechanical Engineers and the American Physical Society. His recognition extends beyond the western world; he was invited to be the guest of the USSR Academy of Sciences and the Romanian Academy of Sciences because of his leadership in fluid mechanics research.

It is with affection for the deep impact that he has had on our lives and with respect for his accomplishments as a teacher and a scholar that we dedicate this symposium to Professor Maurice Holt. We only hope that our interaction with him will continue for many years to come.

W. F. Ballhaus, Jr.
M. Y. Hussaini

Editors' Preface

The present volume of Lecture Notes in Physics covers the proceedings of the Eleventh International Conference on Numerical Methods in Fluid Dynamics, held in Williamsburg, Virginia, June 27-July 1, 1988. It contains 103 papers. Seven of these papers are based on invited lectures, and the rest were selected on the basis of abstracts submitted from all over the world by four paper selection groups, one in the U.S.A. headed by Maurice Holt, another in Europe headed by Roger Temam, the third in U.S.S.R. headed by Victor Rusanov, and the fourth in Japan (representing the Pacific Rim Countries) headed by K. Oshima. Following the usual tradition, the invited papers appear first, and then the contributed papers in alphabetical order by first author.

The conference co-chairmen were D. L. Dwoyer and Robert G. Voigt; they are indebted to many who helped with the detailed organization of the meeting; they thank all of them, and in particular, Ms. Mary Adams who was in charge of the computer graphics displays and Ms. Emily Todd, the conference secretary.

The financial support was provided by NASA Langley Research Center.

We thank Dr. W. Beiglböck and the editorial staff of Springer-Verlag for assistance in preparing the proceedings.

November 1988.

D. L. DWOYER M. Y. HUSSAINI R. G. VOIGT
(Editors)

CONTENTS

INAUGURAL TALK

Computational Fluid Dynamics – A Personal View

M.Y. Hussaini
Institute for Computer Applications in Science and Engineering
NASA Langley Research Center
Hampton, VA 23665

I. Abstract

This paper provides a personal view of computational fluid dynamics. The main theme is divided into two categories – one dealing with algorithms and engineering applications and the other with scientific investigations. The former category may be termed computational aerodynamics with the objective of providing reliable aerodynamic or engineering predictions. The latter category is essentially basic research where the algorithmic tools are used to unravel and elucidate fluid dynamic phenomena hard to obtain in a laboratory. While dealing with the algorithms, only the underlying principles are discussed and the engineering applications are omitted. A personal critique of the numerical solution techniques for both compressible and incompressible flows is included. The discussion on scientific investigations deals in particular with transition and turbulence. In conclusion, some challenges to computational fluid dynamics are mentioned, the grand challenge being turbulence and reacting flows.

II. Introduction

Computational fluid dynamics can broadly be divided into three categories – 1) algorithms, 2) scientific investigations, and 3) engineering applications. Algorithms may be considered as a common foundation for engineering applications and scientific investigations. Engineering applications include many fields such as aerodynamics, meteorology, astrophysics and oceanography. Scientific investigations involve unravelling or elucidating fundamental flow phenomena hard to obtain in a laboratory. In this paper, the principles underlying the algorithms are briefly discussed; next some current engineering applications in aerodynamics are presented; and finally, some representative results of scientific investigations are described in the area of my interest which is stability and transition to turbulence.

The discussion of the algorithms will distinguish between compressible and incompressible flows. For most purposes, the Navier-Stokes equations provide the best description of compressible flows excluding the rarefied gas dynamics regime. These equations do not require further elaboration. From the point of view of numerical discretization of these continuum equations, the inviscid terms have provided the greatest challenge. The algorithms for the inviscid equations, which are usually called the compressible Euler equations, can be extended in a technically straightforward fashion to include viscous terms. Usually, some equivalent of central differencing is employed to discretize the viscous terms. For this reason, this discussion of algorithms for compressible flows will henceforth focus on the Euler equations. They must be viewed, not in isolation, but rather as a limit of the Navier-Stokes equations as the Reynolds number tends to infinity. The physically relevant solutions are those which satisfy additional constraints such as the entropy condition.

The most distinguishing feature of compressible flows is the presence of shocks. Any compressible flow algorithm must be capable of handling a shock wave which means that the Rankine-Hugoniot conditions must be satisfied across the shock. A sufficient condition for this is that the equations be in the conservation form and the numerical scheme be both conservative and dissipative. A shock wave captured in a discrete solution tends to be somewhat diffused, and the solution may have attendant oscillations depending on the details of the scheme and the dimensionality of the problem. Most practitioners rate compressible flow algorithms according to the cosmetic appearance of their solutions to standard test problems; the criteria are how sharp the captured shock is and how well the wiggles are suppressed. A discrete shock will always remain a fertile ground of research for the numerical analysts.

Compressible flow algorithms should also be able to preclude expansion shocks. For instance, the Lax-Wendroff [17] scheme does admit expansion shocks; a remedy for this was provided by Majda and Osher [20] almost a decade later using the entropy condition.

A contact discontinuity poses a different problem for discrete solutions. Although one does not have to satisfy the entropy condition across the discontinuity, one has to deal with the problem of the physical instability of such vortex sheets. The numerical treatments that I am aware of suppress such instabilities through artificial viscosity or compressibility. Again, algorithms are rated according to the sharpness of the contact discontinuity and the smoothness (absence of wiggles) of the solution across the discontinuity. Contact discontinuities are not as ubiquitous as shocks, and seem to have attracted less attention than shocks.

Real gas effects occur in the hypersonic regime and require additional reaction-diffusion equations for the species to be solved. This may add to the stiffness of the problem.

The most obvious examples of singularities are: the intersection of a shock wave with a surface, the triple shock point, and the core of a rolled-up vortex. Discrete Euler solutions in such situations may yield, owing to inherent diffusion, results that tend to agree with experimental observation. The purist will always be uneasy about such agreement.

Let me discuss briefly the principles underlying compressible flow algorithms. The Euler equations are but one example of a hyperbolic system of conservation laws. The fundamental physical principle is the conservation law in the integral form

$$\frac{d}{dt}\int \vec{U}\cdot d\vec{r} = -\int \vec{F}\cdot\hat{n}\,dA. \tag{1}$$

Subject to the uniform continuity of $\vec{U}$ and the differentiability of $\vec{F}$, this reduces to the partial differential equation

$$\vec{U}_t = -\nabla\cdot\vec{F}. \tag{2}$$

Consider now the one-dimensional scalar case and integrate the conservation law across a single cell. This yields the semi-discrete form

$$\frac{\partial \bar{U}_j}{\partial t} = \frac{F_{j+1/2} - F_{j-1/2}}{\Delta x} \tag{3}$$

where the cell average quantity is

$$\bar{U}_j = \int_{x_{j-1/2}}^{x_{j+1/2}} U_j(\xi)\,d\xi \tag{4}$$

and the interface flux is

$$F_{j+1/2} = F(U_{j+1/2}). \tag{5}$$

Note the distinction between pointwise and average values. A historical perspective is useful for appreciating modern techniques for solving Equation 1, which is the first building block of compressible flow algorithms.

III. Compressible Flow Algorithms

A. The Sixties

The two representative methods of the 1960's are Godunov [8] and Lax-Wendroff (L-W) [17] schemes.

Godunov's method exploits the local physics of the hyperbolic problem, i.e., the local characteristics of the problem. It solves the Riemann problem

$$W_t + F(W)_x = 0 \tag{6}$$

where

$$\begin{aligned} W(x; U_L; U_R) &= U_L, \quad x < 0 \\ &= U_R, \quad x > 0 \end{aligned} \tag{7}$$

with piecewise constant initial conditions, and uses this solution to compute the flux function at time level n

$$F^n_{j+1/2} = F(W(0; U^n_j, U^n_{j+1}). \tag{8}$$

It is first-order accurate, has sufficient numerical viscosity for stability, and virtually always works in the sense that it is rarely unstable. The Lax-Wendroff method, on the other hand, is based on Taylor series expansion, and is second order accurate. The two-step version of the method,

$$U^*_{j+1/2} = \frac{1}{2}(U^n_{j+1} + U^n_j) - \frac{\Delta t}{\Delta x}(F^n_{n+1} - F^n_j) \tag{9}$$

$$U^{n+1}_j = U^n_j - \frac{\Delta t}{\Delta x}(F^*_{j+1/2} - F^*_{j-1/2}) \tag{10}$$

is due to Richtmyer [28]. It contains a mild amount of implicit dissipation; additional artificial damping must be explicitly added to control oscillations.

The advantages and disadvantages of these two approaches are apparent in their respective solutions to the now standard one-dimensional shock tube test problem. Shortly after the flow is initiated, there is a shock wave, rarefaction wave, and a contact discontinuity. These features are evident in fig. 1 where the exact solution is denoted by a solid line. In comparison with the Godunov's method, the L-W scheme furnishes a sharper shock at the expense of greater dispersion. The L-W results here purposely did not utilize artificial viscosity (to highlight the dispersion errors); in practice explicit damping is always present, but with a magnitude at the discretion of the user.

These two schemes may now be out-dated, but the basic principles and dilemmas of the methods persist to this day.

B. The Seventies

In the 1970's numerous sequels appeared, some of which are discussed below.

MacCormack's scheme [19], an efficient and robust variant of the Lax-Wendroff scheme has been a workhorse of industrial applications. It is the classic example of a good scheme which comes along at the right time, and soon proves its worth on some non-trivial problems of its day.

Flux-Corrected-Transport (FCT) [1] may be considered as the first example of the total variation diminishing schemes (TVD), although the TVD concept itself was introduced a decade later [12]. It consists of two steps; in the first step, a low-order solution is obtained, and is used to determine that fraction of the antidiffusion flux that can be applied at the second step to produce, in some sense, maximum steepening of the solution, while precluding any new extrema. It is a modular, scalar approach suitable for a single conservation law. For systems, some secondary effects of flux limiters are found, and there appears to be room for improvement.

The hybrid scheme, originally due to Harten [9] is a high order scheme away from the shocks and first order monotonic at the shock.

The random choice method [6] is in effect Godunov's method where random sampling of the Riemann solution replaces cell-averages. Although not conservative, the absence of averaging errors makes it the most accurate scheme. Unfortunately, of all the algorithms listed here, it suffers the most deterioration when extended beyond one dimension.

MUSCL [34,35] is a second order sequel to Godunov's method where the piecewise constant initial conditions for the Riemann problem are replaced by piecewise linear initial conditions with proper switching functions to avoid any new extrema. It may appear surprising that nearly two decades elapsed before an improvement to the Godunov's method was developed. However, despite its ingenuity, it was largely ignored, only in part due to the unpleasantness of coding up a general Riemann solver, and to the uncertainty of extending it beyond one dimension. The real hurdle proved to be the difficulty of devising robust flux limiters for the reconstruction phase.

The $\Lambda-$scheme [21] is based on the physics of signal propagation for the hyperbolic laws. It casts the Euler equations in the form of compatibility equations and uses a finite-difference discretization which conforms to the nature of the local characteristic directions. Incorporation of boundary conditions is efficient.

The artificial compression method (ACM) [10,11], in contrast to the artificial viscosity methods (AVM) [22], sharpens the contact discontinuities and shocks. It is the best method that I know of for contact discontinuities.

C. The Eighties

The 1980's have witnessed great advances in numerical techniques for compressible flows. It is still premature to select the methods which will prove to be the most enduring. The following list presents the leading contenders as well as a few long shots.

Flux splitting [31] for the Euler equations may be considered as characteristic differencing in conservation form. The numerical flux function is split into a sum of forward flux and backward flux, and this facilitates upwind differencing. Sometimes the flux differences are split rather than the flux themselves, and some improvements are achieved thereby [29].

The focus in the 1980's has been on high-order methods. High-order FCT [39], piecewise parabolic methods (PPM) [38], essentially non-oscillatory schemes (ENO) [13,14], and spectral methods were developed for hyperbolic problems. PPM and ENO have been applied to some complicated two-dimensional shocked flows. Spectral methods are relatively in the initial stages of development. Multigrid procedures [23] and TVD concepts [3] have been put to use in practical three dimensional applications in many laboratories and in some industries.

Figure 2 demonstrates how well a 1980 technique such as Phil Roe's flux-difference algorithm performs in the case of the standard shock tube test problem. The shock and contact discontinuities captured by this method are relatively sharp.

Total pressure and surface streamlines from the numerical simulation of a flow-field about an F-18 forebody for an incidence angle of 30° and a Reynolds of 740,000 is illustrated in figure 3. In the first phase of the simulation, after intrinsic consistency checks are performed, one examines how well the gross features of the flow-field are captured. In this case, the primary separation and secondary separation on the forebody, leading edge separation and the secondary separation on the strake are important features. These features are found to be in remarkable agreement with oil-flow visualization pictures from the experiment, not shown here.

Further engineering applications can be found in the paper by Ajay Kumar in these proceedings.

Compressible flow algorithms run into difficulties or otherwise become extremely inefficient at very low Mach numbers due to the fact that the equations become time-singular. In the near-zero Mach number limit, most CFD calculations are performed with algorithms designed specifically for incompressible flow. The history of these methods is indeed quite different from that of compressible flows methods.

IV. Incompressible Navier-Stokes Equations

The basic equations for incompressible viscous flow are well-known. The viscous terms introduce an intrinsic algorithmic difficulty in contrast to the compressible case. The divergence-free constraint, and in practice the viscous terms, must be treated implicitly; the simultaneous enforcement of these conditions is quite challenging. The boundary conditions involve only the velocity – there is no boundary condition on the pressure. Although it is common practice to derive a Neumann boundary condition on the pressure from the momentum equation, this is not always a fool-proof approach. The initial conditions are not as innocuous as they appear, for if they are not divergence-free, then the solutions are apt to have a time singularity.

Many algorithms treat the nonlinear advection terms explicitly and the linear terms implicitly. In this case, the most time-consuming part of the algorithm is the solution of the discrete Stokes problem

$$\begin{pmatrix} I+V & G \\ G^* & 0 \end{pmatrix} \begin{pmatrix} \vec{U} \\ p \end{pmatrix} = (A) \begin{pmatrix} \vec{U} \\ p \end{pmatrix} = \begin{pmatrix} \vec{f} \\ 0 \end{pmatrix} \tag{11}$$

where I is the identity operator, V the discrete viscous operator, G the discrete gradient operator (and G* is its adjoint, which is the negative of the divergence operator). The essential difficulty is that A is too large for a direct solution, (remember we are interested in three-dimensional solutions) and its indefiniteness poses severe challenges for conventional iterative methods.

The major efforts on high-speed flow algorithms have focussed on improving the basic discretization. On the other hand, the most significant milestones for incompressible flows have been those which have improved the efficiency of solving the Stokes problem. A few of these are discussed below, and let me forthwith apologize for neglecting such alternatives as discrete vortex methods with random-walk modelling of diffusion.

The marker-and-cell method [15] was the first consistent numerical method for solving the incompressible Navier-Stokes equations with neither spurious modes, nor artificial pressure boundary condition. It did, however, suffer from the limitation that the viscous terms were treated explicitly.

The artificial compressibility method [4] adds to the continuity equation a fictitious term involving the product of a small parameter and the time-derivative of pressure. Thus it removes indefiniteness but sacrifices incompressibility. It is a method for steady-state solution.

The operator splitting methods [5,33], or the projection methods as they are sometimes called, uncouples the advection-diffusion terms from the incompressibility constraint. They involve two steps. In the first step, advection-diffusion terms are advanced, and in the second step, the pressure correction is carried out to enforce a divergence-free condition. Intermediate boundary conditions are important for such methods. They became available in 1971 [7].

Multigrid procedures have been used to obtain grid-independent convergence rates in the case of both finite-difference and spectral discretizations of the Navier-Stokes equations [2].

Although spectral methods have a long history, they were first demonstrated to be practical for high resolution calculations of incompressible flows in 1971 [25]. This, however, was purely for the periodic case where such difficulties as spurious modes and pressure boundary conditions are not present. Moreover, in this special case, the discrete divergence and viscous operators commute, a property which is lost for nonperiodic problems. It was not until 1978 that these difficulties were overcome for wall-bounded flows [26].

By virtue of certain advantages of spectral methods, such as exponential convergence, minimal phase errors, etc., they have become a dominant tool in simulations of instability and transition to turbulence and turbulent shear flows.

Now, I propose to present one example of transition simulation. A direct simulation of laminar breakdown in a flat plate boundary layer was performed and the data postprocessed for qualitative comparison against experimental results. Bubbles (in the experiment) and particles (in the computation) are periodically released from a wire normal to the plate in the vicinity of the peak plane (where the upwash is maximum). The results are shown in fig. 4.

V. Assessment of CFD Status

The following chart (fig. 5) lends some perspective. The horizontal axis measures the flow complexity and the vertical axis the geometric complexity. The ultimate goal of computational aerodynamics lies in the upper right hand corner, and is far beyond any forseeable capability. The Euler solutions to three dimensional flow past wings takes about 5 to 10 minutes on the Cray-2. The Reynolds-averaged Navier-Stokes solutions take about 8 to 10 hours; however, these algorithms have been made efficient through multigrid procedures, and this calculation time now ranges between 1 to 2 hours [30]. The increased computer resources taken by the Reynolds averaged Navier-Stokes calculations are due in part to the increased complexity of the equations themselves and in part to the additional spatial scales that must be resolved. In the direct simulations the range of spatial scales is orders of magnitude larger, and in addition, it is necessary to resolve an equally large range of temporal scales. Hence an order of magnitude increase in computer time is required.

Since we are so far removed from our ultimate goal, a key question is: should we be encouraged or discouraged by our past achievements, our current rate of progress, and our future prospects. Certainly, judging by some of the more extravagant claims that have been made in recent years, discouragement is in order. We should remember however that CFD is barely three decades old, whereas the related experimental and theoretical disciplines have flourished for more than a century. To borrow a phrase from Peter Lax [18], CFD is just "beginning to emerge, like Hercules from his cradle".

A cursory look through the "Current Capabilities and Future Directions in Computational Fluid Dynamics", the findings and recommendations of a NRC committee [24], reveals the following: 1) CFD has become an integral part of the design cycle in the aircraft industry; 2) solution algorithms for the Euler equations are reaching maturity for practical flow configurations, but those for the three-dimensional Navier-Stokes equations are at a relatively primitive stage; 3) in hypersonics, CFD is expected to be the principal source of flow information for vehicle design. 4) In propulsion, CFD has proven to be a valuable design analysis tool in a wide variety of situations. 5) Laminar-turbulent transition and fully developed turbulence are the pacing items of technology.

These achievements and prospects of CFD should be cause for encouragement, especially in view of the sporadic progress which is endemic to any field of science. Here is a classical example from aeronautics. The first attempt to calculate lift of a plate at incidence in an air stream was based on Newton's theory in 1726, and it was grossly underpredicted. More than a century later, Rayleigh proposed a theory which provided a slightly better approximation for lift, but was still not good enough. Three decades later, Rayleigh Joukowski, in 1907, theorized correctly the flow pattern about the plate. The lift formula derived from this theory predicted lift in close agreement with experiment. This, perhaps one of the most important evolutions of a theoretical concept in aeronautics, took about a century and a half.

CFD is still very young, and it does not have the breakthroughs that can be cited which are commensurate in importance to the discovery of the fundamental laws of nature. It has certainly assumed a new dimension equal in importance to theory and experiment in fluids. One hears sometimes the view that CFD usurps the place of theory. CFD has cut out its own place, and helps theory much the same way as physical experiments by providing insights into the physics. Furthermore, CFD may be used to validate the theory; after all, they have the governing equations in common.

VI. Challenges

A. Algorithms

Euler solutions as an asysmptotic limit of Navier-Stokes solutions, and the singularities of such solutions will always be a challenge for computations. Mathematical analysis of the Navier-Stokes equations leading to existence and uniqueness proofs in the case of the nontrivial three-dimensional flow-fields will provide realistic test beds for computational results. Mathematical results on the influence of small scales on large scales and vice versa and mathematical modelling of such interactions are crucial for dealing with turbulent flows.

Analysis of the discrete Navier-Stokes equations involving error bounds, error sources, stability and convergence proofs for truly multidimensional problems, and rigorous results on resolution requirements, artificial boundary conditions, etc., will be synergistic for leaps-and-bounds progress in CFD.

Improvement in efficiency of algorithms through convergence acceleration techniques, adaptive mesh techniques and methods for handling extremely stiff systems of equations, will allow computational solutions of problems with order of magnitude greater complexity than is possible now.

B. Engineering Applications

So far, the capability exists for the simulation of flow past an aircraft in level flight. The immediate challenge will be a full aircraft simulation including take-offs and landings. The simulation of the entire flight envelope of a military aircraft will be a greater challenge. Maneuverability and control aspects will require definition of new problem areas such as fluid dynamic control, fluid-structure interaction, etc. Such advances require an ability to predict turbulent flow for a wide range of conditions much more accurately than is possible today.

As Jameson [16] has pointed out, the integration of the predictive capability into an automatic design method using optimization, will require advances in algorithmic efficiency and computational power. It can possibly be realized in the next decade. Interactive calculations to optimize design appear to be feasible right now only for airfoil evaluation.

As pointed out in the NRC report, the applications and importance of CFD to hypersonic vehicle design is much greater than for the conventional aircraft. Simulations of hypersonic flow pertinent to upper atmospheric flight must take into account vehicle scale, flight velocity and altitude. In other words, the simulation should duplicate the Reynolds number, Mach number and total enthalpy, which is beyond the capability of the ground-based facilities. Thus, the applications and importance of CFD to hypersonic vehicle design is much greater than for the conventional aircrafts. Simulations for hypersonic aircraft will require an integrated approach involving interaction between external aerodynamics and reactive flow-fields of propulsive systems. Again, the ability to deal accurately with transitional and turbulent flows is crucial to success. This presents a great opportunity and challenge to CFD, and I am sure CFD will meet this challenge and convert even the nonbelievers.

C. Scientific investigations

Reactive flows pose yet another challenge to CFD. The scales in micromixing processes are much smaller than the turbulence scales themselves.

Nonlinear wave mechanics dominate laminar or turbulent flows. In fact it is believed that the nonlinear processes leading to turbulent spots in a laminar boundary layer are similar to the processes of turbulent bursts in turbulent boundary layers.

Turbulence is the grand challenge. It is an all pervasive ubiquitous phenomena present in, to cite a few instances, weather patterns, ocean currents, outer layers of the sun, ionosphere, high-temperature plasma, astrophysical jets, chemical reactors, and combustion. If CFD can solve the turbulence problem, the impact will be the same as the discovery of a new law of nature [37].

VII. Acknowledgement

The author is pleased to acknowledge the help of S. Chakravarthy, G. Erlebacher, A. Harten, A. Jameson, S. Osher, J. Thomas, B. van Leer, M. Salas, S. Zalesak and T. Zang in preparing this paper.

References

[1] J. P. Boris and D. L. Book, J. Comput. Phys. 11 (1973), 38.

[2] C. Canuto, M. Y. Hussaini, A. Quarteroni, and T. A. Zang, *Spectral Methods in Fluid Dynamics*, Springer-Verlag (1987).

[3] S. R. Chakravarthy, NASA Contractor Report 4043 (1987).

[4] A. J. Chorin, J. Comput. Phys. 2 (1967), 12-26.

[5] A. J. Chorin, Math. Comput. 22 (1968), 745-762.

[6] A. J. Chorin, J. Comput. Phys. 22 (1976), 517-536.

[7] M. Fortin, R. Peyret and R. Temam, J. Mech., V. 10 (1971), 357-390.

[8] S. K. Godunov, Mat. Sb. 47 (1959), 271-306.

[9] A. Harten and G. Zwas, J. Comput. Phys. 9 (1972), 568.

[10] A. Harten, AEC R and D Report, COO-3077-50, New York University (1974).

[11] A. Harten, Math. Comput. 32 (1978), 363-389.

[12] A. Harten, J. Comput. Phys. 49 (1983), 357.

[13] A. Harten and S. Osher, SIAM J. Numer. Anal., V. 24 (1987), 279.

[14] A. Harten, B. Engquist, S. Osher, S. Chakravarthy, J. Comput. Phys., V. 71 (1987), 231.

[15] F. H. Harlow and J. E. Welsh, Phys. Fluids 8 (1965), 2182-89.

[16] A. Jameson, AIAA Paper 87-1184 (1987).

[17] P. D. Lax and B. Wendroff, Comm. Pure Appl. Math. 13 (1960), 217-237.

[18] P. D. Lax, J. Stat. Phys., V. 43 (1986), 749-756.

[19] R.W. MacCormack, AIAA paper AIAA 69-354 (1969).

[20] A. Majda and S. Osher, Comm. Pure Appl. Math. 32 (1979), 797-838.

[21] G. Moretti, Computers and Fluids, 7 (1979), 191-205.

[22] J. von Neumann and R. D. Richtmyer (1950), vide R. D. Richtmyer and K. W. Morton, *Difference Methods for Initial-Value Problems*, Interscience, New York, 1967.

[23] R.-H. Ni, AIAA J. 20 (1982), 1565-1571.

[24] NRC Report, *Current Capabilities and Future Directions in Computational Fluid Dynamics*, National Academy Press, Washington, D. C. (1986).

[25] S. A. Orzag, J. Fluid Mech., V. 49 (1971), 75-112.

[26] S. A. Orszag and L. C. Kells, J. Fluid Mech. 96 (1980), 159-205.

[27] K. G. Powell, Vortical Solutions of the Conical Euler Equations, Ph.D. thesis, MIT, 1987.

[28] R. D. Richtmyer, vide R. D. Richtmyer and K. W. Morton, *Difference Methods for Initial-Value Problems*, Interscience, New York, 1967.

[29] P. L. Roe, Lecture Notes in Physics 141 (1980), 354-359.

[30] M. D. Salas, private communication.

[31] J. L. Steger and R. F. Warming, J. Comput. Phys. 40 (1981), 263-293.

[32] W. G. Strang, SIAM J. Numer. Anal. 5 (1968), 506.

[33] R. Temam, Bull. Soc. Math. Fr. 96 (1968), 115-152.

[34] B. van Leer, Lecture Notes in Physics 18 (1973), 163-168.

[35] B. van Leer, J. Comput. Phys. V. 32 (1979), 101.

[36] B. van Leer, Lecture Notes in Physics 170 (1982), 507.

[37] K. G. Wilson, summary talk delivered at the Conference on Computational Physics held at the International Centre for Theoretical Physics, Trieste, Oct. 29-31, (1986).

[38] P. R. Woodward and P. L. Colella, J. Comput. Phys. 54 (1984), 115-173.

[39] S. T. Zalesak, in *Advances in Computer Methods for Partial Differential Equations - IV* Eds. R. Vichnevetsky and R. S. Stepleman, Rutgers University (1981).

COMPRESSIBLE FLOW METHODS OF THE 1960'S

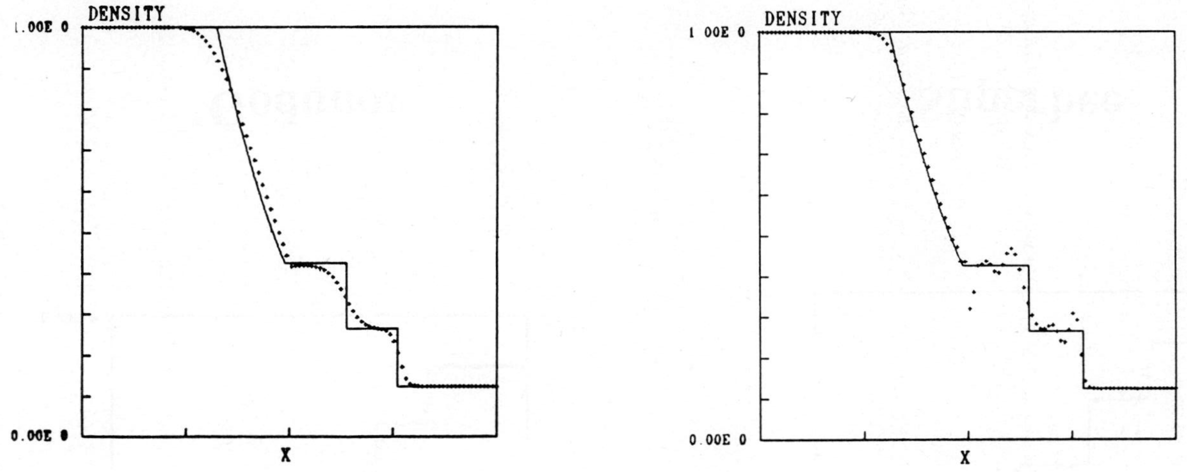

Godunov | **L-W**

Figure 1.

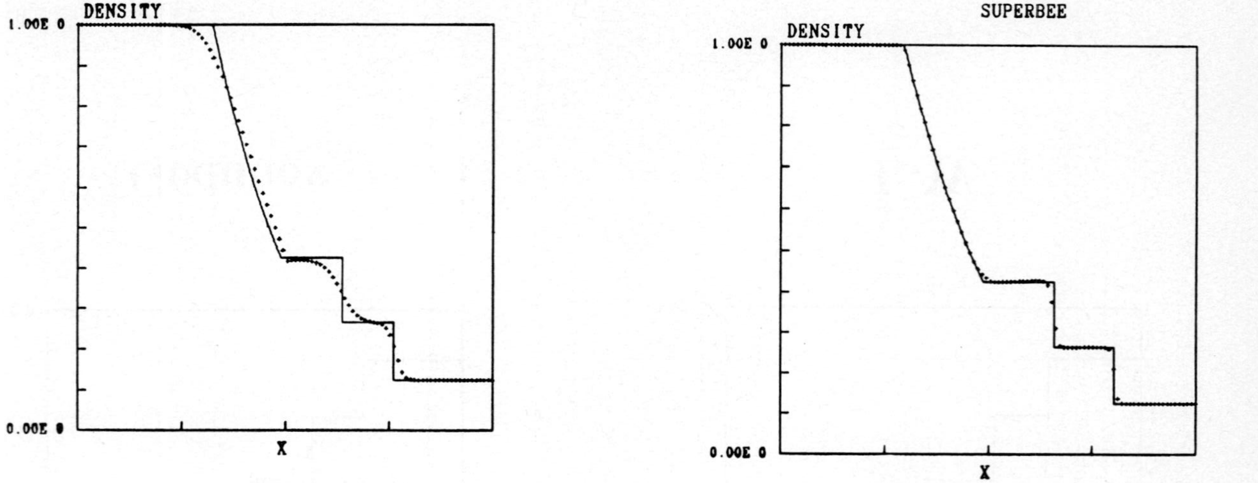

Godunov **Superbee**

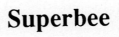

Figure 2.

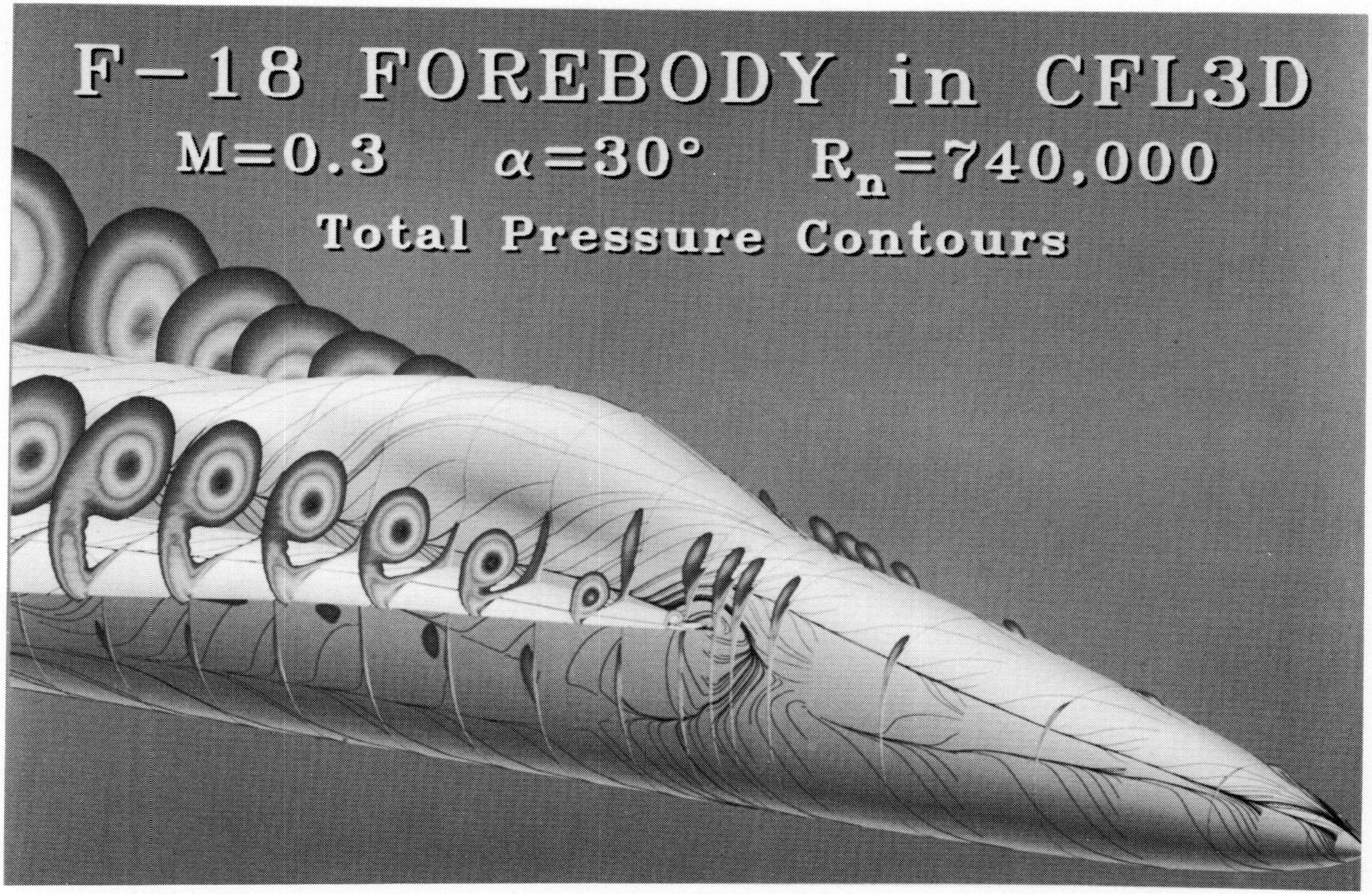

Figure 3. - Total pressures and surface streamlines for F-18 forebody-strake

Figure 4.

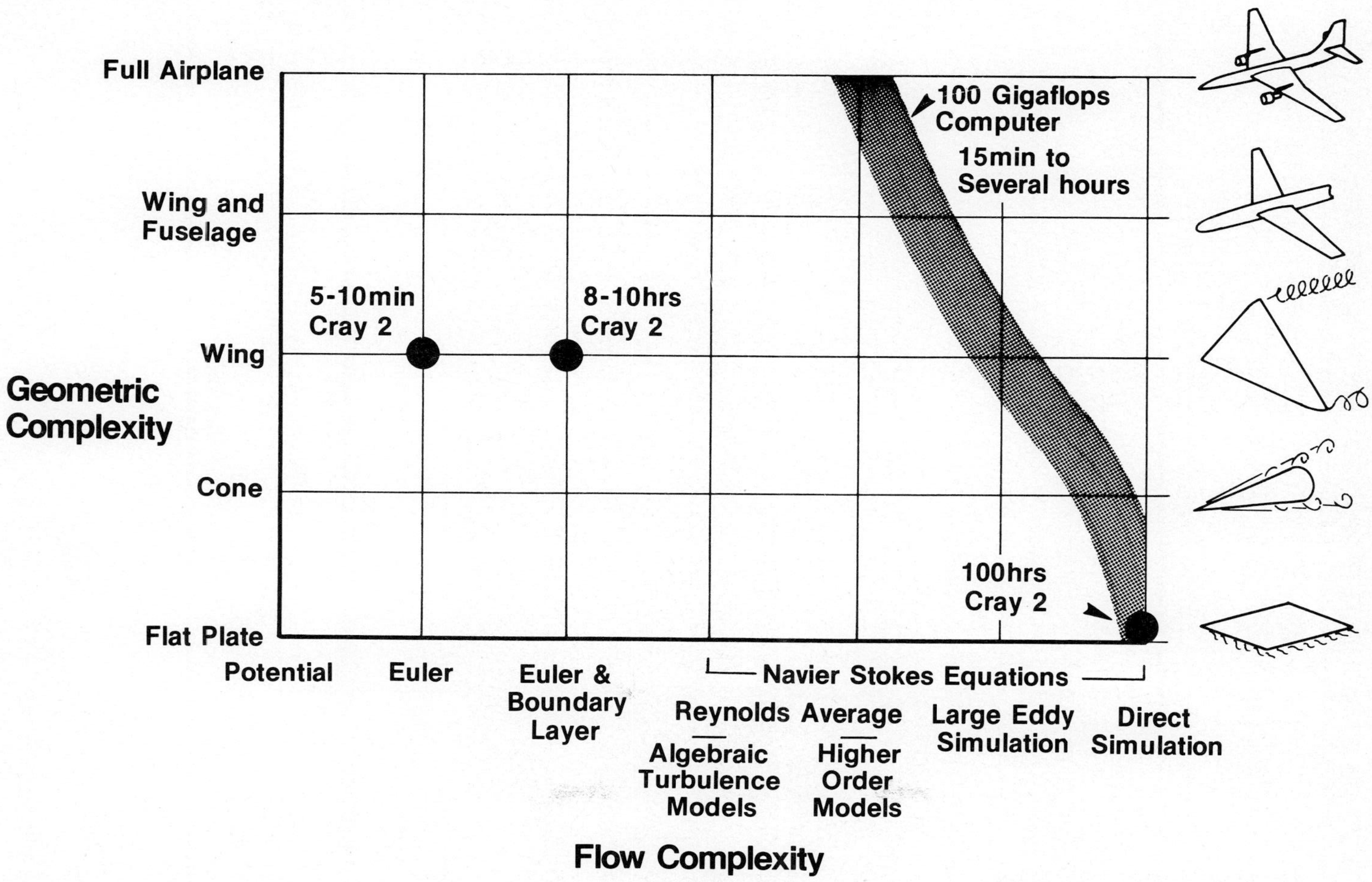
Full Airplane
Wing and Fuselage
Wing
Cone
Flat Plate
Geometric Complexity
5-10min Cray 2
8-10hrs Cray 2
100 Gigaflops Computer
15min to Several hours
100hrs Cray 2
Potential
Euler
Euler & Boundary Layer
Navier Stokes Equations
Reynolds Average
Algebraic Turbulence Models
Higher Order Models
Large Eddy Simulation
Direct Simulation
Flow Complexity

Figure 5.

INVITED LECTURES

COMPUTATIONAL MODELS IN PLASMA DYNAMICS

K.V.Brushlinsky
Keldysh Institute of Applied Mathematics, USSR Acad.Sci., Moscow

Two research branches may be chosen in physics of a dense plasma which is considered as a continuous medium. The first and the most developed one studies formation, characteristics and stability of static, equilibrium plasma configurations. It is mainly due to the interest in the controlled fusion and includes an essential part of works in plasma physics from the 50-es. One can have an idea about them by means of well-known books and reviews, e.g. [1 - 3].

We consider here the second branch, called plasma dynamics, which is less elaborated. It includes investigations of plasma flows in a magnetic field of different configurations, that are of interest in astrophysical problems and in connection with various applications. The both branches correlate with each other as follows. Mathematical models of processes in a dense plasma are based on the magnetogasdynamic equations. In the simpliest case of ideal plasma, i.e. of entirely ionized, infinitely conductive gas without viscosity and heat conductivity, these equations, using the units of measurement avaliable for concrete problems, are:

$$\begin{aligned}
&\frac{\partial \rho}{\partial t} + \vec{\nabla}\cdot\rho\vec{v} = 0\,; \\
&\rho\left(\frac{\partial \vec{v}}{\partial t} + (\vec{v}\cdot\vec{\nabla})\vec{v}\right) = -\vec{\nabla}p + \vec{j}\times\vec{H}\,; \\
&\rho\left(\frac{\partial \mathcal{E}}{\partial t} + \vec{v}\cdot\vec{\nabla}\mathcal{E}\right) + p\,\vec{\nabla}\cdot\vec{v} = 0; \\
&\frac{\partial H}{\partial t} = \vec{\nabla}\times(\vec{v}\times\vec{H})\,; \\
&p = (\gamma-1)\rho\mathcal{E}\,; \qquad \vec{j} = \vec{\nabla}\times\vec{H}\,.
\end{aligned} \tag{1}$$

Equilibrium plasma configurations are described by solutions of the only non-trivial equation that remains of the system (1) if $\frac{\partial}{\partial t}\equiv 0,\ \vec{v}\equiv 0.$

$$\vec{\nabla}p = \vec{j}\times\vec{H}\,. \tag{2}$$

Dynamic plasma configurations are more various. Their description uses all equations (1). Steady-state plasma flows are of particular interest. Their simpliest (e.g. adiabatic or isithermal) models use the equations

$$\vec{\nabla}\cdot(\rho\vec{v})=0, \qquad \rho(\vec{v}\cdot\vec{\nabla})\vec{v}=-\vec{\nabla}p+\vec{j}\times\vec{H}, \qquad \nabla\times(\vec{v}\times\vec{H})=0, \quad (3)$$

that are more rich in solutions in comparison with the equilibrium equations (2). Mathematical and computational modelling plays an essential role in investigation of plasma dynamic processes. It means not only calculations to find some quantitative laws and relations, but also the choice of a suitable model itself. One-dimensional and two-dimensional models call a special attention. They use the symmetry of real plasma devices, simplify the calculations and also may be considered as elements of complicated plasma configurations in general. The one-dimensional models are well-known. First computations were made about 30 years ago [4, 5], a review of one-dimensional ones is in the paper [6]. Among two-dimensional plasma flows two wide-spread classes must be chosen: the flows across a magnetic field and those in the magnetic field plane.

Axisymmetric plasma flows across the proper magnetic field in the channels of plasma accelerators and magnetoplasma compressors [7 - 9, 6] are a typical example of the first class flows. The channel is formed by two coaxial electrodes (fig.I). The electric current running between them produces an azimuthal magnetic field which accelerates the current-carrying plasma down the duct axis. Mathematical models of flows use the magnetogasdynamic equations which, in addition to (I), can take into account finite electric conductivity and, if need be, heat conductivity and viscosity. If problems need distinguish the polarity of electrodes, the equations include also the Hall effect terms. This is a first step to the two-fluid description of plasma as a continuous medium, - the ion and electron velocities are not the same: $\vec{v}_i-\vec{v}_e=\vec{j}/en$ [7]. Time dependent problems are solved numerically, steady-state ones - by means of a stabilizing method.

A quick qualitative analysis of two-dimensional steady-state flows of infinitely conductive plasma may be carried out by means of an analytic technique in the "smooth channel" approximation [IO, 9]. When flow characteristics vary essentially more slow in the channel axis direction than in the radial one ($\frac{\partial f}{\partial z}/\frac{\partial f}{\partial r}\sim\varepsilon\ll 1$ for any function f taking part in the solution), we may rewrite the steady-state MHD-equations in a form without first derivatives in z, i.e. without first order terms in ε. The

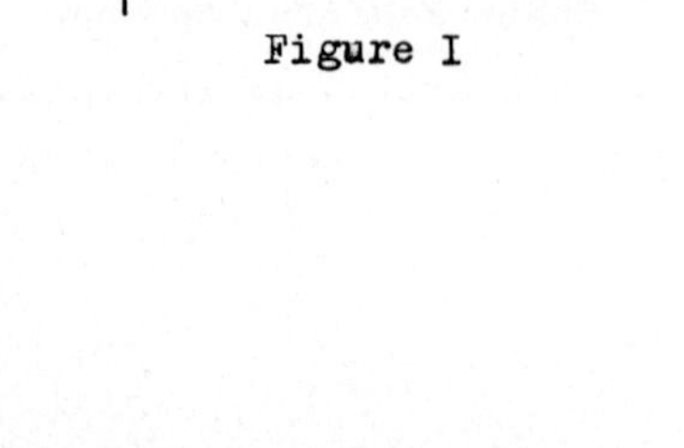

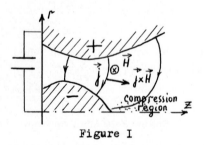

Figure I

terms of order ε^2 and higher being neglected, these equations become ordinary differential equations in r. In some cases we may solve them easily, and the integration constants weakly dependent on z may be chosen according to the statement of problem. In papers [II,I2] there are simple examples of plasma flows described by using this approximation.

To complete the review let us note also flow models including the process of ionization of neutral gas inflowing the channel [7, I3] and a model of the dynamics of solitary impurity ions in the electromagnetic field of the basic MHD-flow. The last model is beeng used in the analysis of some flow details [II, I4].

Numerical analysis of two-dimensional plasma flows in a channel has been carried out by means of the above mathematical models in a set of papers by prof. A.I.Morozov and the author with their colleagues (see reviews [7, 8, I5] and references therein). The result of computations is the investigation of flow characteristics depending on parameters, boundary regimes, Hall effect. One of the properties is the following: near the end of the shortened inner electrode the electric current has an appreciable axial component, and its magnetic field compresses and heats plasma in a narrow region, called the compression region, at the channel axis (fig.I). The idea of the magnetoplasma compressor, offered by A.I.Morozov [I6], is based on this effect.

Steady-state compression plasma flows can be of interest for applications only if they are stable. In this connection the preprint [I2] formulates a problem on the stability of a two-dimensional plasma flow with respect to three-dimensional small perturbations in the linear approach and presents some first results of its solution. Let us dwell on this work for a short while.

An infinitely conductive plasma flow to be investigated is described by a steady-state ($\frac{\partial}{\partial t} \equiv 0$) axisymmetric ($\frac{\partial}{\partial \varphi} \equiv 0$; $v_\varphi \equiv 0$; $H_r \equiv H_z \equiv 0$) solution of the equations (I) in the (z,r)-region corresponding to the channel shape (fig.I - 4). The inflow boundary condition is a given distribution of ρ, p and H at $z=0$, the electrodes $r=r_1(z)$ and $r=r_2(z)$ are solid boundaries - $v_n=0$, and the boundary $r=0$ is the symmetry axis. The flow in a Laval nozzle type channel is usually "transonic", or more exactly, transsignally accelerating: The plasma velocity at the entry is less than the fast magnetic sound speed $C_m = \sqrt{\frac{\gamma p}{\rho} + \frac{H^2}{\rho}}$ and surpasses it at the outlet. Hence there is no outflow boundary condition. In cases of "subsonic" (subsignal, $v_z < C_m$) flows we must have one boundary condition at the outlet $z=1$, for example $\frac{\partial p}{\partial z}=0$.

The steady-state solution may be obtained in previous computation, but here it is given analytically in the smooth channel approximation for the purpose to simplify the problem. Three-dimensional small perturbations are to satisfy the equations (I), linearized on the background of the above steady-state solution. The initial-boundary-value problem is stated in the same (z,r)-region at $0 \leq \varphi \leq 2\pi$ and $t \geq 0$. The channel shape and the boundary conditions are not disturbed, so that the boundary conditions of the linear problem are (the stroke corresponds to perturbations):

$$
\begin{aligned}
&z=0 \quad - \quad \rho' = p' = v_\varphi' = H_z' = H_r' = H_\varphi' = 0, \quad v_r' = \frac{v_r}{v_z} v_z' ; \\
&r = r_1(z),\ r = r_2(z) \quad - \quad v_n' = 0 ; \\
&r = 0 \quad \text{- symmetry conditions,} \\
&z = 1 \ \text{(only if } v_z < C_m) \quad - \quad \frac{\partial p'}{\partial z} = 0 .
\end{aligned}
\tag{4}
$$

The solution must be 2π -periodic in the azimuthal variable φ, and this requirement is also attributed to the boundary conditions.

Let us denote the system of linearized equations (I) with the boundary conditions (4)

$$\frac{\partial U}{\partial t} = \mathcal{L}[U] , \tag{5}$$

where $U = \{\rho', p', v_z', v_r', v_\varphi', H_z', H_r', H_\varphi'\}$ are perturbations dependent on t, z, r, φ ; the linear differential operator $\mathcal{L}$ actes on the space variables z, r, φ and includes the boundary conditions (4). The coefficients of $\mathcal{L}$ are formed of the solution of the steady-state two-dimensional problem. They depend only on z and r, hence the problem (5) admits separation of variables, i.e. has particular solutions

$$U = U_m(t, z, r)\, e^{im\varphi} , \tag{6}$$

where m is any integer. Each Fourier-mode U_m satisfies the system of equations

$$\frac{\partial U_m}{\partial t} = \mathcal{L}_m [U_m] . \tag{7}$$

The in- and outflow boundary conditions and those at the electrodes have the same form (4) for each m, and the symmetry conditions at the axis $r = 0$ are

$$v_{r0}' = v_{\varphi 0}' = H_{r0}' = H_{\varphi 0}' = 0, \quad \frac{\partial \rho_0'}{\partial r} = \frac{\partial p_0'}{\partial r} = \frac{\partial v_{z0}'}{\partial r} = \frac{\partial H_{z0}'}{\partial r} = 0; \ (m=0)$$

$$\rho_1' = p_1' = v_{z1}' = H_{z1}' = 0, \quad v_{r1}' + v_{\varphi 1}' = 0 , \quad H_{r1}' - H_{\varphi 1}' = 0; \ (m=1)$$

$$U_m = 0 \quad (m \geq 2)$$

The coefficients of linear equations (7) do not depend on t, therefore a further separation of variables

$$U_m(t, z, r) = V(z, r)\, e^{\lambda t} \tag{8}$$

is possible, that generates an eigen-value problem

$$\mathcal{L}_m [V] = \lambda V \;. \tag{9}$$

Investigating the flows of infinitely conductive plasma, we must separately consider two classes of problems and two corresponding stability conceptions.

1. If the plasma flow in question is "subsonic" ($v_z < C_m$), there are boundary conditions (4) at the whole (z,r)-region limit, and the problem (9) is a boundary problem with the elliptic operator $\mathcal{L}_m$. The time dependence of solutions of the problem (7) is connected with the senior eigen-value (the spectrum right-hand limit) of the operator $\mathcal{L}_m$ [17]. It can be found by numerical integration of equations (7) with arbitrary initial data until the solution asymptotically takes the form (8). The rate of growth of solutions at $t \to \infty$ and their oscillation frequency determine correspondingly the real and imaginary parts of the senior eigen-value λ. The process under consideration is stable, if $Re\,\lambda \leq 0$, and unstable, if $Re\,\lambda > 0$. Papers [18 - 20] present investigation of stability of some one-dimensional configurations of Z-pinch type, carried out by a similar method.

2. If the plasma flow in the accelerator channel is "transonic" ($v_z > C_m$ at $z = 1$), the outflow boundary condition is omitted, and there is no eigen-value boundary problem. The operator $\mathcal{L}_m$ is hyperbolic in z-direction in the expanding part of the channel, and the equations (9) with zero data at $z = 0$ and at the electrodes have the only zero solution. A physical meaning of this hyperbolicity is that any initial perturbation is being taken away from the channel without reflecting at the outflow cross-section.In a theoretical sense the flow is asymptotically stable in this case. However in practice it is interesting to observe the development of some initial perturbations during a finite time period until they leave the channel. For this purpose the statement of problem may be modified: initial perturbations localized near the entry are fastened by a nonuniform inflow boundary condition, i.e. become a constant source of disturbance. In this case the solution tends to a steady-state regime, when the perturbation "track" can be seen in the whole channel space.

Emphasize, that transonic flows play a special role in this model only in the absence of dissipative processes - resistivity, viscosity and heat conductivity. Even if one of them is taken into account, the problem (9) becomes boundary problem, and the stability investigation scheme is the same as in the previous case.

Computational experiments with both two types of flows of infinitely conductive plasma were carried out in channels of different Laval nozzle shapes. They showed stability of the considered flows with respect to perturbations with Fourier-mode numbers $m = 1, 2, 3$. The evolution of a disturbance of subsonic plasma flow leads in time to undamped oscillations, that corresponds to an imaginary senior eigen-value of the operator $\mathcal{L}_m$. For example, perturbations with $m = 2$ in the channel presented at fig.2, oscillate asymptotically with the period $T \cong 0.9$. That means $\lambda = i\omega$, $\omega = 2\pi/T \cong 7$. Oscillation amplitudes of the density and of the longuitudinal velocity are $\rho' \cong 0.08$, $v_z' \cong 0.02$, when the initial perturbation has had a maximum $\rho'_{max} = 1$. Perturbations of transonic flows decrease fast in time. The fig.3 shows, how the density perturbation ρ' at $m = 1$ spreads in the channel, if it was initially localized near the inflow cross-section. It moves in time along basic flow trajectories, decreases in size, expands in space and leaves the channel with the local maximum value $\rho' \cong 0.06$ (it was initially $\rho'_{max} = 1$). In cases $m > 1$ perturbations decrease still faster. Similar results in axisymmetric problems (m = 0) were obtained earlier in papers [8, 21].

In a set of computations initial perturbations of the density ρ' and

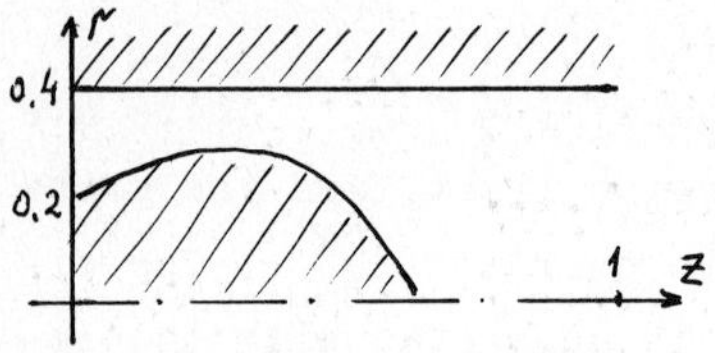

Figure 2

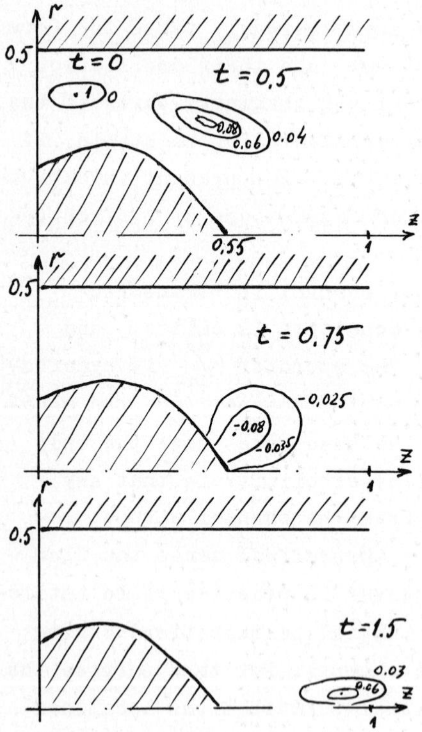

Figure 3

the pressure p' , localized as above near the channel entry, remain constant in time. They are fastened at $z=0$, $t>0$ by the boundary condition. In this case solutions of the problem (7) tend in time to a steady-state regime. Fig.4 presents the "maps" of density perturbations ρ' at m = I, 2. The most appreciable in the compression region and near the channel outlet are again perturbations at m = I. To interprete the presented results in terms of real functions, we must multiply the obtained $\rho'_m, p'_m, v'_{zm}, v'_{rm}, H'_{\varphi m}$ by $\cos m\varphi$ and $v'_{\varphi m}, H'_{zm}, H'_{rm}$ by $\sin m\varphi$ with a small factor and add to the corresponding non-perturbed functions.

Results of the above investigation have shown that the considered plasma flows are stable in the linear approach relative to three-dimensional perturbations given by arbitrary trigonometrical polinomials in φ of the third order. The most essential are perturbations of the snake type (m = I), when the axial symmetry is the most strongly violated, and the channel axis itself moves.

Another typical class of two-dimensional models describes plasma flows in a magnetic field plane. They are more complicated and rich in characteristics in comparison with the preceding ones, and more different from the hydrodynamics. The magnetic field can be produced for example by a system of external currents and then it interacts with currents in plasma. Singular points (or lines) of the field are an essential element of flow. The field size increases infinitely near the wiry conductors and there may be field zero points between them. Flow characteristics depend strongly on the plasma resistivity and therefore it cannot be neglected. As an example of problems on the dynamics of current-carrying plasma in a magnetic field of complicated configuration, the numerical analysis of current sheet formation near the magnetic zero line [22] can be pointed out.

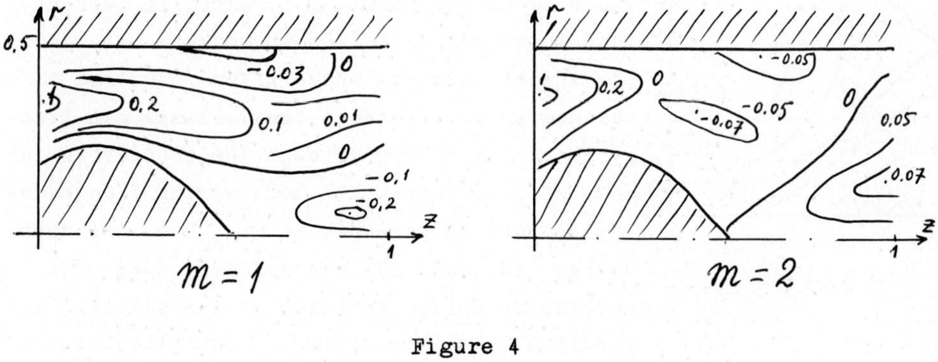

Figure 4

It is convenient to introduce the magnetic field vector-potential $\vec{A}$ in MHD-models of this class of flows: $\vec{H} = \vec{\nabla} \times \vec{A}$. If the vectors $\vec{v}$ and $\vec{H}$ lie in the (x,y)-plane, we have $\vec{A} = (0,0,A)$, $H_x = \frac{\partial A}{\partial y}$, $H_y = -\frac{\partial A}{\partial x}$. As a result the number of unknown functions is less by one, and the Maxwell equation $\vec{\nabla} \cdot \vec{H} = 0$ is automatically satisfied.

In some problems a strong magnetic field plays the role of a frame in the flow picture, and its variation in time is not important. In that case the approximation of a given magnetic field, so called galvanic approximation, may be used. It neglect the magnetic field, induced by electric currents in plasma and simplifies the mathematical model of the problem [23 - 25]. The magnetic field has a form $\vec{H} = \vec{H}_0(\vec{r}) + \vec{H}_1(t,\vec{r})$, $\vec{\nabla} \times \vec{H}_0 = 0$, where $\vec{H}_0$ is a given potential field of external currents and $\vec{H}_1$ is a small addition to it. This addition takes part only in the electric current definition $\vec{j} = \vec{\nabla} \times \vec{H}_1$, and it is neglected in the motion and induction equations. Therefore the electric field has a potential: $\vec{E} = -\vec{\nabla}\varphi$. The elliptic equation for the potential φ follows from the equation $\vec{\nabla} \cdot \vec{j} = 0$ and the Ohm's law $\vec{j}/\sigma = \vec{E} + \vec{v} \times \vec{H}_0$ (or its generalisation $\vec{j}/\sigma = \vec{E} + \vec{v} \times \vec{H}_0 + \frac{1}{en} \vec{j} \times \vec{H}_0 + \frac{1}{en} \vec{\nabla} p_e$, if we take into account the Hall effect). The plasma flow is described only by hydrodynamic equations (I), whose right-hand side includes given magnetic field $\vec{H}_0$ and the current $\vec{j}$, that determined by Ohm's law after the potential φ has been found.

As an example, consider a model of two-dimensional plasma flow in a plane of a given quadrupole type magnetic field. The field configuration (fig.5) contains both two singularities, mentioned above: a seddle type zero point O and a wiry current conductor, that intersect the figure plane in the point K . The plasma flow is due to a steady-state source in a vicinity of the point O . Electric current runs through plasma: the boundaries AB and CD are electrodes, and a given potential difference is kept between them. Plasma can freely leave the region through the boundaries AB and CD , imitating its loss at the electrodes. The picture is periodic in x , i.e. the boundaries AD and BC are symmetry axes. The problem is solved by means of the galvanic approximation with additional assumption of con-

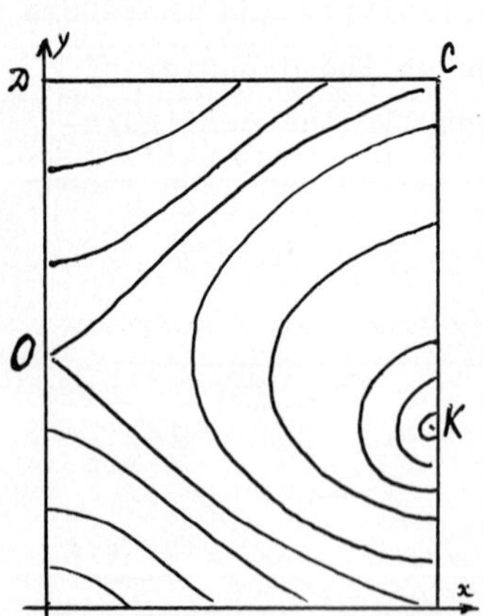

Figure 5

stant temperature. Computatios,,made by the author together with A.I. Morozov and A.M.Zaborov, showed, that the solution of problem quickly reaches a quasi-steady-state regime with following characteristics. Plasma moves from the source O mainly along the magnetic field lines and leaves the region through the boundaries AB and CD near the field separatricies. Flow across the field is of diffusion type with the diffusion coefficient $\mathcal{D} = \beta / 2 Re_m$, where β is the ratio of typical gas and magnetic pressures, and Re_m is the magnetic Reinolds number [7, 25]. The electric current in (x,y)-plane is also directed along magnetic field. It is concentrated near the separatricies, hence a considerable part of the current runs through a vicinity of the zero point O. Fig.6 presents the distribution of: a) velocity field (v_x, v_y) b) current (j_x, j_y), c) density ρ and d) potential φ. There are also velocity and current in z-direction. The velocity v_z is that of movement of a conductor with the current j_y in the magnetic field H_x. Its value is comparable with the electromagnetic drift velocity E/H. The current $j_z \simeq (\vec{v} \times \vec{H})_z$ is small, its maximum is situated at the boundary AD where $\vec{v} \perp \vec{H}$. The Hall effect breaks the symmetry with respect to electrodes: it damps the plasma flow towards the anode and accelerates it towards the cathode. It also diminishes the value of total electric current through the plasma.

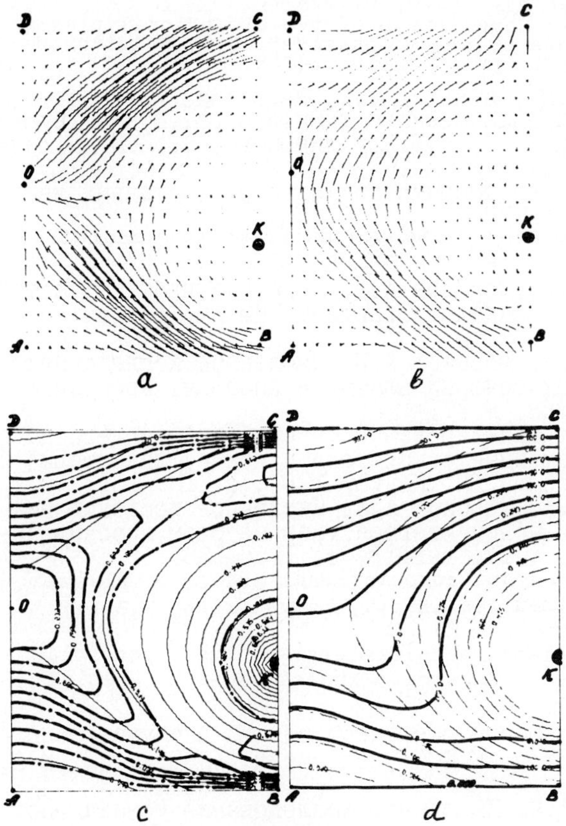

Figure 6

References

1. Арцимович Л.А. Управляемые термоядерные реакции. М.Физматгиз,1963
2. Rose D.J.,Clark M. Plasmas and Controlled Fusion. N.Y.,London, 1961.
3. Основы физики плазмы, п/р А.А.Галеева и Р.Судана, т. 1-2, М. Энергоатомиздат, 1983

4. Брагинский С.И.,Гельфанд И.М.,Федоренко Р.П. Теория сжатия и пульсаций плазменного столба в мощном импульсном разряде. В кн."Физика плазмы и проблема управляемых термоядерных реакций" изд.АН СССР, 1958, т.IУ, с. 201-221.
5. Hain K.,Hain G.,Roberts K.V.,Roberts S.J.,Koeppendorfer W. Zs.Naturforsch., 1960, bd.15a(12), s.1039.
6. Roberts K.V.,Potter D.E. Magnetohydrodynamic calculations. In"Meth. in Comp.Phys.", vol.9, Acad.Press, 1970, p.339-420.
7. Brushlinsky K.V.,Morozov A.I. Computation of two-dimensional plasma flows in channels. In "Reviews of plasma physics" ed.by M.A.Leontovich, Consultants Bureau, N.Y., 1980, vol.8, p.105.
8. Брушлинский К.В.,Морозов А.И.,Савельев В.В. Некоторые вопросы течений плазмы в канале магнитоплазменного компрессора. В кн."Двумерные численные модели плазмы" п/р К.В.Брушлинского, М.,ИПМ им.Келдыша АН СССР, 1979, с. 7-66.
9. Морозов А.И. Физические основы космических электрореактивных двигателей, т.I, М.,Атомиздат, 1978.
10. Morozov A.I.,Solov'ev L.S. Steady-state plasma flows in a magnetic field. In "Reviews of plasma physics", vol.8, p.I
11. Брушлинский К.В.,Козлов А.Н.,Морозов А.И. Динамика пробных частиц в двумерном потоке плазмы в канале. Препринт ИПМ им.Келдыша АН СССР, 1980, № 156.
12. Белова И.В.,Брушлинский К.В.,Морозов А.И. Численное исследование устойчивости компрессионных течений плазмы в каналах. Препринт ИПМ им.Келдыша АН СССР, 1988, № 17.
13. Брушлинский К.В. Численное моделирование течений ионизующегося газа в каналах. В кн."Плазменные ускорители и ионные инжекторы" п/р Н.П.Козлова и А.И.Морозова, М.,Наука, 1984, с. 139-151.
14. Брушлинский К.В.,Козлов А.Н.,Морозов А.И. Численное исследование двумерных течений плазмы и ионизующегося газа методом пробных частиц. Физика плазмы, 1985, т.11, вып.11, с. 1358-1367.
15. Brushlinsky K.V. Mathematical and computational models of plasma flows. Inst.Space and Astronaut.Sci.Rept., Tokyo, 1986, N.Sp.4, p.239-247.
16. Morozov A.I. Steady-state plasma accelerators and their possible applications in thermonuclear research. Nuclear Fusion, Special Supplement, 1969, p. 111-119.
17. Брушлинский К.В. О росте решения смешанной задачи в случае неполноты собственных функций. Изв.АН СССР,сер.матем.,1959, т.23, № 6, с. 893-912.
18. Брушлинский К.В.,Шатанов А.П. Численное исследование устойчивости z-пинча при конечной проводимости плазмы. Препринт ИПМ им.Келдыша АН СССР, 1980, № 150.
19. Белова И.В.,Брушлинский К.В. Численная модель неустойчивости z-пинча в плазме конечной проводимости. Журн.вычислит.матем. и мат.физики, 1988, т.28, № 1, с. 72-79.
20. Dibiase J.A.,Killeen J. A numerical model for resistive magnetohydrodynamic instabilities. J.Comp.Phys.,1977,v.24,N.2, p.158-185.
21. Бадьин Л.В. Численное исследование течений плазмы в канале МПК с дивертором. Препринт ИПМ им.Келдыша АН СССР, 1983, № 131.
22. Брушлинский К.В.,Заборов А.М.,Сыроватский С.И. Численный анализ токового слоя в окрестности магнитной нулевой линии. Физика плазмы, 1980, т.6, вып.2, с. 297-311.
23. Брагинский С.И. К магнитной гидродинамике слабо проводящих жидкостей, ЖЭТФ, 1959, т.37, вып.5/11/, с. 1417-1430.
24. Ватажин А.Б.,Любимов Г.А.,Регирер С.А. Магнитогидродинамические течения в каналах. М.,Наука, 1970.
25. Брушлинский К.В.,Заборов А.М.,Морозов А.И.,Цветков А.В. Двумерные численные модели плазменных течений диффузионного типа во внешнем магнитном поле. Препринт ИПМ им.Келдыша АН СССР, 1984, № 84.

PARALLEL COMPUTERS AND PARALLEL COMPUTING IN SCIENTIFIC SIMULATIONS

Tsutomu Hoshino
Institute of Engineering Mechanics, University of Tsukuba
Tsukuba-shi, Ibaraki-ken, 305, Japan

1. Introduction

The emergence of VLSI since 1970's has been and still is changing all aspects of computer technology. Not only the processors are speeded up and the memory capacity is increased, but also various kinds of computer architectures are made technically and economically feasible. For scientific applications, parallel architecture is one that promises the great possibility of enhancing the performance.

Many parallel computers have been studied in the academic world, and some of them have been put into the market. This short article describes the author's view on this parallel computer architecture in relation to their application to scientific computing.

Parallelism refers to a nature of process, in which multiple tasks can be executed simultaneously. The parallelism can be found in the scientific model itself (inherent parallelism), in the algorithm (algorithmic parallelism), and in the parallel computer (machine parallelism). The number of tasks that can be processed in parallel is called degree of parallelism. The degree of parallelism finally achieved is the minimum of those of inherent, algorithmic and machine parallelisms.

Another key concept in parallel processing is overhead. Because of the uneven workload among processors, some processors may be left in an idle state. Any data transfer among the partitioned tasks is inevitable, since the all tasks have to cooperate with each other to solve the original job. In order to transfer the valid data, as well as to transfer the valid control among the tasks, any synchronization has to be taken. These three extra-tasks or wastes of resources, i.e. idling of processors, data transfer, and synchronization are most common overheads associated with parallel computing.

Granularity is defined as a ratio (unlikely the conventional definition) of the net computation taken place in a single processor, over the data communication overhead associated with that net part[1].

Here, the net means the fraction of computation that is essentially needed to get the result. Granularity will be discussed later in more detail.

2. Parallel computer architectures

Several taxonomies of basic constructs in the contemporary parallel computers are as follows:

1) Instruction-flow vs. Data-flow,
2) SIMD(Single-Instruction Multi-Data) vs. MIMD(Multi-Instruction Multi-Data),
3) Distributed local memory vs. Shared common memory,
4) Arithmetic pipeline vs. Processor array,
5) Small number of fast processors (Low degree of parallelism) vs. Large number of slow processors (High degree of parallelism),
6) General purpose vs. Special purpose,
7) Network topology, or how the multiple processors and memories are interconnected: Nearest-neighbor-mesh(NNM), Hypercubes, Multistage Switching Networks, and other topologies.

Some of these are fundamental, a priori commitments by architects, while others are very superficial and designer's a posteriori decisions.

3. Scientific models

Boundless needs in large-scale scientific and engineering calculations (often called simulation) are the major incentive for the development of parallel supercomputers. What are typical structures in the scientific models? It is difficult to make clear classification on these variety of calculational models. Followings are also superficial, and a single simulation model sometimes belongs to several groups.

1) Continuum models, describing the motion of continuum (fluid, field, etc.) by partial differential equations (PDE's). Physical actions are mostly of nearest-neighbor(NN) type, though the implicit integration schemes frequently require the remote interaction expressed by linear equations.

2) Particle models, describing the motion of many bodies (plasma, molecules, stars, etc.) that interact each other through the field they create, or do not interact at all. The models are usually expressed by ordinary differential equations(ODE's), or ODE's combined

with PDE's. The type of interaction could be either NN or remote.

3) System models, describing the complex systems (generally artificial) with sparse matrix and linear algebra. Interaction among system elements is mostly remote and random, or the rules of interaction in the models cannot be fully exploited by computer architects.

4. Parallel supercomputing

Successful parallel supercomputing should make a good match between these structures of models and computer architectures. In other words, we must investigate the topology of data dependence; whether it can be partitioned into parallel subtasks, and how they can be mapped into memories and processors. We must also study the granularity in the model and the numerical algorithm; whether they can keep processors busy and communication overheads small most of the time.

Performance evaluation is another area where users' expertise is needed. Performance is expressed in terms of the execution time, the efficiency, or whatever performance figures that the user may choose. Let us limit our discussion within a certain type of architecture, such as NNM, hypercube, etc. Obtained performance is represented by a scaling law, a function of hardware and model parameters , such as the total number of space points, number of processors, average time to execute floating point calculation, etc.

Once we establish the scaling law on the parallel machine with that type of architecture, then we can predict the performance of the non-existent new machine with the same architecture, by substituting the new design parameters in the scaling law.

Although the performance is finally evaluated by executing users' benchmark problems on the machine, it cannot be generalized to that will be obtained on other types of benchmark problems.

On the other hand, the hardware performance, such as MFLOPS or MIPS, cannot directly be interpreted to the performance that would be obtained in the end users' applications. Several intermediate performance figures can be obtained by executing the typical set of statements such as Livermore loops, or the subroutine packages such as LINPACK.

The author would like to propose the unit-granularity measure, as one reflecting the hardware potential in parallel processing, that is defined by, unit-granularity:

$$g = \frac{\{\, t_i,\ \text{time to execute the unit floating point operation in a processor} \}}{\{\, t_b,\ \text{time to transfer the unit floating point data from a processor (or a memory directly accessible from the processor) to another processor (or a memory directly accessible from the processor)} \}}$$

It is straightforward to interpret this unit-granularity into the total-granularity defined by (total time for net calculation)/(total time for communication) in their applications. In the continuum type models, it is simply derived by,

total-granularity:

$$g_{tot} = g \, \frac{\{\, f_i,\ \text{total floating point operations} \}}{\{\, f_b,\ \text{total floating point transfers} \}}$$

The smaller the total-granularity, the lower the performance.

The multiprocessor-version of supercomputers with a few vector processors, connected via slow speed large capacity shared memory, have very small unit-granularity, when we assume the transfer of large amount of data via the memory, as it is often needed in fluid dynamics. Hypercube machines have surprisingly small granularity because of the very slow data transfer via interprocessor link, often of single bit width. The access to global shared common memory through the multistage switching network suffers from slow communication speed of the network, and consequently, displays small unit-granularity, unless one intentionally slows down the processor speed to match with the communication speed.

One of the methods to decrease the data transfer overhead is a software effort to overlap the internal calculation with the data transfer (Fig. 1). For example, suppose that the data on the boundary of the subregion allocated to each processing unit are requested before the calculation starts for the internal data that do not interact with any boundary data. After the requested data have arrived, the boundary calculation is made on those data (Fig. 1(b)).

This overlapping scheme could result in considerable improvement in execution time, as illustrated in the comparison between Fig. 1(a) and Fig. 1(b). The complete overlapping as shown in Fig. 1, however, is not always expected; it depends not only on the structure of and the algorithm for individual problem, but also on the particular set of parameters, such as the number of space points laid on the boundary relative to the number of internal space points.

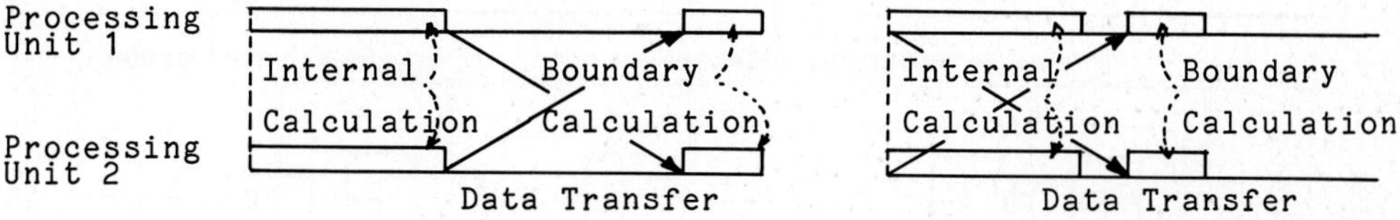

(a) Non-overlapping scheme (b) Overlapping scheme

Fig. 1 Data Transfer and Internal Calculation

It is not always an easy task even for an intelligent compiler to generate the object program doing such an overlapping, where the compiler has to identify automatically which part of the data transfer should be overlapped with which internal processing. Even for the end-users, if they are not experts in the control of parallel processing, it would be a difficult work to instruct explicitly the overlapped operation in their source programs.

For the continuum type simulations, the communication overhead must be so small that the unit-granularity becomes greater than or equal to 1. Assuming the contemporary hardware technology of silicon, the nearest-neighbor-mesh(NNM) connected processor array would be one candidate that realizes this unit-granularity figure, if the machine has a bus link for the NN connection as broad as the internal system bus. Another possibility in the future is the massive optical links among the processors, that could be interconnected via hypercube or other topologies of network.

5. PAX computer

An example of high performance parallel computer with reasonable unit-granularity is the PAX computer developed by the author.

The PAX computer development started in 1977, aiming at the realization of a parallel computer, which can be used in broad range of scientific and engineering calculations [2][3]. The past 4 prototype machines, PAX-9, PAX-32, PAX-128, and PAX-64J were all too slow (0.5 ~ 3 MFLOPS) and too small (1 ~ 4 MB) to match with the users' demands. The next model QCDPAX, however, will exceed the commercial supercomputers in speed and capacity. (Figs. 2 and 3).

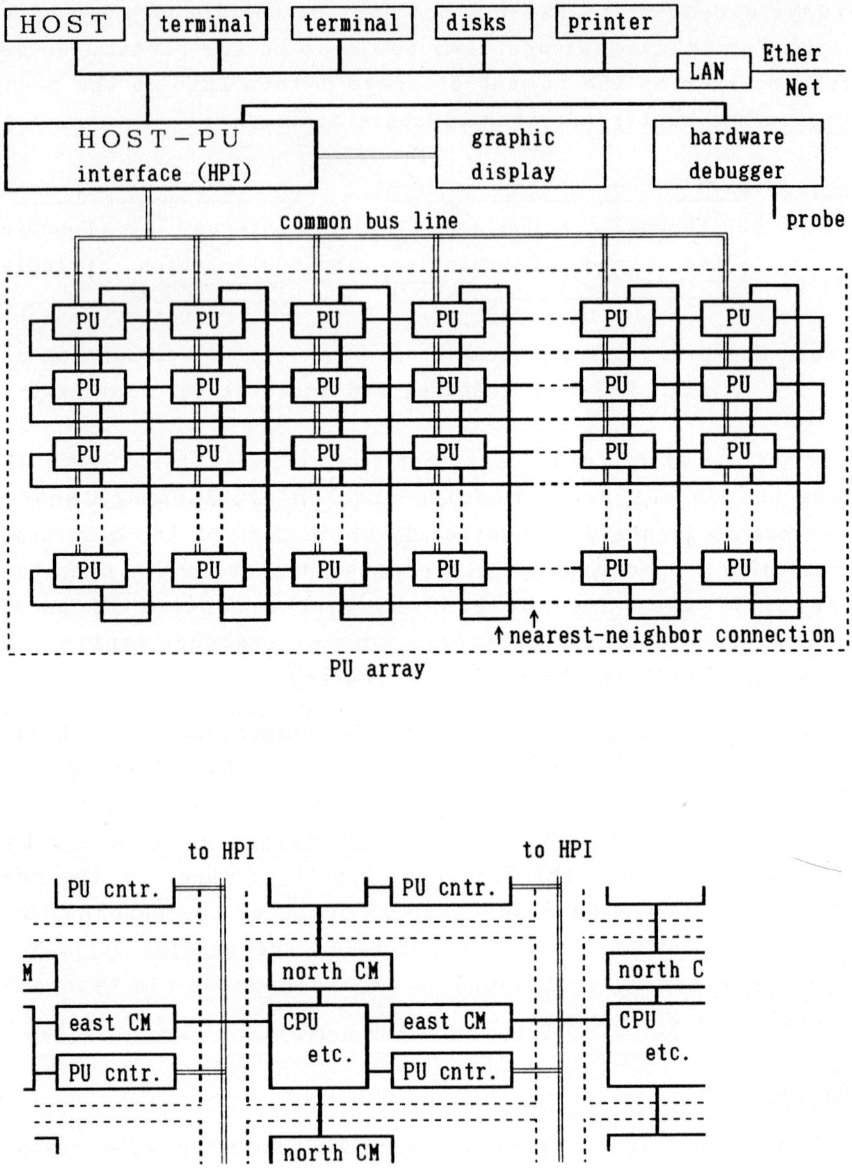

Fig. 2 System Configuration of QCDPAX.

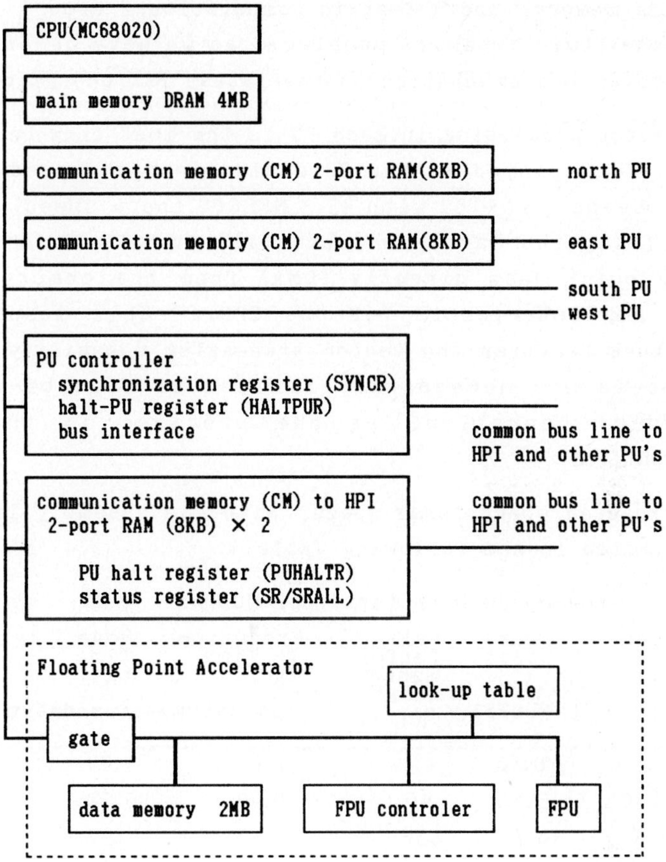

Fig. 3 Configuration of Processing Unit of QCDPAX.

The PAX architecture can be characterized by:

(1) Two-dimensional NNM torus connection,
(2) Broad communication channel between adjacent PU's(Processing Units),
(3) Homogeneous structure of PU array,
(4) Global synchronization,
(5) Global control and data communication (broadcasting) among PU's,
(6) Nodal description in source program for PU,
(7) Debugging of parallel program at synchronization barriers,
(8) Multi-purpose applicability in scientific applications, such as fluid- [4] and fluid-particle-hybrid models [5].

The MIMD architecture enables us to employ various kinds of algorithms, such as incomplete-LU decomposition scheme in solving linear equations for quark propagators in QCD(Quantum Chromo Dynamics) simulation, to which the machine will primarily be devoted. Requirements to the hardware performance are Giga FLOPS average speed,

Giga bytes memory, and 1 year in computational time. It can also be used in the fluid dynamics problems, as we have demonstrated in the Navier-Stokes implementation on the latest PAX prototype machines [4].

The vector processing in each PU is the most enhanced capability in QCDPAX. We employed in the PU design, a high-speed CMOS floating point processor, L64132 with 33.3 MFLOPS (peak speed). In order to exploit fully the hardware potential, it is necessary to feed the floating point data directly (DMA) from the memory to the chips without any intervention of the CPU. This means that the PU architecture provides the vector processing capability in a sense of DMA, but is not necessarily limited to that of the pipelined arithmetic operation, as one provided in the commercial supercomputers.

The expected operational speed, with the assembly language coding, is illustrated in the following Table 1.

Table 1. Speed and Granularity of QCDPAX

Expressions	Peak Speed (MFLOPS)	Half-Performance Length $n_{(1/2)}$	Execution Time t_c (μs/datum)	Move Time t_b (μs/datum)	Unit Granularity g
A = B + C	0.86	---	1.16	1.44	0.81
A(I)=B(I)+C(I)	5.56	40	0.90	0.92	0.98
S=S+A(I)*B(I)	16.7	38	0.29	0.92	0.32
S=S+A(I)*A(I)	33.3	76	0.26	0.92	0.28

We measured, on the prototype board, that the time to transfer a single floating point data is about 1.44 μsec, and that to transfer a vector data with length 10 is about 0.92 μsec/data. The unit-granularity of QCDPAX machine is, therefore, estimated to be about 0.8 for scalar, and from 0.3 to 1 for vector (length 10) operations.

We also enhanced the host-PU interface(HPI) with the video image display, where the 2-port memory between each PU and the host is used for the real-time output of data during the execution of parallel program. Thus very fast real-time display of the scientific simulations will be realized.

Software system must exploit the machine parallelism, as well as provide the comfortable programming environment. Host OS is UNIX, in which several language processors are developed, such as PSC(parallel scientific C for vector processing) and QFA(quick floating-point assembler for optimization of object program by users) for PU programming, and PREC(preprocessor c) for host programming.

6. Conclusion

Users are recommended to know more about the computer architectures and to benchmark the machines before falling in love with a particular machine, unless he/she is willing to die for that love. Even the users' participation in the designing phase of the machine is encouraged, in order to make the end product practical in their applications, otherwise they have to be patient in using the unfriendly machine.

Will future be a peaceful world in which several types of parallel machines coexist, or a battle field where they fight for a share in the market? Somewhat between the both extremes could be anticipated. Decisive factor for the future world will again come from the basic technology of logic devices (silicon and its successors), and of integrating them.

Superconduction may be an important but not decisive factor in speeding-up the machines, if it is solely used for the signal transmission in the computer hardware. The delay in the present computer circuit is caused by the high impedance in the logic device and the stray capacity along the transmission line, where the superconducting line helps little in speeding-up. It is still difficult to predict what the superconductive logic device and its integration will be.

Optical technology may be more important to breakthrough the pin-number-limitation, often hindering some of the parallel architectures with massive connection. The introduction of optical technology in inter-board or inter-logic device connection will push the computer architecture toward more parallelism and higher efficiency.

References

[1] Harold S. Stone, High-Performance Computer Architecture, Addison-Wesley, 1987.
[2] T. Hoshino, An invitation to the world of PAX, IEEE Computer 19(1986) 68-79.
[3] T. Hoshino, T. Shirakawa and K. Tsuboi, Mesh-connected parallel computer PAX for scientific applications, Parallel Computing 5 (1987) 363-371.
[4] T. Hoshino, T. Kamimura, T. Iida and T. Shirakawa, Parallelized ADI scheme using GECR (Gauss-elimination-cyclic-reduction) method and implementation of Navier-Stokes equation in the PAX computer, Proc. 1985 International Conference on Parallel Processing (IEEE Computer Society Press, Silversprings, 1985) 426-433.
[5] T. Hoshino, R. Hiromoto, S. Sekiguchi and S. Majima, Mapping schemes of the particle-in-cell method implemented on the PAX computer, to be published in Parallel Computing.

CFD FOR HYPERSONIC AIRBREATHING AIRCRAFT

Ajay Kumar
NASA Langley Research Center
Hampton, Virginia, 23665
U.S.A.

Abstract

A general discussion is given on the use of advanced computational fluid dynamics (CFD) in analyzing the hypersonic flow field around an airbreathing aircraft. Unique features of the hypersonic flow physics are presented and an assessment is given of the current algorithms in terms of their capability to model hypersonic flows. Several examples of advanced CFD applications are then presented.

Introduction

Research in the general area of hypersonics was almost nonexistent in the period from early 1970's to mid 1980's.[1] During this period, analytical research directed specifically at hypersonics virtually disappeared. Many hypersonic wind tunnels and shock tubes were either dismantled or not used at all. University programs in hypersonics were more or less dropped. There were only a few pockets of research activities located primarily at NASA's Langley and Ames Research Centers. However, in the past few years there has been an explosion of activities in this area all over the country including government research laboratories, universities, and industries. This substantially increased interest in hypersonics has been caused in part by the need for developing a new class of vehicle, called the aero-space plane, powered by an airbreathing propulsion system. A hypersonic airplane powered by airbreathing engines can gain a performance advantage over a rocket powered vehicle by using the atmosphere as the oxidizer. In order to achieve or realize this advantage, however, these airbreathing engines require large capture areas and must be closely integrated with the airframe to avoid excessive drag penalties. This requirement has resulted in designs in which almost the entire undersurface becomes part of the propulsion system (Fig. 1). The forebody precompresses the flow before it enters the engine inlets and the flared afterbody acts as part of the nozzle by further expanding the exhaust products from the engine. The engine itself may have features such as cut-back cowl, spill windows or doors, and swept compression surfaces to operate it over a wide Mach number range. Struts or centerbodies may be necessary to provide additional compression in the inlets and locations for fuel injection in the combustors. Integration of the propulsion system with the airframe is extremely important at hypersonic Mach numbers. Some analyses have shown that the vehicle forebody and afterbody may be responsible for up to 70% of the net thrust.[2]

Due to lack of available ground facilities for testing at high Mach numbers (the ones that are available do not simulate the flight enthalpies at these high Mach numbers) and due to the complex and integrated nature of the flow field, traditional wind-tunnel based design techniques are not adequate. Modern CFD technology offers a promising alternative means of analyzing, designing, and optimizing the hypersonic vehicles. It will be used to extend wind-tunnel model results to full-scale flight conditions where tests can be done and will be the dominant analysis technique at the high Mach numbers where hardly any testing can be done in ground-based facilities.

Through the 1970's and early 1980's research in hypersonic internal flows focused on the development of the individual module of a supersonic combustion ramjet (scramjet) engine for airbreathing propulsion. Since this research took place when CFD was rapidly maturing, computational techniques were developed for analyzing scramjet components. Reference 3 provides a comprehensive survey of the CFD techniques for analyzing hypersonic internal flows. According to this reference, computational codes are fairly mature for inlets but somewhat less so for combustors and nozzles. However, for all the components of the scramjet engine, Reynolds-averaged Navier-Stokes (RANS) and parabolized Navier-Stokes (PNS) codes exist today that account for at least some of the relevant physics.

The same advancements in CFD technology that led to the development of codes for high-speed internal flows have also led to much improved capability to predict hypersonic external flows. Throughout the 1970's interest in hypersonic external aerodynamic predictions focused on entry and reentry vehicles and the shuttle orbiter. A number of sophisticated codes based on the viscous shock layer (VSL) approximation as well as PNS and RANS equations were developed that included real gas, radiation, ablation, and wall catalysis effects relevant to hypersonic entry body flows (see Ref. 4, for example). Thus, it is apparent that CFD codes based on PNS and RANS equations exist today that can analyze flow over hypersonic configuration and through the engine components. However, it is necessary to extend the current CFD technology to one of the most crucial problems in the development of the hypersonic airbreathing airplane--propulsion/airframe integration. To integrate the airframe and propulsion system at high Mach numbers, CFD must treat flow fields substantially more complex than those around the entry vehicles. Reference 5 gives a good discussion on the CFD requirements for hypersonic airplanes versus entry vehicles. This paper provides the status of CFD for hypersonic flows in terms of algorithms, their capability to model the relevant physics, and application to analysis/design/optimization of hypersonic configurations. Some examples of advanced CFD applications are then presented.

Code Development for Hypersonic Flows

The use of CFD as a tool in the analysis and design requires development of advanced codes which can be relied upon. The basic issues that need to be addressed in the code development are

- Proper modeling of flow physics and geometric complexity
- Robust algorithm (accurate, stable, efficient, and compatible with the current day computer architecture)
- Suitable grid for complex geometries
- Validation of code

Once a code is developed, proper strategy must be used for its efficient application in the analysis, design, and optimization of configurations. A brief discussion follows on each of these issues.

Flow Physics and Geometric Complexity

The flow around a hypersonic airplane is predominantly three-dimensional and is dominated by viscous effects. Due to the highly integrated nature of the vehicle design, there is a strong coupling between the vehicle and propulsion system flow fields. An accurate prediction of the state of the boundary layer developed by the forebody is important not only for calculating surface quantities such as skin-friction, heat transfer, and pressure but also for the inlet performance. In addition to the accurate prediction of the boundary-layer, it is also necessary to predict the entire shock-layer profile accurately in order to be able to determine the mass and momentum flux entering the inlets. Of course, real gas effects become important for flight Mach numbers beyond about M = 10.

After the inlet face, the geometric complexity of the flow-field boundaries increases dramatically. At off-design conditions, the inlets will spill a substantial amount of flow thereby setting up a complex multiple inlet forebody flow-field interaction which can be further complicated by the presence of wings. The aft end of the hypersonic airplane is dominated by the interaction of the multiple internal/external nozzle system flow with the vehicle wing/body flow. The nozzle flow includes real gas effects throughout the operating envelope of the vehicle, and at off-design conditions, possible flow separation on the external nozzle surface may occur. Requirements for accurate prediction of the complete three-dimensional flow field are again important on the external nozzle in order to estimate the nozzle thrust coefficients and the direction of the net thrust vector. Prediction of surface properties such as skin-friction, heat transfer, and pressure are also vitally important. The real-gas models incorporated in the nozzle flow analysis must include the chemistry of the combustor products as well as the air chemistry.

The combustor flow field presents perhaps the most difficult situation for the CFD codes both in terms of physical modeling as well as flow complexity. Important physical/chemical processes in the combustor include fuel/air mixing, ignition, combustion, and shock/turbulence/kinetics interactions. One of the major concerns in the high-speed combustors is to enhance the mixing of fuel and air which seems to get adversely affected with increasing Mach numbers.[6] Apart from predicting the overall combustor flow field, there are a number of locations in a typical combustor where highly detailed analysis of very localized processes is required. Such regions include the immediate flow fields in the neighborhood of fuel injectors and flameholders. Accurate and detailed prediction of such highly localized phenomenon is required if CFD codes are to be used in the analysis and design of combustors. References 7 through 10 describe some of the codes developed for combustor flow-field analysis.

A major issue in CFD code development for a hypersonic airplane is transition and turbulence modeling. It is extremely important to know the transition location on the vehicle and its extent. Transition ambiguity can significantly increase the gross weight of the vehicle due to the use of a turbulent rather than laminar boundary layer. So far as turbulence modeling for high Mach numbers is concerned, it appears that the existing models are adequate for the attached boundary layers but significant developmental work is required for predicting turbulent mixing in free-shear layers, shock-turbulent boundary-layer interaction, and three-dimensional separation. Real-gas effects and compressibility effects at high Mach numbers need to be incorporated in the models. Modeling of kinetics/turbulence interaction is of crucial importance for combustor flow. At high altitudes, the flow over the vehicle goes from turbulent to laminar and low-density effects may become important. Transition and turbulence models must, therefore, account for this relaminarization process.

It is apparent from the preceding discussion that there are a number of shortcomings in the physical/chemical modeling and significant effort is necessary to model the physics of high Mach number flow. Not only that, the physics that need to be modeled changes significantly over the flight envelope; the CFD codes should be flexible enough to account for the changing physics if these codes have to be used for the entire range of flight conditions.

Algorithms

As mentioned earlier, the flow field around the hypersonic airplane is three-dimensional and is dominated by viscous effects. It is, therefore, necessary to include viscous terms in the governing equations. In general, the codes that are being developed or used today are based on either the parabolized Navier-Stokes (PNS) equations or the Reynolds averaged Navier-Stokes (RANS) equations (full or under thin-layer approximation). The equations are solved by either central-difference algorithms or upwind-difference algorithms. Of these, the codes that use the central-difference algorithms are the most mature and have been extended to include the most complete physical/chemical models (see, for example, Refs. 9, 11-15 for PNS codes and 7, 8, 10, 16-20 for RANS codes based on central differencing). The upwind-difference based codes[21-23] are relatively new, and to date mostly perfect gas codes are available for the three-dimensional flows. Extensive efforts are underway to include more advanced physical and chemical models in upwind codes.

So far as the applications of the codes are concerned, the PNS codes are the least expensive of the currently available methods for computing hypersonic airplane flow fields. Due to their inability to predict streamwise separation, their application is mostly restricted to external flows with limited use to internal flows.

The central-difference RANS methods are the most widely used today. The popular explicit and implict solution algorithms have shortcomings in stability and computational complexity, respectively, which has led to the search for alternatives. Additionally, the central-difference methods are limited in shock capturing capability. Despite these problems, they are currently the algorithms most widely used in the production codes and have successfully predicted complex internal and external flow fields for hypersonic configurations.

Grid

An integral part of the code development is a grid generation package that can provide suitable grids for complex configurations. A number of techniques are being developed and used for generating grids in these packages. Important issues related to grids are the analysis and reduction of error introduced by the coordinate system, the automation of control of coordinate line spacing to achieve both concentration and smoothness, the dynamic coupling of the coordinate system with the physical solution, and the patching together of regions to represent general configurations with sufficient continuity. Reference 24 provides a good introduction to the state-of-the-art in the area of numerical grid generation.

Validation of Code

Before using a CFD code for an applied problem, a necessary step is to validate it. Code validation requires detailed and/or global experimental data of high quality taken in well calibrated wind tunnels. The model geometry should be well defined, error bounds on the data should be specified, and the type of flow (laminar, transitional, or turbulent) should be known.

Code validation over the entire operating flight envelope requires validating the code over small ranges of the flight envelope. As the vehicle goes through the flight envelope, the physics that needs to be modeled changes, the characteristics of the algorithm change, and the grid requirements for resolving the flow change. Anytime a code is used too far beyond the range of its validation, the uncertainty factor associated with the calculations increases. There are three steps involved in the validation process:

(a) Consistency and internal checks. - The code should be checked for mass, momentum, and energy conservation and preservation of freestream on an arbitrary grid in the absence of a body. It should be compared against available exact solutions for inviscid and viscous problems. Frequently, there are internal numerical control parameters in a code which can affect calculated results. A study of the effects of small perturbations of these parameters on the computed results should also be made.

(b) Code to code comparison.- A newly developed code should be checked against other established codes using similar or different algorithms and also against boundary-layer and parabolized Navier-Stokes codes whenever possible.

(c) Code to experiment comparison.- The final step in validating a code consists of comparing the calculated results against some well defined and reliable experimental data. The experimental data may be from an existing experiment or may be from a new experiment specifically designed for code validation. In the second case, a code developer can make specific inputs in instrumenting the model for proper data acquisition.

Reference 25 provides an example of code validation. It is important to realize that due to a lack of activities in the area of hypersonics for over a decade, a vacuum exists in the available experimental data that can be used for code validation. The experimental facilities and required high-quality instrumentation need to be developed so that a reliable experimental data base can be generated. In fact, there seems to be a general consensus amongst experimentalists that in the coming few years, most of the high-speed testing will be done to produce extensive data base necessary for code validation. Considering the potential use of CFD codes, this is certainly a step in the right direction.

Code Application

The CFD codes can be used either for analysis or design/optimization of a configuration. At the present time, use of the RANS codes is primarily restricted to the analysis of flow. Due to large computer resource requirements, even a 2-D Navier-Stokes code is considered fairly expensive and time consuming in design. However, codes based on the PNS method are certainly finding their use in the design process. Reference 26 provides a good study of CFD code applications to practical problems. A word of caution is necessary at this point on the application of modern CFD codes. These codes cannot be treated as black boxes. The user of the code should have a reasonable background in CFD, both from an algorithm as well as programming point of view. In addition, a good understanding of flow physics is necessary to be able to make sense out of the large amount of data generated by the computers. Use of sophisticated graphics/post processing tools is required to extract meaningful information about the flow from the large data base.

Advanced Applications

This section presents examples of some advanced applications of CFD in analysis and design studies of high-speed flows. In the first application, flow over a hypersonic lifting body is studied; whereas, in the second application, integration and interaction of multiple scramjet inlet modules is studied. The final example demonstrates the use of CFD codes in hypersonic nozzle design subject to some external constraints which make the conventional nozzle design procedure impractical.

Hypersonic Lifting Body

Figure 2 shows the geometry and grid for a hypersonic lifting body tested over a Mach number range of 11 to 19 and Reynolds number range of 11,000/ft to 10^7/ft. A number of test conditions were simulated numerically using an upwind thin-layer Navier-Stokes code, CFL3D, described in Reference 21. A grid consisting of approximately 200,000 points (46 points axially and 65 x 65 points in the cross-plane) was used in the analysis. Numerical results are presented here for laminar flow at Mach 19.2 and Reynolds number 360,000/ft. The angle of attack is zero.

Figure 3 shows the particle traces around the lifting body. It is seen that for this configuration, the flow tends to pile up on the lower surface along the symmetry plane (not a very desirable feature as this low momentum flow is supposed to enter the inlets of the propulsion system). Figure 4 and 5 show the comparison of predicted surface heating rates with the experimental results. Axial heating rates in the upper and lower symmetry planes are compared in Figure 4 whereas cross-section heating rates are compared in Figure 5. It is seen from these figures that the analysis has predicted heating rates that compare well with the experimental data. More detailed results for the preceding and other flow conditions are presented in Reference 28.

Multiple Module Scramjet Inlet Integration

One of the major requirements in the development of hypersonic vehicles is to closely integrate the vehicle airframe and the propulsion system. In an effort to investigate this problem, an experimental as well as analytical program has been devised. Figure 6 shows the schematic of an experimental model with three inlet modules mounted on a flat plate that is used to generate a simulated forebody boundary layer. The compression surfaces of each module are swept wedges. The aft body expansion is simulated by an expansion on the plate. The numerical results presented here were obtained from a 3-D RANS code, SCRAMIN, described in Reference 17. The flow conditions were for a freestream Mach number of 4.03, static pressure of 8724 N/m^2, static temperature of 70°K, and zero angle of attack and yaw. Each inlet module had a geometric contraction ratio of 4.0 and the cowl closure began at the throat of the inlets. A grid of approximately 340,000 points was used in the calculations. Only half of the configuration was analyzed due to flow symmetry. Figure 7 shows the computational grid in one of the cross planes and the symmetry plane of the configuration. Out of a total of 61 grid planes in the vertical direction, 25 planes were below the cowl plane to account for the interaction between the high-pressure internal flow and the low-pressure external flow caused by the aft placement of the cowl.

A 2-D version of SCRAMIN was used on the front part of the flat plate to calculate the flow up to the face of the modules. The 3-D code was then used for the flow from the face of the modules to the end of the configuration. Figure 8 shows the velocity vector field and the pressure contours in a streamwise plane located slightly above the cowl plane. The pressure contours show the shock and expansion waves and their interactions. Since it is a cold flow with no fuel injection, the flow expands back to low pressure behind the inlet throat. The velocity vector plot shows relatively small regions of separated flow caused by the shock/boundary-layer interactions.

Figure 9 shows the pressure contours and velocity vector field in the plane of symmetry. The velocity vector plot shows a significant downturn in flow direction ahead of the cowl resulting in some flow spillage. The downturn is caused by the sidewall sweep and the interaction between the internal and external flow. Once the inlet flow passes behind the cowl leading edge, it is turned back parallel to the cowl plane, and this turning results in a cowl shock which is evident in the pressure contour plot.

Multiple Inlet Interactions

In the preceding study, one of the concerns that needs to be investigated both numerically as well as experimentally is the potential for interactions between closely mounted multiple inlets. In order to examine such interactions, a two-strut scramjet inlet shown in Figure 10 and analyzed in Reference 27 was used as a model problem for a three-inlet system. As is seen, the inlet has three separate passages. The two center struts have an initial compression angle of 9°, and the initial cowl closure begins at the throat for which $x/x_T = 1$. To study the interactions, an attempt was made to unstart the center passage and see the impact of this unstart on the two side passages. In the initial attempts to unstart the center passage, the cowl location was fixed at the throat but the geometric contraction ratio of the center passage was increased substantially increasing the strut compression angle from 9° to 10.75°. This increase in center passage contraction ratio did not cause it to choke and resulted in no interaction with the side passages as is evident from the pressure contours in Figure 11 which remain unchanged with increasing contraction ratio. However, the increased contraction ratio resulted in much higher pressure in the center passage as well as an increased downturn of the flow ahead of the cowl as is seen from Figures 12 and 13. A capture plot of the inlet is shown in Figure 14. It is seen that as the center passage is gradually closed, the total inlet capture goes down, but the capture plots of individual passages show that all the decrease in capture is due to the increased spillage from the center passage. The capture of the side passages remains constant. This again confirms that there is no interaction between the center and side passages.

The second attempt to choke the center passage was made by moving the cowl forward from its initial location of $x/x_T = 1$. Two cowl locations of $x/x_T = .85$ and $.67$ were tried with the strut compression angle remaining at 9°. For both cowl locations, the center passage still did not choke but the inlet capture increased significantly as is seen from Figure 15. But for $x/x_T = .67$, when the strut compression angle was increased to 10°, choking or unstart of the center passage occurred. Figure 16 shows the pressure contours in the symmetry plane of the inlet. The results of 9° strut compression angle are used as the starting solution. Pressure contours at 5,500, 10,000, 13,000, and 17,500 time-steps clearly show the development and formation of a bow shock ahead of the cowl. This bow shock stands in front of the cowl producing a region of subsonic flow between the shock and the cowl and resulting in significantly increased spillage from the center passage. The pressure contours in the cross plane located slightly above the cowl plane are shown in Figure 17. It is seen that the flow in the side passages has also been modified due to the unstart of the center passage. Obviously, once the subsonic flow ahead of the cowl is established, it interacts with the flow in the side passages and modifies it.

Nozzle Design

The numerical analysis codes are still not widely used in the design of hypersonic configuration components. A major exception is, of course, in the design of wind-tunnel and rocket nozzles where techniques based on method of characteristics with boundary-layer correction have been in use for many years. These design techniques are restricted to either 2-D or axisymmetric flows. Three-dimensional designs or designs whose constraints force the relaxation of the requirement of shockless flow, will require more sophisticated CFD tools. The effort to design the nozzle contours for the NASA Langley 8' High-Temperature Tunnel (HTT) provides an example of the use of advanced CFD codes in design projects. Under this effort, two nozzles for Mach 4 and 5 were designed such that they smoothly blend with the existing Mach 7 nozzle about 200 inches upstream of the test section. The specified constraints were: (1) the axial position and radius of the entrance to the subsonic region; (2) the throat axial location; and (3) the axial station where current and new nozzle walls must smoothly blend together. The Mach number variation was required to remain below ± 0.1 about its mean value across 60% of the core flow. The high temperature flow in the tunnel required the possibility of foreign gas injection for transpiration cooling in the nozzle throats. Furthermore, the large static temperature variation in the nozzles resulted in a significant variation in gas properties, and those variations had to be properly modeled.

Due to the imposed constraints, it was found that the conventional shockless nozzle design procedure was not applicable. A new iterative design procedure was developed under this effort that coupled an Euler code, a method of characteristics code, and a boundary-layer code. A Navier-Stokes code was used to check the overall flow quality of the final design. All codes included consistent real gas chemistry packages for a H-C-O-N gas system. In addition, the Navier-Stokes and boundary-layer codes had the capability to account for foreign gas injection for transpiration cooling. A detailed discussion of this iterative design procedure is given in Reference 29.

Figures 18 and 19 show some results obtained from the design of the Mach 5 nozzle. Figure 18 shows the Mach number profiles in the exit plane of the nozzle calculated by the Navier-Stokes and Euler codes. The profile has a mean value of 4.96 with a variation of ± 0.06 over more than 70% of the test section radius, thus satisfying the Mach number variation constraint. However, a weak shock forms near the nozzle throat and intersects the exit plane, as is clearly seen from the Mach number contours in Figure 19. It appears that the weak shock cannot be avoided under the specified geometric constraints. Figures 18 and 19 also illustrate the qualitative similarities in the flow solution obtained from the iterative design procedure and the Navier-Stokes code.

Conclusions

The purpose of this paper has been to discuss CFD technology as it relates to the computation of flow fields associated with hypersonic airbreathing aircrafts. It appears at this point that CFD is the only available means for analyzing the flow over the entire flight trajectory. However, advances in algorithm robustness and speed, geometric flexibility, and inclusion of more complete flow physics are required in the CFD codes if they are to be widely used in the development of hypersonic airbreathing aircrafts.

References

1. Anderson, J. D., Jr.: A Survey of Modern Research in Hypersonic Aerodynamics. AIAA Paper 84-1578, 1984.

2. Henry, J. R.; and Anderson, G. Y.: Design Considerations for the Airframe Integrated Scramjet. NASA TM-X-2895, 1973.

3. White, M. E.; Drummond, J. P.; and Kumar, A.: Evolution and Status of CFD Techniques for Scramjet Applications. J. of Propulsion and Power, vol. 3, no. 5, Sept.-Oct. 1987, pp. 423-439.

4. Kumar, A.; Graves, R. A., Jr.; Weilmuenster, K. J.; and Tiwari, S. N.: Laminar and Turbulent Flow Solutions with Radiation and Ablation Injection for Jovian Entry. Aerothermodynamics & Planetary Entry, Progress in Astronautics and Aeronautics, vol. 77, edited by A. L. Crosbie, AIAA, New York, 1981, pp. 351-373.

5. Dwoyer, D. L.; and Kumar, A.: Computational Analysis of Hypersonic Airbreathing Aircraft Flow Fields. AIAA Paper 87-0279, 1987.

6. Kumar, A.; Bushnell, D. M.; and Hussaini, M. Y.: A Mixing Augmentation Technique for Hypervelocity Scramjets. AIAA Paper 87-1882, 1987.

7. Drummond, J. P.; Rogers, R. C.; and Hussaini, M. Y.: A Detailed Numerical Study of a Supersonic Reacting Mixing Layer. AIAA Paper 86-1427, 1986.

8. Uneshi, K.; Rogers, R. C.; and Northam, G. B.: Three-Dimensional Computations of Transverse Hydrogen Jet Combustion in a Supersonic Airstream. AIAA Paper 87-0089, 1987.

9. Sinha, N.; and Dash, S. M.: Parabolized Navier-Stokes Analysis of Ducted Turbulent Mixing Problems with Finite-Rate Chemistry. AIAA Paper 86-0004, 1986.

10. Chitsomboon, T.; Kumar, A.; Drummond, J. P.; and Tiwari, S. N.: Numerical Study of Supersonic Combustion Using a Finite-Rate Chemistry Model. AIAA Paper 86-0309, 1986.

11. Schiff, L. B.; and Steger, J. L.: Numerical Simulation of Steady Supersonic Viscous Flow. AIAA Paper 79-0130, 1979.

12. Kaul, U. K.; and Chaussee, D. S.: AFWAL Parabolized Navier-Stokes Code: 1983 AFWAL/NASA Merged Baseline Version. AFWAL-TR-83-3118, May 1984.

13. Gnoffo, P. A.: Hypersonic Flows Over Biconics Using a Variable-Effective Gamma, Parabolized Navier-Stokes Code. AIAA Paper 83-1666, 1983.

14. Gielda, T.; and McRae, D.: An Accurate, Stable, Explicit, Parabolized Navier-Stokes Solver for High-Speed Flows. AIAA Paper 86-1116, 1986.

15. Chitsomboon, T.; Kumar, A.; and Tiwari, S. N.: A Parabolized Navier-Stokes Algorithm for Separated Supersonic Internal Flows. AIAA Paper 85-1411, 1985.

16. MacCormack, R. W.: The Effect of Viscosity in Hyper-Velocity Impact Cratering. AIAA Paper 69-354, 1969.

17. Kumar, A.: Numerical Simulation of Scramjet Inlet Flow Fields. NASA TP-2517, May 1986.

18. Beam, R.; and Warming, R. F.: An Implicit Factored Scheme for the Compressible Navier-Stokes Equations. AIAA J., vol. 16, April 1978, pp. 393-402.

19. Jameson, A.; Schmidt, W.; and Turkel, E.: Numerical Solutions of the Euler Equations by Finite-Volume Methods Using Runge-Kutta Time-Stepping Schemes. AIAA Paper 81-1259, 1981.

20. Vatsa, V. N.: Accurate Numerical Solutions for Transonic Viscous Flow Over Finite Wings. J. of Aircraft, vol. 24, no. 6, June 1987, pp. 377-385.

21. Thomas, J. L.; and Walters, R. W.: Upwind Relaxation Algorithms for the Navier-Stokes Equations. AIAA Paper 85-1501-CP, 1985.

22. Chakravarthy, S. R.; Szema, K. Y.; Goldberg, U. C.; and Gorski, J. L.: Application of a New Class of High Accuracy TVD Schemes to the Navier-Stokes Equations. AIAA Paper 85-0165, 1985.

23. Gnoffo, P. A.; McCandless, R. S.; and Yee, H. C.: Enhancements to Program LAURA for Computation of 3-D Hypersonic Flow. AIAA Paper 87-0280, 1987.

24. Thompson, J. F.: Numerical Grid Generation. North-Holland, New York, 1982.

25. Rudy, D. H.; Kumar, A.; Thomas, J. L.; Gnoffo, P. A.; and Chakravarthy, S. R.: A Comparative Study and Validation of Upwind and Central-Difference Navier-Stokes Codes for High-Speed Flows. AGARD Symposium on Validation of Computational Fluid Dynamics, Lisbon, Portugal, May 1988.

26. Barber, T. J.; and Cox, G. B., Jr.: Hypersonic Vehicle Propulsion: A CFD Application Case Study. AIAA Paper 88-0475, 1988.

27. Kumar, A.: Numerical Simulation of Flow Through Scramjet Inlets Using a Three-Dimensional Navier-Stokes Code. AIAA Paper 85-1664, 1985.

28. Richardson, P. F.; Parlette, E. B.; Morrison, J. H.; Switzer, G. F.; Dilley, A. D.; and Eppard, W. M.: Data Comparison Between a 3-D Navier-Stokes Analysis and Experiment for the McDonnell Douglas Lifting Body: Generic Option #2. Paper No. 52, Fourth National Aero-Space Plane Symposium, Feb. 1988.

29. Erlebacher, G.; Kumar, A.; Anderson, E. C.; Rogers, R. C.; Dwoyer, D. L.; Salas, M. D.; and Harris, J. E.: A Computational Design Procedure for Actively Cooled Hypersonic Wind-Tunnel Nozzles Subject to Wall Shape Constraints. Proceedings of Computational Fluid Dynamics in Aerospace Design Workshop, University of Tennessee Space Institute, UTSI Pub. E02-4005-029-85, June 1985.

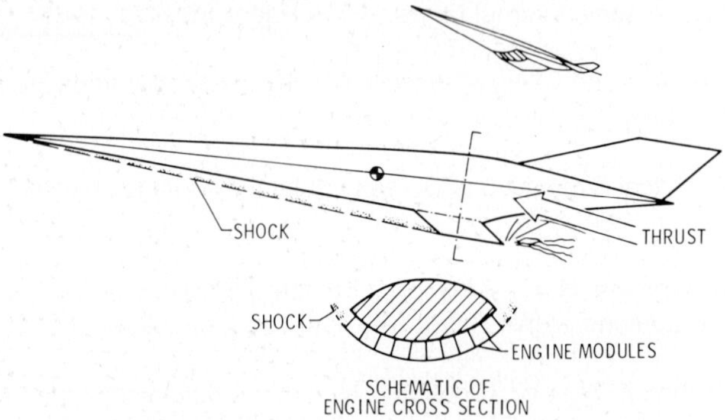

Fig. 1 - Schematic of a hypersonic aircraft with airframe-integrated scramjet-engine concept.

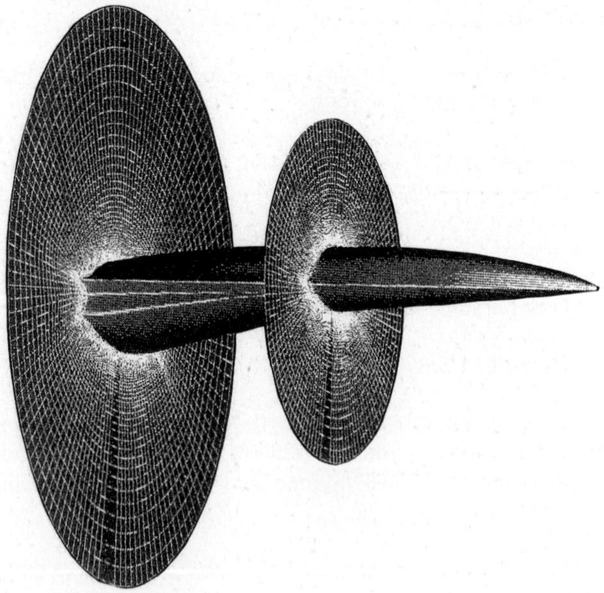

Fig. 2 - Geometry and grid for a hypersonic lifting body.

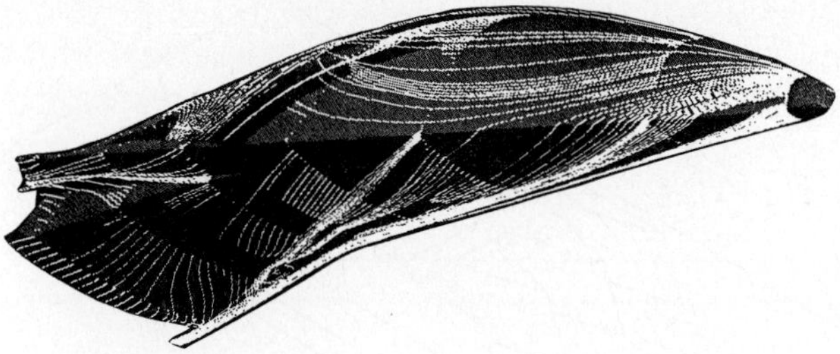

Fig. 3 - Particle traces on the liting body.

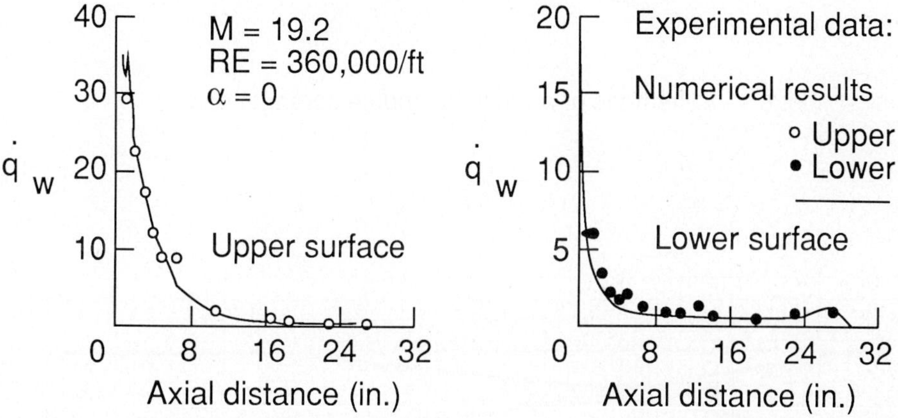

Fig. 4 - Axial heat transfer rate $\dot{q}_w$ (btu/ft^2 sec) in symmetry plane on upper and lower surfaces.

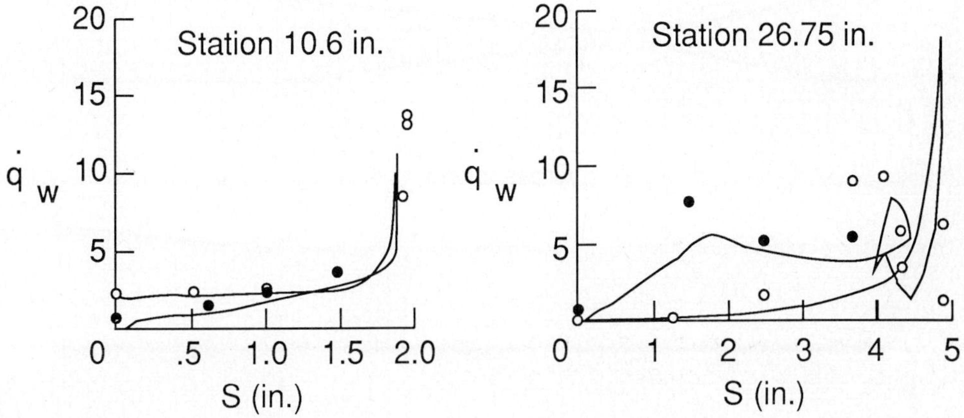

Fig. 5 - Surface heat transfer rate $\dot{q}_w$ (btu/ft^2 sec) in cross planes.

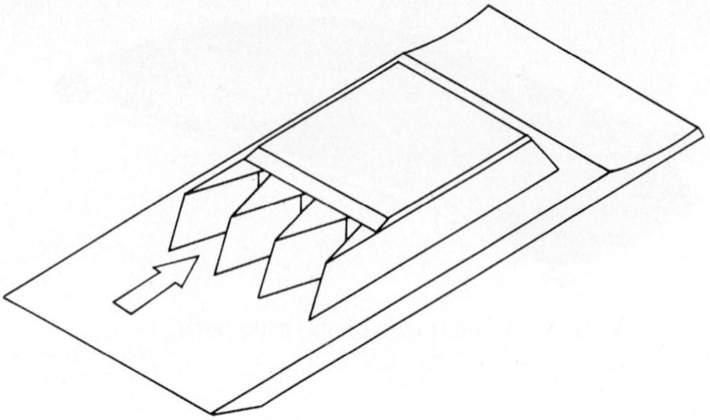

Fig. 6 - Schematic of a multiple module scramjet engine.

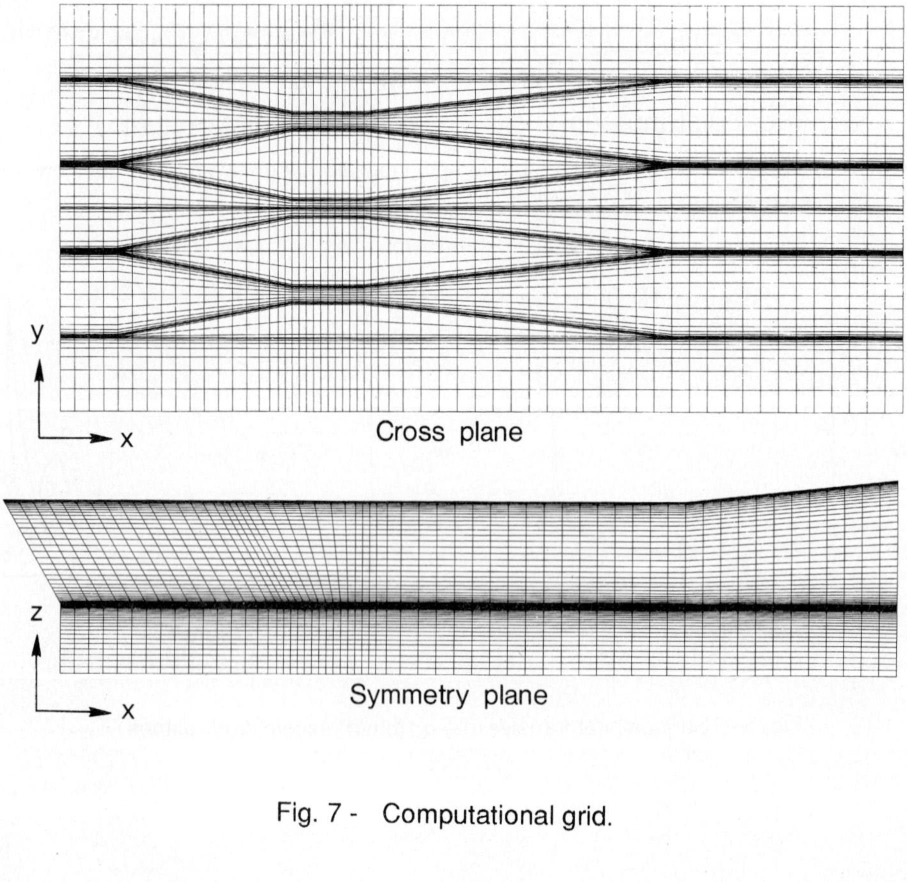

Fig. 7 - Computational grid.

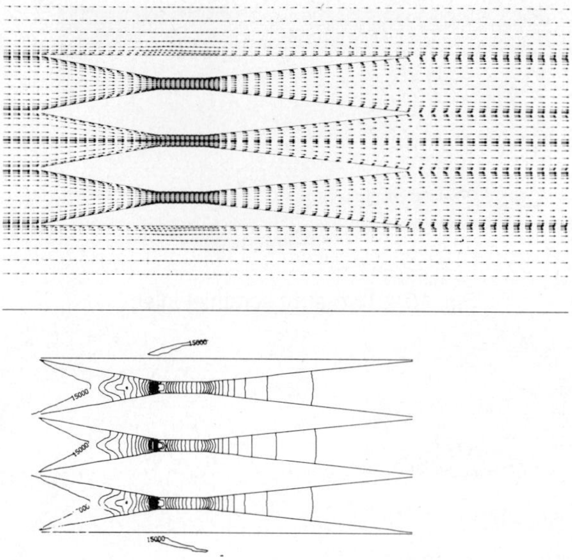

Fig. 8 - Velocity vector field and pressure controus in a cross plane slightly above the cowl plane.

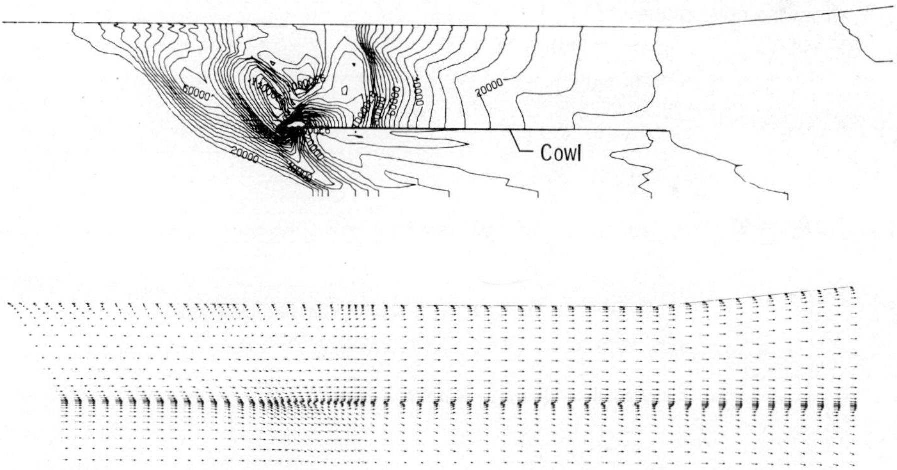

Fig. 9 - Pressure contours and velocity vector field in the symmetry plane.

Fig. 10 - Two-strut scramjet inlet.

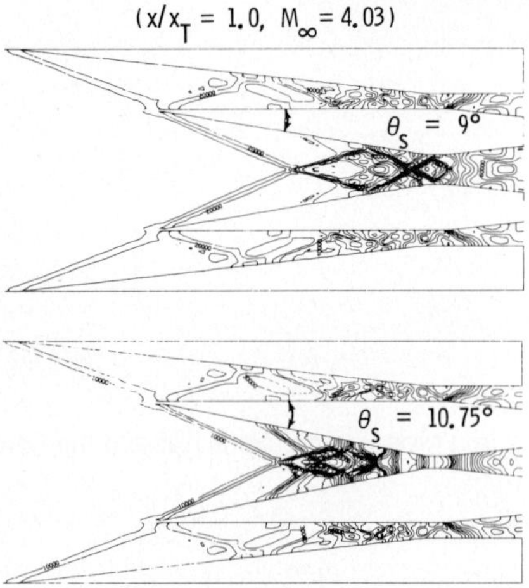

Fig. 11 - Pressure contours in a cross plane located at 12% of the inlet height from the cowl plane.

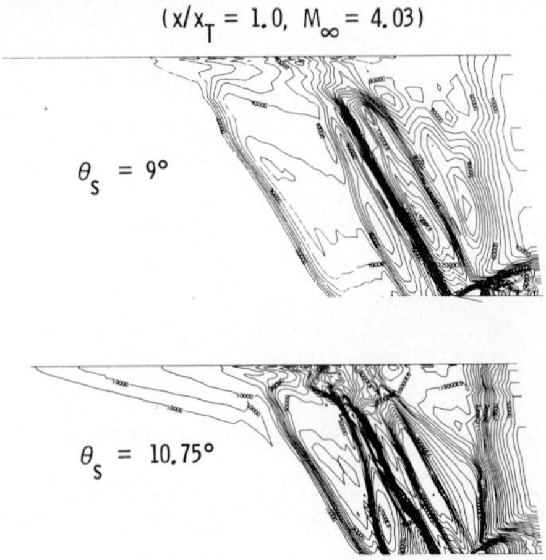

Fig. 12 - Pressure contours in the symmetry plane.

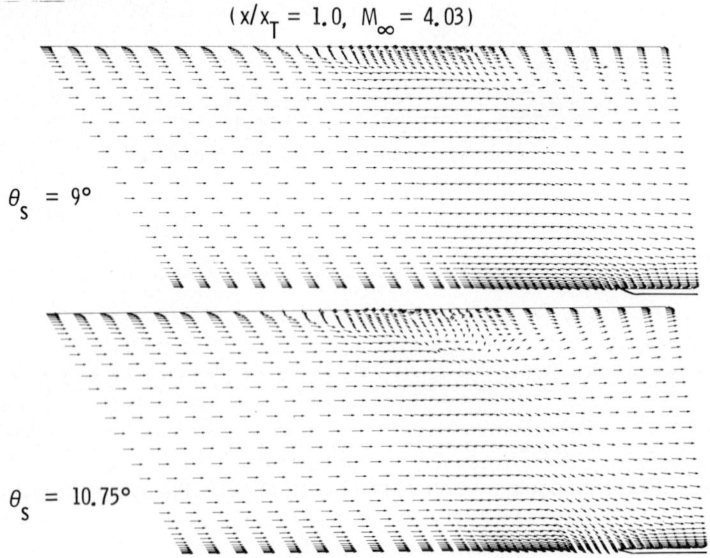

Fig. 13 - Velocity vector field in the symmetry plane.

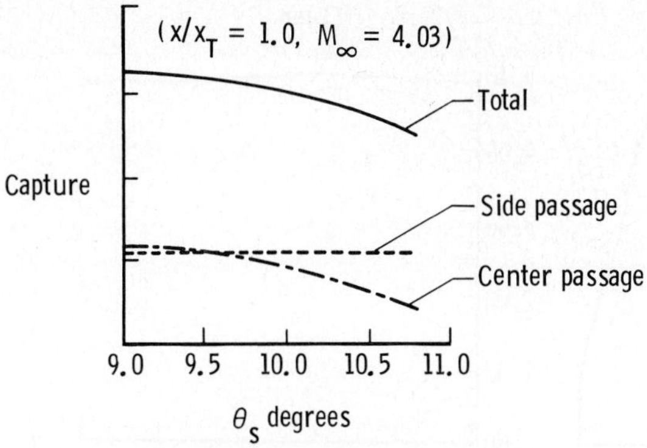

Fig. 14 - Inlet capture as a function of strut compression angle, θ_s.

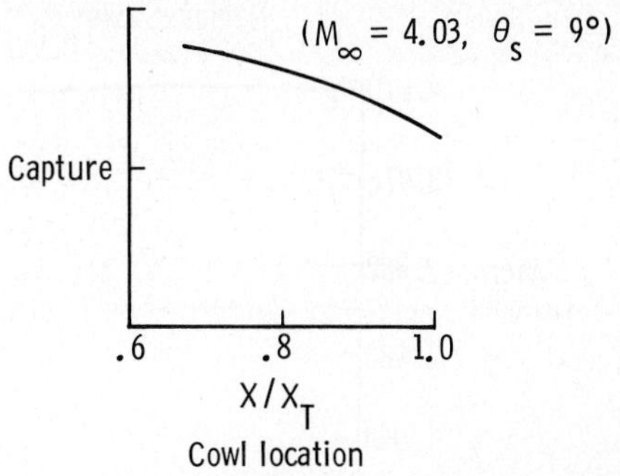

Fig. 15- Inlet capture as a function of cowl location for strut compression angle, $\theta_s = 9^\circ$.

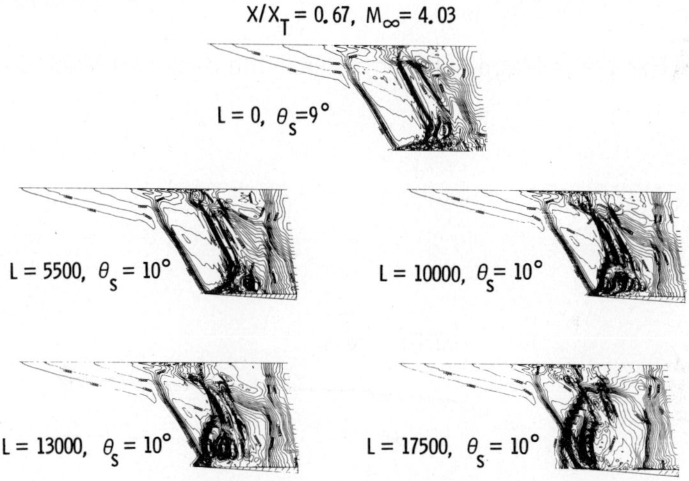

Fig. 16 - Pressure contours in the symmetry plane of the inlet.

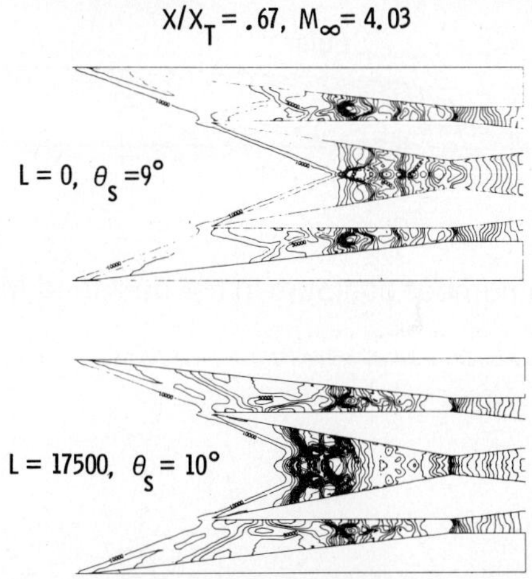

Fig. 17 - Pressure contours in a cross plane located at 12% of the inlet height from the cowl plane.

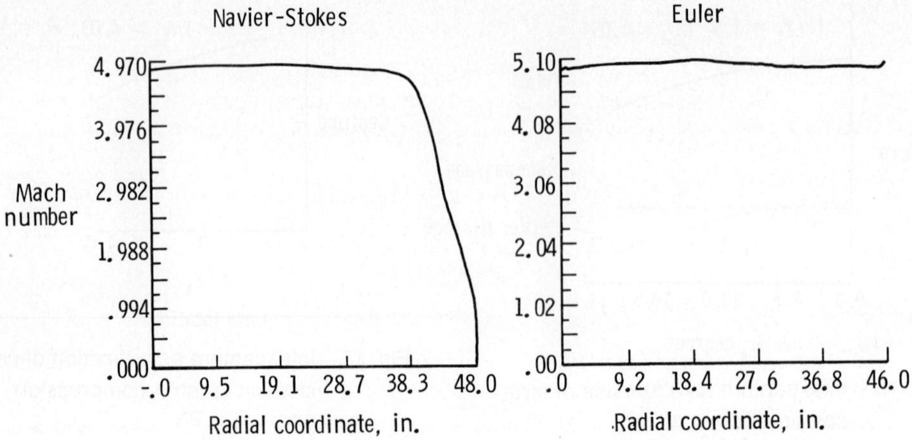

Fig. 18 - Exit plane Mach number profile in the designed Mach 5 nozzle.

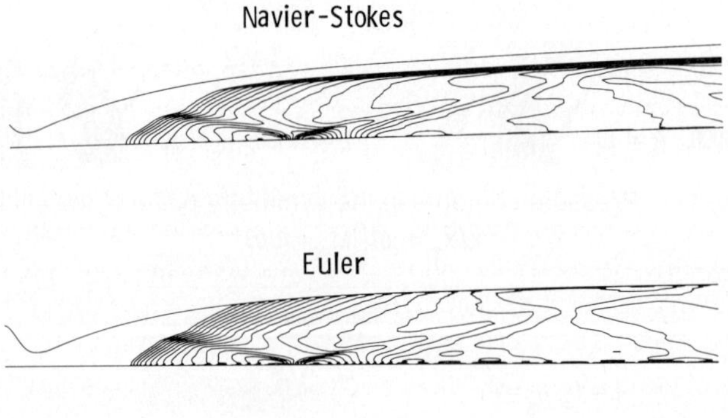

Fig. 19 - Mach number contours in the designed Mach 5 nozzle.

Multigrid for the Steady-State Incompressible Navier-Stokes Equations: a Survey

J. Linden, G. Lonsdale, B. Steckel, K. Stüben

GMD-F1T, Postfach 1240, D-5205 St. Augustin 1, West Germany

1. Introduction

In the last decade, a lot of work has been invested in the development of multigrid solvers for Navier-Stokes equations and considerable progress has been achieved in terms of computing times, when comparing the resulting methods with classical single-grid solvers. In this paper we review the current state of both multigrid methods and related discretization aspects for the steady-state Navier-Stokes equations. Due to the tremendous amount of related literature, we consider only the incompressible equations in primitive variable formulation

$$L\begin{pmatrix} \mathbf{u} \\ p \end{pmatrix} := \begin{pmatrix} Q[\mathbf{u}] & \nabla \\ \nabla^t & 0 \end{pmatrix}\begin{pmatrix} \mathbf{u} \\ p \end{pmatrix} = \begin{pmatrix} \mathbf{f} \\ 0 \end{pmatrix}, \quad Q[\mathbf{u}] = -\Delta + Re(\mathbf{u}\cdot\nabla), \tag{1}$$

discretized by means of finite differences or finite volumes. We aim to highlight approaches which, in our opinion, appear to provide a way forward towards the treatment of more general flow domains.

We discuss stable discretizations of the Navier-Stokes equations. An emphasis is laid on *non-staggered* grid discretizations. In multigrid development, such discretizations are far less common compared to their staggered analogue but they are, in our opinion, much more convenient in connection with general non-uniform and non-orthogonal grids (boundary fitted grids) which are the current norm when solving flow problems on complex regions. Various relaxation schemes (including ***pressure correction based methods***, ***distributive*** and ***box schemes***), which are used as smoothers in multigrid methods, are discussed. We focus on analysis and algorithmic aspects of construction of smoothing schemes, as these form the most important component of any multigrid method.

In particular, we propose a stabilized discretization for non-staggered grids as well as a generalized form of box-relaxation which turns out to be quite an effective and simple smoother and is easily extendable to general boundary fitted grids. As box-relaxation has not been analyzed in the literature before, we show the basic principle of such an analysis.

Because of space limitations, we have to confine ourselves to the most fundamental aspects, assuming that the reader is familiar with the basics of multigrid (see, e.g., [16]) as well as with the general peculiarities of the Navier-Stokes discretization. At places, we give references to papers where more details can be found. Moreover, for technical simplicity, we consider only the 2D case; an extension to three dimensions is straightforward. For the same reason, in the description of algorithms and analysis below we assume square, cartesian grids.

2. Stable discretizations

The basic requirement of any multigrid method in solving elliptic problems is the availability of a relaxation process with sufficient error smoothing properties, a necessary condition being the *stability* of the underlying discretization scheme. Moreover, as multigrid solvers employ a hierarchy of grids of different resolution, we need *mesh independent* smoothing properties. The mesh independent convergence of multigrid methods, which is their major advantage, arises from exactly this mesh independent smoothing. To obtain this, a sufficient, mesh independent "amount" of stability is needed; stability alone is not enough.

Essentially, interior stability of a difference scheme is a matter of high frequency error components. Therefore, in order to obtain an approximate insight into stability properties, one may ignore boundary conditions and investigate a given discretization on an infinite grid. A basic theoretical tool for this is *local Fourier analysis*. In [3], the so-called *h-ellipticity* measure has been introduced which represents a quantitative measure of stability. A system of difference equations is called h-elliptic if no high frequency Fourier components are annihilated. The h-ellipticity measure E_h itself is defined as

$$E_h = min\{|\lambda_h(\Theta)| \ : \ \pi/2 \leq |\Theta| \leq \pi\} \ / \ max\{|\lambda_h(\Theta)| \ : \ |\Theta| \leq \pi\} \tag{2}$$

where $\lambda_h(\Theta)$ denotes the *symbol* of the difference operator [3] with respect to the Fourier component $e^{i\Theta \cdot x/h}$ ($\Theta = (\Theta_1, \Theta_2)$, $-\pi \leq \Theta_i \leq \pi$, $|\Theta| = max\{|\Theta_1|, |\Theta_2|\}$) (for an example, see below).

In the Navier-Stokes equations there are two sources of instability. The first one is caused by the convection-diffusion part, $Q[\mathrm{u}]$, of the momentum equations (singular perturbation): for high Reynolds numbers the ellipticity measure of standard (central) discretization schemes decreases and one has to introduce additional ellipticity to keep the discrete equations stable. With respect to multigrid development, this kind of instability has been investigated rather extensively in the past, mostly by means of simple (scalar) model equations. Stability can be achieved in various well-known ways (most of them being essentially equivalent), e.g., by using upwind or hybrid schemes for the convection terms or by explicitly adding h-dependent artificial viscosity.

The second source of instability is of a quite different nature. It does not depend on the size of the Reynolds number and it arises because pressure occurs only in terms of its first derivatives. For instance, if the Navier-Stokes equations are discretized on a non-staggered grid (see Figure 1b) by means of central differencing, pressure values are directly coupled only between grid points which are at a distance of $2h$. Thus, the entire grid can be subdivided into four subgrids (in a four-colour fashion) between which the pressure values are decoupled, i.e., three highly oscillating pressure frequencies (e.g., the "checkerboard" frequency) are annihilated.

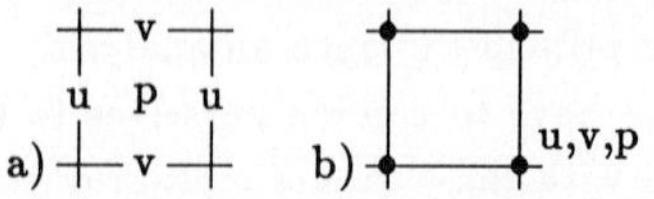

Figure 1: Staggered and non-staggered location of unknowns

There are several remedies to cure this instability. The most common one is to use a staggered distribution of unknowns (MAC-type grid, see Figure 1a) instead of a non-staggered one. The corresponding central differences are stable with respect to pressure oscillations, essentially because pressure differencing is now done with a distance of h rather than $2h$.

Another possibility is to keep the non-staggered distribution of unknowns but to add an elliptic artificial pressure term to the discrete continuity equation, e.g. $-\omega h^2 \Delta_h p_h$, giving the modified continuity equation

$$\partial_{2h}^x u_h + \partial_{2h}^y v_h - \omega h^2 \Delta_h p_h = 0 \tag{3}$$

where ∂_{2h}^x and ∂_{2h}^y denote central differences (stepsize $2h$) with respect to x and y, respectively, and Δ_h is the discrete Laplacian. The parameter ω has to be chosen small enough to maintain good accuracy but large enough that the discrete Navier-Stokes system becomes sufficiently elliptic. Note that the additional term does not change the order of consistency.

In order to obtain some guess about a proper choice of ω, we compute E_h. For simplicity, we do this only for the discrete Stokes equations

$$\begin{pmatrix} -\Delta_h & 0 & \partial_{2h}^x \\ 0 & -\Delta_h & \partial_{2h}^y \\ \partial_{2h}^x & \partial_{2h}^y & -\omega h^2 \Delta_h \end{pmatrix} \begin{pmatrix} u_h \\ v_h \\ p_h \end{pmatrix} = \begin{pmatrix} f_h^u \\ f_h^v \\ 0 \end{pmatrix} . \tag{4}$$

The symbol of the discrete Stokes operator, $\lambda_h(\Theta)$, is given by

$$\lambda_h(\Theta) = det \begin{pmatrix} -\hat{\Delta}_h(\Theta) & 0 & \hat{\partial}_{2h}^x(\Theta) \\ 0 & -\hat{\Delta}_h(\Theta) & \hat{\partial}_{2h}^y(\Theta) \\ \hat{\partial}_{2h}^x(\Theta) & \hat{\partial}_{2h}^y(\Theta) & -\omega h^2 \hat{\Delta}_h(\Theta) \end{pmatrix} = -\hat{\Delta}_h(\omega h^2 \hat{\Delta}_h \hat{\Delta}_h - \hat{\Delta}_{2h}) \tag{5}$$

where $\hat{\Delta}_h$, $\hat{\Delta}_{2h}$, $\hat{\partial}_{2h}^x$ and $\hat{\partial}_{2h}^y$ denote the symbols (formal eigenvalues on the infinite grid) of the respective scalar difference operators. For instance,

$$\begin{aligned} \Delta_h e^{i\Theta\cdot x/h} &= \hat{\Delta}_h(\Theta) e^{i\Theta\cdot x/h} \quad \text{with} \quad \hat{\Delta}_h(\Theta) = 2(cos\Theta_1 + cos\Theta_2 - 2)/h^2 , \\ \Delta_{2h} e^{i\Theta\cdot x/h} &= \hat{\Delta}_{2h}(\Theta) e^{i\Theta\cdot x/h} \quad \text{with} \quad \hat{\Delta}_{2h}(\Theta) = (cos2\Theta_1 + cos2\Theta_2 - 2)/2h^2 . \end{aligned}$$

Inserting (5) into (2), one can compute $E_h = E_h(\omega)$. Note first that $E_h(\omega)$ does *not* depend on h which means that the ellipticity of the discrete Stokes equations, stabilized by using (3), is indeed *h-independent*. To simplify the explicit computation of $E_h(\omega)$, note that only the second term of λ_h (the term in parentheses in (5)) is critical, i.e., may become small for high frequencies. Ignoring the (fully elliptic) factor $\hat{\Delta}_h$, one obtains

$$E_h(\omega) \equiv E(\omega) \sim \begin{cases} 8\omega(1-8\omega) & (\omega \le 1/16) \\ 1/4 & (1/16 \le \omega \le 1/12) \\ (1+1/4\omega)/16 & (\omega \ge 1/12) . \end{cases} \tag{6}$$

As expected, for $\omega = 0$ the ellipticity measure becomes zero. Due to (6), the smallest ω for which the measure is a maximum is $\omega = 1/16$. This value has been tested in practice and turns out to be a reasonable choice both with respect to accuracy and stability. Compared to the staggered dicretization – assuming a cartesian grid and the same number of grid cells – global discretization errors of the non-staggered version typically are larger by a factor of around 2 to 3. This is mainly due to the long-distance differencing of first order terms. At this point, we want to note that central differencing on non-staggered grids formally requires pressure values along physical boundaries. They can most easily be obtained by means of extrapolation from interior pressure values. However, in order to asymptotically obtain the same order of accuracy as for the staggered discretization, extrapolation has to be sufficiently accurate. For instance, in case of rectangular domains, linear extrapolation is sufficient to retain second order accuracy for velocity and pressure (with respect to the L_2-norm).

For the Navier-Stokes equations with low Reynolds numbers, the same ω can be used. This changes, however, if a significant amount of artificial viscosity is introduced. A more detailed analysis then shows that an appropriate choice of ω depends on the actual amount of artificial viscosity. Roughly speaking, ω should be inversely proportional to the Peclet number. This has been confirmed also by numerical tests.

The choice of staggered or non-staggered discretization depends somewhat on the application at hand. As long as one is dealing with rather simple geometries and orthogonal grids, the staggered discretization is actually superior to the non-staggered one, the major reason being the slightly better numerical accuracy mentioned above. For complex flow regions, however, discretization of the Navier-Stokes equations is usually done on non-orthogonal meshes (boundary fitted grids, see, e.g., [15], [8] and the references given there). By explicitly transforming the Navier-Stokes equations along with the velocities to the corresponding new coordinates, the staggered discretization can, in principle, be generalized also to such cases. However, long-distance differencing can no longer be avoided (either in the momentum equations or in the continuity equation, depending on whether co- or contravariant velocities are introduced as new dependent unknowns). In fact, the numerical accuracy of the staggered discretization turns out to depend more sensitively on the non-orthogonality of a grid than its non-staggered analogue. In addition, the non-staggered discretization is considerably easier to implement: the equations can be discretized directly in terms of the original *cartesian* velocity components and do not have to be transformed explicitly to the new coordinates.

Based on the above non-staggered approach, we are currently developing a parallel multigrid solver for the Navier-Stokes equations on general $2D$-domains [8]. To our knowledge, explicit artificial pressure terms as in (3) have not – in the framework of finite differences or finite volumes – been used before. We want to point out, however, that the use of such artificial terms is not new. Similar approaches are used, mostly indirectly, in several other Stokes or Navier-Stokes solvers. For instance, the SIMPLE-method for non-staggered grids as discussed in [12], can be seen to implicitly solve a discrete continuity equation of the form (3) (with $\omega = 1/8$). Also in the framework of (stabilized) finite-element approximations, artificial pressure terms are quite common: such terms are, e.g., used explicitly in [5] (with $\omega = 1$). Another popular finite-element approach uses so-called "bubble functions" (mini-elements [1], [5]). It can easily be seen that, eliminating the bubble functions from the finite-element equations (on cartesian grids) again leads to a discrete continuity equation with an artificial pressure term (and $\omega = 1/80$).

To conclude this section, we want to mention that interior stability of the non-staggered discretization can also be obtained without introducing an artificial pressure term by simply using non-central discretizations, e.g., forward differences for p and backward differences for $\mathbf{u}$ (in the continuity equation). This follows from the above analysis by replacing the first order differences in (4) by the respective one-sided ones (and setting $\omega = 0$). Ellipticity immediately follows from $\lambda_h = \hat{\Delta}_h \hat{\Delta}_h$. In contrast to the above approach, however, this discretization is always only of first order. Some results are reported in [6].

3. Overview of current smoothing schemes

Even if stability is ensured, the construction of efficient and robust smoothing schemes for the discrete Navier-Stokes equations in the primitive variable formulation is not easy. This is mainly due to the fact that the pressure field is determined by the entire system and not just by one of the equations, i.e., for the relaxation of the pressure field the coupling between momentum and continuity is essential. Several schemes have been proposed in the literature most of them belonging to the very general class of so-called *distributive relaxations*. An essentially different approach is given by *box-relaxation*. Both approaches are described briefly below.

A major step in developing smoothing schemes is the investigation of their smoothing properties. This is important not only for obtaining a rough a priori comparison of different smoothers, but also for judging the final multigrid convergence behavior (tuning and debugging). The easiest way is by local Fourier analysis as proposed by Brandt [3]. Although this analysis is not rigorous (as it assumes frozen coefficients and infinite grids) it has proved to be a useful tool in that it gives realistic estimates for multigrid smoothing and convergence (as long as perturbations from the boundaries are small and coefficients are smoothly varying). Using this technique, the (*interior*) *smoothing factor*, a measure of the reduction of high frequency error components per relaxation sweep (usually denoted by μ), can be computed. Approaches for performing the analysis will be outlined below.

3.1 Distributive smoothing

In this section, we first review the most often used distributive schemes in the form in which they are usually presented and physically motivated. Though having been developed and tested mainly for staggered grids, they can be modified for use on non-staggered grids, at least formally. For ease of presentation, the description is done in terms of the linearized momentum equations (nonlinear terms frozen to current values).

Each step of these distributive relaxations, calculating $\mathbf{u}_h^{n+1}$, p_h^{n+1} from $\mathbf{u}_h^n$, p_h^n, is of *predictor-corrector* type. In the *prediction step*, a new velocity $\mathbf{u}_h^\star$ is computed by approximately solving the momentum equations

$$Q_h \mathbf{u}_h^\star + \nabla_h p_h^n = \mathbf{f}_h \quad \text{with} \quad Q_h = Q_h[\mathbf{u}_h^n] \tag{7}$$

by performing a few standard relaxation sweeps (e.g., Gauss-Seidel). In the *correction step*, both velocity and pressure are updated

$$\mathbf{u}_h^{n+1} = \mathbf{u}_h^\star + \delta \mathbf{u}_h\,, \quad p_h^{n+1} = p_h^n + \delta p_h\,. \tag{8}$$

The aim is to compute these changes such that (i) the continuity equation is satisfied and (ii) the current momentum defect, i.e., $\mathbf{d}_h^\star = \mathbf{f}_h - Q_h \mathbf{u}_h^\star - \nabla_h p_h^n$, remains unchanged.

The various schemes proposed differ in the way in which these changes are actually computed. Fuchs and Zhao [6] use changes of the form

$$\delta \mathbf{u}_h = \nabla_h \chi_h\,, \quad \delta p_h = -Q_h \chi_h \tag{9}$$

where χ_h is a grid function defined at pressure locations. In order to satisfy (i), χ_h is determined by a Poisson equation, namely, $\Delta_h \chi_h = -\nabla_h \cdot \mathbf{u}_h^\star$. Thus, the new momentum defect is given by $\mathbf{d}_h^{n+1} = \mathbf{d}_h^\star - (Q_h \nabla_h - \nabla_h Q_h)\chi_h$. In particular, the momentum defect is maintained exactly only if Q_h and ∇_h commute which, generally, is not true for the discrete operators. For small *Re*-numbers or sufficiently fine grids, however, commutativity is nearly fulfilled (at least in grid points not adjacent to boundaries) and the momentum defects are maintained approximately which, in practice, turns out to be sufficient for good (interior) smoothing. If, however, commutativity is violated strongly, the above approach does no longer yield good smoothers and corresponding multigrid methods slow down and may even diverge.

It should be noted that, in the context of smoothing, the above Poisson equation for χ_h does not have to be solved exactly; some sweeps of a simple relaxation method like plain Gauss-Seidel are sufficient in practice [6].

The DGS (distributed Gauss-Seidel) scheme introduced by Brandt and Dinar [4] (which actually lead to the very first Navier-Stokes multigrid solver in primitive variables) can be regarded as a special case of the above approach: in DGS, while sweeping over the grid cell by cell (in the correction step), pressure and velocity changes are computed locally by using a grid function χ_h which is non-vanishing only at the center of the current grid cell (where pressure is located). This smoothing process is particularly cheap, but its effectiveness also depends sensitively on the above commutativity.

Several attempts have been made to overcome the difficulties arising from non-commutativity. Boundary influences, e.g., can be damped by using additional boundary relaxations [10]. Another approach is the PGA (pressure gradient average) scheme proposed by Fuchs [7] (in which pressure derivatives are used as dependent variables instead of the pressure itself) details of which together with some comparisons to DGS can be found in [7].

An alternative to the above χ_h-approach (9) are *pressure-correction* based schemes. There, the velocity correction $\delta\mathbf{u}_h$ is defined directly in terms of pressure changes δp_h which, due to continuity, leads to a Poisson-like equation for δp_h (the *pressure correction equation*)

$$\delta\mathbf{u}_h = D_h^{-1}\nabla_h\delta p_h \quad \text{and} \quad \nabla_h\cdot(D_h^{-1}\nabla_h\delta p_h) = -\nabla_h\cdot\mathbf{u}_h^\star \tag{10}$$

where D_h is an easy-to-invert approximation to Q_h. The new momentum defect is $\mathbf{d}_h^{n+1} = \mathbf{d}_h^\star + (Q_hD_h^{-1} - I_h)\nabla_h\delta p_h$. Thus, the better D_h approximates Q_h, the better the momentum defect is maintained. Multigrid methods based on pressure correction type smoothers (where exact solution of the pressure correction equation is replaced by a few suitably chosen relaxation sweeps) were used in [9], [2], [13].

In their most general form, distributive relaxations have been considered first by Brandt [3]; a more rigorous theoretical description can be found in [18]. The starting point for the investigation of distributive relaxations from a more general point of view was the following observation by Brandt: DGS for the Navier-Stokes equations as outlined above can *equivalently* be described as a *standard* pointwise Gauss-Seidel relaxation applied to the transformed system $L_hM_h(\widehat{\mathbf{u}}_h,\widehat{p}_h)^t = (\mathbf{f}_h,0)^t$ with

$$M_h = \begin{pmatrix} I_h & \nabla_h \\ 0 & -Q_h \end{pmatrix}. \tag{11}$$

This observation gave rise to the definition of a general class of distributive relaxations, applicable to any elliptic system $L_h\mathbf{U}_h = \mathbf{F}_h$, defined as follows:

1. Change the system $L_h\mathbf{U}_h = \mathbf{F}_h$ to a "simpler" one by a transformation of the dependent variables $\mathbf{U}_h \to \widehat{\mathbf{U}}_h$ of the form $\mathbf{U}_h = M_h\widehat{\mathbf{U}}_h$ with some suitably choosen operator M_h

(the so-called *distributor*). The resulting transformed system is just the product system $L_h M_h \widehat{\mathbf{U}}_h = \mathbf{F}_h$;

2. Choose a classical, "simple" relaxation process for the transformed system of the form $\widehat{\mathbf{U}}_h^{i+1} = \widehat{\mathbf{U}}_h^i + A_h(\mathbf{F}_h - L_h M_h \widehat{\mathbf{U}}_h^i)$ with A_h being some approximate inverse of the product operator $L_h M_h$;

3. Re-formulate this relaxation scheme in terms of the original operator and unknowns by using $\mathbf{U}_h = M_h \widehat{\mathbf{U}}_h$: $\mathbf{U}_h^{i+1} = \mathbf{U}_h^i + M_h A_h(\mathbf{F}_h - L_h \mathbf{U}_h^i)$.

This new definition covers a wide range of popular distributive relaxations for Navier-Stokes equations, in particular, the above mentioned PGA and pressure correction schemes [18]. The latter ones, e.g., are obtained by using the distributor

$$M_h = \begin{pmatrix} I_h & -D_h^{-1}\nabla_h \\ 0 & I_h \end{pmatrix}$$

(with D_h being the approximation of Q_h as in (10)) along with suitably choosen relaxation schemes for the resulting product system $L_h M_h$ (e.g. line Gauss-Seidel successively for velocity and pressure).

This unifying description of distributive relaxations (as "normal" relaxations for a properly transformed system) may serve as a general basis for the construction of smoothing schemes for Navier-Stokes equations. Moreover, it also leads to a very elegant approach for analyzing smoothing properties (by means of local Fourier analysis). For DGS, e.g., the smoothing analysis can be performed in terms of pointwise Gauss-Seidel applied to the product operator

$$L_h M_h = \begin{pmatrix} Q_h & C_h \\ -\nabla_h^t & -\Delta_h \end{pmatrix} \tag{12}$$

where M_h as in (11) and $C_h = Q_h\nabla_h - \nabla_h Q_h$ is the "commutator" whose influence has already been discussed above. Now, one easily sees that in cases in which it is justified to neglect C_h (essentially for low Re-numbers), the smoothing properties only depend on the diagonal terms Q_h, Δ_h of (12). Thus, e.g., for Stokes flow ($Q_h = -\Delta_h$), the DGS smoothing factor μ is just that of pointwise Gauss-Seidel for the Laplacian $-\Delta_h$, i.e., $\mu = 0.5$ [3].

Pressure correction based smoothing schemes have not yet been analyzed in this fashion. Smoothing results are, however, available in [13]. There, a more direct approach in applying local Fourier techniques is used which turns out to be technically considerably more complicated. As a result of this analysis, one gets, e.g., a smoothing factor of $\mu = 0.6$ for low Re-numbers (using alternating line-SOR for momentum equations and assuming the pressure correction equation to be solved exactly which is actually not necessary in practice).

3.2 Box smoothing

Compared to the distributive relaxations, the so-called *box relaxation* is conceptually much simpler. The basic idea of one sweep of box-relaxation is to solve the linearized discrete Navier-Stokes equations *locally* cell by cell ("box").

For staggered grids, Vanka [17] proposed to successively update *simultaneously* the 5 unknowns as indicated in Figure 2a, using the respective 4 momentum equations along the edges and the

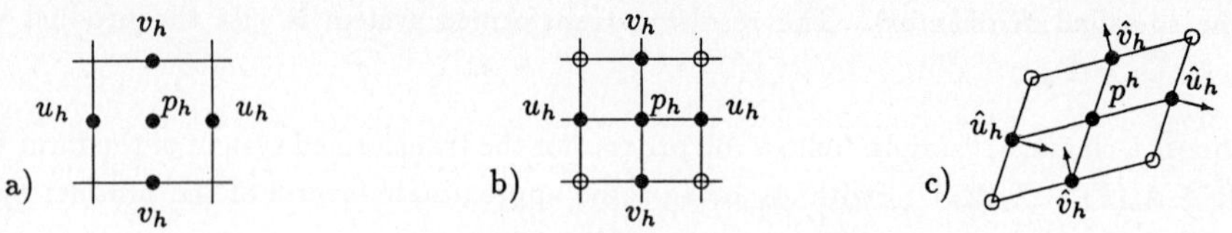

Figure 2: Unknowns updated simultaneously by box-relaxation

continuity equation in the center of the box. More precisely, for any fixed box, in the momentum equations all off-diagonal velocity components as well as the nonlinear velocity components are replaced by their most recent old values. Thus, for any box, one has to solve a 5×5 linear system of equations, with the corresponding matrix being of the very simple form

$$\begin{pmatrix} \star & 0 & 0 & 0 & \star \\ 0 & \star & 0 & 0 & \star \\ 0 & 0 & \star & 0 & \star \\ 0 & 0 & 0 & \star & \star \\ \star & \star & \star & \star & 0 \end{pmatrix} \tag{13}$$

where the $\star$'s denote non-vanishing entries.

In [8], we extended box-relaxation for non-staggered discretizations. For cartesian coordinates, it is defined in essentially the same way as above except that the edge lengths of the boxes are now twice as large (see Figure 2b). The generalization of box-relaxation to the non-staggered case mainly served as a first step towards solution on arbitrary, non-orthogonal grids for which we already pointed out that non-staggered discretizations are more convenient than staggered ones. On non-orthogonal discretization grids, we found from numerical investigations that the best way to proceed is, roughly, to update the velocities by correcting only their component *normal* to the boundary of a box (see Figure 2c). Consequently, for each box, one again has to solve only 5×5 systems of the same structure as indicated in (13), except that the bottom right element of the matrix is now non-zero due to the artificial pressure term in (3).

In case of significant anisotropies it is well known, that smoothing properties of "pointwise" oriented relaxation processes deteriorate. Such anisotropies occur, e.g., in case of strongly non-uniform grids. To still obtain good smoothing, one has to simultaneously resolve all strong couplings between unknowns. In such cases, box-relaxation should be replaced by a corresponding box-line-version. For instance, in case of x-line relaxation, this means that all unknowns marked in Figure 3 are updated simultaneously (which results in the solution of one block-tridiagonal system of equations per line).

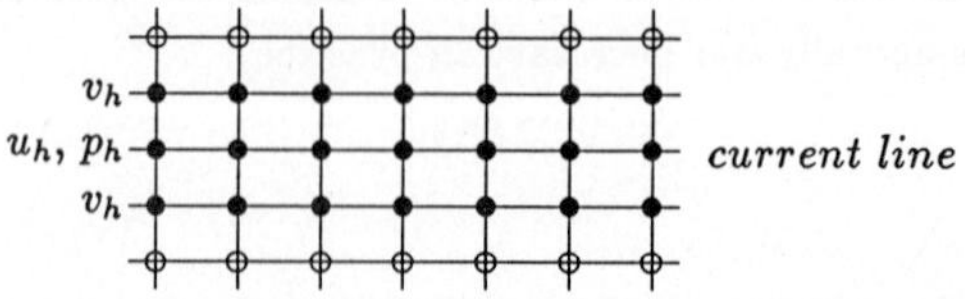

Figure 3: Unknowns updated simultaneously by box-line-relaxation

While conceptually simpler than distributive relaxation, in practice, box-relaxation turns out to have quite robust smoothing properties even for higher Re-numbers. Amazingly enough that no systematic investigation of its smoothing properties is found in the literature. Therefore, we want to show how box-relaxation can be analyzed in the framework of local Fourier analysis. Because of space limitations, we confine ourselves to the basic case of Stokes equations on non-staggered grids. Staggered grids can be treated similarly. An extension to Navier-Stokes equations can be done by treating worst (frozen coefficient) cases.

According to the definition of box-relaxation, the velocity components are updated twice, the pressure is updated only once. Therefore, one sweep of box-relaxation, starting from some first approximations u^h, v^h, p^h yields new approximations $\overline{u}^h$, $\overline{v}^h$, $\overline{p}^h$ using some intermediate velocity components $\widehat{u}^h$, $\widehat{v}^h$.

The relation between the corresponding *error* quantities (also denoted by u, v, p and omitting the subscript h) in relaxing any fixed box is given by the discrete continuity equation

$$h^2\omega\,(4\,\overline{p}_C - \overline{p}_W - p_E - p_N - \overline{p}_S) + \frac{h}{2}\,(\widehat{v}_N - \overline{v}_S) + \frac{h}{2}\,(\widehat{u}_E - \overline{u}_W) = 0$$

and the four momentum equations

$$\begin{aligned}
(4\,\widehat{v}_N - \overline{v}_{NW} - v_{NE} - v_{NN} - \widehat{v}_C) &+ h\,(p_{NN} - \overline{p}_C)/2 &= 0\,,\\
(4\,\overline{v}_S - \overline{v}_{SW} - v_{SE} - \overline{v}_{SS} - \widehat{v}_C) &+ h\,(\overline{p}_C - \overline{p}_{SS})/2 &= 0\,,\\
(4\,\overline{u}_W - \overline{u}_{WW} - \widehat{u}_C - \widehat{u}_{NW} - \overline{u}_{SW}) &+ h\,(\overline{p}_C - \overline{p}_{WW})/2 &= 0\,,\\
(4\,\widehat{u}_E - \widehat{u}_C - u_{EE} - u_{NE} - \widehat{u}_{SE}) &+ h\,(p_{EE} - \overline{p}_C)/2 &= 0\,.
\end{aligned}$$

Here, for any fixed box, the subscripts denote the point locations according to the Figure below. Furthermore, we assume the order of relaxing the boxes to be columnwise (from bottom to top) with the columns being ordered from left to right.

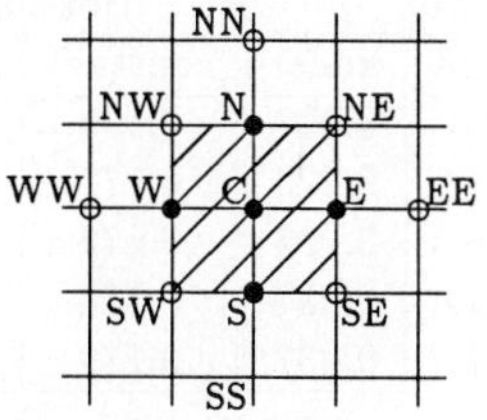

Assuming that $(u,v,p)^t = e^{i\Theta x/h}(a,b,c)^t = \varphi(\Theta)$ is some Fourier frequency, one easily sees that the above infinite system of equations has solutions of the same form, namely, $(\overline{u},\overline{v},\overline{p})^t = e^{i\Theta x/h}(\overline{a},\overline{b},\overline{c})^t$ and $(\widehat{u},\widehat{v})^t = e^{i\Theta x/h}(\widehat{a},\widehat{b})^t$. Inserting these frequencies into the above equations and eliminating the resulting relations for $\widehat{a}$, $\widehat{b}$, one obtains

$$\begin{pmatrix}\overline{a}\\ \overline{b}\\ \overline{c}\end{pmatrix} = M(\Theta,h)\begin{pmatrix}a\\ b\\ c\end{pmatrix}$$

with some 3×3-matrix M. The amplification factor $\mu(\Theta)$ of $\varphi(\Theta)$ is defined as the spectral radius of $M(\Theta,h)$ (which turns out not to be independent of h). Using $\omega = 1/16$ in (3), one computes a smoothing factor of $\mu = \max\{\mu(\Theta) : \pi/2 \le |\Theta| \le \pi\} = 0.625$. A similar analysis for box-line-relaxation gives $\mu = 0.564$.

Smoothing properties can be improved further by using underrelaxation for the pressure. For a corresponding analysis replace $\overline{p}_C$ by $(\overline{p}_C - (1-\sigma)p_C)/\sigma$ in the above five equations where σ

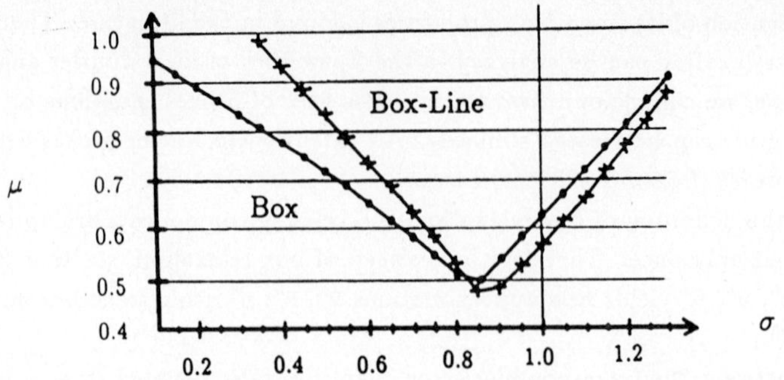

Figure 4: Smoothing factors for box-relaxation and box-line-relaxation

is some relaxation parameter. Figure (4) shows the behavior of the smoothing factor as a function of σ both for the box-relaxation and its linewise counterpart.

As a demonstration of the real multigrid behavior, Table 1 contains average convergence factors of multigrid $W(2,1)$-cycles for Stokes equations, observed in practice (on a cartesian $N \times N$ grid using $\omega = 1/16$, $\sigma = 1$ and standard inter-grid transfers) both for box- and box-line-relaxation as smoother and for constant and linear pressure extrapolation at physical boundaries. Ideally, according to the smoothing factors given above, one would expect a convergence factor per cycle of around 0.244 ($\approx 0.625^3$) and 0.179 ($\approx 0.564^3$) in case of box-smoothing and box-line-smoothing, respectively. This is in perfect agreement with the results shown in the table which indicates that the multigrid iteration works optimally; there is neither a negative influence from the boundaries nor from the coarser grids.

	Box		Box-line	
N	constant	linear	constant	linear
16	0.215	0.237	0.173	0.140
32	0.217	0.223	0.177	0.131
64	0.218	0.235	0.179	0.130
128	0.241	0.237	0.178	0.138
256	0.241	0.237	0.179	0.142

Table 1: Average convergence factors: $W(2,1)$-cycle for solving Stokes problem.

4. Concluding remarks

The multigrid algorithms described above all fall into the category of what one might call *direct* multigrid for the Navier-Stokes system, in that a multigrid method is constructed for the system as a whole. Quite often, single grid codes are already in existence and it might be easier to replace only a time-consuming component by an efficient multigrid solver. For example, this "acceleration" approach has been reported by several authors (see [9], [14] and the references therein) in conjuction with the SIMPLE method [11] which is a single-grid iterative method based on the pressure correction scheme described in Section 3.1. Within this iteration, accurate solution of the pressure correction equation (10) is required. Thus a standard, efficient multigrid method (for Poisson-like equations) is used instead of many single-grid relaxation sweeps for the solution of this equation, to achieve the required accuracy at reduced cost. Although this approach does indeed give speed-ups compared with the original SIMPLE method, the convergence of the *global* iteration – which, in fact, is extremely slow – will be unchanged. The increase in efficiency is, therefore, very limited.

Full multigrid efficiency, characterized by mesh-independent convergence factors, can only be achieved using the direct approach. However, at present, there is no method which is both robust *and* efficient in all circumstances. On the other hand, in particular situations one can make definite recommendations. For simple domains, for which it is natural to use staggered, cartesian grids, the DGS smoother leads to a highly efficient multigrid method – provided that the *Re*-number is not too large. Box smoothers are more robust, particularly for higher *Re*-numbers, but are approximately twice as expensive. Even more expensive are the generalized distributive schemes which are conceptually more complicated and, moreover, are somewhat sensitive to the details of implementation. For complex geometries the choice of method is less clear. In particular, it should be suited for non-orthogonal, boundary-fitted coordinates and, as previously motivated, for non-staggered grids. Our experience indicates that the generalized (non-staggered) box scheme, in combination with the artificial pressure stabilization, is a promising possibility.

An alternative approach for dealing with complex geometries has been proposed [3], in which cartesian grids are used in the interior of the domain combined with local refinement and orthogonal, boundary fitted grids close to the boundaries. While this approach would lead (theoretically) to an improvement in efficiency and would also allow the use of staggered grids, it has only been used for simple model problems to date. Its effectiveness for realistic flow simulation has yet to be proved.

Finally, we at least want to mention the aspect of parallelizability. With the availability of new multiprocessor machines (MIMD, local memory), with hundreds of autonomous computing units working together, there is a good chance to drastically reduce the tremendous amount of computing time needed for solving the Navier-Stokes equations for "real-life" problems. However, in order to exploit the power of such machines, the parallelizability of numerical methods becomes a major question in algorithm development. Here, multigrid methods play an outstanding role because they are not only "optimally efficient" on sequential computers (i.e., the numerical work to solve a problem is proportional to the number of unknowns) but also their components (relaxation, interpolation, restriction) are by their nature fully parallelizable. For some discussion on parallel aspects of multigrid methods and further references, see, e.g., [8].

References

[1] Abdalass, E.M.: *Resolution performante du probleme de Stokes par mini-elements, maillages auto-adaptatifs et methodes multigrilles – applications.* Ph.D. Thesis, Ecole Centrale de Lyon, 1987

[2] Barcus, M.; Peric, M.; Scheuerer, G.: *A control volume based full multigrid procedure for the prediction of two-dimensional, laminar, incompressible flows.* Proceedings of the 7th GAMM-Conference on Numerical Methods in Fluid Mechanics, Louvain-la-Neuve, Belgium, Vieweg-Verlag, 1987

[3] Brandt, A.: *Multigrid Techniques: 1984 Guide with Applications to Fluid Dynamics.* Monograph, GMD-Studien 85, 1984

[4] Brandt, A.; Dinar, N.: *Multigrid solutions to elliptic flow problems.* In "Numerical Methods for PDE" (Parter, S.V., ed.), Academic Press, 1984

[5] Brezzi, F.; Pitkäranta, J.: *On the stabilization of finite-element approximations of the Stokes equations.* In "Efficient Solutions of Elliptic Systems" (Hackbusch, W., ed.), Notes on Numerical Fluid Mechanics 10, Vieweg-Verlag, 1984, pp.11-19

[6] Fuchs, L.; Zhao, H.S.: *Solution of three-dimensional viscous incompressible flows by a multigrid method.* Int. J. Numer. Methods in Fluids 4, 1984, pp. 539-555

[7] Fuchs, L.: *Multigrid schemes for incompressible flows.* In "Efficient Solution of Elliptic Systems" (Hackbusch, W., ed.), Notes on Numerical Fluid Mechanics 10, Vieweg Verlag, 1984, pp. 38-51

[8] Linden, J.; Steckel, B.; Stüben, K.: *Parallel multigrid solution of the Navier-Stokes equations on general 2D-domains.* To appear in: Parallel Computing 1988, North Holland (in press)

[9] Lonsdale, G.: *Solution of a rotating Navier-Stokes problem by a nonlinear multigrid algorithm.* J. Comp. Phys. 74, 1988, pp. 177-190

[10] Niestegge, A.; Witsch, K.: *Analysis of a multigrid Stokes solver.* To be submitted to AMC

[11] Patankar, S.V.: *Numerical heat transfer and fluid flow.* Hemisphere, Washington DC - New York, 1980

[12] Peric, M.: *A finite volume method for the prediction of three-dimensional fluid flow in complex ducts.* Ph.D. Thesis, Mech. Eng. Dept., Imperial College, London, 1985

[13] Shaw, G.J.; Sivalagonathan, S.: *On the smoothing properties of the SIMPLE pressure-correction algorithm.* Int. J. Numer. Methods Fluids 8, 1988, pp.441-461

[14] Solchenbach, K.; Trottenberg, U.: *On the multigrid acceleration approach in computational fluid dynamics.* GMD-Arbeitspapier 251, 1987

[15] Stüben, K.; Linden, J.: *Multigrid methods: An overview with emphasis on grid generation processes.* Proceedings "Numerical Grid Generation in Computational Fluid Dynamics" (Häuser, J.; Taylor, C., eds.), Pineridge Press, Swansea, 1986, pp. 483-509

[16] Stüben, K; Trottenberg, U.: *Multigrid methods: Fundamental algorithms, model problem analysis and applications.* In "Multigrid Methods" (Hackbusch, W.; Trottenberg, U., eds.), Lecture Notes in Mathematics 960, Springer-Verlag, 1982, pp. 1-176

[17] Vanka, S.P.: *Block-implicit multigrid solution of Navier-Stokes equations in primitive variables.* J. Comp. Phys. 65, 1986, pp.138-156

[18] Wittum, G.: *Distributive Iterationen für indefinite Systeme als Glätter in Mehrgitterverfahren am Beispiel der Stokes- und Navier-Stokes-Gleichungen mit Schwerpunkt auf unvollständigen Zerlegungen.* Dissertation, University of Kiel, 1986

A SURVEY OF UPWIND DIFFERENCING TECHNIQUES

P.L. Roe
College of Aeronautics
Cranfield Institute of Technology
Bedford MK43 OAL, U.K.

1. PREAMBLE

Algorithm improvement is no less important today than when Dean Chapman demonstrated to this Conference in 1980 the equal roles played by hardware and software in developing the utility of CFD [1]. Where do good algorithms come from? The sources of inspiration may lie in mathematics, or in computer science, or in physics. Respectively these disciplines might direct our thoughts to spectral methods, multi-grid or upwinding. Of course, whatever the primary inspiration we cannot neglect the others, but in CFD there is one over-riding requirement. The "computational fluid" that we create must have "dynamics" as close as possible to those of the real fluid.

2. NUMERICAL ADVECTION

Modelling the dynamics is easier for some problems than for others. Evolution problems in which the spatial derivatives are of even order merely damp out or amplify any waves present in the initial data. Numerical schemes based on central differences share this property, and the only error to be eliminated is an amplification error. Odd order derivatives translate the waves, preserving their amplitude in simple cases. A numerical scheme will suffer both an amplification and a phase error (dissipation and dispersion). It is easy to devise schemes without dissipation errors, but this is not a good idea. The classical example is the leapfrog scheme, where any high frequency component of the data propagates undamped with a totally incorrect velocity, completely destroying, after sufficient time, any initial waveform except a pure sine wave [2,3].

The same disease is suffered to some extent by any linear scheme for solving the simple advection equation

$$u_t + au_x = 0 \qquad (1)$$

which has even order truncation error. For such schemes, dispersion always dominates dissipation [2], leading to pronounced 'wiggles' in the solution. The art of solving equ. (1) lies in balancing dissipation with dispersion, in which respect odd-order truncation is more desirable. The superiority of odd-order schemes emerges clearly from numerical experiments reported by Leonard [4].

To be more specific, consider schemes for solving (1) on a regular grid, of the form

$$u_i^{n+1} = \Sigma c_k u_{i+k}^n \qquad (2)$$

where u_i^n approximates $u(i\Delta x, n\Delta t)$, and the c_k are coefficients satisfying the accuracy conditions.

$$\sum_{k=r}^{k=s} k^p c_k = (-\upsilon)^p \qquad (3)$$

for $p = 0,1,\ldots q$, where q is the order of the truncation error and $\upsilon = a\Delta t/\Delta x$ is Courant number. In what follows we assume a, hence υ, positive and treat the contrary case by symmetry. Schemes which offer the most accuracy for a given amount of computation are those for which $q = r+s$. If q is even, such schemes are stable [5] only if $r = s$ (central differencing) or $r = s+2$ (two points with upwind bias). If q is odd, they are stable only if $r = s+1$. Fig 1 shows how some of these schemes perform on simple step data, simulating a shockwave. Note that $r = 1$, $s = 1$ is the Lax Wendroff scheme; $r = 2$, $s = 0$ is second-order upwinding.

These results are typical of every advection study known to the writer. Any attempt to derive an optimum scheme for solving (1), under any constraints, leads to a preference for odd-order methods and biassed stencils. I do not know of one which leads to a symmetrical scheme having the same properties for each direction of flow. In such a scheme the coefficients c_k would be polynomial functions of ν satisfying $c_k(\nu) = c_{-k}(-\nu)$. The most famous example of an asymmetric optimisation was Godunov's [6] result concerning the monotonicity preserving scheme with smallest truncation error. It is the first order upwind scheme shown on Table 1 (also the r = 1, s = 0 scheme in Fig 1). Godunov's paper is also epoch-making in that it marks the beginning of a quest for adequate definitions of 'quality' in CFD. This still elusive property is only partially revealed by truncation errors and Fourier analysis. Godunov's original objective was for a scheme that could be guaranteed to produce a monotone solution from monotone data, but many quite different desiderata could be proposed.

For example, Wesseling sought, among the class of schemes with r = s = 2, the one that most closely reproduces a step function after one iteration [7]. It turned out to be Fromm's scheme (see Table 1). Note that this optimisation rejected one of its offered degrees of freedom, by putting $c_2 = 0$. Fromm's scheme gives results very close to those of the (r = 2, s = 1) third-order scheme; also shown in Table 1.

TABLE 1

	c_{-2}	c_{-1}	c_0	c_1
Upwind	0	ν	$1-\nu$	0
Fromm	$-\nu(1-\nu)/4$	$\nu(5-\nu)/4$	$1-\nu(3+\nu)/4$	$-\nu(1-\nu)/4$
Third-order	$-\nu(1-\nu^2)/6$	$\nu(2-\nu)(1+\nu)/2$	$1-(1-\nu^2)(2-\nu)/2$	$-\nu(2-\nu)(1-\nu)/6$

These three linear schemes are the chief examples to occur in practice. However, Godunov's analysis also shows that no linear scheme of second-order accuracy or better can preserve monotonicity. This theorem stands like a dragon protecting the realm of hyperbolic problems. Whoever would solve them accurately must first slay it in battle. Harten [8] has reviewed the emphasis this places on schemes that are non-linear, even when applied to (1).

Typical of these are flux-limiter schemes [9,10,11] in which (1) is solved in conservation form,

$$u_i^{n+1} = u_i^n - \frac{\Delta t}{\Delta x}[f^n_{i+\frac{1}{2}} - f^n_{i-\frac{1}{2}}]$$

with

$$f^n_{i+\frac{1}{2}} = a[u_i^n + \tfrac{1}{2}(1-\nu)B(\Delta^n_{i-\frac{1}{2}}, \Delta^n_{i+\frac{1}{2}})] \tag{4}$$

where B is some (generally non-linear) average of the two differences indicated ($\Delta^n_{i+\frac{1}{2}} = u^n_{i+1} - u^n_i$). Choosing the first argument leads to the r = 2, s = 0 upwind scheme, and choosing the second gives Lax-Wendroff. Choosing the arithmetic mean gives the Fromm scheme. More general averages are able to overcome Godunov's theorem. Suppose we insist that u_i^{n+1} lies between u_i^n and u_i^{n+1} (call this the interpolation property). This is a stronger condition than the often quoted TVD constraint [11,12], and avoids certain anomalies. This places certain constraints on the averaging function B, most easily expressed by noting that on dimensional grounds B must be a homogenous function of its arguments, and can therefore be written as

$$B(\Delta i-\tfrac{1}{2}, \Delta i+\tfrac{1}{2}) = \Delta_{i+\frac{1}{2}}\phi\,\frac{(\Delta i-\frac{1}{2})}{(\Delta i+\frac{1}{2})} = \Delta i+\tfrac{1}{2}\,\phi(r_{i+\frac{1}{2}}) \tag{5}$$

It is easy to show that the interpolation property implies that the function (r) satisfies.

$$-\frac{2}{\nu} \leq \phi(r) - \frac{\phi(r)}{r} \leq \frac{2}{1-\nu} \tag{6}$$

This is a necessary and sufficient condition. It is also frequently quoted as a sufficient condition for the TVD property. It implies further necessary and sufficient conditions

$$-k \leq \phi \leq \frac{2}{1-\nu} - k \tag{7a}$$

$$-k \leq \frac{\phi}{r} \leq \frac{2}{\nu} - k \tag{7b}$$

In the earliest studies it was not realised that taking $k \neq 0$ was a significant generalisation. Fig 2a, 2b show 'limiter diagrams'. Fig 2a, in which $k = 0$, has been most explored. Taking $\phi(r)$ close to the permitted boundary gives excellent resolution of discontinuities; in fact, if $\phi(r)$ lies on the boundary, they remain optimally resolved for any number of iterations. However, such limiter functions turn smooth data into a series of step functions (staircasing). See the examples in [4]. They may, however, be used as interface trackers in multi-media problems. A limiter scheme is second-order accurate if and only if

$$\phi(1 + \varepsilon) = 1 + O(\varepsilon) \tag{8}$$

for example van Leers 'harmonic mean' limiter

$$\phi = \frac{r + |r|}{1 + r} \tag{9}$$

Such limiters are found experimentally to diffuse discontinuities over a width proportional to $t^{1/3}$ [10] but do a reasonable job of transporting smooth data. An unexplained observation is that they show remarkably little dispersion; the main errors are a slow loss of peak amplititde. An attempt to combine sharp discontinuities with good continuous transport is Roe's 'Superbee' limiter [14]. This combines second-order accuracy in smooth regions with discontinuities that remain of finite width for all time. However, it does seem very difficult to find all desirable properties simultaneously within this class of scheme. There is a local loss of accuracy near extreme values, and this tends to accumulate.

Taking $k \neq 0$ introduces room to manoeuvre when $r < 0$ (precisely the condition that obtains near extrema). The limiter H shown in Fig 26 represents Harten's second-order ENO scheme (of which more later) which greatly improves performance on smoothly oscillating data. A possible drawback here (and to other limiters) is that introducing non-differentiable factions like this will invalidate many techniques for rapidly progressing to the steady state, such as Newton's method. Van Albada [15] introduced the limiter V, defined by

$$\phi = \frac{r + r^2}{1 + r^2} \tag{10}$$

Although only very simple ideas are involved, more research is still needed to clarify the properties of limiter functions.

3. RIEMANN SOLVERS

Any improvement in techniques for solving linear advection can be applied fairly directly to one-dimensional conservation laws of the form

$$\underset{\sim}{u}_t + \underset{\sim}{F}(\underset{\sim}{u})_x = 0 \tag{11}$$

Godunov's original scheme [6] followed a line of reasoning that has since been generalised profitably. He supposed that the values u^n represent average values of the conserved variables u (mass, momentum and energy in fluid dynamics) in the cell at time $n\Delta t$. At the beginning of each time step, these average values are represented by constant states within each cell. He then solved exactly the problem of how two such constant states interact; i.e. he solved the Riemann problem, which is the initial-value problem for (11) with $u(x,0) \equiv u_L$ if $x \leq 0$, or u_R if $x>0$.

Whatever the states $\underset{\sim}{u}_L, \underset{\sim}{u}_R$, and whatever the "flux function" $\underset{\sim}{F}(\underset{\sim}{u})$, there is no characteristic length to this problem. The solution, if one exists, is therefore self-similar $\underset{\sim}{u}(x,t) = \underset{\sim}{u}(x/t)$ (see Fig 3a). Existence and uniqueness of the solution are non-trivial questions, but for the well-studied case of ideal gas dynamics, there is always a unique solution [16]. Finding a solution is intimately linked with the study of centred simple waves. These are given implicity by

$$x/t = \xi = \lambda(\underset{\sim}{u}) \tag{12}$$

where λ is an eigenvalue of the Jacobian matrix $A(\underset{\sim}{u}) = \partial F/\partial u$ equal to a local wave-speed. Also involved are shockwaves, across which the jump relationship

$$\xi = \Delta F/\Delta u \qquad (13)$$

holds with ξ now the shock speed, and constant states. One way to view the Riemann problem is as a search for a path in 'phase space' $(\underset{\sim}{u})$ linking $\underset{\sim}{u}_L$ to $\underset{\sim}{u}_R$ by subpaths which obey either (12) or (13), and such that a unique value of u corresponds to each ξ. For present purposes only the solution at $\xi = 0$ is needed. Denote the flux there by $\underset{\sim}{F}^*(\underset{\sim}{u}_L,\underset{\sim}{u}_R)$. Then Godunov's method is obtained by integrating round 1234 in Fig 3b.

$$\underset{\sim}{u}_i^{n+1} = \underset{\sim}{u}_i^n - \frac{\Delta t}{\Delta x}[\underset{\sim}{F}^*(\underset{\sim}{u}_i^n, \underset{\sim}{u}_{i+1}^n) - \underset{\sim}{F}^*(\underset{\sim}{u}_{i-1}^n,\underset{\sim}{u}_i^n)] \qquad (14)$$

For linear advection this is indeed the optimum monotonicity preserving scheme.

The computation of F^* for ideal gasdynamics is generally an iterative process involving the shockwave and simple wave equations. By reputation this is an expensive business, but maybe not so dear as commonly supposed. A thorough study by Gottlieb et al [17] recommends an initial guess based on assuming simple waves for both the outer disturbances. This gives a closed form solution for the inner region which is exact if the assumption is true, and a second-order estimate otherwise, with a built-in error indication. The expense involved in this first iteration is one logarithm and two exponentials, which is all that will be needed in the majority of any flow. A fast Newton iteration will give the solution elsewhere. However, for more general equations of state, analytic forms of the jump and wave relationships may not be available, and then the expense of an exact solution becomes very great.

Alternative ways to tackle the Riemann problem become essential in such cases, and are of interest in the ideal case also. We will mention first Osher's solution, which passes in phase space from the left state $\underset{\sim}{u}_L$ to the right state $\underset{\sim}{u}_R$ by paths composed entirely of simple waves, (12). Along such paths

$$dF = \lambda(\underset{\sim}{u})\, du \qquad (15)$$

so the flux difference $\underset{\sim}{F}_R - \underset{\sim}{F}_L$ may be expressed as

$$\Delta\underset{\sim}{F} = \underset{\sim}{F}_R - \underset{\sim}{F}_L = \int_{\underset{\sim}{u}_L}^{\underset{\sim}{u}_R} \lambda(\underset{\sim}{u})\, d\underset{\sim}{u} \qquad (16)$$

However the simple wave paths are chosen, (16) is an identity. Osher and Solomon [18] split the total path into two disjoint sets, depending on the sign of λ. Positive wavespeed is used to indicate a right-going disturbance, so the right going part of the flux-difference is

$$\Delta\vec{\underset{\sim}{F}} = \tfrac{1}{2}\int_{\underset{\sim}{u}_L}^{\underset{\sim}{u}_R}(\lambda(\underset{\sim}{u}) + |\lambda(\underset{\sim}{u})|)\, d\underset{\sim}{u} \qquad (17)$$

with a corresponding expression for the left-going part. The resulting expression for the interface flux is

$$\underset{\sim}{F}^*(\underset{\sim}{u}_L,\underset{\sim}{u}_R) = \tfrac{1}{2}(\underset{\sim}{F}_L + \underset{\sim}{F}_R) - \tfrac{1}{2}\int_{\underset{\sim}{u}_L}^{\underset{\sim}{u}_R} |\lambda(\underset{\sim}{u})|\, d\underset{\sim}{u} \qquad (18)$$

This manipulation can be carried out for any choice of wavepath, but the integral in (18) is path-dependent. The original choice of Osher and Solomon involves reversing the natural order (Fig 4a). With that choice (O-variant) they could prove that the numerical scheme (14),(18) always gives monotone change through a steady shock. Although the integral can be found analytically, there is considerable expense involved. Speckreijse [19] finds the typical computational cost to be six logarithms and seven exponentials, considerably more than the exact solution. He prefers the natural path (Fig 4b) described as the P-variant, which involves the same algebraic expense as Gotlieb's first approximation to the exact solution.

The O-variant cannot be justified as a cheap replacement for the exact solution; its interest lies in providing a scheme with better numerical properties than the exact solution. The flux formula to which it leads (18) is a differentiable function of u_L, u_R (Speckreijse lists the derivatives) whereas the Godunov flux has a discontinuous derivative if a shockwave lies along $\xi = 0$. As mentioned earlier, differentiability is an important issue for implicit schemes. Also, Roberts [20] has investigated the phenomenon of noise radiated from slowly-moving shockwaves, first observed by Woodward and Collella [21]. The minimum width of a numerical shock transition is either Δx or $2\Delta x$, depending on the phase of the shock relative to the mesh. Slowly moving shocks therefore pulsate. If the problem is a nonlinear system, waves may radiate along the other characteristics (Fig.5). The phenomenon depends very little on Courant number, and only slightly on Mach number, (H.C.Lin, private communication). When the exact Riemann solution is used, the maximum amplitude is about 3% of the shock jump, for shocks that move with one or two percent of the characteristic speed. Presumably this has rather a damaging effect on the late stages of convergence to a steady state. According to Roberts [20] the phenomenon almost disappears for the O-variant of Osher's scheme, but is strongly present in the P-variant. Here we have another example of an approximate Riemann solver having better properties than the exact solution. Maybe it would be more exact to speak of alternative Riemann solvers.

An alternative for which economy is the chief motivation is my own Riemann solver [22]. This exploits the fact that we can easily solve the Riemann problem for any linear system of equations; we therefore replace (11) by the local linearisation

$$\underset{\sim}{u}_t + A_{i+\frac{1}{2}}(\underset{\sim}{u}_i, \underset{\sim}{u}_{i+1})\underset{\sim}{u}_x = 0 \tag{19}$$

where $A_{i+\frac{1}{2}}$ is a locally constant Jacobian, constructed to have the crucial property

$$A_{i+\frac{1}{2}}(\underset{\sim}{u}_i, \underset{\sim}{u}_{i+1})(\underset{\sim}{u}_{i+1} - \underset{\sim}{u}_i) = \underset{\sim}{F}_{i+1} - \underset{\sim}{F}_i \tag{20}$$

The key property bestowed by (20) is that the solution to the linearised problem (19) coincides with the solution to the exact problem whenever this involves merely a single shock or a single contact. Computational experience is that this property makes the associated numerical scheme very robust.

An apparent drawback of the method is that because of the linearisation no allowance is made for the spreading of expansion waves. This can lead to non-physical 'expansion shocks' appearing in the computed flows. Two cures are available, each of them still inexpensive. Harten [12] adds dissipation to slowly moving waves. (H.C.Lin reports that this also cures the noise problem, from which the Roe method otherwise suffers). My own preference [23,14] is to add dissipation proportional to the strength of any expansion. This in fact spreads expansion waves more strongly, and more correctly, than using the exact solution, reinforcing earlier comments. As CFD is applied more and more ambitiously it becomes increasingly important (because increasingly less obvious) that the numerical schemes can be relied upon to converge to the proper phsyical (entropy-satisfying) solutions. Rigorous treatment of this point for scalar problems [24,25] is being extended to systems [26].

The only expensive operation involved in Roe's linearisation applied to unsteady ideal gasdynamics is one square root. This is a useful, but perhaps not decisive advantage. What is more significant is that similar inexpensive linearisations are available for many other types of problem, although no completely unified way of carrying through the algebra has emerged. The original paper introduced a parameter vector $\underset{\sim}{w}$ such that $\underset{\sim}{u}, \underset{\sim}{F}$ are both quadratic functions of $\underset{\sim}{w}$. This approach has led to linearisations for steady ideal gasdynamics [22] and unsteady general relativistic gas dynamics [27] Assuming that $A_{i+\frac{1}{2}} = A(\underset{\sim}{u}_{i+\frac{1}{2}})$ (the exact Jaocbian evaluated at some average state) leads to linearisations for the shallow water equations [28] and for ideal magnetohydrodynamics [29]. Direct approximation of the eigenvalues and eigenvectors of $A_{i+\frac{1}{2}}$ was introduced for ideal gasdynamics in [30], and can also be used for reactive flows (L.G.Simmonds, private communication). It has been used to provide linearisations valid for arbitrary equations of state. Subtly different solutions are to be found in [31,32,33,34]. See also [35]. Practical testing will perhaps discriminate between the alternatives.

In all linearised cases the interface flux is given by

$$F^*(\underset{\sim}{u}_L, \underset{\sim}{u}_R) = \frac{1}{2}(\underset{\sim}{F}_L + \underset{\sim}{F}_R) - \frac{1}{2}\sum \alpha^{(k)}|\lambda^{(k)}|\underset{\sim}{r}^{(k)} \tag{21}$$

Other linearisations are due to Huang [36], and Lombard [37]. The first of these does not have the ability to recognise strong waves; the second is obtained by factorising A into a series of simpler matrices. The two-dimensional steady version has been given by Dick [38].

What I like to call Very Approximate Riemann solvers were introduced in [39], and developed in [40,41]. If it is assumed that only two waves are present in the Riemann solution (Fig.6) and that the speeds of these are a_L, a_R, then an interface flux can be derived as

$$F^*(u_L, u_R) = \frac{F_L a_R - F_R a_L}{a_R - a_L} + \frac{a_L a_R}{a_R - a_L}(u_R - u_L) \tag{22}$$

If we take $a_R = -a_L = \Delta x/\Delta t$ this returns the Lax-Friedrichs scheme (very diffusive) but improved estimates for a_L, a_R result in better schemes. The cost of such solvers is very small, but in their simple form they cannot resolve contact discontinuities.

This disadvantage is shared by the Flux-Vector Splitting approach [42,43]. Here the function $F(u)$ is divided into 'positive' and 'negative' parts

$$F(u) = F^-(u) + F^+(u) \tag{23}$$

such that the Jacobian $\partial F^+/\partial u$ has no negative eigenvalues, and vice versa. The associated Riemann solution is

$$F^*(u_L, u_R) = F^+(u_L) + F^-(u_R) \tag{24}$$

The earliest versions [42] were insufficiently smooth at sonic points. This was corrected by van Leer [43], and further improved by Hanel [44]. However, the inability to recognise shear waves (in multidimensional applications) seems to disqualify these simpler Riemann solution as foundations for Navier-Stokes codes [45]. The perfect Riemann solver for generating numerical schemes probably does not yet exist. Properties of cheapness,robustness,range of application and (in some not yet fully defined sense) smoothness are all important. There is an irritating problem of coping with near vacuum conditions [46]. Goodman and Leveque [47] replace Roe's locally constant Jacobian with a piecewise constant Jacobian that permits spreading expansions. Bell et al [48] begin by solving the Riemann problem as a series of jumps, then uses interpolation to detect their structure. Perhaps a sythesis of these diverse approaches will take place in the near future.

4. MORE GENERAL SCHEMES FOR SYSTEMS

Advection schemes and Riemann solvers are the twin pillars upon which practical upwind schemes are built. There are three main techniques for putting them together. The first is purely formal, and is most readily applied when combined with a linearised Riemann solution [49]. The interface flux (21) is replaced by

$$F^* = \tfrac{1}{2}(F_L + F_R) - \tfrac{1}{2}\sum \alpha^{(k)} |\lambda^{(k)}| r^{(k)} + \tfrac{1}{2}\sum \phi^{(k)} \alpha^{(k)} (1 - \tfrac{\Delta t}{\Delta x} |\lambda^{(k)}|) \lambda^{(k)} r^{(k)}$$

The first two terms give the first-order correction. The limiter function ϕ is the same function of $\alpha^{(k)}$ and the neighbouring wavestrengths that is involved in some chosen advection scheme. The neighbours are chosen to left or right according to the sign of $\lambda^{(k)}$.

The other technique has a more physical basis. The keywords are Reconstruction, Evolution, Projection. The idea is, given average values of the conserved variables within each cell, to reconstruct the underlying data that would give rise to those averages. The reconstruction should be simple enough to permit its evolution to be computed; usually the reconstruction within each cell is polynomial. So far we have been considering piecewise constant reconstruction. The first attempts to go beyond this were the piecewise linear MUSCL scheme of van Leer [50] and the piecewise parabolic method (PPM) of Woodward and Colella [21]. Lack of space restricts us to discussion of the MUSCL scheme. The reconstructed function describing the conserved cariables in the ith cell is (see Fig.7)

$$\underset{\sim}{R}_i^n(x) = \underset{\sim}{u}_i + \underset{\sim}{S}_i^n(x - x_i) \tag{26}$$

where S_i^n is some slope determined inside the ith cell. For example, we can take the arithmetic mean of $(\underset{\sim}{u}_{i+1}-\underset{\sim}{u}_i)$ with $(\underset{\sim}{u}_i-\underset{\sim}{u}_{i-1})$ leading to central differencing, or we can take some non-linear average, as discussed under flux limiters. More elaborate reconstructions [21,51] use higher order polynomials. The reconstructions in adjacent cells will not in general be continuous across the interface (although they will tend to be more so as the order of the reconstructions increases), and one way to evolve the solution is to solve generalised Riemann problems, in which the classical data is extended to include non-uniform left and right states [52]. (Other generalisations incorporate source terms [53,54] or reactive terms [55].

Alternatively, the reconstructed flow in each cell can be advanced independently of its neighbours. For a linear reconstruction, this is trivial; for the more general case, the Cauchy-Kowalewsky construction is used [51]. At each interface, there is now a different function $\underset{\sim}{u}(t)$ defined on either side. For a second-order scheme, it is sufficient to evaluate these half-way through the time-step, and then find the interface fluxes in (14) by solving the Riemann problem (by any preferred method) for this data. See [51] for the more general case. The description "projection" is justified by the fact that in every case what is computed by (14) is a new average value in each cell. The details of the old reconstruction are lost, and a new reconstruction will be made.

The third method is to compute the interaction between the reconstructions at the beginning of each time step, yielding a value of $\partial\underset{\sim}{u}/\partial t$ (the semi-discrete residual) within each cell. Runge-Kutta methods may be used to advance in time. Chosen carefully they do not destroy monotonicity properties [56]. For finding steady states, the residual becomes the right-hand side of an implicit operator [57,58,59].

5. UPWINDING VIA FINITE ELEMENTS

Finite element methods, for the present purpose, will be distinguished as those in which the flow is represented (everywhere, not just at nodes) as the sum of a finite number of trial functions, and in which the weak form of the governing equations is satisfied exactly when the solution is tested against some class of test functions. For example, in linear advection, the data may be represented as

$$u(x,n\Delta t) = \Sigma \; c_i \phi_i(x) \tag{27}$$

where each $\phi_i(x)$ is a hat function (Fig 8a). The weak form of (1) is

$$\int_{-\infty}^{\infty} w_j(x) \; [u_t + \Sigma \; c_i \; \phi_i'(x)]dx = 0 \tag{28}$$

Equation (28) is required to be satisfied for every value of i, and every test function $w_j(x)$ in a certain class. If the $w_j(x)$ are chosen also to be hat functions, we have the classical Galerkin method, leading in this case to an unstable central differencing scheme. Upwind bias can be introduced by choosing alternative test functions, a procedure known generically as the Petrov-Galerkin technique. The choice

$$w_j(x) = \phi_j(x) + \tau a \phi_j'(x) \tag{29}$$

has been developed by Hughes and collaborators [60,61]. Here τ is a constant with the dimensions of time, serving to bias the integral in (28) in favour of the upwind half of the trial function. A similar bias is obtained by choosing

$$w_j(x) = \phi_j(x + a\tau) \tag{30}$$

This leads to the characteristic Galerkin method [62,63] in which the test function is the trial function displaced along the characteristics of the problem. Note that (29) is an expansion of (30), to first order in τ. These are the only two techniques known to the author in which upwinding is incorporated fundamentally into the finite element machinery.

6. MULTIDIMENSIONAL CALCULATIONS

These are either very hard, or very easy, according to the viewpoint adopted. Most multidimensional calculations make use of a quadrilateral mesh aligned with the surface of the obstacle. One approach then is to adopt an operator-splitting (or dimension-by-dimension) technique [64], where the individual operators act along rows and columns of the mesh. Interface flux formulae may be used that are slight modifications of (25), where the quantities $\alpha,\lambda,\underset{\sim}{r}$, now derive from solving Riemann problems for the flow normal to each cell interface. Algorithm details, with impressive results, are given in [66,67]. Any other one-dimensional technique can be used in the same way. Yee [66] also uses upwinding as a 'postprocessor', to a MacCormack-type scheme. The difference between the flux formula (25) and a Lax-Wendroff flux is

$$\Delta \underset{\sim}{F} = \tfrac{1}{2}\sum(\phi^{(k)}-1)\left(1-\frac{\Delta t}{\Delta x}|\lambda^{(k)}|\right)\alpha^{(k)}\underset{\sim}{r}^{(k)} \tag{31}$$

which is a third-order diffusion term. Equ (31) and a counterpart in the other direction are used to replace the artificial diffusion.

A more directly multidimensional version of the MUSCL approach [67] is to estimate inside each cell both x and y-gradients, limiting them if need be. Then each cell is updated by $\tfrac{1}{2}\Delta t$, and the average fluxes across the boundary are found for the whole timestep solving Riemann problems normal to the cell faces. Similar ideas can be used on unstructured (triangular or irregular) grids [68]. At first glance, such schemes avoid the possible errors inherent in operator-splitting, but in fact there is a crucial assumption still made that only wave motion normal to the faces need be considered. However, since oblique waves will not be correctly recognised by such considerations [69], the physical basis of the numerical scheme is undermined. Indeed no entirely adequate explanation has been given of why the methods work as well as they undoubtedly do. Part of the explanation appears to be [69] that the 'splitting error' is largely a high frequency affair, and is averaged out by finite-volume methods. Where the one-dimensional methods are too successful, in preserving high frequencies, the multidimensional results deteriorate [70]. These observations throw into doubt the Reconstruction-Evolution-Advection approach as a basis for multi-dimensional calculations if the evolution stage uses only one-dimensional physics.

The potential benefits of a properly multidimensional approach were shown by Davis [71], who computed a steady shock reflection problem using velocity gradients as an indicator of wave orientation. He found greatly improved results compared with the dimension-by-dimension approach (which for some reason seems to do better on unsteady problems). More truly multidimensional wave recognition techniques have been proposed [72,73]. The development of algorithms based on them presents some distinctive problems [74]. The finite-element method of Hughes [61] does not directly attempt to recognise wave patterns, but introduces diffusion terms related to the flow gradients.

7. STEADY STATE CONVERGENCE

Fast convergence to the steady state is a desirable property in many applications. Studies of idealised problems usually show that upwind schemes do very well in this regard, rapidly removing errors to the domain boundary [36,75]. Practical experience is often the reverse [65,66], and possibly the 'noise' phenomenom [20] is responsible. Another possibility is that limit cycles occur; they have been found in model problems [76]. Accelerated convergence via multigrid techniques is frequently attempted; the particular link between multigrid and upwinding is most closely explored in [19].

8. CONCLUSIONS AND APOLOGIES

The number of theoretically significant papers on this topic is at least twice the number that I have been able to mention. I have tried to select material that contributes to what I see as the mainstream of upwind methodology, but I am guiltily aware of many omissions. From the material presented I believe the following conclusions can be drawn.

The case for upwind schemes is overwhelming when simple model problems are studied. The question is whether the expenditure remains worthwhile for more complex flows. For moderately nonlinear steady problems the case is probably not yet made. For highly non-linear transient flows I believe that upwind schemes are establishing themselves as superior, although the perfect scheme surely remains to be found. Progress is still needed in the design of limiter functions, finding Riemann solvers with desirable properties, efficient multidimensional techniques, and convergence acceleration. Because upwind schemes stay so close to the physics of the underlying problem, making that progress should be considerably eased.

REFERENCES

1. Chapman, D. R., 1981, Lecture Notes in Physics, 141, pp. 1-11, Springer.
2. Trefethen, L. N., 1986, in Numerical Analysis, eds. Griffiths and Watson, pp. 200-219, Longman.
3. Iserles, A., 1986, IMA J. Num. Anal., 6, pp. 381-392.
4. Leonard, B. P., 1988, NASA Lewis, ICOMP Report, to appear.
5. Strang, G., Iserles, A., SIAM J. Num. Anal., 20, pp. 1251-1257.
6. Godunov, S. K., 1959, Mat Sbornik, 47, pp. 357-393.
7. Wesseling, P., 1973, J. Eng. Math., 7, pp. 1-31.
8. Harten, A., 1987, in Wave Motion, Theory, Modelling, and Computations, ed. Chorin, Springer.
9. Roe, P. L., Baines, M. J., 1982, Notes on Numerical Fluid Mechanics, 5, ed. Viviand, pp. 281-290, Vieweg.
10. Roe, P. L., Baines, M. J., 1984, Notes on Numerical Fluid Mechanics, 7, eds. Pandolfi, Piva, pp. 283-290, Vieweg.
11. Sweby, P. K., 1984, SIAM J. Numer. Anal., 21, pp. 995-1011.
12. Harten, A., 1983, J. Comp. Phys., 49, pp. 357-393.
13. Osher, S., Chakravarthy, S., 1983, J. Comp. Phys., 50, pp. 447-481.
14. Roe, P. L., 1985, Lectures on Applied Mathematics, 22, ed. Sommerville et al., pp. 163-194, Am. Math. Soc.
15. van Albada, G. D., et al., 1982, Astron. Astrophysics, 108, pp. 76-84.
16. Smoller, J., 1983, Shock Waves and Reaction-Diffusion Equations, Springer.
17. Gottlieb, J. J., et al., 1987, preprint, Univ. of Toronto.
18. Osher, S., Solomon, F., 1982, Math. Comput., 38, pp. 339-374.
19. Hemker, P. W., Spekreijse, S., 1986, Appl. Num Math., 2, pp. 475-493.
20. Roberts, T. W., 1988, Proc. ICFD Conference, Oxford, to appear.
21. Colella, P., Woodward, P., 1984, J. Comp. Phys., 54, pp. 174-201.
22. Roe, P. L., 1981, J. Comp. Phys., 43, pp. 357-372.
23. van Leer, B., 1981, SIAM J. Sci. Stat. Comp., 5, pp. 1-20.
24. Osher, S., 1984, SIAM J. Num. Anal., 21, pp. 217-235.
25. Tadmor, E., 1984, Math. Comp., 43, pp. 369-381.
26. Tadmor, E.. 1987, Math. Comp., 49, pp. 91-103.
27. Eulderink,F,1988, Astronomy and Astrophysics, submitted.
28. Glaister, P., 1987, Univ. of Reading, Num. Anal. Report, 9/87.
29. Brio, M. Wu, C. C., 1988, J. Comp. Phys., 75, pp. 400-422.
30. Roe, P. L., Pike, J., 1984, Computing Methods in Applied Science and Engineering, North-Holland.
31. Vinokur, M., Liu, Y., 1988, AIAA Paper 88-0127.
32. Glaister, P., 1988, J. Comp. Phys., 74, pp. 382-408.
33. Grossman, B., Walters, R. W., 1987, AIAA Paper 87-1117 CP.
34. Liou, M-S., van Leer, B., Shuen, J-S., 1988, ICOMP Report 88-7.
35. Colella, P., Glaz, H., 1985, J. Comp. Phys., 59, p. 264.
36. Huang, L. C., 1981, J. Comp. Phys., 42, p. 195-211.
37. Lombard, C., Oliger, J., Yang, J. Y., 1982, AIAA Paper 82-0976.
38. Dick, E., 1988, J. Comp. Phys., 26, pp. 19-32.
39. Harten, A., Lax, P. D., van Leer, B., 1983, SIAM Review, 25, pp. 35-61.
40. Davis, S. F., 1988, SIAM J. Sci. Stat. Comp., 9, pp. 445-473.
41. Einfeldt, B., 1988, SIAM J. Num. Anal., 23, p. 294.
42. Steger, J. L., Warming, R. F., 1981, J. Comp. Phys., 40, p. 395.
43. van Leer, B., 1982, Lecture Notes in Physics, 170, p. 507.

44. Hanel, P., Schwane, R., Seider, G., 1987, AIAA Paper 87-1105 CP.
45. van Leer, B., Thomas, J. L., Roe, P. L., Newsome, R. W., 1987, AIAA Paper 1104 CP.
46. Sjorgreen, B., 1988, preprint, Univeristy of Uppsala
47. Goodman, J. B., Leveque, R., 1988, SIAM J. Num. Anal., 23, p. 268.
48. Bell, J. B., Colella, P., Trangenstein, J. A., 1987, preprint, UCRL-97258.
49. Roe, P. L., 1986, Annual Review of Fluid Mechanics, pp. 337-365, Annual Reviews.
50. van Leer, B., 1977, J. Comp. Phys., 23, pp. 276-293.
51. Harten, S., et al., 1987, J. Comp. Phys., 71, pp. 231-303.
52. Bourgeade, A., Le Floch, Ph., Raviart, P. A., 1987, Lecture Notes in Mathematics, 1270, eds. Carasso, et al., Springer.
53. Glimm, J., Marshall, G., Plohr, B., 1984, Adv. App. Math., 5, pp. 1-30.
54. Glaz, H., Liu, T-P., 1984, Adv. App. Math., 5, pp. 111-146.
55. Ben-Artzi, M., 1986, preprint UCLA.
56. Shu, C-W., Osher, S., 1987, ICASE Report 87-33.
57. Mulder, W. A., van Leer, B., 1985, J. Comp. Phys., 59, p. 232.
58. Yee, H. C., Warming, R. F., Harten, A., 1985, J. Comp. Phys., 57, pp. 327-360.
59. Thomas, J. L., van Leer, B., Walters, R. W., 1985, AIAA Paper 85-1680.
60. Hughes, T. J. R., Mallet, M., Mizukami, A., 1986, Comp. Meth. in App. Mech. and Eng., 54, pp. 341-355.
61. Hughes, T. J. R., Mallet, M., 1986, Comp. Meth. in App. Mech. and Eng., 58, pp. 329-336.
62. Morton, K. W., 1984, Oxford Univ. Comp. Lab. Rep. 8411.
63. Morton, K. W., Sweby, P. K., 1987, J. Comp. Phys., 73, pp. 203-230.
64. Strang, G., 1968, SIAM J. Num. Anal., 5, pp. 506-517.
65. Lytton, C. C., 1987, J. Comp. Phys., 73, pp. 395-431.
66. Yee, H. C., 1986, Seminar on Computational Aerodynamics, ed. M. Hafez, AIAA, also NASA TM 89464.
67. Borrel, M., Montagné, J. L., 1985, AIAA Paper 85-1497.
68. Billey, V., et al., 1987, Lecture Notes in Mathematics, 1270, eds. Carrasso, et al., Springer.
69. Roe, P. L., 1987, ICASE Report 87-64.
70. Toro, E. F., Roe, P. L., 1988, ICFD Conference, Oxford, to appear.
71. Davis, S. F., 1984, J. Comp. Phys., 56, pp. 65-92.
72. Roe, P. L., 1986, J. Comp. Phys., 63, pp. 458-476.
73. Deconinck, H., Hirsch, C., Peuteman, J., 1986, Lecture Notes in Physics, 264, pp. 216-221.
74. Roe, P. L., 1987, Cranfield, COA Report 8720.
75. Engquist, B., Gustafsson, B., 1987, Math. Comp., 49, pp. 39-64.
76. Roe, P. L., van Leer, B., 1988, ICFD Conference, Oxford, to appear.

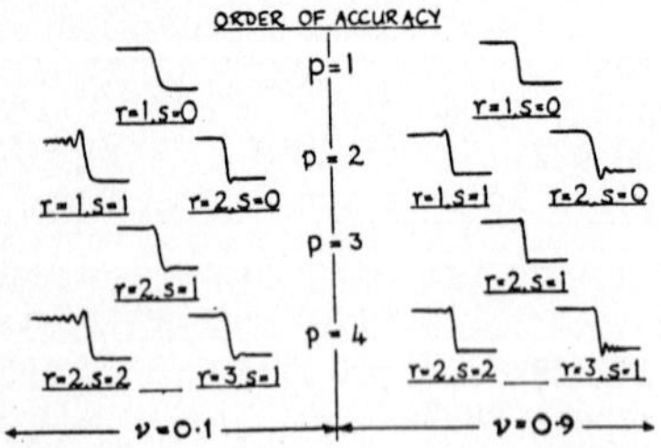

Fig. 1

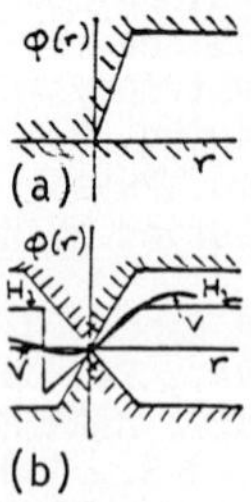

Fig. 2

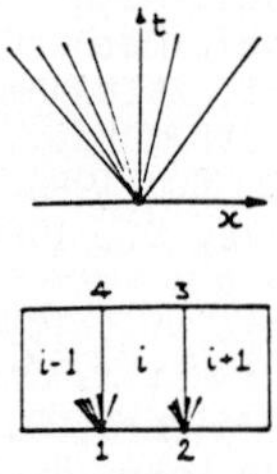

Fig. 3

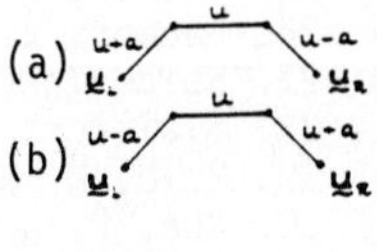

Fig. 4

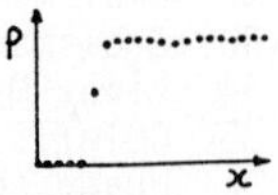
Fig. 5

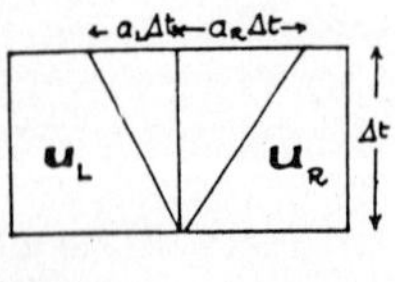
Fig. 6

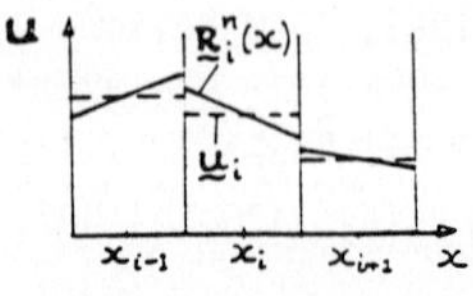
Fig. 7

Fig. 8

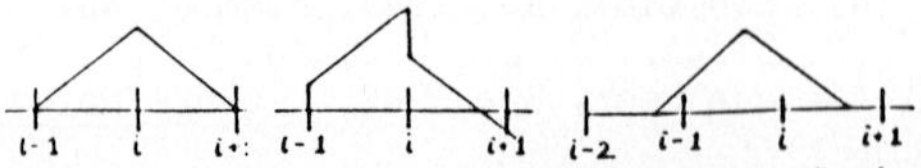
(a) $\phi_i(x)$ (b) eqn. (29) (c) eqn. (30)

DYNAMICAL SYSTEMS, TURBULENCE AND THE NUMERICAL SOLUTION OF THE NAVIER-STOKES EQUATIONS

R. Temam[1]

INTRODUCTION

Our aim in this article is to describe new numerical algorithms that are well suited for the solution of the Navier-Stokes equations over large intervals of time. Although we investigate here the case of incompressible flows, the methods can be extended to other flows such as thermohydraulics and reactive flows or, as well, compressible flows.

The understanding and the numerical solution of the Navier-Stokes equations are major problems of fluid mechanics with important implications in engineering and in fundamental research : the study of compressible or incompressible flows with or without reactive, electromagnetic or thermal phenomena necessitates among other things the understanding of the Navier-Stokes equations.

On the other hand the considerable increase of the computing power due to the appearance of supercomputers has made possible the solution of numerical problems that were unthinkable a few years ago. In the case of the Navier-Stokes equations we can now consider ranges of values of the physical parameters that are close to values physically relevant (at least in two space dimensions) and we can consider the onset of turbulence. In particular we can study turbulent flows that are time dependent, the laminar stationary solution being unstable ; and the computation of such flows necessitates clearly a large (theoretically infinite) time integration of the Navier-Stokes equations.

One of the first difficulties encountered in the solution of these equations and which appears even when the flow is laminar, is the treatment of the incompressibility condition

$$\text{div } u = 0. \tag{0.1}$$

There are now several methods that are available for the treatment of (0.1) and that are well suited for computing stationary solutions and laminar flows. In particular, among the many important contributions of N.N. Yanenko to numerical analysis and computational fluid dynamics, the *fractional step method* (or splitting-up method) that he has introduced and contributed to develop is one of the classical available methods (see [RY][Y1,Y2]).

[1]Laboratoire d'Analyse Numérique, Université Paris-Sud, Bâtiment 425, 91405 Orsay (France), and Institute for Applied Mathematics and Scientific Computing, Bloomington, Indiana (U.S.A.)

The fractional step method has also laid A.J. Chorin [C2] and R. Temam [T3] to introduce the *projection method.* Also although this is far from the preoccupations of this article we should like to mention a recent application of the fractional step method to liquid crystal problems (see R. Cohen *et al.* [CHKL], R. Cohen [C0]) connected to the extension of the fractional step method to constrained optimization (see J.L. Lions and R. Temam [LT1][LT2]).

When we reach turbulent regimes new difficulties arise. In particular an essential aspect of turbulent flows is the relation/interaction between small and large eddies. All the frequencies of the spectrum, up to the Kolmogorov dissipation frequency k_d interact ; large eddies break into small eddies and those, in turn, feed the large eddies. Besides the usual difficulties (incompressibility, nonlinearity, large Reynolds number), a new difficulty occuring in computing high Reynolds fluid flows is the interaction of small and large eddies : small eddies are negligible at each given time but their cumulative effect is not negligible on a large interval of time. Thus *a proper and economical treatment of small eddies is necessary for large time computations.* The algorithms that we propose here, called nonlinear Galerkin methods, stem from recent developments in dynamical systems theory and are motivated by this preoccupation.

This article is organized as follows. In Section 1 we study the interaction of small and large eddies in a turbulent flow. In Section 2 we present the simplest nonlinear Galerkin method while Section 3 contains another (more involved) version of the method. Section 4 gives some theoretical justifications of the methods related to recent developments in the dynamical system approach to turbulence. Finally Section 5 presents the results of some *spectral* numerical computations based on the nonlinear Galerkin method in Section 2 ; these results show for a given accuracy, a significant gain in computing time (of the order of 20 % to 40 %).

CONTENTS.

1. INTERACTION OF SMALL AND LARGE EDDIES.

We consider in space dimension $n = 2$ or 3 the incompressible Navier-Stokes equations with density 1 :

$$\frac{\partial u}{\partial t} - \nu \Delta u + (u.\nabla) + \nabla p = f, \tag{1.1}$$

$$\nabla \cdot u = 0. \tag{1.2}$$

Here u is the velocity vector, p the pressure ; $\nu > 0$ is the kinematic viscosity and f represents volume forces.

For the sake of simplicity we restrict ourselves to space periodic flows and assume that the average flow vanishes. Assuming that the period is 2π in each direction we expand u in Fourier series

$$u = \sum_{j \in Z^n} u_j e^{ij \cdot x}, \tag{1.3}$$

where $n = 2$ or 3, $j = \{j_1, j_2\}$or $\{j_1, j_2, j_3\}$and $u \in C^n$, $u_{-j} = \bar{u}_j$. Of course $u = u(x,t)$ and $u_j = u_j(t)$.

At each time t the kinetic energy of the flow $u(t)$ is

$$\frac{1}{2} \| u(t) \|^2 = \frac{1}{2} \int_\Omega | u(x,t) |^2 \, dx = \frac{(2\pi)^n}{2} \sum_{j \in Z^n} | u_j(t) |^2, \tag{1.4}$$

where Ω is the cube $(0, 2\pi)^n$. The enstrophy is

$$\| \nabla u(t) \|^2 = \int_\Omega | \text{grad } u(x,t) |^2 \, dx = (2\pi)^n \sum_{j \in Z^n} | j |^2 | u_j(t) |^2, \tag{1.5}$$

$| j |^2 = | j_1 |^2 + | j_2 |^2$ or $| j_1 |^2 + | j_2 |^2 + | j_3 |^2$.

We denote by λ_r, $r \in \mathcal{N}$, the eigenvalues of the Stokes operator, i.e. $| j |^2, j \in Z^n$, ordered in a sequential increasing order. We select a cut-off value m and decompose u into a finite part

$$y_m = P_m u = \sum_{|j|^2 \leq \lambda_m} u_j(t) e^{ijx}, \tag{1.6}$$

and a tail

$$z_m = Q_m u = \sum_{|j|^2 \geq \lambda_{m+1}} u_j(t) e^{ijx}. \tag{1.7}$$

Of course y_m corresponding to the small eigenvalues represent the *large eddies* of the flow, while z_m corresponding to the large eigenvalues of the flow represent the *small structures.*

By projection of (1.1) onto the space of divergence free functions, the pressure gradient disappears and we obtain the functional form of the equation, namely

$$\frac{\partial u}{\partial t} - \nu \Delta u + B(u,u) = f \tag{1.8}$$

where for two vector fields v, w,

$$B(v,w) = \Pi\{(v.\nabla)w\}, \tag{1.9}$$

$\Pi\varphi$ corresponding to the divergence free component of φ in the Helmotz decomposition of a vector field φ.

Now we write

$$u = y_m + z_m$$

and project equation (1.8) (or (1.1)) onto the first m modes (small wavelengths) and onto the other modes (large wavelengths). This yields a coupled system of equations for y_m and z_m :

$$\frac{\partial y_m}{\partial t} - \nu\Delta y_m + P_m B(y_m + z_m, y_m + z_m) = P_m f \tag{1.10}$$

$$\frac{\partial z_m}{\partial t} - \nu\Delta z_m + Q_m B(y_m + z_m, y_m + z_m) = Q_m f. \tag{1.11}$$

Here P_m and Q_m are the Fourier series truncations corresponding to the modes $|\ j\ |^2 \leq \lambda_m$ and $|\ j\ |^2 \geq \lambda_{m+1}$ (see (1.6),(1.7)).

Behavior of small eddies.

Some computations which will not be reproduced here show that if m is sufficiently large, the small eddies component z_m carries only a small part of the kinetic energy (and of the enstrophy). More precisely after a transient time of the order of $\kappa(\nu\lambda_{m+1})^{-1}$, we have

$$\| z_m(t) \| \leq \left(\begin{array}{l} K\delta L^{1/2} \text{ in dimension 2} \\ \\ K\delta^{2/3} \text{ in dimension 3,} \end{array} \right. \tag{1.12}$$

where

$$\delta = \frac{\lambda_1}{\lambda_{m+1}}, \quad L = 1 + \log \frac{1}{\delta} \tag{1.13}$$

and K is a nondimensional number depending only on the Reynolds number Re, and of upper bounds M_0, M_1 of the kinetic energy and the enstrophy :

$$\begin{aligned} &\frac{1}{2} \| u(t) \|^2 \leq M_0, \ \forall t, \\ &\| \nabla u(t) \|^2 \leq M_1, \ \forall t. \end{aligned} \tag{1.14}$$

The Reynolds number is that based on the norm of $f, \| f \|$, which has the dimension of the square of the velocity, and the typical length 1 (or 2Π) :

$$Re = \frac{\| f \|^{1/2}}{\nu} \tag{1.15}$$

In fact, in the space-periodic case considered here the norm of $\| z_m(t) \|$ decays much faster than indicated in (1.12), namely at an exponential rate

$$\| z_m(t) \| \leq K_1 \exp\left(-K_2 m^{1/n}\right), \quad n = 2 \text{ or } 3, \tag{1.16}$$

This follows readily from the properties of Fourier series coefficient of an analytic function. Explicit values of K_1 and K_2 can be derived from [FT2].

A simplified interaction law.

A simplified interaction law of small and large eddies is obtained by taking into account the following facts :

(i) As recalled before z_m is small compared to y_m (which is of the same order as u itself) :

$$\| z_m \| << \| y_m \| \simeq \| u \| = \quad \| y_m + z_m \| .$$

(ii) The relaxation time for the small eddies of the order of $(\nu \lambda_{m+1})^{-1}$, is much smaller than that of the large eddies (of the order of $(\nu \lambda_1)^{-1}$).

Taking into account these observations, we can simplify (1.10),(1.11). Due to (ii) it seems legitimate to neglect $\partial z_m / \partial t$ in (1.11) and, due to (i), to approximate $B(y_m + z_m, y_m + z_m)$ by $B(y_m, y_m)$. Hence (1.1) yields

$$-\nu \Delta z_m + Q_m B(y_m, y_m) \simeq Q_m f. \tag{1.17}$$

Relation (1.17) appears as an adiabatic (or quasistatic) law of interaction between small and large eddies.

Returning then to (1.10), we observe that

$$B(y_m + z_m, y_m + z_m) = B(y_m, y_m) + B(y_m, z_m) + B(z_m, y_m) + B(z_m, z_m),$$

and

$$\| B(y_m + z_m, y_m + z_m) \| \simeq$$
$$\simeq \| B(y_m, y_m) \| >> \| B(y_m, z_m) \| + \| B(z_m, y_m)) \| >> \| B(z_m, z_m) \| .$$

We could of course approximate again $B(y_m + z_m, y_m + z_m)$ by $B(y_m, y_m)$, but this has the undesirable effect of destroying the interaction of small eddies on the large one. Hence we can either leave (1.10) unchanged or replace it by

$$\frac{\partial y_m}{\partial t} - \nu \Delta y_m + P_m\{B(y_m, z_m) + B(z_m, y_m)\} + P_m B(y_m, y_m) \simeq P_m f. \tag{1.18}$$

More rigorous justifications of (1.17),(1.18) (or (1.18),(1.10)) were derived in [FMT]; the arguments will be sketched in Section 4. Also Section 5 contains some numerical evidences supporting this model. For the moment we shall show how one can infer from (1.17), (1.18) a new type of Galerkin approximation of the Navier-Stokes equations.

2. THE NONLINEAR GALERKIN METHOD.

The eigenfunctions for the Stokes operator are just the exponential functions

$$\alpha_j e^{ijx}, \quad j \in Z^n, \ \alpha_j \in \mathcal{R}^n, \quad | \alpha_j | = 1.$$

In space dimension 2 we simply have

$$\alpha_j = \frac{\tilde{j}}{| j |}, \ \tilde{j} = \{j_2, -j_1\}, \ j = \{j_1, j_2\}.$$

The corresponding eigenvalue is $| j |^2$ and, as indicated before, we number in sequential order the eigenvalues and the eigenfunctions ; for r integer we denote by w_r the eigenfunction corresponding to the eigenvalue λ_r.

For m fixed the Galerkin method based on $w_1, ..., w_m$, reduces to one of the spectral methods advocated in [GO][CHQZ]. With the notations above the approximate functions u_m is of the form

$$u_m(x,t) = \sum_{k=1}^{m} g_{km}(t) w_k(x) \tag{2.1}$$

and satisfies the approximate equation

$$\frac{\partial u_m}{\partial t} - \nu \Delta u_m + P_m B(u_m, u_m) = P_m f. \tag{2.2}$$

The nonlinear Galerkin method based on the simplified model (1.10),(1.12) proceeds as follows : we consider $2m$ eigenfunctions w_k and write (see (2.1)) :

$$u_{2m} = y_m + z_m \ ^{(1)} \tag{2.3}$$

$$y_m(x,t) = \sum_{k=1}^{m} g_{km}(t) w_k(x), \quad z_m(x,t) = \sum_{k=m+1}^{2m} g_{km}(t) w_k(x). \tag{2.4}$$

The equations satisfied by y_m, z_m are (compare to (1.10),(1.11)) :

$$\frac{\partial y_m}{\partial t} - \nu \Delta y_m + P_m\{B(y_m) + B(y_m, z_m) + B(z_m, y_m)\} = P_m f \tag{2.5}$$

---

(1) Note that the quantities denoted here y_m, z_m are not the same as in Section 1.

$$-\nu\Delta z_m + (P_{2m} - P_m)B(y_m, y_m) = (P_{2m} - P_m)f. \tag{2.6}$$

At this point we emphasize the fact that (2.3),(2.4) is by no mean a modeling of turbulence : we are working with the full Navier-Stokes equations ; and *the point of view is that some terms have been indentified as (very) small and we just set them equal to 0 instead of ordering the computer to handle very small quantities.* The situation is reminescent of the uncomplete Cholesky factorization in linear algebra.

With an estimate similar to (1.12) one can show that at each given time, after a transient period, z_m is small compared to y_m. Hence $u_m(t)$ and $y_m(t)$ are very close at each time t. However, on a large interval of time, the effects of the z_m add up and produce significant changes ; this is one of the aspects of sensitive dependence to initial data which is well known in nonlinear dynamics.

We call the algorithm (2.5),(2.6) a nonlinear Galerkin method : if we set $z_m = 0$ in (2.5), then $u_m = y_m$ and (2.5) reduces to (2.2). Thus (2.5) appears as a modification of (2.2) with some nonlinear correction terms $z_m = z_m(y_m)$ which are "explicitly" determined in terms of y_m by resolution of the Stokes problem (2.6). Hence the terminology.

The convergence of the nonlinear spectral-Galerkin method (2.5),(2.6) is proved in [MT]. As far as computing time is concerned, for a spectral method with m modes, it is known that the computing time is proportional to m^2 (in space dimension 2). Hence the computing time for a usual Galerkin method with $2m$ modes is 4 times that of the computing time for m modes. A significant part of this computing time (about 1/3) is spent in the FFT computations related to the nonlinear terms. Since the nonlinear terms in (2.5),(2.6) involve only m modes, it is about 3/4 of the computing time on the nonlinear terms which is saved with the utilization of the nonlinear Galerkin method.

The results of numerical computations performed with this nonlinear Galerkin method are reported in Section 5.

3. ANOTHER NONLINEAR GALERKIN METHOD.

The nonlinear Galerkin method (2.5),(2.6) is based on the simplified form (1.17),(1.18) of the Navier-Stokes equations. This simplified form has been derived by observing that $\| z_m(t) \|$ is small compared to $\| y_m(t) \|$ like a power of

$$\delta = \frac{\lambda_1}{\lambda_{m+1}}$$

(see (1.12)) and, a sort of small perturbation method has been used ; (2.5),(2.6) correspond to the first order perturbation terms. Now, using asymptotic expansion technics we can envisage higher order perturbation terms leading to higher order perturbations ; this program has been carried out in [T5] and our aim now is to describe another simplified interaction law between small and large eddies and the corresponding nonlinear Galerkin method.

A second order perturbation equation consists in leaving (1.10) unchanged and replacing (1.11) by

$$-\nu\Delta z_m + Q_m\{B(y_m, y_m) + B(y_m, z_m) + B(z_m, y_m)\} \cong Q_m f. \tag{3.1}$$

By comparison with (1.11), the term $\partial z_m/\partial t$ has been dropped and, as well, $Q_m B(z_m, z_m)$.

For a nonlinear Galerkin (spectral) method, we change the notations and consider $u_{2m} = y_m + z_m$ as in (2.3),(2.4). The function y_m, z_m satisfy

$$\frac{\partial y_m}{\partial t} - \nu\Delta y_m + P_m B(y_m + z_m, y_m + z_m) = P_m f, \tag{3.2}$$

$$\begin{aligned} -\nu\Delta z_m + (P_{2m} - P_m)\{B(y_m, y_m) + B(y_m, z_m) + \\ + B(z_m, y_m)\} = (P_{2m} - P_m)f. \end{aligned} \tag{3.3}$$

Note that (3.3) is a linear equation for z_m, which gives z_m in terms of y_m by resolution of a linearized steady Navier-Stokes equation. We then insert $z_m = z_m(y_m)$ in (3.2) and obtain a nonlinear perturbation of the $m-$modes spectral method.

The convergence of this method as $m \to \infty$, has been proved in [MT]. In theory this method is more accurate than that of Section 3 (see [T5] and Section 4) ; it would be interesting to test the method and see if computational efficiency confirms the theoretical result.

Remark 3.1.

(i) Higher order methods corresponding to higher order perturbations can be also derived; they necessitate a proper approximation of $\partial z_m/\partial t$ (and $B(z_m, z_m)$) in (1.11).

(ii) For (3.2),(3.3) as well as for (2.5),(2.6) we have chosen to approximate z_m with as many modes (m) as y_m. However we could approximate z_m with $(d-1)m$ modes, $d > 2$, so that the approximation involves a total of dm modes. The choice $d = 2$ is convenient in FFTs and seems also appropriate.

4. THEORETICAL JUSTIFICATION : ATTRACTORS AND INERTIAL MANIFOLDS.

Although the practical value of a numerical method is decided by the quality of the actual computational tests, we would like here to give some indications on the theoretical aspects and motivations of the nonlinear Galerkin methods.

We are considering the Navier-Stokes equations (1.1),(1.2) with space periodic boundary condition and a driving force f which maintains the motion ; the force f is time independent. For simplicity we restrict ourselves to space dimension 2. If the Reynolds number Re defined in Section 1, based on f, is small then there exists a unique stationary (laminar) solution u_S to (1.1),(1.2) and each solution of these equations converges to the steady state as $t \to \infty$. It is also believed that if the Reynolds number is larger than a critical Reynolds number, then the stationary solution u_S loses its stability. In this case the solutions of (1.1),(1.2) never converge to a steady state and the flow remains constantly time-dependent (and turbulent).

The natural phase space for the problem is an infinite dimensional Hilbert space H, spanned by the eigenfunction w_k. A stationary solution is a point in H, while a time dependent soluton of (1.1),(1.2) is represented by a curve in H. In the laminar situation the orbit $u(t)$ converges to the unique stable stationary solution u_S. In the turbulent regime *the flow becomes time dependent* : even if the data are time independent, the flow never converges to a stationary state and remains time dependent. It was shown by Foias-Temam [FT1] that there exists a compact attractor $\mathcal{A}$ to which all the orbits converge. The attractor $\mathcal{A}$ appears as a substitute of u_S when u_S loses its stability; it is the mathematical object that represents the turbulent permanent regime.

It was shown in [FT1] that $\mathcal{A}$ has finite dimension. An explicit bound on this dimension was derived in [CFT1] for space dimension 3 and in [CFT2] for space dimension 2 ; in the latter case it reads :

$$\dim \mathcal{A} \leq Re^{4/3} \log Re. \tag{4.1}$$

Furthermore the estimate (4.1) is essentially optimal, since Babin-Vishik [BV] proved that the dimension can be $\geq Re^{4/3}$. Hence *the number of degrees of freedom of such a turbulent flow can be as large as* $Re^{4/3}$ and this agrees with Batchelor and Kraichnan results on two-dimensional turbulence [B][K].

For the computation of a permanent turbulent regime we may need a number of parameters of the same order as the number of degrees of freedom, i.e. the dimension of the corresponding attractor.

Since the attractor can be a complicated set (of fractal type), the idea in [FST] was to imbed it in a smooth finite dimensional manifold called an inertial manifold. This manifold which is positively invariant for the semigroup associated to the equation attracts all the orbits at an exponential rate. Using a decomposition of the space H similar to that used in the previous sections, namely $H = P_m H \oplus Q_m H$, $u = y + z$, $y = P_m u$, $z = Q_m u$, the inertial manifold $\mathcal{M}$ is a manifold of equation

$$z = \Phi(y). \tag{4.2}$$

When the manifold $\mathcal{M}$ exists, its equation (4.2) provides a quasistatic interaction law between small and large eddies. And this law applies in the permanent regime i.e. as soon as the orbits are close enough from the manifold $\mathcal{M}$.

Unfortunately the existence of inertial manifolds for the Navier-Stokes equations is not proved. A substitute and perhaps computationally more convenient concept is that of approximate inertial manifolds (AIM) : these are manifolds that contains the attractor and attract all the orbits in a thin neighborhod.

For example a restatement of (1.12) is that the linear space $P_m H$ is an approximate inertial manifold for m sufficiently large. Less trivial AIMs were found in [FMT] and [T5]. The approximate inertial manifold $\mathcal{M}_0$ of [FMT] is precisely that appearing in (1.13) and of equation

$$-\nu\Delta z + Q_m B(y,y) = Q_m f. \tag{4.3}$$

The approximate inertial manifold $\mathcal{M}_1$ of [T] is that appearing in (3.1) and of equation

$$-\nu\Delta z + Q_m\{B(y,y) + B(y,z) + B(z,y)\} = Q_m f. \tag{4.4}$$

The thickness of the neighborhod of $\mathcal{M}_0$ and $\mathcal{M}_1$ attracting the orbits are determined in [T5] and are respectively in space dimension 2

$$\delta^{3/2}L \text{ and } \delta^2 L^{3/2}, \quad L = 1 + \log\frac{1}{\delta}.$$

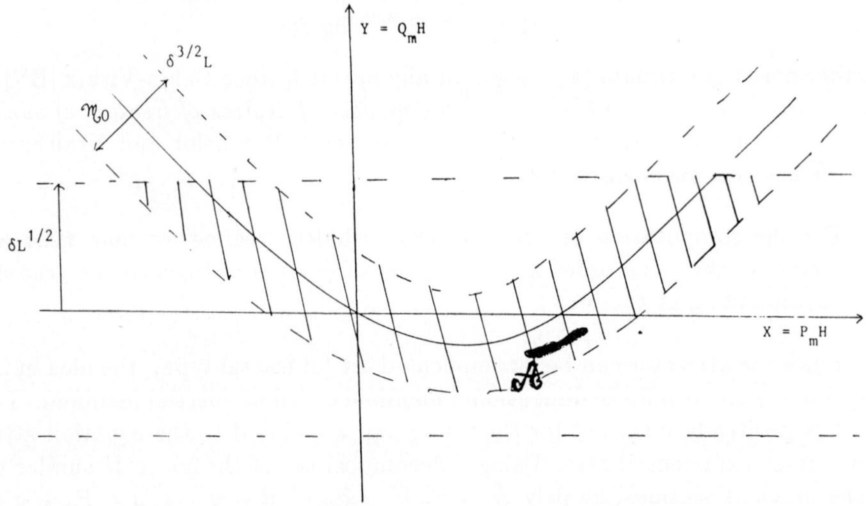

Figure **4.1.**

Localization of the attractor $\mathcal{A}$ in the phase space : $\mathcal{A}$ lies in the dashed region

In conclusion, the attractor $\mathcal{A}$ is closer to $\mathcal{M}_1$ than it is to $\mathcal{M}_0$ and it is closer to $\mathcal{M}_0$ than it is to $P_m H$. Projecting the Navier-Stokes equations onto $P_m H$ we obtain the usual Galerkin method. Projecting the equations onto $\mathcal{M}_0$ or $\mathcal{M}_1$, we obtain the nonlinear Galerkin methods of Section 2 or 3.

5. NUMERICAL COMPUTATIONS.

We describe here the results of computational tests performed with the nonlinear Galerkin method (2.5),(2.6). Comparisons are also made with the usual Galerkin method (i.e. (2.2)). In both cases the two-dimensional Navier-Stokes equations with space periodicity boundary conditions are considered. The numerical results are due to C. Rosier [R].

The time discretization of the equations is performed in both cases with a predictor-corrector scheme. The predictor step is done by solving :

$$\tilde{y}^{n+1} = e^{-A\Delta t}y^n + \Delta t \sum_{j=0}^{\rho-1} \alpha_0^j P[f^{n-j} - b(u^{n-j}, u^{n-j})] - \nabla\tilde{\Pi}_y^{n+1}$$

$$\nu A\tilde{z}^{n+1} = Q[b(y^n, y^n) - f^n] - \nabla\tilde{\Pi}_z^{n+1}$$

$$\tilde{u}^{n+1} = \tilde{y}^{n+1} + \tilde{z}^{n+1}.$$

The corrector step is the following

$$\tilde{y}^{n+1} = e^{-A\Delta t}y^n + \Delta t\alpha_1^0 P_m[f^{n+1} - b(\tilde{u}^{n+1}, \tilde{u}^{n+1}])$$

$$+\sum_{j=1}^{\rho-1} \alpha_1^j P[f^{n+1-j} - b(u^{n+1-j}, u^{n+1-j})] - \nabla\Pi_y^{n+1}$$

$$\nu Az^{n+1} = Q[b(y^{n+1}, y^{n+1}) - f^{n+1}] - \nabla\Pi_z^{n+1}$$

$$u^{n+1} = y^{n+1} + z^{n+1}$$

where n is the time step, ρ is the order of the scheme, $\alpha_r^j = \int_0^1 e^{A(f-1)\Delta t}\Phi_r^j(f)df$, and Φ_r^j is a polynomial of degree $\rho - 1$ such that $\Phi_r^j(1-k) = \delta_{jk},\ j,k = 0,...,\rho-1,\ r = 0,1$.

The nonlinear terms are treated as usual by collocation and FFT.

Implicit in (2.5),(2.6) (and in the notation B) is the existence of a pressure like function associated to y_m and z_m; we denote by p_m and q_m the two pressure like functions, $p_m + q_m$ approximating the actual pressure function. Hence (2.5) and (2.6) are equivalent to

$$\frac{\partial y_m}{\partial t} - \nu\Delta y_m + P_m\{(y_m\cdot\nabla)(y_m + z_m) + (z_m\cdot\nabla)y_m + \nabla p_m\} = P_m f \tag{5.1}$$

$$\nabla\cdot y_m = 0 \tag{5.2}$$

$$-\nu\Delta z_m + (P_{2m} - P_m)\{(y_m\cdot\nabla)y_m + \nabla q_m\} = (P_{2m} - P_m)f \tag{5.3}$$

$$\nabla \cdot z_m = 0 \tag{5.4}$$

Two examples have been considered. In both cases the solution u is a priori chosen and $p = 0$, and f is determined from (1.1). Hence the exact solution of the equation is known and it is easy to test safely the accuracy.

In the first example (see Figures 5.1a,5.1b,5.1c), $m = 8$ and $u = \{u_1, u_2\}$,

$$u_1(x_1, x_2, t) = (\cos t)\ \varphi(x_2)$$

$$u_2(x_1, x_2, t) = (\cos t)\ \varphi(x_1)$$

$$\varphi(x_1) = 10^{-2}\left(x_1^2(x_1^2 - 4\Pi x_1 + 4\Pi^2) - \frac{8\Pi^4}{15}\right)$$

$$Re = 48.30$$

Figure 5.1a shows that the errors for both schemes are about the same as t evolves. Figures 5.1b and 5.1c show the absolute and relative gain in computing time between the usual and the present nonlinear Galerkin method ; here the relative gain of computing time is approximately 18% .

In the second example (see Figures 5.2a, 5.2b, 5.2c), $m = 32$, $Re = 1{,}704$ and $u = \{u_1, u_2\}$ is as in the first example with now $\varphi(x_1) = 10^{-1}(x_1^2(x_1^2 - 4\pi x_1 + 4\pi^2) - \frac{8\pi^4}{15})$. Figure 5.2a shows that the errors are about the same for both schemes as t evolves. Figures 5.2b and 5.2c show the absolute and relative gain in computing time between the usual and the present nonlinear Galerkin method ; here the relative gain in computing time is approximately 40% .

Finally, Figures 5.3a,b,c, are numerical evidences supporting the small-large eddies model described in Section 1. They give also a posteriori justifications of the nonlinear Galerking method proposed here. The computations were made for the first example above. Figure 5.3a shows the evolution of the ratio

$$|\ B(y^{n+1}, y^{n+1})\ |_{L^2}\ /\ |\ (z^{n+1} - z^n/\Delta t)\ |_{L^2}$$

along the iterations ; the ratio is of the order of 10^5. Figure 5.3b shows the evolution of the ratio

$$|\ B(y^n, y^n)\ |_{L^2}\ /\ |\ B(y^n, y^n) + B(z^n, y^n) + B(z^n, z^n)\ |_{L^2}$$

Finally Figure 5.3c shows the evolution along the iterations of the ratio

$$|\ (y^{n+1} - y^n)/\Delta t\ |_{L^2}\ /\ |\ (z^{n+1} - z^n)/\Delta t\ |_{L^2}$$

The ratio is of order 10^4 ($\Delta t = 10^{-3}$ in the three figures).

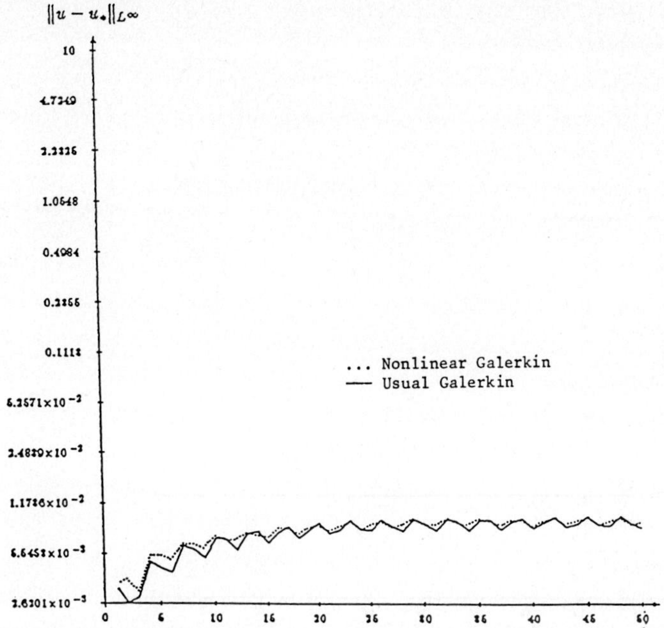

Figure 5.1a. Comparison of L norm of the error for the nonlinear Galerkin scheme and the usual Galerkin scheme

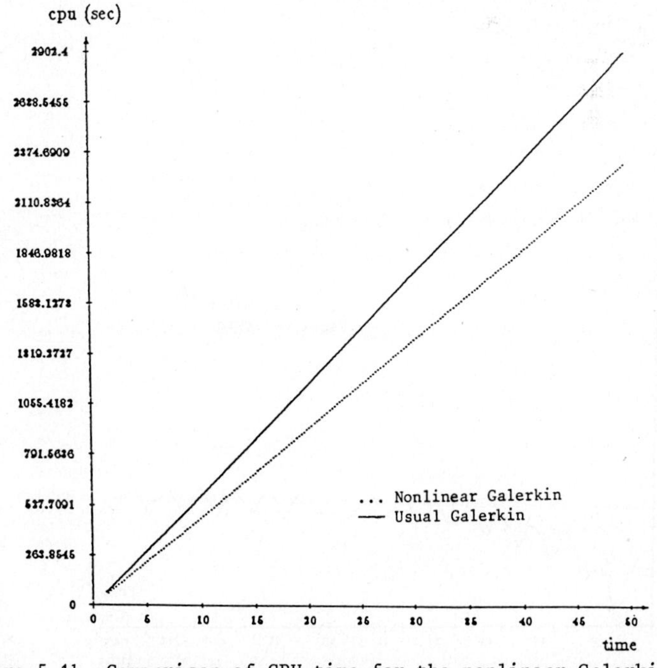

Figure 5.1b. Comparison of CPU time for the nonlinear Galerkin scheme and the usual Galerkin scheme

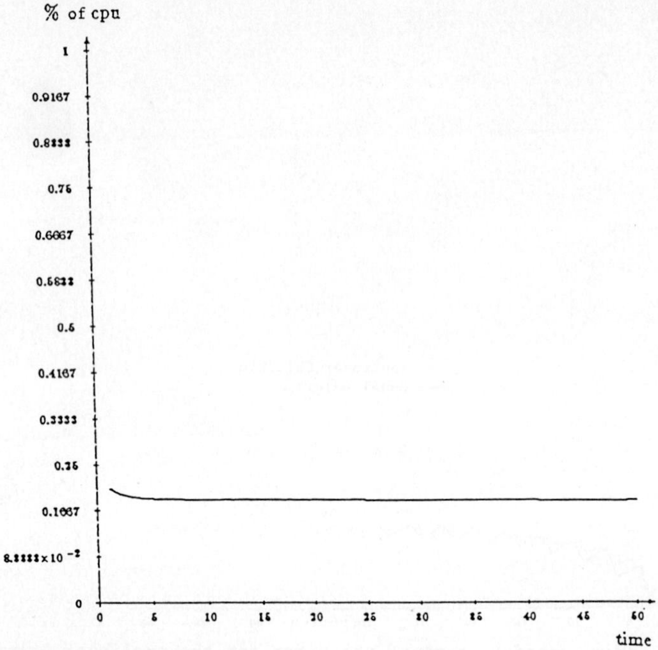

Figure 5.1c. Gain of CPU time for the nonlinear Galerkin scheme over the usual Galerkin scheme.

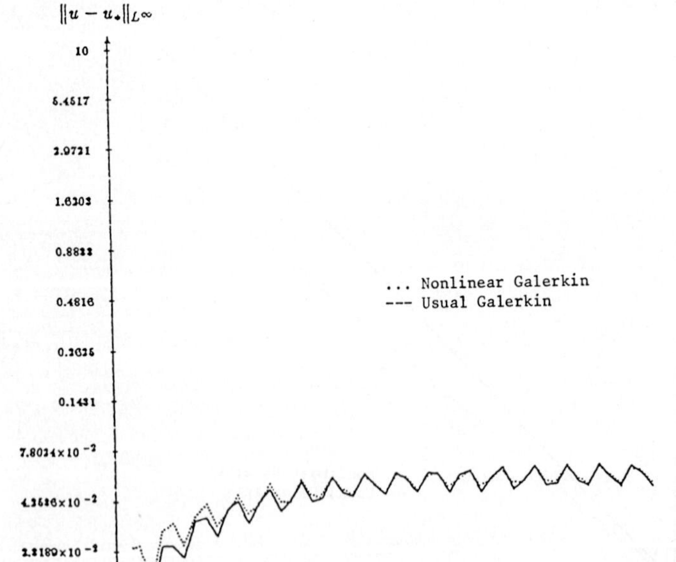

Figure 5.2a. Comparison of L norm of the error for the nonlinear Galerkin scheme and the usual Galerkin scheme.

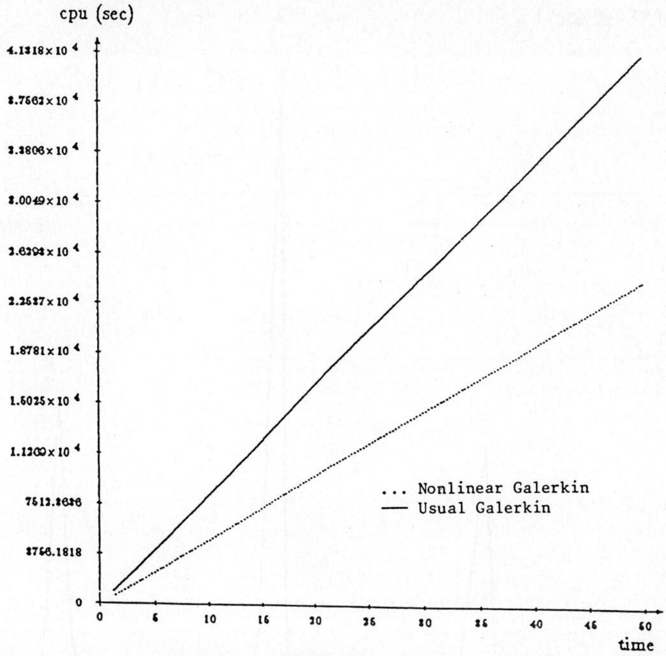

Figure 5.2b. Comparison of CPU time for the nonlinear Galerkin scheme and the usual Galerkin scheme.

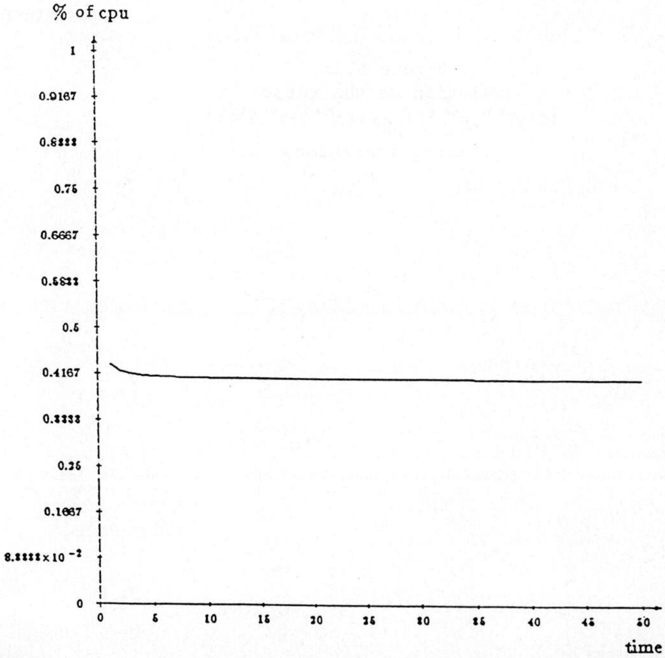

Figure 5.2c. Gain of CPU time for the nonlinear Galerkin scheme and the usual Galerkin scheme

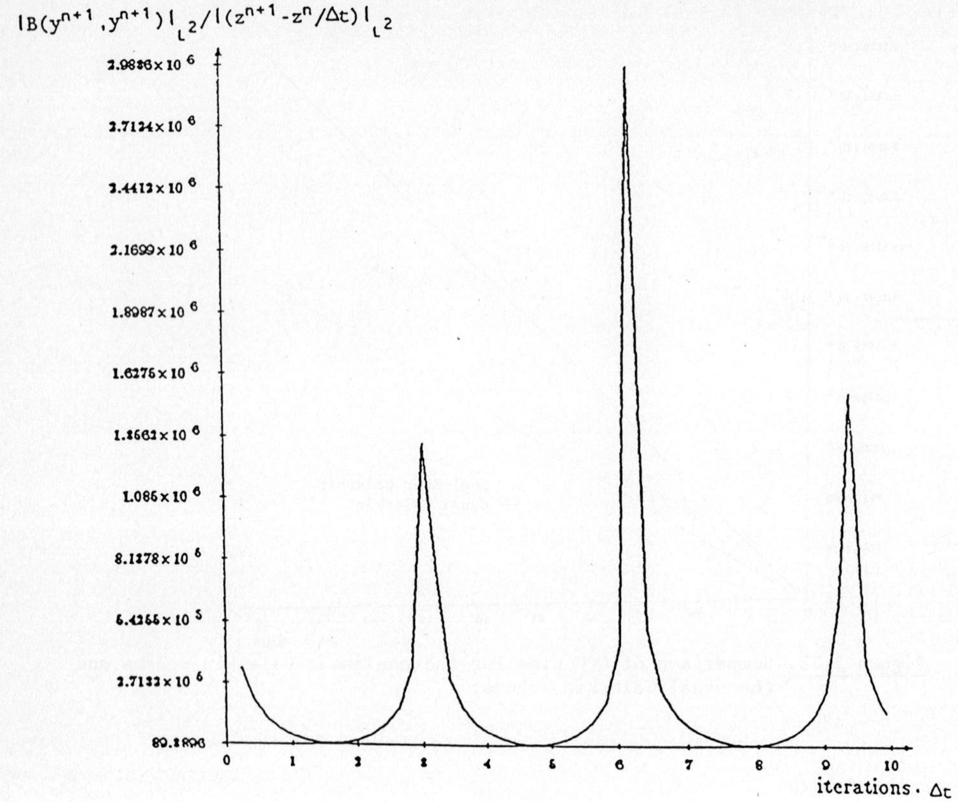

Figure 5.3a
Evolution of the ratio
$|B(y^{n+1}, y^{n+1})|_{L^2} / |(z^{n+1} - z^n/\Delta t)|_{L^2}$
along iterations

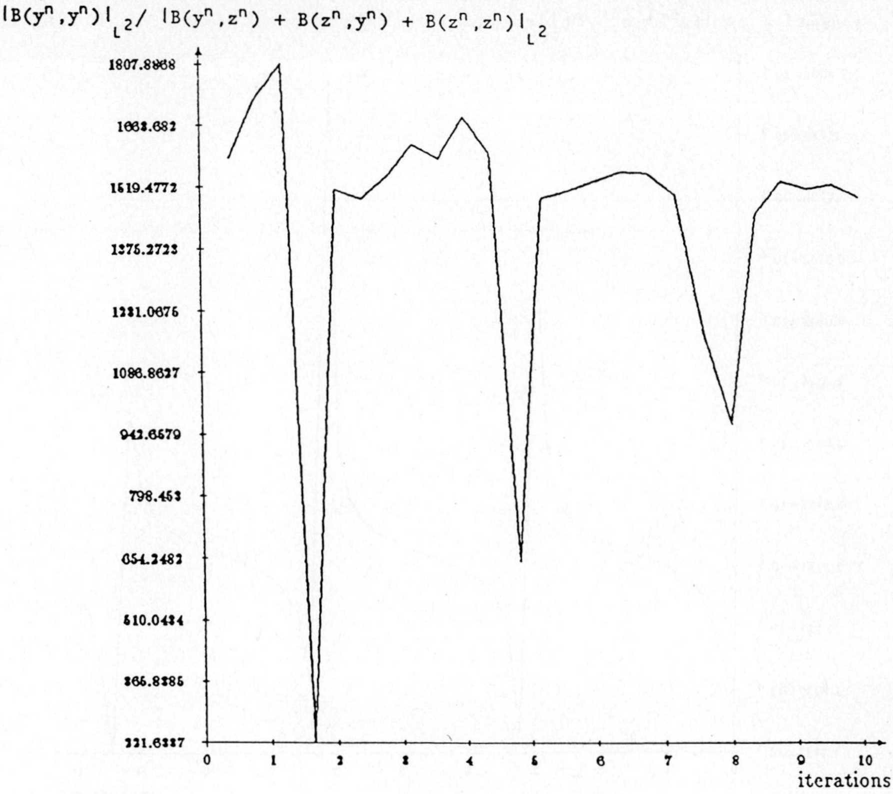

Figure 5.3b

Evolution of the ratio

$|B(y^n,y^n)|_{L^2} / |B(y^n,z^n) + B(z^n,y^n) + B(z^n,z^n)|_{L^2}$

along iterations

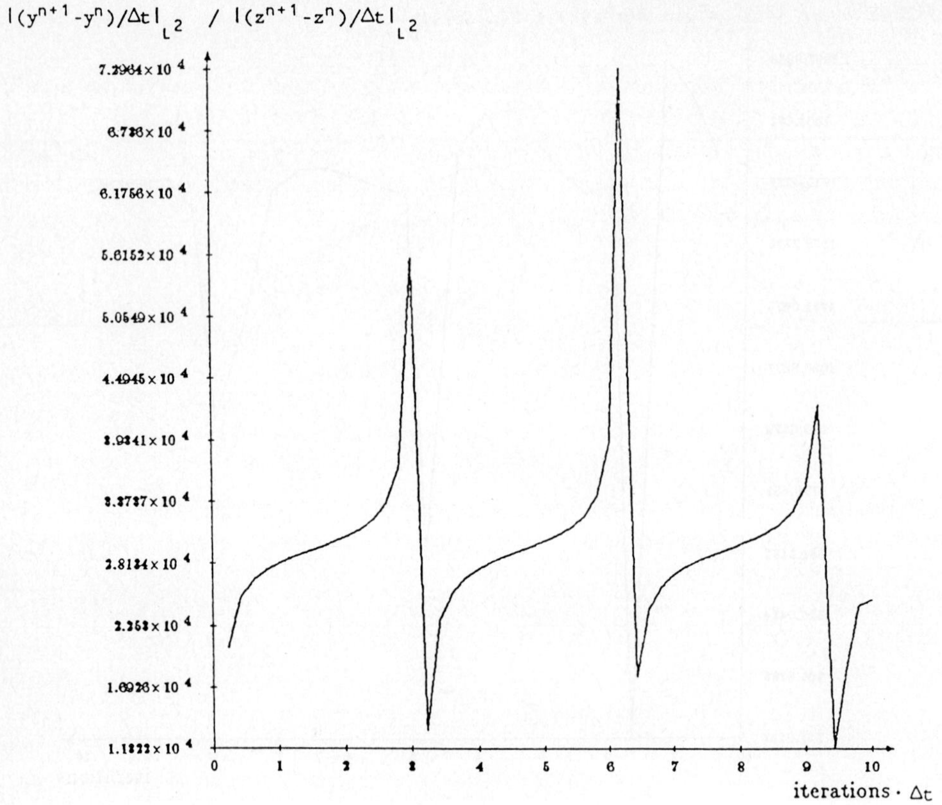

Figure 5.3c
Evolution of the ratio
$|(y^{n+1}-y^n)/\Delta t|_{L^2} \;/\; |(z^{n+1}-z^n)/\Delta t|_{L^2}$
along iterations

CONCLUSIONS.

In this article we have presented a new discretization algorithm called the nonlinear Galerkin method. The algorithm seems robust and well suited for large time integration of the Navier-Stokes equations. The preliminary numerical tests show a substantial gain in computing time. Further numerical experiments and the extension of the method to other equations and to other forms of discretization (finite elements, finite differences...) will be presented elsewhere.

Acknowledgements. This research was partially supported by AFOSR, Contract n° 88-103, and by the Research Fund of Indiana University. Parts of the computations were made on the Cray 2 of the Centre de Calcul Vectoriel pour la Recherche at Palaiseau, France.

REFERENCES

[B] G.K. Batchelor, Computation of the energy spectrum in homogeneous, two dimensional turbulence, *Phys. Fluids,* 12, suppl. 2 (1969) 233-239.

[BV] A.V. Babin and M.I. Vishik, Attractors of partial differential equations and estimate of their dimension, *Uspekhi Mat. Nauk,* 38 (1983) 133-187 (in Russian), *Russian Math. Surveys,* 38 (1983) 151-213 (in english).

[C1] A.J. Chorin, A numerical method for solving incompressible viscous flow problems, *J. Comp. Phys.,* 2 (1967) 12-26.

[C2] A.J. Chorin, Numerical solution of the Navier-Stokes equations, *Math. Comp.,* 23 (1968) 341-354.

[CFT1] P. Constantin, C. Foias and R. Temam, Attractors representing turbulent flows, *Memoirs of A.M.S.,* 53, 114 (1985), 67+vii pages.

[CFT2] P. Constantin, C. Foias and R. Temam, On the dimension of the attractors in two-dimensional turbulence, *Physica D,* 30 (1988) 284-296.

[CO] R. Cohen, Ph. Dissertation, University of Minnesota, Minneapolis, 1988.

[CHKL] R. Cohen, R. Hardt, D. Kinderlehrer, S.-Y. Lin and M. Luskin, Minimum energy configurations for liquid crystals : computational results, to appear.

[CHQZ] C. Canuto, M.Y. Hussaini, A. Quarteroni and T.A. Zang, *Spectral methods in fluid dynamics,* Springer Series in Computational Physics, Springer-Verlag, New York, 1988

[FMT] C. Foias, O. Manley and R. Temam, Modelling of the interaction of small and large eddies in two dimensional turbulent flows, *Math. Mod. and Num. Anal., (M2AN),* 22 (1988) 93-114.

[FST] C. Foias, G. Sell and R. Temam, Inertial manifolds for nonlinear evolutionary equations, *J. Diff. Equ.,* 73 (1988) 309-353.

[FT1] C. Foias and R. Temam, Some analytic and geometric properties of the evolution Navier-Stokes equations, *J. Math. Pures Appl.,* 58 (1979) 339-368.

[FT2] C. Foias and R. Temam, Gevrey class regularity for the solutions of the Navier-Stokes equations, to appear.

[GO] D. Gottlieb and S. Orszag, *Numerical Analysis of Spectral Mehtods : Theory and Applications,* SIAM-CBMS, Philadelphia, 1977.

[K] R.H. Kraichnan, Inertial ranges in two-dimensional turbulence, *Phys. Fluids,* 10 (1967) 1417-1423.

[LT1] J.L. Lions et R. Temam, Une méthode d'éclatement des opérateurs et des contraintes en calcul des variations, *C.R. Acad Sci. Paris,* Série A, 263 (1966) 563-565.

[LT2] J.L. Lions et R. Temam, Eclatement et décentralisation en calcul des variations, *Symposium on Optimization,* Lecture Notes in Mathematics, vol. 132, Springer-Verlag, 1970, 196-217.

[MT] M. Marion and R. Temam, Nonlinear Galerkin methods, submitted to *SIAM J. Num. Anal.*

[R] C. Rosier, Thesis in preparation, Université de Paris-Sud, Orsay, 1989.

[RY] B.L. Rohdezestvensky and N.N. Yanenko, *Systems of quasilinear equations and their applications to gas dynamics,* Mir, Moscow, 1978.

[T1] R. Temam, Sur la stabilité et la convergence de la méthode des pas fractionnaires, *Ann. Mat.Pura Appl.,* LXXIV (1968) 191-380.

[T2] R. Temam, Sur l'approximation de la solution des équations de Navier-Stokes par la méthode des pas fractionnaires (I), *Arch. Rat. Mech. Anal.,* 32,n° 2 (1969) 135-153.

[T3] R. Temam, Sur l'approximation de la solution des équations de Navier-Stokes par la méthode des pas fractionnaires (II), *Arch. Rat. Mech. Anal.,* 33,n° 5 (1969) 377-385.

[T4] R. Temam, *Navier-Stokes Equations,* 3rd revised edition, North-Holland, 1984. Russian translation edited by N.N. Yanenko.

[T5] R. Temam, Variétés inertielles approximatives pour les équations de Navier-Stokes bidimensionnelles, *C.R. Acad. Sci. Paris,* 306, Série II, 1988, 399-402, and Induced trajectories and approximate inertial manifolds, to appear.

CONTRIBUTED PAPERS

A Comparative Study of TV Stable Schemes for Shock Interacting Flows

Takayuki AKI
National Aerospace Laboratory, Chofu, Tokyo 182, Japan

1. Introduction Over the past several years, substantial advances have been made in the numerical analysis of the Euler equations. A variety of total variation (TV) stable schemes for solving the equations have been developed. TV diminishing (TVD) schemes which appeared as the first class among the TV stable ones can be constructed to be of very higher order accuracy greater than the second. At critical points in the solution such as those of local extrema and sonic points, all the existing TVD schemes automatically reduce to being low order discretization. The maximum grobal accuracy attainable by the TVD schemes is restricted to first order at most, no matter how their formal accuracy is high enough in monotone smooth regions. Even presently available TVD scheme resulting to 15th order accuracy its high accuracy can be maintained only in monotone smooth regions, it reduces to that of the basis scheme (E-scheme, necessarily only first order) at the critical points mentioned.

To overcome this difficulty, schemes which are of globally high-order accuracy in smooth regions have been constructed recently. Although a complete mathematical theory for such schemes is still unavailable, numerical experiments to prove their potentiality have been conducted by scheme exploring people themselves. However, these experiments were restricted mostly to scalar conservation laws or the one-dimensional (1-D) Euler equations in the Cartesian coordinate. Little information is now available about performance of these new TV stable schemes, particularly for computation of multi-dimensional flows under a generalized coordinate transformation, except TVD schemes. In this paper, we will display a comparative study of such schemes with emphasis on computing complex shock interacting flows.

2. Numerical Method The Euler eauations were discretized at the second order accuracy, and all the computations have been implemented with explicit method. This is straightforward to 1-D computation, and 2-D computation was executed by using the fractional step method of Strang type which is known as a simple method attaining the second order accuracy. In this manner we may write the basic difference equations in a unified form as

$$U_j^{n+1} = U_j^n - \lambda\,(\tilde{F}_{j+1/2} - \tilde{F}_{j-1/2}) \qquad (1)$$

where U_j is the column vector whose elements are conservative variables and scaled with J_j (see below), $\tilde{F}_{j+1/2}$ is a numerical flux function in the direction being considered, and λ is a mesh ratio of time to space. The numerical flux fuction can be written as

$$\tilde{F}_{j+1/2} = (1/2)\,(F_j + F_{j+1} - R_{j+1/2}\,\Phi_{j+1/2} \,/\, J_{j+1/2}) \qquad (2)$$

where F_j is the physical flux appeared in the Euler equations and scaled with appropriate metrics of the transformation, $R_{j+1/2}$ is the 4x4 matrix whose colums are composed of the right eigenvectors of Jacobian $\partial U/\partial F$, and $J_{j+1/2}$ is Jacobian of the coordinate transformation. For a generalized coordinate, (2) can be expressed in other ways. The present expression is a form of the pseudo finite volume ones [1].

Let $\phi^m_{j+1/2}$ be the m-th element of $\Phi_{j+1/2}$, it can be written as

$$\phi^m_{j+1/2} = \sigma(a^m_{j+1/2})(g^m_j + g^m_{j+1}) - \phi(a^m_{j+1/2} + \beta^m_{j+1/2})\ \alpha^m_{j+1/2} \tag{3a}$$

where $a^m_{j+1/2}$ is an eigenvalue of Jacobian F/ U and $\phi(z)$ is a function introduced as entropy correction [2]. The fuction $\sigma(z)=0.5[\phi(z)-\lambda z]$ and

$$\beta^m_{j+1/2} = \begin{cases} (g^m_{j+1} - g^m_j)/\ \alpha^m_{j+1/2} & \alpha^m_{j+1/2} \neq 0 \\ 0 & \alpha^m_{j+1/2} = 0 \end{cases} \tag{3b}$$

Here,

$$\alpha^m_{j+1/2} = \hat{U}^m_{j+1} - \hat{U}^m_j \tag{3c}$$

where $\hat{U}_j = J\ U_j$, or the unscaled solution vector.

Equations (1 - 3) are shaered among schemes tested. The function g has its own expression for respective scheme. A basic code originally written for testing upwind type TVD scheme of Harten-Yee [3] was modified to include TVB [4] and UNO [5] schemes. Formulas required for the modifications are as folows.

For TVD scheme, a family of the flux limiter function can be used for g. Here the ultra bee in Roe's terminology [6] was applied to the nonlinear fields and the super bee to the linear ones.

For TVB modification, the function g should be read as

$$g_j = (1/2)[mc(\alpha_{j-1/2},\ b\alpha_{j+1/2}) + mc(\alpha_{j+1/2},\ b\alpha_{j-1/2})] \tag{4a}$$

where

$$mc(x,y) = m[x, y+\Delta^2 M\ sgn(x)] \tag{4b}$$

where m is the conventional minmod function and $1 \leq b \leq 3$ and M are constants. In the present test cases, b=3 and M=10 were used. Δ is the prescribed space mesh width.

On the other hand the function g for UNO scheme can be given as

$$g_j = m(\tilde{g}_{j+1/2}, \tilde{g}_{j-1/2}) \tag{5a}$$

$$\begin{aligned} g_{j+1/2} = (1/2)[\phi\ (\nu_{j+1/2}) - |\nu_{j+1/2}|\ \alpha_{j+1/2} \\ + \max(0,\ \nu_{j+1/2})(1-\nu_{j-1/2})\hat{S}_j - \max(0,\ \nu_{j+1/2})(1-\nu_{j+3/2})\hat{S}_{j+1} \end{aligned} \tag{5b}$$

where

$$\hat{S}_j = m(S^-_j, S^+_j)/(1+\nu_{j+1/2} - \nu_{j-1/2}) \tag{5c}$$

$$S^{\pm}_j = \alpha_{j+1/2} \mp (1/2)m(\alpha_{j+3/2} - \alpha_{j+1/2},\ \alpha_{j+1/2} - \alpha_{j-1/2}) \tag{5d}$$

where $\nu_{j+1/2} = \lambda\ a_{j+1/2}$. Using (5), let $\sigma(z)=1$ in (3a).

Feature to be noted is treatment of the difference stencil. Both the TVB and UNO is five points scheme at the the second order. The former uses a fixed stencil with symmetry to attain the required uniform accuracy whereas the latter uses an adaptive stencil with asymmetry or ENO interpolation everywhere.

3. Results Several 1-D shock flow problems were computed by using respective scheme mentioned above and compared respectively with the exact solution of given problem for obtaining definite accuracies in some norms. TV at every time step was also traced and compared with each scheme. Results obtained for the standard shock tube problem (Sod's one) are shown in Fig.1. Cartesian equipartitioned 101 space grid and CFL=0.99 were used in these computations.

Test results for 2-D and generalized coordinates frame work are shown in Fig. 2. Problem in this figure is of "reflection of initially planar shock around a smoothly courved bend in a two-dimensional duct" which was discussed in [7]. Results in Fig. 2 were obtained under the same computational conditions on grid (131 across the width x 4 points per one degree around the bend), shock and gas parameters (M_S =2.2 and γ =1.4), CFL=0.99, and etc. Key feature of this problem from the shock dynamical point of view is the Mach shock system evoluted at the propagating front. Shock transitions in the Mach system are crisp and the shock system is captured well by any scheme. However, as is anticipated from the treatment of the difference stencil, TVB result shows a largest oscillatory behavior among the solutions. Moreover, a closer inspection of the figure shows TVB result fails to capture or smear heftily a contact discontinuity following the frontal Mach shock system. UNO gives the smoothest result. Although TVD result is inferior in the smoothness of solution compared to UNO result, it captures a kink in the slip surface emanated from the triple point of the Mach shock system. The kink in the slip surface has been observed in experiment. Computer time required increases in order of TVD, TVB, and UNO. We may conclude tentatively that TVD can give a reliable result in all aspects of shock dynamics with minimum cost.

4. Conclusions New TV stable schemes for solving the Euler equations of compressible flow were applied to compute time-dependent shock interacting flows. Results obtained showed that the schemes tested will equally applicable to muti-dimensional shock interacting flows under a generalized coordinate framework as the TVD schemes were the case, though within the second order accuracy. The present results showed that key features of shock dynamics will be captured by using any scheme tested. However, extended use of the TVB scheme at further higher order might be criticized due to its accuracy correcting method using the difference stencil symmetrically fixed, which result mostly source of oscillation in the solution due to recovery of unlimited numerical flux as was in classical difference schemes. Merit of using the UNO scheme will be realizable at some higher order accuracy than the second one or by possible improvements of the present numerical flux function to ones upwind biased at the second order. Because the UNO scheme at higher order requires more computational effort compared to its second order one, study upcoming shall be directed to this. Considering the existing TVD schemes, even at higher order, can produce an efficient code for vector computers available today. They will continue to preserve their superiority in the sense stated above until a higher order upwind biased UNO scheme with efficiency competitive with the TVD ones can be available in future study.

References

[1] Yee,H.C. et al., NASA TM100097, 1988.

[2] Harten,A., JCP. Vol. 71, 1987.

[3] Yee,H.C., NASA TM8946, 1987.

[4] Shu,C.-W., Math. Comp. Vol. 49, 1987.

[5] Harten,A. and Osher,S., SINUM. Vol. 24, 1987.

[6] Roe,P., Lectures in Applied Mathematics, Vol. 22-Part 1, 1985.

[7] Aki.T., Lecture Notes in Physics, Vol. 264, 1986.

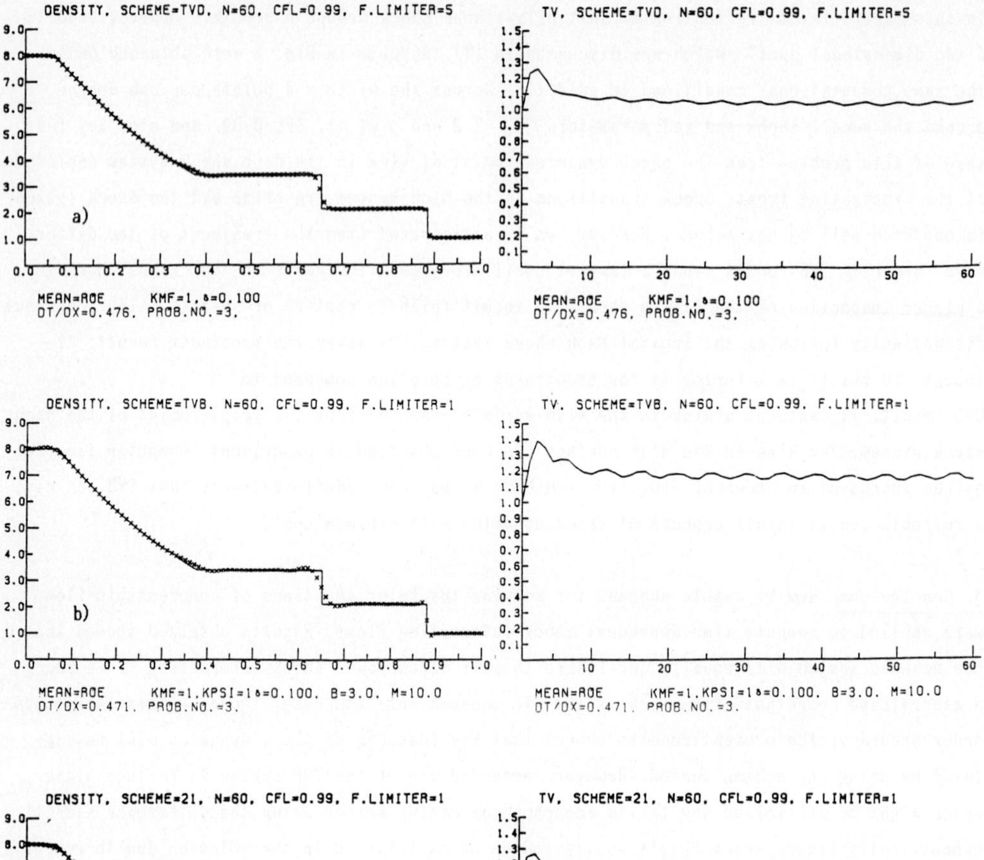

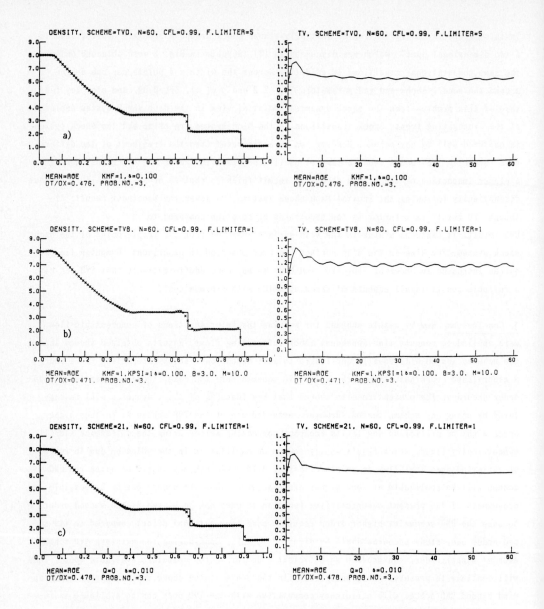

Figure 1. 1-D results for Sod's problem. Left;density, Right;TV profiles: a) TVD, b) TVB, and c) UNO.

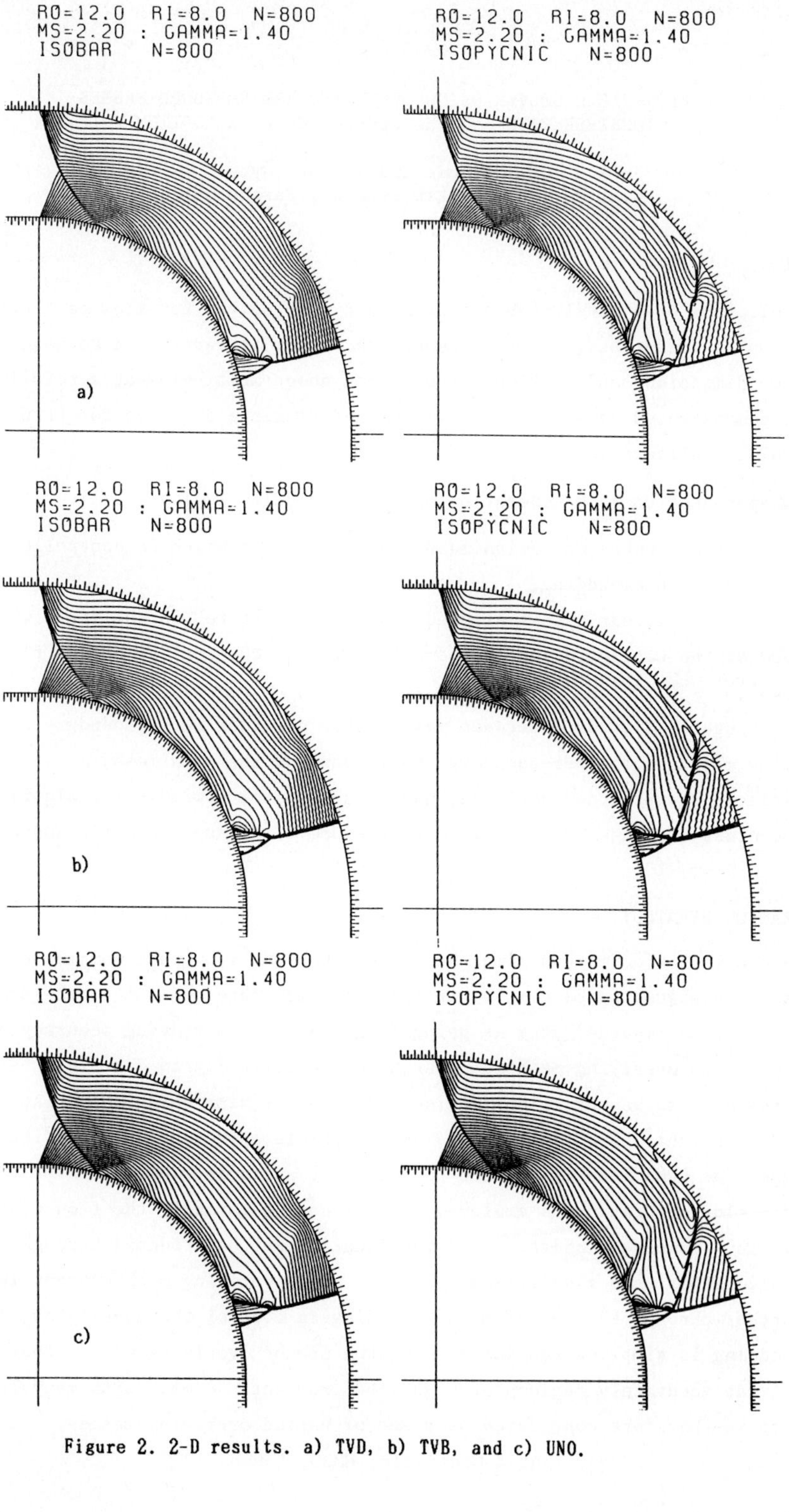

Figure 2. 2-D results. a) TVD, b) TVB, and c) UNO.

A FLOW-FIELD SOLVER USING OVERLYING AND EMBEDDED MESHES TOGETHER WITH A NOVEL COMPACT EULER ALGORITHM

C.M. Albone and Gaynor Joyce
Royal Aerospace Establishment, Farnborough, UK

INTRODUCTION

The development of a novel mesh generator and flow solver for flow past complex configurations is described. In this paper, the study is restricted to mesh generation in two dimensions and to flow calculations about multi-element aerofoils. The techniques demonstrated constitute a powerful and flexible approach for three-dimensional complex configurations.

The new approach[1] to mesh generation and flow solution employs:

i) a main (or background) Cartesian mesh, the density of which is controlled by the extensive use of mesh embedding,
ii) overlying curvilinear meshes aligned with each "feature" in the field whether it is a feature of the configuration (component surfaces, corners) or of the flow (shock waves[2], jets),
iii) an exchange of flow data between meshes that provides outer-boundary conditions for overlying meshes and inner-boundary conditions for the main mesh,
iv) an explicit, first-order accurate, non-conservative, upwind Euler algorithm[3], made second-order accurate in space through the use of truncation error source terms[4,5].

MESH GENERATION STRATEGY

Any field calculation method aimed at calculating the flow past complex configurations must be designed to be simple yet flexible, accurate and fast. The use of overlying meshes is regarded here as being fundamental to achieving accuracy and flexibility. Each overlying mesh is constructed specifically to deal in a natural way with each feature, so as not to compromise the accuracy with which that feature is represented. This has been demonstrated as a flexible approach to shock fitting[2], and in this paper the same idea is applied to component surfaces, where an overlying mesh of limited field penetration is employed to represent accurately the flow-tangency condition. Thus component addition and modification may be effected through essentially local, rather than global, mesh changes. The overlying mesh concept is taken a stage further through the use of mesh embedding to control the density of the main mesh. Embedding is simply a special case of the use of overlying meshes that is designed to treat accurately regions of high flow gradient. A mesh with many levels of embedding is therefore considered as a set of nested overlying meshes.

THE FLOW ALGORITHM

The Euler algorithm adopted here combines accuracy with simplicity. It consists of two parts. In the first part, the equations are discretized using a first-order accurate upwind algorithm, which is essentially the split-coefficient-matrix method of Chakravarthy[3], and steady-flow conditions are achieved through explicit time-marching. If N_1 is defined as the first-order accurate upwind discretization operator for the steady-state Euler equations, then the corresponding solution vector, $\underline{V}_1$, subject to given boundary conditions satisfies $N_1\underline{V}_1 = 0$. In the second part of the algorithm, a solution vector, $\underline{V}_2$, is obtained by solving the equation $N_1\underline{V}_2 = -N_2\underline{V}_1$, subject to the same boundary conditions, where N_2 is any second-order accurate discretization operator for the steady-state Euler equations. Whilst $\underline{V}_2$ is not precisely the solution of $N_2\underline{V} = 0$, it is shown in Ref 5 that it is the solution corresponding to <u>some</u> second-order accurate discretization operator. Second-order accuracy arises because the source term, $-N_2\underline{V}_1$, closely approximates the discretization error associated with the use of N_1. The algorithm has two desirable features. Firstly, in both parts of the algorithm, the operator on the left-hand side is N_1. Thus simple forward differencing may be used for the time-marching and boundary conditions are treated naturally. Secondly, since N_1 is a first-order accurate upwind operator, a compact computational molecule (having just five points in two dimensions and seven points in three dimensions) can be constructed by choosing N_2 to be the central-difference operator.

IMPLEMENTATION OF THE EMBEDDING STRATEGY IN TWO DIMENSIONS

The main objective of the work reported here is to devise and test an embedding scheme for the main mesh that allows for the use of very fine meshes locally and coarser meshes in the mid-to-far field, whilst at the same time retaining some mesh structure. This is achieved by generating patches of locally uniform Cartesian mesh; each patch being synthesised from computationally identical sub-patches called blocks. Embedding is driven by the requirement that the main mesh is locally at least as fine as each overlying mesh where the overlapping occurs. Any number of levels of embedding is allowed. The number of cells per block controls the rate of decay of mesh density away from the configuration surfaces. The outer (far-field) boundary is a square in two dimensions and the main mesh extends inside each component of the configuration. Main and overlying meshes for a three-element aerofoil with 64 surface mesh points on each element are shown in Fig 1, where, for clarity, the main mesh points inside component surfaces are omitted.

Implementation of the flow algorithm on the embedded/overlying mesh structure takes the following form:

i) the flow algorithm is used to advance in time the flow variables at each point of all blocks, even if a block (or part of it) lies inside a component and even if a block is overlaid by a finer block or by any part of the component-orientated overlying meshes,

ii) outer boundary conditions are supplied to the main mesh by imposing undisturbed stream conditions on the far-field boundary,

iii) where part of all of a block is overlaid by a finer block, the flow data there are replaced by those from the finer block, thus providing inner boundary conditions to the coarser blocks,

iv) outer boundary conditions for each finer level of blocks are provided by interpolation of flow data from adjacent coarser blocks,

v) flow data at main mesh points lying too close to the configuration surface to permit correct use of the main mesh Euler algorithm (marked by triangles in Fig 2) are replaced by values interpolated from within the appropriate overlying mesh, so providing inner boundary conditions to the main mesh,

vi) outer boundary conditions for each overlying mesh (applied at points marked by circles in Fig 2) are obtained by interpolation from within the appropriate block of the main mesh,

vii) inner boundary conditions for each overlying mesh are obtained by using the Euler algorithm coupled with the flow-tangency condition at mesh points on the configuration surface,

viii) where adjacent blocks are on the same embedding level, their common boundary is made transparent to the Euler algorithm.

DISCUSSION AND RESULTS

Validation of the method has been carried out for single and multi-element aerofoils. The overlying meshes around each element are of C-type, truncated two mesh intervals downstream of the trailing edge. Each overlying mesh has a field penetration, normal to the surface of the element, of just one mesh interval. Main and overlying mesh generation for an aerofoil with any number of elements requires no user expertise, since it follows automatically from the specified surface point distribution and from three simple control parameters. The time-marching Euler algorithm, which employs local time steps, does not require the addition of any explicit dissipation terms either to stabilise the marching process or to capture shocks. The only data required (apart from the mesh generation output) are the undisturbed stream Mach number and angle of attack, together with a Courant number for each mesh.

Calculations on single-element aerofoils show that the method has an accuracy comparable to that of more conventional methods. The flow past the three-element aerofoil of Fig 1 has been calculated at a Mach number of 0.2 and an angle of attack of 20° with meshes having 64, 96 and 128 points on the surface of each element. Pressure distributions for these three meshes are shown in Fig 3 together with the exact incompressible result.

CURRENT AND FUTURE DEVELOPMENT

The success in two dimensions of the present novel approach to mesh generation has led to a programme of work to extend the ideas to mesh generation in three dimensions[6].

This work, which is currently in progress, adopts the same philosophy of main mesh embedding but will employ overlying meshes having a field penetration of two or more mesh intervals, so further decoupling the two interpolation processes between main and overlying meshes. The two-dimensional method is also being used as the basis for a zonal approach to solutions of the Navier-Stokes equations, where each overlying mesh covers the shear layers and the Euler equations are solved on the main mesh.

REFERENCES

1 C.M. Albone — Euler methods for complex geometries - are we, or you, losing the way? In Issue 2/85 of CFD News, Ed. F. Walkden and P. Laidler, Pub. Univ. of Salford, June 1985.

2 C.M. Albone — A shock-fitting scheme for the Euler equations using dynamically overlying meshes. In Numerical Methods for Fluid Dynamics II, pp 427-437, Pub. OUP, 1986.

3 S.R. Chakravarthy, D.A. Anderson, M.D. Salas — The split coefficient matrix method for hyperbolic systems of gas dynamic equations. AIAA Paper 80-0268, 1980.

4 F. Walkden — Private communication, 1985.

5 C.M. Albone — A second-order accurate Euler algorithm using a source-term approach. RAE Technical Report to be published.

6 C.M. Albone — An approach to geometric and flow complexity using feature-associated mesh embedding (FAME): strategy and first results. In Numerical Methods for Fluid Dynamics III, to be published by Oxford University Press in 1988.

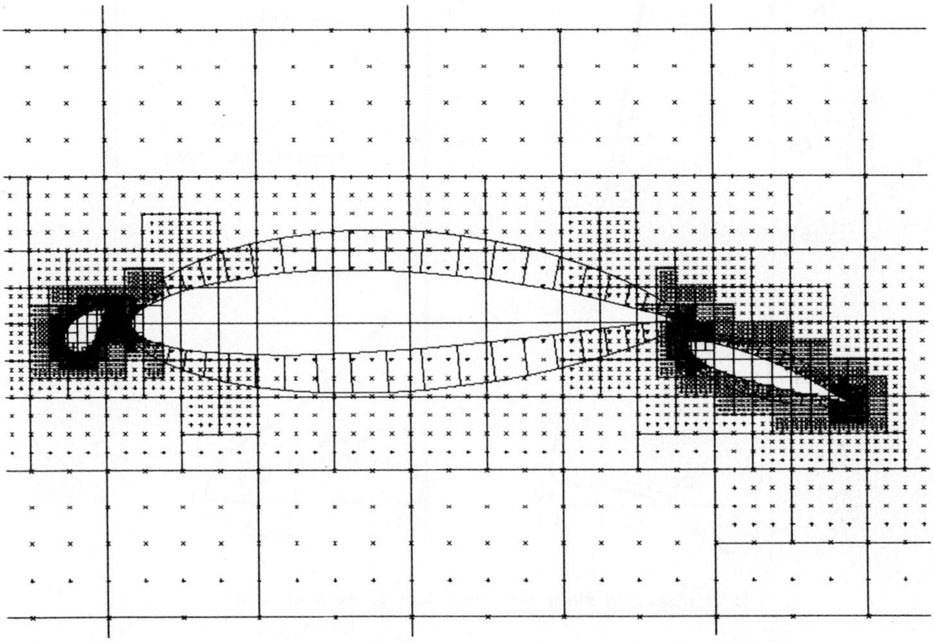

Fig 1 Main and overlying meshes for a three-element aerofoil

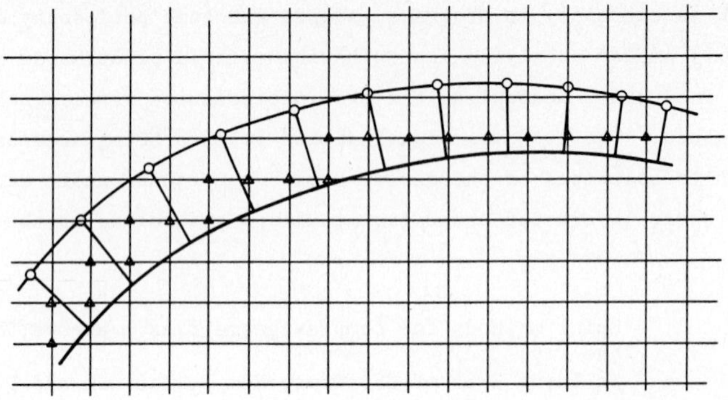

Fig 2 Transfer of flow data between main and overlying meshes

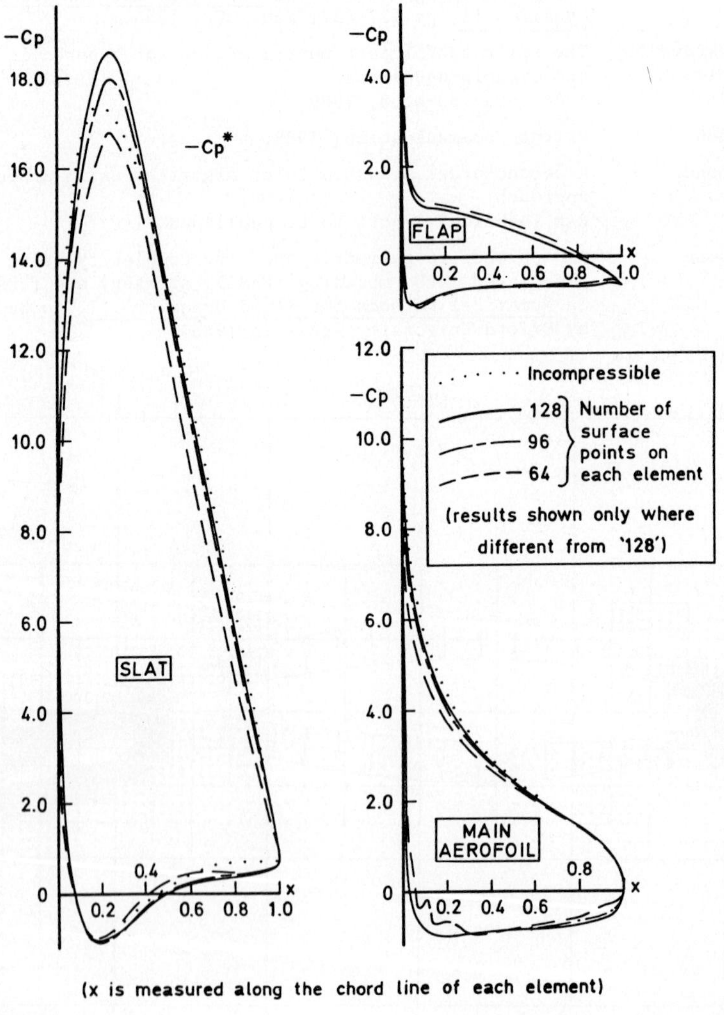

Fig 3 Three-element aerofoil: Mach number 0.2; Incidence 20 deg

MULTIDIMENSIONAL ADAPTIVE EULER SOLVER

Renzo Arina
CNR-CSDF, Politecnico di Torino
24, C.so Duca degli Abruzzi, I-10129 Torino

Bernardo Favini
Università 'la Sapienza'
19, Via Eudossiana, Roma

Luca Zannetti
Politecnico di Torino
24, C.so Duca degli Abruzzi, I-10129 Torino

Abstract.

The difficulties to be faced when attempting the numerical description of an inviscid compressible flow are mainly related to the treatment of discontinuities and to the discretization of complicated geometries. Different ideas and approaches are available: briefly, shock capturing or shock fitting as regards the treatment of discontinuities, numerical or algebraic or analytical body fitted grid generation as regards the geometrical discretization.

The present work is based on a peculiar use of a shock capturing technique and on an adaptive grid generator, each other related. The main questions addressed in the paper are the following: 1) what is the capability of upwind numerical processes on modelling multidimensional flow fields; 2) how the grid affects point (1) and, as a consequence, what kind of grid has to be used.

Introduction.

In ref.[1] it has been shown that it is possible to conceive a non-linear difference scheme that approximates the quasi-linear form of the equations where the solution is continuous and that approximates the divergence form where the flow exhibites discontinuities, exploiting the strict relationship between the Lambda-type schemes [2,3,4] and the Flux-Vector-Splitting-type schemes [5]. In this way, we can exploit the good qualities of both formulations, obtaining good accuracy, simplicity, correct boundary treatment in smooth regions and monotonic capturing of discontinuities. However, a general difficulty common to any shock capturing process remains unresolved: how to deal with the true multidimensional nature of the flow field with shock geometries which are independent of the grid shape, while the numerical results strongly depend on the grid shape. Many upwind formulations which aims to be independent of the grid have been attempted. In ref.[6] the doubt that such aiming is illusory has been cast. We believe that the grid plays an important role in a numerical process, and our conclusion is that the main improvement in the description of multidimensional flow fields can be obtained by considering the locations of the grid points dependent of the solution, and related to it by an adaptive process. For these reasons, an adaptive orthogonal grid generator [7] has been coupled with the present technique in order to calculate two-dimensional inviscid flows.

The numerical scheme.

The adopted numerical scheme is based on the ideas developed in ref.[1,6]. Denoting by x^i the cartesian coordinates, and by ξ^i a curvilinear body-fitted coordinate system, where $i = 0$ denotes time and $i = 1,2,3$ space directions, the coordinate transformation has the form

$$x^0 = \xi^0 \quad , \quad x^i = f(\xi^j) \quad i = 1,3 \;\; j = 0,3 \tag{1}$$

Defining a vector $\mathbf{Q}$ of the space-time domain, with space components coinciding with the components of the flow velocity $\mathbf{q}$, and with time component equal to one, $Q^0 = 1$, the Euler equations can be written as, with $i = 0,3$ and $m,k = 1,3$,

$$\begin{aligned} Q^i a \,|_i \; + \delta a Q^i \,|_i &= 0 \\ Q^i Q^k \,|_i \; + a\left(\frac{a\,|_m}{\delta} - \kappa s\,|_m\right) g^{mk} &= 0 \\ Q^i s\,|_i &= 0 \end{aligned} \tag{2}$$

where '|' denotes covariant derivative, a is the speed of sound, s the entropy, $\delta = \frac{\gamma-1}{2}, \kappa = \frac{a}{2\gamma\delta}$ and g^{mk} the contravariant metric tensor components with respect to the curvilinear coordinates ξ^i.

System (2) can be rewritten into the matrix form, with $\mathbf{u} = (a, Q^1, Q^2, Q^3, s)^T$,

$$\mathbf{u}\,|_0 \; + \mathbf{A}^m \mathbf{u}\,|_m = 0 \tag{3}$$

The matrices $\mathbf{A}^m$ can be splitted according to the Lambda technique [2,3,4], and we can define an upwind scheme which approximates the equations written in quasi-linear form (3). In ref.(2) it has been shown, according to the model of wave fronts, that the Lambda-type methods have the same theoretical background as the more sophisticated multidimensional upwind methods [8,9]. The only difference consists into a different interpretation of the initial data known at the grid points, which is in any case arbitrary, being in general the exact solution of the problem unknown. Moreover the Lambda-type methods enable the enforcement of the boundary conditions consistent with the wave model assumed in the inner region, without the need for additional boundary conditions.

For the correct capture of discontinuities, the present scheme is switched from the non-conservative to the conservative form depending on the local nature of the flow. As remarked in ref.[1], this is accomplished by writing eqn.s(3) in divergence form

$$\mathbf{v}\,|_0 \; + \mathbf{F}^m \,|_m = 0 \quad m = 1,3 \tag{4}$$

where $\mathbf{v}$ denotes conserved variables and $\mathbf{F}^m$ the fluxes. Exploiting the property that fluxes are homogeneous functions of degree one of the conserved variables, $\mathbf{F}^m = \mathbf{B}^m\mathbf{v}$, eqn.s(4) can be recast in the form

$$\mathbf{v}\,|_0 \; + \mathbf{B}^m \mathbf{v}\,|_m \; + (\mathbf{B}^m)\,|_m \,\mathbf{v} \; = \; 0 \tag{5}$$

For the definition of Jacobian matrix, the third term in (5) is identically zero in the absence of discontinuities. A scheme in non-conservative form neglects this term. However in the presence of a discontinuity, this term is not negligible anymore: in fact, it provides the property of shock-capturing for a numerical scheme written in conservative form.

With this correction across a discontinuity, the present numerical technique corresponds to the FVS scheme proposed by Steger and Warming [5].

Grid generation.

As stated in ref.[6], it is our opinion that the grid plays a fundamental role, and that it is illusory to conceive a grid-independent numerical scheme. An improvement of the numerical solution can be obtained mainly improving the grid point distribution by an adaptive control.

The major problems that result in generating boundary-fitted coordinates with adaptive control, are the occurrence of folded regions, and an excessive grid skewness, which can lead to inaccuracies for finite-difference calculations. In the present work an adaptive grid generation technique for two-dimensional domains has been applied [7]. Imposing the constraints of orthogonality, that is zero off-diagonal metric tensor components, the functions $x^i(\xi^j)$ must satisfy the relations

$$F\frac{\partial x^1}{\partial \xi} = \frac{\partial x^2}{\partial \eta} \quad , \quad \frac{\partial x^1}{\partial \eta} = -F\frac{\partial x^2}{\partial \xi} \tag{6}$$

where $F = \sqrt{g_{22}/g_{11}}$. From eqn.s(6) it is possible to obtain a mapping system formed by the elliptic self-adjoint equations

$$\frac{\partial}{\partial \xi}\left(F\frac{\partial x^i}{\partial \xi}\right) + \frac{\partial}{\partial \eta}\left(\frac{1}{F}\frac{\partial x^i}{\partial \eta}\right) = 0 \qquad i = 1,2 \tag{7}$$

Eqn.s(7) are coupled with a boundary-point relocation procedure [7] in order to satisfy at the same time the conditions of orthogonality (6) and the shape correspondence between the given boundary of the physical domain and the image of the boundary of the computational one. In ref.[7] it is shown that the class of curvilinear coordinates solution of this mapping system, belongs to the family of quasiconformal mappings. It is then possible to state the existence of regular unfolded coordinates. Moreover, under the constraint of adaptation with respect to discontinuities of the associated inviscid flow solution, unfolded orthogonal grids are obtained, capable of aligning and clustering along shock waves one family of the coordinate lines, while the other crosses the discontinuity orthogonally.

Computed results.

The numerical performance of the proposed approach is tested by considering two different problems, computed by solving the time-dependent Euler equations (2).

The first example is the regular shock reflection at a wall, with a free-stream Mach number $M_\infty = 2.9$ and an incident shock angle of 29°. The computational domain is $0 \leq x \leq 4$ and $0 \leq y \leq 1$. This standard test problem has been widely discussed in the literature [10]. In fig.1-a the iso-Mach lines of the solution computed on a uniform grid of size 41×21, are shown. In fig.1-b the grid obtained by solving eqn.s(7) is displayed. The adaptive control of the dilatation function

F is based on the flow field computed on the uniform grid. Fig.1-c illustrates the iso-Mach lines of the solution computed on the adapted grid. It can be seen that the clustering and alignement of the coordinate lines along the discontinuities lead to a sharper shock. However the conflicting constraints of adaptation and orthogonality with respect to the boundaries, make the grid unable to correctly fit the solution near the upper corners. Moreover the requirement of orthogonality has a double effect: it pushes the grid to align one family of coordinate lines along the discontinuity, but also tends to cluster the second family in regions of uniform flow.

The second problem is the supersonic flow past a circular cylinder, with free-stream Mach number $M_\infty = 3$. In fig.2-a it is shown the bow shock (iso-Mach lines) computed on the conformal grid with 41×21 points, generated with eqn.s(7) by setting $F = 1$. Fig.2-b,c show the grid adapted with respect to the previous flow field and the computed shock respectively. It is interesting to note the improvement of the solution, in particular in the region where the shock is weak.

Conclusions.

In this work a shock-capturing technique based on a combination of a Lambda-type scheme and of the FSV technique, has been presented. It has been shown that the capability of upwind schemes to model multidimensional flow fiels is affected by the grid. For this reason the present technique is combined with an adaptive orthogonal grid generator to solve 2-D inviscid flows.

References.

[1] B.Favini, L.Zannetti, *A Difference Scheme for Weak Solutions of Gasdynamic Equations.* AIAA Paper, AIAA-87-0537.

[2] G.Moretti, *The λ-scheme*, Computers and Fluids, 7, 1979, 191-205.

[3] L.Zannetti, G.Colasurdo, *Unsteady Compressible Flow: A Computational Method Consistent with the Physical Phenomena*, AIAA J., 19, July 1981, 852-856.

[4] S.R.Chakravarthy, D.A.Anderson, M.D.Salas, *The Split-Coefficient-Matrix Method for Hyperbolic Systems of Gasdynamic Equations*, AIAA Paper, AIAA-80-0286.

[5] J.L.Steger, R.F.Warming, *Flux Vector Splitting of the Inviscid Gasdynamic Equations with Application to Finite Difference Methods*, J.Comput.Phys., 40, 1981, 263.

[4] S.R.Chakravarthy, D.A.Anderson, M.D.Salas, *The Split-Coefficient-Matrix Method for Hyperbolic Systems of Gasdynamic Equations*, AIAA Paper, AIAA-80-0286.

[6] L.Zannetti, B.Favini, *About the numerical Modelling of Multidimensional Unsteady Compressible Flows.* To appear in Computers and Fluids.

[7] R.Arina, *Adaptive Orthogonal Curvilinear Coordinates*, ICFD 1988 Conference on Numerical Methods for Fluid Dynamics, IMA Series, Oxford Univ. Press

[8] P.L.Roe, *Approximate Riemann Solvers, Parameter Vectors and Difference Schemes*, J.Comput.Phys., 43, 1981, 357-372.

[9] H.Deconick, C.Hirsch, J.Peutman, *Characteristic Decomposition Methods for the Multidimensional Euler Equations*, 10th ICNMFD ,Beijing, China, 1986.

[10] H.C.Yee, R.F.Warming, A.Harten, *Implicit Total Variation Diminishing (TVD) Schemes for Steady-State Calculations*, J.Comput.Phys., 57, 1985, 327-360.

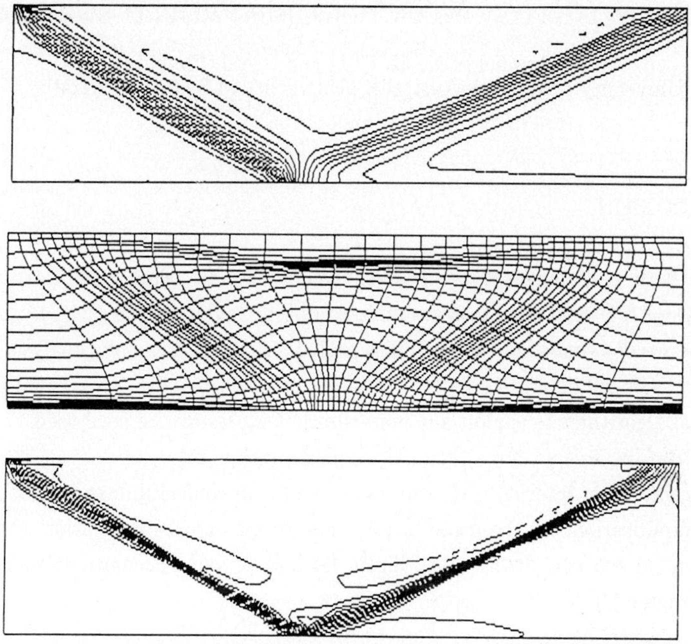

1-a

1-b

1-c

fig.1 - Regular shock reflection at a wall ($M_\infty = 2.9, 41 \times 21$)

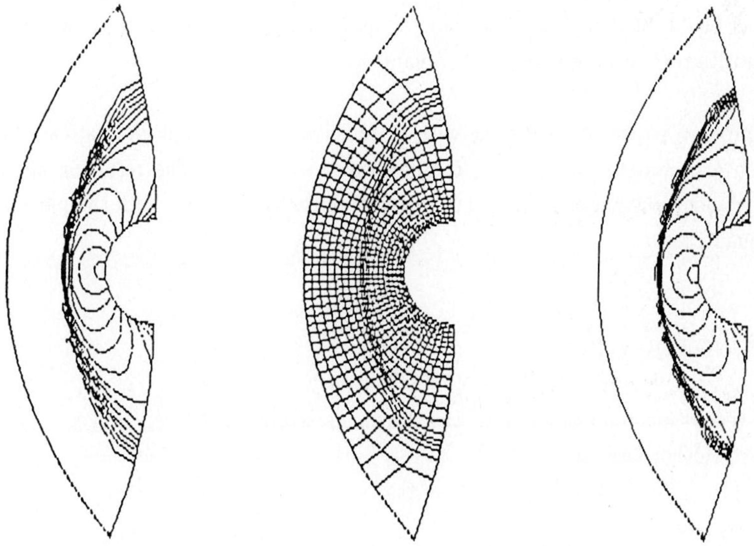

2-a 2-b 2-c

fig.2 - Supersonic flow past a circular cylinder ($M_\infty = 3, 41 \times 21$)

INTERNAL SWIRLING FLOW PREDICTIONS USING A MULTI–SWEEP SCHEME

S.W. Armfield[1], N.–H. Cho[2] and C.A.J. Fletcher[2]
[1]University of Western Australia, [2]University of Sydney, Australia

1. INTRODUCTION

Flows with a dominant flow direction usually permit some simplification of the full Navier–Stokes (FNS) equations to a system of reduced Navier–Stokes (RNS) equations via an order–of–magnitude removal of the axial dissipative terms [1]. Symbolic analysis [1,2] provides an a priori means of determining the character of the resulting RNS equations. In turn computational algorithms to exploit any non–elliptic features can be developed.

For example, for internal turbulent swirling flow in conical diffusers up to about 10 deg, additional simplifications can be made to the transverse momentum equation which render the RNS equation system completely non–elliptic [3] and produce accurate solutions in a single downstream march [4].

For larger diffuser angles (≈ 20 deg), but with sufficient swirl to prevent axial separation at the wall, the additional simplifications to the transverse momentum equation cannot be justified. Symbolic analysis indicates that appropriate RNS equations have an elliptic character associated with the pressure, continuity interaction. But the velocity field, if the pressure field could be temporarily frozen, is non–elliptic and could be determined by a marching algorithm coinciding with the dominant flow direction.

The present paper is concerned with a multisweep (repeated downstream marches) scheme which exploits the character of the governing equations. The resulting algorithm is applied to turbulent swirling flow in a 20 deg conical diffuser to predict both the mean flow and turbulence quantities.

2. GOVERNING EQUATIONS

For steady incompressible flow the Reynolds–averaged Navier–Stokes equations in coordinate–free tensor form are

$$u^i_{;i} = 0 \tag{1}$$

$$\rho u^j u^i_{;j} + \rho \tau^{ij}_{;j} + g^{ij} p_{,j} - \mu g^{\ell j} u^i_{,\ell j} = 0, \tag{2}$$

where u is the mean velocity vector, τ_{ij} is the Reynolds stress tensor, p is the pressure and g is the metric tensor. Accurate computational solutions have been obtained using both the k–ϵ and algebraic Reynolds stress (ASM) turbulence models to relate τ^{ij} to the mean flow and k, ϵ. The specific form of the equations, the choice of constants and a comparison of the ability of the two turbulence models to predict the mean flow and turbulence quantities are provided in companion papers [5,6]. In this paper only k–ϵ turbulence model results are presented.

For the test problem of axisymmetric swirling flow in a conical diffuser the above coordinate–free equations are specialised to spherical coordinates. An order–of–magnitude analysis of the governing equations indicates that terms associated with downstream (x) diffusion, both laminar and turbulent, are negligible in comparison with transverse (θ) diffusion terms. Removal of the downstream diffusion terms produces a system of reduced Navier–Stokes (RNS) equations.

3. COMPUTATIONAL ALGORITHM

The application of symbolic analysis leads to the following solution strategy. If the pressure field is frozen the velocity solution can be obtained from a downstream spatial march. However the pressure solution algorithm must reflect the elliptic influence of the pressure. Consequently the solution is upgraded sequentially.

The sequence of solution is as follows. The domain is marched once in the axial direction. At each axial location all the momentum equations and the k and ϵ equations are inverted to obtain new values at that location. The resulting axial velocity approximation is corrected to enforce constant mass flow [4], and simultaneously the axial pressure component is obtained. The axial pressure component is identified with the wall pressure. Since forward differencing is used to discretise the axial pressure gradient in the x–momentum equation the elliptic influence of the pressure is partially catered for.

To obtain the radially–dependent pressure component a discrete Poisson equation is constructed to satisfy local continuity and solved with the velocity field frozen. The sequence of downstream velocity march and pressure Poisson solver is repeated until a pre–set convergence criterion is obtained.

Second–order differencing on a regular (non–staggered) exponentially–stretched mesh is used for discretisation in the radial (θ) direction. A hybrid of two–point upwind and central differencing is used to discretise axial convective terms. For the present flow conditions a 40 x 40 grid is sufficiently refined [6] to avoid truncation error contamination from the hybrid discretisation. However for flows with local axial flow reversal it is expected that generalised four–point upwind discretisation [1] for the convective terms will be desirable.

At every downstream station each velocity component is upgraded sequentially via a scalar tridiagonal system spanning the radial (θ) domain (Fig. 1). The discrete Poisson equation for the radial pressure component is iterated using an ADI/approximate factorisation construction.

It has been found, particularly with strongly diffusing flows, that it is necessary to under–relax all the variables. Additionally the Poisson equation is not solved fully at each iteration, rather the result is used to correct the stored pressure field, and typically only around five sweeps of the domain in each direction are made to obtain the pressure correction. The pressure solution will then converge with the overall convergence of all the dependent variables.

4. RESULTS AND DISCUSSION

Typical convergence data is shown in Fig. 2. The solid line denotes the axial momentum equation and the dotted line denotes the k equation. These results have been obtained with a k–ϵ turbulence model on a 25(θ) x 20(x) axial grid. Computation of the u,v and w solutions from the momentum equations uses an under–relaxation factor of 0.1 for the first 80 iterations and 0.4 thereafter. Generally the iterative convergence is more robust with the k–ϵ turbulence model than with the ASM turbulence model. This is essentially due to the greater diagonal dominance in the tridiagonal systems of equations formed from the momentum equations when a k–ϵ turbulence model is used [6]. It is of interest that the present algorithm converges more rapidly with ASM turbulence modelling when the full Navier–Stokes equations are used in place of the reduced Navier–Stokes equations.

Computed solutions generated by the present algorithm are shown in Figs. 3 to 6. These results have been obtained with a k–ϵ turbulence model on 40(θ) x 40(x) grid with a two–layer wall function adjacent to the diffuser wall. For the present results the experimental data just downstream of the diffuser entrance (Station 2, Fig. 1) is used as an upstream boundary condition. Convergence is assumed when the maximum residual error in the x–momentum equation is less than 1×10^{-3}.

The development of the axial flow distribution in the diffuser is shown in Fig. 3. The influence of the swirl, causing a reduction in the axial flow close to the axis and an increase near the wall, is clearly evident. Generally good agreement is indicated with the experimental data of Claussen and Wood [7]. The corresponding decay in the swirl velocity is shown in Fig. 4.

The downstream development of the turbulent kinetic energy, k, is indicated in Fig. 5. The flowfield is characterised by relatively low levels of k close to the axis and a local peak close to the wall. The local peak initially grows in the downstream direction and then decays. This behaviour, which is discussed elsewhere [6], is predicted closely. The Reynolds stress, $\overline{u'v'}$, behaviour (Fig. 6) indicates a corresponding peak adjacent to the wall which grows and then decays. To obtain accurate predictions of the normal stresses the use of ASM turbulence models is preferred [6].

5. CONCLUSIONS

A sequential multisweep algorithm has been constructed to exploit the non–elliptic nature of the velocity field and the elliptic nature of the pressure field for swirling turbulent diffuser flows. The iterative convergence behaviour of the algorithm is indicated. It is found that accurate predictions of the mean and major turbulence quantities are obtained when the reduced Navier–Stokes equations are used with a k–ϵ turbulence model.

REFERENCES

1. C.A.J. Fletcher, *Computational Techniques for Fluid Dynamics*, Springer–Verlag, 1988, Vol. 2
2. C.A.J. Fletcher and S.W. Armfield, *Comm Appl Numer Analysis*, 4 (1988), 219–226
3. S.W. Armfield and C.A.J. Fletcher, *Comm Appl Numer Analysis*, 2 (1986), 377–383
4. S.W. Armfield and C.A.J. Fletcher, *Int.J.Num.Meth. Fluids*, 6 (1986), 541–556
5. S.W. Armfield and C.A.J. Fletcher, "Comparison of k–ϵ and Algebraic Reynolds Stress Models for Swirling Diffuser Flow", *Int.J.Num.Meth. Fluids*, (1988, to appear)
6. S.W. Armfield, N.–H. Cho and C.A.J. Fletcher, "Prediction of Turbulence Quantities for Swirling Flow in a Conical Diffuser", (1988, submitted)
7. P.D. Clausen and D.H. Wood, 6th Symp. Turb. Shear Flows, Toulouse, 1987, 1.3.1–5

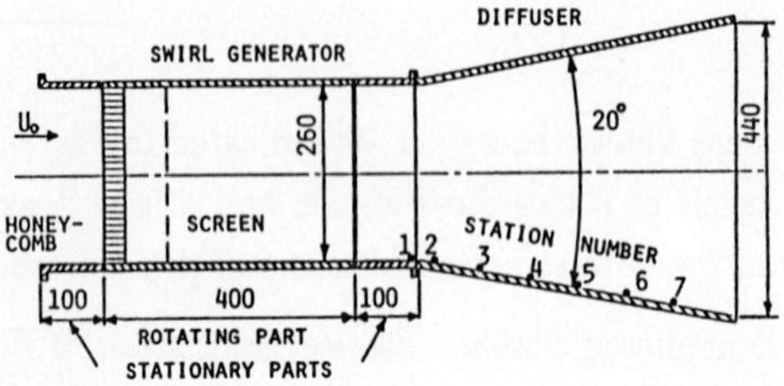

Fig. 1 Computational domain for swirling diffuser flow.

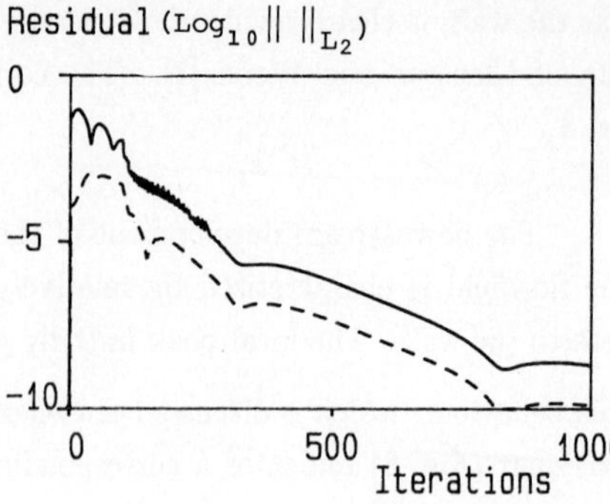

Fig. 2 Iteration convergence.

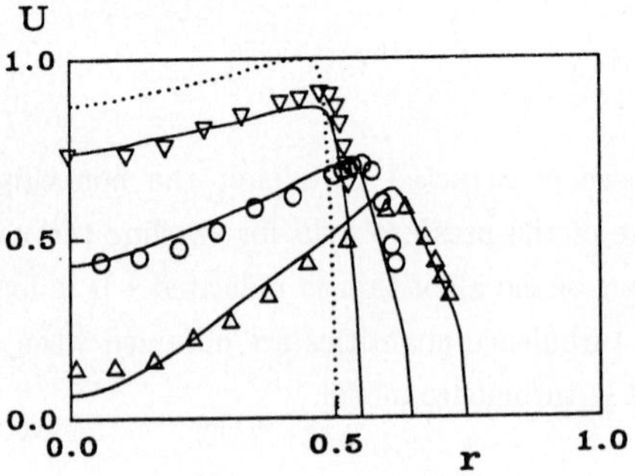

Fig. 3 Radial distribution of axial velocity.
.....initial (upstream) profile
▽,○,△experimental data [7].

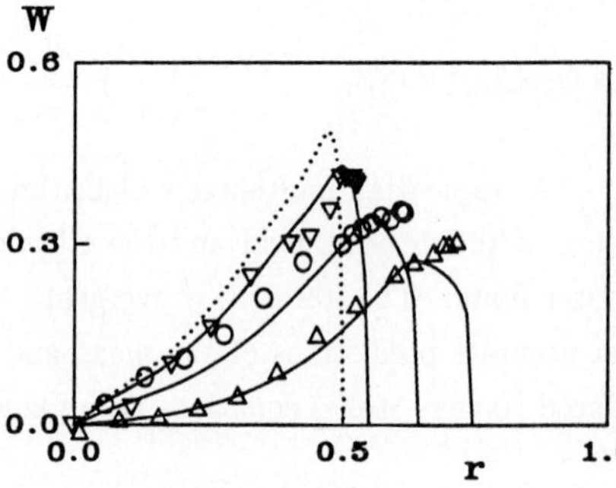

Fig. 4 Radial distribution of circumferential velocity.

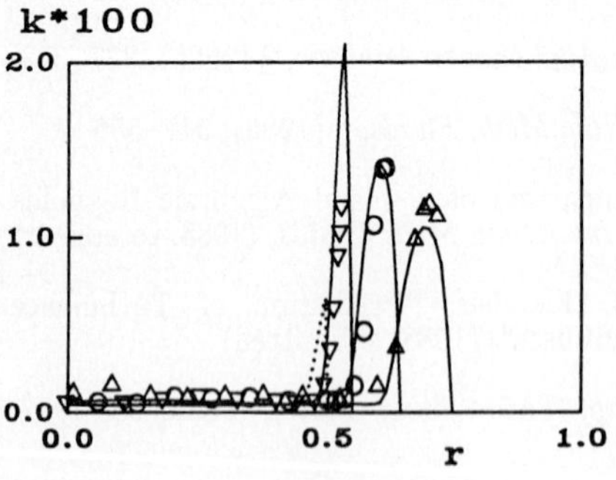

Fig. 5 Radial distribution of turbulent kinetic energy.

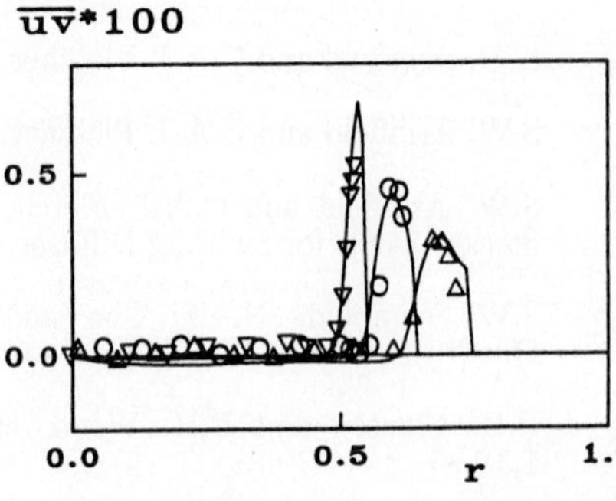

Fig. 6 Radial distribution of Reynolds stress $\overline{uv}$.

A PRESSURE GRADIENT FIELD SPECTRAL COLLOCATION EVALUATION FOR 3-D NUMERICAL EXPERIMENTS IN INCOMPRESSIBLE FLUID DYNAMICS

by

M. AZAIEZ ++-+, G. LABROSSE ++, H. VANDEVEN +++

++ Laboratoire FAST, UA CNRS
Dept. Hydrodynamique, Université de Paris Sud
Bat 502, 91405 ORSAY Cedex, FRANCE

+ Laboratoire d'Analyse Numérique, Université Paris XI, UA CNRS

+++ Laboratoire d'Analyse Numérique, Université Paris VI, UA CNRS

Three dimensional numerical experiments in Fluid Dynamics should be more and more feasible, owing to the increasing of computer performances and memory capacities. Many physical situations of flows confined in simple geometries are, then, worth being studied in some details with the help of accurate 3-D numerical schemes, formulated in primitive variables. Spectral methods are well designed for this goal (1). To evaluate a velocity field with a spectral accuracy is not a very hard task (2), providing the fluid properties are held constant (with temperature for example) and the advective terms are treated explicity in time. But the difficulty arises for the determination of the pressure field to get a residual to the constraint equation div($\underline{v}$)=0 of the same magnitude as the one obtained for the momentum conservation equation. This is clearly a criterion of coherence for the numerical scheme and of confidence for the numerical results physical reality.

An iterative spectral collocation Stokes solver has been implemented for mixed boundary conditions on the velocity. We have used an unique grid made of the Gauss-Lobatto points on which the incompressibility condition is imposed.

Both, div($\underline{v}$)=0 and $\underline{\nabla}p$ are obtained, with the spectral accuracy, in a reasonable number of iterations as compared to the collocation points number.

This code has been tested in 2-D orthogonal geometries, cartesian and bounded cylindrical annulus, its extension to 3-D being straightforward.

1 - NUMERICAL FORMULATION OF THE PROBLEM

Let us consider the Stokes system to be solved within a domain Ω bounded by Γ :

$$(\underline{\nabla}^2 - h^2)\underline{v} - \underline{\nabla}p = \underline{F} \quad \text{in } \Omega$$

boundary conditions for $\underline{v}$ on Γ

$$\text{div}(\underline{v}) = 0 \quad \text{in } \bar{\Omega}$$

where h^2 and $\underline{F}$ are given Helmholtz constant and source term respectively.

The first two equation lines define a linear non singular operator, H, acting on $\underline{v}$, and can be writen formally :

$$H\,\underline{v} - \underline{\nabla}'p = \underline{F}'$$

where $\underline{\nabla}'$ and $\underline{F}'$ are deduced from the corresponding quantities by properly taking into account the boundary conditions on $\underline{v}$.

The operator H and $\underline{\nabla}'$ are not anymore pure differential operators and do not commute with the divergence operator. The constraint $\mathrm{div}(\underline{v}) = 0$ gives, then, the following operator acting on p :

$$Ap = (\mathrm{div}.H^{-1}.\underline{\nabla}')p = -\,(\mathrm{div}.H^{-1})\underline{F}'$$

Noting that the differential order of A being zero, no extra conditions are required on p to solve it, as it should be in this problem.

From a practical point of view : 1) The A operator mixes all the space directions and its matricial representation is full. For the time being, there is no efficient direct solver for it. 2) Some redundancies appear between the imposed conditions on $\underline{v}$: $\underline{v}|_\Gamma = \underline{v}_0$ (for example) and $\mathrm{div}(\underline{v})=0$ (4). The pressure and velocity mathematical representations must then be properly chosen if one needs an unique and accurate determination of the pressure field (3). This amounts, for example, to use staggered grids in a spectral collocation code. The theoritical analysis of such a procedure has been performed for a Legendre spectral decomposition of the p, $\underline{v}$ fields in a 2-D cartesian geometry (3). Otherwise, spurious pressure modes are present in the linear system. They are well localised in the spectral space (4) when a Tau method is chosen to solve the differential system, while some of them are distributed on the whole physical and spectral spaces in the case of a collocation method (3). In a 2-D bounded cylindrical annulus, we have the same spurious modes with an unique grid as in the 2-D cartesian case. And, from a theoritical analysis, it can be shown that the ($\underline{v}$,p) numerical solution is unique if those modes are excluded from the p representation space.

2 - THEORITICAL STATEMENTS ABOUT THE 2-D BOUNDED CYLINDRICAL ANNULUS CASE

The problem to be solved is :

$$-\mu[((r+\alpha)\underline{v}_{,r})_{,r}/(r+\alpha) + \underline{v}_{,zz}] + \underline{\nabla}p = \underline{F} \quad \text{in } \Omega$$

$$\underline{v} = 0 \quad \text{in } \Gamma$$

$$\mathrm{div}(\underline{v}(r+\alpha)) = 0 \quad \text{in } \bar{\Omega}$$

where the constant α comes from the transform [R1,R2] to [-1,1]

Let us define the functional spaces X_N and M_N

$$X_N = (H^1_0(\Omega))^2 \cap (P_N(\Omega))^2$$

$$M_N \text{ in } L^2_0(\Omega) \cap P_N(\bar{\Omega})$$

and the collocation point set $E_N = \{ \underline{x}=(x_i,x_j)\ 0\le i\le N,\ 0\le j\le N \}$ where the x_i are solution of $(1-x^2)L'_N$, L_N being the Nth degree Legendre polynomial. One has : $x_0 = -1 < x_1 < x_2 \ldots\ldots < x_N = 1$.

The discretized formulation of the problem is then : find $(\underline{v}_N, p_N)$ in $X_N \times M_N$ such that

$$-\mu[((x_i+\alpha)\underline{v}_{,r})_{,r}/(x_i+\alpha) + \underline{v}_{,zz}](\underline{x})+\underline{\nabla}p(\underline{x}) = \underline{F}(\underline{x}),\ \underline{x} \text{ in } \Omega\cap E_N$$

$$\underline{v}(\underline{x}) = 0,\ \underline{x} \text{ in } \Gamma\cap E_N$$

$$[(1/(x_i+\alpha))[(x_i+\alpha)v_r]_{,r} + v_{z,z}](\underline{x}) = 0,\ \underline{x} \text{ in } \bar{\Omega}\cap E_N$$

Making use of the discretized Legendre scalar product $(f,g)_N$ one has the following steps :

a) <u>Proposition</u> : The problem is equivalent to the following variational problem : find $(\underline{u}_N, p_N)$ such that :

for any $\underline{v}_N$ in X_N, $\mu(\underline{\nabla}\ \underline{u}_N, (r+\alpha)\underline{\nabla}\ \underline{v}_N)_N + (\underline{\nabla}p_N, (r+\alpha)\underline{v}_N)_N = (\underline{F}, (r+\alpha)\underline{v}_N)_N$

for any q_N in $P_N(\bar{\Omega})$, $(\mathrm{div}(\underline{u}_N(r+\alpha)), q_N)_N = 0$

b) <u>Proposition</u> : One reduces M_N to the complement of the spurious mode space, the hypothesis of the theorem given in (6) being verified, one has :

for any q_N in M_N, $\exists\ \underline{v}_N \neq 0$ in X_N such that

$$(\underline{\nabla}q_N, \underline{v}_N)_N \ge c(N)\| \underline{v}_N \|_X \| q_N \|_M$$

c) <u>Theorem</u> : For any $\underline{F}$ in $(C^0(\bar{\Omega}))^2$, the collocation spectral problem admits an unique solution in $X_N \times M_N$. Moreover, if the solution $(\underline{v},p)$ belongs to $(H^\tau(\Omega))^2 \times H^{\tau-1}(\Omega)$, $\tau\ge 0$, and if $\underline{F}$ belongs to $(H^\sigma(\Omega))^2$ with the integer $\sigma > 1$, then it exists a constant independant of N such that :

$$\| \underline{v}-\underline{v}_N \|_X \le C_1\{ N^{1-\tau}\| \underline{v} \|_{\tau,\Omega} + N^{1-\sigma}\| \underline{F} \|_{\sigma,\Omega} \}$$

$$\| p-p_N \|_M \le C_2\{ N.C(N)(\| \underline{v} \|_{\tau,\Omega} + \| p \|_{\tau-1,\Omega}) + N^{3-\sigma}\| \underline{F} \|_{\sigma,\Omega} \}$$

3 - <u>NUMERICAL FORMULATION OF THE PROBLEM</u>

The Stokes system has been solved in the following configurations :

- inhomogenous Dirichlet or Neumann boundary conditions for $\underline{v}$.
- 2-D cartesian or meridian plan of a bounded cylndrical annulus.
- Legendre or Chebyshev collocation method with an unique grid for the pressure and velocity fields. 8 spurious pressure modes are then expected (4) to be present in our linear problem. Their gradient cancels on the Ω collocation points.

The Stokes solver proceeds as follows :

- in a preprocessing stage, the H operator eigenspaces are built up following the well known sequential diagonalisation procedure (2).

-the A operator is then solved for the pressure gradient field evaluation. An extended conjugate gradient method, as proposed in (5), has been used for this unsymetric operator.

-then comes the velocity field determination.

Convergence rates and accuracy have been measured by running the code on polynomial solutions (whose higher degree is given by the spectral cut-off) and on large spectral band solutions.

The essential features are :

- there is no significant difference between our results in Legendre or Chebyshev decompositions. We present therefore only data obtained with the Chebyshev collocation method.

- the iterative procedure converges to the expected spectral accuracy of the $\mathrm{div}(\underline{v})$ and $\underline{\nabla}p$ fields. A moderate number of iterations, as compared to the number of collocation points, allows to get very satisfactory results. Figure 1 presents, as function of the iteration number n, the euclidian norm of $\mathrm{div}(\underline{v}^{(n)})$ and $\underline{\nabla}(p^{(n)}-p_{exact})$ in the case of a polynomial solution corresponding to an inhomogeneous Dirichlet boundary conditions on $\underline{v}$. Choosing two spectral cut-off, 10^2 and 19^2, we have reported only results with $\| \mathrm{div}(\underline{v}^{(n)}) \|_2 \geq 10^{-16}$: 30 and 44 iterations, respectively, have been necessary to go down 14 decades in $\mathrm{div}(\underline{v})$. On figure 2 are given similar results for bounded cylindrical annulus, with inhomogeneous Dirichlet or Neumann boundary conditions on $\underline{v}$. Curve (c) on this figure corresponds to a non limited spectral band solution.

The $(\underline{v}, \underline{\nabla}p)$ fields can thus be obtained at any given accuracy, just by choosing adequate spectral cut-off. The pressure field situation is somewhat different because of the spurious modes. On table 1 is shown, with an arbitrary scale, the map of $\delta p=p_N-p_{exact}$ on the grid points, excluding the four corners of our domain where there is no access to the pressure. Of course since $\underline{\nabla}\delta p = 0$ on the internal points, these pressure modes have no effect, neither on the iterative procedure, nor on the velocity field evaluation. It is only a matter of filtering them out if one needs an accurate pressure field !

4 - CONCLUSION

The Stokes system can be solved within the framework of spectral methods to get accurate solenoidal velocity fields. The present drawback of the numerical procedure lies in the pressure iterative loop. Although its convergence is surprisingly good, it will not be a refinement to conceive an accelerated algorithm, particularly for unsteady 3-D incompressible fluid flows.

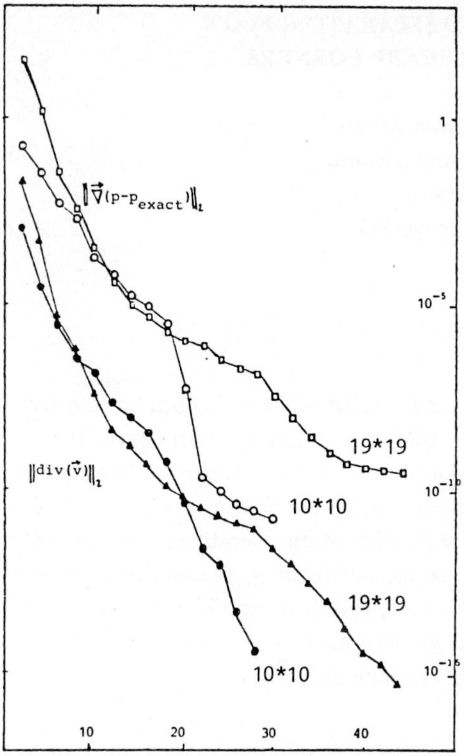

Figure 1

2-D Cartesian geometry, collocation Chebychev method
Inhomogeneous Dirichlet boundary conditions on velocity.
Are presented, as function of the iteration number n,
the Euclidian norms of $\mathrm{div}(\vec{v}^{(n)})$ and $\vec{\nabla}(p^{(n)}-p_{exact})$.

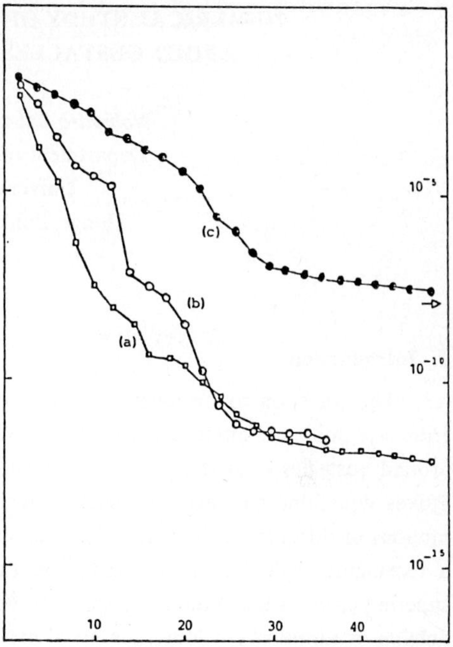

Figure 2

Meridian plan of a bounded cylindrical annulus. The Euclidian norm of $\mathrm{div}(v^{(n)})$ is presented, as function of the iteration number n, for the following cases:

curve (a) : polynomial solution, with inhomogeneous Dirichlet b.c., spectral cut-off are 11*11

curve (b) : polynomial solution, with inhomogeneous Neumann b.c., spectral cut-off are 20*20

curve (c) : non limited spectral band solution, same b.c. as in (b), spectral cut-off are 13*13.

→ indicates the expected spectral value.

.	17	14	17	14	17	.
17	10	17	10	17	10	17
14	17	14	17	14	17	14
17	10	17	10	17	10	17
14	17	14	17	14	17	14
17	10	17	10	17	10	17
.	17	14	17	14	17	.

Table 1.

REFERENCES

(1)-D.GOTTLIEB, S.A.ORZAG (1977), "Numerical analysis of spectral methods : theory and applications. " SIAM, Philadephia.

(2)-D.B.HAIDVOGEL, T.ZANG (1979), "The accurate solution of Poisson's equation by expansion in Chebychev polynomials.", J.Comp.Phys, 30, 167-180

-P.HALDENWANG, G.LABROSSE, S.ABBOUDI, M.DEVILLE (1984), "Chebyshev 3-D spectral and 2-D pseudospectral solvers for the Helmholtz equation." J.Comp.Phys., 55, 115-128.

(3)-C.BERNARDI, Y.MADAY, B.METIVET (1987), "Calcul de la pression dans la résolution spectrale du problème de Stokes." La Recherche Aérospatiale, n° 1987-1, 1-21.

(4)-P.HALDENWANG (1984), "Résolution tridimensionnelle des équations de Navier-Stokes par methodes spectrales Tchebycheff : application à la convection naturelle." Ph.D.Thèse, Université de Provence.

-P.HALDENWANG, G.LABROSSE (1986), proceedings of the 6th Int.Symp. on Finite Element Methods in Flow Problems, ed. by M.O.Bristeau, R.Glowinski, A.Hausel, J.Periaux, 261-266

(5)-O.AXELSON (1980), "Conjugate gradient type methods for unsymetric and inconsistent systems of linear equations." Linear algebra and its applications, 29, 1-16

(6)-F.BREZZI (1974), "On the existence, uniqueness and approximation of saddle-point problems arising from Lagrange multipliers.", RAIRO Anal.Numer. 8/R2, 129-151.

NUMERICAL STUDY OF THE 3D SEPARATING FLOW ABOUT OBSTACLES WITH SHARP CORNERS

Nobuhiro Baba and Hideaki Miyata
Department of Naval Architecture
University of Tokyo
Hongo,Bunkyo-ku,Tokyo 113

1. Introduction

The transiton to turbulence in the separating and free shear flows is of great interest from scientific and engineering points of views. We study the vortical structures of the flows around obstacles with sharp corners by numerical integration of the incompressible Navier-Stokes equations and explain some features of the nonlinear interaction between vortical motions of different scales. The three-dimensional simulation of the transitional flow around a rectangular cylinder demonstrates the existence of counter-rotating, streamwise vortices superimposed on the spanwise-organized, vortical structure and their contribution to the production of turbulence. Some numerical experiments are also made to examine the effects of numerical errors arising from the nonlinear terms on the finite difference solution.

2. Numerical method

The computational procedure used here is common to our previous works (Baba & Miyata 1987 and Miyata et al. 1987). The Navier-Stokes equations and the continuity equation are solved by the finite difference method, of which the fundamental algorithm is based on the MAC method by Harlow & Welch (1965).

We use the three forms for the advective terms of the Navier-Stokes equations, namely the gradient form $u^j\partial_j u^i$, the divergence form $\partial_j(u^i u^j)$, and the rotation form $\partial_i(\frac{1}{2}u^j u^j) - u^j(\partial_i u^j - \partial_j u^i)$. These forms have different conservative properties from each other. Many works have emphasized the importance of conservative properties in the finite differnce solution (Arakawa 1966, Piacsek & Williams 1970, Mansour et al. 1979 and others). The divergence form unconditionally conserves momentum while the other two forms conditionally conserve it only if the continuity constraint is exactly satisfied. The rotation form has the conditionally conservative property for kinetic energy, but the other forms do not have this property. These three forms with different conservative properties as well as different truncation errors are supposed to give rise to noticeable influences on the solution.

The time advancement is made explicitly using the second-order Adams-Bashforth method for the advective terms and the Euler method for another terms. The continuity constraint is satisfied by solving the Poisson equation for pressure at each time step by the relaxation method. We employ the standard Cartesian grid system fixed on the rectangular obstacles with uniform grid spacing. The centered differences are employed for the spatial derivatives : the fourth-order accurate centered differences for the advective terms and the second-order accurate ones for another terms. To obtain a stable solution at higher Reynolds numbers, we incorporate the nonlinear dissipation model as $-\frac{\alpha_4}{24}|u^j|\Delta_j^3 \overset{c}{\underset{2}{D}}{}_j^4 u^i$ into the Navier-Stokes equation. Here $\overset{c}{\underset{2}{D}}{}_j^4$ denotes the second-order accurate centered difference operator of the fourth-derivative with respect to x^j. We control the magnitude of the fourth-derivative dissipation with the parameter α_4 so that the diffusion effect may be sufficiently small.

3. Results and discussion

3.1. Flow over a double step

First we examine the effects of the above forms of advective terms on the solution of the vortical flow over a double step. Solution procedure is exactly the same in the three cases. The computations are carried out for the two-dimensional flows at the lower Reynolds numbers R of 200, 500 and 3000 based on the height of a double step. The difference of the form hardly gives effect on the solution at R=200 but meaningful effect on the solutions at R=500 and 3000. The contour maps of vorticity at R=500 are shown in Fig.1. In the case of the gradient form, the primary vortex shed from a double step is not transferred downstream but enlarged excessively until the solution breaks down at the dimensionless time t=74. In the case of the divergence form, unfavorable oscillations of variables take place above the top surface of the step and it causes the solution break at the earlier time of t=11. Only the case of the rotation form shows successive vortex shedding. It is shown that the difference of the form of the advective terms gives noticeable influences on the simulated flow. This suggests the real significance of numerical errors arising from the nonlinear terms. It seems clear that the rotational form with the conservative property of kinetic energy as well as momentum is most rubust for the simulation of these vortical flows.

The time sequence of the vortex development at R=3000 computed by the rotation form is shown in Fig.2. The separated flow from the leading edge rolles into the small-scale vortex B above the step, while the separated flow from the trailing edge forms the primary downstream vortex A behind the step. The former vortex B merges into the latter one A, and the two vortices rotate around each other until they amalgamate. The vortex A+B shed downstream induces a secondary vortex C with the opposite rotation in the vicinity of the rear corner. This secondary vortex C interacts not only with the downstream vortex A+B but also with another leading-edge vortex D in the subsequent cycle. The time variation of the location of the shed vortices are shown in Fig 3, where the location is determined as the x^1 coordinate of the vortex core with the minimum value of pressure. This plot shows the nonuniformity of the speed of vortex transfer, which is attributable to the above nonlinear interaction between vortices.

An example of computational results of the three-dimensional transitional flow at the higher Reynolds number of 13000 is compared with the visualization experiment in Fig.4. It is shown that the flow past the double step is disturbed by the small-scale, three-dimensional motion, resulting in the turbulence like appearance.

3.2. Flow around a rectangular cylinder

The flow around a rectangular cylinder at Reynolds number of 6000 ,which is visualized with the hydrogen-bubble technique as Fig.5, is chosen to study the three-dimensional structure of separated shear flows. The flow is accelerated from the rest state, while a very weak three-dimensional disturbance is given all over the accelerating flow. Fig.6 shows the formation of the small-scale vortices in the separated shear layer emerging from the leading edge as well as the large-scale Strouhal vortices behind the cylinder although the periodically oscillating wake has not yet been fully developed. The several leading-edge vortices merge into the Strouhal vortex in the formation region until it is shed downstream. More than two vortices rotate around each other until they amalgamate. This process of the nonlinear interaction between two-dimensional vortices of different scales is very similar to those observed in the visualization experiment.

In this early stage of vortex development, the flow is subjected to three-dimensional instability. Fig 7 is the contour map of streamwise vorticity at the same time as Fig 6. Counter-rotating,streamwise vortices develop in the separating shear flow around a cylinder. By the side of the cylinder, the multiple-row structure of streamwise vortices is formed between the cylinder wall and the outer freestream although their streamwise length is relatively short. They are located on both sides of the inflexion point of the streamwise velocity profile as shown in Fig.8. The double-row structure is similar to that predicted by the weak nonlinear instability theory by Benney (1961). Behind the cylinder, the streamwise vortices are highly elongated from the vicinity of the trailing edge into the Strouhal vortex, so that their structure of the pairs of counter-rotating, streamwise vortices become much clearer. This structure is very similar to that of the mashroom-shaped vortex pairs observed experimentally by Bernal & Roshko (1986) in the plane mixing layer and by Wei & Smith (1986) in the near wake of a circular cylinder. Fig.9 shows the relation of this three-dimensional structure to the fluctuated velocity field. The counter-rotating, streamwise elongated vortices exist in a layer of strong mean gradient of streamwise velocity, so that they make the periodically oscillating velocity field where the fluid of low speed is moved outwards from the wake of the cylinder and the fluid of high speed is moved in from the outer freestream. The measurement with a hot-film anemometer indicates that the separating flow from the leading edge rapidly grows downstream and degenerates to turbulence in the vicinity of the trailing edge. These results suggest that the pairs of counter-rotating, streamwise vortex contribute to the production of turbulence and that they may play an important role in the transition to turbulence and its development in the separating shear flow.

References

Arakawa,A. (1966) *J.Comput.Phys.1*, 119.
Baba,N. and Miyata,H. (1987) *J.Comput.Phys.69* 362.
Benny,D.J. (1961) *J.Fluid Mech.10*, 209.
Bernal,L.P. and Roshko,A. (1986) *J.Fluid Mech.170*, 499.
Harlow,F.H. and Welch,J.E. (1965) *Phys.Fluids 8*, 2182.
Mansour, N.N,Moin,P., Reynolds,W.C., and Ferziger,J.H. (1979) "Improved Methods for Large Eddy Simulations of Turbulence," in *Turbulent Shear Flows I*, edited by Durst,F.et al. (Springer,Berlin,1979),386
Miyata,H., Sato,T., and Baba,N. (1987) *J.Comput.Phys.72*, 393.
Piacsek,S.A. and Williams,G.P. (1970) *J.Comput.Phys.6*, 392.
Wei,T. and Smith,C.R. (1986) *J.Fluid Mech.169*, 513.

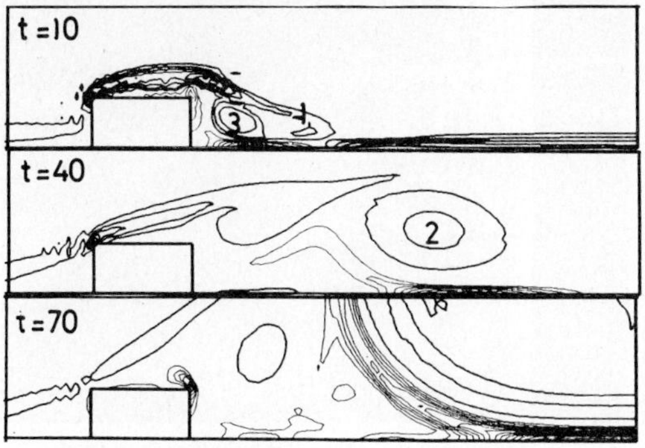

(a) Gradient form

(b) Divergence form

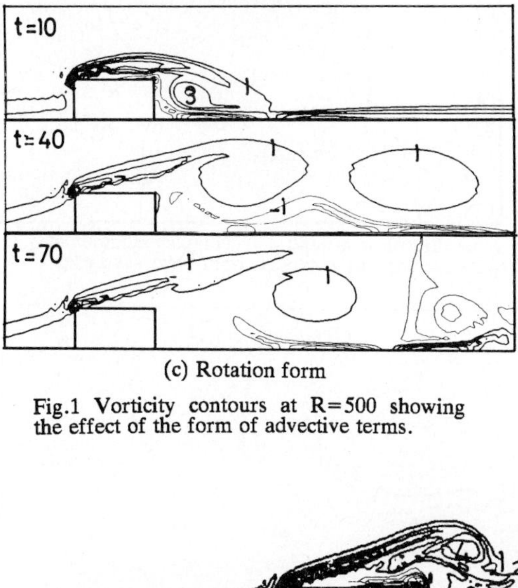

(c) Rotation form

Fig.1 Vorticity contours at R=500 showing the effect of the form of advective terms.

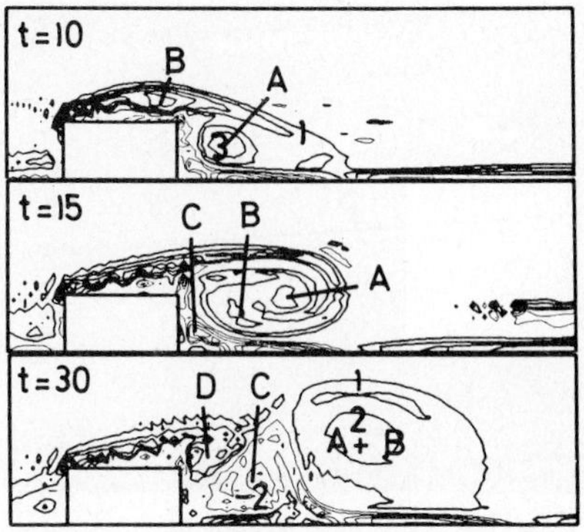

Fig.2 Time sequence of vorticity contours at R=3000 in the case of the rotation form showing the nonlinear interaction between the two-dimensional vortices over a double step.

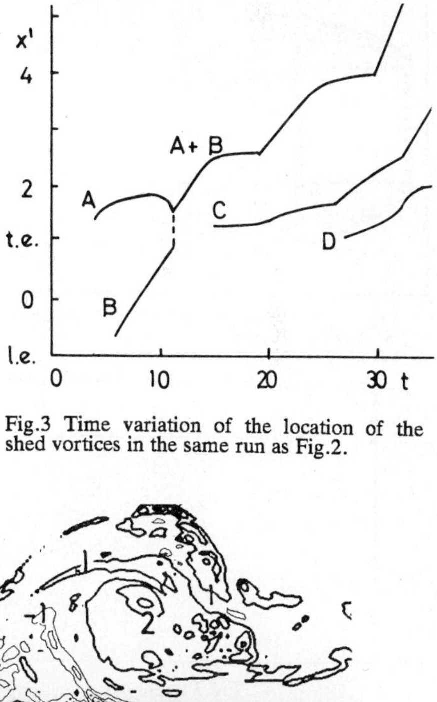

Fig.3 Time variation of the location of the shed vortices in the same run as Fig.2.

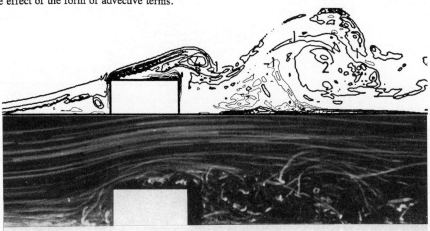

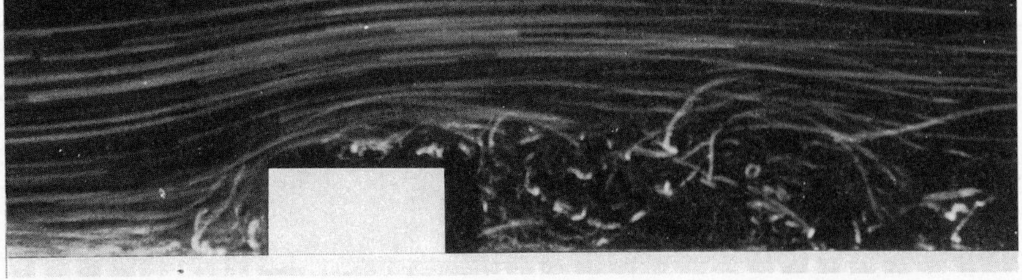

Fig.4 Vorticity contours of the three-dimensional transitional flow at R=13000 (above) and a visualization photograph at R=13170 (below).

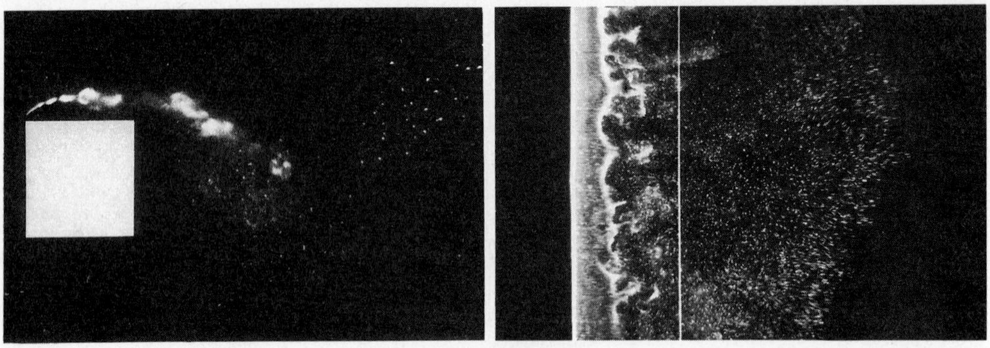

Fig.5 A set of photographs of the separated flow around a rectangular cylinder at R=6000, visualized with the hydrogen-bubble technique, left ; side view, right ; top view.

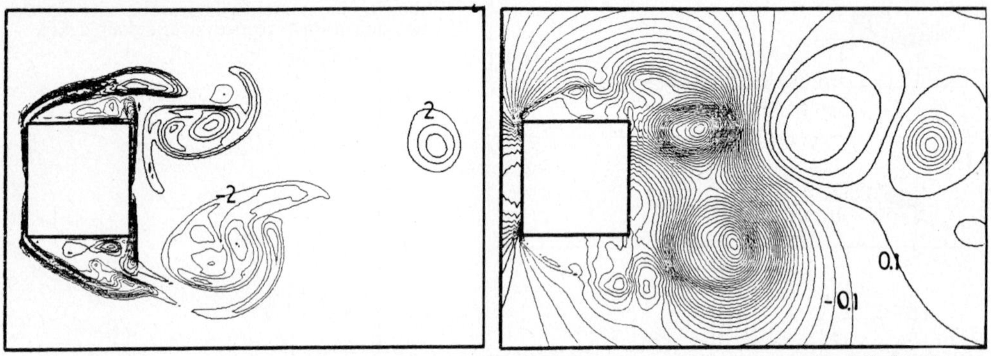

Fig.6 Spanwise-organized vortical structure of the separating flow around a rectangular cylinder at R=6000 at the early stage of t=10, left ; averaged spanwise vorticity contours, right ; averaged pressure contours.

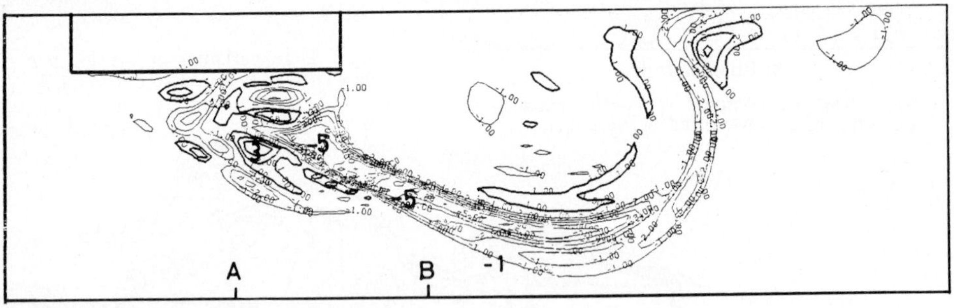

Fig.7 Streamwise vorticity contours at the same time as Fig.6.

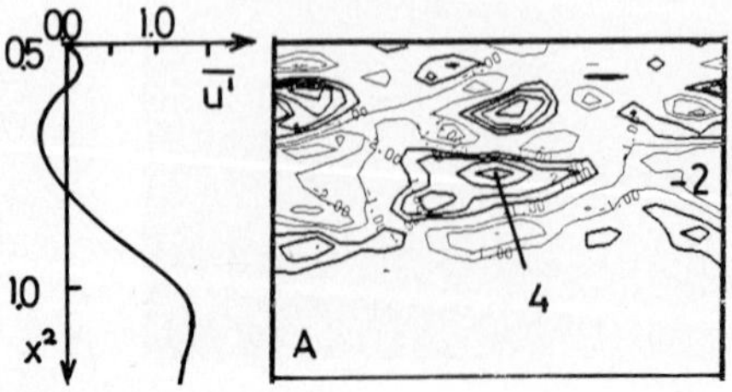

Fig.8 Streamwise vorticity contours at the cross section A and averaged streamwise velocity profile.

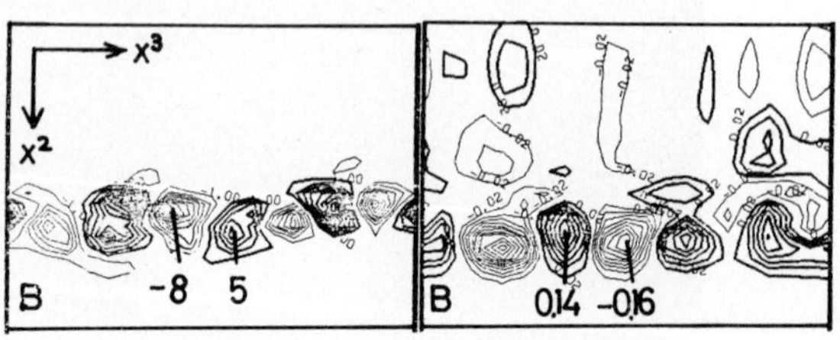

Fig.9 Streamwise vorticity contours (left) and fluctuated streamwise velocity contours (right) at the cross section B.

FINITE VOLUME TVD RUNGE KUTTA SCHEME
FOR NAVIER STOKES COMPUTATIONS

F. Bassi*, F. Grasso**, M. Savini***

* Universita' di Catania, Istituto di Macchine, via A. Doria 6, 95125 Catania, Italia

** Universita' di Roma "La Sapienza", Dip. di Mecc. e Aeronautica, via Eudossiana 18, 00100 Roma, Italia

*** CNR/CNPM, Viale Baracca 69, 20068 Peschiera Borromeo (Mi), Italia

INTRODUCTION

In the present work a finite volume algorithm that exploits the properties of both total variation diminishing schemes and Runge Kutta ones is developed for the solution of the compressible Navier Stokes equations. To date the approach followed by most of the researchers for the solution of the compressible Navier Stokes equations has been that of treating separately the inviscid and viscous parts [1-2]. Thus many upwind and/or centered schemes developed for the Euler equations have been extended to solve high Reynolds number flows. Among the many algorithms that have recently appeared, total variation diminishing (TVD) schemes have properties, such as monotonicity, high order accuracy, entropy preserving, etc., that make them very attractive when high resolution of discontinuities is needed [3-8]. However very little has been published on their use for Navier Stokes equations [9]. Here is proposed a finite volume method that treats the inviscid (Euler) part by means of an upwind biased second order TVD scheme that follows the one originally developed by Harten [3], whilst the viscous contribution is central differenced. Time integration of the system of ordinary differential equations, obtained after spatial discretization, is performed by a Runge Kutta algorithm. Enforcement of the TVD property requires careful analysis for the selection of the coefficients of the Runge Kutta algorithm. In the next sections the numerical solution is described and three applications are presented to demonstrate the ability of the method to solve complex viscous flows by comparison with experiments.

NUMERICAL SOLUTION

The governing equations are the compressible Navier Stokes equations, that in conservation form become

$$\frac{\partial}{\partial t}\int_V W\,dV = -\oint_{\partial V}\left(F_E - F_V\right)\cdot\vec{n}\,dS \qquad (1)$$

where

$$W = \left[\rho, \rho\vec{u}, \rho E\right]^T$$

$$F_E = \left[\rho\vec{u}, \rho\vec{u}\vec{u} + p, \rho\vec{u}\left(E + \frac{p}{\rho}\right)\right]^T$$

$$F_V = \left[0, \vec{\sigma}, -\left(\vec{q} - \vec{u}\cdot\vec{\sigma}\right)\right]^T$$

For the computed turbulent flows the Baldwin-Lomax model of turbulence is used [12]. The different nature of convective and viscous contributions can be effectively accounted for by using different discretization schemes. Hence a TVD scheme is employed for the Euler terms due to the propagation character of the inviscid contribution, whilst central differencing is used for the viscous terms. Several applications (mainly for Euler equations) of TVD schemes have shown the good properties of TVD approaches in the presence of shoks and/or contact discontinuities. In the present paper TVD is implemented in the framework of an explicit finite volume approach. For an arbitrary computational cell, as the one depicted in Fig. 1, the discretized (Euler) flux at cell interface i+1/2 is evaluated as follows:

$$\left(\vec{F}_E \cdot \vec{n}\Delta S\right)_{i+\frac{1}{2}} = [\frac{1}{2}(F_i + F_{i+1})\cos\theta_{i+\frac{1}{2}} + \frac{1}{2}(G_i + G_{i+1})\sin\theta_{i+\frac{1}{2}} + \Phi_{i+\frac{1}{2}}]\Delta S_{i+\frac{1}{2}} \tag{2}$$

where the flux function Φ modifies the fluxes F, G to make the scheme upwind biased and second order accurate through an antidiffusive flux contribution. Its expression is obtained via local characteristic decomposition along the normal to cell face in order to detect wind direction for each characteristic variable.

Let $U_{i+1/2}$ and $V_{i+1/2}$ be respectively the velocity components normal and tangential to cell face i+1/2 (which are equivalent to controvariant and covariant velocities), i.e.

$$U_{i+\frac{1}{2}} = (\ u\cos\theta + v\sin\theta)_{i+\frac{1}{2}}$$

$$V_{i+\frac{1}{2}} = (-u\sin\theta + v\cos\theta)_{i+\frac{1}{2}}$$

The eigenvalues (a) of the Jacobian of the unmodified numerical flux and the associated right eigenvector matrix (R) are given by the following

$$a^l_{i+\frac{1}{2}} = (U \pm c\qquad)_{i+\frac{1}{2}} \qquad\qquad l = 1,3$$

$$a^l_{i+\frac{1}{2}} = U_{i+\frac{1}{2}} \qquad\qquad l = 2,4$$

$$R_{i+\frac{1}{2}} = \begin{pmatrix} 1 & 1 & 1 & 0 \\ u - c\cos\theta & u & u + c\cos\theta & -\sin\theta \\ v - c\sin\theta & v & v + c\sin\theta & \cos\theta \\ H - Uc & \frac{q^2}{2} & H + Uc & V \end{pmatrix}_{i+\frac{1}{2}} \tag{3}$$

where Roe's averaging is used to evaluate u, v, c, H at i+1/2.
Hence the following formula for $\Phi_{i+1/2}$ is obtained [3]

$$\Phi_{i+\frac{1}{2}} = \frac{1}{2}\sum_{l=1}^{4}\left[g^l_i + g^l_{i+1} - Q\left(a^l_{i+\frac{1}{2}} + \gamma^l_{i+\frac{1}{2}}\right)\alpha^l_{i+\frac{1}{2}}\right]R^l_{i+\frac{1}{2}} \tag{4}$$

where Q, g^l, γ^l, α^l are, respectively, the numerical viscosity function (that renders the three-point scheme upwind biased), the antidiffusive flux (that makes the scheme second order accurate), the characteristic speed of the antidiffusive flux and the differences of the characteristic variables. The function Q here adopted is equal to

$$Q(z) = \begin{pmatrix} \frac{1}{2}\left(\frac{z^2}{\epsilon} + \epsilon\right) & z \le \epsilon \\ |z| & z > \epsilon \end{pmatrix} \tag{5}$$

The parameter ϵ has the effect of enforcing entropy condition for vanishing eigenvalues. A value of ϵ = .1 has here been used. The definition of the dissipation function as $Q(a) + Q(\gamma)$, as used by Jameson, introduces more dissipation.

Several formulas for the antidiffusive flux (or else the flux limiter) g can be constructed as function of successive differences of characteristic variables which satisfy TVD constraints, as indicated by Sweby [10]. In the present work Van Leer limiter is employed for the non linear fields and Roe's "superbee" limiter is used for the linear ones, i.e.

$$g_i^l = \frac{\tilde{g}^l_{i-\frac{1}{2}}\tilde{g}^l_{i+\frac{1}{2}} + |\tilde{g}^l_{i-\frac{1}{2}}\tilde{g}^l_{i+\frac{1}{2}}|}{\tilde{g}^l_{i-\frac{1}{2}} + \tilde{g}^l_{i+\frac{1}{2}}} \qquad l = 1,3$$

$$g_i^l = s\max\left[0, \min\left(2|\tilde{g}^l_{i+\frac{1}{2}}|, s\tilde{g}^l_{i-\frac{1}{2}}\right), \min\left(|\tilde{g}^l_{i+\frac{1}{2}}|, 2s\tilde{g}^l_{i-\frac{1}{2}}\right)\right] \qquad l = 2,4 \tag{6}$$

$$s = sign\left(\tilde{g}^l_{i+\frac{1}{2}}\right)$$

The characteristic speed $\gamma^l_{i+1/2}$ is given by

$$\gamma^l_{i+\frac{1}{2}} = \begin{cases} \dfrac{g^l_{i+1} - g^l_i}{\alpha^l_{i+\frac{1}{2}}} & \alpha^l_{i+\frac{1}{2}} \neq 0 \\ 0 & \alpha^l_{i+\frac{1}{2}} = 0 \end{cases}$$

The expression of the vector $\alpha^l_{i+1/2}$ is

$$\alpha^l_{i+\frac{1}{2}} = R^{-1}_{i+\frac{1}{2}}\Delta_{i+\frac{1}{2}}W$$

It must be pointed out that the second order TVD scheme described above is a five-point one. Hence it requires boundary conditions for the antidiffusive flux limiters g. In the present work at all boundaries the normal derivative of g is set equal to zero; moreover for all cells along the walls there is no contribution in the normal direction due to the modified flux Φ.

Time integration of the system of ordinary differential equation obtained after spatial discretization is performed by the following three stage Runge Kutta algorithm:

$$W^{(0)} = W^n$$

$$W^{(k)} = W^{(0)} - \alpha_k \frac{\Delta t}{V} P\left(F_E^{(k-1)}, F_V^{(0)}\right) \qquad k = 1,3 \tag{7}$$

$$W^{n+1} = W^{(3)}$$

where P (F_E, F_V) is the spatial discretization operator. When applying such an algorithm to a scalar hyperbolic equation with upwind spatial discretization, one finds that selecting $\alpha_1 = 1/5$, $\alpha_2 = 1/2$, $\alpha_3 = 1$ makes the scheme TVD if the modified flux contribution (Φ) is evaluated at least the first two stages, with a CFL number ≤ 1.5. Note that TVD formulation, being upwind biased, is favourably affected by enlargement of stability bounds on the real axis, which is guaranteed by the above choice of the Runge Kutta coefficients.

RESULTS AND DISCUSSION

Three numerical experiments are presented and the results are compared with the ones obtained using a numerical scheme which employes an adaptive dissipation, RKAD [14].

A. Double throat nozzle (laminar; $Re = 1600$)

This numerical test case has been selected because it presents strong viscous and inviscid phenomena. The flow is assumed to be laminar and undergoes a transition from subsonic to supersonic regime. Computed results obtained on a 152x32 grid show that the method accurately resolves the flow details after the interaction between shock and the boundary layer. Note that such a feature cannot be detected when using the RKAD.

B. RAE2822 airfoil (turbulent; $M_\infty = .75$; $\alpha = 2.70$; $Re = 6.2\ 10^6$)

This test case is a rather critical one since it requires a suitable mesh refinement near the wall. Having used for this test case a rather coarse mesh (100x28), the high degree of stretching makes the test a very severe one for the method. The results compare well with the experiments, [15], and the ones obtained by the RKAD with adaptive mesh embedding developed by the authors, [14]. The comparison shows a better resolution of the shock (captured within two cells); however the overall performance of RKTVD is not dramatically different from RKAD.

C. LS59TG cascade (turbulent; $M_{2is} = 1.31$; $\alpha_1 = 30$; $Re_1 = .6\ 10^6$)

Prediction of turbulent flow in high turning cascades is a difficult task due to geometrical complexities as well as to physical flow features involved (shocks, wake, shock boundary-layer interaction with possible separation, flow separation at blunt trailing edge). The comparison between experimental, [16], and computed results shows that for this test case many of the essential flow features are adequately resolved: flow separates near the trailing edge where the shock inpinges on the boundary layer and a large region of stagnation flow is formed before the confluence of suction and pressure side streams.

CONCLUSIONS

TVD concepts applied to inviscid (Euler) terms of fully Navier Stokes equations seem to improve the accuracy of the numerical solutions; this is obtained at the expense of greater computational effort, due both to code complexity and to slow rate of convergence. Further investigation is needed to asses the superiority of the present approach versus others.

REFERENCES

[1] GAMM Workshop on Numerical Simulation of Compressible Navier Stokes Equations, Notes in Numerical Fluid Mechanics, vol. 18, Vieweg Verlag, 1986.

[2] Holst, T.L., AIAA paper no. 87-1460, 1987.

[3] Harten, A., J. Comp. Phys., vol. 49, pp. 357-393, 1983.

[4] Yee, H.C., Warming, R.F., Harten, A., J. Comp. Phys., vol. 57, pp. 327-360, 1985.

[5] Van Leer, B., J. Comp. Phys., vol. 32, pp. 101-136, 1979.

[6] Davis, S.F., ICASE report no. 84-20, 1984.

[7] Roe, P.L., ICASE report no. 84-53, 1984.

[8] Yee, H.C., NASA TM-89464, 1987.

[9] Yee, H.C., J. Comp. & Maths. with Appls., vol. 12, pp. 413-432, 1986.

[10] Sweby, P.K., SIAM J. Numer. Anal., vol. 21, pp. 995-1011, 1984.

[11] Jameson, A., Lax, P.D., ICASE report no. 86-18, 1986.

[12] Baldwin, B.S., Lomax, H., AIAA paper 78-257, 1978.

[13] Bassi, F., Grasso, F., Savini, M., Lect. Notes in Phys., vol. 264, pp. 113-119, 1986.

[14] Bassi, F., Grasso, F., Savini, M., Lect. Notes in Num. Fluid Mech., 1987.

[15] Cook, P.H., McDonald, M.A., Firmin, M.C., AGARD AR-138, 1979.

[16] Sieverding, C., Von Karman Institute, LS 1973-59, 1973.

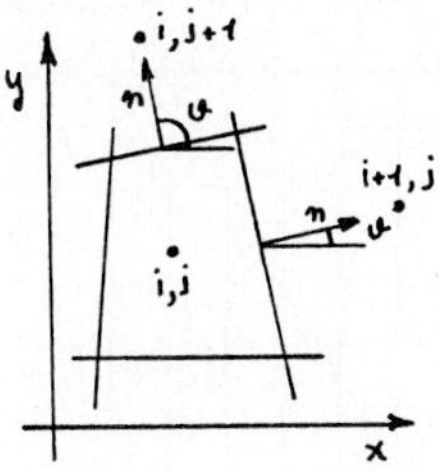

Fig. 1 Computational cell

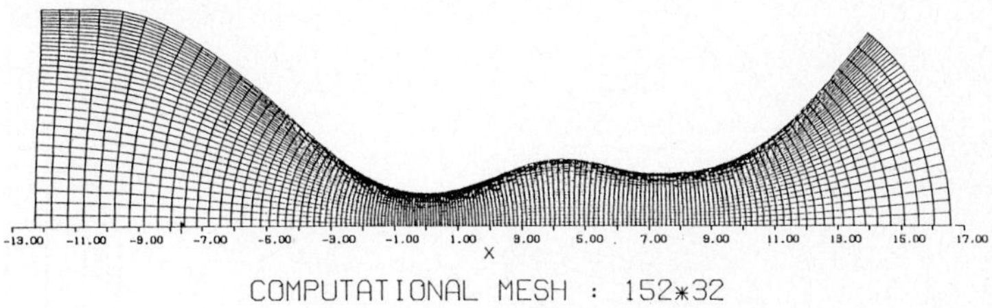

Fig. 2a Case A: computational grid

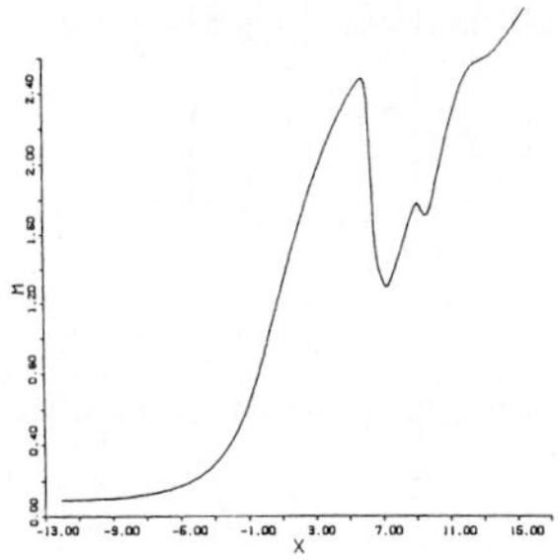

Fig. 2b Case A: Centerline Mach number

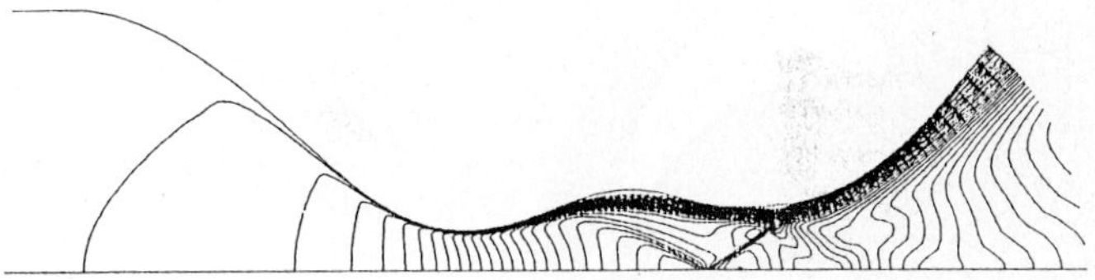

Fig. 2c Case A: isoMach contours

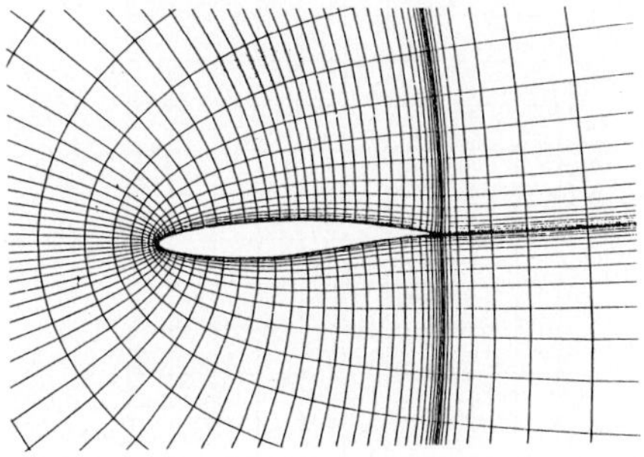

Fig. 3a Case B: computational grid

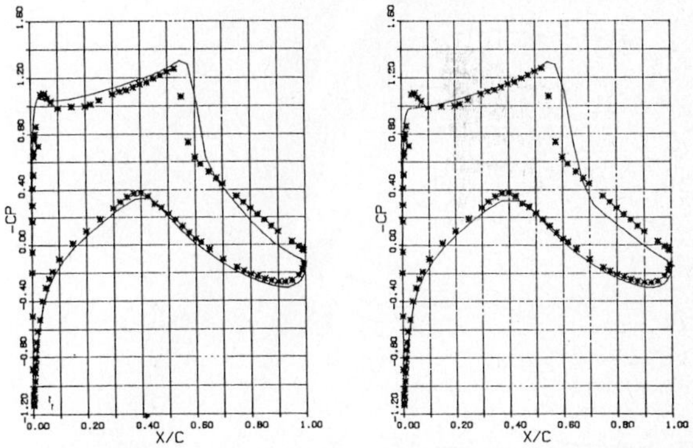

Fig. 3b Case B: CP vs chord (TVD - embedding)

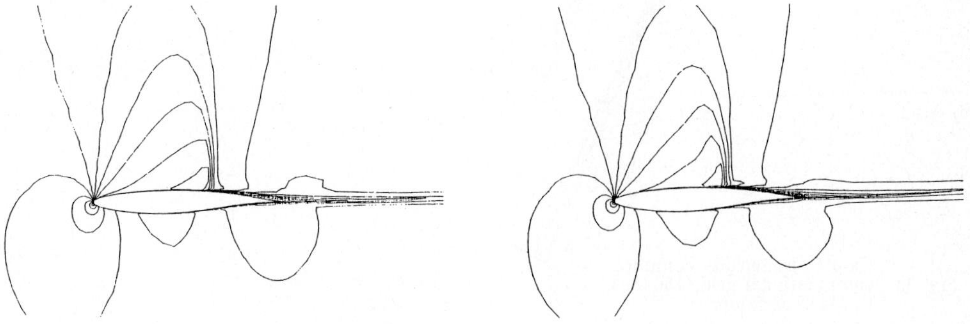

Fig. 3c Case B: isoMach contours (TVD - embedding)

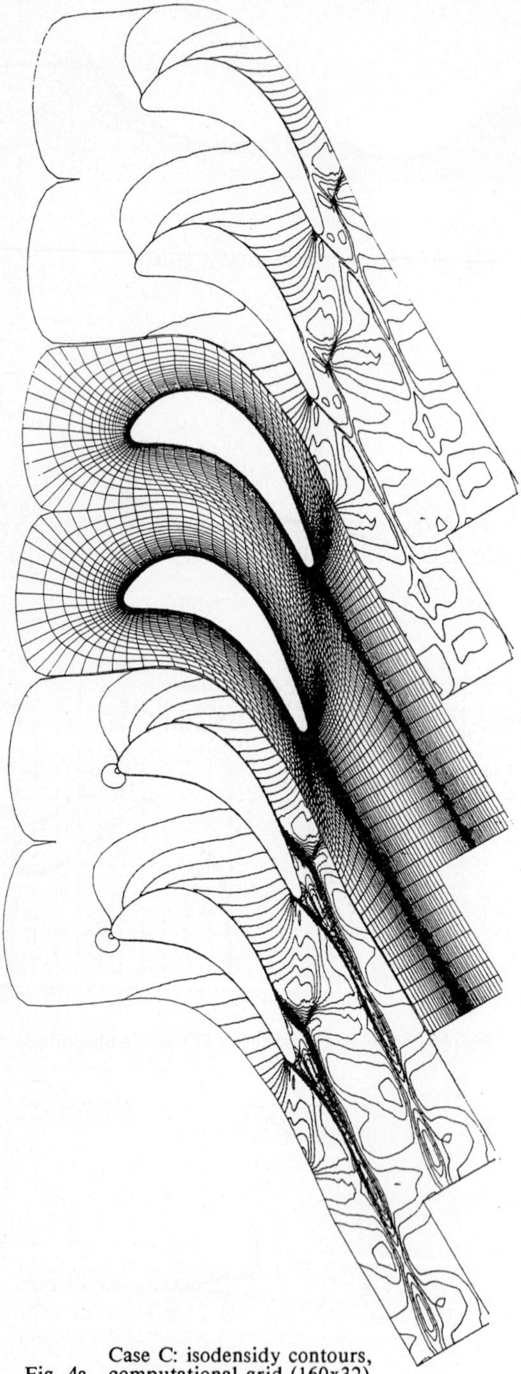

Fig. 4a Case C: isodensidy contours, computational grid (160x32), isoMach contours

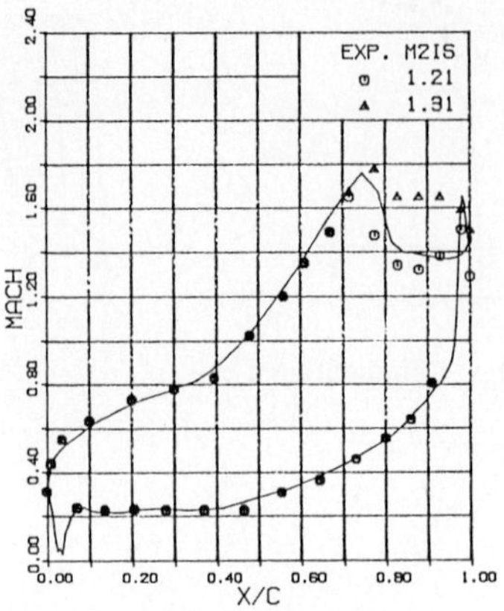

Fig. 4b Case C: surface isentropic Mach number

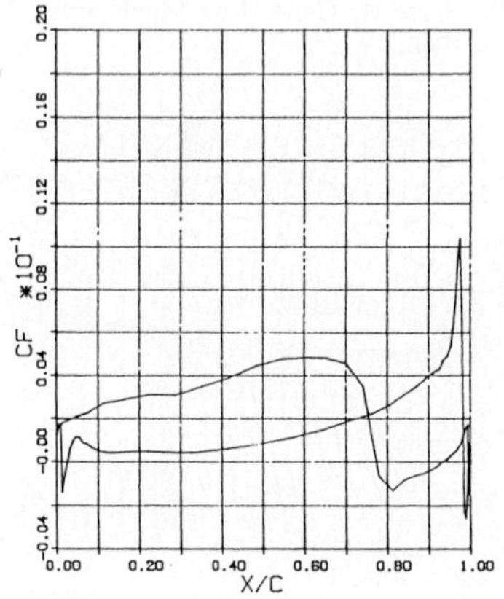

Fig. 4c Case C: surface skin friction coefficient

Godunov Methods and Adaptive Algorithms for Unsteady Fluid Dynamics†

John Bell, Phillip Colella, John Trangenstein, Michael Welcome
Lawrence Livermore National Laboratory
Livermore, CA 94550

Higher-order versions of Godunov's method have proven highly successful for high-Mach-number compressible flow. One goal of the research being described in this paper is to extend the range of applicability of these methods to more general systems of hyperbolic conservation laws such as magnetohydrodynamics, flow in porous media and finite deformations of elastic-plastic solids. A second goal is to apply Godunov methods to problems involving more complex physical and solution geometries than can be treated on a simple rectangular grid. This requires the introduction of various adaptive methodologies: global moving and body-fitted meshes, local adaptive mesh refinement, and front tracking.

Extension of the Godunov methodology to other systems of equations is more difficult due to the presence of pathologies in the local wave structure which do not appear in gas dynamics. In the applications listed above, one finds that the wave speeds are nonmonotonic functions along their associated wave curves in phase space, requiring more complicated entropy conditions to ensure physically realizable solutions. In addition, there are surfaces in phase space along which the linearized coefficient matrices fail to have a complete set of eigenvectors. We have developed an approximate Riemann problem solver based on a generalization of the Engquist-Osher flux that is suitable for these types of problems [1]. The initial test of the method has been for flow of a multiphase mixture of gas, oil and water in a porous medium. The underlying hyperbolic system consists of three conservation laws for the component densities and exhibits the degeneracies cited above. An additional constraint that the fluid fills the available pore volume leads to an associated parabolic system for pressure and total volumetric flow rate, which appear in the hyperbolic flux as spatially dependent terms. The first example, Fig. 1, shows gravity inversion in a core saturated with pure water in the top third, pure oil in the middle third and pure gas in the bottom third. In addition to the complex frontal structure shown in the figure, there is also a substantial drop in pressure (40%) arising from the mixing of the pure components. The second example, Fig. 2, shows a two-dimensional flow in which water is being injected along the left edge, and oil and gas are being withdrawn along the right edge. Gravity imposes a force in the (-y) direction, so we see some vertical structure in the solution, since the various phases in the problem have different mass densities. In addition to porous media flow this method is also being used to model shock-waves in solids [2] and for compressible magnetohydrodynamics [3].

The other facet of our development of Godunov methods is the introduction of adaptive methodology. To treat more complex flow and problem geometry we have extended the second-order unsplit Godunov method first introduced by Colella [4] to a logically-rectangular moving quadrilateral mesh [5]. This algorithm is based on an upstream-centered predictor-corrector formalism, with the conservative corrector step using finite-volume differencing. The method is second-order accurate for smooth solutions on smooth grids, has a robust treatment of discontinuities with minimal numerical diffusion, and is freestream-preserving. Figure 3 shows a comparison of a numerical solution with experimental data of Zhang and Glass [6] for shock diffraction. A body-fitted quadrilateral grid is generated using a biharmonic solver, so that it is smooth, though not orthogonal. Nonetheless, there is no difficulty resolving this complicated time-dependent flow

† This work was performed under the auspices of the U.S. Department of Energy by the Lawrence Livermore National Laboratory under contract No. W-7405-Eng-48. Partial support under contract No. W-7405-Eng-48 was provided by the Applied Mathematical Sciences Program of the Office of Energy Research, the Defense Nuclear Agency under IACRO 88-873, and the Air Force Office of Scientific Research under contract number AFOSR-ISSA-870016.

pattern. (See Glass et al [7] for a more detailed comparison.) In figure 4, we show an example of a moving grid calculation, namely, a self-similar ramp shock reflection. At the initial time, an algebraic grid is generated. Then each corner of the grid is given a velocity proportional to its initial location, so that the self-similar reflection pattern approaches a discrete steady state on the moving grid in the limit of long times.

We are also developing other adaptive methods aimed at focusing computational effort where it is needed. In figure 5, we show a result obtained by coupling the adaptive mesh refinement algorithm of Berger and Colella [8] for shock hydrodynamics to a volume-of-fluid type algorithm for tracking an interface between two materials. In the adaptive mesh refinement algorithm, the mesh is locally refined in space and time in response to the appearance of large errors or features in the solution. Refined regions may appear or disappear as a function of time; also, there may be multiple levels of refinement. The multifluid algorithm is coupled to an operator split Godunov method, although operator splitting is not essential. In the approach taken here, only the thermodynamic discontinuity is tracked. There are multiple values for the density and energy for each cell, but only a single velocity per cell, so that any slip that forms along the interface is captured. The figure shows a comparison to experimental data of L. F. Henderson [9] for shock refraction by an oblique material interface. This methodology is being used in a more detailed study of slow-fast refraction [10]. A second application of the multifluid-local refinement algorithm is shown in Fig. 6. The figure contains the result of a calculation of the interaction of a shock with a dense spherical cloud [11]. The calculation was done in cylindrical geometry, with the axis of symmetry located along the left edge of the domain. There are two levels of refinement, each by a factor of four. The finest grids are outlined with boxes in the figure; thus it is apparent that at late times, the finest grids are reserved for the cloud, which in this problem is the region of greatest interest to us. The multifluid tracking eliminates a large class of diffusive errors from the transport of the cloud material. To have performed this calculation on an equivalent uniform grid would have taken 10 times as much CPU time, and 5 times as much memory.

References

1. J. B. Bell, P. Colella, and J. A. Trangenstein, "Higher-Order Godunov Methods for General Systems of Hyperbolic Conservation Laws," UCRL-97258, LLNL, to appear in J. Comp. Phys..
2. J. A. Trangenstein and P. Colella, *A Higher-Order Godunov Method for Finite Deformation in Elastic-Plastic Solids*, submitted for publication.
3. A. Zachary and P. Colella, *Higher-Order Godunov Methods for Compressible MHD*, in preparation.
4. P. Colella, "A Multidimensional Second Order Godunov Scheme for Conservation Laws," LBL-17023, Lawrence Berkeley Laboratory, to appear in J. Comp. Phys..
5. J. A. Trangenstein and P. Colella, *An Unsplit, Second-Order Godunov Method for Moving Quadrilateral Grids*, in preparation.
6. D. L. Zhang and I. I. Glass, "An Interferometric Investigation of the Diffraction of Planar Shock Waves over a Half-Diamond Cylinder in Air," UTIAS Report No. 322, March 1988.
7. I. I. Glass, J. A. Trangenstein, D. L. Zhang, J. B. Bell, M. Welcome, P. Collins, and H. M. Glaz, *Numerical and Experimental Study of High-Speed Flow over a Half-Diamond Cylinder*, in preparation.
8. M. J. Berger and P. Colella, "Local Adaptive Mesh Refinement for Shock Hydrodynamics," UCRL-97196, LLNL, to appear in J. Comp. Phys..
9. A. M. Abd-el-Fattah and L. F. Henderson, "Shock waves at a slow-fast gas interface," *J. Fluid Mech.*, vol. 89, pp. 79-85, 1978.
10. P. Colella, L. F. Henderson, and E. G. Puckett, *Numerical and Experimental Analysis of Refraction of a Shock by a Material Interface, Part I: The Slow-Fast Case*, in preparation.
11. P. Colella, R. Klein, and C. McKee, *The Nonlinear Interaction of Supernova Shocks with Galactic Molecular Clouds*, in preparation.

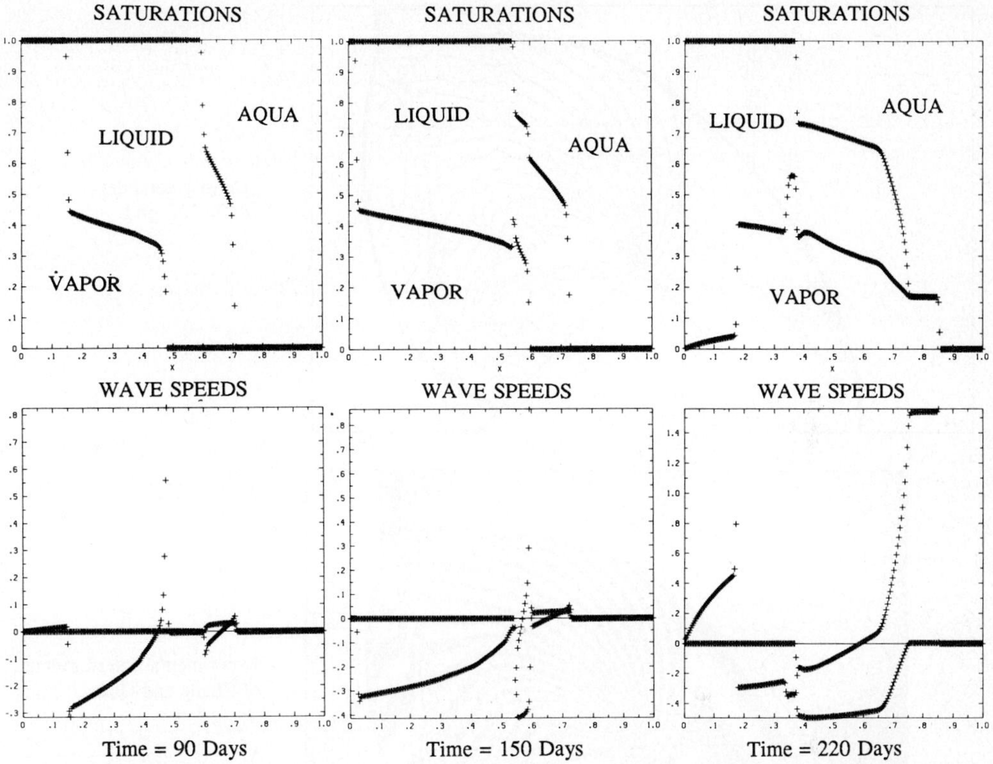

Fig.1. Time history of three component gravity inversion using 240 grid points. Left edge is the bottom; right edge is the top.

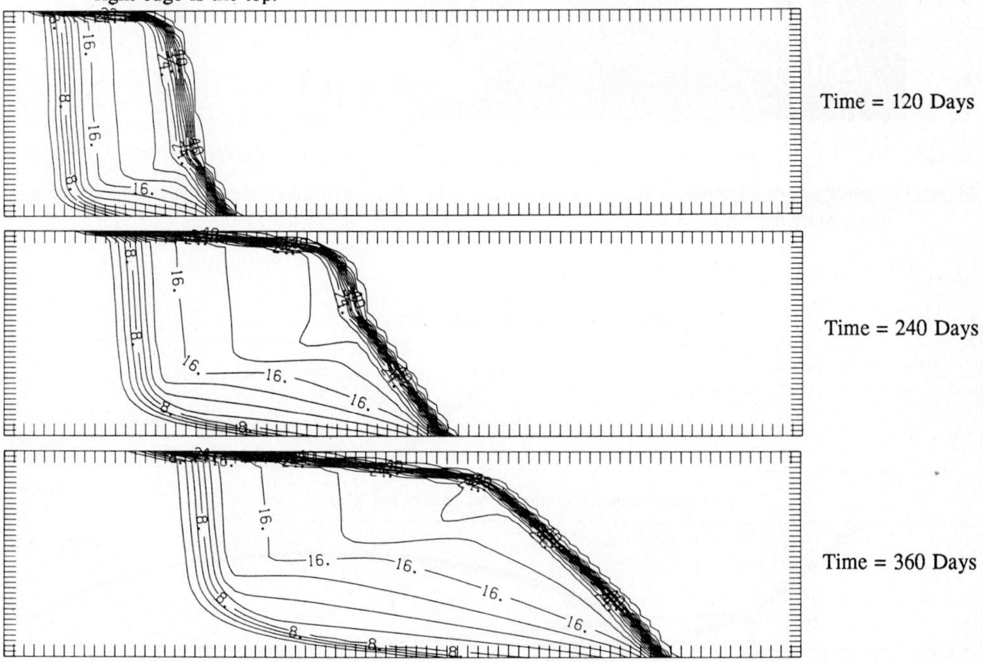

Fig.2. Contour of gas component density for two-dimensional waterflood on 240×80 grid.

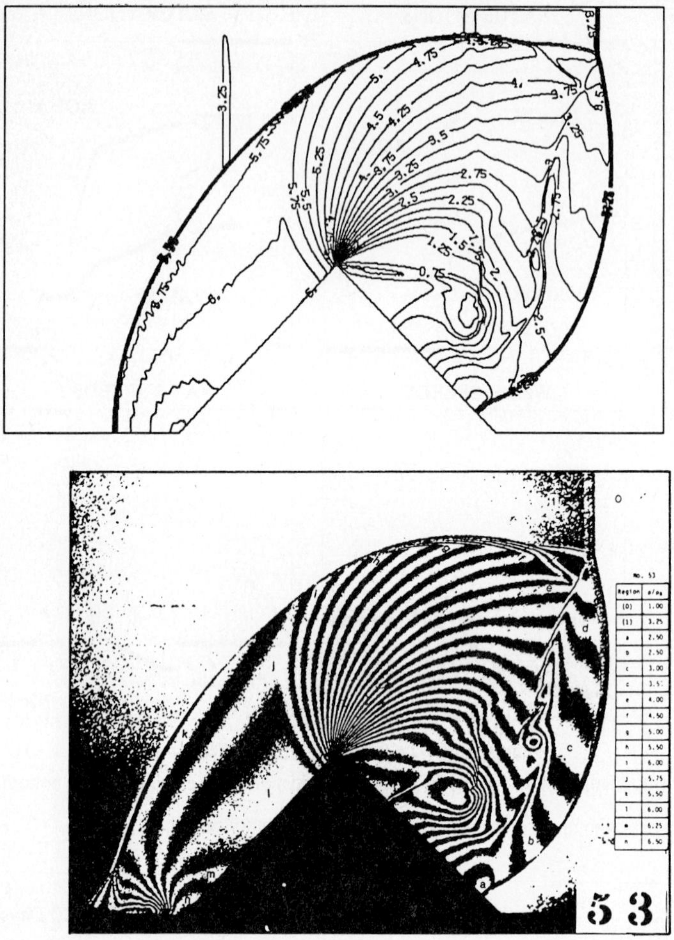

Density contours
440×220 grid

Experimental Interferogram
of Zhang and Glass

Fig.3. Comparison of numerical solution with experiment for shock diffraction over a half-diamond cylinder. Shock Mach number is 2.45.

Density contours at early time

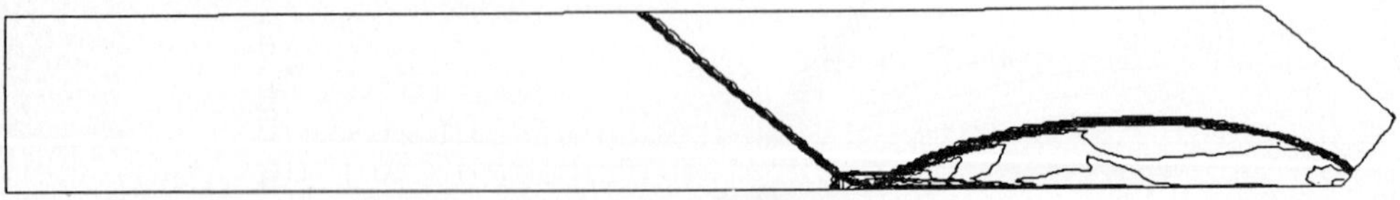

Stationary density contours on expanding grid

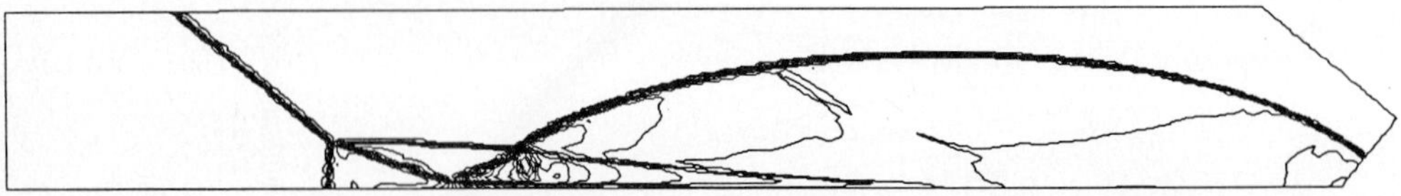

Fig. 4. Self-similar ramp reflection on a moving quadrilateral grid.

Logarithmically spaced density contours　　Schlieren photograph (courtesy of L. F. Henderson)

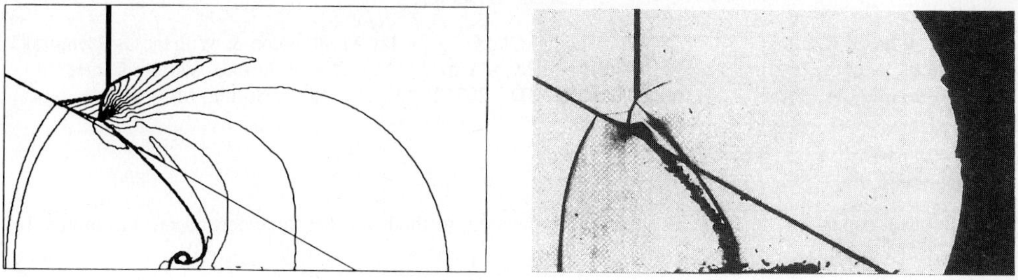

Figure 5. Shock refraction from an oblique material interface.

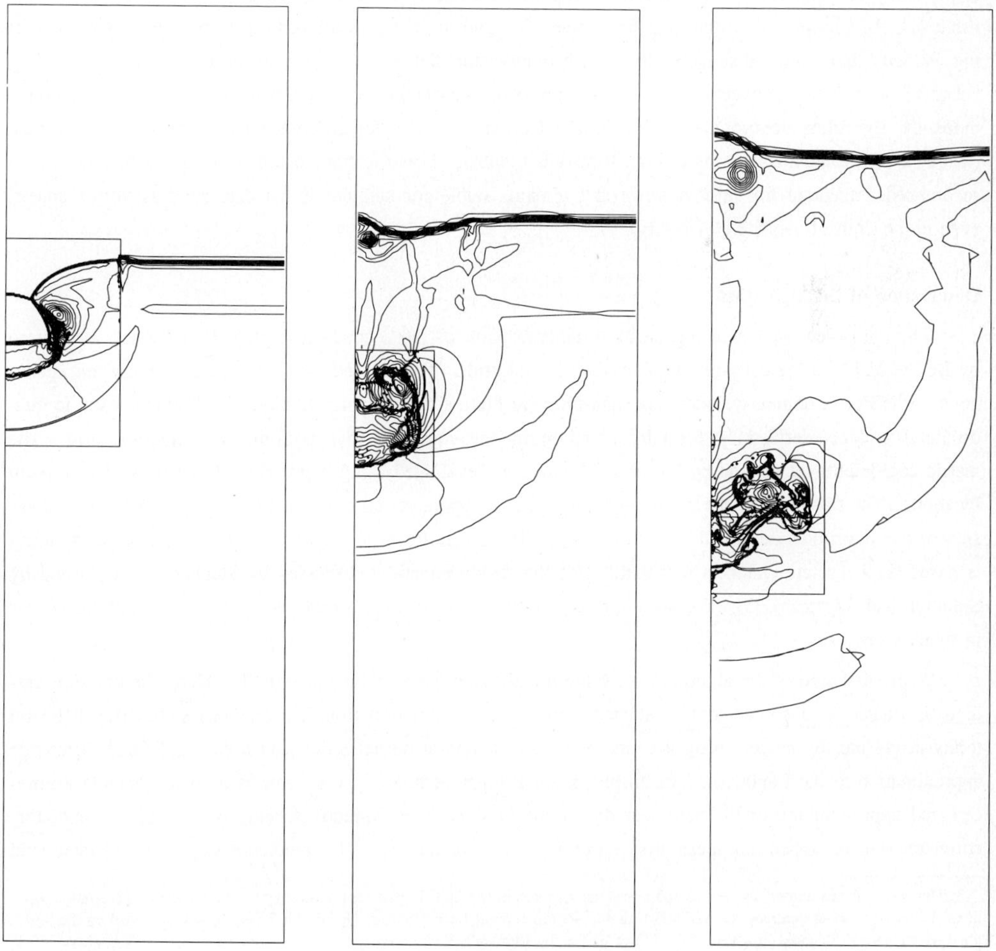

Time = .004　　Time = .008　　Time = .014

Fig.6. Refraction of a planar shock by a dense spherical cloud

Application of a Second-Order Projection Method to the Study of Shear Layers

John B. Bell† Lawrence Livermore Laboratory Livermore, CA 94550

Harland M. Glaz§ University of Maryland College Park, MD 20742

Jay M. Solomon & William G. Szymczak‡ Naval Surface Warfare Center Silver Spring, MD 20903

Introduction

In this paper we apply a second-order projection method for the time-dependent, incompressible Navier-Stokes equations

$$u_t + (u \cdot \nabla)\, u = \frac{1}{\text{Re}} \Delta u - \nabla p \qquad (1)$$

$$\nabla \cdot u = 0 \qquad (2)$$

to the study of shear layers. The algorithm represents a higher-order extension of Chorin's projection algorithm [1]. In Chorin's algorithm one first solves (1) ignoring the pressure term and then projects the resulting velocity field onto discretely divergence-free vector fields. Our method introduces more coupling between the diffusion-convection step and the projection to obtain second-order temporal accuracy. Furthermore, the algorithm incorporates a second-order Godunov method for differencing (1) that provides a robust treatment of the nonlinear terms at high Reynolds number. These features combine to give a method that is second-order accurate for smooth flows and remains stable for singular initial data such as vortex sheets, even in the limit of vanishing viscosity.

Description of the algorithm

The algorithm used here represents a generalization of the second-order projection method introduced by Bell et al [2] to logically-rectangular quadrilateral grids. The method uses a staggered grid system analogous to the Fortin bilinear-velocity, constant-pressure finite element approximation [3]. The extension to quadrilateral grids is derived by formally transforming the equations to a uniform computational grid. The metric coefficients introduced by the transformation are evaluated using appropriate differences of grid-point locations. The treatment of the convection and diffusion operators on the transformed grid is a routine extension of the algorithms described by Bell et al [2]. The projection onto discretely divergence-free vector fields is based on a discrete Galerkin formulation for the steady equations discussed by Stephens et al [4] and by Solomon and Szymczak [5]. We now briefly describe the algorithm; a more detailed description will appear in future work.

In the first step of the algorithm we solve the diffusion-convection equation (1). Here, the pressure gradient is treated as a source term evaluated using its approximation from the previous step. The diffusion terms in (1) are discretized using a Crank-Nicholson temporal discretization and a standard finite difference approximation to the Laplacian. An unsplit, second-order Godunov-type scheme is used to compute a time-centered approximation to the nonlinear flux terms $(u \cdot \nabla)u$. The Godunov scheme is an explicit, predictor-corrector scheme based on ideas first introduced by Colella [6]. The predictor step uses characteristic

† This work of this author was performed under the auspices of the U.S. Department of Energy by the Lawrence Livermore National Laboratory under contract No. W-7405-Eng-48. Partial support under contract No. W-7405-Eng-48 was provided by the Applied Mathematical Sciences Program of the Office of Energy Research.

§ The work of this author was partially supported by the National Science Foundation under grant DMS-87-3971.

‡ The work of these authors was supported by the NSWC Independent Research Fund.

information to extrapolate velocities to $t^{n+½}$ from the solution at time t^n. These predicted velocities are then used to compute a second-order upwind approximation to the nonlinear convection terms. The scheme incorporates a limiting mechanism that selectively adds dissipation near discontinuities or under-resolved gradients providing a second-order spatial discretization that is robust for nonsmooth data. Since the Godunov scheme computes a second-order approximation to $(u \cdot \nabla)u$ without any implicit coupling to time t^{n+1}, the linear algebra for solving (1) reduces to the solution of the symmetric positive-definite system associated with Crank-Nicholson discretization of the heat equation.

Because the approximation to the pressure in the discretization of (1) is lagged, the result of the first step of the algorithm does not satisfy the incompressibility condition (2). In the second step of the algorithm we apply a discrete projection of the vector field computed in the first step that decomposes it into a discretely divergence-free vector field and the discrete gradient of some scalar. These two components of the decomposition define updates for the velocities and for the pressure gradient. The discrete-Galerkin approach for approximating the projection requires only the solution of a symmetric, positive-definite matrix with spectral properties analagous to a discrete Laplacian. We have treated the system using a conjugate gradient algorithm preconditioned by a modified incomplete Cholesky factorization of the matrix. (For periodic calculations we have also used Fourier techniques.)

Initial tests of the second-order projection formulation combined with Godunov-type treatment of the nonlinear convection terms were presented by Bell et al [2]. These tests demonstrate second-order convergence of the method for smooth flow for a variety of Reynolds numbers including the incompressible Euler equations and provide some indication of the overall performance of the method. They also show that the method is stable for discontinuous initial velocity data, even in the zero viscosity limit.

Numerical results

The first computational example illustrates the behavior of the method for nonsmooth initial data. For this example the initial data consists of two patches of constant vorticity, each of radius 0.08 centered at (.38,.5) and (.62,.5); thus, the initial velocities are continuous but have discontinuous derivatives. The computation was performed on a uniform grid for the incompressible Euler equations using a noflow condition on the boundary of the unit square. (No boundary condition was specified for the tangential velocity component.) Vorticity contours for various times are shown in Figure 1 for 64×64, 128×128 and 256×256 grids. For this example we observe a dramatic increase in resolution as the grid is refined. The separate integrity of the patches is preserved for a longer time and the tails extending from the central core are better resolved on the finer grids. Nevertheless, even the coarse grid effectively captures the gross feature of the merger.

The second set of calculations are a direct comparison with the experimental study of Ho & Huang [7] on the nonlinear instability of a forced shear layer. The forcing is introduced using an upstream perturbation based on the stability analysis of Monkewitz & Huerre [8]. We note that a number of other authors have also treated this problem; see, for example, Kuhl et al [9] which includes a survey of the other literature.

Our computations were performed on a 300×60 uniform grid with a highly stretched grid 25 zones wide around the top, bottom and right boundaries to provide a buffer to eliminate boundary effects. The Reynolds number for the flow was 30,000, based on the length of the uniform grid region. In Figure 2 we present a time history of vorticity contours on the uniform grid region for three cases corresponding to the fundamental perturbation frequency (ω_1) plus a subharmonic. The cases correspond to the first (ω_2), second (ω_3) and third (ω_4) subharmonics, respectively. The largest perturbation amplitude was 1% for the fundamental with reduced amplitudes for the subharmonics as given by Monkewitz and Huerre [8]. The numerical results show the instability of the shear layer and the downstream vortex merging pattern. The first case consist of a simple pairing. In the second case the vortices merge in groups of three with the leading two vortices merging first. The third case shows the merger of groups of four. Here the two interior vortices merge first, followed by the trailing vortex and finally (downstream of the window plotted) the leading vortex. The prediction of the correct merging patterns and the near periodicity of the results serve to validate the numerical method on a realistic flow problem.

References

1. A. J. Chorin, "On the Convergence of Discrete Approximations to the Navier-Stokes Equations," *Math. Comp.*, vol. 23, pp. 341-353, April 1969.
2. J. B. Bell, P. Colella, and H. M. Glaz, *A Second-Order Projection Method for the Incompressible Navier-Stokes Equations*, submitted for publication.
3. M. Fortin, "Numerical Solutions of the Steady State Navier-Stokes Equations," in *Numerical Methods in Fluid Dynamics*, ed. J. J. Smolderen, AGARD-LS-48, 1972.
4. A. B. Stephens, J. B. Bell, J. M. Solomon, and L. B. Hackerman, "A Finite Difference Galerkin Formulation of the Incompressible Navier-Stokes Equations," *J. Comp. Phys.*, vol. 53, pp. 152-172, Jan. 1984.
5. J. M. Solomon and W. G. Szymczak, "Finite Difference Solutions for the Incompressible Navier-Stokes Equations using Galerkin Techniques," *Fifth IMACS International Symposium on Computer Methods for Partial Differential Equations*, Lehigh University, June 19-21, 1984.
6. P. Colella, "A Multidimensional Second Order Godunov Scheme for Conservation Laws," LBL-17023, Lawrence Berkeley Laboratory, to appear in J. Comp. Phys..
7. C.M. Ho and L.S. Huang, "Subharmonics and Vortex Merging in Mixing Layers," *J. Fluid Mech.*, vol. 119, pp. 443-473, 1982.
8. P.A. Monkewitz and P. Huerre, "Influence of the Velocity Ratio on the Spatial Instability of Mixing Layers," *Physics of Fluids*, vol. 25, pp. 1137-1143, 1982.
9. A.L. Kuhl, K.-Y. Chien, R.E. Ferguson, H.M. Glaz, and P. Colella , "Inviscid Dynamics of Unstable Shear Layers," RDA-TR-161604-004, R&D Associates, Marina del Rey, April 1988.

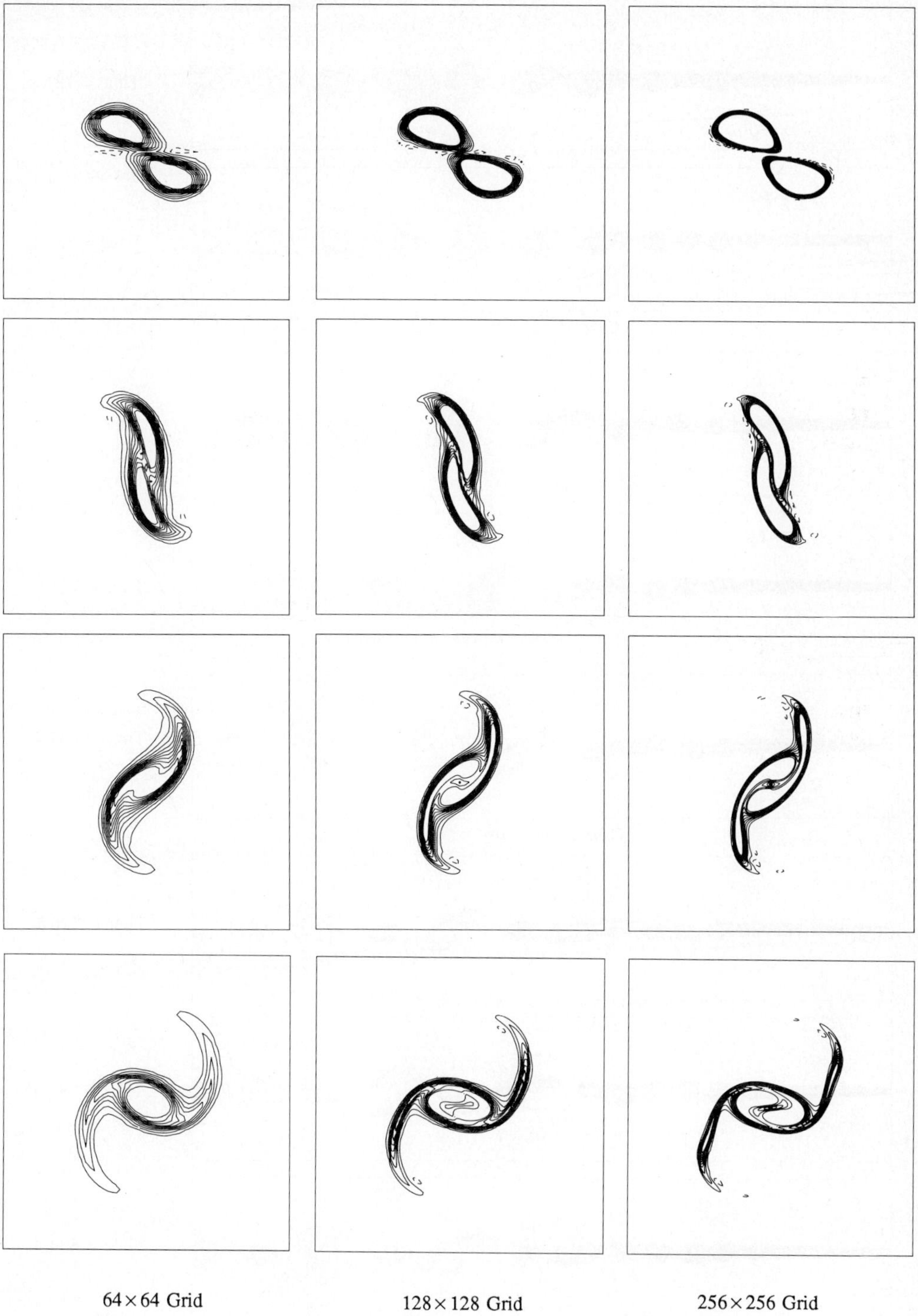

Fig. 1. Vorticity contours for vortex merger in a finite domain.

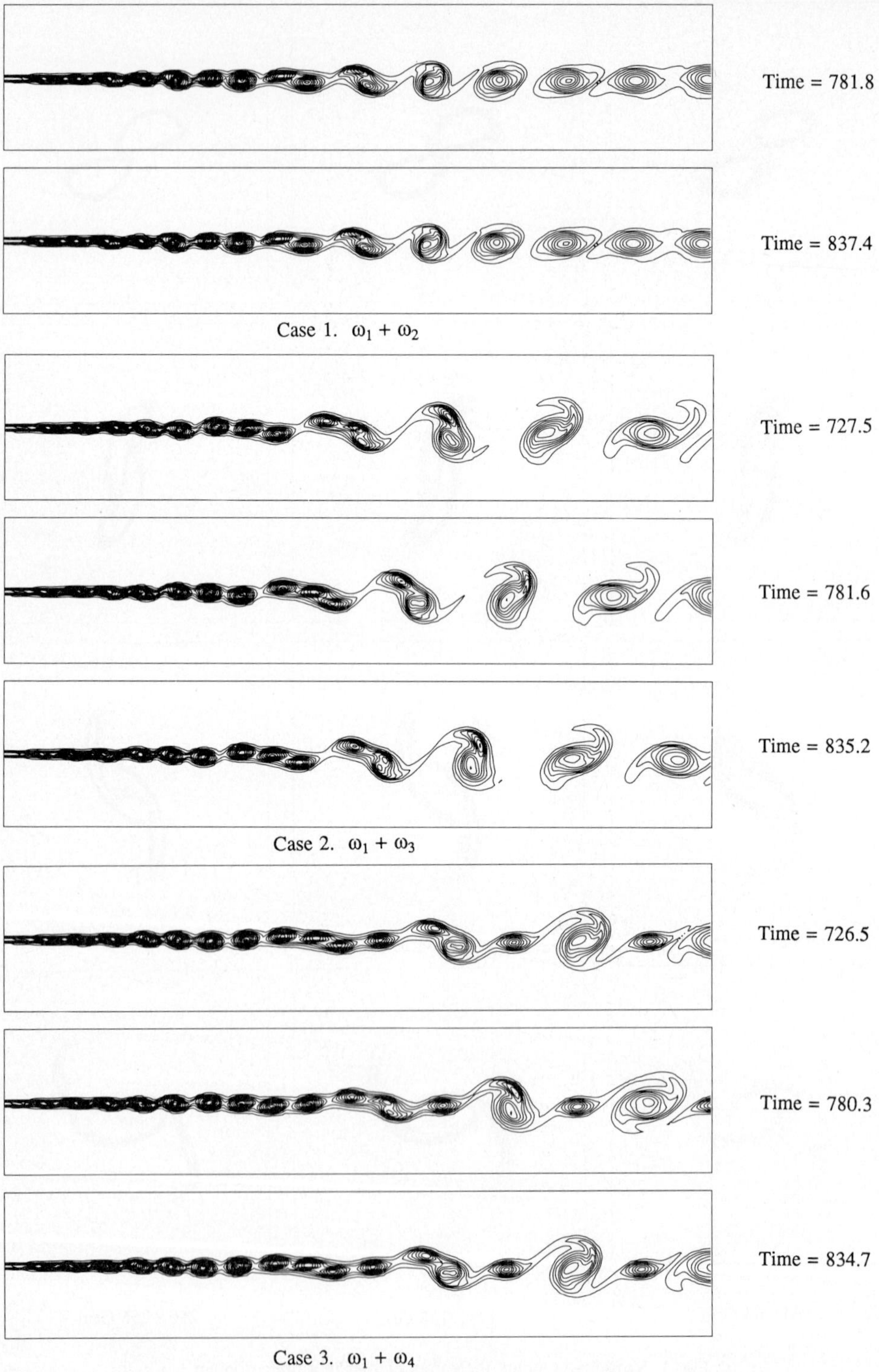

Fig. 2. Vorticity contours for nonlinear instability of a shear layer.

A GRP-SCHEME FOR REACTIVE DUCT FLOWS IN EXTERNAL FIELDS

M.Ben-Artzi
Department of Mathematics

A. Birman
Department of Physics

Technion, Israel Institute of Technology
Haifa 32000, Israel

1. INTRODUCTION

Consider the Euler equations that model the time dependent flow of an inviscid, compressible, reactive fluid through a duct of smoothly varying cross section. In addition to the hydrodynamical pressure force we allow an external conservative field which does not vary with time. We are using here the quasi one-dimensional approximation, namely, the hypothesis that all flow variables are uniform across a fixed cross section. Note that our treatment applies in particular to all problems with planar, cylindrical or spherical symmetry. Such problems arise, e.g., in astrophysics. Denoting by r the spatial coordinate and by A(r) the area of the cross section at r, our equations are,

$$A\frac{\partial}{\partial t}U + \frac{\partial}{\partial t}\big(AF(U)\big) + A\frac{\partial}{\partial r}G(U) + AH(U) = 0,$$

$$U = \begin{bmatrix}\rho\\ \rho u\\ \rho E\\ \rho z\end{bmatrix},\quad F(U) = \begin{bmatrix}\rho u\\ \rho u^2\\ (\rho E+p)u\\ \rho zu\end{bmatrix},\quad G(U) = \begin{bmatrix}0\\ p\\ 0\\ 0\end{bmatrix},\quad H(U) = \begin{bmatrix}0\\ \rho\phi'(r)\\ 0\\ k\rho\end{bmatrix}, \tag{1}$$

where ρ,p,u are, respectively, density, pressure and velocity. We denote by z the mass fraction of the unburnt fluid, that is, z = 1 (resp. z = 0) represents the completely unburnt (resp. burnt) fluid. The total specific energy E is given by $E = e + \frac{1}{2}u^2 + \phi$, where $\phi = \phi(r)$ is the external potential and e is the specific internal energy (including chemical energy). We are assuming an equation-of-state of the form $p = p(e,\rho,z)$. The reaction rate $k = k(e,\rho,z)$ is assumed to be a positive function. Along a particle path we have $\frac{dz}{dt} = -k$.

Our purpose in this work is to present a robust, high resolution numerical scheme for the time integration of the equations (1). We work within the general framework of the GRP (Generalized Riemann Problem) Scheme [1,2,3]. An outline of the method is given in Section 2. We discuss three numerical examples in Section 3.

2. OUTLINE OF THE METHOD

The GRP (Generalized Riemann Problem) is the following:

Let U(r,t) be the (entropy) solution of (1), where the initial values are linearly distributed with a jump at r = 0,

$$U(r,0) = U_\pm + U'_\pm \cdot r \quad \text{for } \pm r > 0. \tag{2}$$

Determine $\left(\frac{\partial U}{\partial t}\right)_0 = \lim_{t\to 0+} \frac{\partial U}{\partial t}(r=0,t)$.

Note that $U_0 = \lim_{t\to 0+} U(r=0,t)$ is just the solution to the Riemann problem involving the values $U_\pm$ in (2).

It turns out that the solution to the GRP can be obtained in a *fully analytic* manner. We shall not attempt to indicate the proof here, but just make the following observation.

Recall that u,p are continuous across a contact discontinuity. Denoting by $\left(\frac{du}{dt}\right)^*$, $\left(\frac{dp}{dt}\right)^*$, their (total) time derivatives along that streamline, it can be shown that they satisfy a pair of linear equations

$$a_\pm\left(\frac{du}{dt}\right)^* + b_\pm\left(\frac{dp}{dt}\right)^* = d_\pm \,, \tag{3}$$

where the coefficients $a_\pm$, $b_\pm$, $d_\pm$ can be determined from the initial conditions and the solution U_0 of the above mentioned Riemann problem. This leads (see [1]) to a number of significant simplifications of the scheme.

Now the numerical solution is discretized at $t_n = n\Delta t$ as a piecewise linear function, leading to jump discontinuities at cell boundaries, designated as $r_{i+½}$. The difference scheme for (1) is now,

$$U_i^{n+1} - U_i^n = -\frac{\Delta t}{\Delta V_i}\left[A(r_{i+½})F(U)_{i+½}^{n+½} - A(r_{i-½})F(U)_{i-½}^{n+½}\right] \tag{4}$$

$$- \frac{\Delta t}{\Delta r_i}\left[G(U)_{i+½}^{n+½} - G(U)_{i-½}^{n+½}\right] - \Delta t\cdot H\left(U_i^{n+½}\right) ,$$

where $\Delta r_i = r_{i+½} - r_{i-½}$, $\Delta V_i = \int_{r_{i-½}}^{r_{i+½}} A(r)dr$.

The values $U_{i+½}^{n+½}$ are obtained by a first order Taylor expansion (in t), using the solution of the GRP at the jump $r = r_{i+½}$.

3. NUMERICAL EXAMPLES

(a) *Infinite reflected shock*. This is a test problem proposed by Noh [4] in spherical geometry. An infinite sphere of (initially) cold gas is uniformly imploding at u = -1. The resulting flow can be calculated analytically. The results are depicted in Fig.1, where the solid line represents the exact solution and the dots give the GRP numerical solution.

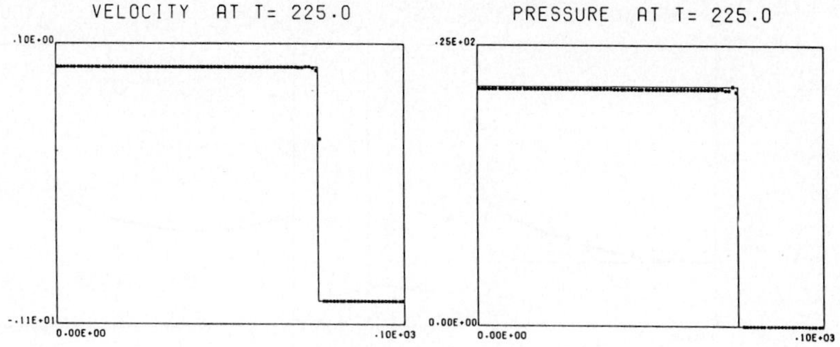

Fig.1 Infinite Reflected Shock

(b) *Taylor blast wave.* Here we calculate the flow profile behind a spherical detonation wave initiated at the origin r = 0. The solid line in Fig.2 here represents Taylor's self-similar solution [5], obtained by using a C-J model (reaction zone of zero width). The GRP solution (given by dots) does not assume such a model, of course.

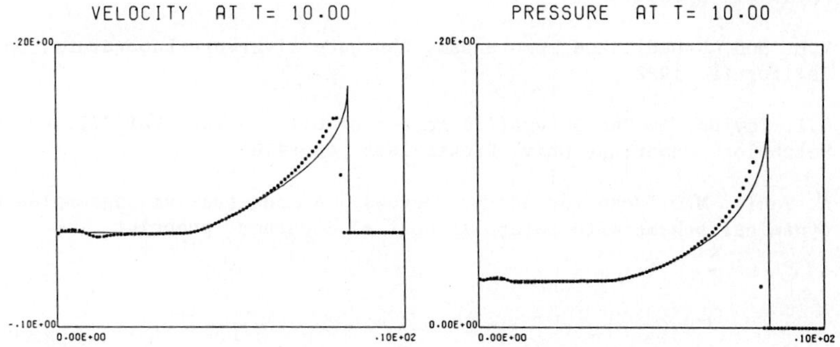

Fig.2 Taylor Blast Wave

(c) *Reactive flow of gas in cluster.* This is an astrophysical example, taken from [6]. A uniform gas, initially at rest, is subject to an external potential simulating self-gravity. As it collapses, the pressure gradient builds up, heating the gas, which is eventually ignited at the center. The resulting blast wave reverses the flow as the gas is burned. The numerical results are shown in Fig. 3.

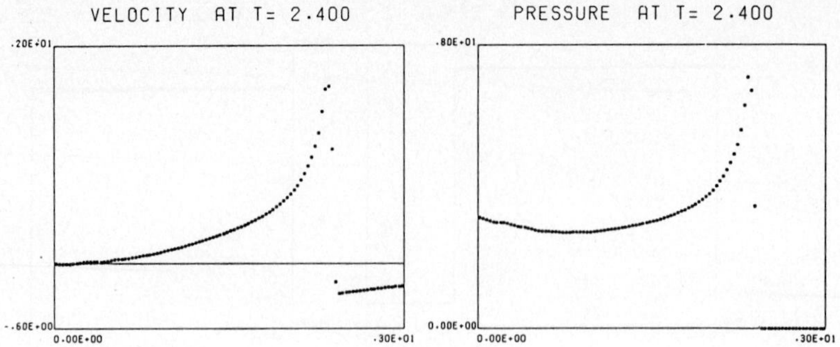

Fig.3 Reactive Flow of Gas in Cluster

REFERENCES

[1] M. Ben-Artzi and J. Falcovitz, in *Lecture Notes in Physics*, *218*, Ed. Soubbaramayer and J.P. Boujot, Spriner-Verlag 1985, 87-91.

[2] M. Ben-Artzi and J. Falcovitz, *J. Comp. Phys.* 55 (1984), 1-32.

[3] M. Ben-Artzi, The generalized Riemann problem for reactive flows, *J. Comp. Phys.* (in press).

[4] W.F. Noh, Unpublished memorandum, Lawrence Livermore Laboratory, Univ. of California, 1982.

[5] G.I. Taylor, in *The Scientific Papers of G.I. Taylor*, *Vol.III*, Ed. G.K. Batchelor, Cambridge Univ. Press, 1963, 465-478.

[6] A. Yahil, M.D. Johnston and A. Burrows, A conservative Lagrangian hydrodynamical scheme with parabolic spatial accuracy, preprint.

NUMERICAL SOLUTION OF THE NAVIER-STOKES EQUATIONS USING ORTHOGONAL BOUNDARY-FITTED COORDINATES

J. S. Bramley and D. M. Sloan
University of Strathclyde
Glasgow, Scotland

A boundary-fitted coordinate method is used to transform an irregularly shaped region onto a rectangular region so that finite difference methods can readily be used to solve the Navier-Stokes equations over an irregular region. The boundary-fitted coordinate transformation introduced by Thompson et al [5] has the disadvantage that the transformed coordinate scheme may not be orthogonal. Bramley and Sloan [1,2] use Thompson's method to transform a divided channel onto a rectangular region with equally spaced grids across the channel and obtain the numerical solution for viscous flow in a bifurcating channel. Thompson's method will allow grid lines to be clustered but it is difficult to relate the grid spacing required to the input parameters. In the present work we use the method of Ryskin and Leal [4] to generate orthogonal grids with grid lines clustered in the boundary layers of the fluid flow.

The numerical solution of the Navier-Stokes equations is obtained for laminar, viscous flow in a bifurcating channel, the geometry of which is given in fig. 1. The channel is symmetric about the centre line and the boundary-fitted coordinate method of Ryskin and Leal [4] is used to transform the domain of the solution from the (x,y) plane to a rectangle in the (ξ,η) plane. The governing equations of the transformation are

$$(fx_\xi)_\xi+(\tfrac{1}{f}x_\eta)_\eta = 0, \qquad (1)$$

$$(fy_\xi)_\xi+(\tfrac{1}{f}y_\eta)_\eta = 0, \qquad (2)$$

where $f(\xi,\eta)$ is the ratio of the grid scale factors in the x and y directions defined by

$$f(\xi,\eta) = \sqrt{\frac{(x_\eta)^2+(y_\eta)^2}{(x_\xi)^2+(y_\xi)^2}}. \qquad (3)$$

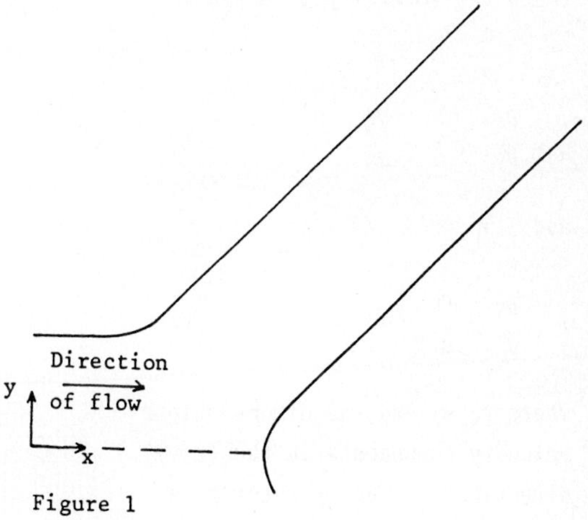

Figure 1

The equations of the boundaries are fixed, and the equations of the boundary curves are as given in Bramley and Sloan [1]. The function $f(\xi,\eta)$ is fixed and then equations (1) and (2) are solved numerically using central differences. The iteration scheme is such that one SOR iteration is performed and a simple geometrical technique is then used to move the boundary nodes along the boundary to achieve boundary othogonality. The process is then repeated until the average residual is less than 10^{-4} for both equations. Numerical results are obtained using a function f which depends only on η, and also a function f which depends on both ξ and η.

The equation for f with variation only in the η direction is

$$f(\xi,\eta) = 4(k_1-1)(\eta-1)\eta+k_1. \tag{4}$$

At the centre of the channel f=1 and with $k_1<1$ the grid lines are clustered at the boundary. In the case of the variation in the ξ and η directions, the dependence on η is as given in (4) but f is also varied linearly with ξ up and down stream of the bifurcation point.

Two typical grids used are shown in figure 2. Figure 2(a) shows grid I with f only varying in the η direction. Grid II is similar to this grid but only half the number of grid points in each direction. Figure 2(b) shows grid III. While the grids are mostly orthogonal there are regions where the grids are clearly not orthogonal. The grids shown do not extend far enough downstream. The grids are extended downstream in an orthogonal manner by algebraic means. This has two advantages: the grid is truly orthogonal and computer time is saved in computing the grid.

The viscous flow problem is solved in terms of the dimensionless stream function ψ and the vorticity ζ, defined by

$$u = \frac{\partial\psi}{\partial y},\quad v = -\frac{\partial\psi}{\partial x} \tag{5}$$

and

$$\zeta = \frac{\partial v}{\partial x} - \frac{\partial u}{\partial y}, \tag{6}$$

where (u,v) are the dimensionless velocity components in the (x,y) directions. The operator D is defined by

$$D \equiv \alpha\frac{\partial^2}{\partial\xi^2} - 2\beta\frac{\partial^2}{\partial\xi\partial\eta} + \gamma\frac{\partial^2}{\partial\eta^2} \tag{7}$$

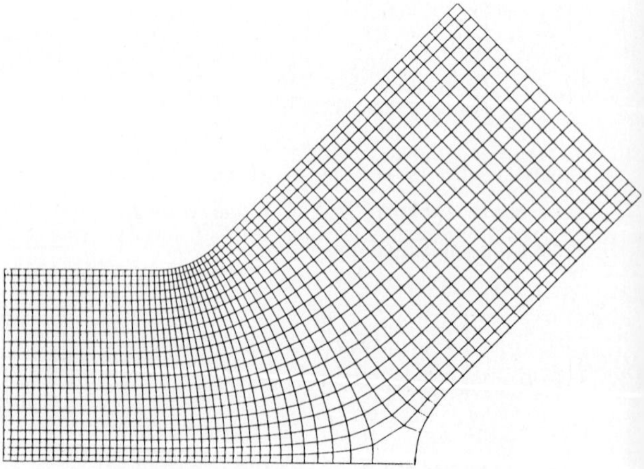

Figure 2(a) Grid I 20 x 68 f varying only with η

where $\alpha = x_\eta^2+y_\eta^2, \quad \beta = x_\xi x_\eta+y_\xi y_\eta, \quad \gamma = x_\xi^2+y_\xi^2.$ (8)

The Navier-Stokes equations in the transformed (ξ,η) coordinates are

$$D\psi+\sigma\psi_\eta+\tau\psi_\xi = -J^2\zeta, \tag{9}$$

$$D\zeta+\sigma\zeta_\eta+\tau\zeta_\xi = ReJ[\psi_\eta\zeta_\xi-\psi_\xi\zeta_\eta], \tag{10}$$

where $J = x_\xi y_\eta-x_\eta y_\xi$, $\sigma = [y_\xi Dx-x_\xi Dy]/J$, $\tau = [x_\eta Dy-y_\eta Dx]/J$ and Re is the Reynolds number based on the upstream channel width. Poiseuille flow values for ψ and ζ are applied upstream and downstream as described in Bramley and Sloan [1]. On the lower boundary of the channel ψ is taken to be zero and ψ=1 on the upper boundary, and on the axis of symmetry ζ=0. On the solid boundary, instead of using the second order Woods [6] boundary condition we use a first order boundary condition. The reason for this is that the mesh is finer at the wall and also the transformation of the Woods boundary condition gave problems with fine meshes at the wall. The grid at the boundary is always orthogonal and the first order boundary condition is then

$$\zeta_b = \frac{2(\psi_b-\psi_i)}{(x_b-x_i)^2+(y_b-y_i)^2} \tag{11}$$

where the subscript b refers to a grid point on the boundary and the subscript i refers to the interior point immediatly next to the boundary point.

At the interior points a central difference scheme is used on equation (9), while the Dennis-Hudson [3] difference scheme is used on equation (10). This choice ensures that the iterative matrix in the Gauss-Seidel scheme is diagonally dominant. The method of solution is fully explained in Bramley and Sloan [1]. The Gauss-Seidel iteration process is carried out until the residuals in both equations are less than 10^{-4}.

The results are calculated for 3 grids and Reynolds numbers 50, 100, 500. It is not possible to present all the results in detail. The results are broadly in agreement with previous work but are obtained with fewer grid points. Figure 3 gives the streamlines for Re = 500 on grid I, while figure 4 is the same problem using grid III. The separation and recirculation region(s) can be calculated in detail with the number of grid points in each direction reduced to half that used in Bramley and Sloan [1]. This leads to an associated reduction in computer time. Clustering the grid points around the point of smallest vorticity on the upper curve causes the flow to separate at Reynolds number 50, where in previous work the flow has not separated at this Reynolds number. It is clear that the vorticity boundary condition needs to be improved in order to use smaller grids at the solid boundaries.

<u>References</u>

1. J. S. Bramley and D. M. Sloan, Tenth International Conference on Numerical Methods in Fluid Dynamics, Lecture Notes in Physics, Springer-Verlag, 1986.

2. J. S. Bramley and D. M. Sloan, J. Comput. Phys. <u>15</u>, 297-311 (1987).

3. S. C. R. Dennis and J. D. Hudson, Proc. 1st Conf. Num. Meth. Laminar and Turbulent Flows, p.69, Pentech Press, London (1978).

4. G. Ryskin and L. G. Leal, J. Comput. Phys. <u>50</u>, 71-100 (1983).

5. J. F. Thompson, F. C. Thames and C. W. Mastin, J. Comput. Phys. <u>24</u>, 274-302 (1977).

6. L. C. Woods, Aeronaut. Q. <u>5</u>, 176-184 (1954).

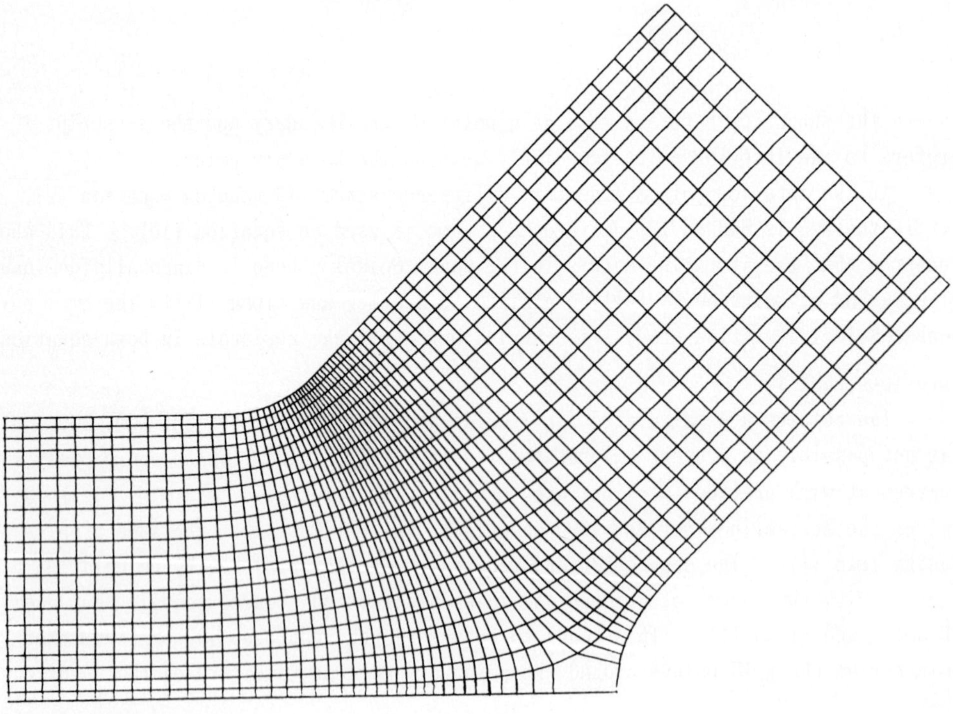

Figure 2(b) Grid III 20 x 65 f varying with ξ and η

Figure 3
Streamlines for Re = 500
using grid I

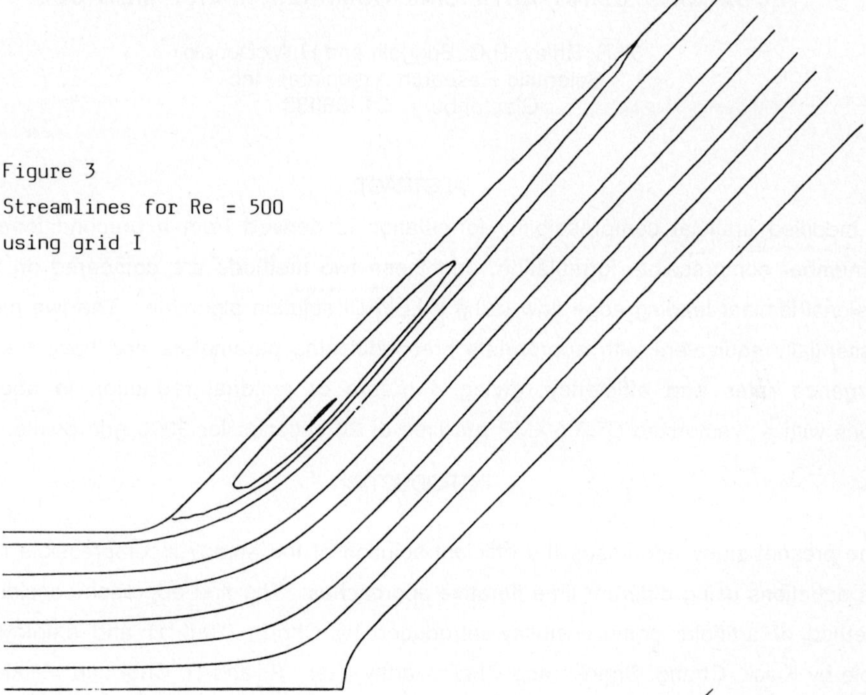

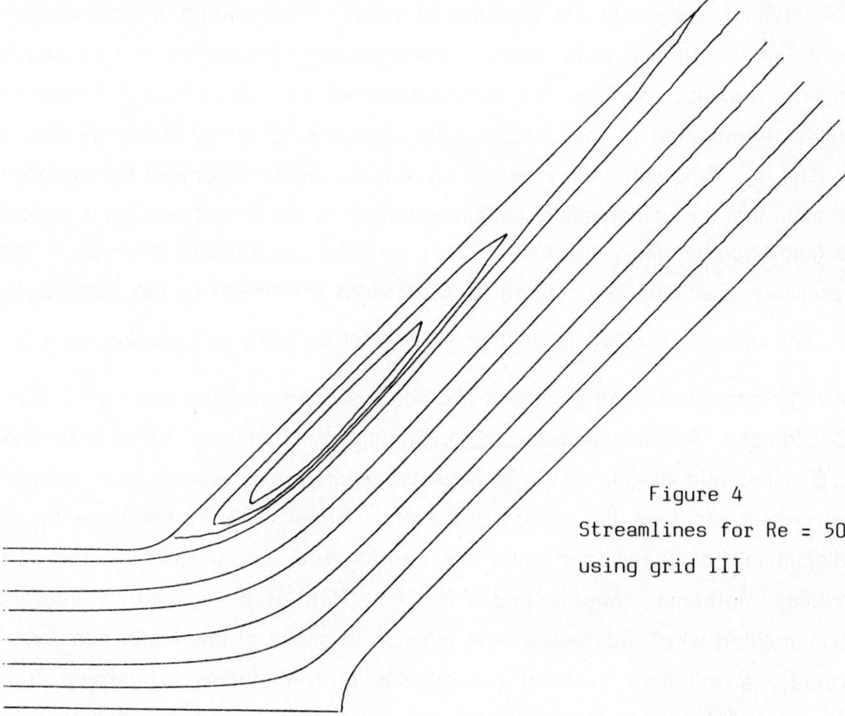

Figure 4
Streamlines for Re = 500
using grid III

SOLUTION OF THE INCOMPRESSIBLE NAVIER-STOKES EQUATIONS USING ARTIFICIAL COMPRESSIBILITY METHODS

W.R. Briley, R.C. Buggeln and H. McDonald
Scientific Research Associates, Inc.
Glastonbury, CT 06033

ABSTRACT

A modified artificial compressibility formulation is derived from a preconditioned low Mach number compressible formulation, and these two methods are compared on a two-dimensional laminar leading edge flow using a LBI/ADI solution algorithm. The two methods are essentially equivalent with appropriate preconditioning parameters and have the same convergence rates and efficiency, giving 4 orders of residual reduction in about 75 iterations with a vectorized CRAY-XMP run time of 20 seconds for 3000 grid points.

INTRODUCTION

The present study addresses the efficient solution of the steady incompressible Navier-Stokes equations using different time iterative approaches. The first approach considered is the method of artificial compressibility introduced by Chorin (Ref. 1) and employed for example by Kwak, Chang, Shanks and Chakravarthy (Ref. 2) and D. Choi and Merkle (Ref. 3). In this method, the continuity equation is modified by adding a time derivative of pressure, $\beta^{-2}\partial p/\partial t$, where β is an artificial compressibility parameter chosen to promote convergence to a steady solution. The second approach considered is a preconditioned low Mach number formulation of the compressible equations given by Briley, McDonald and Shamroth (Ref. 4). This approach is based on a nondimensional form of the equations and the observation that the compressible equations reduce to the incompressible equations as a reference Mach number M_{ref} approaches zero, provided (a) the total enthalpy is constant, and (b) pressure gradients are related to a pressure coefficient of the form $C_p/2 = p - 1/\gamma M_{ref}^2$. The energy equation is omitted since total enthalpy is constant, and thus the system has the same number of equations and hence (in principle) the same efficiency as the incompressible case. A (decoupled) incompressible energy equation is added to the system if needed. D. Choi and Merkle (Ref. 3) have considered a preconditioned compressible formulation which includes the energy equation. Turkel (Ref. 5) has recently given a unified discussion of these and other preconditioned compressible and artificial compressibility methods. Merkle and Y-H Choi (Ref. 6) have recently introduced a perturbation method which addresses heat addition to a gas at low Mach number. In the present study, a modified artificial compressibility formulation is derived from the preconditioned low Mach number compressible formulation, and these two methods are compared on a two-dimensional test problem.

GOVERNING EQUATIONS

The preconditioned compressible low Mach number formulation (Ref. 4) is given in nondimensional form by

$$a \frac{\partial \rho}{\partial t} + \nabla \cdot \rho \mathbf{U} = 0 \qquad (1)$$

$$a \mathbf{U} \frac{\partial \rho}{\partial t} + \rho \frac{\partial \mathbf{U}}{\partial t} + \nabla \left(\rho \mathbf{U U} + C_p/2 \right) = Re^{-1} \tau \qquad (2)$$

$$\rho h_0 = 1 + \gamma M_{ref}^2 \left(C_p + \frac{\gamma - 1}{\gamma} \rho \mathbf{U} \cdot \mathbf{U} \right) / 2 \qquad (3)$$

This system is solved in terms of ρ and components of $\mathbf{U}$ as dependent variables. The total enthalpy h_o is a constant which can be taken as unity, and the quantity $a = h_o/\gamma M_{ref}^2$ is a preconditioning parameter (see Ref. 4). The unsteady equations would have a = 1.

The modified artificial compressibility formulation is derived from Eqs. (1-3) by taking γ=1 in Eq. (3), eliminating density, and taking M_{ref}=0, h_o=1. This formulation is written as

$$\frac{1}{\beta^2} \frac{\partial}{\partial t} \frac{C_p}{2} + \nabla \cdot \mathbf{U} = 0 \qquad (4)$$

$$\frac{\alpha}{\beta^2} \mathbf{U} \frac{\partial}{\partial t} \frac{C_p}{2} + \frac{\partial \mathbf{U}}{\partial t} + \nabla \left(\mathbf{U U} + \frac{C_p}{2} \right) = Re^{-1} \tau \qquad (5)$$

with added preconditioning parameters α and β. This modified formulation is also considered by Turkel (Ref. 5) and differs from that of Chorin (α=0) in that a time derivative of Cp appears in the momentum equation. The choice $\alpha=\beta=1$ is equivalent to the compressible preconditioning $a=h_o/\gamma M_{ref}^2$ used here and in Ref. 4. The value β=1 was found to be optimal and was used for all calculations reported here. Computed results are given here for α=0,1, and 2.

The present solution procedure combines (a) a general nonorthogonal coordinate transformation, (b) a linearized block implicit scheme with Douglas-Gunn splitting, (c) three-point central spatial differences with a small amount of added second-order dissipation, (d) local Δt for convergence acceleration, and (e) implicit boundary conditions. The inflow/outflow conditions consisted of specified total pressure and flow angle at inflow, and static pressure at outflow; other variables were extrapolated. Further details of the solution procedure are given in Ref. 7.

COMPUTED RESULTS

A comparison of artificial compressibility and low Mach number compressible formulations was made for a test problem consisting of laminar flow (Re_H=500) about a blunt leading edge at zero incidence. The (60 x 50) C-grid coordinate system and contours of computed velocity magnitude are shown in Fig. 1. The mesh is stretched to give $\Delta r_{min}/(r_{max}-r_{min})=10^{-3}$ along a radial coordinate line. A local CFL number of 15 was used for all calculations. Convergence rates for this problem are shown in Figs.(2-4). Physical maximum residuals are shown and these are normalized by the initial maximum residual but are not scaled by the time step or coordinate Jacobian. In Fig. 2, the compressible formulation is solved for Mach numbers of 0.2 and 0.5 without any matrix preconditioning (a=1). These calculations demonstrate the degradation of convergence rate which occurs as the Mach number is reduced, a result of increasing stiffness in the ADI splitting error term. In Fig. 3, matrix preconditioning is introduced to remove this stiffness, and this results in rapid convergence (4 orders of residual reduction in about 75 iterations) which is independent of Mach number for Mach numbers as low as 10^{-5}. At $M=10^{-5}$ convergence stops at 4 orders of residual reduction due to round-off error. The solution for M=0.05 was repeated for 1000 iterations and converged to machine accuracy (10^{-12}) in less than 500 iterations. In Fig. 4, convergence using the modified artificial compressibility formulation is shown. The modified method (α=1,2) performs slightly better than the conventional method (α=0) and has the same convergence behavior as the preconditioned compressible formulation (compare Figs. 3 and 4). The analysis of Turkel (Ref. 5) indicates that α=2 is optimal.

Acknowledgment This work was supported by NASA/MSFC.

REFERENCES

1. Chorin, A.J.: A Numerical Method for Solving Incompressible Viscous Flow Problems, J. Comp. Physics, Vol. 2, pp. 12-26, 1967.

2. Kwak, D., Chang, J.L.C., Shanks, S.P., and S.R. Chakravarthy: An Incompressible Navier-Stokes Flow Solver in Three-Dimensional Curvilinear Coordinate System Using Primitive Variables. AIAA Paper No. 84-0253, 1984.

3. Choi, D. and C.L. Merkle: Application of Time-Iterative Schemes to Incompressible Flow. AIAA J. Vol. 23, No. 10, pp. 1518-1524.

4. Briley, W.R., McDonald, H. and Shamroth, S.J.: A Low Mach Number Euler Formulation and Application to Time Iterative LBI Schemes, AIAA Journal, Vol. 21, No. 10, p. 1467-1469, October 1983.

5. Turkel, E.: Preconditioned Methods for Solving the Incompressible and Low Speed Compressible Equations. J. Comp. Physics, Vol. 72, pp. 277-298, 1987.

6. Merkle, C.L. and Y-H Choi: Computation of Low Speed Flow with Heat Addition, AIAA J., Vol. 25, No. 6, pp. 831-839, 1987.

7. Briley, W.R. and McDonald, H.: Solution of the Three-Dimensional Navier-Stokes Equations for a Steady Laminar Horseshoe Vortex Flow. AIAA Paper No. 85-1520-CP, 1985.

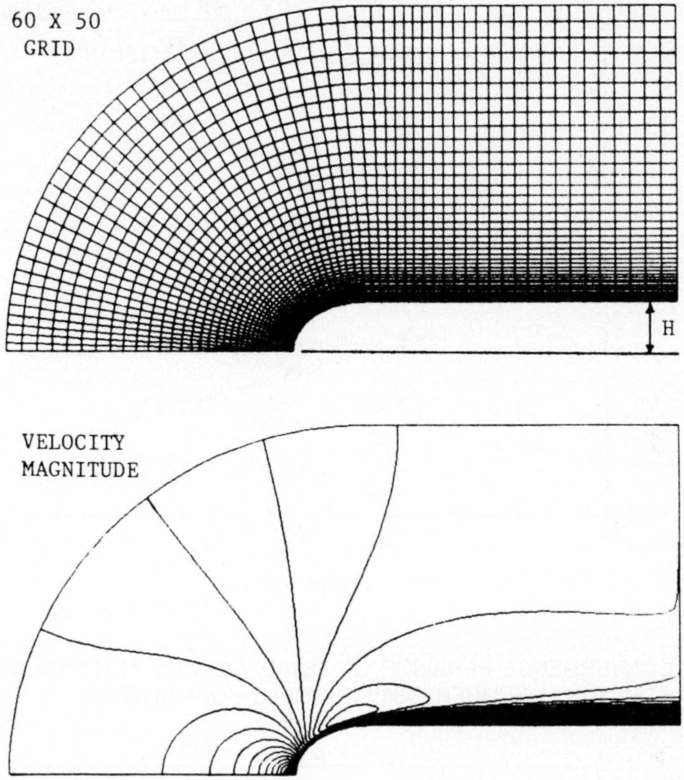

FIG. 1 - TEST PROBLEM : LAMINAR FLOW PAST A BLUNT LEADING EDGE (Re_H = 500)

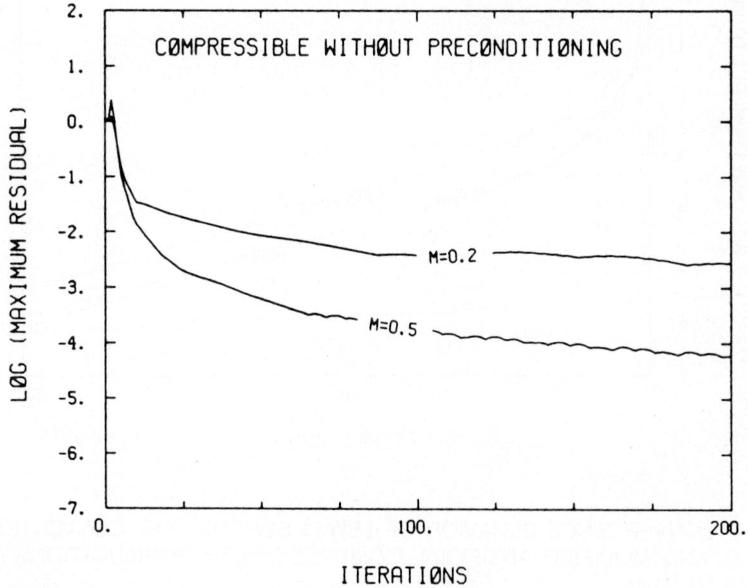

FIG. 2 - CONVERGENCE BEHAVIOR OF LBI/ADI SCHEME FOR COMPRESSIBLE FORMULATION WITHOUT PRECONDITIONING

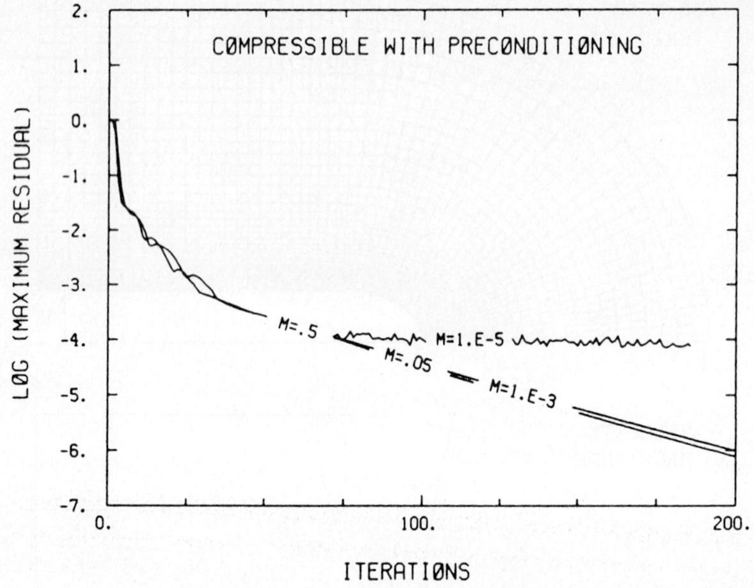

FIG. 3 - CONVERGENCE BEHAVIOR OF LBI/ADI SCHEME FOR PRECONDITIONED LOW MACH NUMBER COMPRESSIBLE FORMULATION

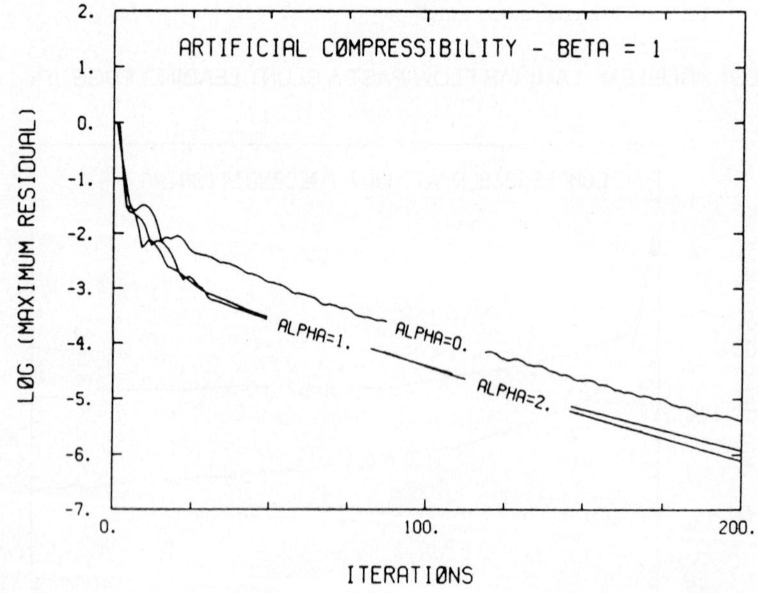

FIG. 4 - CONVERGENCE BEHAVIOR OF LBI/ADI SCHEME FOR CONVENTIONAL AND MODIFIED ARTIFICIAL COMPRESSIBILITY FORMULATIONS

Adaptive Finite Element Methods for Three Dimensional Compressible Viscous Flow Simulation in Aerospace Engineering

----

M.O. Bristeau, INRIA, B.P. 105, 78153 Le Chesnay Cedex, France
R. Glowinski, University of Houston, Houston, Texas, U.S.A. and INRIA
B. Mantel, J. Périaux, G. Rogé, AMD/BA, B.P. 300, 92214 St Cloud Cedex, France.

----

1. Introduction

The numerical solution of complicated two and three dimensional viscous flows modelled by the Navier-Stokes equations can be achieved by various time discretization schemes combined to convenient space discretizations (cf. e.g. [1]). If we consider, for example, operator splitting time discretization, the resulting subproblems require for their solution the use of advanced computational methods (mostly iterative) ; such methods are precisely described in this paper, together with the results of numerical experiments.

2. The compressible Navier-Stokes equations

Let $\rho, \mathbf{u}, T$ be the *density, velocity* and *temperature* variables, respectively. Taking $\sigma = \ln\rho$ as new variable the *Navier-Stokes equations* take the following (*non conservative*) non dimensionalized form

$$(\partial\sigma/\partial t) + \nabla.\mathbf{u} + \mathbf{u}.\nabla\sigma = 0, \tag{2.1}$$

$$(\partial\mathbf{u}/\partial t) + (\mathbf{u}.\nabla)\mathbf{u} + (\gamma-1)(T\nabla\sigma+\nabla T) = (e^{-\sigma}/Re)(\Delta\mathbf{u} + 1/3\ \nabla(\nabla.\mathbf{u})), \tag{2.2}$$

$$(\partial T/\partial t) + \mathbf{u}.\nabla T + (\gamma-1)T\nabla.\mathbf{u} = (e^{-\sigma}/Re)((\gamma/Pr)\ \Delta T + F(\nabla\mathbf{u})), \tag{2.3}$$

where Re, Pr and γ are the *Reynolds* number, the *Prandt* number and the *ratio of specific heats,* respectively.
The term F(.) in (2.3) is given (for two dimensional flows) by

$$F(\nabla\mathbf{u}) = 4/3\{|\partial u_1/\partial x_1|^2 + |\partial u_2/\partial x_2|^2 - (\partial u_1/\partial x_1)(\partial u_2/\partial x_2)\} + |\partial u_1/\partial x_2 + \partial u_2/\partial x_1|^2 \tag{2.4}$$

with $\mathbf{u} = \{u_i\}^2$. *Initial* and *boundary* conditions have to be added to the above equations (see [2] for more details). For the simplicity of the notation we shall rewrite (2.2)-(2.4) as follows

$$(\partial \underset{\sim}{u}/\partial t) - \mu\Delta \underset{\sim}{u} + \beta\underset{\sim}{\nabla}\sigma - \underset{\sim}{\psi}(\sigma,\underset{\sim}{u},T) = 0, \tag{2.5}$$

$$(\partial T/\partial t) - \Pi\,\Delta T - \chi(\sigma,\underset{\sim}{u},T) = 0\,; \tag{2.6}$$

the quantities and functions $\mu, \beta, \underset{\sim}{\psi}(.), \Pi, \chi(.)$ are fully defined in [2].

3. Time discretization of the Navier-Stokes equations

We present here a *time discretization* of the Navier-Stokes equations (2.1), (2.5), (2.6) *based on operator splitting :*

Let $\Delta t(>0)$ be a *time discretization* step and $\theta \in (0,1/2)$; we have then (with $\rho_0, \underset{\sim}{u}_0, T_0$) the *initial* density, velocity and temperature)

$$\sigma^0 = \sigma_0 = \ln \rho_0,\ \underset{\sim}{u}^0 = \underset{\sim}{u}_0,\ T^0 = T_0\,; \tag{3.1}$$

then for $n \geq 0$, *starting from* $\{\sigma^n, \underset{\sim}{u}^n, T^n\}$, *we solve*

$$\begin{cases} (\sigma^{n+\theta} - \sigma^n)/\theta\Delta t + \underset{\sim}{\nabla}.\underset{\sim}{u}^{n+\theta} = -\underset{\sim}{u}^n.\underset{\sim}{\nabla}\,\sigma^n, \\ (\underset{\sim}{u}^{n+\theta} - \underset{\sim}{u}^n)/\,\theta\Delta t - a\,\mu\Delta\underset{\sim}{u}^{n+\theta} + \beta\,\underset{\sim}{\nabla}\sigma^{n+\theta} = \underset{\sim}{\psi}(\sigma^n,\underset{\sim}{u}^n,T^n) + b\,\mu\Delta\underset{\sim}{u}^n, \\ (T^{n+\theta} - T^n)/\,\theta\Delta t - a\,\Pi\,\Delta T^{n+\theta} = \chi(\sigma^n,\underset{\sim}{u}^n,T^n) + b\,\Pi\,\Delta T^n, \end{cases} \tag{3.2}$$

$$\begin{cases} (\sigma^{n+1-\theta} - \sigma^{n+\theta})\,/(1-2\theta)\Delta t + \underset{\sim}{u}^{n+1-\theta}.\underset{\sim}{\nabla}\sigma^{n+1-\theta} = -\underset{\sim}{\nabla}.\underset{\sim}{u}^{n+\theta}, \\ (\underset{\sim}{u}^{n+1-\theta} - \underset{\sim}{u}^{n+\theta})\,/(1-2\theta)\Delta t - b\,\mu\Delta\underset{\sim}{u}^{n+\theta-1} - \underset{\sim}{\psi}(\sigma^{n+1-\theta}, \underset{\sim}{u}^{n+1-\theta}, T^{n+1-\theta}) = \\ \qquad a\,\mu\,\Delta\underset{\sim}{u}^{n+\theta} - \beta\,\underset{\sim}{\nabla}\sigma^{n+\theta}, \\ (T^{n+1-\theta} - T^{n+\theta})\,/(1-2\theta)\Delta t - b\,\Pi\,\Delta T^{n+1-\theta} - \chi\,(\sigma^{n+1-\theta}, \underset{\sim}{u}^{n+1-\theta}, T^{n+1-\theta}) = \\ \qquad a\,\Pi\,\Delta T^{n+\theta}, \end{cases} \tag{3.3}$$

$$\begin{cases} (\sigma^{n+1} - \sigma^{n+1-\theta})/\,\theta\Delta t + \underset{\sim}{\nabla}.\underset{\sim}{u}^{n+1} = -\underset{\sim}{u}^{n+1-\theta}.\underset{\sim}{\nabla}\,\sigma^{n+1-\theta}, \\ (\underset{\sim}{u}^{n+1} - \underset{\sim}{u}^{n+1-\theta})/\,\theta\Delta t - a\mu\,\Delta\underset{\sim}{u}^{n+1} + \beta\underset{\sim}{\nabla}\,\sigma^{n+1} = \\ \qquad \underset{\sim}{\psi}(\sigma^{n+1-\theta}, \underset{\sim}{u}^{n+1-\theta}, T^{n+1-\theta}) + b\,\mu\,\Delta\underset{\sim}{u}^{n+\theta-1} \\ (T^{n+1} - T^{n+1-\theta})/\,\theta\Delta t - a\,\Pi\,\Delta T^{n+1} = \chi(\sigma^{n+1-\theta}, \underset{\sim}{u}^{n+1-\theta}, T^{n+1-\theta}) + b\,\Pi\,\Delta T^{n+1-\theta}, \end{cases} \tag{3.4}$$

with $0 < a,b < 1$, $a+b = 1$, satisfying (cf. [3])

$$a = (1-2\theta)/(1-\theta),\ b = \theta/(1-\theta). \tag{3.5}$$

Other time discretization schemes are presently under investigations, such as the following *fully implicit* one

$$(\sigma^{n+1} - \sigma^{n})/\Delta t + \underset{\sim}{u}^{n+1}.\underset{\sim}{\nabla}\, \sigma^{n+1} + \underset{\sim}{\nabla}.\underset{\sim}{u}^{n+1} = 0, \tag{3.6}$$

$$\begin{cases} (\underset{\sim}{u}^{n+1} - \underset{\sim}{u}^{n})/\Delta t + (\underset{\sim}{u}^{n+1}.\underset{\sim}{\nabla})\, \underset{\sim}{u}^{n+1} + (\gamma-1)(T^{n+1}\, \underset{\sim}{\nabla}\sigma^{n+1} + \underset{\sim}{\nabla} T^{n+1}) \\ - (e^{-\sigma^{n+1}}/Re)[\Delta \underset{\sim}{u}^{n+1} + 1/3\, \underset{\sim}{\nabla}(\underset{\sim}{\nabla}.\underset{\sim}{u}^{n+1})] = \underset{\sim}{0}. \end{cases} \tag{3.7}$$

$$\begin{cases} (T^{n+1} - T^{n})/\Delta t + \underset{\sim}{u}^{n+1}.\underset{\sim}{\nabla}\, T^{n+1} + (\gamma-1)\, T^{n+1}\, \underset{\sim}{\nabla}.\underset{\sim}{u}^{n+1} \\ - (e^{-\sigma^{n+1}}/Re)[(\gamma/Pr)\, \Delta T^{n+1} + F(\underset{\sim}{\nabla}\underset{\sim}{u}^{n+1})] = 0. \end{cases} \tag{3.8}$$

4. Solving the Generalized Stokes Problems

Each time step of scheme (3.1)-(3.4) requires the solution of the two *generalized Stokes problems* (3.2) and (3.4). Both problems are particular cases of

$$\alpha\sigma + \underset{\sim}{\nabla}.\underset{\sim}{u} = g \textit{ in } \Omega, \tag{4.1}$$

$$\alpha\underset{\sim}{u} - a\mu\, \Delta \underset{\sim}{u} + \beta \underset{\sim}{\nabla}\sigma = \underset{\sim}{f} \textit{ in } \Omega, \tag{4.2}$$

where Ω is the *computational flow region*. In order to render (4.1)-(4.2) *well posed*, boundary conditions have to be added ; an appropriate set of such boundary conditions is given in e.g. [2]. Using Fourier Analysis (as in [4], [5] and [6]), we can construct efficient preconditioners which can be used for the *conjugate gradient* solution of the above generalized Stokes problems, reducing the solution of (4.1), (4.2) to a sequence of ordinary Dirichlet and/or Neumann problems (see [4]-[7] for more details).

5. Solving the Nonlinear Problems .

Each time step of both schemes (3.1)-(3.4) and (3.1), (3.6)-(3.8) leads to the solution of a system of *nonlinear equations* ; for example scheme (3.1)-(3.4) requires the solution of the nonlinear problem (3.3) which is a particular case of

$$\alpha\sigma + \underset{\sim}{u}.\underset{\sim}{\nabla}\sigma = g \textit{ in } \Omega, \tag{5.1$_1$}$$

$$\alpha\underset{\sim}{u} - b\mu\, \Delta \underset{\sim}{u} - \underset{\sim}{\psi}(\sigma,\underset{\sim}{u},T) = \underset{\sim}{f} \textit{ in } \Omega, \tag{5.1$_2$}$$

$$\alpha T - b\Pi\, \Delta T - \chi(\sigma,\underset{\sim}{u},T) = h \textit{ in } \Omega, \tag{5.1$_3$}$$

with appropriate boundary conditions (some of them being provided by the previous substep (3.2)).

At the present moment our group of investigators is extending the basic least squares/conjugate gradient methodology described in, e.g. [3] and [8], in order to include those GMRES algorithms advocated in [9], [10] which are based on the construction of *Krylov subspaces*. From the importance of these new developments, we shall give just below the general description of such a GMRES algorithm in the general context of *Hilbert spaces :*

Let's V be a real Hilbert space for the scalar product (.,.) and the corresponding norm ||.||. We denote by V' the dual space of V, by <.,.> the duality pairing between V' and V and finally by S the duality isomorphism between V and V', i.e. the isomorphism from V onto V' such that

$$\langle Sv,w\rangle = (v,w),\ \forall v,w \in V,$$

$$\langle Sv,w\rangle = \langle Sw,v\rangle,\ \forall v,w \in V,$$

$$\langle f,S^{-1}g\rangle = (f,g)_*,\ \forall f,g \in V'.$$

Now with F a (possibly nonlinear) operator from V into V', we consider the problem

$$F(u) = 0. \tag{5.2}$$

Description of the GMRES algorithm for solving (5.2) :

$$u^o \in V, \textit{ given }; \tag{5.3}$$

then, for $n \geq 0$, u^n *being known we obtain* u^{n+1} *as follows* :

$$r_1^n = S^{-1}F(u^n), \tag{5.4}$$

$$w_1^n = r_1^n/\|r_1^n\|. \tag{5.5}$$

Then for j=2,...,k, *we compute* r_j^n *and* w_j^n *by*

$$r_j^n = S^{-1}\, DF(u^n;w_{j-1}^n) - \sum_{i=1}^{j-1} b_{ij-1}^n w_i^n, \tag{5.6}$$

$$w_j^n = r_j^n/(\|r_j^n\|; \tag{5.7}$$

in (5.6), $DF(u^n;w)$ *is defined by either*

$$DF(u^n;w) = F'(u^n).w \tag{5.8}$$

(*where* $F'(u^n)$ *is the differential of* F *at* u^n, *or if the calculation of* F' *is too costly*

$$DF(u^n;w) = (F(u^n+ \varepsilon w)-F(u^n))/\varepsilon \tag{5.9}$$

with ε *sufficiently small ; we define then* b_{il}^n *by*

$$b_{il}^n = \langle DF(u^n;w_l^n),w_i^n\rangle\ (=(S^{-1}DF(u^n;w_l^n),w_i^n))). \tag{5.10}$$

Next, we solve

Find $a^n = \{a_j^n\}_{j=1}^k \in \mathbb{R}^k$ *such that*

$$\forall b = \{b_j\}_{j=1}^k \in \mathbb{R}^k \textit{ we have} \tag{5.11}$$

$$\|F(u^n + \sum_{j=1}^k a_j^n w_j^n)\|_* \leq \| F(u^n + \sum_{j=1}^k b_j w_j^n)\|_*$$

and obtain u^{n+1} *by*

$$u^{n+1} = u^n + \sum_{j=1}^k a_j^n w_j^n . \tag{5.12}$$

Do n = n+1 *and go to* (5.4).
In algorithm (5.3)-(5.12), k is the dimension of the so-called *Krylov space.*

Remark 5.1. : We should easily prove that

$$(w^n_l, w^n_j) = 0, \forall 1 \le l, j \le k, j \ne l. \tag{5.13}$$

Remark 5.2. : To compute $DF(u^n;w)$, we can use instead of the second order accurate discrete derivative

$$DF(u^n;w) = (F(u^n + \varepsilon w) - F(u^n - \varepsilon w))/2\varepsilon \tag{5.14}$$

Remark 5.3. : Norm $\|.\|_*$ in (5.11) satisfies the following relations

$$\|f\|_* = \|S^{-1}f\|, \forall f \in V', \tag{5.15}$$

$$\|f\|_* = \langle f, S^{-1}f\rangle^{1/2}, \forall f \in V', \tag{5.16}$$

indeed we have used (5.15) to evaluate the various norms $\|.\|_*$ occuring in (5.11).

Remark 5.4. : In order to evaluate a^n via the solution of (5.11) it is sufficient to approximate, in the neighborhood of b = 0, the functional

$$b \dashrightarrow \|F(u^n + \sum_{j=1}^{k} b^n_j w^n_j)\|^2_* \tag{5.17}$$

by the *quadratic* one defined by

$$b \dashrightarrow \|F(u^n + \sum_{j=1}^{k} b_j DF(u^n;w^n_j)\|^2_* \tag{5.18}$$

Minimizing the functional (5.18) is clearly equivalent to solving a linear system whose matrix k x k is symmetric and positive definite (this approach is equivalent to taking for a^n the first iterate provided by Newton's method applied to the solution of (5.11) and initialized with b = 0).

Remark 5.5. : In order to control the systematic error introduced by the approximation of the functional (5.17) by the quadratic one in (5.18), *backtracking* techniques are described in [10], together with numerical experiments showing the efficiency of these methods.

Remark 5.6. : The practical implementation of algorithm (5.3)-(5.12) for solving (5.1) requires a convenient choice for the (preconditioning) operator S. At the moment, we are testing *incomplete block* LU *factorization* preconditioners (see [7] for details).

6. Finite Element Approximations. Compatibility between Velocity and Density Approximations.

It is well known that for the incompressible Navier-Stokes equations, *pressure* and *velocity* cannot be approximated independently (cf. e.g. [11] and the references therein). Typical choices are, for example,

i) *continuous* and *piecewise linear* approximations for the *pressure*, associated to a *continuous* and *piecewise quadratic* approximation for the *velocity.*

ii) Same approximation than above for the pressure associated to a *continuous* and *piecewise linear* approximation for the *velocity* on a grid *twice finer* that the pressure one (see [8, Chapt. 7], [3, Sect. 5] for more details).

iii) The so-called Arnold-Brezzi-Fortin mini element (cf. [12] and [3, Sect. 5]).

Concerning now the compressible Navier-Stokes equations there has been a natural tendancy to approximate velocity and density (or logarithmic density) using the same finite element spaces for both quantities (except [17]).

Indeed such a strategy can produce spurious oscillations, particularly for density when using Uzawa algorithm for the generalized Stokes problem (the Glowinski-Pironneau Stokes solver behaves more regularly in terms of density), (see [6]-[7]). Once more, a most efficient cure is again to use approximations of type (i), (ii), (iii), with the role of the pressure played by the density (or the logarithmic one) and the temperature. As in the incompressible case, approximations of type (ii) and (iii) are particularly attractive, and therefore have been used for the subsequent calculations.

7. Self Adaptive Mesh Refinement Techniques

The development of various singularities in the Navier-Stokes flows gives a strong motivation to use a mesh adaptive strategy ; this is particularly important for three dimensional problems for which limiting the number of grid point is crucial in order to fit the size of the core memory.

Actually, with non structured meshes, local enrichment algorithms are rather easy to implement, for finite element approximations, as shown in [3, Sect. 15].

Various physical criteria can be chosen in order to decide those regions where the density of grid points has to be increased. In these regions, each triangle, or tetrahedron, will be splitted into 4 or 8 sub-elements.

Remark 7.1. : With the techniques discussed in [3], the existing grid points are not moved, but extra ones are added. Also, starting from a structured grid, we obtain after refinement a non structured one, on which the method will provide a more accurate solution.

Remark 7.2. : Other mesh refinement strategies are described in [13], [14].

To conclude this section, let's mention that investigations are under way, in order to extend from elliptic problems to hyperbolic ones, the a posteriori error estimates employed to refine the mesh (see e.g. [15], [16] for the elliptic theory).

8. Numerical Experiments

We present some results obtained using the methods (briefly) described in the above sections. The first series of numerical experiments concern a two-dimensional compressible viscous flow around an elliptic body, using adaptivity. We took here $M_\infty = .8$, $Re = 10^3$, $\gamma = 1.4$, $Pr = .72$ and a zero angle of attack. We have shown on Figures 8.1 to 8.3, the isobar contours obtained with : the *same linear* approximation for velocity and density in scheme (3.2)-(3.4) (Figure 8.1), approximation (ii) of Section 6 in the same scheme (Figure 8.2), approximation (iii) of Section 6 in scheme (3.6)-(3.8) (Figure 8.3). We observe that the spurious oscillations-associated to the same approximation for velocity and density and Uzawa algorithm- disappear with the more sophisticated approximations (ii) and (iii) of Section 6. Similar phenomena are observed for three dimensional flow around spheres (see Figures 8.4 and 8.5) and close to the forebody of the space vehicle Hermes at $M_\infty = 2$ and $Re = 10^3$ (see Figures 8.6 and 8.7) ; for these three dimensional calculations mesh adaptive techniques have also been used. We observe that Figures 8.4 and 8.5 correspond to simulations close to a stagnation point where very strong compressions take place.

9. Conclusion :

We have briefly discussed here the numerical solution of the compressible Navier-Stokes equations written in non conservative form. We have shown by these first results that approximations satisfying some "Inf-sup" condition lead, for moderate Reynolds and Mach numbers, to accurate results without upwinding or artificial viscosity. The solution techniques which have been discussed are extended presently to *conservative* formulations in view of genuine hypersonic simulations including real gas effects.We can hope from the first results that, for higher Mach and Reynolds numbers, we will obtain accurate solutions while introducing less dissipation in the schemes; the accuracy, particularly of pressure, being a critical point in view of further calculations in hypersonic design.

Acknowledgment : This work is partly supported by a grant from DRET (Grant n° 88.103).

A part of computational work was performed on CRAY 2 of CCVR.

We would like to thank P. Leyland, F. Brezzi and M. Fortin for fruitful discussions on the compatible approximations and also Mrs. C. Demars for the typing of the manuscript.

References :

[1] R. Peyret and T.D. Taylor, *Computational Methods for Fluid Flow* (Springer-Verlag, New-York, 1982).

[2] M.O. Bristeau, R. Glowinski, B. Mantel, J. Périaux and G. Rogé, Acceleration of Compressible Navier-Stokes calculations, Proceedings of ICFD 88 Conference on Numerical Methods for Fluid Dynamics, Oxford, March 21-24, 1988. K.W. Morton ed.

[3] M.O. Bristeau, R. Glowinski and J. Périaux, Numerical Methods for the Navier-Stokes equations. Applications to the simulation of compressible and incompressible viscous flows. *Computer Physics Report* 6 (1987), North-Holland, Amsterdam, pp. 73-187.

[4] J. Cahouet and J.P. Chabard, Multi-Domains and Multi-Solvers Finite Element Approach for the Stokes problem, in : *Innovative Numerical Methods in Engineering*, eds., R.P. Shaw, J. Périaux, A. Chadouet, J. Wu, C. Marino, C.A. Brebbia (Springer-Verlag, Berlin, 1986), p. 317-322.

[5] C. Bègue, R. Glowinski and J. Périaux, Détermination d'un opérateur de préconditionnement pour la résolution itérative du problème de Stokes dans la formulation d'Helmholtz, *C.R. Acad. Sci.* Paris, t. 306, Série I, p. 247-252, 1988.

[6] C. Bègue, M.O. Bristeau, R. Glowinski, B. Mantel, J. Périaux and G. Rogé, Accélération de la convergence dans le calcul des écoulements de fluides visqueux. 1ère partie : cas incompressible et cas compressible non conservatif. Rapport INRIA n° 861, 1988.

[7] G. Rogé, Approximation mixte et accélération de la convergence pour la résolution des équations de Navier-Stokes compressible en éléments finis, Thèse de 3ème cycle, Université Pierre et Marie Curie, Paris 6, à paraître.

[8] R. Glowinski, *Numerical Methods for Nonlinear Variational Problems* (Springer-Verlag, New-York, 1984).

[9] Y. Saad and M.H. Schultz, GMRES : A generalized minimal residual algorithm for solving nonsymmetric linear systems, *SIAM J. Sci. Stat. Comp.*, 7 (1986), pp. 856-869.

[10] P.N. Brown, Y. Saad, Hybrid Krylov Methods for Nonlinear Systems of Equations, *Lawrence Livermore National Laboratory Research Report UCLR-97645,* Nov. 1987.

[11] V. Girault and P.A. Raviart, *Finite Element Methods for Navier-Stokes Equations* (Springer-Verlag, Berlin, 1986).

[12] D.N. Arnold, F. Brezzi and M. Fortin, A stable finite element for the Stokes equations, *Calcolo* 21 (1984) 337.

[13] J. Peraire, M. Vahdati, K. Morgan and O.C. Zienkiewicz, Adaptive remeshing for compressible flow computations, *J. Comp. Phys.* 72,2, 1987.

[14] M.C. Rivara, Algorithms for refining triangular grids suitable for adaptive and multigrid techniques, *Int. J. Num. Meth. in Eng.*, Vol. 20, pp. 745-756 (1984).

[15] I. Babuska and W.C. Rheinbold, A posteriori error estimates for the Finite Element Method, *Intern. J. Numer. Meth. Eng.* 112 (1978) 1597.

[16] R.E. Bank , Analysis of a local a posteriori error estimate for elliptic equations in *Accuracy Estimates and Adaptivity for Finite Elements,* J. Wiley & Sons, New-York, 1986, pp. 119-128.

[17] Y. Secretan, G. Dhatt, D. Nguyen, Compressible Viscous Flow around a NACA0012 airfoil, *Numerical Simulation of Compressible Navier-Stokes flows,* Notes on Num. Fluid Mech., Vieweg, Vol. 18, 1987.

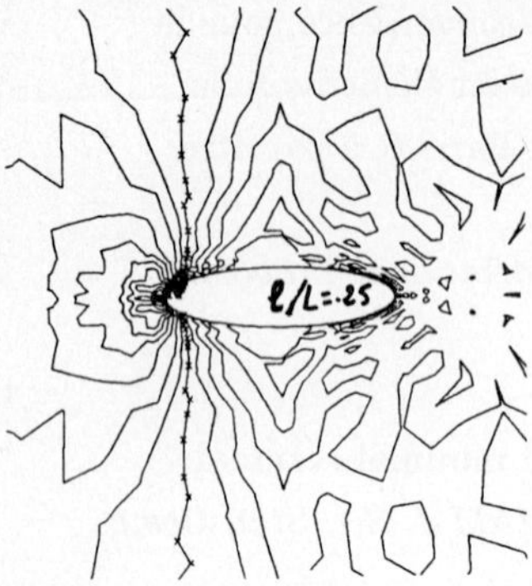

Figure 8.1.
P_1 approximations

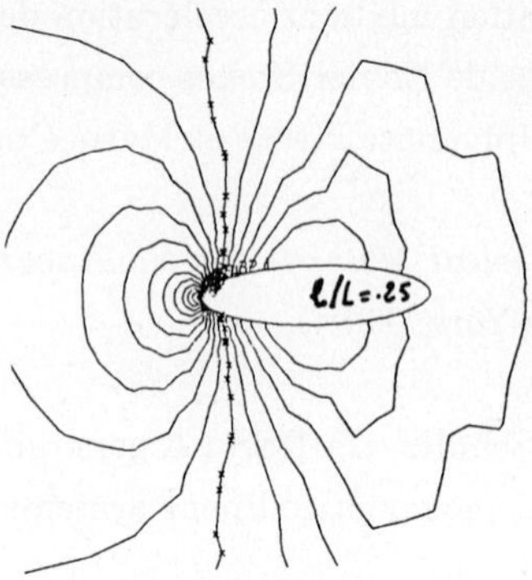

Figure 8.2.
P_1, mini-element approximations

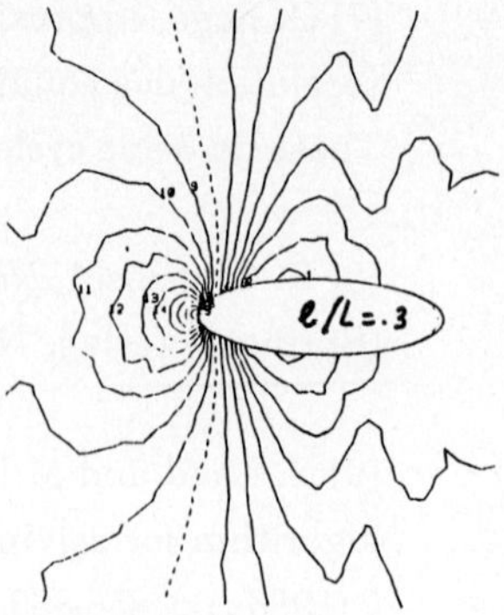

Figure 8.3.
P_1, P_1 iso-P_2 approximations

Flow around an ellipse, $M_\infty = 0.8$, Re = 1000
Iso-C_p lines

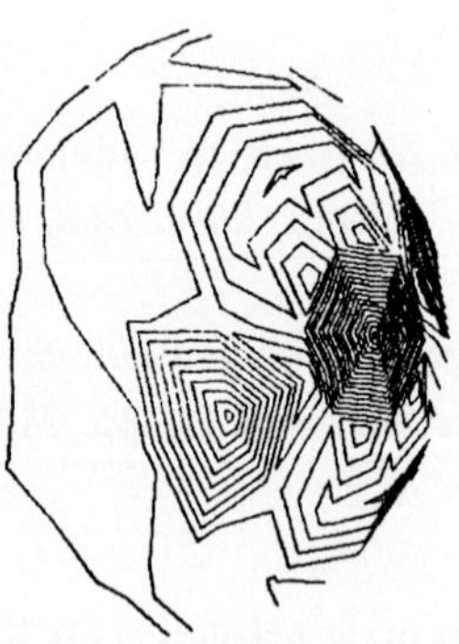

Figure 8.4.
P_1 approximations

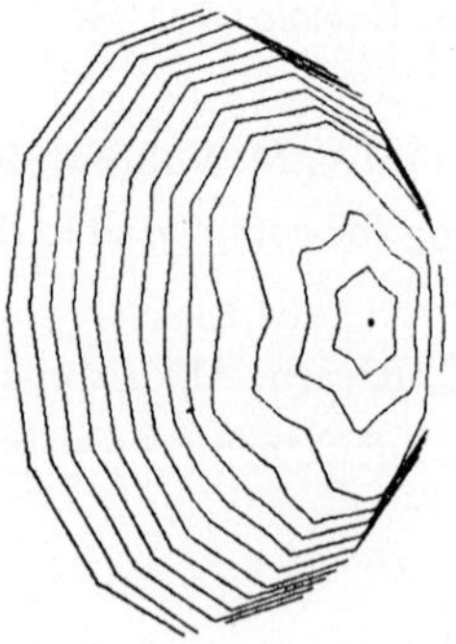

Figure 8.5.
P_1, mini-element approximations

Flow around a sphere, $M_\infty = 0.9$, Re = 100.

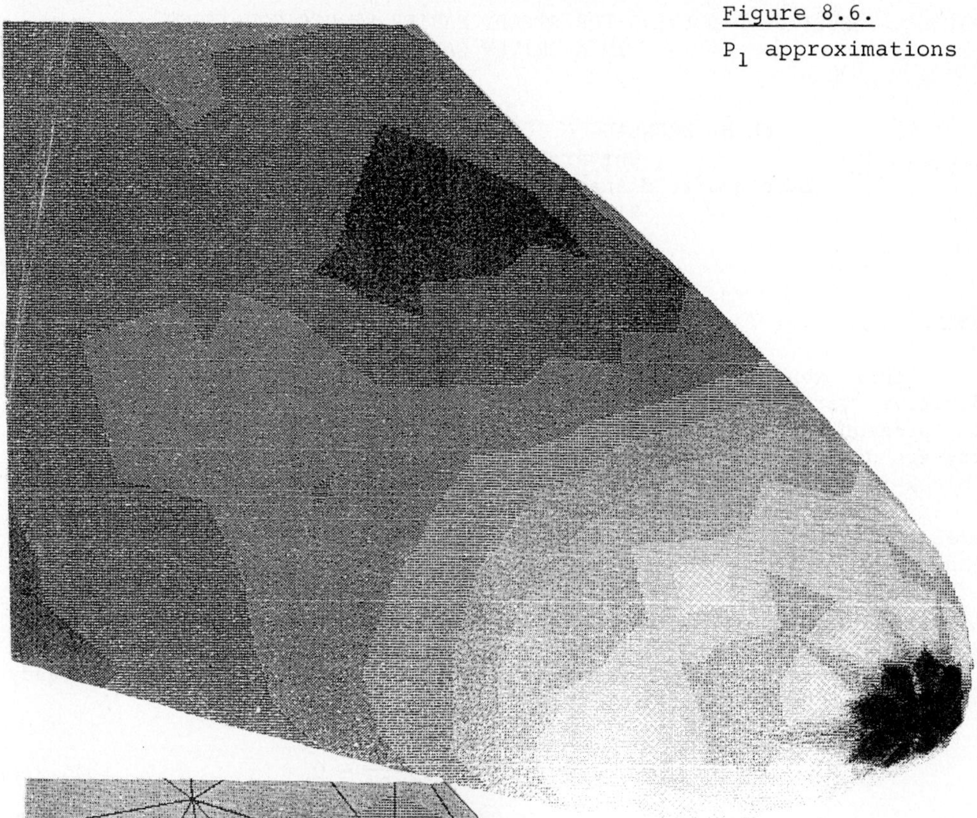

Figure 8.6.
P_1 approximations

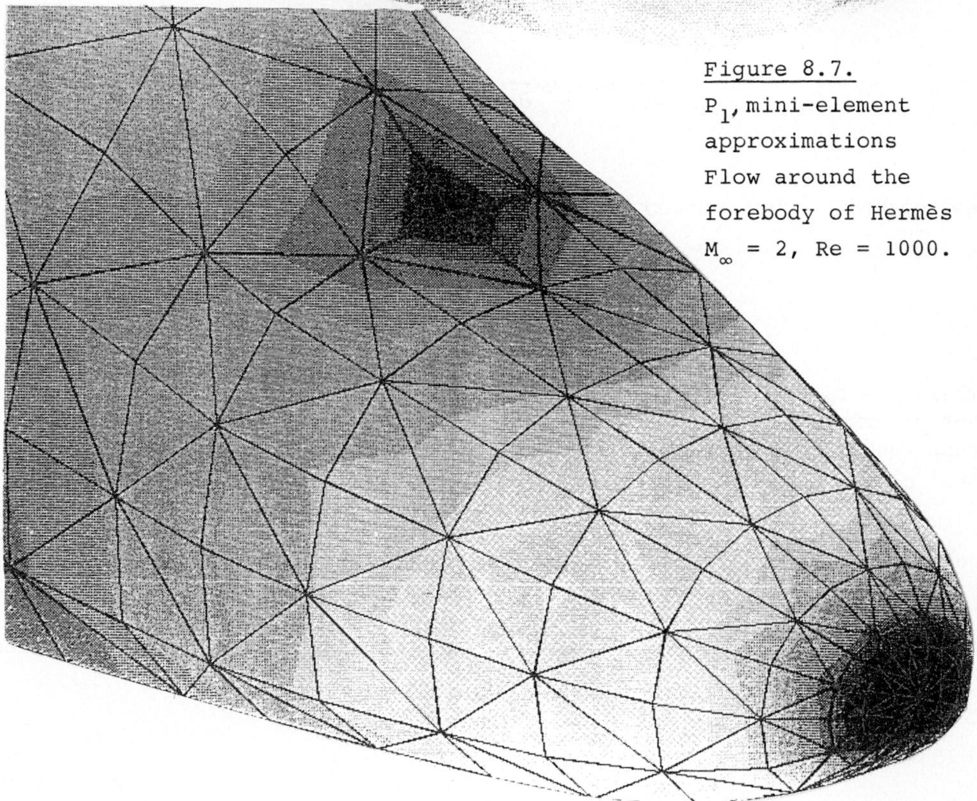

Figure 8.7.
P_1, mini-element approximations
Flow around the forebody of Hermès
$M_\infty = 2$, Re = 1000.

MULTIGRID SOLVERS FOR STEADY NAVIER-STOKES EQUATIONS IN A DRIVEN CAVITY

Ch.H. BRUNEAU, C. JOURON and L. B. ZHANG
Université Paris-Sud
Laboratoire d'Analyse Numérique, Bâtiment 425
91405 Orsay, France.

ABSTRACT.

This work is devoted to solving steady Navier-Stokes equations in primitive variables in a 2D driven cavity. Two new robust upwind schemes are presented to approach the convection terms. The pressure and the velocity are discretized on staggered grids, the FMG-FAS algorithm is used with a block-implicit relaxation procedure to solve the whole system.

These schemes are stable and the obtained solutions are in good agreement with earlier numerical works. In addition with one of the two schemes for Re=10,000 the transition to turbulence is observed numerically.

1. THE DRIVEN CAVITY PROBLEM.

The driven cavity problem is to solve the adimensioned steady Navier-Stokes equations in the 2D domain $\Omega = (0,1)x(0,1)$:

$$\begin{cases} -\frac{1}{Re}\Delta u + u\frac{\partial u}{\partial x} + v\frac{\partial u}{\partial y} + \frac{\partial p}{\partial x} = 0 \\ -\frac{1}{Re}\Delta v + u\frac{\partial v}{\partial x} + v\frac{\partial v}{\partial y} + \frac{\partial p}{\partial y} = 0 \end{cases} \tag{1}$$

$$\operatorname{div} \vec{q} = 0 \tag{2}$$

associated to the boundary conditions :

$$\begin{cases} \vec{q} = (1,0) \text{ on } I = (0,1)x\{1\} \\ \vec{q} = (0,0) \text{ on } \partial\Omega\backslash I \end{cases} \tag{3}$$

where $\vec{q} = (u,v)$ is the velocity, p the pressure and Re the Reynolds number. Let us denote by D the diffusion operator and by C the convection operator, then equations (1) read :

$$D(\vec{q}) + C(\vec{q}) + \nabla p = 0 \tag{4}$$

For large Reynolds numbers this problem is quite difficult to solve, in particular the approximation of the nonlinear operator C. Here we propose two new uncentered schemes which give a good balance between D and C.

The driven cavity problem is a test problem very convenient to make comparisons with previous works like [1], [2] and [3].

2. THE MULTIGRID PROCEDURE AND THE APPROXIMATION.

The Navier-Stokes equations are solved on a set of staggered grids (figure 1) by a FMG-FAS (Full Multigrid - Full Approximation Storage) algorithm.

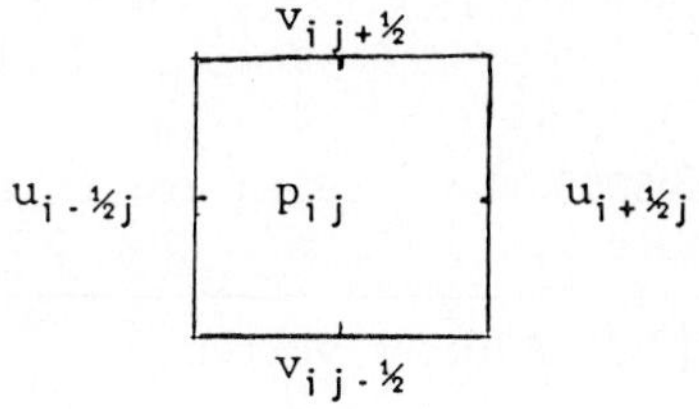

Unknowns on the cell (i,j)
Figure 1

On each grid the discretized equations are solved by the cell-by-cell underrelaxation procedure proposed in [3]. This leads to solve implicitly a 5x5 linear sytem on the cell (i,j) for the five unknowns.

Equations (1), (2) are approximated by centered finite differences at the current point except for the convection terms where second order uncentered differences are used. We describe here the term $u\frac{\partial u}{\partial x}$ at point (i-1/2, j) :

First scheme [5]

$$\begin{cases} \dfrac{u^{n-1}_{i-1/2,j}}{2\Delta x}\left(3u_{i-1/2,j} - 4u^{n-1}_{i-3/2,j} + u^{n-1}_{i-5/2,j}\right) & \text{if } u^{n-1}_{i-1/2,j} > 0 \\[2ex] -\dfrac{u^{n-1}_{i-1/2,j}}{2\Delta x}\left(3u_{i-1/2,j} - 4u^{n-1}_{i+1/2,j} + u^{n-1}_{i+3/2,j}\right) & \text{otherwise} \end{cases} \tag{5}$$

Second scheme [6]

$$\begin{cases} \dfrac{u^{n-1}_{i-1,j}}{3\Delta x}\left(4u_{i-1/2,j} - 5u^{n-1}_{i-3/2,j} + u^{n-1}_{i-5/2,j}\right) & \text{if } u^{n-1}_{i-1,j} > 0 \\[2ex] -\dfrac{u^{n-1}_{i,j}}{3\Delta x}\left(4u_{i-1/2,j} - 5u^{n-1}_{i+1/2,j} + u^{n-1}_{i+3/2,j}\right) & \text{if } u^{n-1}_{i,j} < 0 \end{cases} \tag{6}$$

where Δx is the size of the mesh, n the index of relaxation iterations and $u^{n-1}_{i-1,j}$, $u^{n-1}_{i,j}$ are computed by linear interpolation of the neighbours.

In the first scheme only the derivative $\frac{\partial u}{\partial x}$ is approximated by second order uncentered finite differences at point (i-1/2,j). On the contrary, in the second scheme, the whole term $u\frac{\partial u}{\partial x}$ is a sum (sometimes reduced to zero) of uncentered finite differences at points (i-1,j) and (i,j).

The two schemes add positive terms on the main diagonal of the linear 5x5 system; which improves the stability of the whole process.

3. NUMERICAL RESULTS.

For Re=1000 and Re=5000 the results are in good agreement with [1], [2] and [3] as shown in table 1.

	u_{min} y_{min} on vertical centerline	Primary vortex Φ (location)	Secondary vortex bottom right Φ (location)
Re=1000			
[1]	-0.3829 0.1719	-0.1179 (0.5313,0.5625)	0.175E-2 (0.8594,0.1094)
[2]		-0.1160 (0.5286,0.5643)	0.170E-2 (0.8643,0.1071)
[3]	-0.3870 0.1734	-0.1173 (0.5438,0.5625)	0.174E-2 (0.8625,0.1063)
scheme (5)	-0.3901 0.1699	-0.1193 (0.5313,0.5664)	0.174E-2 (0.8633,0.1133)
scheme (6)	-0.3764 0.1602	-0.1163 (0.5313,0.5586)	0.191E-2 (0.8711,0.1094)
Re=5000			
[1]	-0.4364 0.0703	-0.1190 (0.5117,0.5352)	0.308E-2 (0.8086,0.0742)
scheme(5)	-0.4535 0.0762	-0.1240 (0.5156,0.5352)	0.306E-2 (0.8008,0.0742)
scheme(6)	-0.4359 0.0664	-0.1142 (0.5156,0.5313)	0.465E-2 (0.8301,0.0703)

Comparison of solutions
Table 1

The two schemes were compared to the one used by Vanka in [3] and the second scheme appeared to the more stable and efficient in term of CPU time. It gives for Re=5000 on a 512x512 grid a steady solution which good representation of secondary vortices (figure 2).

Morover, with this second scheme, we observed numerically the transition to turbulence. Indeed for Re = 10000 we get a stable steady solution on the 256x256 grid as the boundary layer is well represented (figure 3). Then on the 512x512 grid there is no convergence any more to a steady solution as turbulence can be captured by the fine mesh. We see on figure 4 the evolution of the solution with iterations in this last case.

CONCLUSION.

The results are in good agreement with other results in the square driven cavity for Reynolds numbers less or equal to 5000. It appears that the solution is laminar for Re $\leqslant$ 5000 and becomes unstable for the steady model with Re = 10000 on the fine 512x512 grid when small eddies develop along the walls in the corners.

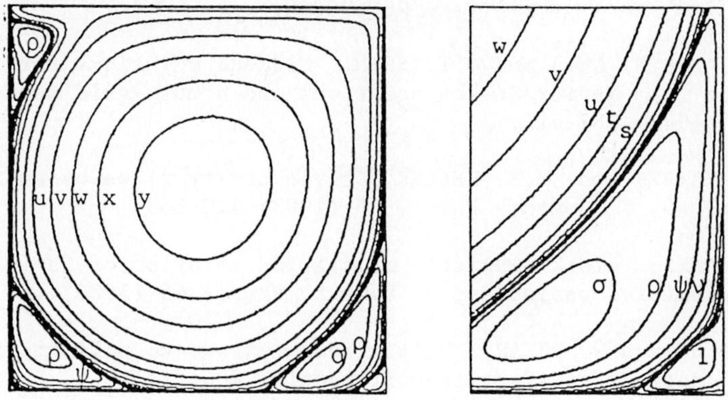

Streamline for Re = 5000

Figure 2

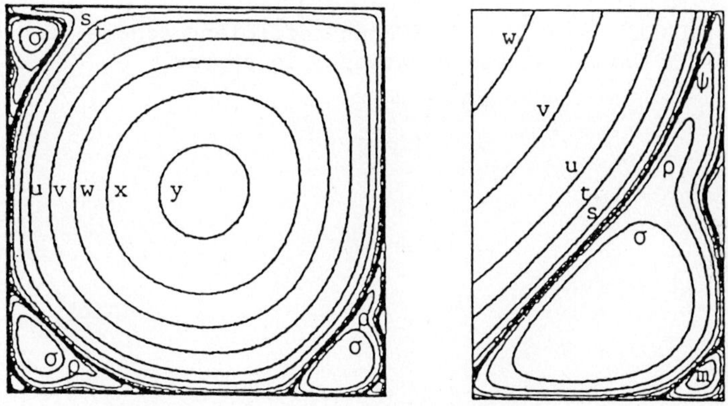

Streamline for Re = 10000 on 256x256 grid

Converged solutions

Figure 3

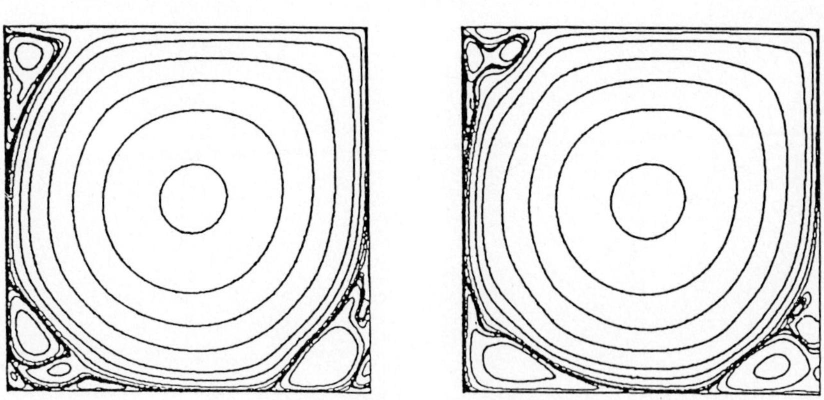

Evolution of solution during iterations
for Re = 10000 on 512x512 grid

Figure 4

REFERENCES

[1] U. GHIA, K.N. GHIA and C.T. SHIN, High-Re solutions for incompressible flow using the Navier-Stokes equations and a multigrid method, J. Comput. Phys., 48 (1982) 387-411.

[2] R. SCHREIBER and H.B. KELLER, Driven cavity flows by efficient numerical techniques, J. Comput. Phys., 49 (1983) 310-333.

[3] S.P. VANKA, Block-implicit multigrid solution of Navier-Stokes equations in primitive variables, J. Comput. Phys., 65 (1986) 138-158.

[4] A. BRANDT and N. DINAR, Multigrid solutions to elliptic flow problems, Numerical Methods for Partial Differential Equations, V. Parter Ed., Madison, 1978, 53-147.

[5] L.B. ZHANG, Résolution numérique des équations de Navier-Stokes stationnaires par la méthode multigrille, thesis, University of Paris-Sud, Orsay, 1987.

[6] Ch.H. BRUNEAU and C. JOURON An efficient scheme for solving steady incompressible Navier-Stokes equations, to appear.

COMPUTATION OF HYPERSONIC VORTEX FLOWS WITH AN EULER MODEL *

Charles-Henri Bruneau, Jacques Laminie
Laboratoire d'Analyse Numérique, Bâtiment 425
Université Paris-Sud, 91405 Orsay, France

Jean-Jacques Chattot
Aérospatiale, 78133 Les Mureaux, France

* Work performed with financial support of DRET and under grants from the CCVR Ecole Polytechnique, France.

ABSTRACT

The variational approach of the steady Euler equations presented at the 10th ICNMFD [1] is extended to the treatment of supersonic and hypersonic flows by introducing the energy equation in the least-squares formulation. The approximation is made with cubic or prismatic linear finite elements and the results are presented for flows around a rectangular flat plate or a thin delta wing for various Mach numbers and angles of attack. They show the occurrence of vortical flows on the upper surface of the wings due to the sharp edges.

The simulation code is run on a CRAY2 and the computing time is kept to a minimum by using the large storage capacity available, an efficient vectorization and a full multigrid procedure.

INTRODUCTION

The purpose of this work is to capture the vortex structure developing around wings with sharp leading edges for a wide range of Mach numbers and angles of attack, to see how the vortex sheet is related to the Mach number at infinity, the angle of attack and the geometry.

This study concerns only simplified geometries without thickness : a rectangular flat plate of small aspect ratio and a thin delta wing with sweep angle $\beta = 70°$. A structured mesh with cubic or prismatic cells is used to improve the efficiency of the code and due to the symmetry only half of the flow is computed.

The flow at infinity is given at the entrance sections of the domain and the tangency condition is imposed on both sides of the wings as done in [2].

DESCRIPTION OF THE METHOD

The full steady Euler equations are written in conservative partial differential form in a 3D domain Ω as follows :

$$
\begin{aligned}
A &= \frac{\partial \rho u}{\partial x} + \frac{\partial \rho v}{\partial y} + \frac{\partial \rho w}{\partial z} = 0 \\
B &= \frac{\partial \rho u^2}{\partial x} + \frac{\partial \rho uv}{\partial y} + \frac{\partial \rho uw}{\partial z} + \frac{\partial (\gamma-1)\rho e}{\partial x} = 0 \\
C &= \frac{\partial \rho uv}{\partial x} + \frac{\partial \rho v^2}{\partial y} + \frac{\partial \rho vw}{\partial z} + \frac{\partial (\gamma-1)\rho e}{\partial y} = 0 \qquad (1) \\
D &= \frac{\partial \rho uw}{\partial x} + \frac{\partial \rho vw}{\partial y} + \frac{\partial \rho w^2}{\partial z} + \frac{\partial (\gamma-1)\rho e}{\partial y} = 0 \\
E &= \frac{\partial (\gamma \rho e + \frac{1}{2}\rho q^2)u}{\partial x} + \frac{\partial (\gamma \rho e + \frac{1}{2}\rho q^2)v}{\partial y} + \frac{\partial (\gamma \rho e + \frac{1}{2}\rho q^2)w}{\partial z} = 0
\end{aligned}
$$

where the density ρ, the velocity $\vec{q} = (u,v,w)^T$ and the internal energy e are the unknowns ; γ is the ratio of specific heats and $q = |\vec{q}|$. The main difference with [1] is that the last equation, the conservation of energy, is solved as a partial differential equation instead of using the algebraic Bernoulli's equation. Consequently e is chosen as an unknown and the pressure p is replaced by $(\gamma-1)\rho e$ on one hand, and the whole system is linearized by Newton's method for the five unknowns on the other hand. This also means that the total enthalpy H is not imposed everywhere and that its value $H = \gamma e + \frac{1}{2} q^2$ is computed at each point. We show the variation of this quantity in the numerical results.

Starting with the non linear problem (1) for the unknowns $U=(\rho,u,v,w,e)^T$ we solve at each iteration of Newton's method a first order linear system for the unknowns $\tilde{U} = (\tilde{\rho},\tilde{u},\tilde{v},\tilde{w},\tilde{e})^T$ and we update the solution in the formula $U^{n+1} = U^n - \tilde{U}$ where n is the index of Newton step and the sequence U^n converges to the solution of (1).

This linear system is solved through the minimization of a least-squares functional which transforms the first order hyperbolic system into a second order parabolic one. If $A_\ell, B_\ell, C_\ell, D_\ell$ and E_ℓ are the five linear equations the functional reads

$$J(\tilde{U}) = \int_\Omega [A_\ell^2 + B_\ell^2 + C_\ell^2 + D_\ell^2 + E_\ell^2]dxdydz.$$

Then the resulting variational problem is symmetric and also the linear system we need to inverse at each Newton iteration.

For both geometries the domain is a rectangular hexaedron and a block-successive over-relaxation (BSOR) method is used to solve the linear system where each block represents the unknowns of a vertical plane (y,z) with the same number of points. One block is inverted by an incomplete Cholesky preconditioned conjugate gradient (ICCG) method.

APPROXIMATION

The groups of conservatives variables are approximated by a finite element method, for instance the approximation of ρu is given by

$$\rho_h u_h = \sum_{i=1}^{I} (\rho u)_i \phi_i$$

where ϕ_i is the basis of the approximate space. The finite element is either a linear cubic element spanned by the set $\{1,x,y,z,xy,xz,yz,xyz\}$ or a linear prismatic element spanned by the set $\{1,x,y,z,xz,yz\}$. That means in this last case that the triangular faces are in the planes (x,y) and that the other faces are rectangular. This is very convenient to approximate the flow around the delta wing. Nevertheless, a test with the rectangular flat plate shows that the results are about the same with both approximations.

NUMERICAL ASPECTS AND EFFICIENCY

To improve the efficiency of the whole process we use a full multigrid procedure (FMG) on a set of successive grids (for instance $14\times16\times12$, $28\times32\times24$ and $56\times64\times48$ around the wing). This procedure consists in interpolating the results from a coarse grid to the finer one, which divides the computing time by a factor two.

Then we use the large memory of the CRAY2 to store the matrices, independent of the solution, representing terms like

$$\int_\Omega \frac{\partial\phi_i}{\partial x} \cdot \frac{\partial\phi_j}{\partial x} \, dxdydz, \quad 1 \leqslant i,j \leqslant I.$$

So at each iteration of Newton's linearization the assembling of the linear system is reduced to a product of these matrices by some vectors. This last operation is highly vectorizable, and this part of the calculus is kept to a minimum.

Finally 80% of the inversion of the linear system is also vectorized and a very good efficiency is reached for the whole computation in terms of CPU time. Indeed the solution for 80,000 unknowns at Mach number at infinity $M_\infty = 0.7$ and angle of attack $\alpha = 10°$ converges in 50 iterations and is obtained in about half an hour on one processor of the CRAY2.

NUMERICAL RESULTS

The results show the occurrence of a vortex sheet that rolls up at the tip of the plate or at the leading edge of the delta wing with a conical shape. For a given Mach number at infinity this vortex grows with the angle of attack. On the contrary, for a given angle of attack, the vortex tightens as the Mach at infinity increases.

The results on the thin delta wing for Mach numbers at infinity $M_\infty = 0.7$ or $M_\infty = 2$ and angle of attack $\alpha = 10°$ show the same characteristics than those available in the literature (for instance [3] to [5]) as shown on the figure.

The accuracy of the resolution of the system including the energy equation is indicated by the variation on the total enthalpy that does not exceed 3% ; moreover the maximum is reached in the vortex core.

CONCLUSION

The extended method presented above is able to capture vortex flows around wings with sharp edges in subsonic as well as in supersonic or hypersonic regime (the tests include a computation at $M_\infty = 5$).

Nevertheless the computation with 200,000 points requires about 100 megawords of storage and 8 hours of CPU time.

REFERENCES

[1] Ch.H. Bruneau, J.J. Chattot, J. Laminie and R. Temam, Computation of vortex flow past a flat plate at high angle of attack, 10th ICNMFD, Beijing, 1986, Lecture Notes in Physics 264.
[2] Ch.H. Bruneau, J. Laminie and J.J. Chattot, Computation of 3D vortex flows past a flat plate at incidence through a variational approach of the full steady Euler equations, to appear in Int. J. Num. Meth. in Fluids.
[3] K. Powell, E. Murman, E. Perez and J. Baron, Total pressure loss in vortical solutions of the conical Euler equations, AIAA paper 85-1701.
[4] C. Koeck, Computation of three-dimensional flow using the Euler equations and a multiple-grid scheme, Int. J. Num. Meth. in Fluids, 5, 1985.
[5] L.E. Eriksson and A. Rizzi, Computation of vortex flow around wings using the Euler equations, 4th GAMM Conf. on Num. Meth. in Fluid Mech., 1981.

50%

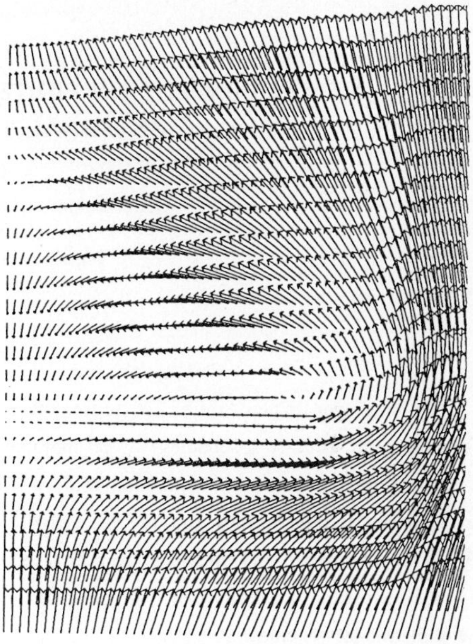

75%

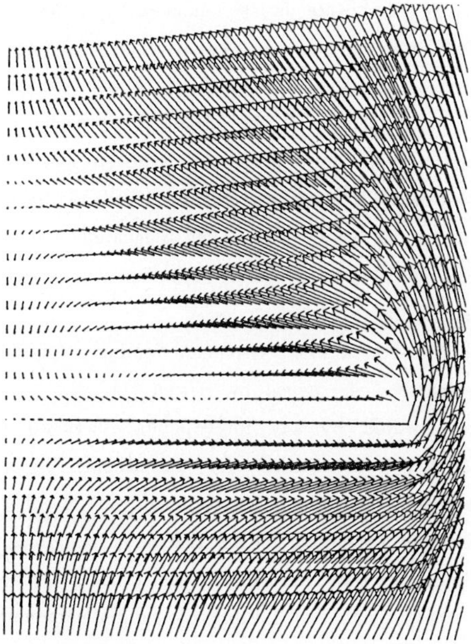

100%

Pressure contours

75%

$M_{\infty} = 2.0 \quad \alpha = 10°$

Solution around the delta wing with the 56x64x48 grid

A HIGH RESOLUTION FINITE VOLUME SCHEME FOR STEADY EXTERNAL TRANSONIC FLOW

D. M. Causon
Department of Mathematics & Physics
Manchester Polytechnic
Manchester, M1 5GD, UK

SUMMARY

An improved Euler solver is presented for computing steady external transonic flows around projectiles and aircraft forebodies. The method solves the unsteady Euler equations by time-marching using operator-splitting in conjunction with a finite volume formulation. The resolution of captured shock waves is improved by the use of a total variation diminishing (TVD) version of the well-known MacCormack scheme and artificial compression techniques. Existing production code implementations of MacCormack's method can be updated easily, as described, to reflect recent advances in one-dimensional schemes.

INTRODUCTION

Over the last fifteen years, substantial advances have been made in the numerical solution of hyperbolic partial differential equations, especially in those arising in computational external aerodynamics. Numerical solutions obtained using the early schemes of the 1970's were characterised by the appearance of spurious oscillations around the shock profile. The classical approach for damping oscillations and removing non-linear instabilities was to introduce extra artificial viscosity terms explicitly. Unfortunately, this approach tends to cause a spurious entropy layer to form, emanating particularly from regions of high spatial gradient like shock waves and stagnation points. The problem was that the added dissipative terms had a global and problem dependent coefficient. What was needed was to apply artificial viscosity more selectively.

In the 1980's various workers began to put the procedures on a more systematic footing. Out of this work emerged the total variation diminishing (TVD) schemes, so-named by Harten [1], who gave a set of theorems, leading to conditions, which when satisfied by any three-point, conservative, second-order accurate finite-difference scheme, are necessary and sufficient for the scheme to be TVD. As a sequel to this work, Davis [2] showed that it is possible to re-formulate the classical Lax-Wendroff scheme in TVD form. This is accomplished by appending to the scheme a non-linear term which applies precisely the correct amount of artificial viscosity needed at each mesh point to limit overshoots and undershoots. This work opened the way to updating existing production computer codes which employ the well-known MacCormack scheme, a variant of Lax-Wendroff.

FORMULATION

Since the use of a body-fitted mesh is anticipated, the equations of motion are written in integral form as a prelude to discretisation by the finite volume method.

$$\frac{\partial}{\partial t}\iiint_{vol} \underset{\sim}{U}\, d\,vol + \iint_{s} \underline{\underline{H}} \cdot \hat{\underline{n}}\, ds = 0 \tag{1}$$

where $\underset{\sim}{U} = [\rho, \rho w_1, \rho w_2, \rho w_3, e]^T$ $\quad \underline{\underline{H}} = \begin{bmatrix} \rho \underline{q} \\ \rho w_1\, \underline{q} + p\underline{a}_1 \\ \rho w_2\, \underline{q} + p\underline{a}_2 \\ \rho w_3\, \underline{q} + p\underline{a}_3 \\ (e+p)\underline{q} \end{bmatrix}$,

and the flow velocity $\underline{q} = w_\ell \underline{a}_\ell$, where $\underline{a}_\ell$ are the Cartesian unit base vectors. We solve equations (1) using a factored sequence of one-dimensional difference operators, where each component operator relates to its respective co-ordinate direction. Further details may be found in [3].

TVD MACCORMACK FINITE VOLUME SCHEME (TVDM)

The MacCormack finite volume operator $L_1(\Delta t)$ is:

$$\overline{\underset{\sim}{U}^{n+1}_{ijk}} = \underset{\sim}{U}^n_{ijk} - \Delta t \left[\underline{\underline{H}}^n_{ijk} \underline{S}_{i+\frac{1}{2}} + \underline{\underline{H}}^n_{i-1jk} \underline{S}_{i-\frac{1}{2}} \right] \tag{2a}$$

$$\underset{\sim}{U}^{n+1}_{ijk} = \tfrac{1}{2}\left[\underset{\sim}{U}^n_{ijk} + \overline{\underset{\sim}{U}^{n+1}_{ijk}} - \Delta t \left[\overline{\underline{\underline{H}}^{n+1}_{i+1jk}} \underline{S}_{i+\frac{1}{2}} + \overline{\underline{\underline{H}}^{n+1}_{ijk}} \underline{S}_{i-\frac{1}{2}} \right]\right] \tag{2b}$$

where $\underset{\sim}{U}_{ijk} = vol_{ijk}[\rho, \rho w_1, \rho w_2, \rho w_3, e]^T_{ijk}$ and $\underline{S}_{i\pm\frac{1}{2}}$ are the area vectors on opposite faces of the cell, normal to the surface x^1 = constant. Scheme (2) can be updated easily to total variation diminishing (TVD) form by appending to the right hand side of the "corrector" step (2b) the term:

$$+ \left[G^+_i + G^-_{i+1}\right] \Delta \underset{\sim}{U}^n_{i+\frac{1}{2}} - \left[G^+_{i-1} + G^-_i\right] \Delta \underset{\sim}{U}^n_{i-\frac{1}{2}} \tag{3}$$

where for clarity we have suppressed subscripts j, k and:

$$\Delta \underset{\sim}{U}^n_{i+\frac{1}{2}} = \underset{\sim}{U}^n_{i+1} - \underset{\sim}{U}^n_i \,, \quad \Delta \underset{\sim}{U}^n_{i-\frac{1}{2}} = \underset{\sim}{U}^n_i - \underset{\sim}{U}^n_{i-1} \tag{4}$$

$$G^\pm_i = G\left[r^\pm_i\right] = 0.5\, C(\upsilon) \left[1 - \varphi\left[r^\pm_i\right]\right] , \tag{5}$$

$$C(\upsilon) = \begin{cases} \upsilon(1-\upsilon) & , \ \upsilon \leqslant 0.5 \\ 0.25 & , \ \upsilon > 0.5 \end{cases} , \tag{6}$$

$$\upsilon = \upsilon_i = \frac{\left[\underline{q}^n_{ijk} \cdot \underline{S}_{i+\frac{1}{2}} \right] + c_{ijk} S_{i+\frac{1}{2}}}{Vol_{ijk}} \cdot \Delta t_{ijk} \tag{7}$$

where c_{ijk} is the local speed of sound,

$$r_i^+ = \frac{\sum_{\ell} \Delta^{\ell}_{i-\frac{1}{2}} \cdot \Delta U^{\ell}_{i+\frac{1}{2}}}{\sum_{\ell} \Delta U^{\ell}_{i+\frac{1}{2}} \cdot \Delta U^{\ell}_{i+\frac{1}{2}}} \;, \quad r_i^- = \frac{\sum_{\ell} \Delta U^{\ell}_{i-\frac{1}{2}} \cdot \Delta U^{\ell}_{i+\frac{1}{2}}}{\sum_{\ell} \Delta U^{\ell}_{i-\frac{1}{2}} \cdot \Delta U^{\ell}_{i-\frac{1}{2}}} \tag{8}$$

where the superscript ℓ indicates the ℓth component of the vector $\Delta \underset{\sim}{U}^n_{i\pm\frac{1}{2}}$.

$$\varphi(r) = \begin{cases} \min(2r,1) & , \; r > 0 \\ 0 & , \; r \leq 0 \end{cases} \tag{9}$$

If required, the resolution of contact discontinuities can be improved further by the use of artificial compression techniques. Further discussion of the theoretical details of the schemes can be found in [3,4].

APPLICATIONS

Fig 1a illustrates an aircraft forebody geometry which is typical of the type of problem which we wish to solve. Here, we wish to predict the disturbance field around the forebody at proposed intake locations, at transonic Mach numbers and various angles of attack. The forebody has an elliptical nose dipped at 5°, sharp corners and flat sides to accommodate engine intakes. A body fitted H–O type grid was used which had $57(x) \times 16(\theta) \times 10(r)$ cells and a converged solution was obtained after 800 time steps, commencing the calculation with an impulsive start at Mach 1.40 and −5° angle of attack. Fig 1b shows the isobar contours on the surface of the body. Unlike similar calculations using the MacCormack method, there were no parameters to be adjusted to control the amount of added dissipation for shock capturing. Fig 2 shows a similar calculation for a forebody typical of a civil aircraft. An analytical description of this geometry is given in the reference cited.

Fig 3a,b,c illustrates a secant–ogive cylinder boat–tail (SOCBT) projectile geometry and the computed surface Cp distribution and velocity vector field in the base region. These results relate to $\alpha = 0°$ and $M = 0.9$. The initial conditions in the field region are that the flow variables are assumed uniformly to be those consistent with the freestream Mach number. In the base region the fluid is assumed to be at rest. The use of scheme TVDM in each split operator prevents the appearance of any numerical oscillations in the base region when mixing begins to occur. The results shown in Figs 3b,c agree closely with those given in [5].

CONCLUSIONS

A high resolution version of the MacCormack method has been presented, which is more robust and has an improved shock capturing capability. The required modifications to the standard method are minor and will enable many existing production computer codes to take advantage of recent advances in one–dimensional schemes.

ACKNOWLEDGEMENTS

This work has been carried out with the support of the Procurement Executive, Ministry of Defence.

REFERENCES

1. Harten A. High Resolution Schemes for Hyperbolic Conservation Laws. *J Comp Phys*, vol 49, 357–392, 1983.

2. Davis SF. TVD Finite Difference Schemes and Artificial Viscosity. NASA CR 172373, 1984.

3. Causon DM. High Resolution Finite Volume Schemes and Computational Aerodynamics. In *Proc 2nd Int Conf on Hyperbolic Problems*, Spring-Verlag, 1988.

4. Yee HC. Upwind and Symmetric Shock Capturing Schemes. NASA TM 89464, 1984.

5. Sturek WB. Applications of CFD to the Aerodynamics of a Spinning Shell. AIAA Paper No 84-0323, 1984.

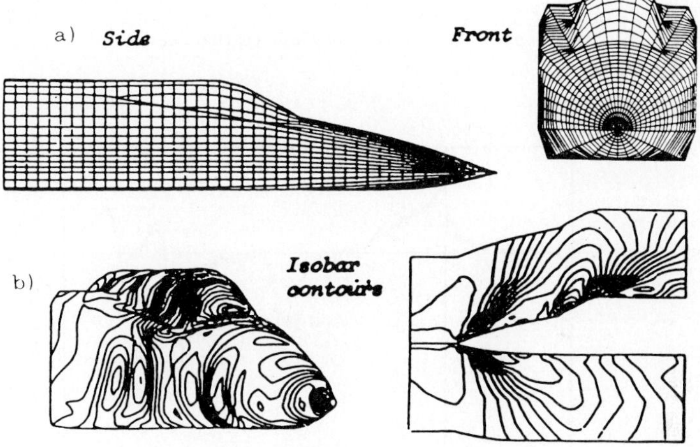

Fig. 1 Combat Aircraft Forebody at Mach 1.40 and $\alpha = -5^\circ$

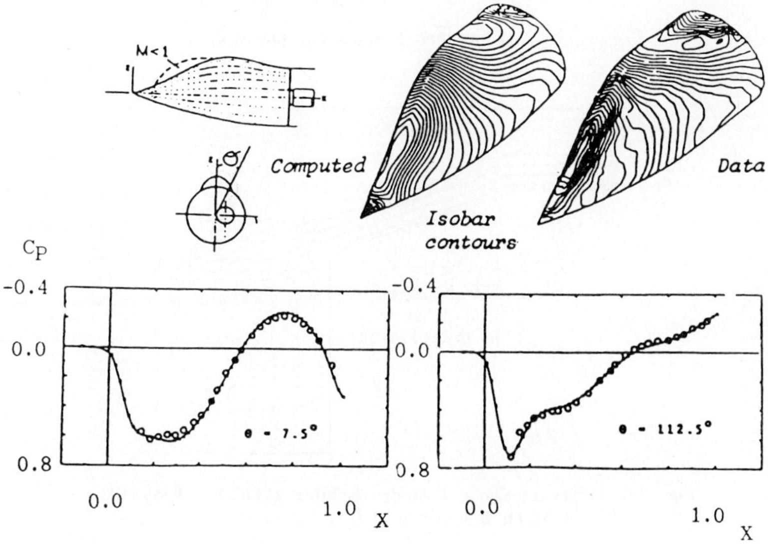

Fig. 2 Civil Aircraft Forebody at Mach 1.70 and $\alpha = -5^\circ$ from NASA TM 80062

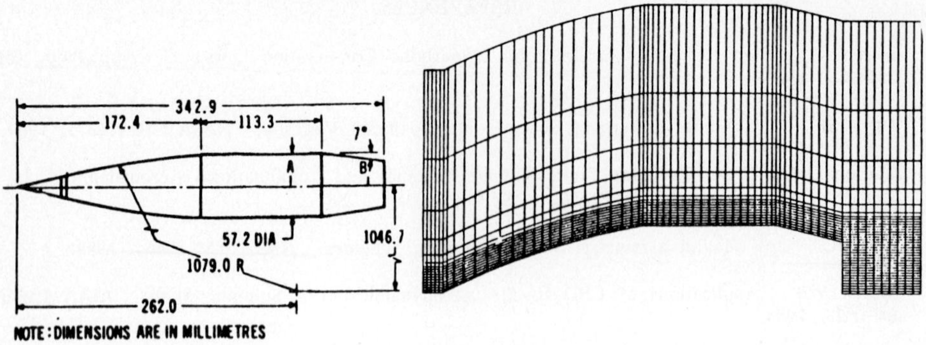

a) Geometry and Mesh (partial view)

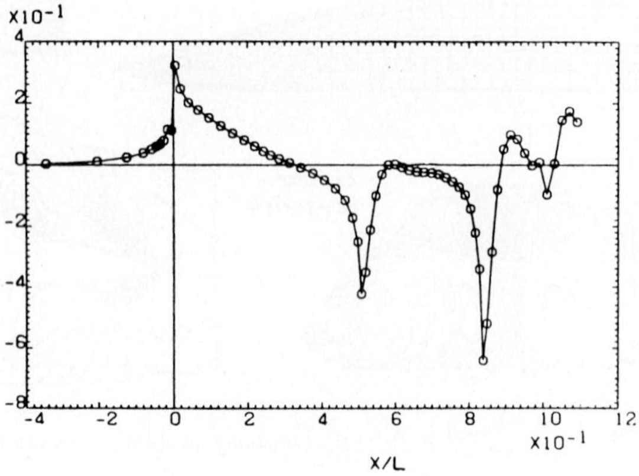

b) Computed Surface Cp Distribution

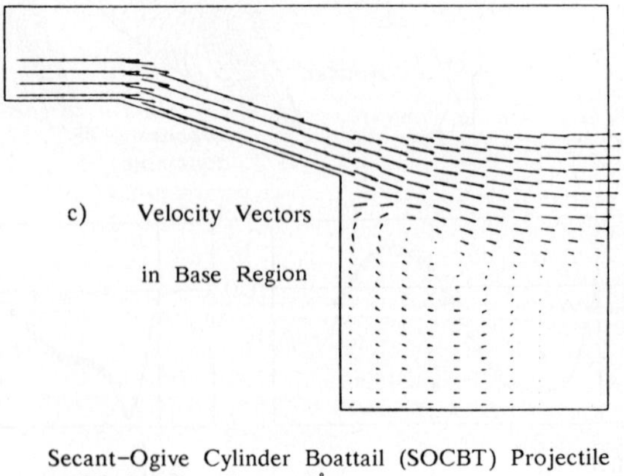

c) Velocity Vectors in Base Region

Fig. 3 Secant–Ogive Cylinder Boattail (SOCBT) Projectile at Mach 0.9 and $\alpha = 0^\circ$

Development of a Highly Efficient and Accurate 3D Euler Flow Solver

H. C. Chen and N. J. Yu
Boeing Commercial Airplanes, Seattle, Washington

Introduction

An Euler flow solver has been developed employing a highly efficient multigrid scheme. A successive mesh refinement procedure has been incorporated to further improve convergence of the solution. A new dissipation model has also been implemented to ensure that the solution is grid insensitive. Results presented include the numerical assessment of the solution consistency through comparison of Euler solutions using different dissipation parameters and grids with varying quality. Convergence of the solution with respect to mesh refinement is demonstrated using a 2D version of the code. An improved numerical boundary condition method is studied to further improve the solution accuracy.

Flow Solver

The present Euler analysis method is based on a cell-centered out-of-core multigrid Euler solver for a complete airplane analysis (ref. 1). A finite-volume scheme is used for spatial discretization and a five-stage Runge-Kutta scheme is used for time integration (ref. 2). The complete flowfield data are stored on solid state disk, and the computation is done in a block-by-block manner through the use of highly efficient input/output data management. Fluxes are fully conserved across the interblock boundaries to ensure that the multi-block solution agrees with the single-block solution.

Important improvements have been made to the dissipation formulation to provide more accurate and reliable solutions for complex geometry. The spectral radius scaling of dissipation terms (ref. 3, 4) has been adapted to our Euler code. The new formulation improves the consistency of the solution by decreasing sensitivity to the dissipation parameters and to the quality of the grid. This improvement over conventional dissipation formulations (ref. 5) is especially pronounced on grids with large aspect ratio or with nonsmooth mesh distributions.

Results

Employment of both successive mesh refinement and the multigrid acceleration scheme is effective in speeding up the convergence of the solutions to steady state. For a typical flow solution of a wing-body-winglet configuration containing 200,000 grid points, three levels of mesh refinement are used in the calculations. On the first two mesh levels, 200 multigrid cycles are used. The fine mesh results for a wing-body-winglet analysis at 25 and 50 multi-grid cycles are compared with the results at 200 cycles in Fig. 1. The differences in solutions are hardly discernable after 50 multigrid cycles on the fine mesh. To obtain a similarly converged solution without using

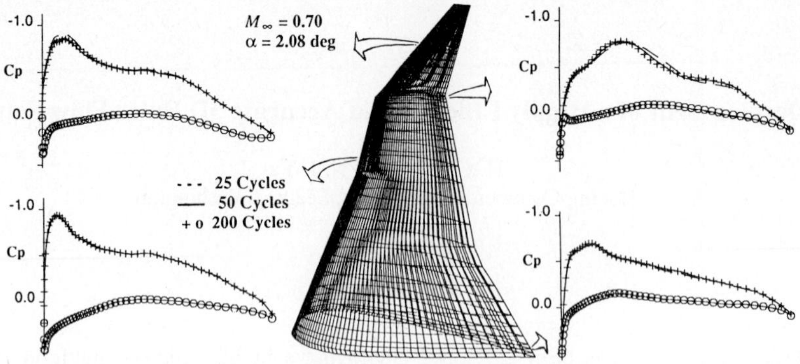

Fig. 1. Convergence of Euler Solution on the Fine Mesh

successive mesh refinement requires more than 100 multigrid cycles in the fine mesh. For transonic cases with strong shocks, successive mesh refinement reduces the multigrid cycles from 400 to 200 on the fine mesh. The shock capturing scheme in the present Euler solver typically smears out the shock discontinuities over three to five grid points. The convergence of the shock discontinuities is more efficient when successive mesh refinement is used.

For the consistency test with respect to dissipation parameters, Euler analyses for a wing-body-winglet configuration with two different sets of dissipation parameters were conducted. In Fig. 2, symbols show a solution using dissipation parameters twice as large as those for the solution given by the solid line. The two results are very close although increasing the values of the dissipation parameters appears to smooth out some of the local variations in the solution. Important flow features, such as the peaks and locations of the leading edge expansions, are unchanged from the wing-root to the tip of the winglet.

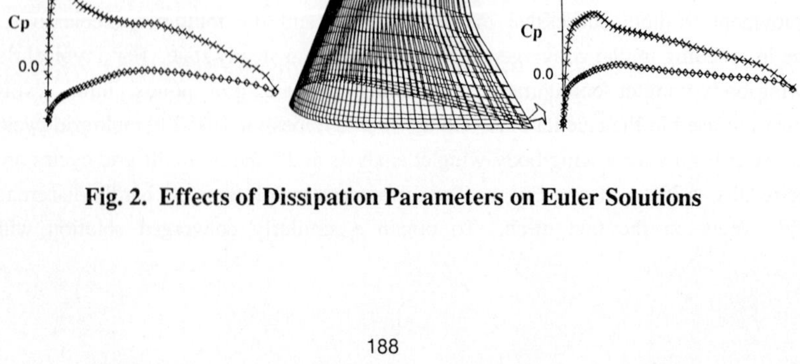

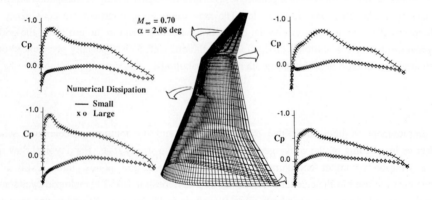

Fig. 2. Effects of Dissipation Parameters on Euler Solutions

For the grid sensitivity test, the Euler solutions on two grids with significantly different grid quality were obtained for comparison. Both grids contain 200,000 points and have identical surface grid distributions. However, Grid-1 has a strong kink at the tip of the winglet, whereas Grid-2 is smooth near the tip of the winglet. The difference in solutions due to these grid effects are small and localized (Fig. 3). Away from the tip of the winglet, the differences are essentially non-discernable. Again, important flow features such as shock locations are unchanged throughout the entire wing.

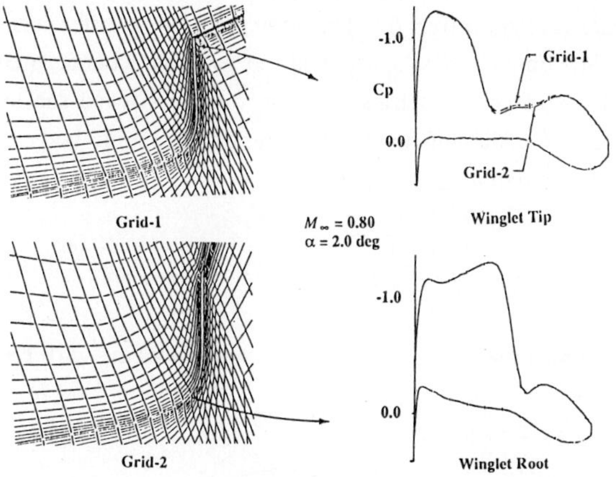

Fig. 3. Effects of Grid Quality on Euler Solutions

A convergence study of the Euler solutions with respect to mesh refinement requires the use of extreme fine mesh. This becomes prohibitively expansive in 3D. Therefore, a 2D version of the Euler code was employeed for such studies. We chose the NLR7301 shockfree airfoil as a test case. Four successive level of grids were used (Fig. 4) starting from grid-1 of 40x8, then grid-2 of 80x16,

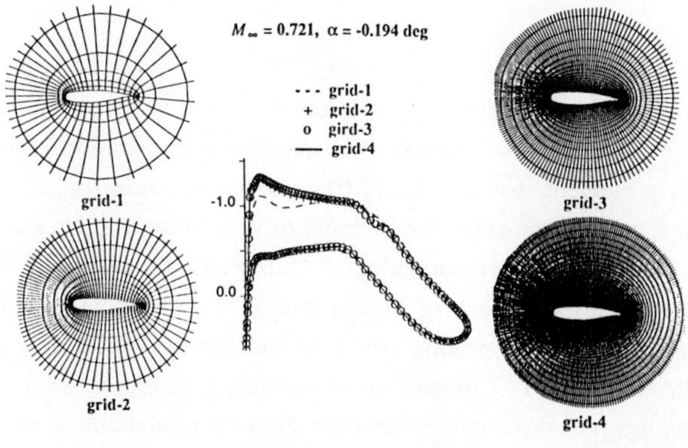

Fig. 4. Convergence of Euler Solutions with respect to Grid Refinement

grid-3 of 160x32 and ending with grid-4 of 320x64. The Euler solution in grid-1 is noticeably different from the finer grid solutions. The airfoil geometry is not very well defined in this grid. As the grid is refined, it is seen that the solutions on grid-2 and -3 are reasonably close to the finest grid solution on grid-4. This suggest that the 160x32 grid is adequate for this type of calculation.

We also compared the cell-centered scheme with a nodal scheme. Fig. 5a sketches a volume cell adjacent to a solid body boundary. Along side AB, the only contribution to momentum flux is due to the pressure integral alone. In the cell-centered scheme, the pressure along side AB is estimated using the normal momentum relation (ref. 6). This requires a finite difference estimate of the cell face projection areas in the computational space, and the solution accuracy may be affected when dealing with very complex surface geometries involving highly skewed surface grids. The nodal scheme, however, solves for nodal unknowns directly on the solid body boundary using the conservation laws and does not require such differences.

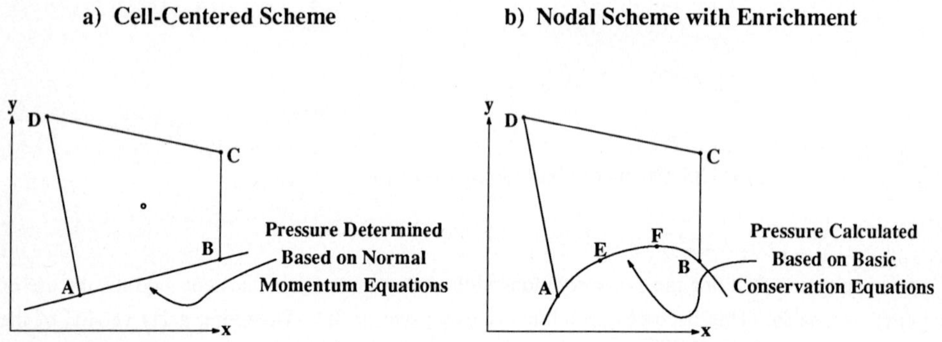

Fig. 5. A Flux Cell Adjacent to a Solid Body Boundary

In addition, the nodal scheme offers opportunities to enrich the surface geometries to provide a more accurate boundary condition method. To improve a conventional nodal scheme using a bilinear isoparametric element, one can insert (Fig. 5b), along side AB, more geometry defining points (e. g., points E and F), and allow the pressure to vary linearly along curve AEFB to facilitate a more accurate evaluation of the pressure integral. A test case of subcritical flow around a circular cylinder was chosen to study the effects of surface grid point enrichment. The flow was analyzed on a 160x32 grid. Case-1 used 321 defining points on the surface, while case-2 picked every other point on the surface from case-1. Comparison of converged results showed that differences in the two Cp distributions are not discernable, but false entropy production is reduced (Fig. 6) due to surface enrichment.

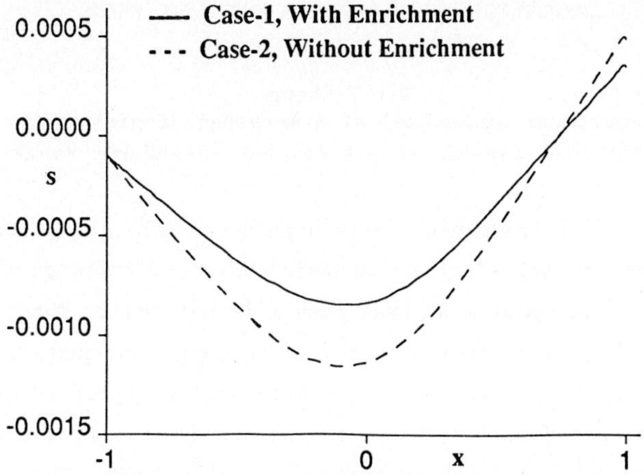

Fig. 6. Effects of Surface Enrichment on False Entropy Production

Conclusion

In summary, a highly efficient and accurate Euler solver for the analysis of transonic flow over a complete airplane configuration with winglet has been developed. The solution method is insensitive to grid quality and dissipation parameters. Convergence of the solution with respect to mesh refinement was demonstrated in 2D to further assess the numerical accuracy of the Euler solutions. The improved numerical boundary condition method, based on a nodal scheme, reduced the false entropy production in the flow.

References

1. Yu, N. J., Chen, H. C., Chen, A. W. and K. R. Wittenberg, "Grid Generation and Flow Code Development for Winglet Analysis," AIAA Paper 88-2548, presented at AIAA Sixth Applied Aerodynamic Conference, June 1988.
2. Jameson, A. and Baker, T. J.,, "Multigrid Solution of the Euler Equations for Aircraft Configurations," AIAA Paper 84-0093, 1984.
3. Jameson, A., "Successes and Challenges in Computational Aerodynamics," AIAA Paper 87-1184, 1987.
4. Swanson, R. C. and Turkel, E., "Artificial Dissipation and Central Difference Schemes for the Euler and Navier-Stokes Equations," AIAA Paper 87-1107, 1987.
5. Jameson, A., Schmidt, W., and Turkel, E., "Numerical Solutions of the Euler Equations by Finite Volume Methods Using Runge-Kutta Time-Stepping Scheme," AIAA 81-1259, 1981.
6. Chen, H. C., Yu, N. J. and Rubbert, P. E., "Flow Simulations for General Nacelle Configurations Using Euler Equations," AIAA 83-0539, 1983.

COMPUTATION OF RAREFIED HYPERSONIC FLOWS

Sin-I Cheng
Department of Mechanical & Aerospace Engineering
Princeton University, Princeton, New Jersey 08544 U.S.A.

Our interest in lifting maneuverable hypersonic vehicles in orbital transfer or cruise calls for the lift-drag aerodynamic characteristics at altitudes much above 40 Km where the gaseous mean free path λ is not small. The "merged layer" formulation[1] of the Navier-Stokes system is no longer adequate. For free molecular flows with the gas-gas molecular interaction neglected entirely, we have the Newtonian approximation,[2] where the aerodynamic pressure on a surface is locally determined by the surface angle ϕ as $\sim\rho U^2 \sin^2\phi$. This turns out to be an excellent approximation in the Continuum Limit, especially after some minor empirical corrections. With little difference of surface pressure in both the free molecular and the continuum limits, it is widely presumed that the pressure and the force coefficients on a surface vary smoothly and "monotonically" through the transition range. This is adopted in the analysis of the vehicle acceleration data of some early Shuttle flights[3][4]. An anomalous (~50%) deviation of :mospheric density from the standard atmosphere arou i 120Km was revealed. Thus SUMS program was established to study it. Air sample was taken from Shuttle surface through a short straight tube, and led to an instrument package with a Mass Spectrometer which has been calibrated with some quasi-steady gas reservoir conditions. For interpreting the data, the "reservoir" condition has to be related to the gas states entering the tube (Internal Flow problem) and then to the undisturbed atmosphere flowing over the Shuttle (External Flow Problem). This is analogous to the Rayleigh probe problem, fundamental to the measurements in continuum supersonic gas dynamics. The rarefied gas in both the internal and the external flow problems is highly "Nonequilibrium" with molecular distribution function governed by the Boltzmann equation whose solution is difficult. The popular Direct Simulation Monte Carlo Method (DSMC), does not pretend to solve the Boltzmann Equation[6,7] but to simulate the evolution of the distribution function with a much smaller ensemble of Monte Carlo Particles. A smooth, monotonic variation of the drag coefficient from the continuum to the free molecular state for the external flow was obtained with double peaked distribution functions over SUMS inlet; but no meaningful internal flow solution[7][8] can be generated despite free adjustments of many numerical and physical parameters.

We cannot help recalling the fundamental deficiencies of the DSMC that repeated sampling in a small computational cell is incompatible with the physics of a hypersonic rarefied flow where $M\ K_N >> 1$ (not $<< 1$). For such a non-equilibrium system, the central limit theorem and the principle of detailed balance are inapplicable. The DSMC computations let many collisions among the randomly chosen pair of Monte Carlo Particles (MCP) to proceed for some fixed, small convective $\Delta t_c < < \Delta x/U$, chosen in such a manner that the molecules within each computational cell will not cross the cell boundary. When each such MCP is identified as an individual molecule[9], the free flight time of such molecules will be λ/c. In order to have the many collisions as the central limit theorem calls for, we should have $\lambda/c < < \Delta t_c < < \Delta x/U < < D/U$; i.e. $MK_N < < 1$ contrary to the condition of a hypersonic rarefied flow with $MK_N > > 1$. A plausible alternative to avoid such a contradiction is to recognize each MAC as a micro-ensemble of molecules, (denied by the popular notion of DSMC[9]) whose collective interaction can be compatible with $MK_N > > 1$. Moreover, the collision cross section of such collective interactions can involve many "internal degrees" of freedom, accompanied by many physical constants, desired and proposed[6] for DSMC to reproduce better the physical results. Such DSMC will surely provide some reasonable external flow solution once we know what it should be. For the internal flow problems, we have to consider some alternative approach, however.

The double peaked molecular distribution function $f = f_1 + f_2$ consists of a nearly mono-energetic peak of incoming molecules f_1 and a broad spectrum of scattered molecules f_2 with mole fractions β and $1-\beta$ respectively. The mean values of each beam defined through Maxwellian fits of f_1 and f_2, are governed by the integrated moment equations of Boltzmann over their respective velocity spaces. They are like Navier-Stokes, one set for each beam with stress tensor τ_{ij} and heat transfer vector $\dot{q}_i$ resulting from (i) the departure of f_1 or f_2 from Maxwellian and (ii) the inter-beam collisions. If beam 1 molecules are identified as those undisturbed molecules the equations system for beam 1 gives the rate of change of its population $\rho_1 = \rho\beta$. We then define the mean properties of the mixture as $\rho = \rho_1 + \rho_2$, $p = p_1 + p_2$, $T = \beta\ T_1 + (1-\beta)T_2$ and $u = \beta\ u_1 + (1-\beta)u_2$; and replace $\rho_2 = \rho - \rho_1$ etc. in the N. S. like system for beam 2. For the flow of such a hypersonic rarefied gas mixture through a straight-inlet tube, we can integrate across the tube to obtain:

$$\frac{\partial \rho}{\partial t} + \frac{\partial (\rho u)}{\partial x} = 0 \qquad (1)$$

$$\frac{\partial (\rho u)}{\partial t} + \frac{\partial}{\partial x} (\rho u^2 + p) = -4 \tau_w \qquad (2)$$

$$\frac{\partial (\rho e)}{\partial t} + \frac{\partial}{\partial x} (\rho u e) + p \frac{\partial u}{\partial x} = - 4 \dot{q}_w \qquad (3)$$

where u is the mean velocity along the tube. Our gas-surface interaction model gives:

$$\left(\frac{\partial}{\partial t} + u \frac{\partial}{\partial x} \right) (\rho \beta) = \rho \beta C_1 \qquad (4)$$

$$\tau_w = \frac{1}{4} \rho \sigma_t [\beta C_1 U_1 + (1-\beta) C_2 u_2] \qquad (5)$$

$$\dot{q}_w = \frac{1}{4} \rho \sigma_e [\beta C_1 \{(e_1 + \frac{(u_1)^2}{2}) - (e_w + \frac{(1-\sigma_t)^2 (u_1)^2}{2}) \}$$

$$+ (1-\beta) C_2 \{(e_2 + \frac{(u_2)^2}{2}) - e_w + (\frac{(1-\sigma_t)^2 (u_2)^2}{2})\}] \qquad (6)$$

for the evolution of β, the wall friction τ_w and the wall heat transfer vector q_w. σ_t and σ_e are the accommodation coefficients of the tangential momentum and the total energy respectively.

The gas molecules collide almost exclusively with the molecules in the adsorbed gaseous layer over the tube wall to establish an approximate equilibrium flow with varying mean flow properties along the tube. Their evolution is analogous to that of one dimensional continuum gas flow in a tube with friction and heat transfer. As an initial value problem, the evolution can be treated first with a vacuum reservoir condition in the downstream. The residual gas at some low pressure in the reservoir will generate a counter flow from the exit of the tube (reservoir) toward the inlet, without interfering with the free molecular inflow. Therefore, we solve equations system (1)-(6) for flows into a "vacuum reservoir", with the back flow from the reservoir to be superposed separately.

We used explicit, forward time, centered space differencing in terms of the latest available data to discretize (1) - (4). Figures 1, 2 and 3 give some typical results of the variations of different mean flow properties along the tube at 5000 time steps Δt for different fixed, constant values of the accomodation coefficients σ_t and σ_e. $\Delta t = D/U \sim 10^{-7}$ sec. for the reference velocity U to travel the tube diameter D. The tube was filled with the inlet gas before opening the tube exit. As illustrated in Fig. 4 different initial conditions in the tube give reproducible results after the first few hundred time steps ($\delta t < 10^{-4}$ sec). A steady state is reached at $10 \sim 5 \times 10^{-3}$ sec. with zero mean flow velocity and some maximum pressure at the tube exit (Figs. 5,6), expanding freely into the vacuum reservoir with an efflux equal to the influx at the tube inlet. This steady flow condition of the gas at the tube exit expanding freely into a "vacuum" reservoir and the evolution of the gaseous flow within the tube toward this steady state, depend on the accommodation coefficients σ_t and σ_e and the temperature distribution along the tube wall. They are independent of the rarefied reservoir conditions so long as a "Steady State" is reached.

There is a maximum limit of the steady state mass influx into a given tube, beyond which the tube will become "choked" before reaching a steady state of lower mass flux into the vacuum reservoir. The reduction of the influx is accomplished by the generation of upstream propagating pressure waves which will eventually pop out of the tube to alter the external flow over SUMS inlet. This phenomenon is preceded by the occurrence of negative local mean velocity somewhere in the tube. We adopted this latter criterion of imminent choking to define a choking boundary in the σ_t and σ_e space from our computed results with a given wall temperature Tw distribution and a given tube inlet condition. This boundary defines the maximum steady mass flux that can be pushed through the tube into a vacuum reserver. When the reserver contains some low pressure gas it will generate a counter flow out of the inlet tube, with the maximum steady state mass influx correspondingly reduced. Thus the matching of this internal flow solution with the external flow solution over SUMS inlet and with the dynamic characteristics of the instrument package in reducing the SUMS data is somewhat more complicated. This work is sponsored by NASA Langley under Contract No. NAS1 17234. The author thanks Mr. Y. W. Ma and Ms. Ann Bourlioux for carrying out the computations.

REFERENCES

1. Cheng, S. I. and Chen, J. H., "Slips, Friction and Heat Transfer Laws in a Merged Regime", Phys. Fluid, 17, 9, (1974).

2. Hayes, W. D., Probstein, R. F., "Hypersonic Flow Theory", Academic Press, 1959.

3. Blanchard, R. C. and Larman, K. T.,"Rarefied Aerodynamics and Upper Atmospheric Flight Results from the Orbital High Resolution Accelerometer Package Experiment". Presented at AIAA Atmospheric Flight Mechanics Conference. Paper No. 87-2366, (1987).

4. Blanchard, R. C., Duckett, R. J. and Hinson, E. W., "A Shuttle Upper Atmospheric Mass Spectrometer (SUMS) Experiment," AIAA Paper 82-1334, August 1982.

5. Bird, G. A., "Monte Carlo Simulation of Gas Flows", Annual Reviews of Fluid Mechanics, Vol. 10, Van Dyke Ed., Annual Review Inc., Palo Alto (1979).

6. Bird, G. A., "Low Density Aerothermodynamics", AIAA Paper 85-0994. (1985).

7. Bienkowski, G. K., "Inference of Free Stream Properties from SUMS Experiment," Rarefied Gas Dynamics, Vol. 1, Oguchu, Ed. (1984).

8. Moss, J. M. and Bird, G. A., "Direct Simulation on Transitional Flow for Hypersonic Reentry Conditions", Progress in Astronautics and Aeronautics: Thermal Design of Aeroassisted Orbital Transfer Vehicles, Nelson Ed. Vol. 96, (1985).

9. Bird, G. A., "Private Communication" (1987).

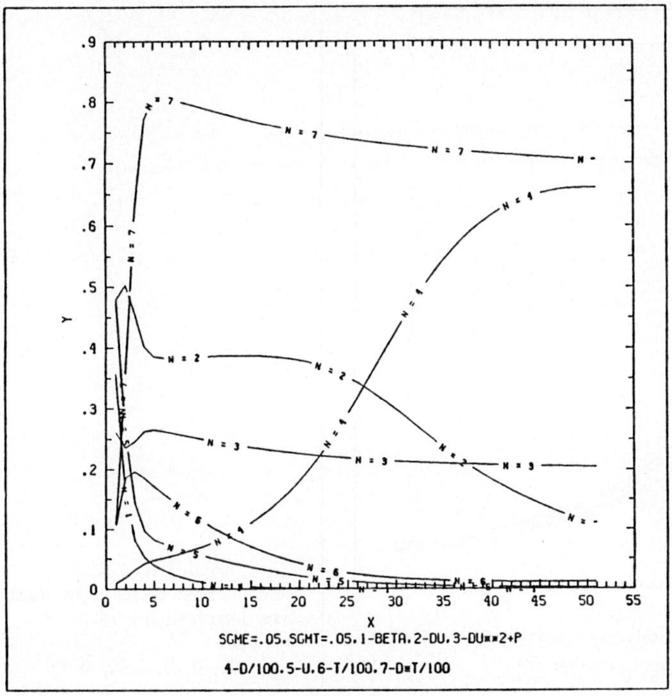

Fig.1. Variation of Mean Flow Properties at 500Δt along Inlet tube length.

Mole fraction	β	(N=1)
Mass flux	DU	(N=2)
Stream Thrust	$p+DU^2$	(N=3)
Density	D/100	(N=4)
Velocity	U	(N=5)
Temperature	T/100	(N=6)
Pressure	DT/100	(N=7)

With $\sigma_e = 0.05$ $\sigma_t = 0.05$

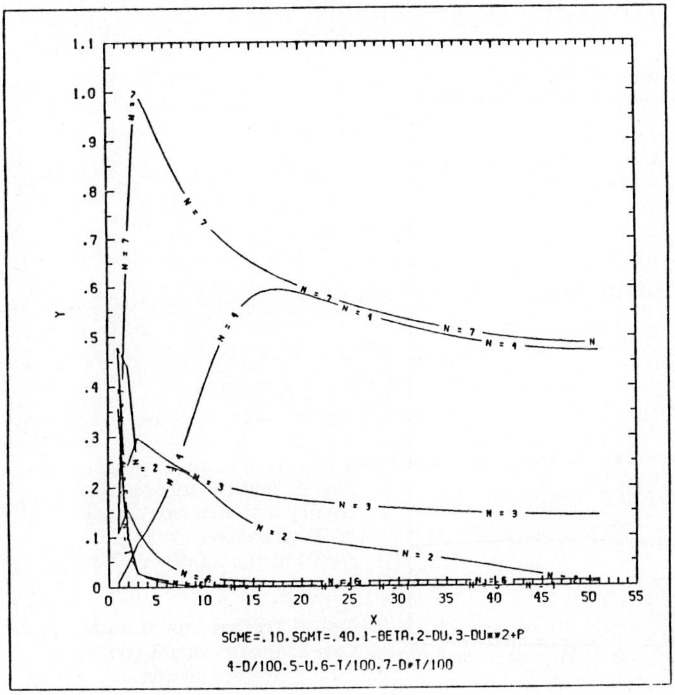

Fig.2. Variation of Mean Flow Properties at 5000Δt along Inlet tube length with $\sigma_e = 0.10$ $\sigma_t = 0.40$

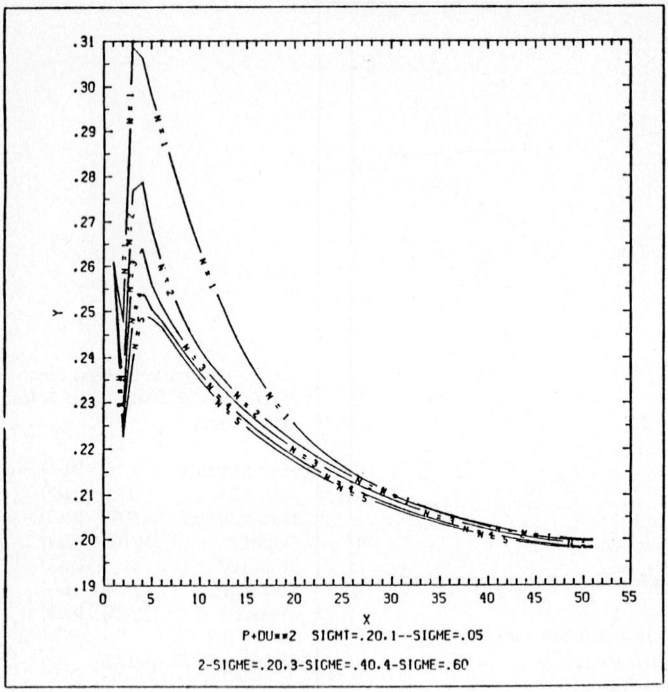

Fig.3. Variation of Mean Flow Stream Thrust $p+DU^2$ along Inlet tube length with
$\sigma_t = 0.20$
$\sigma_e = 0.05,\ 0.20,\ 0.40,\ 0.60$

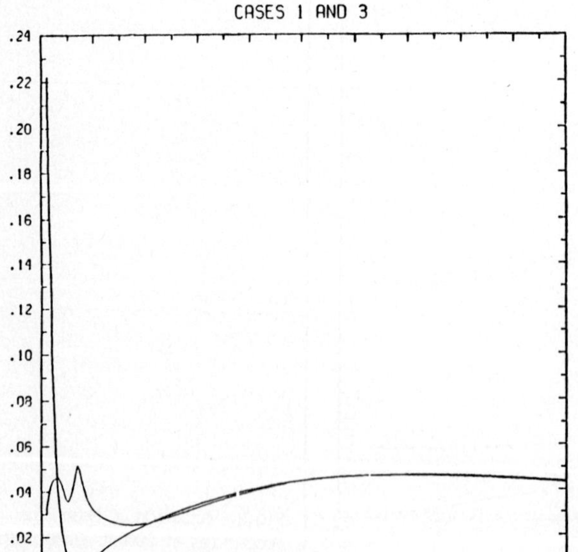

Fig. 4. Temporal variations of Density D at tube exit in unit of 20 time steps. (N/20) with $\sigma_e = 0.05$, $\sigma_t = 0.2$ $\beta_o = 0.35$.
Case 1. Uniform initial study.
Case 2. Linear initial density to zero at exit.

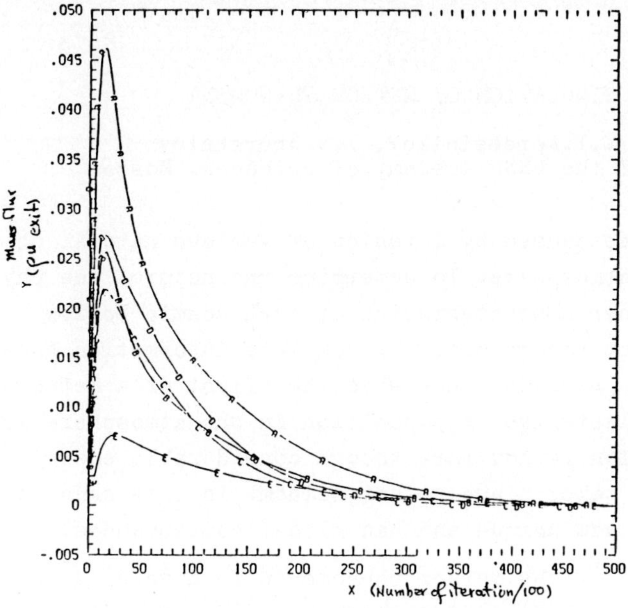

Fig.5. Temporal variations of Mean Flow mass efflux (DU) at tube exit in unit of 100 time steps (N/100).

Cases	σ_e	σ_t	σ_Q
A	0.05	0.20	0.35
B	0.05	0.20	0.10
C	0.20	0.20	0.10
D	0.20	0.20	0.20
E	0.05	0.40	0.10

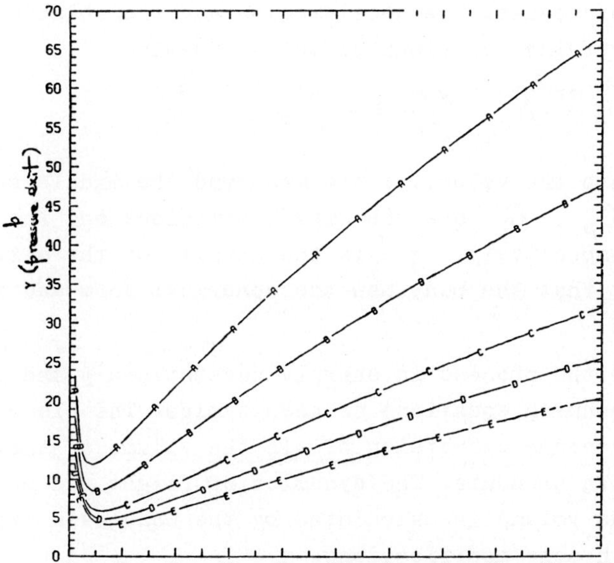

Fig.6. Temporal variations of pressure at tube exit in unit of 50 time steps cases same as Fig.5.

GASDYNAMICAL SIMULATION OF METEOR PHENOMENA

P.I.Chushkin, V.P.Korobeinikov, L.V.Shurshalov
Computing Centre of the USSR Academy of Sciences, Moscow

Meteor phenomena are caused by invasion of various natural objects into the Earth's atmosphere. To determine the nature, the physical properties and other characteristics of such cosmic bodies using registered indirect and generally incomplete information gasdynamical simulation is used. The case when the flight of a meteoroid is finished by its blast-like decomposition in the atmosphere is considered. Such situation is not rare though considerable energy release in the atmosphere takes place rather seldom. In this sense the Tunguska meteorite fall was unique and had global consequences.

A new complex model of the meteor phenomenon is developed in the paper. The entire process is divided into two stages - the flight stage and the blast stage. At the first stage the flight and the interaction of the meteoroid with the atmosphere are assumed to proceed as the quasistationary processes. The deceleration and the ablation are described by the governing equations of meteor physics

$$m\frac{dV}{dt} = -\frac{1}{2}C_D \rho V^2 S, \quad \frac{dm}{dt} = -\frac{1}{2}\sigma \rho V^3 S,$$

where V, m, S are the velocity, the mass and the middle section area of the body, C_D, σ are the drag coefficient and the ablation coefficient, respectively, ρ is the density of the air. For simplicity we assume that the body has the spherical form and its density ρ_b is constant.

At the second stage the process is sharply nonstationary and is described by the corresponding equations of gasdynamics. The blast is simulated by the instantaneous conversion of all the volume of the body into a gas under high pressure. The dynamics of flight and expansion of such a gaseous volume is calculated by the Godunov's finite difference method with some modifications.

Some results of these calculations are presented in Fig. 1. The velocity of the centre of mass is shown here as dependence on the distance ℓ. The curves are obtained for different values of the initial body density ρ_b and the ratio of the explosion energy E_0 to the total energy of the body $E_0 + K_0$, i.e. $\varepsilon = E_0/(E_0+K_0)$

where K_0 is the kinetic energy of the body immediately before the explosion.

Determination of the unknown body parameters through consequences of the event leads to an inverse problem: having that or either registered consequences of the meteor phenomenon it is necessary to restore its basic characteristics. Since available information contains only fragmentary data about the meteoroid, its parameters are obtained actually within some range of admissible values.

The possibilities of presented gasdynamical modelling are demonstrated on the example of the Tunguska meteorite. Here the basic registered consequences are the region of flattened forest, the zone of radiation burn of trees and also the corresponding barograms and seismograms.

The flight and the blast of the Tunguska cosmic body are calculated satisfying the condition that the body produced such an energy release at the final part of the trajectory, which is necessary to obtain the real region of flattened forest as the result of the action of the ballistic and explosion waves.

Two essential parameters appear in this model, namely the density of the body ρ_b and the energy ratio ε. The region of admissible values of these parameters was found in the calculations taking into account several restrictions. One such region in the plane ρ_b, ε is shown in Fig. 2a for the case of constant ablation coefficient which approximately corresponds to an icy cosmic body. Analogous diagram for the variable ablation coefficient corresponding to carbonaceous chondrites (its value was taken from [1]) is presented in Fig. 2b. The upper boundary of such a region is determined by the maximum pressure of non-nuclear explosion $p_0 \approx 2 \cdot 10^5$ atm. The lower boundary reflects the fact that the Tunguska body did not reach the Earth's surface. The left boundary corresponds to the lowest possible entry velocity $V_e = 11.2$ km/s. Besides several curves $V_e = const$ are plotted by the dashed lines. Each point on the diagrams uniquely corresponds to certain velocity V_e, the mass m_e and the radius R_e of the cosmic body entering the atmosphere. As it is seen ρ_b, ε-diagram in the case of the carbonaceous chondrite (Fig. 2b) turns out to be broader than for an icy body and the same values of ρ_b correspond to lower values of the entry velocity V_e.

If one supposes the icy-comet nature of the Tunguska body then its entry velocity must exceed 20 km/s from astronomical considerations. This condition immediately narrows the region of admissible values of all the parameters. (It is worth to note that exact knowledge of the entry velocity would reduce this region to a short line.) Taking a middle point of this sufficiently narrow region, i.e. ρ_e = 0.6 g/cm^3, ε =0.2, we determine the following basic characteristics of the Tunguska cosmic body: V_e =22.7 km/s, m_e=233000 t, R_e = 45.4 m, the trajectory inclination angle α =40°, the height of the explosion H_o= 6.5 km, the total energy $E_o + K_o$ =10^{16}J. For these parameters in Fig. 3 the velocity V , the mass m , the radius R and the ballistic energy E_1 are presented as functions of the height H .

Using the obtained energy distribution $E_1(H)$ the computations were carried out for the region of flattened forest with the model of equivalent explosion [2]. In Fig. 4b the arrows show the directions of the fallen trees, the closed curves are the shock wave isochrones. The observed picture of the flattened forest is shown in Fig. 4a. The boundary covering the zone of radiation burn of trees was also calculated with the model of radiation heat action [3,4]. In Fig. 5 registered data on the radiation burn of trees are shown by the different circles corresponding to a weak, medium and strong degree of burn. The computed boundary is drawn as the solid line. As it is seen the good agreement between the theoretical and really observed pictures takes place in both cases.

Application of the described complex model allows to improve essentially the value of the total TNT equivalent of the Tunguska phenomenon. This value is determined here through the total energy of the body at the entry in the atmosphere. It is equal to 15 Mt and agrees with the results of treatment of the barograms and the seismograms.

The presented results maintain, that the gasdynamical simulation is a perspective tool for investigation of the complicated processes of interaction between natural cosmic bodies and the atmosphere.

References

1. Re Velle D. A quasi-simple ablation model for large meteorite entry: theory vs observations. J. Atmosph. Terrest. Phys., 1979, v.41, p.453-473.
2. Korobeinikov V.P., Chushkin P.I., Shurshalov L.V. Mathematical model and computation of the Tunguska meteorite explosion. Acta Astronaut., 1976, v.3, p.615-622.
3. Putyatin B.V. On radiation influence on the Earth during flight of large meteorite bodies in atmosphere. Dokl. AN SSSR, 1980, v.252, p.318-322.
4. Korobeinikov V.P., Chushkin P.I., Shurshalov L.V. Interaction between large cosmic bodies and atmosphere. Acta Astronaut., 1982, v.9, p.641-643.

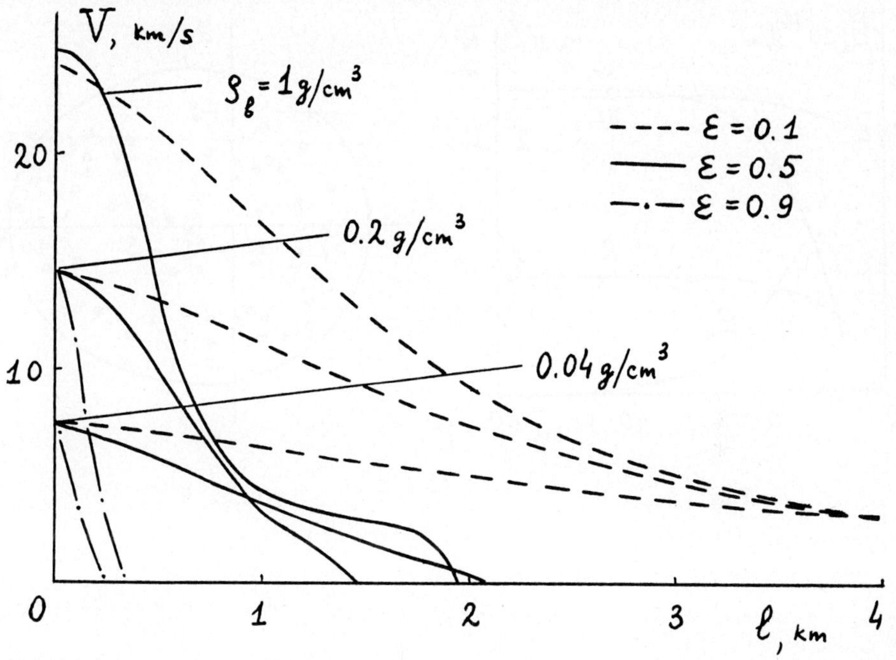
V, km/s
ρ_ℓ = 1 g/cm³
0.2 g/cm³
0.04 g/cm³
---- ε = 0.1
—— ε = 0.5
—·— ε = 0.9
20
10
0
1
2
3
4
ℓ, km

Fig. 1

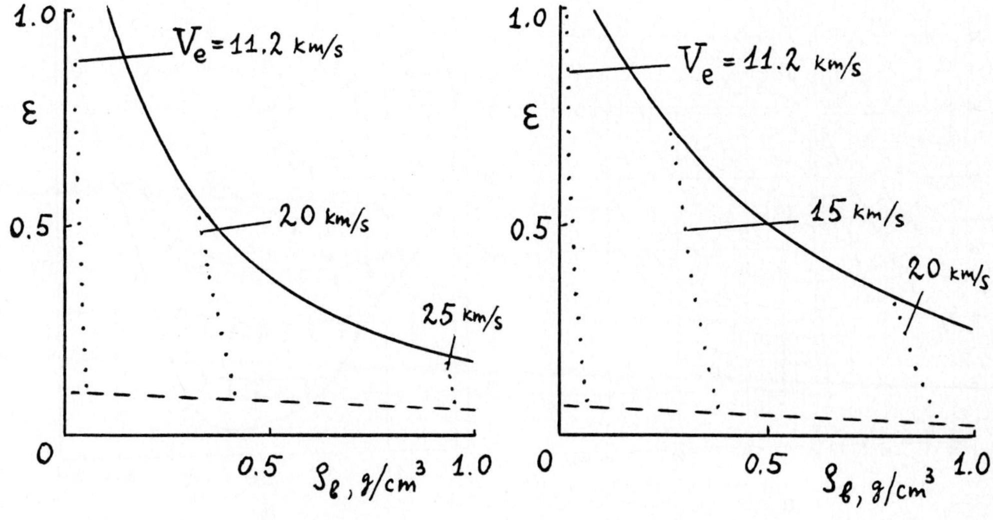
1.0
ε
0.5
0
0.5
ρ_ℓ, g/cm³
1.0
V_e = 11.2 km/s
20 km/s
25 km/s
1.0
ε
0.5
0
0.5
ρ_ℓ, g/cm³
1.0
V_e = 11.2 km/s
15 km/s
20 km/s

a b

Fig. 2

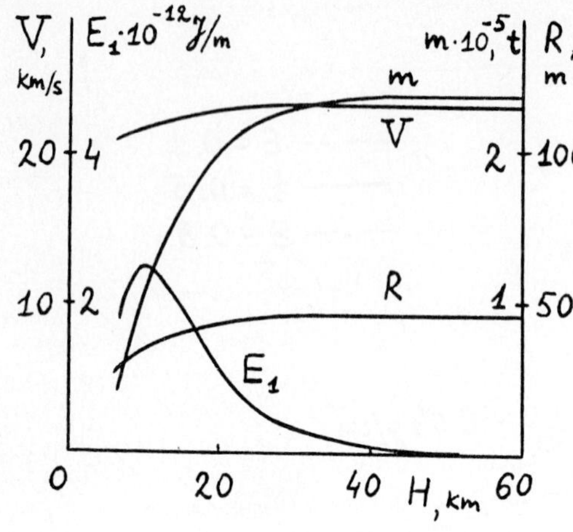

Fig. 3

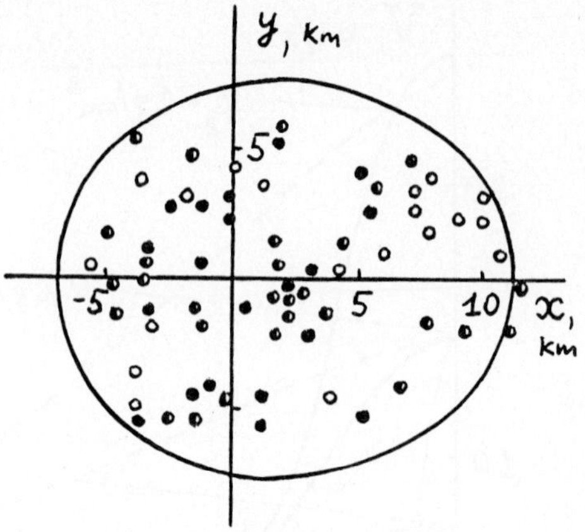

Fig. 5

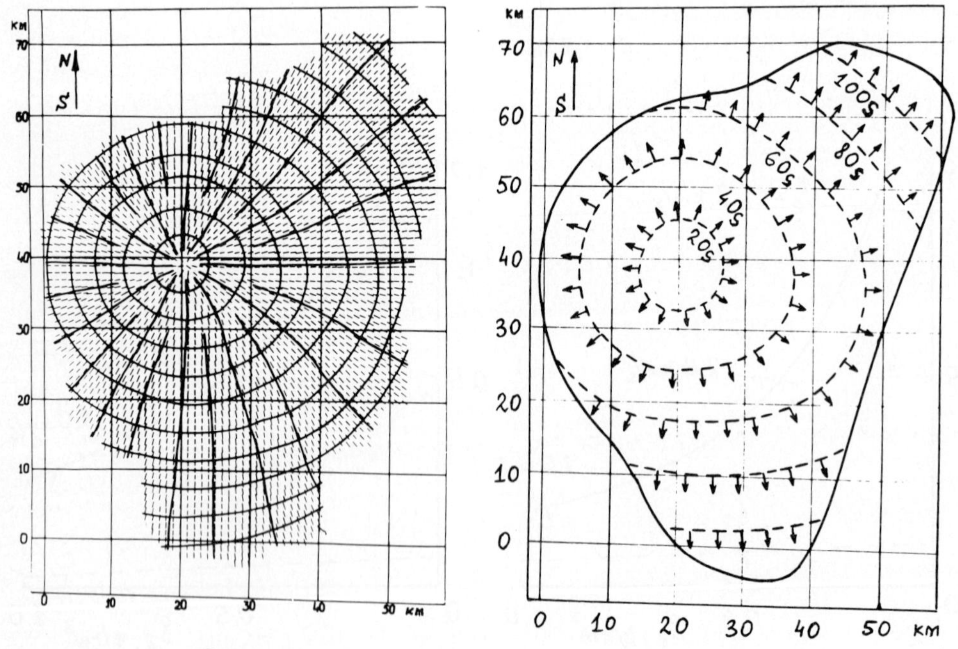

a

b

Fig. 4

A COIN VARIANT OF THE EULER EQUATIONS

Andrea Dadone
Istituto di Macchine ed Energetica, Universita' di Bari
via Re David 200, 70125 Bari, Italy

1. Introduction

The author of the present paper has widely applied the Lambda Formulation, suggested by Moretti /1/ to compute inviscid compressible flows, because of its desirable features and has contributed to improve its accuracy by means of the COIN (Compressible Over INcompressible) variant /2/, which reformulates the Lambda formulation equations in terms of the differences between the variables corresponding to the sought compressible flow and those corresponding to an appropriate known "incompressible" one. In this way, the geometry induced gradients (e.g., around a stagnation point) are mostly accounted for by the incompressible flow solution and the "compressible over incompressible" problem, which is smoother and better behaved than the full compressible one, can be solved very accurately even on a coarse grid.

In Ref. 2 the compressible and incompressible flows in the leading edge region of a circular cylinder are shown to be practically coincident in symmetric flow conditions, whereas minor variations and consequent minor accuracies must be found in unsymmetric flow computations, because of the different leading edge position in incompressible and compressible flow conditions. Some unsymmetric flow computations, performed by the present author as well as by other researchers /3; 4/, seem to prove the COIN technique to retain its superior accuracy, but no extensive study on this matter has ever been performed; moreover, no attempt has ever been made to extend and to check the possible merits of the COIN variant when applied to other types of Euler equations formulations.

The present paper presents a COIN variant of the Euler equations and tests the COIN accuracy, when the leading edge position changes between the reference incompressible flow conditions and the sought compressible ones, by means of some numerical experiments.

2. Governing equations

Let us define an appropriate "incompressible" flow, characterized by the following governing equations:

$$\underline{\nabla}\rho^\circ = 0 \qquad ; \qquad \underline{\nabla}\cdot\underline{V}^\circ = \underline{\nabla}\mathrm{x}\underline{V}^\circ = 0$$
$$\underline{\nabla}H^\circ = 0 \qquad ; \qquad \underline{\nabla}(p^\circ + \rho^\circ V^{\circ 2}/2) = 0 \tag{1}$$

where $\underline{V}$, p, ρ, and H are the velocity vector, pressure, density, and total enthalpy per unit mass, respectively, and $^\circ$ refers to the "incompressible" flow conditions. Accordingly, the COIN Euler equations in vector form can be written as follows:

$$\rho'_t + \underline{\nabla}\cdot(\rho\,\underline{V})' = 0$$
$$(\rho\,\underline{V})'_t + \underline{\nabla}p' + \underline{\nabla}\cdot(\rho\,\underline{V}\,\underline{V})' = 0 \tag{2}$$

$$(\rho\ E)'_t + \underline{\nabla}\cdot(\rho\ H\ \underline{V})' = 0$$

where E is the total internal energy per unit mass and the prime denotes the COIN variables, which are the differences between the variables corresponding to the sought compressible flow and those corresponding to the reference incompressible flow:

$$(\rho\ \underline{V})' = (\rho\ \underline{V}) - (\rho^\circ\ \underline{V}^\circ) \quad ; \quad (\rho\ \underline{V}\ \underline{V})' = (\rho\ \underline{V}\ \underline{V}) - (\rho^\circ\ \underline{V}^\circ\ \underline{V}^\circ)$$

$$(\rho\ E)' = (\rho\ E) - (\rho^\circ\ E^\circ) \quad ; \quad (\rho\ H\ \underline{V})' = (\rho\ H\ \underline{V}) - (\rho^\circ\ H^\circ\ \underline{V}^\circ) \tag{3}$$

With reference to a two dimensional flow and to a curvilinear coordinate system (ξ, η) with metric coefficients ξ_x, ξ_y, η_x, η_y, the following conservative form of the COIN Euler equations can be easily obtained from Eqns. (2):

$$W_t + F_\xi + G_\eta = 0 \tag{4}$$

where the vectors W, F, and G are given by

$$W = \frac{1}{J}\begin{vmatrix} \rho \\ \rho u \\ \rho v \\ \rho E \end{vmatrix} \quad ; \quad F = \frac{1}{J}\begin{vmatrix} (\rho u)'\ \xi_x + (\rho v)'\ \xi_y \\ (p+\rho u^2)'\xi_x + (\rho uv)'\xi_y \\ (\rho uv)'\xi_x + (p+\rho v^2)'\xi_y \\ (\rho uH)'\xi_x + (\rho vH)'\xi_y \end{vmatrix} \quad ; \quad G = \frac{1}{J}\begin{vmatrix} (\rho u)'\ \eta_x + (\rho v)'\ \eta_y \\ (p+\rho u^2)'\eta_x + (\rho uv)'\eta_y \\ (\rho uv)'\eta_x + (p+\rho v^2)'\eta_y \\ (\rho uH)'\eta_x + (\rho vH)'\eta_y \end{vmatrix} \tag{5}$$

being u, v, and J the cartesian components of the velocity $\underline{V}$ and the Jacobian of the coordinate transformation, respectively.

The flux difference splitting method suggested in Ref. 5 has then been applied to split the fluxes; in particular the Riemann problems pertaining to the reference "incompressible" flow have been solved once at the beginning of the computations and the corresponding fluxes have been used to evaluate the COIN (primed) fluxes to be used to integrate Eqn. (4). Moreover a spatially first order accurate numerical scheme has been used.

The COIN Lambda equations are reported in Ref. 2 and are not here repeated for the sake of conciseness; in such a case a spatially second order accurate numerical scheme has been used.

3. Results

In order to test the COIN accuracy, two different test problems are here considered: the flow around a circular cylinder with an undisturbed Mach number equal to .38, suggested in Ref. 4, and the flow around a NACA 0012 airfoil with no incidence at an undisturbed Mach number equal to .72. In order to force the leading edge to be differently located in compressible and incompressible flow conditions, the reference incompressible flow is chosen with an angle of incidence, instead of the more accurate choice of an incompressible flow at no incidence.

Figs. 1 to 4 show the results referring to the flow around the circular cylinder, computed by means of the Flux Difference Splitting method and by using the mesh suggested in Ref. 4: 32 radial and 128 circumferential intervals. Fig. 1 presents the pressure at the uppermost position on the circular cylinder referred to the total pressure of the undisturbed flow (p_u), while Figs. 2 and 3 show the errors in the computed trailing (ε_{te}) and leading edge (ε_{le}) pressure, respectively, and Fig.

4 the mean absolute error of total pressure on the circular cylinder (ε_{tp}). All the results are plotted versus the difference (θ) of the leading edge angular position pertaining to the computed compressible flow case and the one pertaining to the reference incompressible flow used to define the "compressible over incompressible" problem. The continuous lines represent the results obtained by means of the COIN Flux Difference Splitting, while the circles refer to those obtained by means of the classical formulation. Owing to the spatially first order accurate scheme used, the results are not very accurate in comparison with the Lambda Formulation results (see Figs. 5 to 8). However it can be observed the errors to be always smaller when the COIN variant is used; the leading edge movement obviously lowers the COIN variant accuracy, but a superior accuracy is still retained when such a movement is as large as 10°, which corresponds to more than three mesh intervals in the tangential direction.

Figs. 5 to 8 show the results referring to the same test case computed by means of the Lambda Formulation. Owing to the superior accuracy of the spatially second order accurate scheme used in such a case, the computations have been also performed by means of a coarser mesh (16x64), obtained by using only the odd lines of the finer mesh, and the corresponding results have been represented by dashed lines and squares. A close inspection of these figures shows the following features: the pressure at the uppermost position on the circular cylinder and at the leading edge are accurately computed with the COIN and the classical formulation only if the finer mesh is used, while a good accuracy is preserved only by the COIN variant if the coarser mesh is used; the errors in the total pressure on the body and in the trailing edge pressure are always smaller when the COIN formulation is used; the leading edge movement obviously lowers the COIN variant accuracy, but a very good accuracy level is still retained when such movement is as large as 10°.

Figs. 1 to 8 point out the COIN variant, applied either to the Flux Difference Splitting or to the Lambda Formulation, to give always more accurate results, even with a very large leading edge movement; such a superior accuracy is more relevant when coarser meshes are used.

The more general case of the flow around the NACA 0012 airfoil with no incidence at an undisturbed Mach number equal to .72 has then been computed by means of the Lambda Formulation. The corresponding results are plotted in Figs. 9 and 10; the first one presents the pressure at 15% of the chord (p_{15}), while the second one gives the leading edge pressure errors (ε_{le}), plotted as a continuous line and a circle, and the mean absolute errors of total pressure on the airfoil (ε_{tp}), represented by a dashed line and a square, being the lines related to the COIN variant and the symbols to the classical formulation. All the results are plotted versus the angle of incidence (θ) of the reference incompressible flow and have been computed by using a 128x32 intervals C grid, with the trailing edge not pertaining to the computational nodes; accordingly, the pressure at such point has not been reported. The previous conclusions still hold and the superior accuracy of the COIN variant is proved in such flow conditions too, at least for an incidence change as large as 3°, which corresponds to a leading edge movement superior to three tangential mesh intervals. Finally, an important feature of the COIN variant has been outlined by some numerical experiments: the Compressible Over INcompressible formulation is much less sensitive to stretching parameters changes than the classical formulation.

In conclusion, the COIN variant has proven to be applicable to Euler equations in conservation form, at least to the Flux Difference Splitting methods, and the performed numerical experiments have shown the Compressible Over INcompressible problem to be always characterized by more accurate results, even with a very large leading edge movement; such a superior accuracy is more relevant when coarser meshes are used.

Acknowledgements

This research has been supported by M.P.I. and C.N.R.. Thanks are due to prof Pandolfi for having provided the basic code for F. D. S. computations.

References

1. Moretti G., "The λ- Scheme", Computers and Fluids, vol. 7, 1979, pp. 191-205.
2. Dadone A., "A Quasi-Conservative COIN Lambda Formulation", Lecture Notes in Physics, vol. 264, 1986, pp. 200-204.
3. Pandolfi M., Larocca F. and Ayele T., "A Contribution to the Numerical Prediction of Transonic Flows", Notes on Numerical Fluid Mechanics, Vieweg-Verlag, to be published.
4. Moretti G. and Lippolis A., "Airfoil and Intake Calculations", Notes on Numerical Fluid Mechanics, Vieweg-Verlag, to be published.
5. Pandolfi M., "On the flux-difference splitting method in multidimensional unsteady flows", AIAA papr 84-0166.
6. Dervieux A., Periaux J., Rizzi A. and Van Leer B., "Report on a GAMM Workshop on the Numerical Simulation of Compressible Euler Flows, 10-13 June 1986, INRIA-Le Chesnay, France", 7th GAMM C.N.M.F.M., Louvain, Sept. 1987.

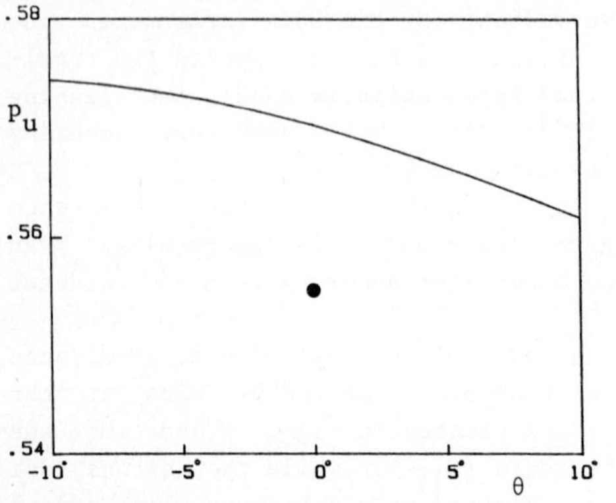

Fig. 1 - F.D.S., pressure at 90°.

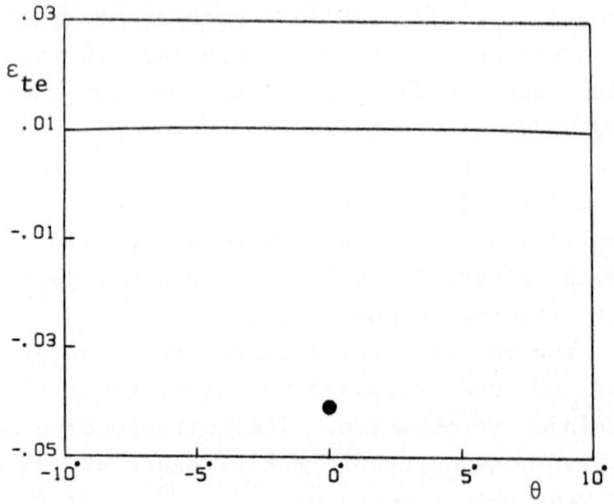

Fig. 2 - F.D.S., T.E. pressure error.

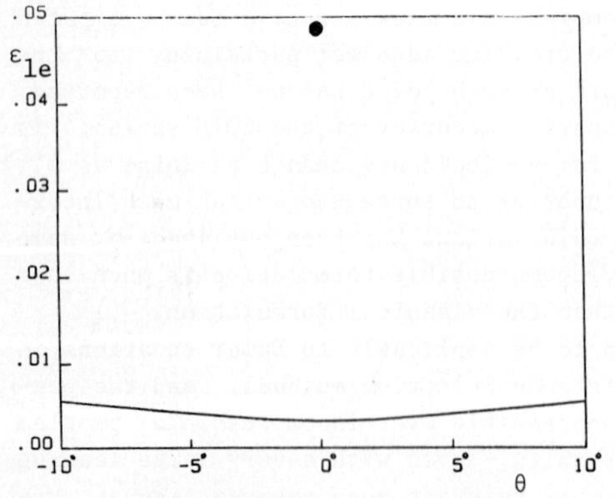

Fig. 3 - F.D.S., L.E. pressure error.

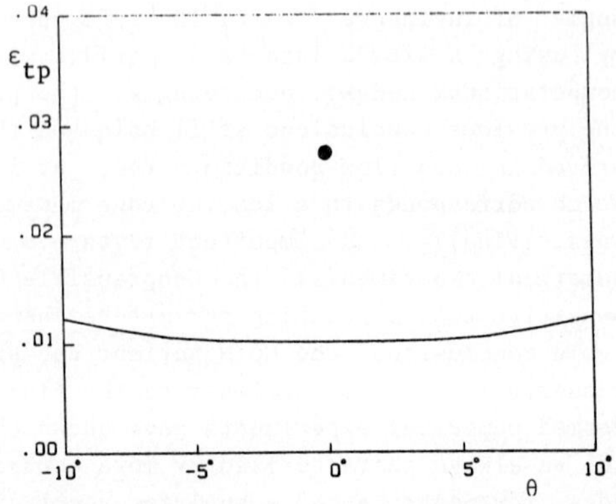

Fig. 4 - F.D.S., total pressure error.

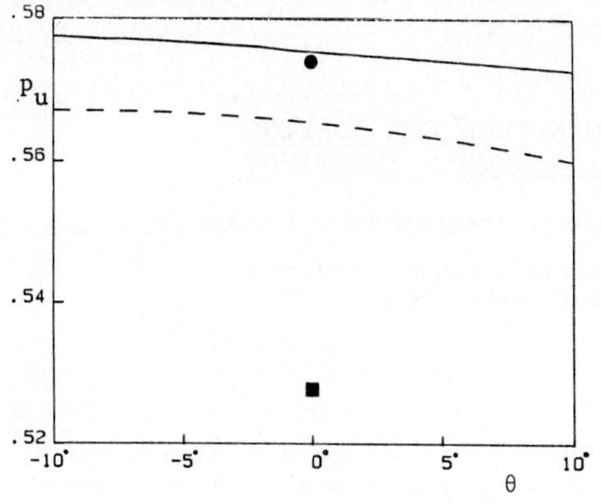

Fig. 5 - λ form., pressure at 90°.

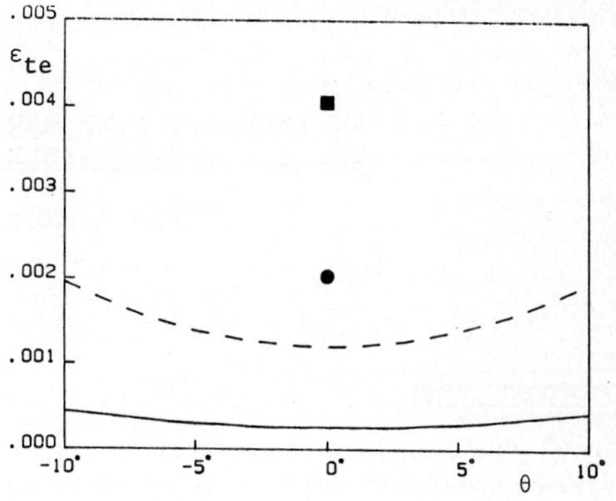

Fig. 6 - λ form., T.E. pressure error.

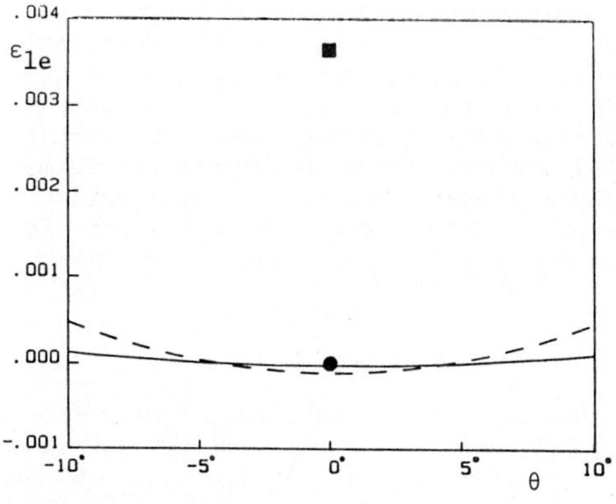

Fig. 7 - λ form., L.E. pressure error.

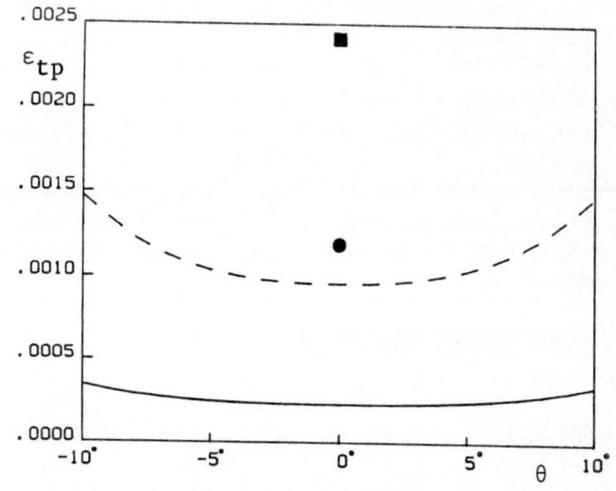

Fig. 8 - λ form., total pressure error.

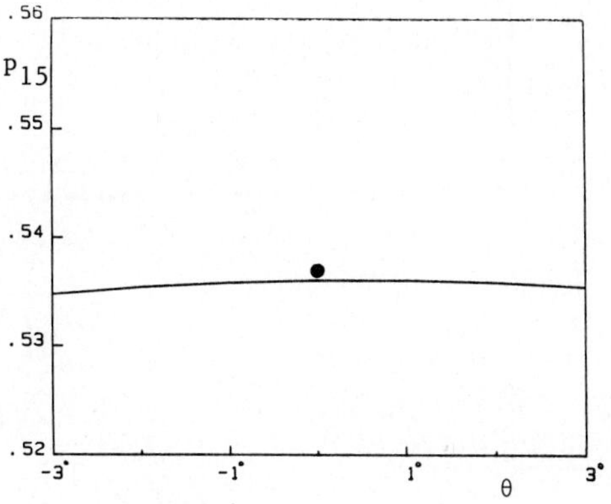

Fig. 9 - NACA 0012, pressure at x/c=.15.

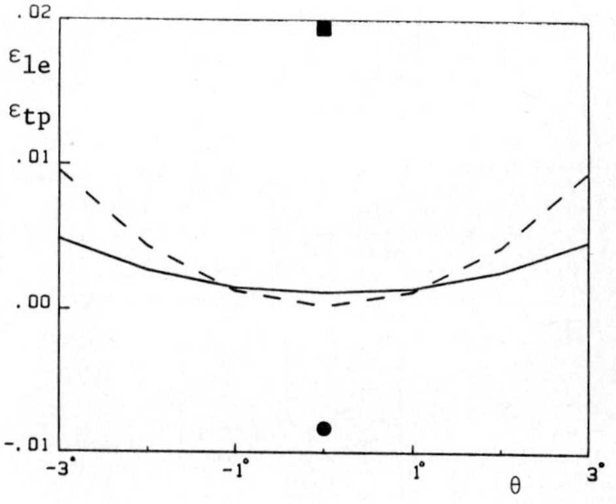

Fig. 10 - NACA 0012, pressure errors.

AN IMPLICIT TIME-MARCHING METHOD FOR SOLVING THE 3-D COMPRESSIBLE NAVIER-STOKES EQUATIONS

Hisaaki DAIGUJI and Satoru YAMAMOTO

Department of Mechanical Engineering, Tohoku University
Aza-Aoba, Aramaki, Sendai 980, JAPAN

INTRODUCTION

The implicit time-marching finite-difference scheme for the steady three-dimensional compressible Euler equations developed by the authors[1] is extended to the Navier-Stokes equations taking account of the diffusion terms. The distinctive features of the previous scheme are to make use of the momentum equations of contravariant velocities instead of the physical velocities, and to be able to treat the solid wall boundary conditions exactly and the periodic boundary condition for the turbomachine impeller flow easily.

The momentum equations of contravariant velocities are originally nonconservative form, but can be written in a divergence form by considering an additional term. The modified equations can be solved implicitly by the same manner as the equations of the existing methods. That is, the computational techniques such as the delta-form approximate-factorization scheme, diagonalization and flux vector splitting can be applied to them. The diffusion, body force and additional terms are treated explicitly. The present method using the NS equations of contravariant velocities is capable of the simple treatments of the boundary conditions for the impeller flow, and seems to be convenient to introduce the anisotropic turbulence models.

FUNDAMENTAL EQUATIONS

The fundamental equations of the relative flow through an impeller of turbomachines in general curvilinear coordinates $\xi_1\xi_2\xi_3$ are

$$\frac{\partial \hat{q}}{\partial t} + \frac{\partial \hat{E}_i}{\partial \xi_i} + \frac{1}{R_e}\hat{S} + \hat{H} = 0 \qquad (1)$$

where

$$\hat{q} = J\begin{pmatrix} \rho \\ \rho w_1 \\ \rho w_2 \\ \rho w_3 \\ e \end{pmatrix}, \quad \hat{E}_i = J\begin{pmatrix} \rho W_i \\ \rho w_1 W_i + \partial\xi_i/\partial x_1 \, p \\ \rho w_2 W_i + \partial\xi_i/\partial x_2 \, p \\ \rho w_3 W_i + \partial\xi_i/\partial x_3 \, p \\ (e+p)W_i \end{pmatrix} \quad (i=1,2,3)$$

$$\hat{S} = -J\frac{\partial \xi_i}{\partial x_j}\frac{\partial}{\partial \xi_i}\begin{pmatrix} 0 \\ \tau_{1j} \\ \tau_{2j} \\ \tau_{3j} \\ \tau_{kj} w_k + K\partial T/\partial x_j \end{pmatrix}, \qquad (2)$$

$$\hat{H} = J\rho\begin{pmatrix} 0 \\ 0 \\ -\omega^2 x_2 - 2\omega w_3 \\ -\omega^2 x_3 + 2\omega w_2 \\ 0 \end{pmatrix}$$

and the internal energy e, Jacobian J, contravariant velocities W_i and components of stress tensor τ_{ij} are

$$e = \rho(\varepsilon + \mathbf{v}^2/2 - \omega r v_u) \qquad (3)$$

$$J = \partial(x_1, x_2, x_3)/\partial(\xi_1, \xi_2, \xi_3) \qquad (4)$$

$$W_i = (\partial\xi_i/\partial x_j) w_j \qquad (i=1,2,3) \qquad (5)$$

$$\tau_{ij} = \mu[(\frac{\partial w_i}{\partial x_j} + \frac{\partial w_j}{\partial x_i}) - \frac{2}{3}\delta_{ij}\frac{\partial w_k}{\partial x_k}] \qquad (i,j=1,2,3) \qquad (6)$$

x_i are cartesian coordinates, w_i cartesian components of relative velocity $\mathbf{w}$, $\boldsymbol{\omega}$ angular velocity of impeller, R_e Reynolds number, $\hat{S}$ diffusion term and $\hat{H}$ body force (centrifugal and coriolis forces) term.

In the present paper we use the momentum equations of contravariant velocities which are derived by taking linear combinations of the momentum equations in Eq. (1). That is[2],

$$\frac{\partial\tilde{q}}{\partial t} + \frac{\partial\tilde{E}_i}{\partial\xi_i} + \tilde{R} + \frac{1}{R_e}\tilde{S} + \tilde{H} = 0 \qquad (7)$$

where

$$\tilde{q} = J\begin{pmatrix}\rho\\ \rho W_1\\ \rho W_2\\ \rho W_3\\ e\end{pmatrix}, \qquad \tilde{E}_i = J\begin{pmatrix}\rho W_i\\ \rho W_1 W_i + g_{1i}\,p\\ \rho W_2 W_i + g_{2i}\,p\\ \rho W_3 W_i + g_{3i}\,p\\ (e+p)W_i\end{pmatrix} \qquad (i=1,2,3)$$

$$\tilde{R} = -J(\rho\mathbf{w}W_i + p\nabla\xi_i)\cdot\frac{\partial}{\partial\xi_i}\begin{pmatrix}0\\ \nabla\xi_1\\ \nabla\xi_2\\ \nabla\xi_3\\ 0\end{pmatrix} \qquad (8)$$

$$\tilde{S} = -J\begin{pmatrix}0\\ (\nabla\cdot\Pi)\cdot\nabla\xi_1\\ (\nabla\cdot\Pi)\cdot\nabla\xi_2\\ (\nabla\cdot\Pi)\cdot\nabla\xi_3\\ \nabla\cdot(\Pi\cdot\mathbf{w}+K\nabla T)\end{pmatrix}, \qquad \tilde{H} = J\rho(-\mathbf{r}\omega^2 + 2\boldsymbol{\omega}\times\mathbf{w})\cdot\begin{pmatrix}0\\ \nabla\xi_1\\ \nabla\xi_2\\ \nabla\xi_3\\ 0\end{pmatrix}$$

$g_{ij} = \nabla\xi_i\cdot\nabla\xi_j$, $\tilde{R}$ is an additional term introduced in order to express the principal part of Eq.(7) in conservative form, and Π the stress tensor whose components are τ_{ij}. To make use of Eq.(7) simplifies the treatments of boundary conditions, especially the periodic condition. In Eqs.(7) and (8), all cartesian components w_i and x_i are included in their vector or tensor forms. Therefore, at the corresponding points on the periodic boundary of the 3-D cascade flow problem, all the variables namely the components of $\tilde{q}$, $\tilde{E}_i$, $\tilde{R}$, $\tilde{S}$ and $\tilde{H}$ have the same values, so the treatments of this boundary condition become very simple.

NUMERICAL METHOD

We here obtain the steady state solutions by solving Eq.(7) using an implicit time-marching finite-difference scheme. In this scheme, first the flux vectors $\tilde{E}_i$ in Eq.(7) are linearized by the expression[1]

$$\tilde{E}_i = \tilde{A}_i\tilde{q} \qquad (i=1,2,3) \qquad (9)$$

Using Eq.(9) in Eq.(7) and applying the Beam-Warming delta-form approximate-factorization scheme, we get

$$(I+\Delta t\theta\delta_1\tilde{A}_1)(I+\Delta t\theta\delta_2\tilde{A}_2)(I+\Delta t\theta\delta_3\tilde{A}_3)\Delta\tilde{q}^n = -\Delta t(\delta_i\tilde{E}_i + \tilde{R} + R_e^{-1}\tilde{S} + \tilde{H})^n \qquad (10)$$

where

$$\Delta\tilde{q}^n = \tilde{q}^{n+1} - \tilde{q}^n \tag{11}$$

$\theta = 1/2$ or 1, and δ_i are the finite-difference operators with respect to ξ_i.

Next, in order to apply the theory of characteristics, Jacobian matrices $\tilde{A}_i$ are put as

$$\tilde{A}_i = \tilde{S}_i^{-1} \Lambda_i \tilde{S}_i \qquad (i=1,2,3) \tag{12}$$

where Λ_i are the diagonal matrices of eigenvalues of $\tilde{A}_i$ and $\tilde{S}_i$ are the matrices composed of eigenvectors of $\tilde{A}_i$. Further Λ_i and $\tilde{A}_i$ are split according to sign of eigenvalues as

$$\begin{aligned} \Lambda_i &= \Lambda_i^+ + \Lambda_i^- , & & (i=1,2,3) \\ \tilde{A}_i &= \tilde{A}_i^+ + \tilde{A}_i^- , & \tilde{E}_i &= \tilde{E}_i^+ + \tilde{E}_i^- \end{aligned} \tag{13}$$

where Λ_i^+ and Λ_i^- are the matrices of only positive or negative eigenvalues, and $A_i^{\pm} = \tilde{S}_i^{-1}\Lambda_i^{\pm}\tilde{S}_i$, $\tilde{E}_i^{\pm} = \tilde{A}_i^{\pm}\tilde{q}$. Using Eqs.(12) and (13) in Eq.(10) and appling the techniques of the diagonalization and the flux vector splitting, we get

$$\begin{aligned} &\tilde{S}_1^{-1}\{I+\Delta t\theta(\Lambda_1^+\nabla_1 + \Lambda_1^-\Delta_1)\}\tilde{M}_1^{-1}\{I+\Delta t\theta(\Lambda_2^+\nabla_2 + \Lambda_2^-\Delta_2)\} \\ &\tilde{M}_2^{-1}\{I+\Delta t\theta(\Lambda_3^+\nabla_3 + \Lambda_3^-\Delta_3)\}\tilde{S}_3\Delta\tilde{q}^n \\ &\qquad = -\Delta t(\nabla_i\tilde{E}_i^+ + \Delta_i\tilde{E}_i^- + \tilde{R} + R_e^{-1}\tilde{S} + \tilde{H})^n \end{aligned} \tag{14}$$

where

$$\tilde{M}_1 = \tilde{S}_2\tilde{S}_1^{-1} , \qquad \tilde{M}_2 = \tilde{S}_3\tilde{S}_2^{-1} \tag{15}$$

∇_i and Δ_i are the backward- and forward-difference operators with respect to ξ_i, respectively, and $\Lambda_i^+\nabla_i + \Lambda_i^-\Delta_i$ mean the upstream-difference operators according to sign of the phase velocity of the wave propagating along a characteristic curve. This upstream-differenc scheme in the previous paper was the fourth-order central-difference scheme having the fourth-order artificial dissipation term, but in the recent papers was replaced by the Chakravarthy-Osher TVD scheme[3].

Eq.(14) is solved dividing into seven steps. The odd steps are products of matrix by vector at each grid points, and the even steps are computations of five sets of linear equations having a penta-diagonal matrix. The diagonalization is effective for the saving of CPU-time, and the TVD-scheme for the stabilization of the solution and the clear shock capturing.

CALCULATION OF TURBULENT FLOW

The accuracy of the steady state solution depends on the RHS of Eq.(14), but is independent of the LHS. For this reason, the diffusion term $\tilde{S}/R_e$ must be completely included in the RHS, but is partially in Λ_i of LHS. We here perform the calculation of turbulent flow using the two-equation k-ε model by Launder-Spalding. The viscosity coefficient μ in $\tilde{S}$ is replaced by $\mu_{eff} = \mu + \mu_t$ under the Boussinesq approximation. The transport equations of k and ε are rewritten to apply to the curvilinear coordinates[4], and solved by an implicit time-marching method. The delta-form AF scheme and the Chakravarthy-Osher TVD scheme are also used here. To reduce the calculating points near the solid wall, the empirical law of the wall together with the k-ε model is used along the wall.

2-D CASCADE FLOWS

We first calculated 2-D cascade flows of a transonic axial-flow compressor using 2-D NS code[5]. The computational grid has 131×31 grid points and was generated by the algebraic method. The Reynolds number based on chord is 1.6×10^6, the inlet relative Mach number 1.02, and the static pressure ratio 1.34. The computational grid and the calculated relative Mach number contours are shown in Figs. 1 and 2. The position and strength of shocks agree well with the experimental results by Schreiber and Starken[6]. The next example is 2-D cascade flows through an axial flow turbine nozzle. The Reynolds number based on chord is 8.5×10^5, the inlet Mach number 0.60, and the static pressure ratio 0.67. The computational grid and the calculated Mach number contours show in Figs.3 and 4. It is found from these figures that the shock waves and the wakes can be obtained clearly by the present method using the nearly orthogonal grid, the Chakravarthy-Osher TVD scheme and k-ε model.

3-D CASCADE FLOWS

We next calculated 3-D cascade flows through a transonic axial-flow compressor rotor with tip clearance using 3-D NS code. The computational grid has $81 \times 25 \times 25$ grid points concentrated toward the hub, casing and blade surfaces as shown in Fig. 5. The Reynolds number based on blade chord at the hub is 1.0×10^6, the inlet relative Mach number 0.911(hub) to 1.291(tip), and the static pressure ratio 1.50. The calculated pressure contours on the blade and hub surfaces are shown in Fig.6, and the Mach number contours on the blade-to-blade surfaces in Fig.7. In spite of the relatively coarse grid, the shock waves in the channel and the leakage vortices near the casing wall are captured clearly.

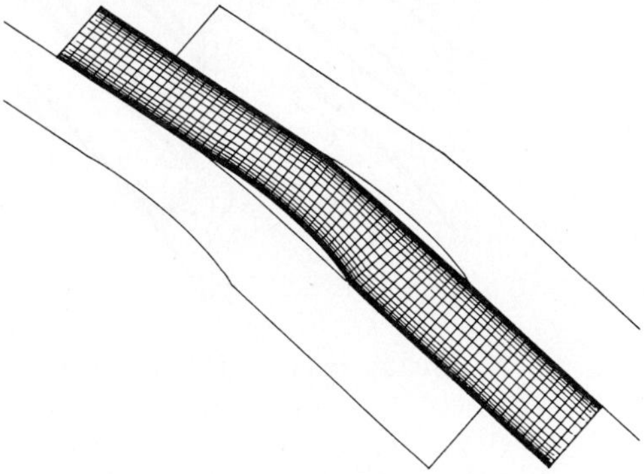

Fig.1 Computational grid of 2-D compressor cascade flow

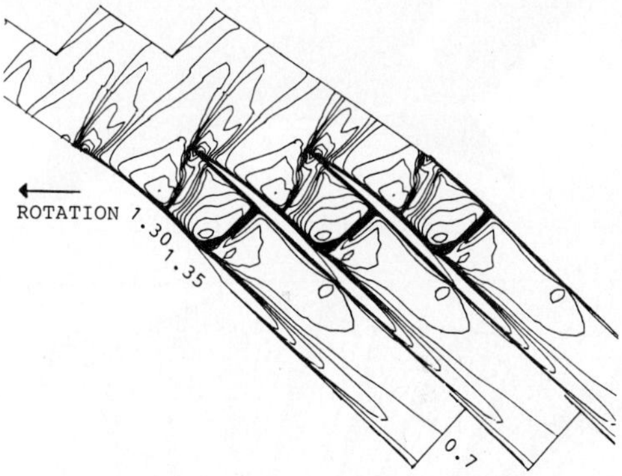

Fig.2 Calculated Mach number contours

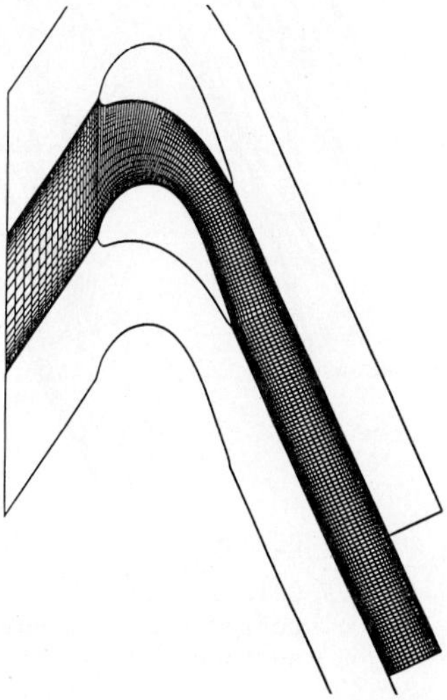

Fig.3 Computational grid of 2-D turbine nozzle flow

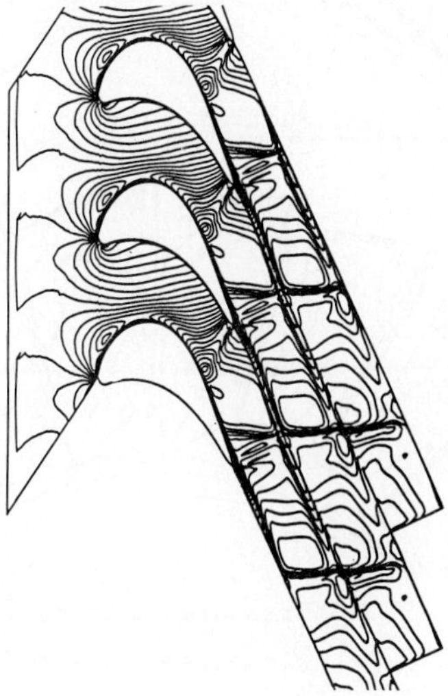

Fig.4 Calculated Mach number contours

REFERENCES

[1] Daiguji, H., Motohashi, Y. and Yamamoto, S., Proc. of 10th Int. Conf. on Numerical Methods in Fluid Dynamics, (1986), 205-210, Springer-Verlag.

[2] Yamamoto, S., Daiguji, H., and Ito, K., Proc. of 1987 Tokyo Int. Gas Turbine Cong. (1987), Vol.II, 295-302.

[3] Chakravarthy, S.R. and Osher, S., AIAA Paper 85-0363, (1985).

[4] Sahu, J. and Danberg, J.E., AIAA Paper 85-0373, (1985).

[5] Yamamoto, S., Daiguji, H. and Ishigaki, H., Proc. of Int. Symp. on Computationl Fluid Dynamics, (1987), to appear.

[6] Schreiber, H.A. and Starken, H., Trans. ASME, J. Engng. Power, 106-2(1984), 288-294.

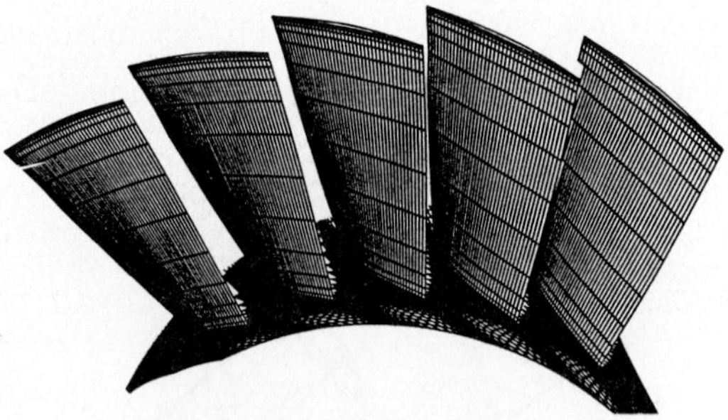

Fig.5 Computational Grid

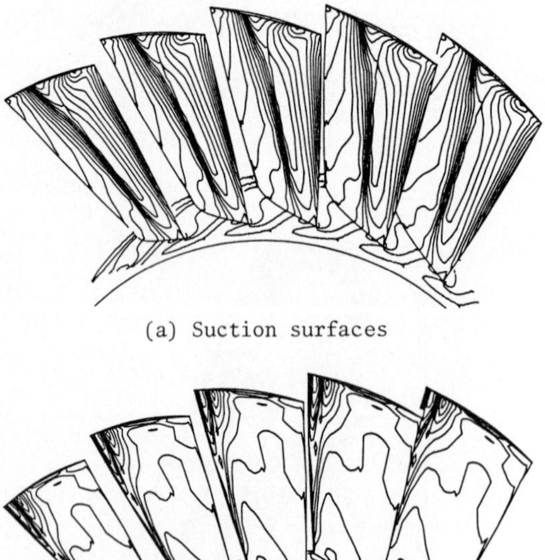

(a) Suction surfaces

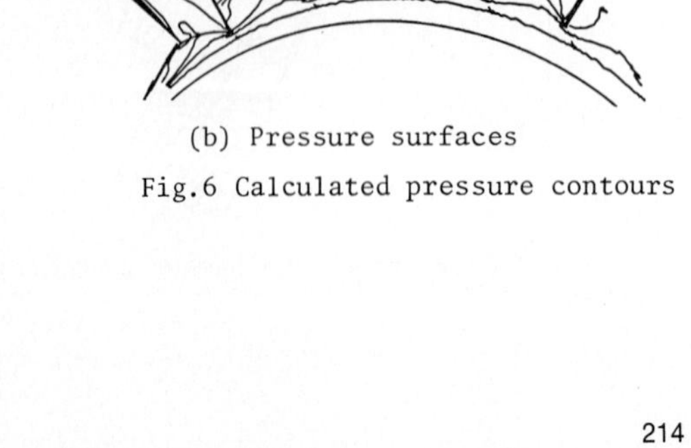

(b) Pressure surfaces

Fig.6 Calculated pressure contours

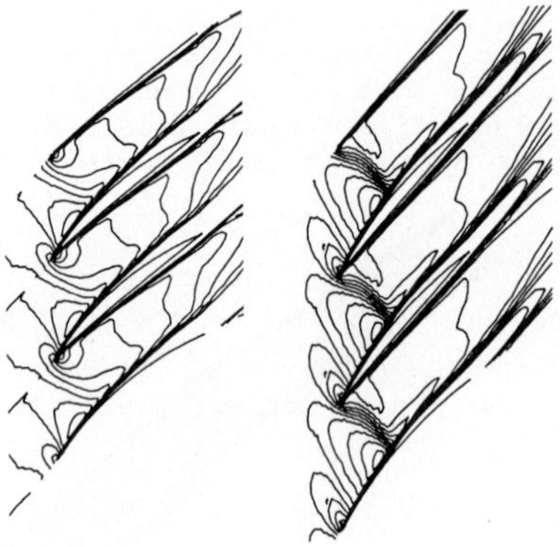

(a) 13% span (b) 58% span

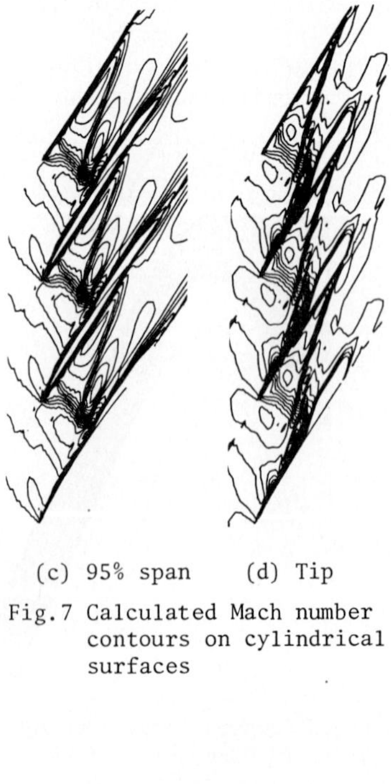

(c) 95% span (d) Tip

Fig.7 Calculated Mach number contours on cylindrical surfaces

LOW-STORAGE IMPLICIT UPWIND-FEM SCHEMES FOR THE EULER EQUATIONS

A. DERVIEUX - L. FEZOUI - H. STEVE (*)
J. PERIAUX - B. STOUFFLET (**)
(*) INRIA Sophia-Antipolis, 2004 Route des Lucioles, 06560 VALBONNE (FRANCE)
(**) AMD-BA DGT-DEA BP 300, 78 Quai Marcel Dassault, 92214 Saint-Cloud (FRANCE)

1. Introduction

During the last few years, several teams in the CFD community have opted for the Finite-Element type 3-D methods applicable to unstructured meshes as an efficient approach for the solution of a complex Euler flow around a realistic geometry. Two typical recent papers have illustrated what can already be done and what is anticipated : In [5], a Galerkin (central differencing) spatial scheme is combined with an explicit-implicit (Runge-Kutta + residual averaging) time integration, and applied to the calculation of the transonic flow around a quadri-reactor aircraft with a mesh of 35,000 nodes. Several studies have shown the very good behavior of second-order upwind schemes for calculations using highly non regular meshes. In [11], a linearized implicit upwind scheme is applied to the calculation of high Mach flows around a spatial aircraft (10-15,000 nodes) ; this method reaches an attractive level of robustness and efficiency, but requests the storage of a large matrix, limiting the maximum possible number of nodes that can be handled for a given computer. The purpose of this paper is to present new versions of the linearized implicit scheme, with, among others, two following main improvements : 1) at most only the block-diagonal terms of the matrix are stored. 2) Non-recurrent solution algorithms are applied, which allows to adapt the schemes to supercomputers.

2. Spatial discrete scheme :

Conservative form : The Euler equations are shortly written in conservative form as follows :

$$W = \begin{pmatrix} \rho \\ \rho\vec{V} \\ E \end{pmatrix} \quad \begin{array}{l} W_t + \vec{\nabla}.\vec{\mathcal{F}}(W) = 0 \\ + \text{ boundary conditions} \\ + \text{ initial condition} \end{array}$$

in which and $\vec{\mathcal{F}}(W)$ is the Euler's flux where ρ is the density, $\vec{V}$ is the velocity and E is the total energy by unit volume.

Spatial discretization : We chose tetrahedra as elements for easier mesh handling. Let S_i the nodes of the mesh, we build the the control volume $\hat{S}_i$ like in [11] .

For the calculation of the solution W_i^n at S_i at time t^n, we need the evaluation of numerical fluxes that are :

- The Osher difference-flux splitting [7] for the explicit stage of the implicit code
- The Steger-Warming difference-flux splitting [9] for the implicit stage of the implicit code. The second-order spatial approximation relies on an extension of *MUSCL-FEM* method [3] derivated from ideas of B. Van Leer [12] : we build interpolated P_1-functions and we take a 1-D Van Albada limiter [2].

3. Implicit time-stepping :

Implicit unfactored relaxation schemes seem to, be more and more attractive ; for unstructured mesh calculation, the use of such schemes seems efficient [4]. We calculate the solution at each time-step in two phases. Firstly, an explicit phase is performed :

$$\widehat{\delta W} = -\Delta t E_2(W^n)$$

where Δt is the time-step and E_2 the second-order numerical flux in which Osher's splitting is combined to the MUSCL-FEM formulation. Secondly, an implicit phase is applied. The implicit terms use a matrix A^n with first-order accurate fluxes, they come from the linearization at W^n of the Steger-Warming decomposition:

$$X = (A^n)^{-1}\,\widehat{\delta W}\ ;\ W^{n+1} = W^n + X.$$

Therefore at each time-step we have to solve a linear problem; the way to solve it efficiently without using too much memory storage is the main point of this paper.

4. Implicit scheme with Jacobi iteration :

Algorithm : When heavy geometries (many nodes) have to be taken into account, the matrix of the linear operator cannot be stored; in this work we present a trade-off that consists in storing diagonal terms and re-computing the extra-diagonal ones [10]. Let D denote the block-diagonal part of A^n

and let E denote the extra-diagonal one, i.e. $E = A - D$; the above linear system is partly solved by a few block-collective Jacobi iterations ($kmax = 4$ here) : $X^0 = (D^n)^{-1}\widehat{\delta W}$; for $k = 1$ to $kmax$, we do $X^{k+1} = (D^n)^{-1}(\widehat{\delta W} - E^n X^k)$ and $W^{n+1} = W^n + X^{kmax}$. The inverse $(D^n)^{-1}$ of the block diagonal terms of the matrix is stored, while the other term E^n is not stored, but calculated at each Jacobi iteration. There are two main reasons for the choice of the Jacobi method : first, the algorithm is not recurrent (unlike the Gauss-Seidel method), therefore independency is preserved for vectorization. Secondly, the low storage version is derived more easily. When a scalar computer is used, this method is less efficient than the implicit full-storage method, but still 5 to 12 times more than an explicit scheme (3-D tests). For vector computers, the loss of efficiency w.r.t. the Jacobi full-storage method is lower since extra calculations do not involve many scatter and gather and are therefore fastly vectorizable.

Memory storage : In 3-D, the total matrix storage requires $350 \times NS$ real numbers (NS is the total number of nodes) with only $25 \times NS$ for the diagonal terms. (We need $100 \times NS$ words for the rest of storage). The Cray-2 (256 Mwords) would allow to use up to 2,000,000 nodes, instead of 500,000 for the full-storage algorithm.

Supercomputing : Since the purpose of this work is to perform massive calculations on supercomputers, the different ressources of this kind of computer must be used efficiently. Best performances are obtained by using the following principles :

- minimizing indirect addressing : one way is to chose algorithms that converge in a few iterations only.
- Vectorize indirect addressing (scatter - gather) by the use of colouring procedures [1].
- Perform rest of the calculation at high rate : 100-130 Mflops on Cray-2.
- Lastly, use tetraedra to allow a conforming refinement, that is a local refinement that transforms an admissible Finite Element mesh in a new admissible one.

Results : We present a 3-D Euler flow around a *delta-wing* (Fig. 1). The Mach at infinity is 1.2 and the angle of attack is 10deg. The initial coarse mesh contains 5,085 nodes, the converged steady solution is calculated with 5 minutes CPU of (monoprocessing) Cray-2. Then, we built by **local refinement** a new mesh involving 22,169 nodes. A steady solution is obtained after 100 time-steps and 20 minutes CPU monoprocessing Cray-2 . For this calculation, the implicit code reaches 50 Mflops.

5. GMRES iteration

Efficient iterative methods relying on optimization have been recently developped [8] and applied to the calculation of compressible flow [6]. We recall the main features of the algorithm.

Algorithm : At each time-step the following operator is introduced:

$$G_n(W) = A^n(W^n)\,(W - W^n) - \Delta t\, E^2(W^n)$$

We want to solve the problem: $G_n(W) = 0$. Let us denote by $DG_n(U,d)$ the Jacobian of G_n at point U in the direction d. At each iteration p , the algorithm will yield $X^{(p+1)} = X^{(p)} + Y$ where $X^{(0)}$ is an initial guess and Y belongs to the "Krylov space" K [8] constructed from successive Jacobians of the functional G_n . Y is chosen such that $||G_n(X^{(p+1)})||$ is minimum. The algorithm first proceeds to find an orthonormal basis of space K. The following calculations are performed :

For $p = 0$ to q:

$$r_1^{(p)} = G_n(X^{(p)})\,,\; v_1^{(p)} = \frac{r_1^{(p)}}{||r_1^{(p)}||}$$

$$\text{For } i = 1 \text{ to } k : r_{i+1}^{(p)} = DG_n(X^{(p)}\,;\,v_i^{(p)}) - \sum_{j=1}^{i} b_{i+1,j}^{(p)}\, v_j^{(p)}\;;\; v_{i+1}^{(p)} = \frac{r_{i+1}^{(p)}}{||r_{i+1}^{(p)}||}$$

where $DG_n(X^{(p)};v)$ is computed exactly or approximated by using the following formulas (ε being a small number):

$$DG_n(X^{(p)};v) = \frac{G_n(X^{(p)} + \varepsilon v) - G_n(X^{(p)})}{\varepsilon}\;;\; b_{il}^{(p)} = (DG_n(X^{(p)};v_i^{(p)})\,,\, v_l^{(p)}).$$

Then $X^{(p)}$ is the solution of the minimization problem :

Find $\lambda^{(p)} = (\lambda^{(p)})_{j=1}^{k}$ such that, for any μ,

$$||G_n(X^{(p)} + \sum_{j=1}^{k} \lambda_j^{(p)} v_j^{(p)})|| \leq ||G_n(X^{(p)} + \sum_{j=1}^{k} \mu_j^{(p)} v_j^{(p)})|| \qquad X^{(p+1)} = X^{(p)} + \sum_{j=1}^{k} \lambda_j^{(p)} v_j^{(p)}$$

and finally $W^{n+1} = X^{(q)}$.

Results : A sample of 2-D experiments have been performed; they are summed up in the two tables below, in which
- EXPL holds for a multi-step explicit code,
- NLGMRES holds for the presented method,
- NLGMRES+1 holds for the same method, but with the use of the diagonal D^n as preconditioner of the iteration,
- LIN.IMPL. holds for full-storage version with Gauss-Seidel as linear solver.

It is observed that, for heavier meshes, the efficiency of the presented method involving the diagonal preconditioner compares well with the full-storage Gauss-Seidel one, that is the fastest of all the investigated approaches; furthermore the presented method requires much less memory space.

3D - Result : We present a 3-D computation of an Euler hypersonic flow past the Hermes space shuttle at Mach = 8 and angle of attack of 20 deg. Fig. 2 shows the Mach number distribution on the surface. The number of nodes in the domain is about 10.000. Another computation has been performed with a crude mesh of about 2600 nodes. It is interesting to compare the performance of the implicit linearized solver combined with relaxation (nodewise Jacobi) and the diagonal preconditionned linear GMRES algorithm for both cases. A typical convergence comparison of one solution of the linear system is presented for each case. One can see that when increasing the number of degrees of freedom, the behaviour of the linear GMRES compared to relaxation is better in terms of number of iterations. One has to note that the CPU cost of both algorithms are practically the same.

6. Conclusion :

We have described two solution methods, applying to the same spatial approximation, that present the following advantages :
– The memory requirements are not much larger than for an explicit method.
– Because the schemes used are parameter free, *only one run* was sufficient for each calculation presented.
– The convergence to the steady solution is fast.
– Mesh adaptation is combined without losing efficiency in the solution.

7. References :

[1] F. ANGRAND - J. EHREL, Vectorized Finite Element codes for compressible Flows, INRIA research report no 622 (1987).
[2] V. BILLEY - A. DERVIEUX - L. FEZOUI - J. PERIAUX - V. SELMIN - B. STOUFFLET, Recent improvements in Galerkin and upwind Euler solvers and appplication to 3-D transonic flow in aircraft design, 8th Int. Conf. on Computing Methods in Applied Sciences and Engineering, Versailles, dec. 14-18, 1987, to be published by North Holland.
[3] L. FEZOUI, Résolution des équations d'Euler par un schéma de Van Leer en éléments finis, INRIA Research Report, no 358 (1985).
[4] L. FEZOUI - B. STOUFFLET, A class of implicit upwind schemes for Euler simulations with unstructured meshes, INRIA Research Report, no 517 (1986).
[5] A. JAMESON - T.J. BAKER, Improvements to fine aircraft method, AIAA paper 87-0452 Reno.
[6] M. MALLET - J. PERIAUX - B. STOUFFLET, Convergence acceleration of finite element methods for the solution of the Euler and Navier-Stokes equations, 8th GAMM Conference on Numerical Methods in Fluid Mechanics, proc. to be published by Vieweg (1988).
[7] S. OSHER - F. SOLOMON, Upwind difference schemes for hyperbolic systems of conservation laws, Journal Math. comp. (1982).
[8] Y. SAAD - M.H. SCHULTZ, "GMRES": a generalized minimal residual algorithm for solving nonsymmetric linear systems, Research Report YALEU/DCS RR-254, (1983)
[9] J. STEGER - R.F. WARMING, Flux-vector splitting for the inviscid gas dynamic equations with applications to finite difference methods, Journal Comp. Physics, vol 40, no 2, pp 263-293 (1981).
[10] H. STEVE, Efficient implicit solvers for the resolution of Euler equation in finite element methods, INRIA Research Report, (Thesis to appear).
[11] B. STOUFFLET - J. PERIAUX - L. FEZOUI - A. DERVIEUX, Numerical simulation of 3-D hypersonic Euler flow around space vehicules using adapted finite elements, AIAA paper 87-0560.
[12] B. VAN LEER, Towards the ultimate conservative difference scheme. The quest of monotonicity. Lecture Notes in Physics, Vol 18 pp 163 (1972).

Fig. 1 : Flow past a delta wing : MACH=1.2, $\alpha = 10$ deg.

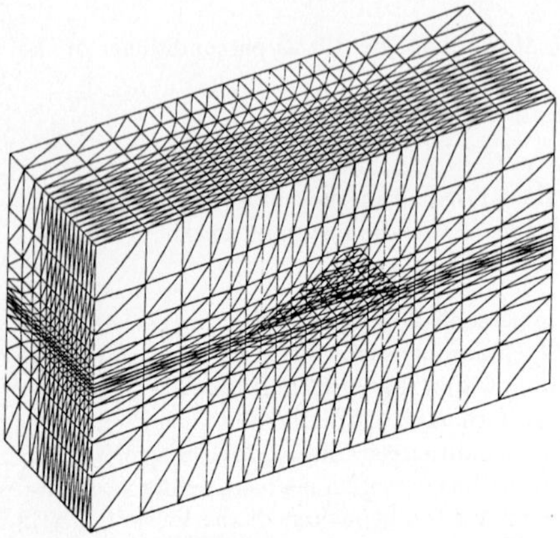

5085 nodes

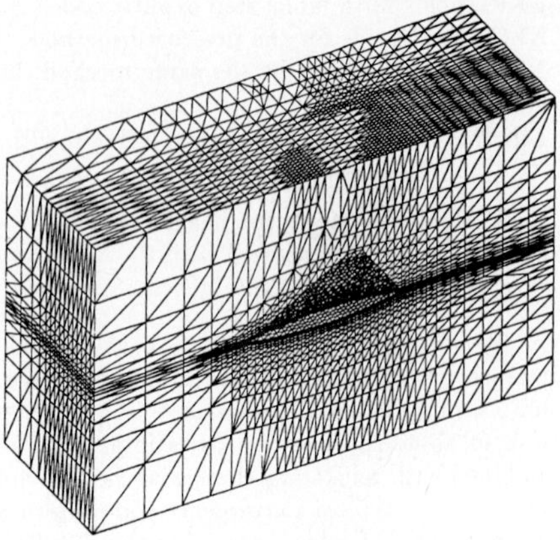

22160 nodes

Convergence over the two successive meshes

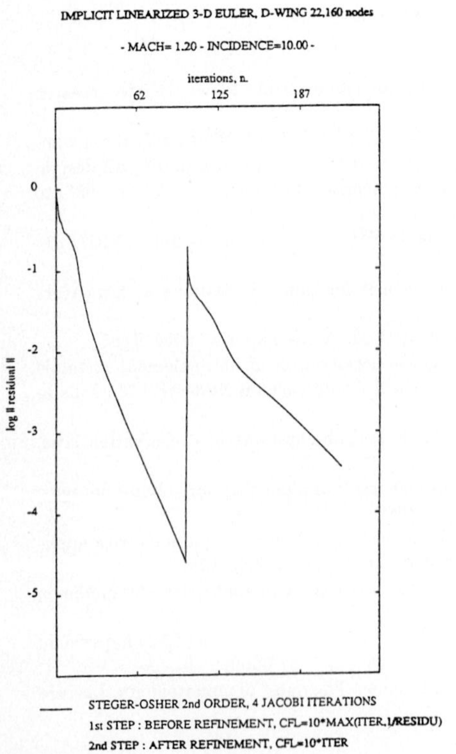

Cp-Lines

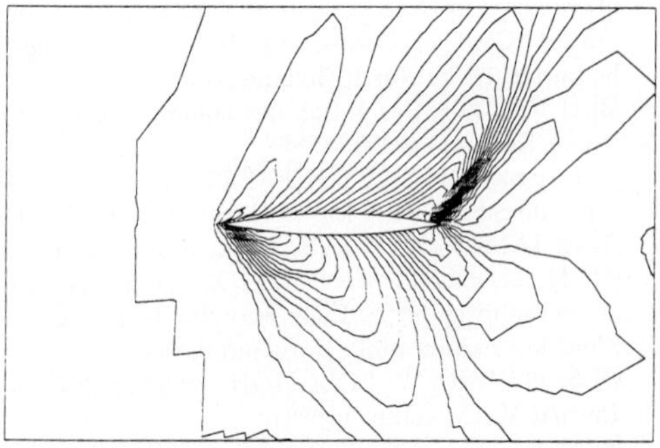

NACA0012	EXPL.	NLGMRES	NLGMRES+1	LIN.IMPL.
355nodes M=.85 ,α=0	1.	.80	.60	.20
1360nodes M=.85 ,α=0	1.	.80	.60	.20
2500nodes M=.85 ,α=0	1.	.43	.25	.19
1360nodes M=2. ,α=10	1.	.88	.60	.20
2500nodes M=2. ,α=10	1.	.80	.54	.20

CPU requirements

	EXPL.	NLGMRES	NLGMRES+1	LIN.IMPL.
NACA12(355nodes)	0	.10	.25	1.
NACA12(1360nodes)	0	.10	.25	1.
NACA12(2500nodes)	0	.10	.25	1.

storage requirements

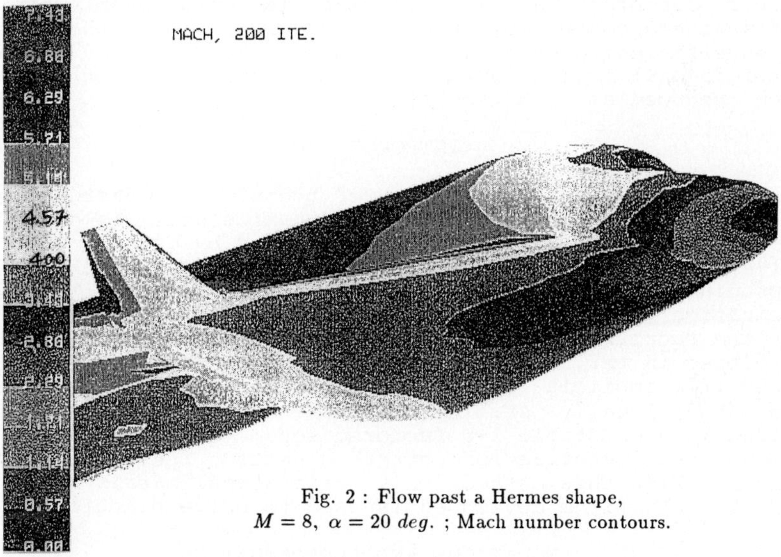

Fig. 2 : Flow past a Hermes shape,
$M = 8$, $\alpha = 20$ *deg.* ; Mach number contours.

Typical convergence of one resolution of the linear system (CFL = 20) Mach = 8

0 GMRES + Diagonal preconditioning
△ GMRES without preconditioning
+ Gauss-Seidel iteration

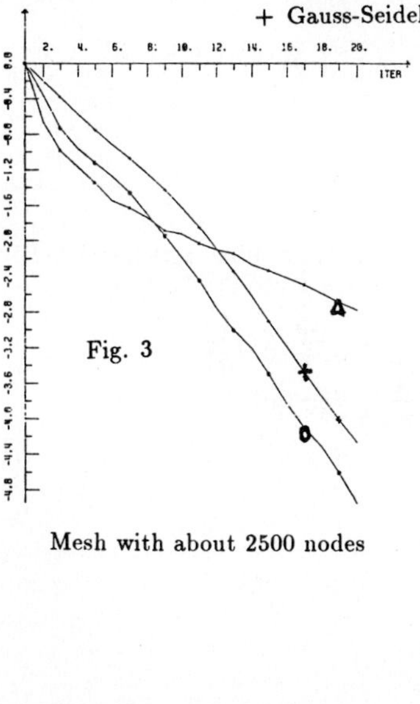

Fig. 3

Mesh with about 2500 nodes

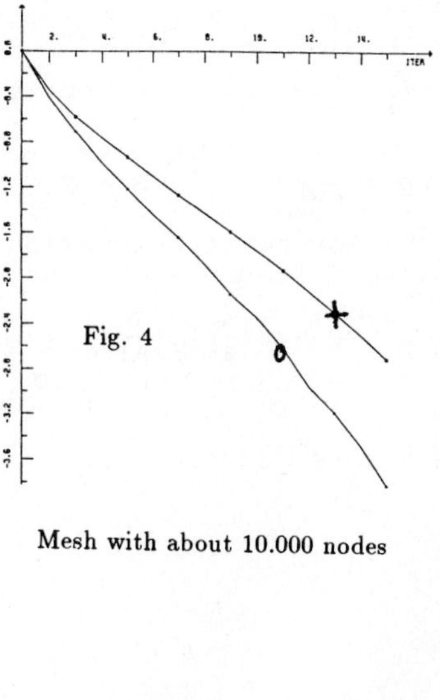

Fig. 4

Mesh with about 10.000 nodes

AN EFFICIENT NESTED ITERATIVE METHOD FOR SOLVING THE AERODYNAMIC EQUATIONS

Fu Dexun and Ma Yanwen
Beijing Institute of Aerodynamics

ABSTRACT

Algorithm for solving the difference equations is considered. The difference eq. obtained with approximate factorization for 3-D stable implicit schemes may become unstable or conditionally stable. If proper Jacobian matrix splitting is used stable approximately factored scheme can be obtained. In order to remove the factorization error en efficient nested iterative method is suggested. In this method only tridiagonal matrix inversions are needed. Amount of computation is much reduced at each time step.

INTRODUCTION

Although development of computer technique is fast how to improve efficiency of solution method in aerodynamics still is one of the most important problems. Efficiency is improved by using implicit schemes (ref.1-4). For 2-D or 3-D implicit schemes complex block pentadiagonal or septadiagonal matrix operators have to be solved. The approximate factorization is one of the simplest methods (ref.3,4). In order to remove the factorization error Line Gauss-Seidel iterative procedure is developed in ref.5 and 6. It was known that for 3-D case the scheme obtained from implicit schemes with approximate factorization may become conditionally stable or unstable. If the Jacobian matrix splitting is used stable 3-D factored scheme can be obtained. In order to remove the factorization error a nested iterative algorithm is suggested. With this method only tridiagonal matrix inversions are needed. The algorithm developed is used to solve practical problems.

EQUATIONS AND APPROXIMATION

Consider the following 2-D NS eq. in vector form

$$\frac{\partial}{\partial t}U + \frac{\partial}{\partial x}f_1 + \frac{\partial}{\partial y}f_2 = F \tag{1}$$

where

$$f_1 = [\rho u,\ \rho u^2 + p,\ \rho uv,\ u(E+p)]^T, \quad U = [\rho,\ \rho u,\ \rho v,\ E]^T$$

$$f_2 = [\rho v,\ \rho uv,\ \rho v^2 + p,\ v(E+p)]^T$$

$$p = \frac{1}{\gamma M_\infty^2}\rho T, \qquad E = \rho\left(C_v T + \frac{u^2+v^2}{2}\right) \tag{2}$$

Using Taylor series expansion and eq. (1) for the inviscid flow the following difference expression can be obtained (ref.7).

$$U^{n+1} = U^n - \Delta t\left(\frac{\partial}{\partial x}f_1 + \frac{\partial}{\partial y}f_2\right) + \frac{\Delta t^2}{2}\Big[\alpha \frac{\partial^2}{\partial t^2}U - \beta\left(\frac{\partial}{\partial x}A_1 + \frac{\partial}{\partial y}A_2\right)\frac{\partial}{\partial t}U + \gamma\left(\frac{\partial}{\partial x}A_1 + \frac{\partial}{\partial y}A_2\right)\left(\frac{\partial}{\partial x}f_1 + \frac{\partial}{\partial y}f_2\right) \tag{3}$$

where A_K are Jacobian matrices

$$A_k = D(f_k)/D(U)$$

and α, β and γ are free parameters. If the first derivatives are approximated with second order accuracy, the higher derivatives are approximated with first order accuracy and $\alpha+\beta+\gamma=1$. the obtained schemes have second order accuracy. A general expression for a class of difference schemes is given in ref.7. The simplest second order scheme from them is

$$\Delta U_{i,j} = -\frac{\Delta t}{\Delta x}\delta^{o}_{x} f_1 + \frac{\Delta t}{\Delta y}\delta^{o}_{y} f_2 \tag{4}$$

$$\left[I + \frac{\beta}{2}(\alpha_1 \delta^{-}_{x} A^{+}_{1} + \alpha_1 \delta^{+}_{x} A^{-}_{1} + \alpha_2 \delta^{-}_{y} A^{+}_{2} + \alpha_2 \delta^{+}_{y} A^{-}_{2} \right) \delta_t U_{i,j} = \Delta U^{n}_{i,j} \tag{5}$$

$$U^{n+1} = U^{n} + \delta_t U \tag{6}$$

where I is a unit matrix, δ^{o}, δ^{+} and δ^{-} is central, forward and backward difference operators respectively, and

$$\alpha_1 = \frac{\Delta t}{\Delta x}, \quad \alpha_2 = \frac{\Delta t}{\Delta y}, \quad A^{\pm}_{k} = S^{-1}_{k} \Lambda^{\pm}_{k} S_k$$

S_k is a matrix consisted of left row eigenvectors of matrix A_k, and $\Lambda^{\pm}_{k}$ is a diagonal matrix with elements $\lambda^{\pm}_{j}(A_k)$

$$\lambda^{+}_{j}(A_k) + \lambda^{-}_{j}(A_k) = \lambda_j(A_k), \quad \lambda^{+}_{j}(A_k) \geq 0, \quad \lambda^{-}_{j}(A_k) \leq 0,$$

λ_j are eigenvalues of matrix A_k. Different splitting technique can be used. The viscous part of the eq. is approximated in standard way with operator addition (ref.8).

SOLUTION METHOD FOR DIFFERENCE EQUATIONS

Approximate Factorization

The factored scheme is one of the simplest method for solving the aerodynamics eq.. For solving 3-D problem with factorization care must be taken. Consider the following model equation

$$\frac{\partial}{\partial t}u + a_1\frac{\partial}{\partial x}u + a_2\frac{\partial}{\partial y}u + a_3\frac{\partial}{\partial z}u = 0, \qquad a_k = \text{const.} \geq 0 \tag{7}$$

Positivity $a_k \geq 0$ can be provided by coordinate transformation. The following three kinds of approximations are considered

$$\left[1+\frac{\beta}{2}(\alpha_1\delta^{o}_{x}+\alpha_2\delta^{o}_{y}+\alpha_3\delta^{o}_{z})\right]\delta_t u_{i,j,k} = -(\alpha_1\delta^{o}_{x}+\alpha_2\delta^{o}_{y}+\alpha_3\delta^{o}_{z})u^{n}_{i,j,k} = \Delta u^{n}_{i,j,k} \tag{8}$$

$$(1+\frac{\beta}{2}\alpha_1\delta^{o}_{x})(1+\frac{\beta}{2}\alpha_2\delta^{o}_{y})(1+\frac{\beta}{2}\alpha_3\delta^{o}_{z})\delta_t u_{i,j,k} = \Delta u^{n}_{i,j,k} \tag{9}$$

$$(1+\frac{\beta}{2}\alpha^{+}_{1}\delta^{+}_{x})(1+\frac{\beta}{2}\alpha^{-}_{1}\delta^{+}_{x})(1+\frac{\beta}{2}\alpha^{+}_{2}\delta^{-}_{y})(1+\frac{\beta}{2}\alpha^{-}_{2}\delta^{+}_{y})\cdot \tag{10}$$

$$(1+\frac{\beta}{2}\alpha^{+}_{3}\delta^{-}_{z})(1+\frac{\beta}{2}\alpha^{-}_{3}\delta^{+}_{z})\,\delta_t u_{i,j,k} = \Delta u^{n}_{i,j,k}$$

where

$$\alpha_k = a_k \Delta t/\Delta x_k \qquad \alpha^{\pm}_{k} = a^{\pm}_{k}\Delta t/\Delta x_k \qquad x_1 = x, \quad x_2 = y, \quad x_3 = z$$

$$a^{+}_{k} + a^{-}_{k} = a_k, \qquad a^{+}_{k} \geq 0, \qquad a^{-}_{k} \leq 0 \tag{11}$$

The scheme (8) is stable if $\beta \geq 1$. The scheme (9) is conditionally stable for $\beta=2$ and is unstable for $\beta=1$. It can be shown the scheme (10) is stable if $\beta \geq 1$ and "contrast" between a^{+}_{k} and a^{-}_{k} is big enough. Big contrast means increasing $|a^{\pm}_{k}|$ under condition (11)

Nested Iterative Method

In order to avoid the factorization error the Line Gouss-Seidel iteration is used in ref. 5 and 6. The difference eq. (5) can be written as follows

$$B\delta_t U_{i,j+1} + A\delta_t U_{i,j} + C\delta_t U_{i,j-1} + D\delta_t U_{i+1,j} + E\delta_t U_{i-1,j} = \Delta U_{i,j} \tag{12}$$

All coefficients are known in process of computation. For convinence we deleted the superscripts on $\delta_t U$ and subscripts on the matrix coefficients. The eq. (12) can be solved by Line Gouss-Seidel iteration with alternating sweeps in the backward and forward x direction. For p=1,3,...

Backward sweep

$$B\delta_t U^{p}_{i,j+1} + A\delta_t U^{p}_{i,j} + C\delta_t U^{p}_{i,j-1} + D\delta_t U^{p}_{i+1,j} + E\delta_t U^{p-1}_{i-1,j} = \Delta U_{i,j} \tag{13}$$

Forward sweep

$$B\delta_t U^{p+1}_{i,j+1} + A\delta_t U^{p+1}_{i,j} + C\delta_t U^{p+1}_{i,j-1} + D\delta_t U^{p}_{i+1,j} + E\delta_t U^{p+1}_{i-1,j} = \Delta U_{i,j} \tag{14}$$

(13-14) are called outer iteration for our calculation. Now the quastion is how to solve the following algebraic eq. with fixed I in both backward and forward sweeps

$$B\delta_t U_{i,j+1} + A\delta_t U_{i,j} + C\delta_t U_{i,j-1} = \Delta U_{i,j} - D\delta_t U_{i+1,j} - E\delta_t U_{i-1,j} = \Delta\overline{U}_{i,j} \tag{15}$$

An iterative procedure is suggested to solve the equation (15). Each of the matrices B, A and C is split as follows

$$B = B_0 + B_1, \qquad A = A_0 + A_1, \qquad C = C_0 + C_1.$$

The iteration procedure suggested is as follows

$$B_0\delta_t U^{q+1}_{i,j+1} + A_0\delta_t U^{q+1}_{i,j} + C_0\delta_t U^{q+1}_{i,j-1} = \Delta\overline{U}_{i,j} - B_1\delta_t U^{q}_{i,j+1} - A_1\delta_t U^{q}_{i,j} - C_1\delta_t U^{q}_{i,j-1} \tag{16}$$

We try to split the matrices B, C and A so that the eq. (16) can be solved easily. According to the splitting technique No.3 in ref.7 the matrices A_0, B_0 and C_0 are defined as

$$B_0 = b_0 I, \qquad A_0 = a_0 I, \qquad C_0 = c_0 I$$

a_0, b_0 and c_0 are scalar functions. The final iterative procedure for solving the equation (16) is as follows

$$b_0\delta_t U^{q+1}_{i,j+1} + a_0\delta_t U^{q+1}_{i,j} + c_0\delta_t U^{q+1}_{i,j-1} = d^{q} \tag{17}$$

where

$$b_0 = \frac{\beta}{2}\alpha_2 v^{-}_{i,j+1}, \quad a_0 = 1 + \frac{\beta}{2}\left[\alpha_1(|u|+c) + \alpha_2(|v|+c)\right], \quad c = -\frac{\beta}{2}\alpha_2 v^{+}_{i,j-1}$$

$$d = \frac{\beta}{2}\alpha_2\left\{\left[cS_2^{-1}I_3S_2\delta_t U^{q}\right]_{i,j+1} + \left[cS_2^{-1}I_0S_2\delta_t U^{q}\right]_{i,j-1}\right\} + \frac{\beta}{2}\alpha_1\left\{c\left[I - S_1^{-1}I_3S_1 - S_1^{-1}I_4S_1\right]\cdot \delta_t U^{q}\right\}_{i,j} + \frac{\beta}{2}\alpha_2\left\{c\left[I - S_2^{-1}I_3S_2 - S_2^{-1}I_4S_2\right]\delta_t U^{q}\right\}_{i,j}$$

$$S_\ell^{-1}I_kS_\ell\delta_t U = \beta_{\ell,k}\left[(S_\ell^{-1})_{1,k}, (S_\ell^{-1})_{2,k}, (S_\ell^{-1})_{3,k}, (S_\ell^{-1})_{4,k}\right]^T$$

$$\beta_{\ell,k} = \sum_i (S_\ell)_{k,i}(\delta_t U)_i$$

where I is a diagonal matrix with all elements equal to zero except the k-th which is equal to one. (S) and (S) are the elements of

the matrices S and S respectively. (U) are the elements of the vector U. The parameter q is the iteration number. The coefficients of the eq. (17) are scalar functions. The vector d is known during the iteration. This process is called inner iteration. For the particular case the convergence of inner iteration can be proved easily. For the general case the convergence is proved by numerical experiment.

Comparison between the direct solution method (15) and the iterative method (17) is given in the following table

	direct solution method		iterative solution method
number	$25/6N^3+9.5N^2-N$		15N+4
of	414	(2-D problem N=4)	62
x & /	750	(3-D problem N=5)	79

The number N=4 for 2-D problem and N=5 for the 3-D problem. In the present paper the iteration number q=1 is used.

NUMERICAL EXPERIMENTS

A. Flow around sphre cone

Two cases are computed:

a). M_∞=7.97 , Re=7000000., σ=10.5

b). M_∞=13.41, Re=1515 , σ=7.5

σ is the half cone angle.Part of computed results is given in the Fig.1.

B. Flow around indented shape body

The parameters used in computation are :

M_∞=5.85, Re=1000000.

Part of computed results are given in the Fig.2.

REFERENCE

1. MacCormack,R.W., AIAA No.69-354(1969).
2. Beam,R.M. and Warming,R.F., J. of Computational Physics, Vol.22, pp 87-110, 1976.
3. Steger,J.L. and Warming,R.F. J. of Computational Physics. Vol.40, No.2(1981.
4. Ma Yanwen And Fu Dexun, Lecture Notes In Physics No.264(1986)
5. MacCormack,R.W., AIAA Paper No.85-032.
6. Chakravarthy,S.R. AIAA Paper No.84-0165(1984).
7. Ma Yanwen and Fu Dexun, AIAA Paper No.87-1123(1987).
8. Fu Dexun and Ma Yanwen, Lecture Notes In Physics Vol.218(1984)

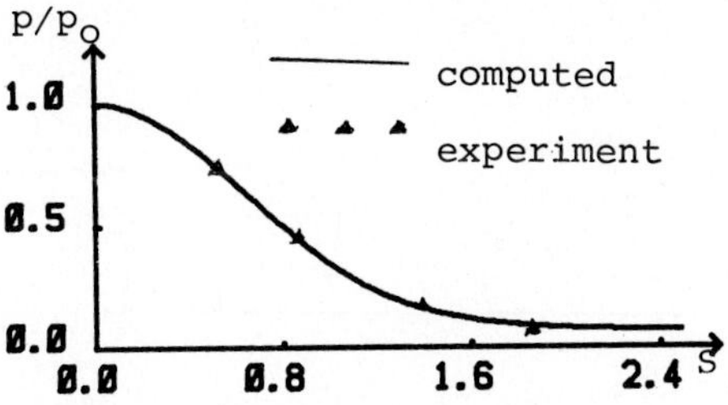

a) Pressure on the surface

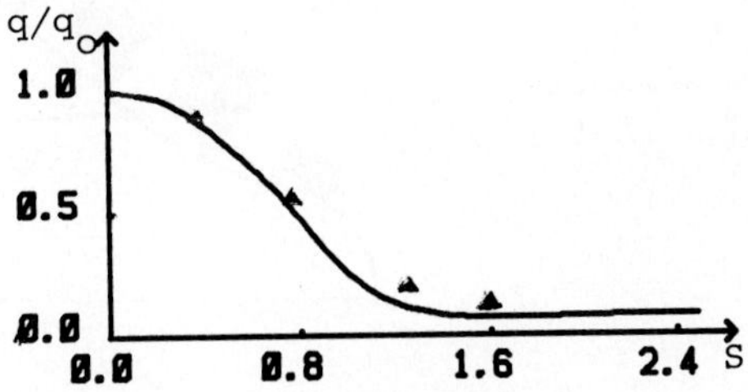

b) Heat transfer on the surface

Fig.1, The pressure and heat transfer on the sphre-cone surface.

a). The pressure on the indented shape body surface

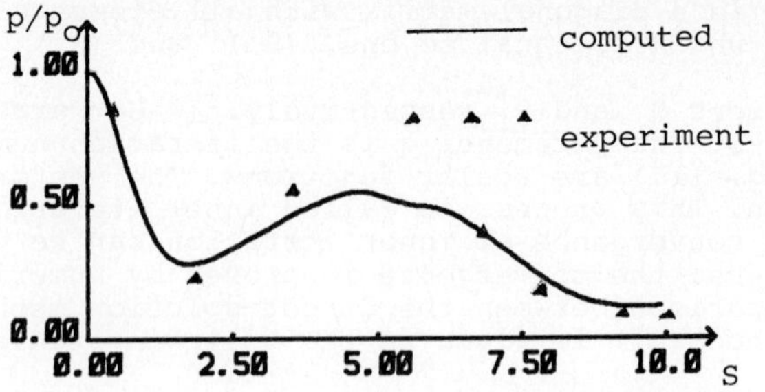

b). The skin friction on the indented shape body surface

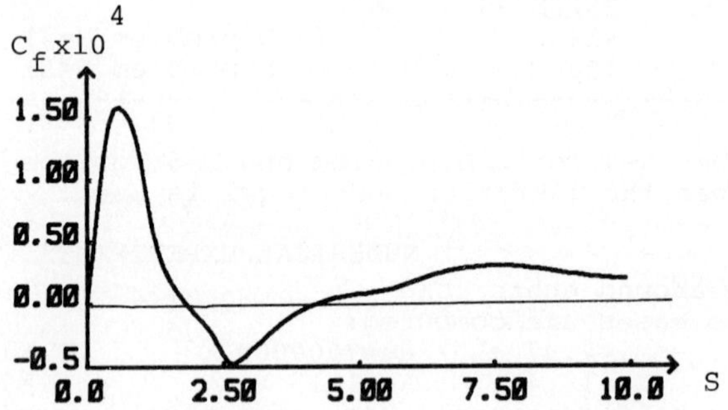

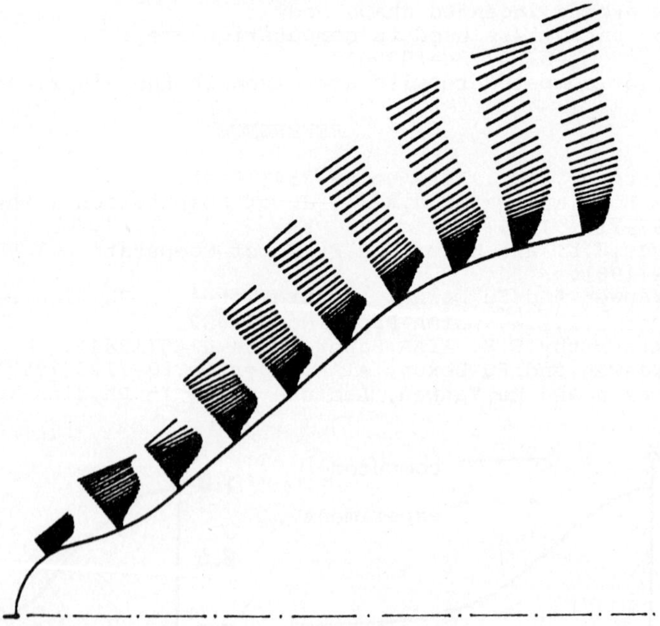

c). The velocity field

Fig.2, The pressure, the skin friction and the velocity field for the indented shape body.

A MULTIGRID METHOD FOR STEADY EULER EQUATIONS BASED ON POLYNOMIAL FLUX-DIFFERENCE SPLITTING

Erik Dick
Department of Machinery, State University of Ghent
Sint Pietersnieuwstraat 41, B-9000 GENT

Abstract

A flux-difference splitting for steady Euler equations based on the polynomial character of the flux vectors is introduced. The splitting is applied to vertex-centered finite volumes. A discrete set of equations is obtained which is both conservative and positive. The splitting is done in an algebraically exact way, so that shocks are represented as sharp discontinuities, without wiggles. Due to the positivity, the set of equations can be solved by collective relaxation methods. A full multigrid method based on symmetric successive relaxation, full weighting, bilinear interpolation and W-cycle is presented. Typical full multigrid efficiency is achieved for the GAMM transonic bump test case since after the nested iteration and one cycle, the solution cannot be distinguished anymore from the fully converged solution.

Flux-difference splitting with respect to primitive variables

Steady Euler equations, in two dimensions, take the form :

$$\partial f/\partial x + \partial g/\partial y = 0 \qquad (1)$$

with flux-vectors :

$$f^T = \{\rho u,\ \rho uu+p,\ \rho uv,\ \rho Hu\} \qquad g^T = \{\rho v,\ \rho uv,\ \rho vv+p,\ \rho Hv\}$$

where $H = \gamma p/(\gamma-1)\rho + \frac{1}{2}u^2 + \frac{1}{2}v^2$ is total enthalpy.

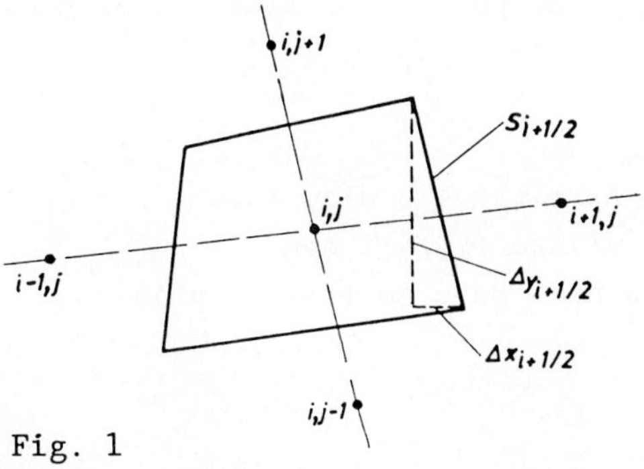

Fig. 1

Figure 1 shows a control volume with the node (i,j) inside it. Also the nodes inside the adjacent elements are indicated.

According to the flux-difference principle, the flux through the surface $S_{i+\frac{1}{2}}$ can be expressed by :

$$F_{i+\frac{1}{2}} = \frac{1}{2}\,\{F_i + F_{i+1} - |\Delta F_{i,i+1}|\} \qquad (2)$$

where F_i and F_{i+1} denote fluxes computed with the values of the variables at the node (i,j) and (i+1,j) respectively. $\Delta F_{i,i+1}$ denotes the flux-difference. In the above, for brevity, the non-varying index is omitted. To give meaning to an expression in the form (2), it is imperative that the flux-difference can be written as :

$$\Delta F_{i,i+1} = F_{i+1} - F_i = \hat{A}_{i,i+1}(\xi_{i+1} - \xi_i) = (\hat{A}^+_{i,i+1} + \hat{A}^-_{i,i+1})\Delta\xi_{i,i+1} \quad (3)$$

where $\hat{A}_{i,i+1}$ is a Jacobian matrix with real eigenvalues and a complete set of eigenvectors, so that it can be split into a positive part, i.e. having non-negative eigenvalues, and a negative part, i.e. having non-positive eigenvalues and where ξ is a vector of dependent variables. The absolute value of the flux-difference is then defined by :

$$|\Delta F_{i,i+1}| = (\hat{A}^+_{i,i+1} - \hat{A}^-_{i,i+1})\Delta\xi_{i,i+1} \quad (4)$$

Well known splitters realizing (3) are the splitters of Roe and of Osher. The splitting of Roe [1] is based on the quadratic character of the flux-vectors with respect to the variables $\sqrt{\rho}$, $\sqrt{\rho}\,u$, $\sqrt{\rho}\,v$, $\sqrt{\rho}\,H$. The splitting of Osher [2] is based on characteristic paths and results in a splitting with respect to the variables $\sqrt{\gamma p/\rho}$, u, v, $\ln(p/\rho^\gamma)$. A very simple splitting based on the polynomial character of the flux-vectors with respect to the primitive variables ρ, u, v, p was proposed by Lombard et al. [3]. It was shown by Hemker and Spekreijse [4] that the Osher scheme can be used directly on steady Euler equations, to form a multigrid method. In this paper, a similar approach is used, but based on the simpler flux-difference splitting of Lombard et al.. In contrast to the original approach of Lombard et al., which used an approximate splitting, the splitting is done here in an algebraically exact way. This is necessary to treat the steady equations directly, avoiding the time marching necessary in the original approach. Also, due to the algebraically exact manipulation, boundary conditions can be introduced in a rigourous way. A detailed description of the splitting technique was given by the author in [5]. In this paper, the principles of the method are summarized and the multigrid formulation is treated.

Since the components of the flux-vectors form polynomials with respect to the primitive variables ρ, u, v and p, components of flux-differences can be written as follows :

$$\Delta\rho u = \bar{u}\,\Delta\rho + \bar{\rho}\,\Delta u$$
$$\Delta(\rho uu+p) = \overline{\rho u}\,\Delta u + \bar{u}\,\Delta\rho u + \Delta p = \bar{u}^2\,\Delta\rho + (\overline{\rho u} + \bar{\rho}\,\bar{u})\Delta u + \Delta p$$
$$\Delta\rho H u = \overline{\rho u}(\tfrac{1}{2}\,\Delta u^2 + \tfrac{1}{2}\,\Delta v^2) + \tfrac{1}{2}\,(\overline{u^2} + \overline{v^2})\Delta\rho u + \gamma/\gamma\text{-}1\,\Delta p u$$
$$= \tfrac{1}{2}\,(\overline{u^2} + \overline{v^2})\bar{u}\,\Delta\rho + \tfrac{1}{2}\,(\overline{u^2} + \overline{v^2})\bar{\rho}\,\Delta u + \overline{\rho u}\,\bar{u}\,\Delta u + \gamma/\gamma\text{-}1\,\bar{p}\,\Delta u$$
$$+ \overline{\rho u}\,\bar{v}\,\Delta v + \gamma/\gamma\text{-}1\,\bar{u}\,\Delta p$$

where the bar denotes mean value.

This leads to : $\Delta f = A_1 \Delta\xi$ and $\Delta g = A_2 \Delta\xi$

As a result, the flux-difference across the surface $S_{i+\frac{1}{2}}$ in figure 1 has the form :

$$\Delta F_{i,i+1} = \Delta s_{i+\frac{1}{2}} (\alpha_1 A_1 + \alpha_2 A_2)\Delta\xi_{i,i+1} = \Delta s_{i+\frac{1}{2}} A \, \Delta\xi_{i,i+1}$$

where A is expressed in function of mean values of combinations of primitive variables.

It is easy to verify that the matrix A has real eigenvalues and a complete set of eigenvectors [5].

Further, the procedure of Steger and Warming [6], is used, i.e. the splitting is based on the splitting of the eigenvalue matrix. This leads to :

$$|\Delta F_{i,i+1}| = \Delta s_{i+\frac{1}{2}} (A^{+}_{i,i+1} - A^{-}_{i,i+1})\Delta\xi_{i,i+1} \qquad (5)$$

With (5), the flux $F_{i+\frac{1}{2}}$ given by (2) can be written in either of the two following ways, which are completely equivalent :

$$F_{i+\frac{1}{2}} = F_i + \frac{1}{2} \Delta F_{i,i+1} - \frac{1}{2} |\Delta F_{i,i+1}| = F_i + \Delta s_{i+\frac{1}{2}} A^{-}_{i,i+1} \Delta\xi_{i,i+1} \qquad (6)$$

$$F_{i+\frac{1}{2}} = F_{i+1} - \frac{1}{2} \Delta F_{i,i+1} - \frac{1}{2} |\Delta F_{i,i+1}| = F_{i+1} - \Delta s_{i+\frac{1}{2}} A^{+}_{i,i+1} \Delta\xi_{i,i+1} \qquad (7)$$

The fluxes on the other surfaces of the control volume $S_{i-\frac{1}{2}}$, $S_{j+\frac{1}{2}}$, $S_{j-\frac{1}{2}}$, can be treated in a similar way as the flux on the surface $S_{i+\frac{1}{2}}$. With (6) and (7), the flux balance on the control volume of figure 1 can be brought into the following form :

$$\Delta s_{i+\frac{1}{2}} A^{-}_{i,i+1} [\xi_{i+1} - \xi_i] + \Delta s_{i-\frac{1}{2}} A^{+}_{i,i-1} [\xi_i - \xi_{i-1}]$$

$$+ \Delta s_{j+\frac{1}{2}} A^{-}_{j,j+1} [\xi_{j+1} - \xi_j] + \Delta s_{j-\frac{1}{2}} A^{+}_{j,j-1} [\xi_j - \xi_{j-1}] = 0 \qquad (8)$$

The set of equations defined by (8) is a so-called positive set. It can be solved by a collective variant of any scalar relaxation method. By a collective variant it is meant that in each node, all components of the vector of dependent variables ξ are relaxed simultaneously. For the treatment of the boundary conditions, the reader is referred to [5].

Multigrid formulation

Figure 2 shows the solution in the form of iso-Machlines for the well known GAMM-test case [7] for transonic flows. The grid shown has 12 x 4 elements. This grid is the coarsest of a series of four. The finest grid on which the solution is shown has 96 x 32 elements. Vertex-centered finite volumes, as indicated in figure 1, were used.

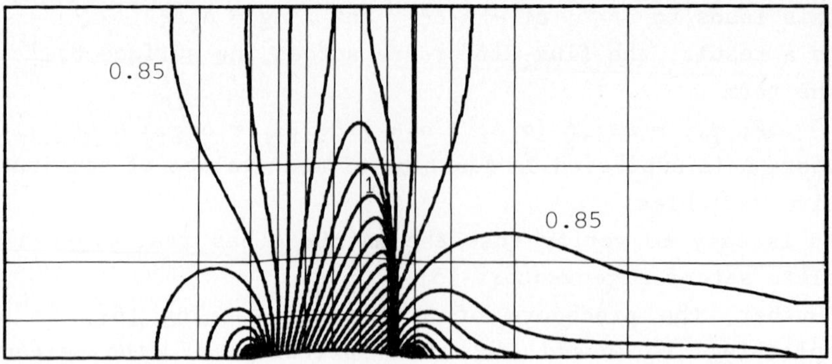

Fig. 2

At inflow, the specification of a horizontal flow direction was used as boundary condition. At outflow, the Mach number was fixed at 0.85. The obtained solution coincides almost with the solution obtained from the most reliable time-marching methods reported in [7]. However, unlike most time-marching solutions, due to the guaranteed positivity everywhere, the solution has no wiggles in the shock region.

Figure 3 shows the geometry of the multigrid method. The method has a full multigrid formulation with a full approximation scheme on the non-linear equations (8). The cycle is of W-form. Also the nested iteration, used as starting cycle, has W-form. The relaxation algorithm is symmetric successive underrelaxation with relaxation factor 0.9. The order of relaxation is lexicographic, i.e. going from the lower left point to the upper right point first varying the row index and then going from the upper right point to the lower left point in the reverse order. In relaxing the set of equations (8), the coefficients are formed with the latest available information. This means that in the first sweep $A^{\dagger}_{i,i-1}$ is evaluated with the function values in node (i,j) on the old level, but with the function values in node (i-1,j) on the new level. After determination of the new values in node (i,j), no updates of coefficients and no extra iterations are done. This means that the set of equations (8) is treated as a quasi-linear set. As restriction operator, full weighting is used within the flow field while injection is used at the boundaries. The prolongation operator is bilinear interpolation. Restriction is not employed for function values. As an approximation for a restricted function value, the last available value on the same grid is taken. The calculation starts from a uniform flow with Mach number 0.85 on the coarsest grid. In figure 3, the operation count is indicated. A relaxation on the current grid is taken as one local work unit. So, the symmetric relaxation is seen as two work units. A residual evaluation plus the associated grid transfer is also taken as one work unit. Hence, the 4 in figure 3, in going down, stands for the construction of the right hand side in the FAS-formulation, two relaxations and one residual evaluation. With this way of evaluating the work, the cost of the cycle is 8.6875 work units on the finest level.

The cost of the nested iteration is about 4.39 work units.

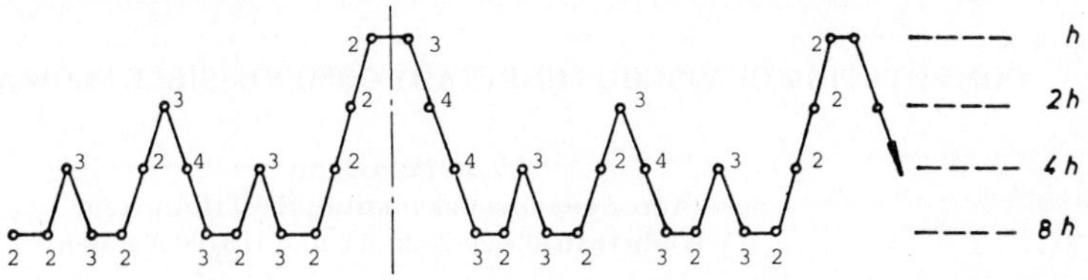

Fig. 3

Figure 4 shows the convergence behaviour of the single grid and the multigrid formulation. The residual shown is the maximum residual of all equations, after normalizing these equations, i.e. bringing the coefficient of ρ, u, v and p on 1 in the mass-, momentum-x-, momentum-y- and energy equation respectively, and dividing the variables by their value in the initial uniform flow.

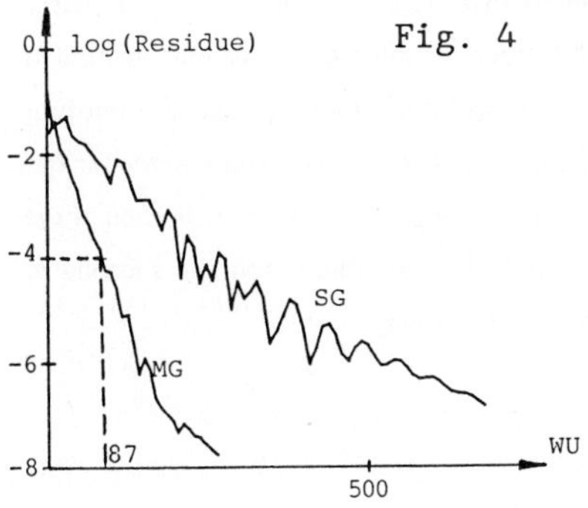

Fig. 4

A maximum residual of 10^{-4} is reached after approximately 87 work units. The convergence factor of the multigrid method, i.e. the residual reduction per work unit is about 0.915. This probably can be considered as being optimal. This is seen by the pressure distribution on the bottom obtained after the nested iteration and one cycle. Up to plotting accuracy this pressure distribution coincides with the distribution obtained after full convergence.

Acknowledgement

The research reported in this paper was granted by the Belgian National Science Foundation (N.F.W.O.).

References

[1] P.L. ROE, J. Comp. Phys. 43 (1981) 357-372.
[2] S. OSHER, in : O. Axelsson et al. (Eds.) Mathematical and numerical approaches to asymptotic problems in analysis, North Holland, 1981, 179-205.
[3] C.K. LOMBARD, J. OLIGER and J.Y. YANG, AIAA paper 82-0976 (1982).
[4] P.W. HEMKER and S.P. SPEKREIJSE, in : D. Braess et al. (Eds.), Notes on numerical fluid dynamics 11, Vieweg, 1985, 33-44.
[5] E. DICK, A flux-difference splitting method for steady Euler equations, J. Comp. Phys. (1988), to appear.
[6] J.L. STEGER and R.F. WARMING, J. Comp. Phys. 40 (1981) 263-293.
[7] A. RIZZI and H. VIVIAND, (Eds.), Notes on numerical fluid dynamics 3, Vieweg, 1981.

COMPUTATION OF VISCOUS UNSTEADY COMPRESSIBLE FLOW ABOUT AIRFOILS

K. Dortmann
Aerodynamisches Institut, RWTH Aachen,
Wüllnerstraße zwischen 5 u. 7, D 5100 Aachen

Introduction

This investigation deals with the computation of unsteady subsonic flows around an airfoil with time- dependent inflow conditions, and with the study of vortex formation in the flow near the trailing edge. Three different time scales have to be considered, the one characteristic for the inflow conditions, the gasdynamic scale, which is given by the speed of sound and the chord length, and a wake scale, for example the periodicity of the vortex street. To resolve the smallest scale, the numerical time step had to be sufficiently small, corresponding to a Courant number of order one. For this reason an explicit five-step Runge-Kutta time stepping scheme was used, which is second-order accurate in time for nonlinear problems. The flow considered is the one around a NACA 4412, as studied in experimental investigations in [1]. Improved resolution, especially near the trailing edge, was achieved using a cell-vertex discretization for the inviscid forces [2] and a node-centred scheme for the viscous forces. Furtheron special care was taken with regard to numerical damping, which can, if chosen too strong, impair spatial and temporal accuracy. In the following the governing equations and the method of solution will be described together with some results for unsteady flows about airfoils.

Governing Equations and Method of Solution

The flow is described by the Navier-Stokes equations for two-dimensional laminar flow, formulated in a curvilinear time-dependent coordinate system $\xi = \xi(x,y,t)\ ;\ \eta = \eta(x,y,t)$:

$$\frac{\partial}{\partial t}\hat{Q} + \frac{\partial}{\partial \xi}\hat{E} + \frac{\partial}{\partial \eta}\hat{F} = \frac{1}{Re}\left[\frac{\partial}{\partial \xi}\hat{E}_v + \frac{\partial}{\partial \eta}\hat{F}_v\right]$$

Therein Q is the vector of the conservative variables and E, F, E_v, F_v are vectors of the convective and diffusive fluxes. Details may be found in [3]. For the time discretization a five-step Runge-Kutta method [4], [5] was used. It is second-order accurate in time for non-linear problems and can be represented by the following sequence:

$$Q^{(o)} = Q^{(n)}$$
$$Q^{(k)} = Q^{(o)} - \alpha_k \Delta t\, Res\,(Q^{(k-1)}) + D(Q^{(o)})$$
$$Q^{(n+1)} = Q^{(k)}$$

For the spatial discretization a combination of two different central approximations, a node- centered and a cell vertex scheme, were used. Both result in a formal second-order accuracy, but the cell vertex scheme was found to be more precise near the trailing edge for C-type meshes with overlapping wake cut. To suppress the odd-even decoupling blended second and fourth order damping terms were added to the steady state operator [7].

Boundary Conditions and Mesh Generation

The flow field around an airfoil is mapped on to a C-type grid, shown in Fig. 1. The typical distance between farfield boundaries and airfoil is about ten chord lengths. The boundary conditions for the outer boundaries are approximated by using the local characteristcs for the given steady-state part of pressure and velocity at infinity. The unsteady part of the inflow is introduced in the computation moving the airfoil together with the grid. The nearly orthogonal mesh is numerically generated by solving a Laplace equation corresponding to an incompressible potential flow. The wake cut is formed by the contour streamline of a flat plate at an angle of attack.

Results

The unsteady flow was calculated for a NACA 4412 airfoil at zero angle of attack for Reynolds numbers between $1 \cdot 10^4$ and $2 \cdot 10^4$. Experiments have shown that even for this angle of attack the flow is separating. This flow is therefore well suited for studying the unsteady behaviour of the solution. For the investigation of accelerated inflow the Mach number was increased sinusoidally from $Ma_\infty = 0.2$,to $Ma_\infty = 0.4$, resulting in a change of Reynolds number from $1 \cdot 10^4$ to $1.9 \cdot 10^4$. Fig. 2 shows the computed lift coefficient as a function of time. The periodic variation of the lift coefficient before ($t < 0$) and after the acceleration ($t > 0,5$) indicates the existance of a Kármán vortex street. During the acceleration the separation point moves downstream which results in a rapid increase of the lift. The reason of the decrease of the lift can be found in the reestablishment of the trailing edge separation at the end of the acceleration. A comparison of computed and experimentally determined streaklines is given in Figs. 7a-c. Good agreement was achieved with respect to the movement of the separation point and the flow structures in Fig. 7c. The differences between the computed results in Fig. 7a and Fig. 7c can be explained by the mesh resolution which is more suitable for the flow with the higher Reynolds number. The above discussed results revealed a rather sensible response to slight changes in numerical damping and they could only be achieved after minimizing the coefficients of the fourth order damping terms. Fig. 3 shows the lift coefficient as a function of time for the same case but with damping coefficients increased by a factor of four. Although the amount of damping ist still within the range used in steady-state calculations the self-induced time-dependent behaviour is completely suppressed. In order to get more detailed information about this effect the flow over a NACA 0012 airfoil with $\alpha = 20°$, $Re_\infty = 20.000$ and $Ma_\infty = 0.3$ was computed varying the coefficients of the fourth order damping terms ε_4. The mesh which is shown in Fig. 4 has a resolution of 171 by 51 grid points. Fig. 5 shows the velocity vectors at one point of time. The characteristics of this flow are trailing edge and leading edge separation and a huge oscillating region of separated flow. Fig. 6 shows the lift coefficient as a function of time and damping. It can be seen, that the damping terms not only affect the range of the lift coefficient but also the modes of the solution resulting in different Strouhal numbers.

Conclusions

The Navier-Stokes equations were solved for calculations of unsteady viscous flows. For an accelerated flow the time history of the lift and corresponding flow patterns were presented for a flow about an airfoil . The results of the unsteady computations indicate a rather sensible reaction of the solution to spatial discretization and numerical damping. Further work is required to gain more insight into this problem.

References

[1] Schweitzer, B., Krause, E., Ehrhardt, G.: Experimentelle Untersuchung instationärer Tragflügelströmungen. Abhandlungen aus dem Aerodynamischen Institut, RWTH Aachen, Heft 27, 1985.

[2] Hall, M. G.: Cell-vertex multigrid schemes for solution of the Euler equations. Numerical Methods for Fluid Dynamics II. Edited by K. W. Morton and M. J. Baines, Clarendon Press, Oxford 1986.

[3] Pulliam, T. H.: Euler and thin-layer Navier-Stokes codes: ARC2D, ARC3D. Notes for Computational Fluid Dynamics User's Workshop, University of Tennessee, Space Institute Tullahoma, March 1984.

[4] Jameson, A., Schmidt, W., Turkel, E.: Numerical solution of the Euler equations by finite-volume methods using Runge-Kutta Time-Stepping Schemes. AIAA-81-1959.

[5] Radespiel, R., Kroll, N.: Progress in the development of an efficient finite-volume code for the three-dimensional Euler equations. DFVLR, April 1985.

[6] Rossow, C.: Comparison of cell centered and cell vertex finite-volume schemes. Held at the 7th GAMM Conference of Numerical Methods in Fluid Mechanics. Dauvain-La-Neuve, 1987.

[7] MacCormack, R. W., Baldwin, B. S.: A numercial method for solving the Navier-Stokes equations with applications to shock-boundary layer interactions. AIAA Paper 75-1, presented at AIAA 13th Aerospace Sciences Meeting, Pasadena, Calif., Jan. 1975

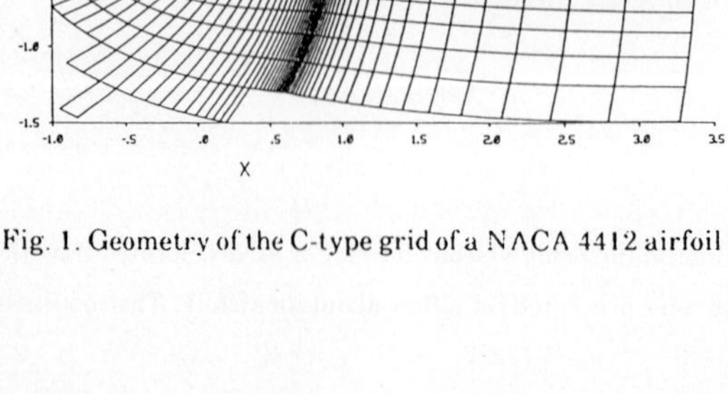

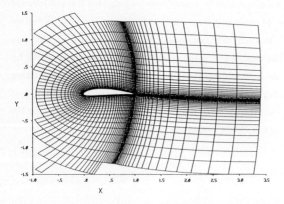

Fig. 1. Geometry of the C-type grid of a NACA 4412 airfoil (141 x 41 grid points).

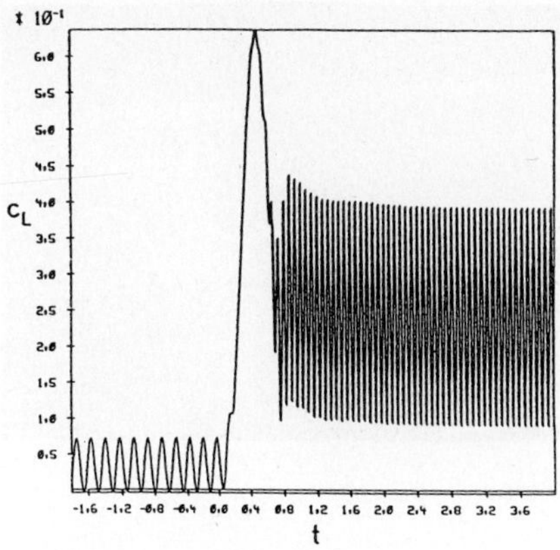

Fig. 2. History of lift coefficient
$t < 0$: $Re_\infty = 10000$, $Ma_\infty = 0{,}2$
$0 \le t \le 0{,}5$: acceleration,
$t > 0$: $Re_\infty = 19000$, $Ma_\infty = 0{,}4$

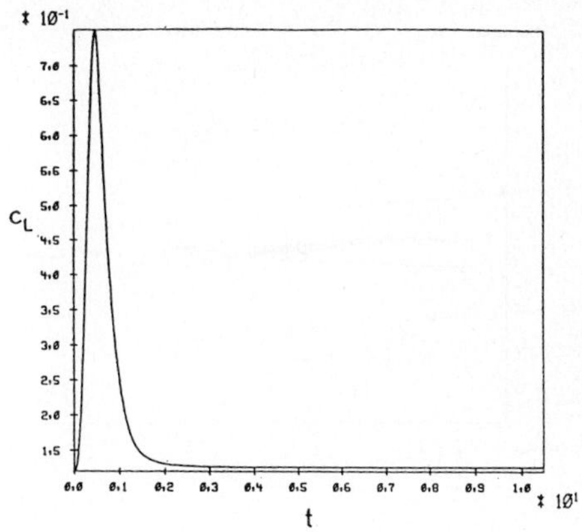

Fig. 3. History of lift coefficient computed with increased damping.

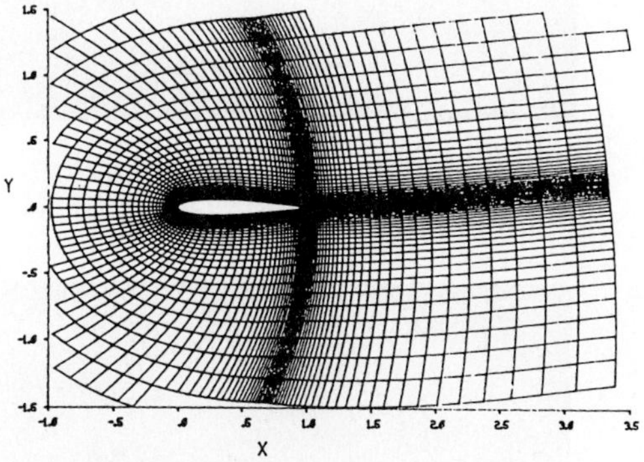

Fi.g 4. Geometry of the C-type grid of a NACA 0012 airfoil (171 x 51 grid points).

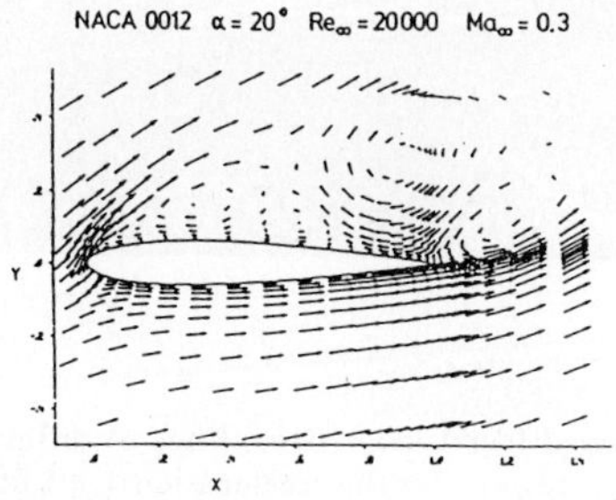

Fig. 5. Velocity vectors for one point of time.

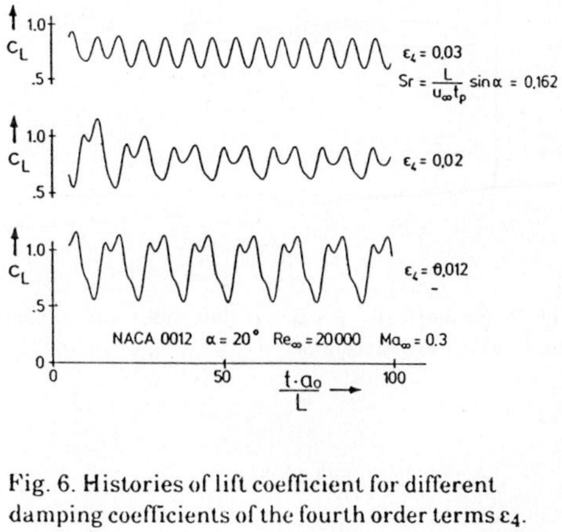

Fig. 6. Histories of lift coefficient for different damping coefficients of the fourth order terms ε_4.

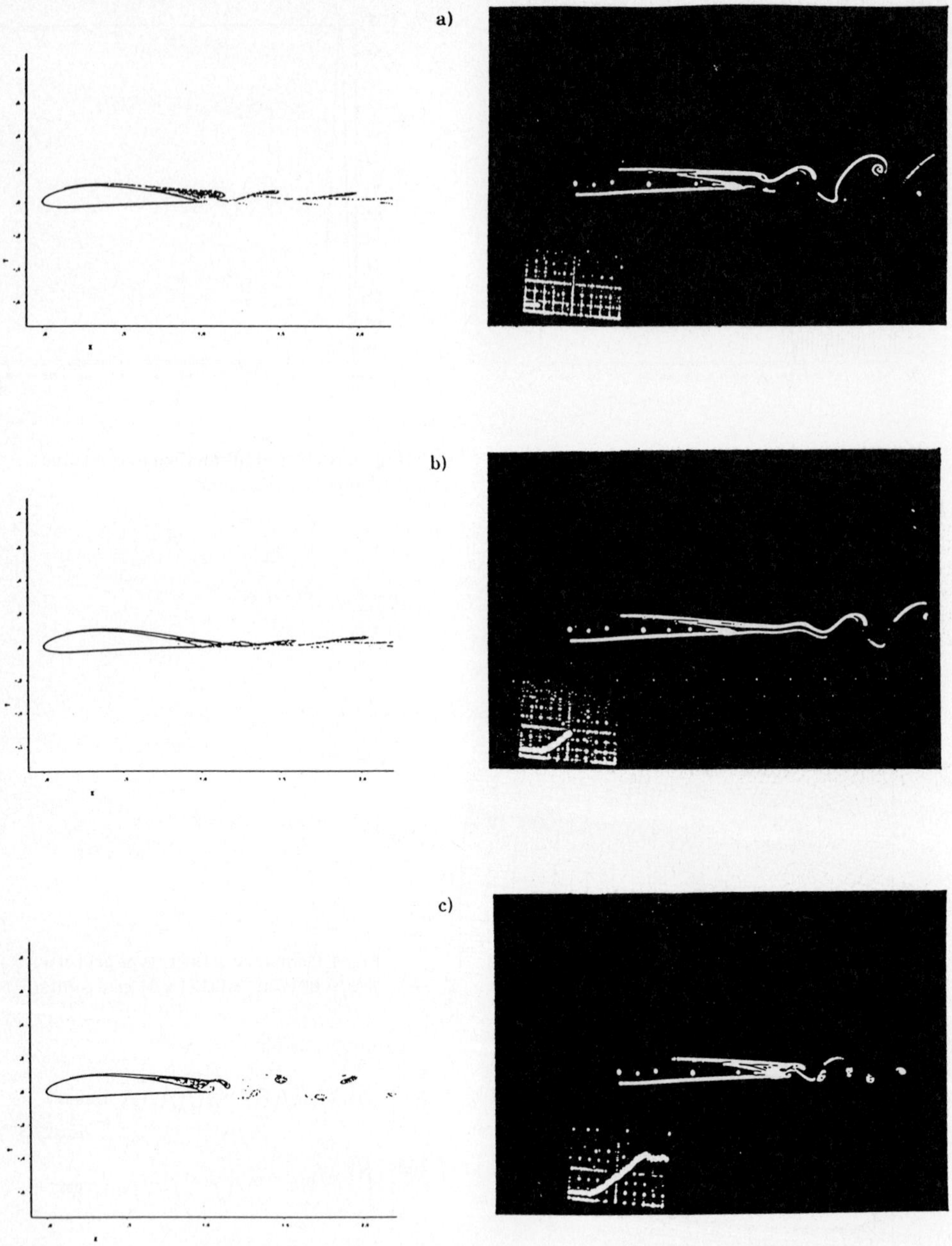

Fig. 7. Numerically predicted (left side) and experimentally determined (right side) streaklines at different times: a) begin of acceleration (t = 0,09), b) during the acceleration (t = 0,33), c) after the acceleration (t = 1,05).

PARALLEL MULTILEVEL ADAPTIVE METHODS*

B. DOWELL, M. GOVETT, S. MCCORMICK, D. QUINLAN

Colorado Research and Development Corporation, Inc.
P. O. Box 9182
Drake Creekside Two, Suite 319
2629 Redwing Road
Fort Collins, Colorado 80525

Abstract.
We report on the progress of a project for design and analysis of a multilevel adaptive algorithm (AFAC [HM]) targeted for the Navier Stokes Computer [NKZ]. We include the results of timing tests for a model AFAC code on a 16-node Intel iPSC/2 hypercube.

1. Introduction.

The asynchronous fast adaptive composite grid method (AFAC [HM]) is a multilevel method for adaptive grid discretization and solution of partial differential equations. AFAC appears to have optimal complexity in a parallel computing environment because it allows for simultaneous processing of all levels of refinement. Coupled with multigrid (MG) processing of each level and nested iteration on the composite grids, AFAC is usually able to solve the composite grid equations in a time proportional to the time it would take to solve the global grid alone. See [HM] for further details.

This report documents initial timing tests of AFAC, coupled with MG and an efficient load balancer, on a 16-node Intel iPSC/2 hypercube. The design of this algorithm is based on the ultimate goal of implementation on the Navier Stokes Computer (NSC [NKZ]).

2. AFAC Algorithm.

We developed a basic code for AFAC to solve Poisson's equation on the unit square, using the finite volume element technique (FVE [LM]) for composite grid discretization and MG [BHMQ] for solution of each grid level problem. This algorithm is currently being coupled with a simple load balancer based on assignment of the processors by levels; that is, instead of assignment by domain decomposition, processors are assigned to levels without regard to the location of these levels. AFAC requires little interlevel communication, so such an effective load assignment and balancing approach is possible. This should be a very effective scheme for the

*This work was supported by the National Aeronautics and Space Administration under grant number NASI–18606.

multi-node NSC which is planned as a fully-connected processor array. We will document tests with a dynamic version of this load balancer in a later report.

AFAC consists of two basic steps: *Step 1.* Given the solution approximation and composite grid residuals on each level, use MG to compute a correction local to that level. *Step 2.* Combine the local corrections with the global solution approximation, compute the global composite grid residual, and transfer the local components of the approximation and residual to each level.

The processing within each step is fully parallelizable. For example, the levels in step 1 are processed simultaneously by a parallelized MG solver [BHMQ]. In the present version of the code, we chose to synchronize between these steps. Asynchronous processing is allowed here, but with our efficient load balancing scheme it would provide little real advantage.

3. Hypercube Implementation.

The AFAC code developed for the iPSC/2 has the following ingredients:

Data Structures. The relationship between the levels of the composite grid is expressed in a tree of arbitrary branching factor at each vertex. Each grid exists as a data structure at one node of this tree. This composite grid tree is replicated in each node. In the case of a change to the composite grid, this change is communicated globally to all processors so the representation of the composite grid in each processor is consistent. The partitioning of all grids is also recorded and is consistent in all processors. Storage of the matrix values uses a one-dimensional array of pointers to row values. These rows are allocated and deallocated dynamically allowing the partitioned matrices to be repartitioned without recopying the entire matrix. This is important to the efficiency of the dynamic load balancer (LB). The effect of noncontiguous matrices is not felt when using this data structure since subscript computation is done by pointer indirection for all but the last subscript, and with the addition of an offset in the case of the last subscript. This organization is particularly effective for partitioning along one dimension.

Grid Manager. The grid manager is responsible for the creation of the grid data structures and the update of the composite grid tree in all processors. Calls to the grid manager allow for the passing of boundary values between processors and the adjustment of the partitions as called for by LB.

MG. A significant change to that described in [BHMQ] is the restriction to one-dimensional decomposition and the allowance of irregular partitionings. A potential additional improvement for greater parallelism is to further partition the grid with each processor, thus making the parallelism finer (although for most applications this may not be needed). We use (2,1) V-cycles that recurse to the coarsest grid, which consists of one interior point. The grid controller is responsible for ordering the execution of the necessary multigrid and intergrid transfer subroutines based on the receipt of messages. For example, multigrid is executed for all grids asynchronously and is driven by the order in which the messages are received.

LB (partitioning strategy). LB is responsible for the dynamic readjustment of the evolving composite grid. As new grids are built, the composite grid is modified, its tree is updated in all processors and the partitions are adjusted to balance the loads. LB can distribute the grids onto the processors, either by domain or by levels. Domains, while making the intergrid transfers cheaper, restrict the partitioning of the grids, thus forcing extra message passing during MG solver execution. This is especially troublesome since most message traffic occurs during the MG solves. Additionally, since we expect that the grid controller will find work while messages are pending, we can better attempt to hide the message latency if message traffic is at a minimum. Thus, we have chosen to distribute the grids to processors by level.

A dynamic version of LB has not yet been implemented on the iPSC/2. The static version tested here is used only by the host for assignment on the initial composite grid.

4. Timing Tests.

To study various properties of our AFAC code, we made several runs on a 16-node Intel iPSC using the seven-level composite grid shown in the figure. We chose each level to be an $n \times n$ grid, counting boundary grid lines, with n varying from 17 to 129. Results of timing these runs are depicted in the table. Here we display the worst case times for one AFAC cycle, including the individual times for the MG solve (step 1) and the interlevel transfers (step 2), for a D dimensional cube with D ranging from 0 to 4. While the data for $n = 65, D = 4$ is missing because of an anomaly, for the case $n = 129$ we show only $D = 4$ data because the iPSC could not support the cases $D < 4$.

Note the reasonably good speed-up exhibited by AFAC, especially for the larger problems. In particular, from a single node to an eight-node cube, speed-up for $n = 17, 33, 65$ were about $5, 6, 7$, respectively. A related observation is that cases with the same ratio of grid points to processors have nearly the same total times. Note also that the sum of the solve and transfer times is often greater than the total time. This is because we report on times that are each worst-case over the processors.

References

[HM] L. Hart and S. McCormick, *Asynchronous multilevel adaptive methods for solving partial differential equations on multiprocessors*, Lecture Notes in Pure and Applied Mathematics/110, Marcel-Dekker, May, 1988.

[NK2] D. Nosenchuck, S. Krist and T. Zang, *On multigrid methods for the Navier Stokes Computer*, Lecture Notes in Pure and Applied Mathematics/110, Marcel-Dekker, May, 1988.

[BHMQ] W. Briggs, L. Hart, S. McCormick and D. Quinlan, *Multigrid methods on a hypercube*, Lecture Notes in Pure and Applied Mathematics/110, Marcel-Dekker, May, 1988.

[LM] C. Liu and S. McCormick, *The finite volume element method (FVE) for planar cavity flow*, these proceedings.

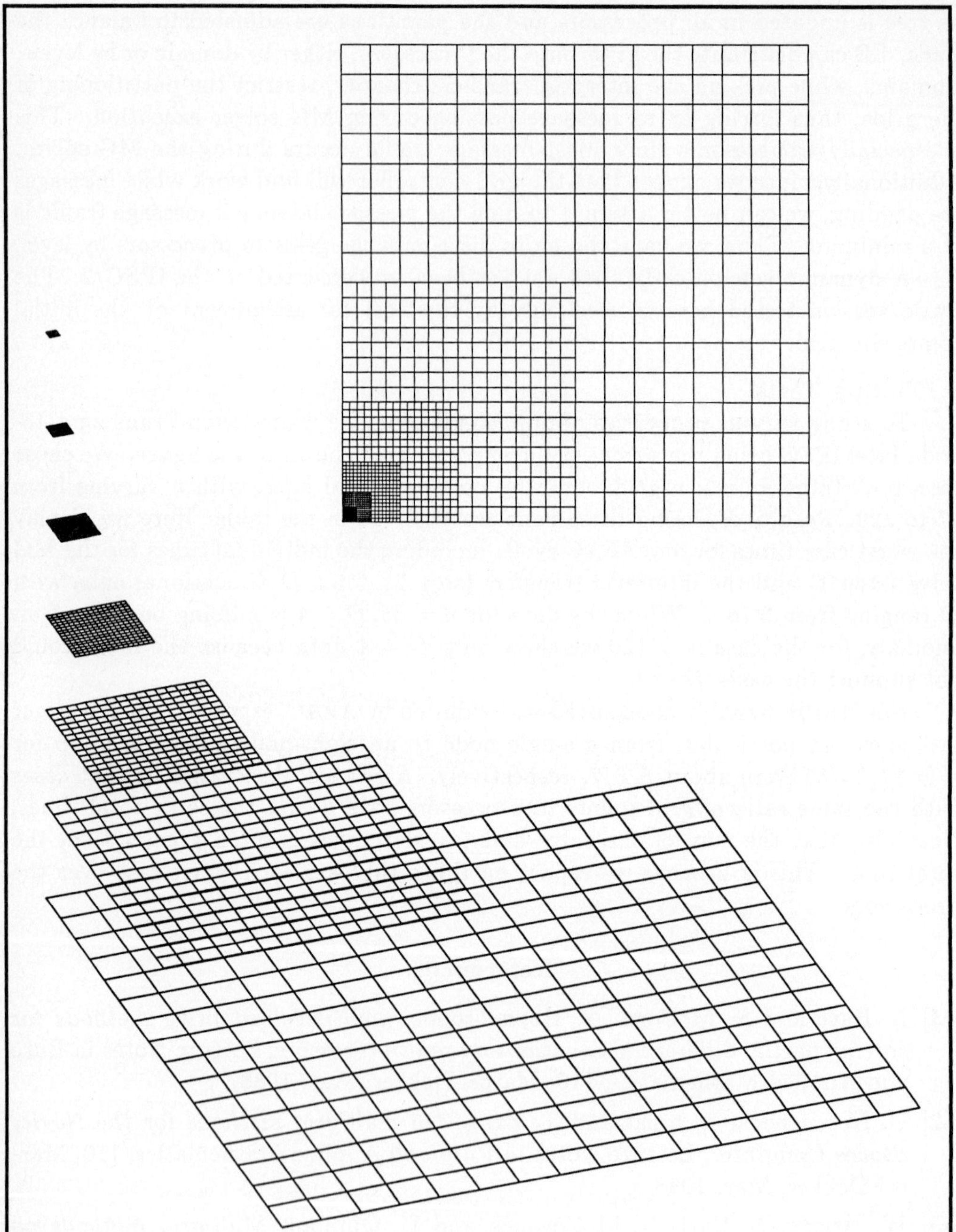

FIGURE

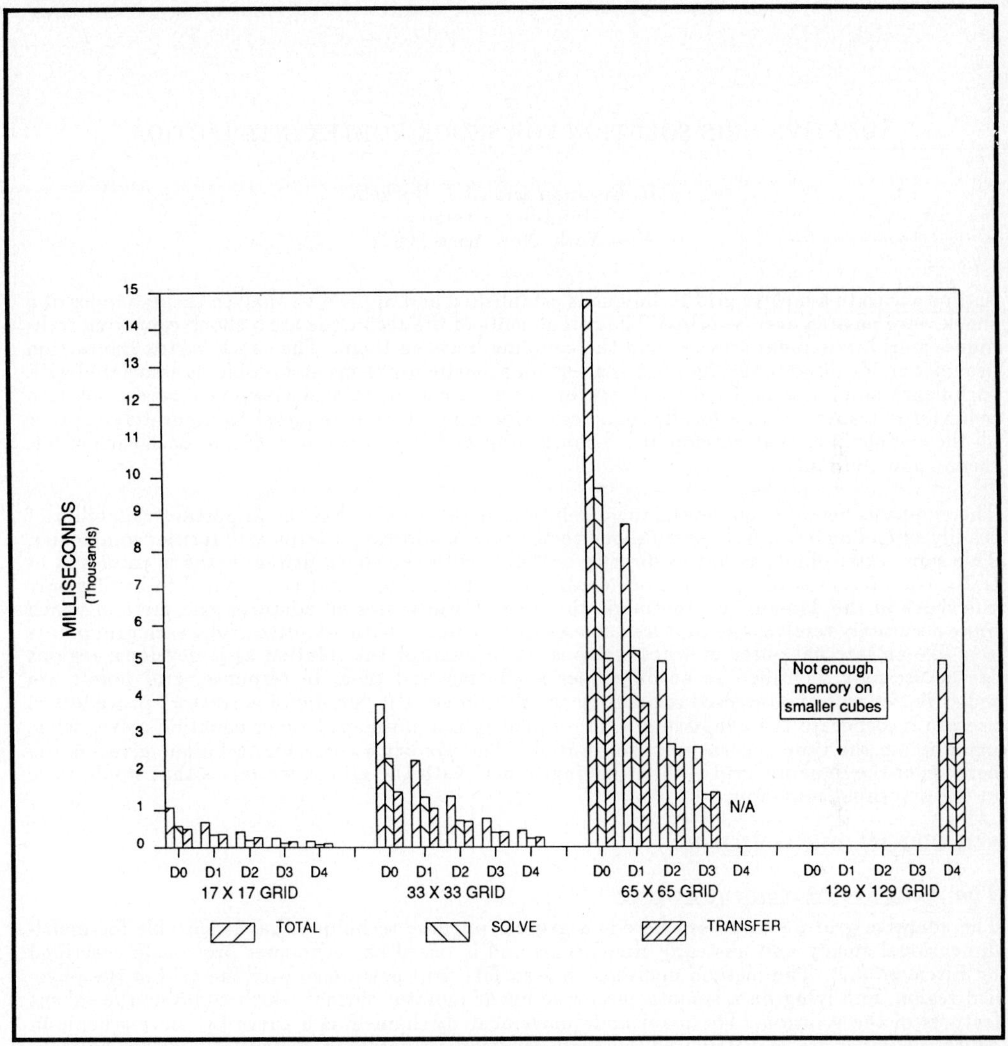
MILLISECONDS (Thousands)
15
14
13
12
11
10
9
8
7
6
5
4
3
2
1
0
Not enough memory on smaller cubes
N/A
D0 D1 D2 D3 D4
17 X 17 GRID
D0 D1 D2 D3 D4
33 X 33 GRID
D0 D1 D2 D3 D4
65 X 65 GRID
D0 D1 D2 D3 D4
129 X 129 GRID
TOTAL
SOLVE
TRANSFER

TABLE

ADAPTIVE GRID SOLUTION FOR SHOCK-VORTEX INTERACTION

P.R. Eiseman and M.J. Bockelie
Columbia University
New York, New York 10027

A time accurate adaptive grid technique is established and applied to analyse the dynamics of a shock wave passing over a vortex. The components of the technique are a shock capturing technique, a grid movement scheme, and the coupling between them. The shock vortex interaction provides an idealized model for studying certain acoustic problems and problems associated with turbulence amplification from shock waves. From a computational viewpoint, severe solution behavior is tracked with a locally high resolution grid. This is required to accurately capture all the complexities that develop in the shock front and the vortex as they approach each other, merge, and continue.

The emphasis herein is on developing a solution method which has the important capability of readily extending beyond the given shock vortex case to address patterns with further complexity. This same extensibility is not as directly accesible with the shock fitting methods employed in [1,2]. The present study uses a shock capturing method that is not constrained to having only one shock in the domain. Furthermore, the present study uses an adaptive grid method which more accurately resolves the captured shocks and vortices. With adaptive grids, each grid points acts like an internal observer whose purpose is to monitor the solution as it develops; regions are accordingly identified as needing finer resolution and then, in response, grid points are redistributed into the targeted regions. A simple and efficent "predictor corrector" procedure is used to incorporate the adaptive grid movement into a *static* grid Euler equation solver while maintaining the time accuracy of the solution. The predictor corrector technique gives us the benefits of the dynamic grid without having to deal with the grid speed terms that would arise in the governing equations.

SOLUTION TECHNIQUE

The Adaptive Movement Scheme

The adaptive grid method used here is a general purpose technique that is suitable for multi-dimensional steady and unsteady simulations and is based on techniques previously described by Eiseman [8,9]. The method operates on a surface grid positioned over the grid in the physical region, and lying on a smooth piecewise linear monitor surface which contains the salient features of the solution. The main node movement mechanism is a curve by curve scheme in which the nodes are re-distributed to satisfy an equidistribution statement of a weight function for each curve on the surface. The weights cause clustering to regions of high normal curvature relative to the uniform conditions of surface arc length. The resulting grid thus provides resolution in regions for which the solution has severe gradients and in the transition regions in which the solution gradients undergo a sharp change. To ensure grid smoothness a low pass filter is applied to the grid after completing the basic adaptive motion. At the completion of all node re-distribution, the surface grid is projected back onto the physical region and the solution is interpolated from the old grid onto the new grid.

The Euler Solver

The flow field is numerically determined with a *static* grid Euler Equation solver which is coupled to a set of modules that perform the the adaptive grid movement. The Euler solver was obtained from the Computational Methods Branch at the NASA Langley Research Center. It uses a finite volume formulation to discretize the Euler equations in generalized coordinates [3] and was written by Joe Morrison. The flux terms are computed with Roe's Approximate Riemann problem method [4] and the spatial variation of the fluid is approximated with the Van Leer MUSCL scheme [5]. The Euler equations are integrated in a time accurate fashion with an explicit four stage Runge-Kutta scheme that is $O(\Delta t^4)$ [6]. The solver is designed for use on *static* grids and when the adaptive grid action is performed the grid will obtain a time dependence; that is, the grid becomes dynamic. From a strict mathematical viewpoint, when computations are performed on a dynamic grid one must account for the changing frame of reference by including additional grid velocity terms in the governing equations of the physical problem. Including the grid speed terms into the governing equations is not a trivial matter because their values are not known a priori. In this study a predictor corrector technique [7] is used that eliminates the need to include the grid speed terms in the Euler equation solver.

The Temporal Coupling

The predictor corrector procedure treats the time integration as a series of initial value problems over short time intervals in which the solution is first advanced to create a new grid and then second, to recompute the solution on the new grid. In effect, the new grid provides a carpet of resolution in the regions over which the solution will evolve. The advanatage gained here is that grid speed terms are not required because the predicted and corrected solutions are computed on static - although different - grids. Thus, the temporal accuracy can be maintained with adaptive grid methods while using only a static grid equation solver. The predictor corrector scheme consists of five basic steps. Given the solution and grid at a time level T, to advance the solution in time we: (1) predict the solution up to a future time level $T+\tau$ by integrating the solution forward p time steps on the given fixed grid; (2) compute a composite monitor surface by averaging intermediate monitor surfaces computed during the forward integration on the fixed grid; (3) adapt the grid to the composite monitor surface to obtain a new grid for the corrector stage; (4) transfer the solution at time level T to the new grid by local linear interpolation; and last, (5) correct the solution by integrating the solution at time level T forward in time on the new grid to a time level $T+\tau'$, where $\tau' \leq \tau$. With the solution and the grid established at the time level $T+\tau'$ the entire predictor corrector procedure is repeated to advance the solution to another time level. By successively repeating the predictor corrector procedure the solution is marched forward through time to the desired final time level.

THE SHOCK VORTEX PROBLEM

The shock vortex problem we model consists of an initially planar shock wave marching toward, and eventually over, a solid core vortex. The shock and vortex are assumed to lie within a channel having solid walls and finite dimensions. The governing equations for the model problem are the unsteady, two dimensional, compressible Euler equations. The Euler equations are well documented in the literature and therefore are not presented here. The initial flow field consists of a planar shock located at $x = 0$ and a counter clockwise rotating solid core vortex located at $x = 0.5$. To the left of the shock wave is the uniform supersonic flow field that would follow behind a shock wave propagating at a relative Mach number of 3 if there is no upstream disturbance. To the right (i.e., ahead) of the shock the initial flow field is obtained by assuming a constant density field, calculating the velocity from the stream function

$$\Psi(x,y) = -\frac{\kappa}{2\pi}\log\left(\left[(x-x_o)^2+(y-y_o)^2+a^2\right]^{1/2}\right)$$

the pressure field from Bernoulli's equation, and the total energy from the equation of state. We assume $\kappa = 0.40$, $a = 0.1$, and $(x_o, y_o) = (0.5, 0)$. At the leading edge of the initial shock front (i.e., $x = 0$) the supersonic flow field behind the shock and the subsonic flow field ahead of the shock are smoothly merged in the space of a few grid cells in order to eliminate a numerically induced "bump" which occurs in the solution if the two flow fields are merged abruptly (i.e., within one grid cell). The additional bump in the solution occurs in the supersonic flow behind the shock front, has a magnitude which is a function of the grid cell size, and occurs when there is a vortex upstream of the shock and when there is no upstream disturbance. The flow field for the smooth shock front is obtained from the solution of a planar shock wave propagating into a region with no upstream disturbance. The modification of the initial flow field only affects the grid points near $x = 0$; away from this region the initial flow field is as described above. On the boundaries of the domain we assume that (a) at the inlet the flow field is that of the initial uniform supersonic flow; (b) at the outlet the flow field is that of the initial vortex; and (c) along the top and bottom of the domain the flow is assumed to be tangent to the boundary (i.e., solid walls).

ADAPTIVE COMPUTATIONS

The Fine Grid Solution

The results for a solution on a static grid with fine grid resolution are presented first to provide a basis of comparison for the adaptive grid solutions. Illustrated in Fig. 1 are contour plots for the pressure field of the solution computed on a static grid with 200 uniformly spaced cells in the axial direction and 50 cells in the transverse direction which have a uniform size in computational space where the transformation from physical to computational space is given by $Y = \tanh\ 2y$. The pressure field for the entire computatinal domain is plotted at four time levels that correspond to the initial condition ($t = 0.00$), before the shock reaches the rim of the vortex core ($t = 0.11$), immediately after the shock exits from the vortex core ($t = 0.20$), and a long time after the interaction has occured ($t = 0.42$).

In this study we only highlight the key features that can be easily discerned from the attached pressure plots; for a thorough review of the physics involved for the dynamic interaction of the shock with the vortex, we refer to [1]. From Fig. 1 it can be seen that before the shock reaches the vortex core that the pressure field behind the shock wave is slightly asymetric but mostly constant. As the shock passes over the vortex core the shock front becomes curved. and the location of the vortex can be determined by the presence of the concentric pressue contours immediately behind the shock. At a much later time, we can see the that shock front is still curved, the vortex has been convected down stream by the flow field following the shock wave, and that the shock wave is separating from the vortex because it is propagating at a faster speed than the vortex. In addition, the presence of an approximately cylindrical sound wave centered on the vortex can be seen moving away from the shock front. With the basic solution established, we can now apply the adaptive grid method to reduce the number of grid points.

Adaptation to only the Shock Wave

In the first adaptive grid solution we adapt an 80×32 cell grid to a monitor surface formed from the density field. Redistributing the grid points based on the density field will cluster the grid points to the shock front because the density field is approximately constant within the vortex and far behind the shock front, but undergoes a large jump in value across the shock front. In the transverse direction, the grid point movement occurs with respect to a background nodal distribution given by a hyperbolic tangent spacing with a damping factor of 2. Therèfore, in the transverse direction the nodes are positioned with the specified hyperbolic tangent spacing which is locally deviated from in regions having a high normal curvature that recieve additional node concentration. Illustrated in Fig. 2 are the grid and contour plots of the pressure field at the same four time levels as presented in Fig. 1. The parameters in the grid movement module that govern the adaptive action are adjusted such that the grid created during each predictor corrector sequence has a grid resolution in the immediate vicinity of the shock front that is comparable to the fine grid resolution in the above static grid case. From Fig. 2 it can be seen that the grid correctly tracks the shock. Comparing Figs. 1 and 2 we can see excellent agreement between the adaptive and static grid solutions. However, in the adaptive grid case significantly fewer grid points have been used.

Adaptation to both the Shock Wave and the Vortex

The results for a second adaptive grid solution are presented in Fig. 3 for a 40×32 cell grid adapted to a monitor surface containing data from both the shock and the vortex. In this case the monitor surface is formed as a linear combination of the density field and the circulation about each point. As described above, the density field will identify the shock front and concentrate points to it. The circulation is defined as $\Gamma = \int \mathbf{V} \cdot dl$ and provides a means of tracking and clustering points to the vortex core because Γ is approximately zero everywhere except within the vortex core. The circulation is computed at each node using a circuit around the node defined by the four adjacent grid points that surround it; on the boundaries of the domain we set $\Gamma = 0$.

The computations of the circulation can contain a large amount of noise in the regions along the shock front. Therefore we apply a smoothing and data clipping scheme to the monitor surface values at the start of the grid movement module. The smoothing is performed with the same relaxation scheme used in the smoothing of the grid positions in the physical domain. The clipping is performed by checking each computed circulation value against the average of the value obtained over a 3×3 patch of nodes centered on the current node; if the value at the center of the patch exceeds the average value by a specified tolerance, then the center value is set equal to the average value. The smoothing is useful for eliminating low amplitude, high frequency variations in the monitor surface values. The clipping is used to eliminate the large amplitude high frequency variations and is followed by the smoothing operation.

As can be seen in Fig. 3, by adapting to the shock and the vortex we can obtain essentially the same results as in the fine static grid case. By looking at the grid plots we can see that the grid is correctly tracking the time dependent locations of the shock and the vortex. However, now the grid points must be positioned to resolve both the shock and the vortex; thus the grid resolution on the shockfront along the centerline of the domain is not as fine as near the boundary where the circulation is zero, nor is the resolution as fine as that illustrated in Fig. 2. If we scrutinize the pressure fields we see that the the curved shock front is adequately defined but that only a hint of the sound wave and the pressure field at the vortex core can be identified.

REFERENCES

1. Pao, S.P. and Salas, M.D., "A Numerical Study Of Two Dimensional Shock-Vortex Interaction," AIAA Paper 81-1205, 1981.
2. Salas, M.E., Zang, T.A. and Hussaini, M.Y., "Shock-Fitted Euler Solutions To Shock-Vortex Interactions," Eighth International Conference on Numerical Methods in Fluid Dynamics, Lecture Notes in Physics, Vol. 170, Springer-Verlag, 1982, pp.461-467.
3. Anderson, W.K., Thomas, J.L. and Whitfield, D.L., "Multigrid Accelerations of the Flux Split Euler Equations," AIAA Paper 86-0274, 1986.
4. Roe, P.L., "The Use of The Riemann Problem in Finite Difference Schemes," Seventh International Conference on Numerical Methods in Fluid Dynamics, Lecture Notes in Physics, Vol. 141, Springer-Verlag, 1981, pp.354-359.
5. Anderson, W.K., Thomas, J.L. and Van Leer, B., "Comparison Of Finite Volume Flux Vector Splitting For The Euler Equations," AIAA Journal, Vol. 24, No. 9, 1986, pp.1453-1460.
6. Jameson, A.,"The Evolution of Computaional Methods in Aerodynamics," ASME Journal of Applied Mechanics, Vol. 50, December, 1983, pp. 1052-1070.
7. Sanz-Serna, J.M., and Christie, I., "A Simple Adaptive Technique for Nonlinear Wave Problems," Journal of Computational Physics, Vol. 67, 1986, pp. 348-360.
8. Eiseman, P.R., "Alternating Direction Grid Generation," AIAA Journal, Vol. 23, No. 4, 1985, pp.551-560.
9. Eiseman, P.R., "Alternating Direction Adaptive Grid Generation For Three Dimensional Regions," Proceeding of The Tenth International Conference on Numerical Methods in Fluid Dynamics, Lecture notes in Physics, No. 264, 1986, pp.258-263.

This work was performed under the Air Force Grant AFOSR-86-0307. Portions of the computational work were performed on the CYBER-205 at the Von Neumann Supercomputing Center, Princeton University.

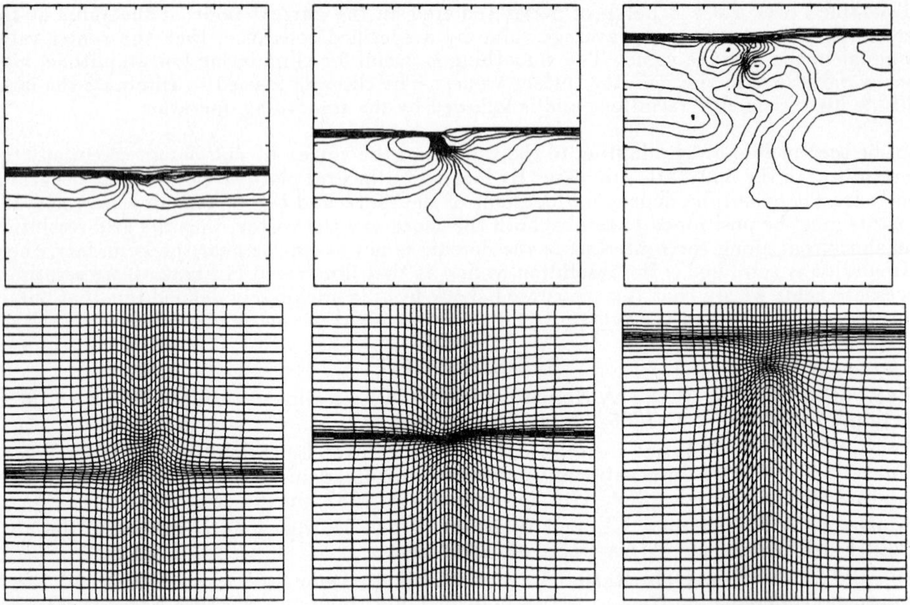

FIGURE 3

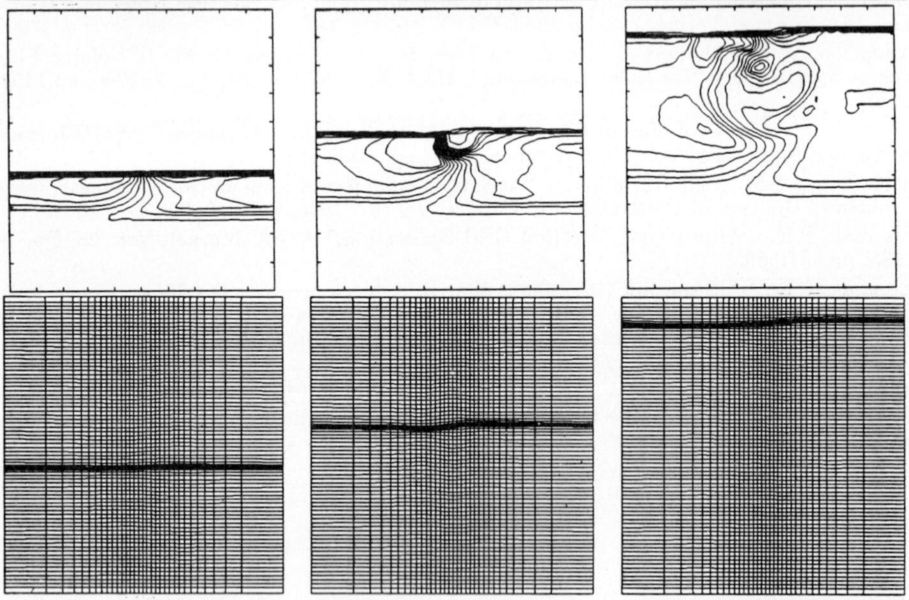

FIGURE 2

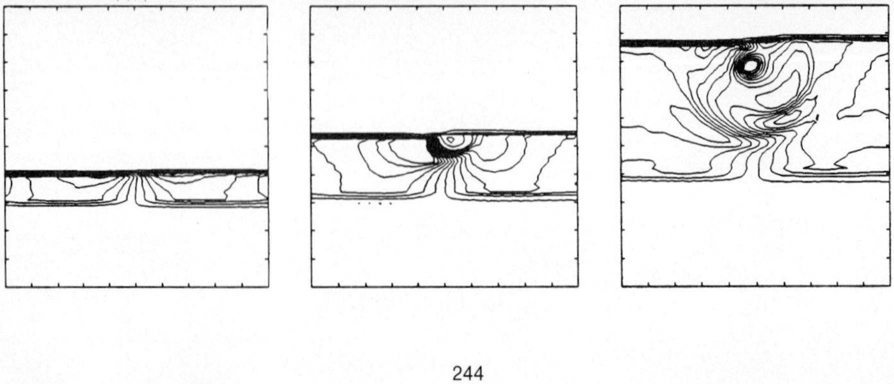

FIGURE 1

COMPUTER SIMULATION OF SOME TYPES OF FLOWS ARISING AT INTERACTIONS BETWEEN A SUPERSONIC FLOW AND A BOUNDARY LAYER

T.G. Elizarova, B.N. Chetverushkin

M.V. Keldysh Institute of Applied Mathematics
Academy of Sciences of the USSR
Miusskaya pl.4, 125047 Moscow

As it is known the gas dynamics Euler and Navier-Stokes equations are obtained by a usual averaging procedure from the integro--differential Boltzmann equation for description of a one-frequancy distribution function. At the difference approximation of the gas dynamics equations we do not use the fact that these equations follow from a more complex transport equation, and macroscopic gasdynamic parameters - the density, the velocity, the temperature - may be obtained as momenta of the distribution function*). Employment of the Boltzmann equation or its simplified kinetic models for describing the behaviour of dense gases appears at first sight to be inexpedient because an amount of computations is considerably larger in this case than for solving the gas dynamics equations.

By basing on kinetic models the authors managed to construct the difference system of equations for determining macroscopic gas parameters (the density, the velocity, the temperature), which depends only on spatial variables and time [1].

Constructing the difference schemes may be treated in the following way: first the difference scheme is written down for the Boltzmann equation and then it is averaged by using the notion about the local Maxwell or local Navier-Stokes form of the distribution function to close the system of macroscopic equations. To a considerable extent this procedure is similar to that which is applied to obtain the Euler or Navier-Stokes equations from the Boltzmann equation. It is natural to call these schemes the kinetic-consistent difference schemes because their form is consistent with a choice of difference approximation for the Boltzman equation. Thus the kinetic

*) Here we speak about describing a rather dense gas for which the Euler or Navier-Stokes equations are valid. For modelling phenomena in rarefied gases kinetic approaches are used sufficiently widely.

consistent difference schemes (k.c.d.s.) differ from other algorithms for solving gas dynamics equations in that the discretization procedure in this case is used first, after which the difference scheme averaging is carried out. We recall that when constructing the difference schemes in the traditional way the order of procedures is opposite: first the transport equation is averaged and then the difference approximation of the obtained equations is constructed [2]. If we consider the kinetic-consistent difference schemes in application to simulation of the nonconducting viscous gas flow they may be treated as a class of schemes with artificial viscosity involved. However, this viscosity is obtained not in the usual phenomenological way but self-consistently with the form of an initial difference approximation of the Boltzmann equation. It has a nontraditional form and exists in all gas dynamics equations; in the motion equation it contains the cubic velocity.

The schemes thus obtained are simple and sufficiently stable, providing at the same time a high accuracy in computations of ideal flows. So the proposed approach is a new trend in extending the numerical techniques to solving inviscid gas dynamics problems.

One of the pressing problems in gas dynamics is calculation of a viscous compressible gas flows for large Reynolds numbers.

It is shown that in the viscous region of the flow the solution obtained by means of k.c.d.s. coincides with that of the Prandtle equation with true viscosity and heat conductivity. Thus k.c.d.s. provide an integral transition from simulation of a nonviscous part of the flow to those regions where consideration of viscosity and heat conductivity is important. It allows using the technique developed by the authors for simulation of a wide range of problems dealing with an analysis of viscous separation flows, including nonstationary or nonsteady, as well as for calculating heat exchange in such processes. Difficulty in solving these problems consists mostly in efficient combined calculations of processes running in the both viscous and nonviscous parts of the flow.

We give a version of k.c.d.s. in the case of a plane two-dimensional geometry as

$$\rho_t + (\rho u)_{\overset{\circ}{x}} + (\rho v)_{\overset{\circ}{y}} = \left[\frac{h_x}{2V}(\rho u^2 + P)_{\bar{x}}\right]_x + \left[\frac{h_y}{2V}(\rho v^2 + P)_{\bar{y}}\right]_y \quad (1)$$

$$(\rho u)_t + (\rho u^2 + P)_{\overset{\circ}{x}} + (\rho u v)_{\overset{\circ}{y}} = \left[\frac{h_x}{2V}(\rho u^3 + 3Pu)_{\bar{x}}\right]_x + \quad (2)$$
$$+ \left[\frac{h_y}{2V}(\rho u v^2)_{\bar{y}}\right]_y + \left[\frac{h_y}{2V}\alpha(T)(Pu)_{\bar{y}}\right]_y$$

$$(\rho v)_t + (\rho u v)_{\overset{\circ}{x}} + (\rho v^2 + p)_{\overset{\circ}{y}} = \left[\frac{h_x}{2V}(\rho u^2 v)_{\bar{x}}\right]_x +$$
$$+\left[\frac{h_x}{2V}\alpha(T)(\rho v)_{\bar{x}}\right]_x + \left[\frac{h_y}{2V}(\rho v^3 + 3pv)_{\bar{y}}\right]_y \quad (3)$$

$$E_t + (u(E+p))_{\overset{\circ}{x}} + (v(E+p))_{\overset{\circ}{y}} =$$
$$= \left[\frac{h_x}{2V}(u^2(E+2p))_{\bar{x}}\right]_x + \left[\frac{h_y}{2V}(v^2(E+2p))_{\bar{y}}\right]_y + \quad (4)$$
$$+\left[\frac{h_x}{2V}\frac{\alpha(T)}{Pr}\frac{p}{\rho}(E+p)_{\bar{x}}\right]_x + \left[\frac{h_y}{2V}\frac{\alpha(T)}{Pr}\frac{p}{\rho}(E+p)_{\bar{y}}\right]_y +$$
$$+\left[\left(1-\frac{\alpha(T)}{Pr}\right)\frac{h_x}{2V}(p\varepsilon)_{\bar{x}}\right]_x + \left[\left(1-\frac{\alpha(T)}{Pr}\right)\frac{h_y}{2V}(p\varepsilon)_{\bar{y}}\right]_y$$

Here ρ is the density, u and v are the velocity components, p is the pressure, $E = \rho(u^2+v^2)/2 + \rho\varepsilon$ is a total energy, V is the characteristic velocity, h_x and h_y are steps of the difference grid.

In the boundary layer approximation the viscosity coefficient for the scheme (1)-(4) has the form

$$\mu = \alpha(T)\,\frac{hp}{2V} \quad (5)$$

where the coefficient $\alpha(T)$ is chosen in correspondence with the true viscosity coefficient for the gas under consideration.

The boundary conditions for the schemes of type (1)-(4) are constructed by averaging the difference scheme for the Boltzmann equation near the boundary which leads to the boundary conditions of the form:

$$u = v = 0$$

$$T = T_w \qquad \text{or} \qquad \partial T/\partial n = W^{\circ} \quad (6)$$

The last condition means that the mass flow through the boundary is absent.

As the first example of calculating viscous flows by using k.c.d.s. an elementary one-parameter problem is considered, where a supersonic flow about a plate is analysed at a zero angle of attack (the Blazius problem). The resulted function of the plate resistence coefficient well agrees with known analytic relations for sufficiently large Reynolds numbers.

The numerical simulation results obtained for nonstationary periodic flows with front separation zones arizing in a supersonic flow around a spiked cylinder are given in Fig. 1. The computing simulation results are in good agreement with physical experiment and correctlydescribe various stages of the flow process. In particular, the both stationary and oscillatory regimes of the first and second kind were calculated depending on the free stream velocity and a geometry of the body. For the oscillatory regime of the second kind the dimensionless frequency of oscillations is Sh = 0.41, while the corresponding experiment value is Sh = 0.42. It is shown that a mechanism of this nonstationary flow has an outlet nature, and the presence of a boundary layer on the spike is a necessary condition for the formation of undamped oscillations [3].

Figures 2 and 3 show the numerical results obtained for a supersonic gasdynamic flow around a cylindrical tube closed from one end and mounted by its open end towards a supersonic free stream (Fig.2). An arising flow picture has a nonstationary character, the shock wave performs oscillations near an average position, and stable pulsations of pressure are observed at the bottom [4]. The pulsation process runs in such a way that the velocity vector along the cylinder vector is always directed inwards, periodically changing its direction nearer to the edges (Fig. 3). The Struhal number in the numerical experiment is Sh = 0.246. By data of the physical experiment Sh = 0.25.

An example of computations of the model problem about a laminar flow in a plane cavern is presented in Fig. 4 [5]. The heat flows on the cavern walls are also computed.

REFERENCES

1. Elizarova, T.G., Chetverushkin B.N. About a computing code for calculation of gasdynamic flows. JVM i MF, 1985, v.25, N 10, p.1526-1533.

2. Elizarova, T.G. On a class of kinetic-consistent difference schemes of gas dynamics. JVM i MF, 1987, v.27, N 11, p.1753-1757.

3. Antonov A.N., Elizarova T.G., Pavlov A.N., Chetverushkin B.N. Direct numerical simulation of oscillatory regimes in a flow about a spiked body. Preprint, Keldysh Inst. Appl. Math., USSR Ac. Sci., 1988, N 56.

4. Elizarova T.G., Chetverushin B.N., Sheretov Yu.V. On some results of calculation of a supersonic flow about an empty cylinder, carried out with kinetic-consistent difference schemes. Preprint, Keldysh Inst. Appl. Math., USSR Ac. Sci., 1988, N 97.

5. Graur I.A., Elizarova, T.G., Chetverushkin B.N. On application of kinetic-consistent difference schemes for calculating heat exchange in supersonic gasdynamic flows. Preprint, Keldysh Inst.

Appl. Math., USSR Ac.Sci., 1988, N 90.

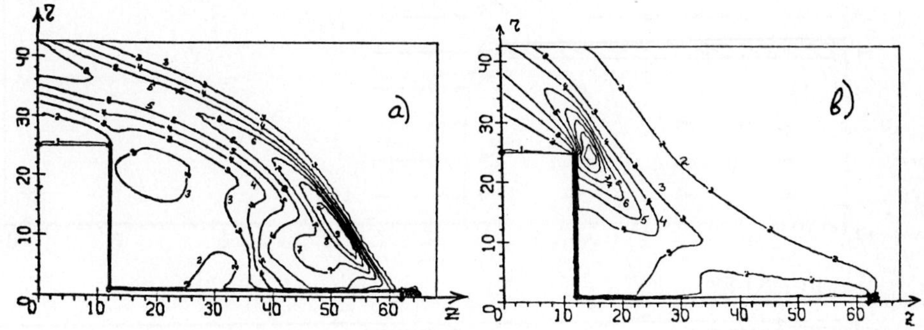

Fig. 1. Isolines of density in the problem of a flow around a spiked body ($M = 6$, $Re = 8.4 \cdot 10^3$, $Pr = 1$) corresponding to a maximal (a) and minimal (b) size of the separation zone.

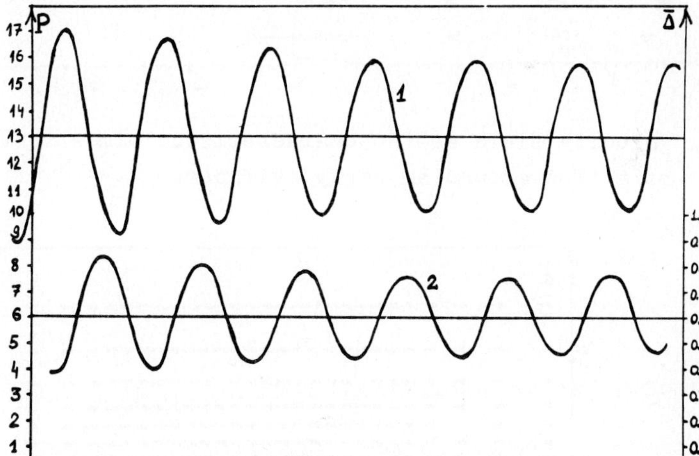

Fig. 2. Pressure at the cavity bottom (curve 1) and separation of the shock wave from the cylinder edge (curve 2) versus time in the problem of a flow around an empty cylinder ($M = 3.7$, $Re = 5 \cdot 10^4$, $Pr = 1$).

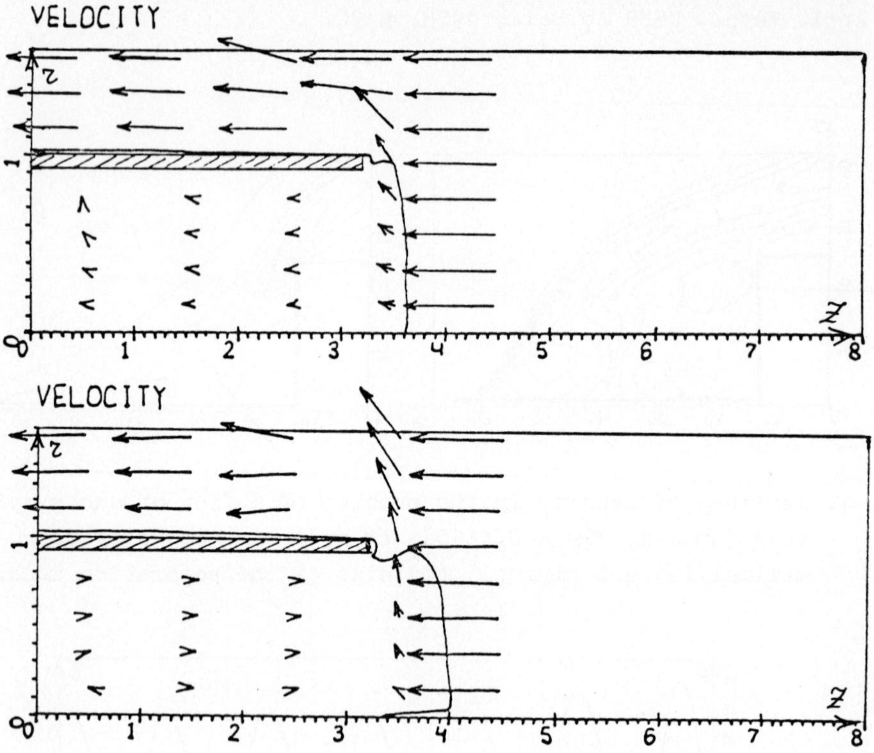

Fig. 3. Velocity field at two characteristic times in the problem of a flow around an empty cylinder.

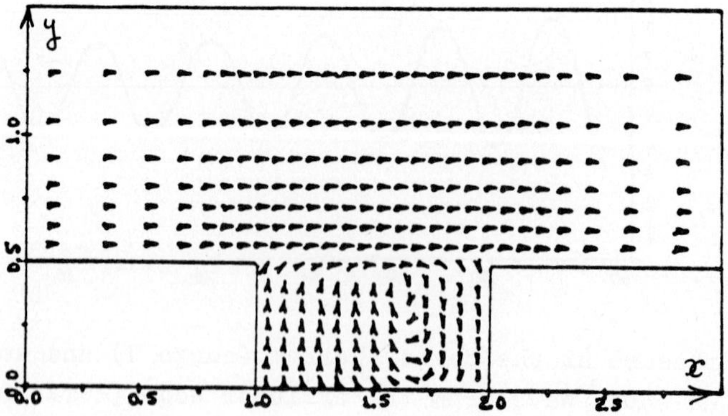

Fig. 4. Gas flow in a cavern (M = 1.05, Re = 1000, Pr = 0.72).

AN IMPLICIT FLUX-VECTOR SPLITTING FINITE-ELEMENT TECHNIQUE FOR AN IMPROVED SOLUTION OF COMPRESSIBLE EULER EQUATIONS ON DISTORTED GRIDS

J.A. Essers
The Florida State University
Mechanical Engineering Dept.
and Supercomputer Computations
Research Inst., P.O. Box 2175
Tallahassee, FL 32316

E. Renard
The University of Liege
Institut de Mecanique
Service d' Aerodynamique Appliquee
rue Ernest Solvay, 21
B-4000 Liege, Belgium

1. A modified finite-element/volume discretization for distorted grids

Consider a linear, time-dependent, 2D convection-diffusion equation:

$$\partial_t u + \sum_{k=1}^{5} a_k \partial_k u = 0 \tag{1}$$

where the a_k are known constants, and the ∂_k correspond to first and second order space derivatives. For $k = 1$ and 2, they denote the convective derivatives ∂_x and ∂_y, whereas the diffusive derivatives ∂_{xx}, ∂_{yy} and ∂_{xy} correspond to k =3, 4 and 5. A centered space discretization of (1) on a structured distorted grid of linear quadrilateral elements can be obtained using a Galerkin or a subdomain finite-element/volume technique. Denoting mesh point by indices (i,j), discretized equation can be written:

$$\underline{M}^{(i,j)}[\partial_t u] + \sum_{k=1}^{5} a_k \underline{L}_k^{(i,j)}[u] = 0 \tag{2}$$

$\underline{L}_k^{(i,j)}[e]$ is a notation for discretization operators applied to any function e and is equivalent to:
$$\sum_{m,n=1}^{3} \left(L_k^{i,j}\right)_{m,n} e_{i+m-2,j+n-2}$$

The $(L_k^{i,j})_{m,n}$ are used to define arrays with dimension (3 x 3) that will be denoted by $\mathbf{L}_k^{i,j}$. Mass arrays $\mathbf{M}^{i,j}$ are used in a similar way to define mass operator $\underline{M}^{i,j}$ [e]. Superscripts (i,j) indicating that the operators are different at the various mesh points will be omitted in the following. To analyze dominant errors related to the use of (2), perform a Taylor series expansion of each term near the "center of mass" G of mesh cell (i,j). For the first term that expansion gives:

$$\underline{M}[\partial_t u] \simeq (\partial_t u)_G + H.O.T.[h^2] \quad ,$$

where $H.O.T.[h^2]$ corresponds to terms of the order of h^2, h denoting a characteristic length of the mesh cell. For each operator of the second term, the expansion yields:

$$\underline{L}_k[u] \simeq \sum_{n=1}^{5} C_{kn}(\partial_n u)_G + H.O.T. \tag{3}$$

The elements of matrix C depend on local mesh topology. We developed an automatic accuracy analyzer based on the evaluation at each mesh point of $R = C - I$, I denoting the unit matrix, and applied it to various distorted grid configurations. The conclusions are similar for different types of discretizations and can be summarized as follows:

- For $k = 1$ and 2, the R_{kn} are of the order of h. Discretized convective terms are therefore first order accurate only. The second order derivatives corresponding to the dominant discretization errors can be responsible for significant dissipative or anti-dissipative numerical effects that can dominate the physical diffusion terms, and can lead to wiggled solutions or to numerical instabilities that would not appear for smooth grids with comparable mesh sizes.
- If a significant amount of skewness is introduced in the grid, the R_{kn} corresponding to $k = 3, 4$ and 5 are of the order of h^o. Diffusion terms are then discretized in an inconsistent manner.

To eliminate the dominant discretization errors mentioned above, we propose to replace all operators $\underline{L}_k$ appearing in (2) by modified operators $\hat{\underline{L}}_k$ coupling the same mesh points. The arrays defining the latter are constructed by using the following formula:

$$\hat{\mathbf{L}}_k = \sum_{n=1}^{5} E_{kn}\mathbf{L}_n \tag{4}$$

where E is the inverse of C. The modified discretized form of the equation can be proved to unconditionally meet the following requirements: second order accuracy for the convective terms, and at least consistency for the diffusion terms. Note that the procedure used to construct the modified operators is general. Its extension to unstructured meshes is straightforward.

To obtain a robust time discretization of the modified version of (2), implicit scheme should be used. A two-level backward Euler formula e.g. gives:

$$\underline{G}[\Delta u] = -\theta \sum_{k=1}^{5} a_k \hat{\underline{L}}_k[u] \quad (5.a), \qquad \text{with:} \quad \underline{G} = \underline{M} + \theta \sum_{k=1}^{5} a_k \hat{\underline{L}}_k \quad (5.b),$$

θ denoting the time-step. R.H.S. term of (5.a) is computed at the known time level. Linear operator $\underline{G}$ couples the values of the unknown time variations Δu at 9 neighboring mesh points.
To improve the computational efficiency, we developed an original approximate factorization procedure that can be used for structured distorted meshes of quadrilaterals. The technique consists in using an approximate form of $\underline{G}$ in eq. (5.a). The 3 x 3 array defining the latter should be equivalent to the "tensor product" of 2 suitable one-dimensional arrays with 3 elements each, i.e.:

$$\mathbf{G} \simeq \mathbf{g}_1 \otimes \mathbf{g}_2 \quad (6) \qquad \text{with:} \quad \underline{g}_q = \underline{\mu}_q + \theta \sum_{k=1}^{5} a_k \underline{\ell}_{k,q} \quad (q = 1 \text{ and } 2) \quad (7)$$

The calculation of the Δu can then be performed using a well-know two steps procedure that requires the solution of tri-diagonal systems.
In (7), the $\underline{\mu}_q$ and $\underline{\ell}_{k,q}$ are 1D operators that are automatically evaluated by the computer using the components of 2D arrays $\mathbf{M}$ and $\mathbf{L}_k$. The procedure used for that evaluation is designed in such a way that the purely transient factorization errors mostly introduce weak dispersive effects of the same order as those of classical finite-difference approximate factorization techniques. In the peculiar situation of a stretched rectangular mesh, our approximate factorization approach can be proven to be equivalent to that devel-

oped by Fletcher et al. [1]. It can, however. also be applied to severely skewed grids of quadrilaterals, and is therefore more general.

The new discretization technique presented here can be extended to non-linear systems of conservation law. Consider a convective 2D system, e.g. corresponding to the compressible Euler equations:

$$\partial_t s + \partial_x f + \partial_y g = 0 \tag{8}$$

where s is a vector of unknown fields, and flux vectors f and g are non-linear functions of the latter. A space discretization of (8) can be obtained from a finite-element technique using a group formulation. Thanks to the latter, a modified discretization with an improved accuracy can be constructed using formula (4). An implicit time-discretization and a classical time linearization then yield:

$$\underline{G}[\Delta s] = -\theta(\hat{\underline{L}}_1[f] + \hat{\underline{L}}_2[g]) \quad (9.a) \; ; \; \underline{G}[\Delta s] \equiv \underline{M}[\Delta s] + \theta(\hat{\underline{L}}_1[A\Delta s] + \hat{\underline{L}}_2[B\Delta s]) \quad (9.b)$$

$A = \dfrac{\partial f}{\partial s}$ and $B = \dfrac{\partial_g}{\partial_s}$ are Jacobian matrices evaluated at the known time level. For structured grids, the solution of linear system (9.a) can be simplified using an approximate factorization of L.H.S. operator constructed from one- dimensional operators defined as:

$$\underline{g}_q[e] = \underline{\mu}_q[e] + \theta(\ell_{1,q}[Ae] + \ell_{2,q}[Be]) \quad \text{for } q = 1 \text{ and } 2 \tag{10}$$

2. A flux-vector splitting finite-element technique for the Euler equations

We have developed a systematic procedure that can be used to transform modified centered finite-element discretizations of the convective terms into "upwind-biased" schemes with the same order of accuracy. Consider the following approximate formula where $x_G^{(i,j)}, y_G^{(i,j)}$ denote the coordinates of the "center of mass" $G_{i,j}$ of cell (i, j):

$$\hat{\underline{L}}_q^{(i,j)} \simeq \hat{\underline{L}}_q^{(i-1,j)} + \left\{x_G^{(i,j)} - x_G^{(i-1,j)}\right\} \hat{\underline{L}}_r^{(i-1,j)} + \left\{y_G^{(i,j)} - y_G^{(i-1,j)}\right\} \hat{\underline{L}}_\ell^{(i-1,j)} \tag{11}$$

for $q = 1$ and 2, with: $r = 3 + 2(q-1)$ and $\ell = 6 - q$.

By performing a Taylor series expansion of left- and right-hand sides of (11) near $G_{i,j}$, this formula can easily be proved to be valid with an error of the order of h^2. It can therefore be used to modify the centered convective operators used in the R.H.S. of (9.a) in order to obtain a second order "upwind" discretization. The L.H.S. operators should also be modified accordingly. To preserve the tri-diagonal character of the algebraic systems, we use simple two points first order upwind differencing for that purpose. It is obviously possible to write 3 different formulas similar to (11), which correspond to upwind-biased discretizations in the positive i, and in the positive and negative j directions. None of the four formulas is in fact applicable to the solution of the Euler equations. For the latter, the R.H.S. term of (9.a) should be split and expressed as the sum of 4 fairly complicated terms according to some appropriate flux-vector splitting rules. Different upwind-biased discretizations should then be used for each term. The flux-vector splitting rules we use are very similar to those developed by Thomas et al. [2].

3. Transonic nozzle flow calculations

To create sufficiently distorted structured meshes, we developed a random grid generation technique. Starting from a fairly regular grid, the different inner mesh points are moved successively and independently into one of the four surrounding quadrilateral elements of the regular grid. The new coordinates of the mesh points depend on the values of random numbers and are used to define a distorted grid with the same number of mesh points. Their motion is limited depending on the value of a user chosen positive parameter α that controls the amount of distortion. The procedure is designed in such a way that mesh cells never overlap if α is smaller than 1.

Starting from a regular grid with 71 x 27 points, we generated more and more distorted meshes in a convergent-divergent channel. One of these meshes is shown on fig. 1. These grids have been used to compute a transonic nozzle flow. Time-dependent Euler equations were discretized using the classical "second order" flux-vector splitting finite-volume method of Thomas et al. [2], and our modified flux-vector splitting finite-element technique. Comparative results obtained with these two techniques using the regular grid and the distorted mesh of fig. 1 are presented on figs. 2-5. The two techniques lead to very similar results on the regular grid. The use of the distorted grid does not significantly modify the results obtained with the new approach, whereas with the finite-volume technique, a severe deterioration is observed. The latter is due to the fact that most of the classical schemes, which are usually claimed to be second order accurate, are in fact not when the grid is significantly distorted.

References

1. Fletcher, C.A.J., "Time-Splitting and the Finite Element Method", Proc. of the 9th Int. Conf. on Numerical Methods in Fluid Dynamics, Lecture Notes in Physics, Vol. 218, pp. 37-47, Springer, 1984

2. Thomas, J.L., Van Leer, B., Walters, R.W., "Implicit Flux- Split Schemes for the Euler Equations", AIAA Paper 85-1680

Acknowledgements

The authors are grateful to the Belgian National Research Foundation (F.N.R.S.) for the financial support provided to one of them (E. Renard) in the frame of the present research.

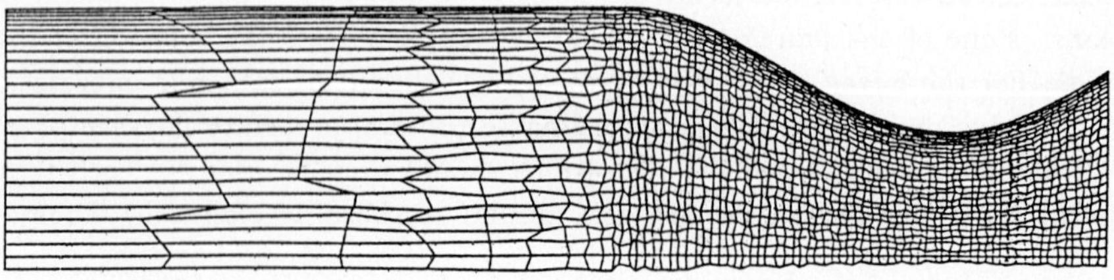

Figure 1. Random grid generation for a convergent-divergent channel (71 X 27 points). $\alpha = 0.6$

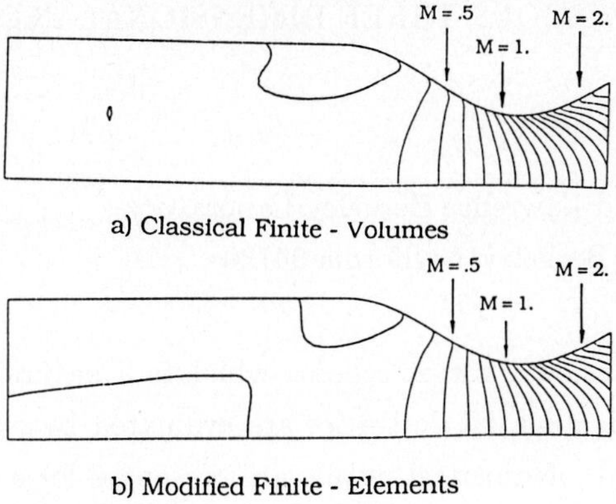

Figure 2. Transonic nozzle flow - isomach lines regular grid (α = 0).

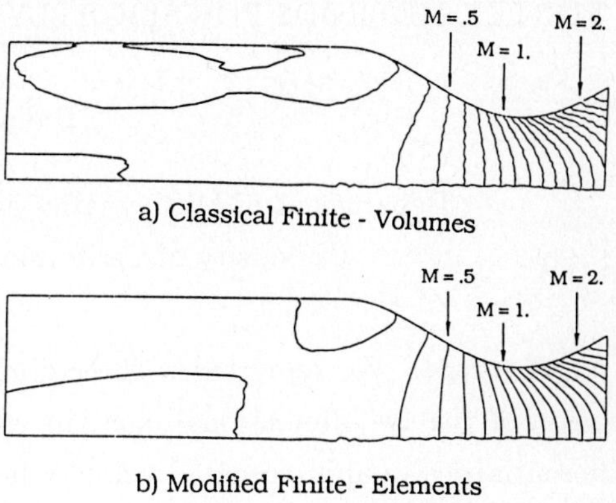

Figure 3. Transonic nozzle flow - isomach lines strongly distorted grid (α = 0.6).

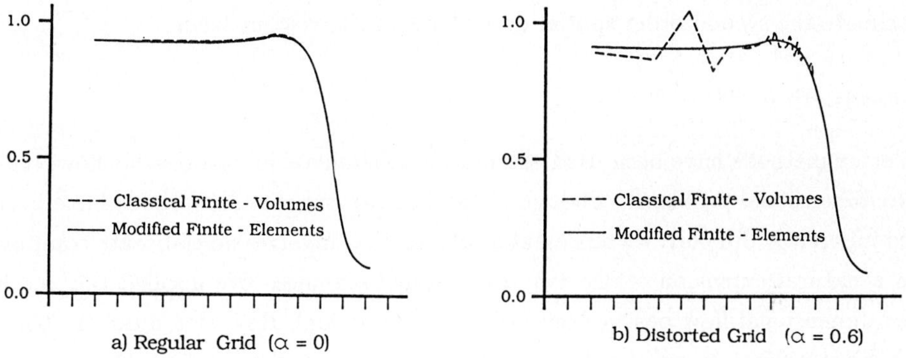

Figure 4. Local Static to Upstream Total Pressure Ratio Along Nozzle Wall.

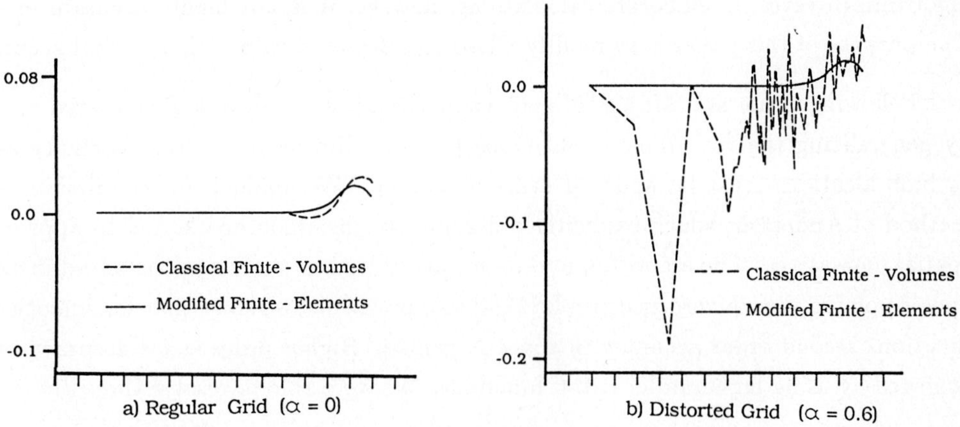

Figure 5. Relative Total Pressure Variation Generated Numerically Along the Nozzle Wall.

VORTEX METHODS FOR SLIGHTLY VISCOUS THREE DIMENSIONAL FLOW

Dalia Fishelov

Department of Mathematics and Lawrence Berkeley Laboratory,
University of California, Berkeley, California 94720

Abstract. We represent a three-dimensional vortex scheme which is a natural extension of the two-dimensional ones, in which spatial derivaties are evaluated by exatly differentiating an approximated velocity field . Numerical results are presented for a flow past a semi-infinite plate, and they demonstrate transition to turbulence. We also suggest a new way to treat the viscous term. The idea is to approximate the vorticity by convolving it with with a cutoff function. We then explicitly differentiate the cutoff function to approximate the second order spatial derivatives in the viscous term.

1. Introduction

Vortex methods have been used extensively to simulate incrmpressible flow, especially for two dimensional problems. Though three dimensional vortex methods have been considered inherently difficult, we represent a scheme that involves no elaborate computations and is a natural extension of the two dimensional schemes. We applied this method to a three dimensional flow past a semi-infinite plate at high Reynolds number. Chorin [5] solved the two dimensional problem numerically; he used computational elements, called blobs, with smoothed kernel. He also intoduced a solution to a three dimensional problem using a filament method. Chorin [5] approximates vortex lines by segments and then, using the Biot-Savart law, he updates the endpoints of the segments every time step. Chorin's algorithm involves no elaborate calculations, however it is not highly accurate in space. The purpose of this paper is to modify Chorin's scheme to gain higher spatial accuracy.

Following Beale and Majda [3] and Anderson [1], we achieve higher spatial accuracy by generalizing the two dimensional blobs to three-dimensional ones. Vorticity as well as blob locations must be updated every time step. We applied, for the first time, the method of Anderson, which explicitly differentiates the smoothed kernel to approximate spatial derivatives. The algorithm and its accuracy is then similar to the two-dimensional one. Applying the convergence proofs [2],[4] to our scheme, we show that for smooth cutoff functions second order accuracy in space is gained. Higher order space accuracy can be achieved by using higher order cutoff functions. We were able to resolve three dimensional features of the flow and transition to turbulence. The numerical results are in agreement with experimental results shown in [6], which suggest that at high Reynolds numbers there exist a large number of small hairpins.

We also suggest a new way to treat the viscous term. The idea is to approximate the vorticity by convolving it with a cutoff function, and then approximate the second order derivatives in the viscous term by explicitly differentiating the cutoff the function. Numerical results performed for the Stokes equation demonstrate the accuracy and convergence of the this scheme.

2. Representation of the Problem

The flow is described by the Navier Stokes equations, formulated for the vorticity ξ:

$$\begin{aligned} &\partial_t \xi + (\mathbf{u} \cdot \nabla)\xi - (\xi \cdot \nabla)\mathbf{u} = R^{-1}\Delta\xi, \\ &\text{div } \mathbf{u} = 0, \end{aligned} \tag{2.1}$$

where $\xi = \text{curl } \mathbf{u}$, $\mathbf{u} = (u, v, w)$ is the velocity vector and Δ is the Laplace operator and R is the Reynolds number. We solve the above equations for a flow past a semi-infinite flat plate. Far away from the plate there is a uniform flow. On the plate we impose the no-leak and the no-slip boundary conditions. As was suggested by physical experiments, we assume that the flow is periodic in the spanwise direction.

The Prandtl equations approximate the Navier-Stokes equations near the plate, and are used therefore in a thin layer near the plate. These equations equations admite the two-dimensional steady state solution - the Blasius solution. However, the three-dimensional Navier-Stokes equations are unstable at high Reynolds numbers($R \geq 1000$), i.e., small perturbations in the Blasius solution may cause large perturbations in the solution as time progresses. The experiments of Head and Bandyapodhyay [6] for high Reynolds numbers indicate the existence of large number of vortex pairs or hairpin vortices, extending through at least a substantial part of the boundary-layer thickness.

3. The Numerical Scheme

We split the Navier-Stokes equations into the Euler equations and the heat equation and apply a Strang-type second order scheme to step the Navier-Stokes equations in time. We first describe the scheme for Euler's equations. These equations can be writen as a set of two ordinary differential equations:

$$\begin{aligned}&\frac{d\mathbf{x}}{dt}(\alpha,t)=\mathbf{u}(\mathbf{x}(\alpha,t),t)\quad,\quad \mathbf{x}(\alpha,0)=\alpha,\\&\frac{d\xi}{dt}(\mathbf{x}(\alpha,t),t)=(\xi(\mathbf{x}(\alpha,t),t)\cdot\nabla)\mathbf{u}(\mathbf{x}(\alpha,t),t),\end{aligned}\tag{3.1}$$

where $\mathbf{x}(\alpha,t)$, where is a particle trajectory. We also use the relation $\mathbf{u}(\mathbf{x},t)=\int K(\mathbf{x}-\mathbf{x}')\xi(\mathbf{x}',t)d\mathbf{x}'$ between vorticity and velocity for incompressible flow to determine the velocity in (3.1). Here K is a matrix valued function given by $K(\mathbf{x})=-\frac{\mathbf{x}}{4\pi|\mathbf{x}|^3}\times$, where $\times$ represent vector product.

To discretize (3.1), we set $\xi=\sum_j\xi_j$, where the ξ_j are functions of small support. Let κ_j be the intensity of the $j-th$ particle, i.e., $\kappa_j=\int\xi_j dxdydz$. Then we obtain the following set of ordinary differential equations for the approximate locations of the particles $\tilde{\mathbf{x}}_i$ and the approximate intensities $\tilde{\kappa}_i$.

$$\begin{aligned}&\frac{d\tilde{\mathbf{x}}_i}{dt}(t)=\sum_{j=1}^{n}K_\delta(\tilde{\mathbf{x}}_i(t)-\tilde{\mathbf{x}}_j(t))\tilde{\kappa}_j(t)\\&\frac{d\tilde{\kappa}_i}{dt}=\sum_{j=1}^{n}(\tilde{\kappa}_i\cdot\nabla\mathbf{x})K_\delta(\tilde{\mathbf{x}}_i(t)-\tilde{\mathbf{x}}_j(t))\tilde{\kappa}_j(t),\end{aligned}\tag{3.2}$$

where $\phi:R^3\to R$, $\phi_\delta=\frac{1}{\delta^3}\phi(\mathbf{x}/\delta)$ is the cutoff function, and $K_\delta=K*\phi_\delta$ is a smoothed kernel. In our calculations we chose $K_\delta(\mathbf{x})=K(\mathbf{x})f_\delta(\mathbf{x})$, where $f_\delta(\mathbf{x})=f(\mathbf{x}/\delta)$. $f(x)$ is radially symmetric and given by $f(r)=1$ for $r\geq 1$ and $f(r)=2.5r^3-1.5r^5$ for $r<1$. This function is continuous with its first derivative at $r=1$.

The second equation to solve is the heat equation $\frac{\partial\xi}{\partial t}=R^{-1}\Delta\xi$. Following Chorin [5] we use the random-walk method to step the heat equation in time, i.e., we move the blobs according to $\tilde{\mathbf{x}}_i^{n+1}=\tilde{\mathbf{x}}_i^n+\eta(\Delta t)$, where $\eta(\Delta t)=(\eta_1(\Delta t),\eta_2(\Delta t),\eta_3(\Delta t))$ and η_1,η_2,η_3 are Gaussian random variables with mean zero and variance $2\Delta t/R$, chosen independently of each other. The Prandtl Equations used in a thin layer $0\leq z\leq z_0$ above the plate, were solved numerically by the tile method, which is the three-dimensional extention of the sheet method (see [5]). This was done to evaluate the boundary conditions on the plate, since blobs did not accurately represrnt the velocity field near the plate.

4. Convergence

Convergence for three-dimensional blob-vortex methods was first proved by Beale[2], and then, using a different approach, by Cottet[4]. Let us first define the Sobolev spaces

$W^{m,p} = \{f, \partial^\alpha f \in L^p(R^n), |\alpha| \leq m\}$ and the norm $\|f\|^p_{m,p} = \sum_{0\leq|\alpha|\leq m} \|\partial^\alpha f\|^p_{0,p}$. Cottet proved that if $\phi \in W^{m,\infty}(R^3) \cap W^{m,1}(R^3)$, $\forall m > 0$,

$$\int_{R^3} \phi(\mathbf{x})d\mathbf{x} = 1, \int_{R^3} \mathbf{x}^\alpha \phi(\mathbf{x})d\mathbf{x} = 0, |\alpha| \leq d-1, \int_{R^3} |\mathbf{x}|^d \phi(\mathbf{x})dx < \infty, \tag{4.1}$$

and if there exist constants C and $\beta > 1$ such that $h \leq C\delta^\beta$, then for h and δ small enough

$$\|\tilde{u} - u\|_{0,p} \leq C\delta^d, \quad p \in (3/2, \infty], \quad t \in [0, \tau].$$

We now apply this theorem to our scheme. Using the relation $\phi(r) = f'(r)/4\pi r^2$, we find that $\phi(r)$ satisfies (4.1) with $d = 2$. In addition, if one chooses the cutoff function ϕ to be infinitely smooth, second order accuracy is achieved. In case that the the cutoff function lies in a Sobolev space for finite m only, Raviat [7, pp. 315] have proved for a two-dimensional problem that if (4.1) holds and $\phi \in W^{m-1,\infty}(R^2)$ for $m \geq 2$ and has compact support, then for all arbitrarily small $s > 0$ there exists a constant C_s, such that $\|\tilde{\mathbf{u}} - \mathbf{u}\|_{0,\infty} \leq C_s\delta^{-s}(\delta^d + h^m/\delta^{m-1})$, provided that $c_2^{-1}\delta^\alpha \leq h \leq c_1\delta^\beta$ with $\alpha \geq \beta > 1$. In our case $\phi \in W^{1,\infty}(R^3)$ and has compact support, then for $\delta = Ch^{2/3}$ the error is at most of order $h^{4/3}$. This can be improved by choosing an infinitely smooth cutoff function. The accuracy of the random-walk algorithm was estimated by Hald for a one-dimensional problem. He proved that the L_2 error decayes like $N^{-1/2}$, where N is the number of tiles.

5. A New Scheme for Viscous Flows

To increase the accuracy of the approximation for the viscous part, we propose the following scheme. For simplicity, we describe the scheme for the two-dimensional case, though it can be easily applied for a three-dimensional problem as well. We follow the charachteristic lines $\frac{d\mathbf{x}}{dt} = \mathbf{u}$, along which the vorticity evolution is given by $\frac{d\xi}{dt} = R^{-1}\Delta\xi$. We approximate the vorticity by convolving it with a cutoff function, and then explicitely differentiate the cutoff function to approximate the second order derivatives appearing in the viscous term. Finally we approximate the integral involved in the convolution by the trapezoid rule and obtain

$$\frac{d\tilde{\mathbf{x}}_i}{dt}(t) = \sum_{j=1}^{n} K_\delta(\tilde{\mathbf{x}}_i(t) - \tilde{\mathbf{x}}_j(t))\tilde{\kappa}_j(t),$$

$$\frac{d\tilde{\kappa}_i}{dt} = R^{-1}\sum_{j=1}^{n} \Delta\phi_\delta(\tilde{\mathbf{x}}_i(t) - \tilde{\mathbf{x}}_j(t)\tilde{\kappa}_j(t).$$

This eliminates the error caused by the statistical process, and yields a scheme which is similar in nature to that applied to the Euler equations. Numerical results performed for the Stokes equation demonstrate the accuracy of the scheme within three significant digits for a coarse initial grid.

6. Numerical Results

We display all the results at $t = 22.5, R = 10000$. In Figure 1 the x, z components of vorticity at the Lagrangian computational points at $y = q/2$ is displayed, where q is the period in y. One can see that for larger x the intensity of the vorticity increases, which is one of the features of transition to turbulence. In Figure 2 we show the y, z components of vorticity at 1.4. Note that for large x the vorticity is no longer directed in one direction. This is in agreement with the results in [6], which indicate the appearance of small hairpins as the flow develops in the streamwise direction. In figure 3 we show the velocity field at $y = q/2$ and in Figure 4 vorticity contours at $x = 1.4$. The computational time to reach $t = 22.5$ on a CRAY X-MP is two hours and thirteen minutes.

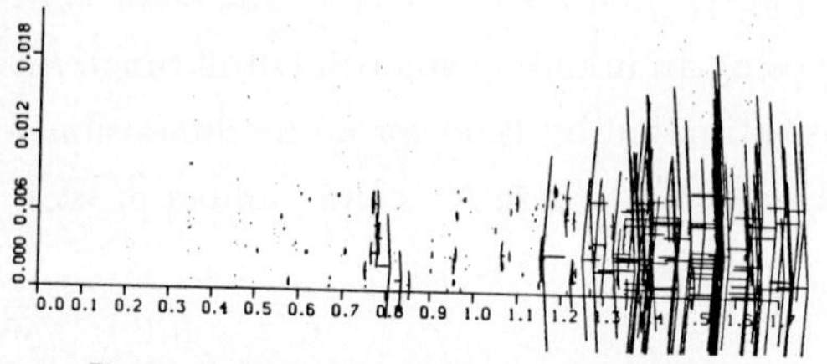

Figure 1. Vorticity in the x, z plane for $y = q/2$, R=10000.

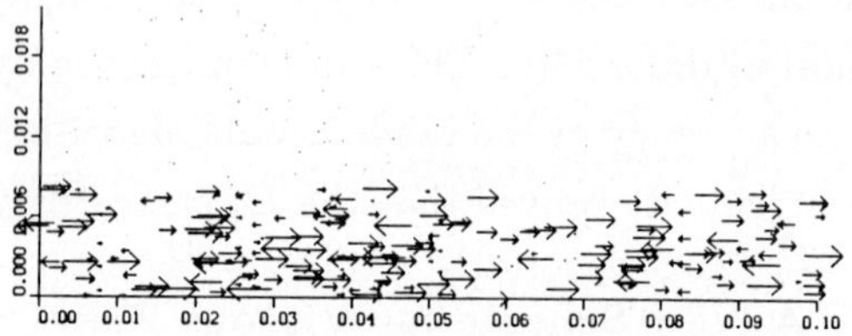

Figure 2. Vorticity in the y, z plane for $x = 1.4$, R=10000.

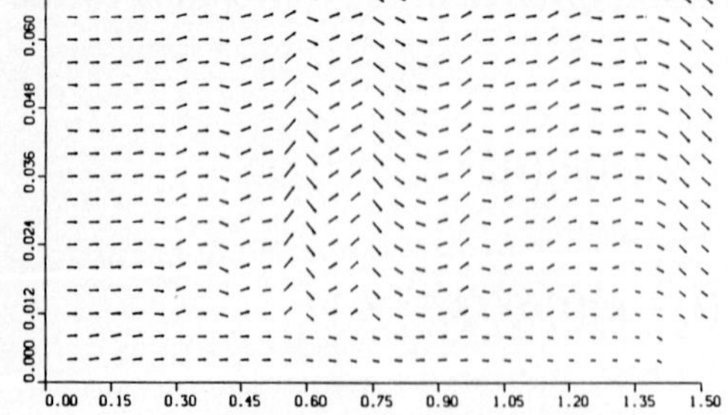

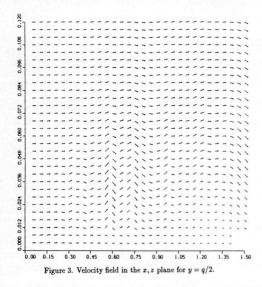

Figure 3. Velocity field in the x, z plane for $y = q/2$.

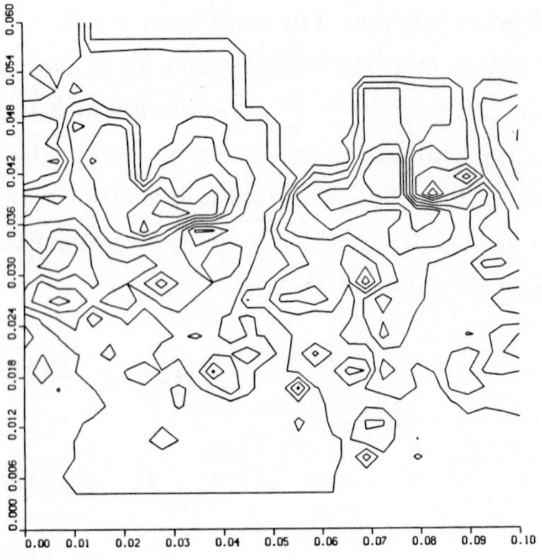

Figure 4. Contours of the z component of vorticity in the y, z plane for $x = 1.4$.

Acknowledgment. I would like to thank Professor Alexandre Chorin for many helpful discussions. I also wish to thank Dr. Scott Baden for vectorizing the code. This work was supported in part at the Lawrence Berkeley Laboratory by the Applied Mathematical Sciences Subprogram of the Office of Energy Research, U.S. Department of Energy, under contract DE-AC03-76SF00098.

References

[1] C. Anderson and C. Greengard, On vortex methods, SIAM J. Numer. Anal., 22 (1985), pp. 413-440.

[2] T. Beale, A convergent 3-D vortex method with grid-free stretching, Math. Comp., 46 (1986), pp. 401-424.

[3] T. Beale and A. Majda, Vortex methods I: Convergence in three dimensions, Math. Comp., 39 (1982), pp.1-27.

[4] G.H. Cottet, A new approach for the analysis of vortex methods in two and three dimensions, To appear in Ann. Inst. Henri Poincare.

[5] A.J. Chorin, Vortex models and boundary layer instability, SIAM, J. Sci. and Stat. Compt., 1 (1980), pp. 1-21.

[6] M.R. Head and P. Bandyapodhyay, New aspects of turbulent boundary layer structure, J. Fluid Mech., 107 (1981), pp. 297-338.

[7] P.A. Raviat, An analysis of particle methods, in Numerical methods in Fluid Dynamics (F. Brezzi ed), Lecture Notes in Mathematics, Vol. 1127, Springer Verlag Berlin (1985).

SECOND ORDER SCHEME IN BIDIMENSIONAL SPACE FOR COMPRESSIBLE GAS WITH ARBITRARY MESH

by

A.J. FORESTIER (*), C. GAUDY (*), & H. BUNG (**)

I - INTRODUCTION

Computational fluid dynamics (C.F.D.) is a large branch of scientific computing that lately undergone explosive growth.

So we want to describe a new bidimensional scheme for compressible flow in any complex geometry.

II - PRESENTATION OF THE SCHEME FOR ONE EQUATION IN MONODIMENSIONNAL SPACE

We describe here an explicit scheme for an hyperbolic equation as

$$\begin{cases} \dfrac{\partial u}{\partial u} + \dfrac{\partial}{\partial x} f(u) = g(u) & \text{where } u : \mathbb{R}^+ \times \mathbb{R} \to \mathbb{R} \\ & f : \mathbb{R} \to \mathbb{R} \\ u(o,x) = u_o(x) & g : \mathbb{R} \to \mathbb{R} \end{cases} \quad (1)$$

A general approach is to obtain a second order scheme (in space and time) which is also entropic.

The contribution of B. VAN LEER [1], A. HARTEN [2], ... is essential in this domain.

We want to describe an approach which is near from B. VAN LEER's one and which can be generalize to Finite Element Method.

The main idea is to use a Riemann exact solver for (1) where if

$$u_o(x) = \begin{cases} u_L & \text{if } x < 0 \\ u_R & \text{if } x > 0 \end{cases}$$, the general solution of (1) is

(2) $u(t,x) = w_R\left(\dfrac{x}{t} ; u_L, u_R\right)$. In particular we remark that if t is positive, u(t,o) is independant of t. So we note

(3) $v_R(u_L, u_R) = w_R\ (0 ; u_L, u_R)$

Now consider a mesh grid where the mesh is centered at $x_{j+1/2}$ and $x_{j+1/2} - x_{j-1/2} = \Delta x$ for simplification.

So we write $x_j = x_{j+1/2} - \dfrac{\Delta x}{2}$ and $x_{j+1} = x_{j+1/2} + \dfrac{\Delta x}{2}$

(*) T.T.M.F./D.E.M.T.
Bat. 91 - C.E.N. SACLAY
91191 Gif sur Yvette - FRANCE
(**) CISI/Ing.
91191 Gif sur Yvette - FRANCE

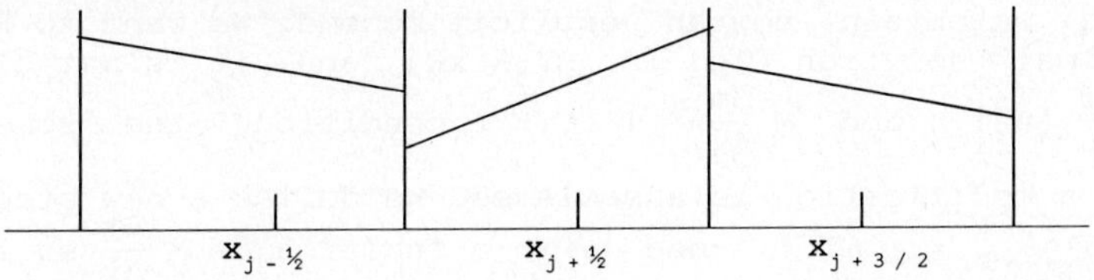

We want to find a solution which is linear in each cell. To know this function, we need to define a mean value and a slope.

In writing $u^n_{j+1/2} \simeq u(n\Delta t, x_{j+1/2})$ (4), we associate in the mesh a slope $\delta^n_{j+1/2}$.

We can now define

$$\begin{cases} u(n\Delta t\ ;\ x_{j+0}) \simeq u^n_{j,+} \equiv u^n_{j+1/2} - \dfrac{1}{2}\,\delta^n_{j+1/2} \\ u(n\Delta t\ ;\ x_{j+1,-0}) \simeq u^n_{j+1,-} \equiv u^n_{j+1/2} + \dfrac{1}{2}\,\delta^n_{j+1/2} \end{cases} \tag{5}$$

The first step in <u>an explicit predictor</u> one as so (we make $g \equiv 0$ but it is easy to extend to $g \neq 0$).

$$u^{n+1/2}_{j,+} \simeq u\left(\left(n+\frac{1}{2}\right)\Delta t; x_j\right) \simeq u(n\Delta t; x_j) + \frac{\Delta t}{2}\left(\frac{\partial u}{\partial t}\right)^n_j \simeq u(n\Delta t; x_j) - \frac{\Delta t}{2}\left(\frac{\partial}{\partial x}(f(u))\right)^n_j$$

and so

$$\begin{cases} u^{n+1/2}_{j,+} = u^n_{j+1/2} - \dfrac{1}{2}\,\delta^n_{j+1/2} - \dfrac{\Delta t}{2\Delta x}\left\{f\left[u^n_{j+1,-}\right] - f\left[u^n_{j,+}\right]\right\} \\ u^{n+1/2}_{j+1,-} = u^n_{j+1/2} + \dfrac{1}{2}\,\delta^n_{j+1/2} - \dfrac{\Delta t}{2\Delta x}\left\{f\left[u^n_{j+1,-}\right] - f\left[u^n_{j,+}\right]\right\} \end{cases} \tag{6}$$

(Notice that the quantities $u^{n+1/2}_{j,+}$ and $u^{n+1/2}_{j+1,-}$ depend only on $u^n_{j+1/2}$ and $\delta^n_{j+1/2}$).

. We use now an <u>explicit Riemann solver</u> and note

$$v^{n+1/2}_j = v_R\left[u^{n+1/2}_{j,-},\ u^{n+1/2}_{j,+}\right] \tag{7}$$

. After this second step, we obtain a conservative explicit solution as :

$$u^{n+1}_{j+1/2} = u^n_{j+1/2} - \frac{\Delta t}{\Delta x}\left[f\left(v^{n+1/2}_{j+1}\right) - f\left(v^{n+1/2}_j\right)\right] \tag{8}$$

. At the end, it is necessary to define <u>the new slope</u> which assome the solution to be monotone (at least, <u>locally</u>) by:

- if $\left(u^{n+1}_{j+3/2} - u^{n+1}_{j+1/2}\right)\left(u^{n+1}_{j+1/2} - u^{n+1}_{j-1/2}\right) > 0$

$$\tilde{\delta}^{n+1}_{j+1/2} = Sg\left(u^{n+1}_{j+1/2} - u^{n+1}_{j/2}\right)\ \mathrm{Min}\left(\left|u^{n+1}_{j+3/2} - u^{n+1}_{j+1/2}\right| - \left|u^{n+1}_{j+1/2} - u^{n+1}_{j-1/2}\right|\right)$$

- in other cases, $\tilde{\delta}^{n+1}_{j+1/2} = 0$

So we obtain now an explicit scheme. We want to know if the numerical solution in in $BV(\mathbb{R}^+ x \mathbb{R})$ and it is not clear if $M = \sup_u \left(\frac{\partial f}{\partial u}(u)\right)$ and $M \frac{\Delta t}{\Delta x} \leqslant 1$ (C.F.L condition) the scheme is stable. So a modification is associated. We define a new step where

(10) $\delta^{n+1}_{j+1/2} = \lambda\ \tilde{\delta}^{n+1}_{j+1/2}$ and λ is a function of $M \frac{\Delta t}{\Delta x}$ so that if $M \frac{\Delta t}{\Delta x} = 1$, we obtain the classical Godounov scheme of first order.

We can then obtain a theorem if by notation $q = \frac{\Delta t}{\Delta x} \sup_u \left|\frac{\partial f}{\partial u}(u)\right|$ (i.e. A. FORESTIER [3]).

<u>THEOREM</u> If λ satisfies $\lambda \leqslant \text{Inf}\left(\frac{1}{q}\,\frac{1-q}{1+q}\ ,\ \frac{1}{1+q}\ ,\ 2\,\frac{1-q}{1+q}\right)$ (11)

the scheme associated to (1) by (6), (7), (8), (9), (10) is convergent in BV $(\mathbb{R}^+ \times \mathbb{R})$ to the entropic solution of (1) and the scheme is of second order.

It is evident then the generalization to a non uniform mesh is easy.

III - <u>GAS DYNAMIC SYSTEM APPLICATION IN MONODIMENSIONNAL SPACE</u>

We have in mind to apply the Riemann solver to the system of compressible fluid where we have mass, momentum and total energy equations of conservation.

$$\left\{\begin{aligned} &\frac{\partial \rho}{\partial t} + \frac{\partial}{\partial x}(\rho u) = 0 \\ &\frac{\partial(\rho u)}{\partial t} + \frac{\partial}{\partial x}(P+\rho u^2) = 0 \\ &\frac{\partial(\rho E)}{\partial t} + \frac{\partial}{\partial x}[(P+\rho E)u] = 0 \end{aligned}\right. \qquad (11)$$

We note : ρ : gas density
ρE : total energy
P : pressure
u : fluid velocity
ρE : internal energy

For closure of this system (11), a state equation is needed : it is the perfect gas where :

$$P = (\gamma-1)\,\rho e \quad \text{or} \quad P = (\gamma-1)\left[\rho E - \frac{1}{2}\frac{(\rho u)^2}{\rho}\right] \qquad (12)$$

If $U = (U_1, U_2, U_3)^T$ with $U_1 = \rho$, $U_2 = \rho u$, $U_3 = \rho E$

and

$F(U) = (V_1, V_2, V_3)^T$ with $V_1 = U_2$, $V_2 = \frac{U_2^2}{U_1} + (\gamma-1)\left[U_3 - \frac{1}{2}\frac{U_2^2}{U_1}\right]$, $V_3 = \frac{U_2}{U_1}[U_3 + V_2]$

we obtain a system

$$\begin{cases} \dfrac{\partial U}{\partial t} + \dfrac{\partial}{\partial X} F(U) = 0 \\ \\ U(o,x) = U_0(x) \end{cases} \tag{13}$$

where $U : \mathbb{R}^+ \times \mathbb{R} \to \mathbb{R}^3$
$F : \mathbb{R}^3 \to \mathbb{R}^3$

In this particular problem, we can use an explicit Riemann solver W_R (see SMOLLER [4]). The procedure scheme is near from the one defined in II.

We describe (in the fundamental variables U_1, U_2, U_3), the solution as a discontinuous linear solution in each cell for U_1, U_2 and U_3. We use three slopes $(\delta_1), (\delta_2), (\delta_3)$ and the vector Δ as $\Delta = (\delta_1, \delta_2, \delta_3)^T$. Then the three steps are :

$$\begin{aligned} \cdot\ U_{j,+}^{n+1/2} &= U_{j+1/2}^n - \frac{1}{2}\Delta_{j+1/2}^n - \frac{\Delta t}{2\Delta x}\left[F\left(U_{j+1,-}^n\right) - F\left(U_{j,+}^n\right)\right] \\ U_{j,-}^{n+1/2} &= U_{j+1/2}^n + \frac{1}{2}\Delta_{j+1/2}^n - \frac{\Delta t}{2\Delta x}\left[F\left(U_{j+1,-}^n\right) - F\left(U_{j,+}^n\right)\right] \end{aligned} \tag{14}$$

$$\cdot\ V_j^{n+1/2} = W_R\ (0;\ U_{j,-}^{n+1/2},\ U_{j,+}^{n+1/2}) \tag{15}$$

$$\cdot\ U_{j+1/2}^{n+1} = U_{j+1/2}^n - \frac{\Delta t}{\Delta x}\left[F\left(V_{j+1}^{n+1/2}\right) - F\left(V_j^{n+1/2}\right)\right] \tag{16}$$

$$\cdot\ \text{If } \left((u_i)_{j+3/2}^{n+1} - (u_i)_{j+1/2}^{n+1}\right)\left((u_i)_{j+1/2}^{n+1} - (u_i)_{j-1/2}^{n+1}\right) > 0$$

$$\begin{aligned} (\delta_i)_{j+1/2}^{n+1} &= Sg\left[(u_i)_{j+1/2}^{n+1} - (u_i)_{j-1/2}^{n+1}\right] \\ &\text{Min}\left(\left|(u_i)_{j+3/2}^{n+1} - (u_i)_{j+1/2}^{n+1}\right|,\ \left|(u_i)_{j+3/2}^{n+1} - (u_i)_{j-1/2}^{n+1}\right|\right) \end{aligned} \tag{17}$$

. In other case $(\tilde{\delta}_i)_{j+1/2}^{n+1} = 0$.

. At the end if $Sg(\tilde{\delta}_1)_{j+1/2}^{n+1} = Sg(\tilde{\delta}_2)_{j+1/2}^{n+1} = Sg(\tilde{\delta}_3)_{j+1/2}^{n+1}$

$$\text{we write} \begin{cases} (\delta_1)_{j+1/2}^{n+1} = (\delta_2)_{j+1/2}^{n+1} = (\delta_3)_{j+1/2}^{n+1} \\ \qquad = \lambda\ Sg\ (\tilde{\delta}_1)_{j+1/2}^{n+1} \ \underset{i}{\text{Min}} \left|(\delta_i)_{j+1/2}^{n+1}\right| \\ \text{and } (\delta_i)_{j+1/2}^{n+1} = 0 \text{ in other cases} \end{cases} \tag{18}$$

So, we obtain an explicit scheme and we can see that it is stable and T.V.D. (Total Variation Diminushing) in fundamental variables for C.F.L. less or equal to 1.

IV - GAS DYNAMIC SYSTEM IN 2D SPACE

It is well-known that the equivalent of the system (11) in two dimensional space is :

$$\begin{cases} \dfrac{\partial \rho}{\partial t} + \dfrac{\partial}{\partial x}(\rho u) + \dfrac{\partial}{\partial y}(\rho v) = 0 \\ \dfrac{\partial(\rho u)}{\partial t} + \dfrac{\partial}{\partial x}(\rho u^2 + P) + \dfrac{\partial}{\partial y}(\rho uv) = 0 \\ \dfrac{\partial(\rho v)}{\partial t} + \dfrac{\partial}{\partial x}(\rho uv) + \dfrac{\partial}{\partial y}(\rho v^2 + P) = 0 \\ \dfrac{\partial(\rho u)}{\partial t} + \dfrac{\partial}{\partial x}[(\rho E+P)u] + \dfrac{\partial}{\partial y}[(\rho E+P)v] = 0 \end{cases}$$

where ρ, P, E, e are defined as in (11) and (u,v) are the two composants of the velocity.

It is possible to write this system as

$$\frac{\partial U}{\partial t} + \frac{\partial}{\partial x} F(U) + \frac{\partial}{\partial y} G(U) = 0$$

where now : $U : \mathbb{R}^+ \times \mathbb{R}^2 \to \mathbb{R}^4$
$F : \mathbb{R}^4 \to \mathbb{R}^4$ and $G : \mathbb{R}^4 \to \mathbb{R}^4$

We use a mesh in $\mathbb{R}^2$ with triangles (on quadrilateral cells) and in each cell, $U = (U_1, U_2, U_3, U_4)^T$ is P_1 - linear - for triangle - (or Q_1 for quadrilateral) function for the four unkowns. It is easy to show that the Riemann problem for compressible gas is equivalent to a monodimensionnal problem in the normal component of a shock propagation.

So we define the cellular fonction by the average value and slopes. We propagate the mean value at step $n+\frac{1}{2}$ as in (14).

The equivalent of (15) is evident and (16) is only changed in a conservative form. The only difficulty is to define the slope in each cell.

For these problem, we cann't use the T.V.D. notion "stricto senso". So we compare the value in the cell with the neighbour cells. For this, we define in the direction of the normal between two cells a slope as in (16).

We can obtain this for each side of the cell (3 for triangle and 4 for quadrilateral) and so a Q_1 fonction that satisfies the monotony for each normal direction (which are only two directions for square cells or in a finite difference scheme).

The scheme, that we have obtained, is equivalent to the 1D case. It is a simple matter to verify that the scheme is stable up to CFL equale 1 (here CFL is $\text{CFL} \equiv \frac{\Delta t}{h} \operatorname{Sup}\left(|u| + |v| + \sqrt{\gamma \frac{P}{\rho}}\right) \leqslant 1$).

VI - CONCLUSIONS

We present a very stable scheme for bidimensionnal problem.

The generalisation to tridimensionnal geometry can be investigate in the same way.

The physical test have been streated without difficulty.

BIBLIOGRAPHY

[1] B. VAN LEER
J.C.P. Vol 32 (1979)

[2] A. HARTEN
On a class of high resolution Total Variation Stable Finite Difference Schemes.
S.I.A.M. Journal of Numerical Analyis (1984), vol. 21

[3] A. FORESTIER
Thesis (University of PARIS VI - 1988)

[4] SMOLLER
Shocks Waves - Reaction Diffusion Equations - Springer-Verlag New-York (1983)

Accurate Simulation of Vortical Flows

Kozo Fujii*
Institute of Space and Astronautical Science,
Yoshinodai 3-1-1, Sagamihara, Kanagawa, 229 Japan

Abstract

Even with recent supercomputers having a large memory, Navier-Stokes simulations for vortical flows do not provide satisfactory results because of the lack of grid resolution to accurately simulate the strength of separation vortices. To overcome this problem, a zonal method is proposed to increase the number of grid points locally. Interface scheme which is critical for an efficient and stable zonal method is based on the Fortified Navier-Stokes concept. Application to both two-dimensional conical and three-dimensional delta wing problems indicates this simple zonal method can improve the accuracy of vortical flow simulations.

Introduction

Recently, there has been progress in the numerical simulation of vortical flow due to three-dimensional flow separation. Euler and/or Navier-Stokes equations were solved numerically to simulate vortical flows such as high-angle-of attack flow over a delta wing[1], over a strake-delta wing[2], over a hemisphere cylinder[3], and so on. Important aspect of these studies compared to earlier computations was the use of much finer grids with the aid of recent supercomputers with large memory such as Cray 2. Thus, more detail of the vortical flow field was revealed. However, the grid distributions used in these studies were the same type as used in the flow simulations without separation vortices. In other words, computational grid is fine only near the body surface. For vortical flows, not only the region near the body but also certain regions away from the body surface are important. Resolution in these regions may be critical for accurate simulations of vortical flows. Vortical flow computations over a strake-delta wing carried out by the present author[4] indicated that the quantitative result was not satisfactory although important phenomenon such as vortex interaction and vortex breakdown were well simulated with fine grid distribution like 850,000 grid points. This indicates that still higher resolution is required for an accurate simulation of vortical flows.

In the present paper, a method to increase the accuracy of the local flow field is proposed. Figure 1 shows the typical cross-sectional grid for the simulation of the flow field of leading-edge separation vortex over a delta wing. Computational region is very large, but the region we are interested in is small (zone 2). What is necessary is to increase the accuracy of the simulation in this small region. Overlap zonal method is proposed for this purpose since it can handle any shape of zone 2 if sophisticated interface scheme such as Chimera[5] method is combined. This fact is important since handling of complicated geometry is now one of the important defects of the finite difference (or volume) method compared to the finite element method. Interface scheme should be simple but accurate and stable, and is based on the Fortified Navier-Stokes concept. In the next section, The Fortified Navier-Stokes equations and the solution method are presented.

Fortified Navier-Stokes Equations and the LU-ADI Algorithm

The 'Fortified Navier-Stokes'(FNS) approach was originally developed by Van Dalsem and Steger[6] to improve the performance of Navier-Stokes algorithms by using fast auxiliary algorithms that solve subsets equations. In this approach, solution to the subset equations is used to add forcing terms to the Navier-Stokes algorithm in the appropriate flow regions. Van Dalsem and Steger used an efficient boundary layer algorithm as subset equations, and accurate drag predictions were obtained with about an order-of-magnitude savings in computer time.

---

* This work was partly carried out when the author was a NRC research associate at NASA Ames Research Center.

In the Fortified Navier-Stokes approach, the "thin-layer" Navier-Stokes equations are modified to include the forcing term as follows (see Ref. 2 for instance for the three-dimensional thin-layer Navier-Stokes equations):

$$\partial_\tau \hat{\mathbf{Q}} + \partial_\xi \hat{\mathbf{F}} + \partial_\eta \hat{\mathbf{G}} + \partial_\zeta \hat{\mathbf{H}} = Re^{-1}\partial_\zeta \hat{\mathbf{S}} + \chi(\hat{\mathbf{Q}}_f - \hat{\mathbf{Q}}) \tag{1}$$

The switching parameter χ is set large compared to all the other terms in the region where the solution $\hat{Q}_f$ of the subset equations is available, and zero outside of the region. For $\chi \gg 1$ the added source term simply forces $\hat{Q} = \hat{Q}_f$. Otherwise it blends $\hat{Q}$ with $\hat{Q}_f$. For a specified $\hat{Q}_f$, this forcing term provides numerical damping, and if treated implicitly, is always stabilizing. For the source term to improve accuracy, $\hat{Q}_f$ must be a better solution than what can be obtained by numerically solving the thin-layer Navier-Stokes equations with the provided grid resolution. The solution algorithm for the thin-layer Navier-Stokes equations should be modified so that the forcing terms are treated implicitly in the solution process. In the present paper LU-ADI solution algorithm[2] is adopted. The essence of this algorithm is that the number of operations in the implicit integration is reduced by using approximate LDU decomposition in addition to the ADI factorization. Each ADI operator is rewritten using the diagonal form, and the flux-vector splitting technique. For example, in the ξ -direction,

$$\begin{aligned} I + h\delta_\xi \hat{A} &\doteq T_\xi (I + h\delta_\xi^b \hat{D}_A^+ + h\delta_\xi^f \hat{D}_A^-) T_\xi^{-1} \\ &\doteq T_\xi (I - \alpha h \hat{D}_{Aj}^- + h\delta_\xi^b \hat{D}_A^+)\ (I + \alpha h |\hat{D}_{Aj}|)^{-1}\ (I + \alpha h \hat{D}_{Aj}^+ + h\delta_\xi^f \hat{D}_A^-) T_\xi^{-1} \end{aligned} \tag{2}$$

This operator is modified by adding the forcing term as follows.

$$T_\xi (I(1+h\chi) - \alpha h \hat{D}_{Aj}^- + h\delta_\xi^b \hat{D}_A^+)\ (I(1+h\chi) + \alpha h |\hat{D}_{Aj}|)^{-1}\ (I(1+h\chi) + \alpha h \hat{D}_{Aj}^+ + h\delta_\xi^f \hat{D}_A^-) T_\xi^{-1} \tag{3}$$

As shown in Eq. 3, this modification is simply to replace identity matrix I with I(1 + hχ). Here, to minimize the factorization error, inversion of I(1 + hχ) is necessary between each ADI operator. Right-hand side is also modified to include the source term. With this implementation, χ can become very large without concern for large factorization errors. As χ becomes large, the algorithm reduces to

$$h\chi \Delta \hat{Q}^n = h\chi(\hat{Q}_f - \hat{Q}^n) \tag{4}$$

or simply $\hat{Q} = \hat{Q}_f$. In other words, the Navier-Stokes algorithm is turned off in regions the where $\chi \gg 1$.

Solution Process

The solution process is explained using Fig. 2. Zone 1 corresponds to the coarse grid and zone 2 corresponds to the local small zone where computational grid is several times as dense as the coarse grid. Currently, zonal interface is simple where zonal boundary coincides as shown in Fig. 2. At each time step, zone 1 is solved first using the Fortified Navier-Stokes equations. The χ function is set large in zone 2 except on the outer boundary of zone 2, and $\hat{Q}_f$ in this zone is specified by the fine grid solution at the previous time step. Namely, the solution in zone 2 is replace by the fine grid solution at the previous time step. Outside of zone 2, the χ function is zero. The χ value can be decreased steeply or gradually at the interface. Next, the physical variables at the interface are interpolated using coarse grid solution and the physical variables for the fine grid are determined at the outer boundary of zone 2. Then, thin-layer Navier-Stokes equations are solved in zone 2, with the χ functions to be zero everywhere. This process is repeated until convergence is obtained for both zones.

Two-dimensional conical flow

First example is a two-dimensional conical flow over a delta wing. The freestream Mach number is 1.4, the angle of attack is 14 degrees, and the Reynolds number is 2 x 10^5. Although the FNS implementation is explained using the three-dimensional equations, the process is almost the same for the conical equations (see Ref. 15 for the conical equations). The grid distribution shown in Fig. 2 is used for this conical computation. The small zone is created locally where the computational grid is twice as dense in the circumferential direction and three time as dense in the radial direction. Figure 3 shows the computed density contour plots. Dashed lines denote the interface boundary. The contour lines are continuous at all the interface boundaries. The FNS works fine as a zonal interface method. Figure 4 shows the regular Navier-Stokes solution for the coarse grid without any local grid increase. Secondary separation near the leading edge seems to be larger in the zonal solution and that changes the location of the primary vortex. The zonal solution clearly improves the accuracy of the vortical flow.

Subsonic flow over a strake-delta wing: 3-D application

Computations for the three-dimensional case were carried out for the flow over a strake-delta wing (see Ref. 2). The freestream Mach number is 0.3, the angle of attack is 12 degrees and the Reynolds number is 1.3×10^6. At this angle of attack, there exist two vortices over the upper surface of the wing; one from the strake leading edge and the other from the main- wing leading edge. However, one flattened vortex is observed when the grid resolution is insufficient. Figure 5 shows the total pressure contour plots for the solution obtained using 61 x 33 x 41 grid points. The solution with a local finer grid zone (added zone 2: 31 x 35 x 67) is shown in Fig. 6. Only the grid points corresponding to the zone-1 grid is used for this plot. The contour plots indicate the existence of two separate vortices after the strake-wing junction, and the solution is obviously improved. In this example, small region like Fig. 2 is created over the wing upper surface. Grid distribution in the chordwise direction is not increased for simplicity.

Recently, many upwind schemes have been proposed for the Euler computation. Some schemes have high resolution for not only non-linear waves but also linear waves, and are less dissipative in the flow field compared to the conventional central differencing schemes with artificial dissipation. In addition, recent study[7] about these schemes showed that the use of proper upwind schemes reduces the unnecssary dissipation in the boundary layer. Thus, the proper upwind schemes may be desirable for the vortical flow simulations. Recent study[8] by the present author indicated that vortex breakdown was observed with much less grid points by the upwind scheme than by the central differencing scheme. This upwind feature was implemented in the present zonal code. Figure 7 shows the total pressure contour plots for the solution obtained by the 'MUSCL' scheme with Roe's average. Figure 8 show the close-up view of the same plots at the 65 % chordwise location. In Fig. 8a representing upwind result, two vortices are more distinct compared to Fig. 8b representing the central difference result.

Summary

A zonal method to improve the accuracy of the simulation of vortical flows is proposed. The application to the two-dimensional and three-dimensional leading-edge separation vortex problem showed that more accurate solutions for the vortical flows were obtained efficiently by the present zonal approach. The use of the present zonal method combined with the proper upwind scheme would result in accurate simulations for vortical flows.

The author would like to thank Dr. W. R. Van Dalsem of NASA Ames Research Center for stimulating discussions during this study.

REFERENCES

1. Thomas, J. L., Taylor, S. L. and Anderson, K. A., "Navier-Stokes Computations of Vortical Flows over Low Aspect Ratio Wing," AIAA Paper 87-0207, 1987.

5. Benek, J. A., Donegan, T. L. and Suhs, N. E., "Extended Chimera Grid Embedding Scheme with Application to Viscous FLows," AIAA Paper 87-1126, 1987.

2. Fujii, K. and Schiff, L. B., "Numerical Simulation of Vortical Flows over a Strake-Delta Wing," AIAA Paper 87-1229, 1987.

3. Ying, X. S., Schiff, L. B. and Steger, J. L., "A Numerical Study of Three-Dimensional Separated Flow Past a Hemisphere Cylinder," AIAA Paper 87-1207, 1987.

4. Fujii, K., Gavali, S. and Holst, T. L., "Evaluation of Navier-Stokes and Euler Solutions for Leading-Edge Separation Vortices," Proc. ICNMLTF, July, 1987, also NASA TM 89458, 1987.

6. Van Dalsem, W. R., and Steger, J. L., "Using the Boundary-Layer Equations in Three-Dimensional Viscous Flow Simulation," Proc. AGARD Fluid Dynamics Panel Symposium on Application of CFD in Aeronautics, Apr. 7-10, 1986.

7. Vatsa, V. N., Thomas, J. L. and Wedan, B. W., "Navier-Stokes Computations of Prolate Spheroids at Angle of Attack," AIAA Paper 87-2627, 1987.

8. Fujii, K and Obayashi, S., "The Use of High-Resolution Upwind Scheme for Vortical Flow Simulations," to appear as NAL TR, Tokyo, Japan.

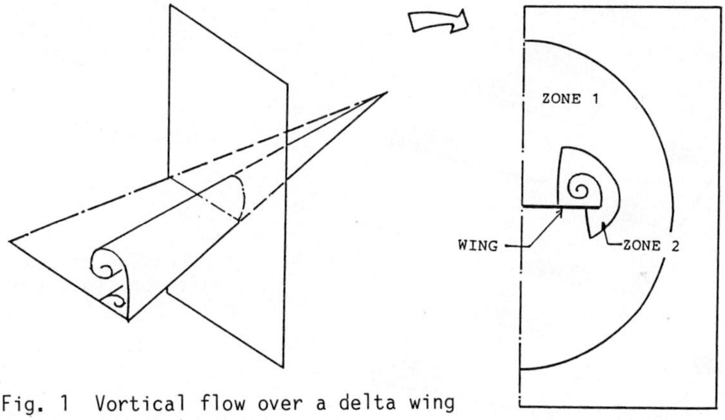

Fig. 1 Vortical flow over a delta wing

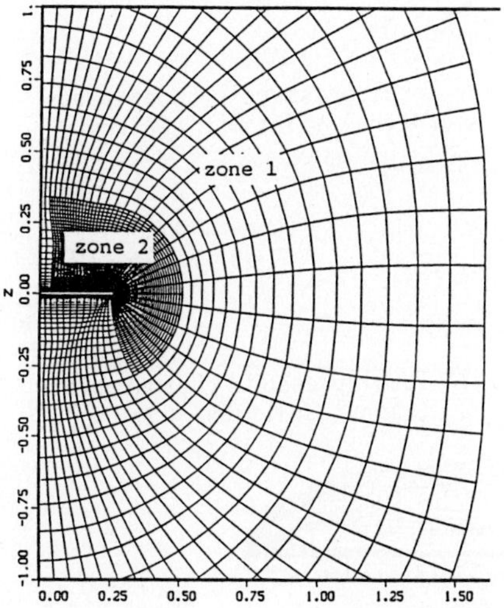

Fig. 2 Zonal grid distributions

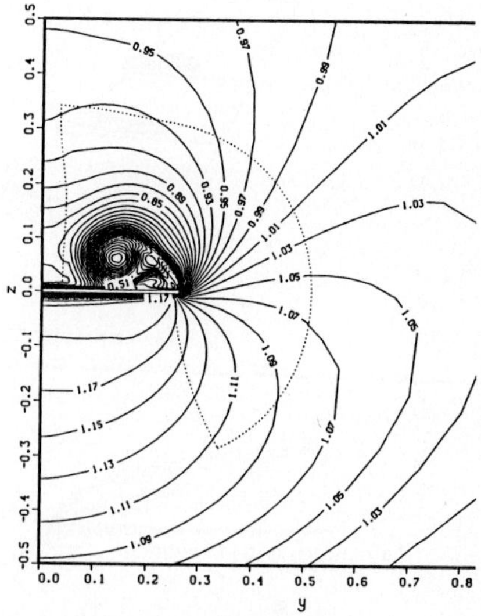

Fig. 3 Computed density contour plots (zonal solution)

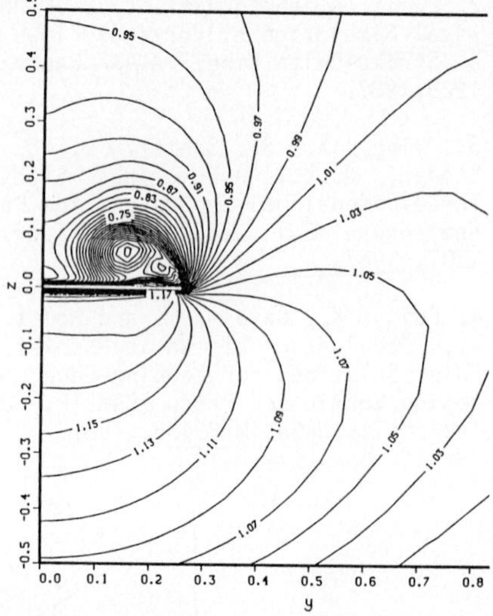

Fig. 4 Computed density contour plots
(single-zone solution)

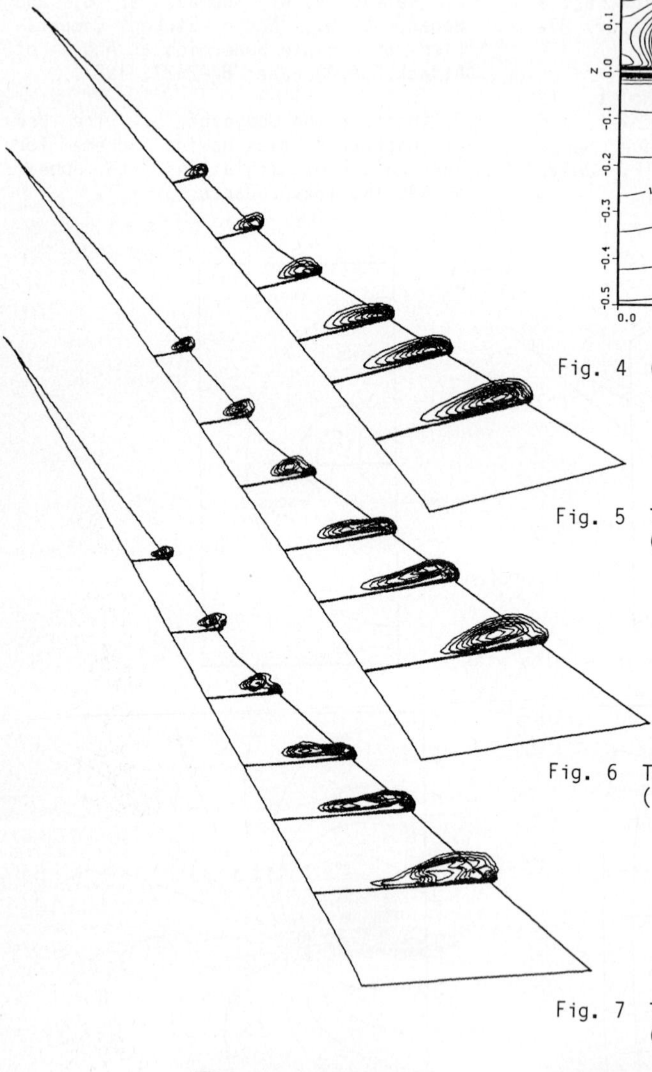

Fig. 5 Total pressure contour plots
(single-zone solution)

Fig. 6 Total pressure contour plots
(zonal solution: central)

Fig. 7 Total pressure contour plots
(zonal solution: upwind)

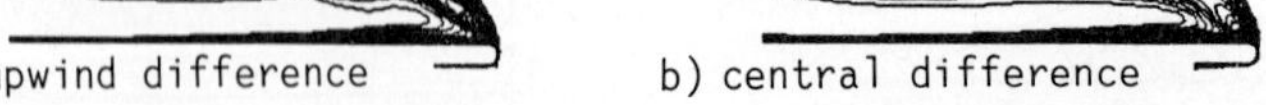

a) upwind difference b) central difference

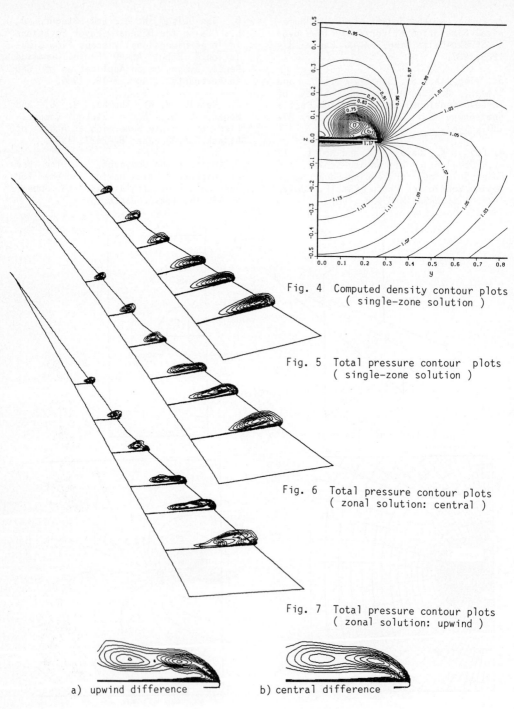

Fig. 8 Spanwise total pressure contour plots
at 65 % chordwise location

Accuracy of Node-Based Solutions on Irregular Meshes

Michael B. Giles*
Department of Aeronautics and Astronautics
Massachusetts Institute of Technology
Cambridge, MA 02139

Introduction

This paper presents an analysis of the order-of-accuracy of discrete steady-state solutions obtained using Jameson's node-based time-marching algorithm [1]. Although this algorithm is usually used in the aeronautical community to solve the Euler or Navier-Stokes equations, in this paper it will be assumed for simplicity that the hyperbolic equation being solved is linear and has constant coefficient matrices.

$$\frac{\partial U}{\partial t} + A\frac{\partial U}{\partial x} + B\frac{\partial U}{\partial y} = 0 \qquad (1)$$

Attention will be limited to irregular computational grids with triangular cells, since these are the grids for which the order of accuracy is most uncertain. Roe has proved [2] that the truncation error of the discrete steady state operator is only first order in these situations, but numerical results presented in a companion paper by Lindquist [3] show that second order accuracy can be achieved if the numerical smoothing is constructed sufficiently carefully. The aim of this paper is to explain this result, but the analysis is not sufficiently rigorous to be considered a proof.

Analysis of local truncation error

If $U(x,y)$ is an analytic solution of the steady equation

$$L\,U(x,y) \equiv A\frac{\partial U}{\partial x} + B\frac{\partial U}{\partial y} = 0, \qquad (2)$$

then integrating over an arbitrary control volume Ω gives

$$\iint_\Omega A\frac{\partial U}{\partial x} + B\frac{\partial U}{\partial y}\,dx\,dy = \oint_{\partial\Omega}(AU\,dy - BU\,dx) = 0 \qquad (3)$$

The control volume Ω for the discrete operator is the union of the triangular cells around one node, and the discrete steady state equation is

$$L^h U^h \equiv \frac{1}{\mathrm{Vol}(\Omega)}\sum_{\partial\Omega}(A\bar{U}\,\Delta y - B\bar{U}\,\Delta x) = 0 \qquad (4)$$

with $(\Delta x, \Delta y)^T$ being the vector face length in the counter-clockwise direction, and $\bar{U}$ being the average value of U on the face obtained by a simple arithmetic average of the values at the two end nodes.

The truncation error T^h is obtained by applying the discrete operator L^h to the analytic solution U.

$$T^h \equiv L^h U = \frac{1}{\mathrm{Vol}(\Omega)}\left(\sum_{\partial\Omega}(A\bar{U}\,\Delta y - B\bar{U}\,\Delta x) - \oint_{\partial\Omega}(AU\,dy - BU\,dx)\right) = \frac{1}{\mathrm{Vol}(\Omega)}\sum_{\partial\Omega}T_f \qquad (5)$$

*Assistant Professor

where T_f is the trapezoidal integration error on each face.

$$\begin{aligned} T_f &= \tfrac{1}{2}A(U_1+U_2)(y_2-y_1) - \tfrac{1}{2}B(U_1+U_2)(x_2-x_1) - \int_1^2 (AU\,dy - BU\,dx) \\ &= (An_x + Bn_y)\left(\tfrac{1}{2}(U_1+U_2)\Delta s - \int_{s_1}^{s_2} U\,ds\right) \\ &= \tfrac{1}{12}(\Delta s)^3(An_x + Bn_y)\frac{\partial^2 U}{\partial s^2} + O\left((\Delta s)^5\right) \end{aligned} \tag{6}$$

In this equation, $(n_x, n_y)^T$ is the outward pointing unit normal vector on the face, and Δs is the length of the face.

There are two important points which arise immediately from this truncation error analysis. The first is that if h is a typical cell dimension then $T_f = O(h^3)$ and $\mathrm{Vol}(\Omega) = O(h^2)$, and hence for a general irregular mesh $T^h = O(h)$, so the local truncation error is first order. (If the mesh is sufficently regular then, as noted by Roe, cancellation of truncation errors T_f on opposing faces leads to the truncation error being second order.) The second point is that if one considers a large control volume Ω whose area is $O(1)$, and sums over all of the cells inside Ω then each of the internal faces contributes equal and opposite amounts to the truncation errors in the two cells on either side. Hence the global integrated truncation error is

$$\sum_{\partial\Omega} T_f = O(h^{-1}) \times O(h^3) = O(h^2), \tag{7}$$

since there are $O(h^{-1})$ faces along the perimeter $\partial\Omega$. These two points suggest that the truncation error has a low-frequency component, with wavelength $O(1)$, whose amplitude is second order, and a high-frequency component with wavelength $O(h)$, whose amplitude is first order. This idea is confirmed in the next section in a different manner.

Spectral content of local truncation error

The truncation error, $T(x,y)$, can be defined to be piecewise linear, with its value on each triangle determined by the values at the three corners. Thus,

$$T(x,y) = \sum_j T_j s_j(x,y) \tag{8}$$

where T_j is the truncation error at node j, and $s_j(x,y)$ is the triangular shape function which varies, in a continuous piecewise-linear manner, from a value of 1 at node j to 0 at its immediate neighbors, and is zero on all other triangles.

The next step is to find the spectral content of $T(x,y)$. For simplicity we now assume that the computational domain is a square of size $\pi \times \pi$, and the boundary conditions are treated perfectly so that the truncation error is zero there. Hence, $T(x,y)$ can be expressed as a sine series expansion.

$$T(x,y) = \sum_{m,n} a_{mn} \sin(mx)\sin(ny) \tag{9}$$

The sum is over positive integer values of m and n. The amplitudes a_{mn} are given by

$$a_{mn} = \frac{4}{\pi^2}\int_0^\pi \int_0^\pi T(x,y)\sin(mx)\sin(ny)\,dx\,dy. \tag{10}$$

The amplitude magnitudes can be bounded by performing an order-of-magnitude analysis. For small, $O(1)$, values of m, n, the integral expression for a_{mn} is first split into a summation over all of the nodes.

$$\int_0^\pi \int_0^\pi T(x,y)\sin(mx)\sin(ny)\,dx\,dy = \sum_j T_j \int_0^\pi \int_0^\pi s_j(x,y)\sin(mx)\sin(ny)\,dx\,dy \tag{11}$$

The contribution from an individual node is then split into two parts, by writing the product of the sine functions as the sum of a constant term and a perturbation.

$$\begin{aligned} &\int_0^\pi \int_0^\pi s_j(x,y)\sin(mx)\sin(ny)\,dx\,dy \\ &\quad= \int_0^\pi \int_0^\pi s_j(x,y)\sin(mx_j)\sin(ny_j)\,dx\,dy \\ &\quad+ \int_0^\pi \int_0^\pi s_j(x,y)\left(\sin(mx)\sin(ny) - \sin(mx_j)\sin(ny_j)\right)dx\,dy \end{aligned} \tag{12}$$

If $m, n = O(1)$, the perturbation term is $O(h)$ and its integral is $O(h^3)$. Since T_j is $O(h)$ and there are $O(h^{-2})$ nodes, the overall contribution to a_{mn} is $O(h^2)$. The other term can be integrated since

$$\int_0^\pi \int_0^\pi s_j(x,y)\,dx\,dy = \tfrac{1}{3}\mathrm{Vol}(\Omega_j) \tag{13}$$

where $\mathrm{Vol}(\Omega_j)$ is the sum of the areas of the triangles which meet at node j, and is the same area that was used earlier in defining the discrete difference operator. Hence, the expression for a_{mn} now becomes

$$a_{mn} = \tfrac{1}{3}\sum_j \mathrm{Vol}(\Omega_j) T_j \sin(mx_j)\sin(ny_j) + O(h^2) \tag{14}$$

Substituting the definition of the truncation error, the area terms cancel and the summation over the nodes can be changed to a summation over faces.

$$a_{mn} = \tfrac{1}{3}\sum_f T_f\left(\sin(mx_{j_-})\sin(ny_{j_-}) - \sin(mx_{j_+})\sin(ny_{j_+})\right) + O(h^2) \tag{15}$$

where T_f is the trapezoidal integration error on an individual face, whose magnitude was estimated earlier, and the nodes j_+ and j_- are the neighboring nodes above and below the face, relative to the direction of the face's unit normal vector $(n_x, n_y)^T$.

If $m, n = O(1)$, then $\sin(mx_{j_-})\sin(ny_{j_-}) - \sin(mx_{j_+})\sin(ny_{j_+}) = O(h)$. $T_f = O(h^3)$ and the number of faces is $O(h^{-2})$ so the sum over all of the faces is $O(h^2)$. Hence, $a_{mn} = O(h^2)$ for $m, n = O(1)$.

The same analysis for $m, n = O(1/h)$ says that $a_{mn} = O(h)$ or smaller. Because of Parseval's theorem,

$$\sum_{m,n} a_{mn}^2 = \frac{4}{\pi^2}\int_0^\pi \int_0^\pi T^2(x,y)\,dx\,dy = O(h^2), \tag{16}$$

there are two possible extremes. If the grid is extremely random then a_{mn} is fairly continuous and so to satisfy Parseval's theorem $a_{mn} = O(h^2)$ for $m, n = O(1/h)$ and decays rapidly for larger m, n. At the other extreme, if the grid is non-uniform in an extremely regular form, then $a_{mn} = O(h)$ at a few values of m, n and is practically zero elsewhere. In either case, the truncation error can be split conceptually into two parts; a low frequency part defined by the amplitudes a_{mn} for

$m, n = O(1)$ and the remaining high frequency part. The low frequency part has a root-mean-square amplitude which is $O(h^2)$ whereas the amplitude of the high frequency part is $O(h)$. The next section determines the corresponding split in the solution error which is produced by this truncation error.

Spectral content of solution error

If U is a solution to the differential equation

$$LU = 0, \tag{17}$$

and U^h is the correponding discrete solution to the difference equation

$$L^h U^h = 0. \tag{18}$$

then the solution error ϵ^h, defined by

$$\epsilon^h = U - U^h, \tag{19}$$

satisfies the discrete equation

$$L^h \epsilon^h = L^h(U - U^h) = L^h U = T^h \tag{20}$$

As shown in the last section, T^h can be split into two parts, a low-frequency part T_1^h and a high-frequency remainder, T_2^h, and the error can be split into two corresponding parts ϵ_1^h and ϵ_2^h. Examing the low-frequency part first, the discrete equation

$$L^h \epsilon_1^h = T_1^h \tag{21}$$

can be accurately approximated by the differential equation

$$L\epsilon_1 = T_1 \tag{22}$$

since L^h is a consistent approximation to L and ϵ_1^h and T_1^h are both smooth. Since this equation is independent of h, it shows that ϵ_1 is of the same order as T_1, and so is second order. (Note: this is the basis of the standard assumption that the solution error is of the same order as the truncation error, because on smooth grids the truncation error is also smooth.)

In calculating ϵ_2^h , the solution error due to the high-frequency truncation error, it is assumed that a fourth difference smoothing is used in the numerical calculation, and that this smoothing has three important properties. The first two are that it is conservative and does not produce a truncation error which is larger than that produced by the physical part of the discrete operator. The third assumption is that it is effective in smoothing all high-frequency oscillations. Given these assumptions, the equation

$$L^h \epsilon_2^h = T_2^h \tag{23}$$

will be approximated in an order-of-magnitude sense by the equation

$$L' \epsilon_2 = T_2 \tag{24}$$

where L' is equal to L plus a term due to the smoothing.

$$L' \equiv A\frac{\partial}{\partial x} + B\frac{\partial}{\partial y} + h^3\left(\frac{\partial^4}{\partial x^4} + \frac{\partial^4}{\partial y^4}\right) \tag{25}$$

Because T_2 is high-frequency, with a typical length scale $O(h)$, boundary effects can be neglected, and the response ϵ_2 can be estimated by considering a solution error which is a constant vector multiplied by a high-frequency sinusoidal function.

$$L' \sin(mx)\sin(ny)\, u_{const} = \Big(m\cos(mx)\sin(ny)A + n\sin(mx)\cos(ny)B + h^3(m^4+n^4)\sin(mx)\sin(ny) \Big)\, u_{const} \qquad (26)$$

This equation shows that if $m,n=O(h^{-1})$, then an $O(1)$ solution error is produced by an $O(h^{-1})$ truncation error, and hence T_2 which is first order produces a solution error ϵ_2 which is second order.

To summarize, the low-frequency component of the truncation error is second order and produces a solution error which is also second order. The high-frequency component of the truncation error is first order, but it produces a solution error which is also second order.

Conclusions

The overall conclusion is that the solution error is second order, even though the local truncation error is first order. The key assumptions in the argument are that boundary condition errors are negligible, and that a fourth difference smoothing is being used which does not increase the order of the truncation error but does smooth high-frequency oscillations.

In discussing the accuracy the root-mean-square error has been used as the indicator of accuracy. It seems likely that the maximum error will behave in a similar manner. In applications one is often most interested in obtaining integral quantities such as lift and drag, and the integration implicit in these will tend to further filter, or suppress, the high frequency component in the error. On the other hand, if the important physical quantity is a derivative variable, such as the surface pressure gradient, then this will amplify the high frequency component.

References

[1] A. Jameson. Current status and future directions of computational transonics. In A. K. Noor, editor, *Computational Mechanics - Advances and Trends*, pages 329–367, ASME, 1986.

[2] P. Roe. *Error Estimates for Cell-Vertex Solutions of the Compressible Euler Equations.* ICASE Report No. 87-6, 1987.

[3] D. R. Lindquist and M. B. Giles. *A Comparison of Numerical Schemes on Triangular and Quadrilateral Meshes.* Proceedings of 11th International Conference on Numerical Methods in Fluid Dynamics, 1988.

SOLUTIONS OF THE INCOMPRESSIBLE NAVIER-STOKES EQUATIONS USING AN UPWIND-DIFFERENCED TVD SCHEME

Joseph J. Gorski
David Taylor Research Center
Bethesda, Maryland

ABSTRACT

The Navier-Stokes equations for incompressible fluid flows have been solved using the pseudo-compressibility concept. The equations are discretized using a third order accurate upwind differenced Total Variational Diminishing (TVD) scheme for the convection terms. The equations are solved implicitly and complex geometries are split into multiple blocks for which structured grids can be generated. This method of solving the Navier-Stokes equations has been implemented in the David Taylor Navier-Stokes (DTNS) series of computer codes. Solutions are provided for several cases including the two-dimensional flow over an airfoil, the axisymmetric flow in a cylindrical container with a rotating lid, and the three-dimensional flow around a circular cylinder mounted on a flat-plate. These results demonstrate the wide applicability of the DTNS computer codes.

INTRODUCTION

For Computational Fluid Dynamics to become more of a design tool it is necessary to develop efficient, reliable, computer codes which can be used effectively with a minimum of effort. The David Taylor Navier-Stokes (DTNS) computer codes, used for solving the Navier-Stokes equations for steady incompressible fluid flows, have been developed with these requirements in mind. With the use of an upwind differenced TVD scheme no numerical parameters need to be changed for different calculations. The equations are solved implicitly and complex geometries can be split into multiple blocks in the grid input for ease of calculation. These aspects of the DTNS codes make them efficient codes which can easily be used for a wide variety of flow calculations. Versions of DTNS are available for two-dimensional (DTNS2D), axisymmetric (DTNSA), and three-dimensional (DTNS3D) flow calculations all of which are demonstrated here.

NAVIER-STOKES EQUATIONS

The Navier-Stokes equations for an incompressible fluid are solved in their primitive variable form using the pseudo-compressibility technique developed by Chorin [1]. A time derivative term for pressure, $\partial(p/\beta)/\partial t$, is added to the continuity equation. The term β controls the convergence rate of the scheme where a value of 1 was used for the present calculations. The addition of a time term to the continuity equation makes the Navier-Stokes equations for incompressible flow hyperbolic in time and thus the equations can be solved using implicit techniques developed for the compressible fluid equations.

DISCRETIZATION TECHNIQUE

The Navier-Stokes equations contain both first derivative convective terms and second derivative viscous terms. The viscous terms are numerically well-behaved diffusion terms and are discretized using central differences. The upwind differenced Total Variational Diminishing (TVD) scheme developed by Chakravarthy et. al.[2,3], for the compressible fluid flow equations, was used for differencing the convective part of the equations. This upwind differenced scheme gives third order accuracy without any artificial dissipation terms being added to the equations. Details of how this discretization method is applied to the Navier-Stokes equations for incompressible fluids can be found in Gorski [4].

SOLUTION PROCEDURE

The equations are solved in an implicit coupled manner using approximate factorization. The convective terms on the implicit side of the equations are discretized with a first-order accurate upwind scheme. This creates a diagonally dominant system which requires the inversion of block tri-diagonal matrices. The implicit side of the equations is only first order accurate but the final converged solution has the high order of accuracy of the explicit part of the equations.

An important quality of any scheme is its convergence rate. The diagonal dominance of the present method allows large time steps to be used for fast convergence. The spatially varying time step of Pulliam and Steger [5] was also used. For even faster convergence the implicit multigrid technique of Jameson and Yoon [6] can be employed.

RESULTS

Results for the two-dimensional flow over an airfoil, the axisymmetric flow in a cylindrical container with a rotating lid, and the three-dimensional flow around a circular cylinder mounted on a flat-plate are presented. No numerical parameters were changed for any of the following calculations.

NACA 4412 AIRFOIL

The two-dimensional turbulent flow over a NACA 4412 airfoil was computed using DTNS2D. The airfoil, investigated experimentally by Coles and Wadcock [7], is at a 13.87

degree angle of attack with a Reynolds number of $1.5X10^6$ based on chord. The turbulence was modelled with the $k-\epsilon$ equations which are also solved using an upwind differenced TVD scheme as described in Ref. [8]. The near-wall turbulence behaviour was modelled using the formulation of Gorski [9]. The airfoil was calculated using a C-type grid with 121 points wrapping around the airfoil and 41 points normal to it. The computed pressure contours for this case are shown in Fig. 1. There is a separation bubble at the trailing edge of the airfoil the extent of which was underpredicted with the present calculation. This may be due to the inadequacy of the $k-\epsilon$ equations in separated flow regions. However, the calculated C_p distribution on the airfoil is in very good agreement with the experiment everywhere but at the trailing edge as can be seen in Fig. 2

CYLINDRICAL CONTAINER FLOW

The axisymmetric flow in a closed circular cylinder produced by a rotating lid was computed using DTNSA. This is a laminar flow with a Reynolds number of 1854 based on angular velocity and cylinder radius. The rotation of the lid creates a meridional circulation in addition to the primary rotating flow. This meridional flow can exhibit the phenomenon of vortex breakdown in the form of a "bubble" on the axis as shown experimentally by Escudier [10]. The calculated streamlines in the total meridional plane for this case are shown in Fig. 3. The cylinder height is twice its radius and the rotating lid is at X = 0. The flow separates on the axis, forming a vortex breakdown bubble, at 79% of the height away from the rotating lid, and reattaches at 62%. The bubble extends outward to 21% of the radius of the cylinder. The calculated shape and location of this vortex breakdown bubble is in good agreement with that observed experimentally by Escudier [10] and also with the computations of Lugt and Abboud [11] based on a completely different algorithm.

CIRCULAR CYLINDER ON A FLAT PLATE

The three-dimensional laminar flow past a circular cylinder mounted on the floor of a wind tunnel with a Reynolds number of 2610, based on cylinder diameter, was computed using DTNS3D. The cylinder height is 1/2 of its diameter and the flow field was measured experimentally by Baker [12]. A horseshoe vortex system is formed in front of the cylinder due to the boundary layer growth on the wind tunnel floor. This vortex system wraps around the cylinder forming a limiting streamline between the free-stream flow and the cylinder. This limiting streamline as well as the flow going over the top of the cylinder can be seen in Fig. 4. This flow was calculated using a three-block grid with approximately 100,000 grid points. Calculated streamwise velocity profiles in the plane of symmetry in front of the cylinder are in good agreement with the experimental data of Baker [12], as shown in Fig. 5, everywhere except at X=-0.73. The calculated vortex is somewhat larger than that measured experimentally but because of the rapid changes the flow is undergoing in this region small discrepancies in the vortex location can lead to such differences in the velocity profiles.

CONCLUSIONS

A set of Navier-Stokes solvers (DTNS) has been developed for steady incompressible fluid flows. The codes are easy to apply to a variety of flow problems, have a high order of accuracy, and utilize solution techniques for rapid convergence. The method is based on an upwind differenced TVD scheme for which no numerical dissipation terms are added to the equations. This paper has presented results for two-dimensional, axisymmetric, and three-dimensional flows for which no numerical parameters were changed for the various calculations. These calculations demonstrate the versatility of the method as well as the DTNS computer codes for a wide range of flow problems.

ACKNOWLEDGEMENT

This work was supported by the ONR under contract N00024-86-WR-10432.

REFERENCES

[1] Chorin, A. J., "A Numerical Method for Solving Incompressible Viscous Flow Problems," *Journal of Computational Physics,* Vol. 2, 1967, pp.12-26.

[2] Chakravarthy, S. R. and Osher, S.,"A New Class of High Accuracy TVD Schemes for Hyperbolic Conservation Laws," AIAA Paper No. 85-0363, 1985.

[3] Chakravarthy, S. R. and Ota, D. K., "Numerical Issues in Computing Inviscid Supersonic Flow Over Conical Delta Wings," AIAA Paper No. 86-0440, 1986.

[4] Gorski, J. J., "TVD Solutions of the Incompressible Navier-Stokes Equations With an Implicit Multigrid Scheme," to be presented at the First National Fluid Dynamics Conference, Cincinnati, Ohio, July 24 - 28, 1988.

[5] Pulliam, T. H. and Steger, J. L., "Recent Improvements in Efficiency, Accuracy, and Convergence for Implicit Factorization Algorithms," AIAA Paper No. 85-0360, 1985.

[6] Jameson, A. and Yoon, S., "Multigrid Solution of the Euler Equations Using Implicit Schemes," *AIAA Journal,* Vol 24, November 1986, pp. 1737 - 1743.

[7] Coles, D. and Wadcock, A. J. ,"Flying Hot Wire Study of Flow Past an NACA 4412 Airfoil at Maximum Lift," *AIAA Journal,* April, 1979, pp. 321-329.

[8] Gorski, J. J., "High Accuracy TVD Schemes for the $k-\epsilon$ Equations of Turbulence," AIAA Paper No. 85-1665, 1985.

[9] Gorski, J. J., "A New Near-Wall Formulation for the $k-\epsilon$ Equations of Turbulence," AIAA Paper No. 86-0556, 1986.

[10] Escudier, M. P., "Observations of the Flow Produced in a Cylindrical Container by a Rotating Endwall," *Experimental Fluids,* Vol 2, 1984, p. 189.

[11] Lugt, H. J. and Abboud, M., "Axisymmetric Vortex Breakdown With and Without Temperature Effects in a Container With a Rotating Lid," *Journal of Fluid Mechanics,* Vol. 179, 1987, pp. 179-200.

[12] Baker, C. J., "The Laminar Horseshoe Vortex," *Journal of Fluid Mechanics,* Vol. 95, part 2, 1979, pp. 347 - 367.

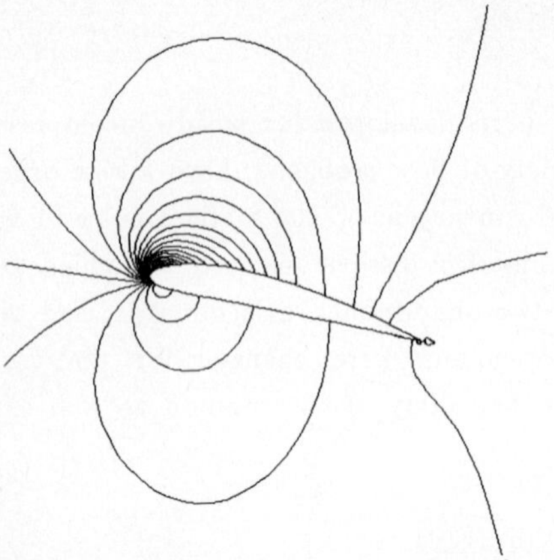

Figure 1. Pressure contours for the NACA 4412 airfoil.

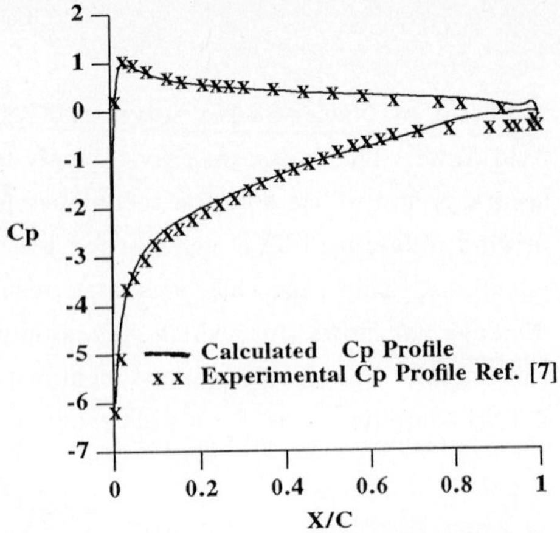

Figure 2. Cp profile for the NACA 4412 airfoil.

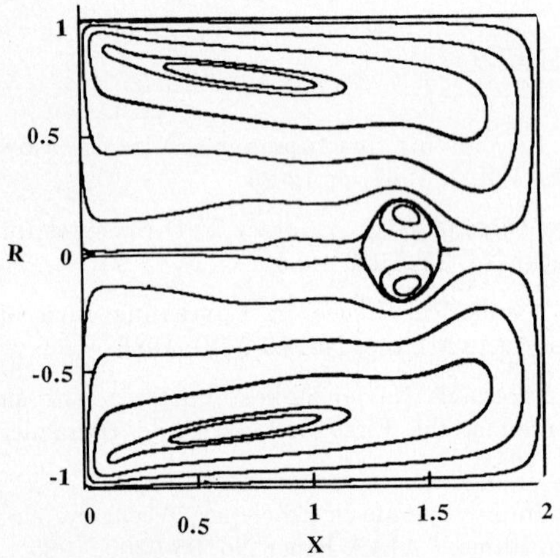

Figure 3. Flow in a cylindrical container with a rotating lid.

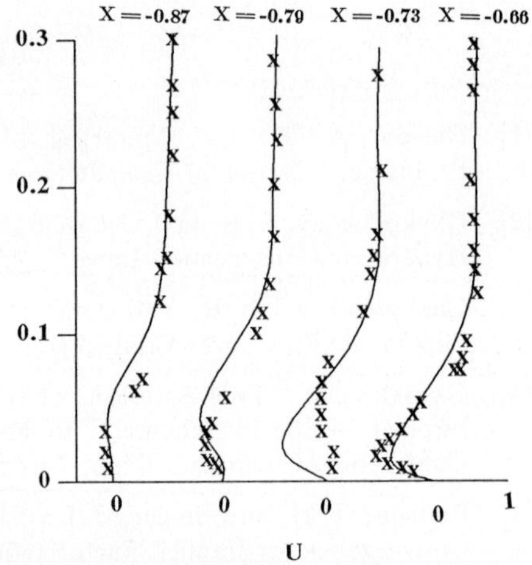

Figure 5. Comparison of computed (-) and experimental (x) streamwise velocity profiles for the 3-D cylinder.

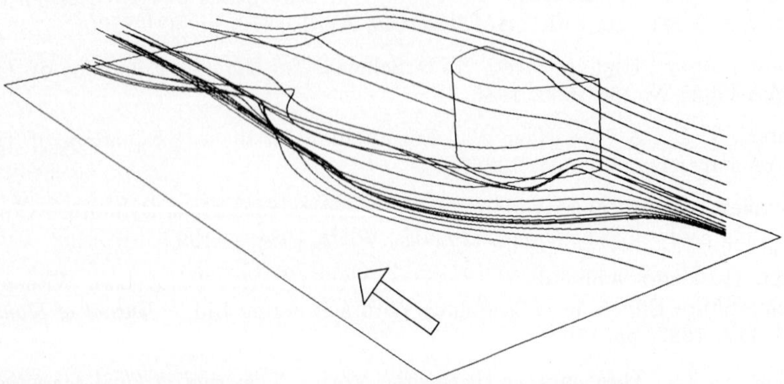

Figure 4. Streamlines past a circular cylinder mounted on a flat plate.

THREE-DIMENSIONAL NUMERICAL SIMULATION OF COMPRESSIBLE, SPATIALLY EVOLVING SHEAR FLOWS

F.F. Grinstein, R.H. Guirguis†, J.P. Dahlburg, and E.S. Oran
Laboratory for Computational Physics & Fluid Dynamics
Naval Research Laboratory
Washington, D.C. 20375

Recent numerical simulations of unforced, subsonic, compressible, spatially evolving two-dimensional shear layers [1,2] have been extended to three dimensions [3] and used to investigate planar shear flows, such as the mixing layer, the plane jet, and the wake behind a rectangular cylinder. In this paper we address the numerical issues of resolution and boundary conditions which are important for the validity of the studies of physical mechanisms in these simulations.

The numerical model solves the time-dependent, inviscid conservation equations for mass, momentum, and energy in three dimensions in order to examine the evolution of large-scale coherent structures. A conserved passive scalar is convected with the velocity field in order to examine entrainment. The equations are solved numerically using a fourth-order phase-accurate Flux-Corrected Transport (FCT) algorithm [4] and direction- and timestep-splitting techniques on structured grids. This approach has been shown to be adequate for simulating the high-Reynolds-number vorticity dynamics in the transition region of free flows and reproduces the basic large-scale features of the flow observed in the laboratory experiments, e.g., the asymmetric entrainment [1], the distribution of merging locations [2], and the spreading rate of the mixing layers [6].

No subgrid modeling other than the natural FCT high-frequency filtering has been included. The nonlinear properties of the FCT algorithm effectively act as a subgrid model by maintaining the large-scale structures and numerically diffusing the wavelengths smaller than a few computational cells. As a consequence, the small residual numerical viscosity of the algorithm mimics the behavior of physical viscosity in the limit of high Reynolds numbers. This was shown, for example, by Grinstein and Guirguis [5], who obtained an upper bound on the effective numerical viscosity by comparing the spread of the mixing layer with that predicted by boundary layer theory for an incompressible laminar mixing layer. In the numerical solution, the mixing layer was forced to remain laminar by constraining the cross stream velocity to be zero; this is a reasonable condition in the limit of high Reynolds number Re, where $v/u = \mathcal{O}(1/Re)$. Figure 1, reproduced from ref. 5, shows the values of the streamwise velocity at several streamwise locations plotted as a function of the similarity variable, $\eta = y/[\nu_e x/U_o]^{\frac{1}{2}}$, where U_o is the velocity of the faster stream. The value of the viscosity, ν_e, was adjusted to obtain the best fit to the laminar mixing layer solution, shown as a solid line. Based on this value of ν_e, an effective Reynolds number, $R_e = \delta\overline{U}/\nu_e$, was evaluated, where $\overline{U}$ is the mean free-stream velocity and δ is the initial mixing layer thickness at $x = 0$. Since numerical diffusion is affected by the grid size, a uniform grid spacing, Δy, was used in the cross stream direction. The results shown in Fig. 1 correspond to a simulation where the initial condition is a step function velocity profile at $x = 0$, in which case, $\delta = \Delta y$, leading to an effective Reynolds number, $R_e \approx 1200$. This is actually a lower bound for the effective Reynolds number, since ν_e becomes smaller when the transition from one velocity to the other in the initial velocity profile takes place over scales of more than a few computational cells. Further details are discussed on ref. 5.

For 3D simulations of shear flows, inflow and outflow boundary conditions are imposed in the streamwise direction, outflow conditions in the cross-stream direction, and periodic boundary conditions are imposed in the spanwise direction. The inflow boundary conditions used in our model were developed and tested in previous simulations for their use in compressible multidimensional shear flow FCT calculations [1,7]. A short inflow plenum is modelled by including a portion of the splitter plate (jet plates or rectangular cylinder) within the computational domain, which provides a realistic transition to the the flow region of interest downstream of the initiation of the mixing layers. The density and velocities are specified at the inflow guard cells and a zero-slope condition on the energy allows the inflow pressure to vary in response to acoustic waves generated by events downstream, thus allowing feedback to occur naturally and with no artificial amplification of reflected acoustic waves. Establishing a reference value for the pressure through the outflow boundary conditions then becomes necessary, to avoid secular errors in the calculations.

Imposing appropriate outflow boundary conditions for subsonic shear flows is an extremely difficult problem. In practice, only a finite portion of the flow can be investigated and we must ensure that the presence of the boundaries that artificially separate the region of interest from the surroundings do not pollute the solution in a significant way. Because of this difficulty, many previous simulations have used well-posed periodic boundary conditions, focusing on temporally (as opposed to spatially) evolving calculations. This can be thought of as describing the time evolution of the flow in compact regions. However, laboratory flows evolve both in space and time and the dynamics of spatially evolving turbulent shear flows cannot be captured by spatially periodic simulations. In previous two-dimensional simulations [1-3,6,7], the approach has been to determine the guard-cell values at the outflow by first or second-order extrapolations of the flow variables and to impose a slow relaxation of the pressure to its ambient value specified at infinity. The flow region of interest (where the gridding is uniform) is surrounded by a buffer region in which the grid gradually stretches towards the outer boundaries where the boundary conditions are actually enforced. The effectiveness of this approach depends on: 1) the dimensions of the buffer region; 2) the time interval over which the simulations are carried out; 3) the nature of the numerical dissipation at the boundaries (due to the gridding) which determines how well the unavoidable spurious reflections are dumped there; and 4) the Mach number of the flow.

To minimize the size of buffer regions in three-dimensional calculations, we have developed new outflow boundary conditions for the shear-flow numerical model. Our guideline is that we would like to define boundary conditions which approximate the time-dependent flow equations as much as possible at the boundaries, while also providing information about the expected flow behavior outside the computational domain. An improved approximation can be obtained by linearizing the inviscid flow equations for the mass and momentum densities, and reducing them to advection equations in the outflow direction (x_{out}),

$$\partial Q/\partial t + V_{out}\partial Q/\partial x_{out} = 0, \tag{1}$$

where V_{out} is the x_{out}-component of the velocity near the boundary. We then dicretize this equation by a first-order upwind scheme and relate the outflow-guard-cell value Q_g^n at the $n-th$ integration cycle as a function of Q_G^{n-1} and the N-th cell value at the boundary, Q_N^{n-1}, from the previous integration cycle by the expression

$$Q_g^n = Q_g^{n-1}(1-\epsilon) + \epsilon Q_N^{n-1}, \tag{2}$$

where

$$\epsilon = V_{out}\Delta t/\Delta x_{out} = c(\Delta x_{min}/\Delta x_{out})M/(1+M), \tag{3}$$

Δt is the integration timestep, c and M are the Courant and Mach numbers, and Δx_{min} and Δx_{out} are the minimum spacing in the grid and the spacing at the outflow boundary, respectively. Using eq. (2) we can assess the quality of the frequently used lower order approximation for the outflow boundary conditions, namely that obtained by defining the guard cell values using a zero-gradient condition. For subsonic flows (say, $M < 0.5$) and Courant numbers $c < 0.5$, from eq. 3 we have $\epsilon < 0.17$; this indicates that the effect of the time derivative cannot be neglected. Thus the zero-gradient condition approximation for the boundary conditions (which corresponds to $\epsilon = 1$) is very unrealistic for these flow regimes.

These concepts were tested in a case study involving a 2D mixing layer problem. The system consisted of two coflowing streams of air with mean free-stream M = 0.3, free-stream velocity ratio 10:1, and normal temperature and pressure conditions. A schematic diagram of the flow configuration is given in Fig. 2a. Fig. 2b and 2c show the grids used for 81 × 94 (run I) and 161 × 100 (run II) computational cells. In both cases, the gridding was uniform in the streamwise direction downstream of the edge of the splitter-plate where 60 or 120 cells were used. The grids were essentially uniform in the cross-stream direction in the region of interest and then gradually stretched towards the outer boundaries.

The outflow conditions specified by eq. 2 were imposed for the mass and momentum densities, and for the passive scalar. The energy density was determined at the guard cells as a function of the other variables. The pressure was relaxed slowly by interpolating between the value at the $N - th$ cell and the value P_{amb} specified at infinity through the expression $P_G = P_N + (P_{amb} - P_N)(r_G - r_N)/r_G$, where r_G and r_N are the distances to the edge of the splitter-plate. The initial shear layer was forced by a planar streamwise momentum perturbation with intensity equal to a few percent of the mean free-stream momentum and frequency in the neighborhood of that of the most amplified mode. The streamwise extent of the computational domains of Figs. 2b and 2c were such that the mean streamwise convection times from the edge of the splitter-plate to the outflow boundary were about $3.5\tau_o$, and $7\tau_o$, respectively, where τ_o is the vortex roll-up period. Contours of the conserved passive scalar for runs I and II at t=$16\tau_o$ are shown in Figs. 3a and 3b. The details of the flow dynamics agree well both qualitatively and quantitatively.

A three-dimensional simulation of the transition region of a subsonic mixing layer evolving both in space and time was performed on the flow configuration shown in Fig. 4. In practice, the number of computational cells ranged from 60 – 240 in the cross-stream direction, 150 – 600 in the streamwise direction, and 40–60 in the spanwise direction. The grid was held fixed in time, and typically used 150 × (60 – 100) × (40 – 60) computational cells in the three-dimensional calculations. The flow was initially constrained to be strictly two-dimensional and allowed to develop in this way. Spanwise vortices roll-up as a result of the nonlinear evolution of the Kelvin-Helmholtz instability. The three-dimensional instabilities were triggered through a spatial (spanwise) sinusoidal perturbation of the cross-stream velocity field. The perturbation was applied during the first computational timestep and then maintained at the inflow boundary. The perturbation level was typically a few percent of the initial mean streamwise velocity, and such that two complete wavelengths of the sinusoidal perturbation were included in the spanwise extent.

Figure 5 shows contours of constant spanwise and streamwise vorticity from three-dimensional simulation of the unforced mixing layer. The contour levels were chosen to be about 30% of the corresponding peak values in the region of the first vortex roll. Higher vorticity values occurred within the areas enclosed by the contours. The figure shows a vortical structure in the process of rolling up just ahead of the edge of the splitter-plate (located just behind the left edge of the vorticity surface in fig. 5a) while a vortex roll and a vortex pairing can be observed further downstream. Although the flow is predominantly two-dimensional, the vortex rolls are distorted due to the presence of the streamwise vortices as shown in fig. 5a. The streamwise vortices in the region of the braids are observed in fig. 5b. A more detailed analysis of the results shows that the vortices occur at the locations of greatest strain rate defined by the inflection points of the spanwise-perturbating function, and appear in counter-rotating pairs as observed in experiments [3]. The distribution of spanwise vorticity in the cores of the vortex rolls is strongly influenced by the presence of the streamwise vortices, and the spread of the mixing layer is considerably enhanced relative to the case in which the mixing layer is constrained to be two-dimensional. Further details on the growth of the simulated mixing layers are discussed in ref. 3, including comparisons with experimental results.

This work was sponsored by the Office of Naval Research and the Naval Research Laboratory. The calculations were partially performed at the computing facilities of the NASA Ames Research Center. We acknowledge useful discussions with Drs. G. Patnaik and K. Kailasanath.

† Science Applications, Inc, McLean, VA.

[1] Grinstein, F.F., Oran, E.S. & Boris, J.P. 1986, *J. Fluid Mech.* **165**, 201.
[2] Grinstein, F.F., Oran, E.S. & Boris, J.P. 1987, *AIAA J.* **25**, 92.
[3] Grinstein, F.F., Oran, E.S. & Hussain, A.K.M.F, Proceedings of the 6th Symposium on Turbulent Shear Flows, Toulouse, France, September 6–9, 1987; AIAA Paper 88-0042, Reno, 1988.
[4] Boris, J.P. & Book, D.L. 1976, in Methods in Computational Physics, Vol. 16, pp. 85-129, Academic Press, New York.
[5] Grinstein, F.F. & Guirguis, R.H. 1988, in preparation, to be submitted to *J. Comp. Phys.*.
[6] Grinstein, F.F., Hussain, F. and Oran, E.S. 1988, Momentum Flux Increases and Coherent-Structure Dynamics in a Subsonic Axisymmetric Free Jet, submitted to *J. Fluid Mech.*.
[7] Boris, J.P., Oran, E.S., Gardner, J.H., Grinstein, F.F., & Oswald, C.E., 1985, in Ninth International Conference on Numerical Methods in Fluid Dynamics, ed. by Soubbaramayer and J.P. Boujot, Springer Verlag, New York, pp. 98-102.

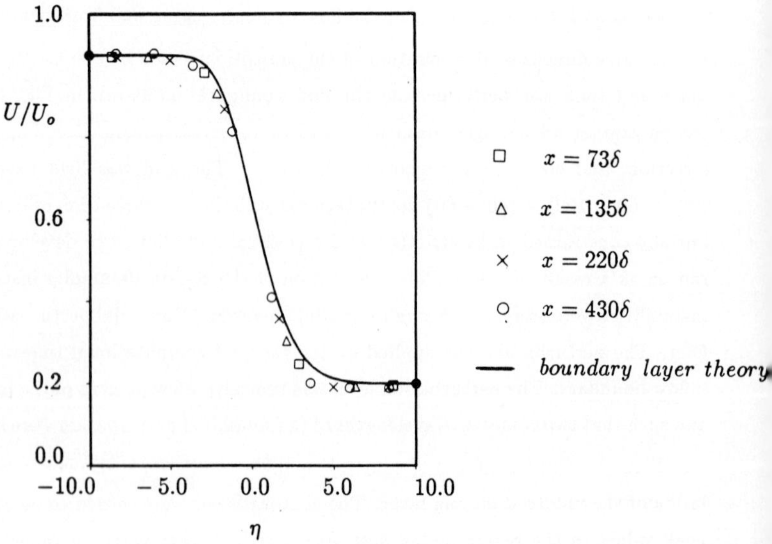

Fig. 1 Initial laminar spread of the mixing layer.

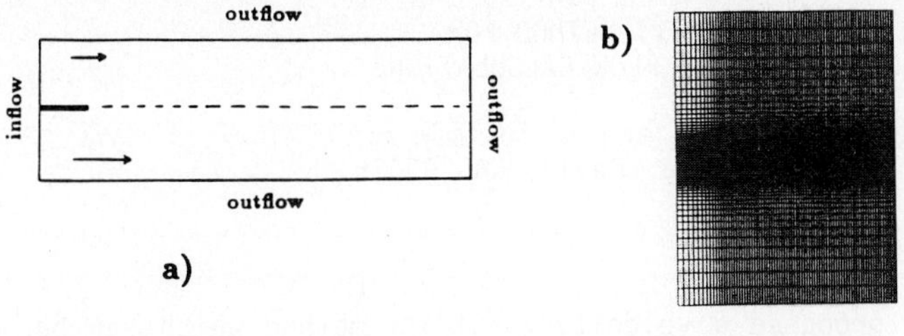

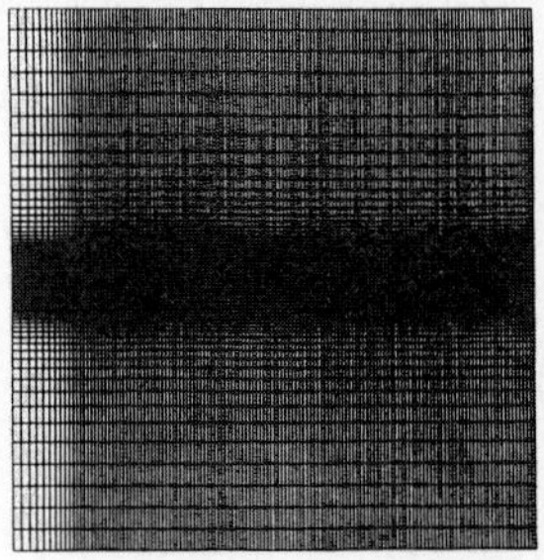

Fig. 2 a) Flow configuration for the 2D mixing layer; b) grid for run I; c) grid for run II.

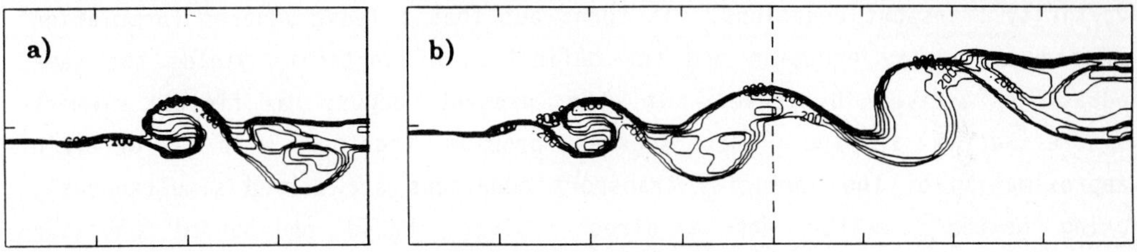

Fig. 3 Contours of the passive scalar, a) run I, b) run II.

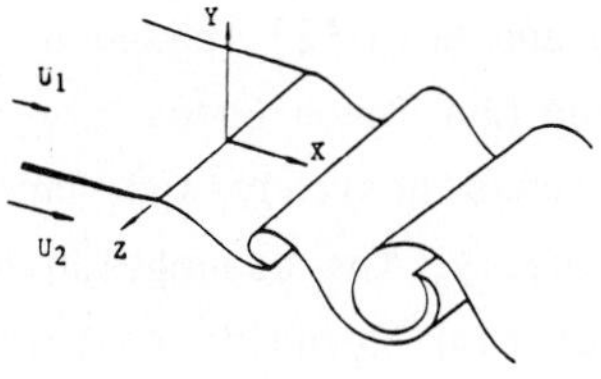

Fig. 4 Flow configuration for the 3D mixing layer.

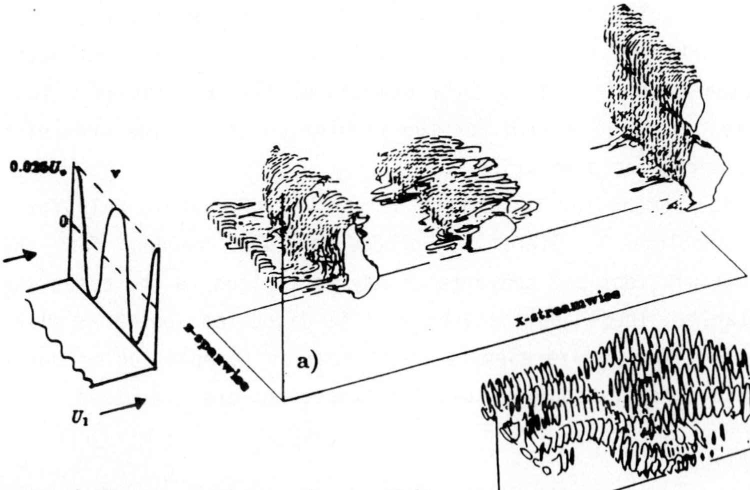

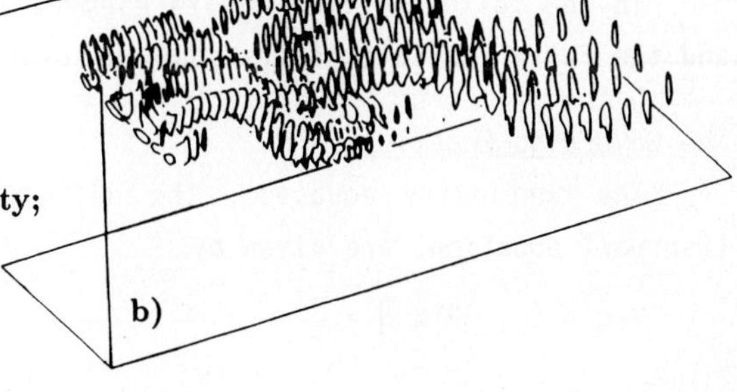

Fig. 5 a) Contours of spanwise vorticity; b) contours of streamwise vorticity.

A VELOCITY/VORTICITY METHOD FOR VISCOUS INCOMPRESSIBLE FLOW CALCULATIONS

M. Hafez, J. Dacles, and M. Soliman
University of California, Davis, CA 95616

Introduction

In 1976, Fasel [1] proposed a velocity/vorticity method which can be used for three dimensional flows. Poisson's equations are solved for the velocity components (with Dirichlet boundary conditions) together with the vorticity transport equations. It turns out that a least squares formulation of the continuity equation and the definition of vorticity yields the same equations derived by Fasel. In the present paper, a finite element discretization of the least squares problem together with a Galerkin approximation of the vorticity transport equations are solved simultaneously using Newton's method and a direct solver (based on banded Gaussian elimination) for a steady laminar flow over a backward facing step. Also, the calculation is repeated using finite differences on a staggered grid. Both solutions are compared to the numerical results of Kim and Moin [2] (based on a fractional step method) and of Armaly [3] et al who used the Teach Code.

There is experimental evidence that the flow becomes unsteady and three dimensional during the transition to turbulence. Obviously, the assumption of a steady laminar two dimensional flow does not hold at high Reynolds numbers; nevertheless numerical solutions of the discrete equations are obtained in this range (for coarse and fine mesh) to demonstrate that the method is robust. Comparison with experimental data is only meaningful for a relatively low Re and provided the mesh is fine enough for the resolution of the features of the physical flow considered for simulation.

The storage requirement for the direct solvers is affordable only for the two dimensional problems. Iterative procedures are needed for three dimensional flow calculations and convergence may be limited to the cases where the solution is stable. These applications will be discussed somewhere else.

In the following, the finite element least squares formulation is derived and the fully implicit, fully coupled solution procedures are discussed.

The Governing Equations

The continuity equation, the definition of vorticity and the vorticity transport equation, are given by

$$\nabla \cdot \vec{q} = s \ , \quad \nabla \times \vec{q} = \vec{\omega} \qquad (1)\text{-}(2)$$

and, $\frac{D\vec{\omega}}{Dt} = \vec{\omega} \cdot \nabla\vec{q} + \nu\nabla^2\vec{\omega}$ (3)

In equation (1), s represents a source distribution in the field. If $s \neq 0$, the continuity equation cannot be automatically satisfied by introducing a stream function for two dimensional and axisymmetric flows. Moreover, the use of stream functions in multiply connected domains requires special treatment of boundary conditions.

To solve equations (1)-(2), with prescribed normal velocity components on the boundary, Rose [4],[5] introduced a compact finite difference scheme which leads to an ill-conditioned system of equations. Osswald, Ghia and Ghia [6] used a direct solver for their discrete velocity equations decoupled from the vorticity equation. Recently Chang and Gunzburger [7] proposed a finite element discretization of equations (1)-(2) which leads to a rectangular algebraic system of equations.

The solution of equations (1)-(2), with prescribed normal velocity component on the boundary, exists only if some compatibility relations hold. For example, using Gauss theorem one obtains

$$\iiint_V \nabla\cdot\vec{q}\ dV = \iiint_V s\ dV = \iint_A \vec{q}\cdot\vec{n}\ dA \quad (4)$$

where $\vec{n}$ is the unit vector perpendicular to A. Similarly, it can be shown that

$$\iiint_V \nabla x\vec{q}\ dV = \iiint_V \vec{\omega}\ dV = \iint_A \vec{n}x\vec{q}\ dA \quad (5)$$

Also, the following vector identity must hold

$$\nabla\cdot\nabla x\vec{q} = \nabla\cdot\vec{\omega} = 0 \quad (6)$$

Fasel proposed to take the curl of equation (2)

$$\nabla x\vec{\omega} = \nabla x(\nabla x\vec{q}) = \nabla(\nabla\cdot\vec{q}) - \nabla^2\vec{q} \quad (7)$$

Substituting equation (1) in equation (7) gives

$$-\nabla^2\vec{q} = \nabla x\vec{\omega} - \nabla s \quad (8)$$

Assuming $\vec{\omega}$ is known, the Poisson's equations with Dirichlet boundary conditions can be solved for the velocity components. In general, equation (8) admits more solutions than those of equations (1)-(2). For the case of "inviscid" flow, in a simply connected domain, Fix and Rose [8] and Phillips [9] showed that the solution of equation (1) and (2) (with prescribed normal velocity components) can be obtained as a solution of equation (8). Beside the normal velocity components, equation (2) is used as boundary conditions for equation (8). With this choice, equation (8) has a unique solution, and extraneous (spurious) solutions are excluded.

To extend the above analysis to the viscous flow case, the prescribed velocity at the boundary is decomposed into two sets; the normal and the tangential components. The normal component together with the definition of vorticity are imposed as boundary conditions for equation (8). The tangential velocity components provide boundary conditions for the vorticity transport equation. If equation (8) and the vorticity transport equation are solved in a coupled manner, there is no need to identify which boundary condition is used for which equation.

<u>A Least Squares Formulation</u>

Consider the functional

$$I = \frac{1}{2}\iiint_V (\nabla\cdot\vec{q}-s)^2 + |\nabla x\vec{q} - \vec{\omega}|^2 \, dV \tag{9}$$

Minimizing I with respect to $\vec{q}$ yields

$$\begin{aligned}\delta I = 0 &= \iiint_V (\nabla\cdot\vec{q} - s)\,\nabla\cdot\delta\vec{q} + (\nabla x\vec{q} - \vec{\omega})\cdot\nabla x\delta\vec{q}\;\, dV\\ &= -\iiint_V \nabla(\nabla\cdot\vec{q} - s)\cdot\delta\vec{q} + \nabla x(\nabla x\vec{q} - \vec{\omega})\cdot\delta\vec{q}\, dV\\ &+ \iint_A (\nabla\cdot\vec{q} - s)\,\delta\vec{q}\cdot\vec{n} + (\nabla x\vec{q} - \vec{\omega})\cdot(\vec{n}x\delta\vec{q})\, dA\end{aligned} \tag{10}$$

Thus, the Euler-Lagrange equation associated with minimizing I is identical to equation (8), provided the boundary terms vanish. The latter condition is satisfied if i) velocity vector $\vec{q}$ is specified at the boundary; ii) the tangential velocity components are specified and equation (1) is imposed at the boundary; or iii) the normal velocity component is specified and the tangential components of equation (2) are imposed at the boundary.

Assume the boundary conditions of iii) are used to determine the velocity field. To solve the vorticity transport equations, three boundary conditions are required; these are the two tangential velocity components prescribed at the boundary as well as the compatibility condition on the vorticity field $\nabla\cdot\vec{\omega} = 0$. It is noticed by Gunzburger and Patterson [10] that the latter is a natural boundary condition associated with the Galerkin formulation of the vorticity transport equation. Moreover, taking the divergence of the vorticity equation, it is shown that $\nabla\cdot\vec{\omega}$ is governed by a (linear) homogeneous equation, hence $\nabla\cdot\vec{\omega}$ vanishes everywhere.

To avoid the coupling between the velocity components at curved boundaries, the boundary condition i) may be used instead, to determine the velocity field. The boundary conditions for the vorticity transport equations are still

$$\vec{\omega}x\vec{n} = \nabla x\vec{q}x\vec{n}. \quad \text{and} \quad \nabla\cdot\vec{\omega} = 0$$

Special treatment may be required, however, to determine accurately the tangential vorticity component at a curved boundary in terms of the derivatives of the velocity components.

Finite Element Discretization

A discrete approximation of the functional I can be easily constructed using finite element techniques. Using standard bilinear shape functions for the velocity components, the functional I can be evaluated in terms of the nodal values. Upon minimization over each element and assembling the contribution from all elements, the nodal equations are readily obtained. A Galerkin method due to Swartz and Wendroff [11] (see also Fletcher [12]) is used for the vorticity equation. The boundary conditions are implemented as discussed before. The resulting nonlinear system of algebraic equations are solved by Newton's method and a direct solver.

For a simple geometry, a finite difference discretization over a staggered grid leads to the five point stencil for the Poisson's equations. Both conservative and nonconservative forms of the vorticity transport equation have been considered using central difference approximations.

Numerical Results

The flow over a backward facing step is simulated using exactly the same geometry of Ref. [2]. The domain of integration is bounded by solid surfaces at the top and the bottom of a rectangular. At the inlet station, a parabolic velocity profile is imposed between the shoulder and the upper surface and it is assumed that the flow is parallel (i.e. the upstream effect of the shoulder is neglected). At the exit, the flow is assumed to be fully developed (i.e. independent of x), hence v = 0. The tangential velocity component (u) is not known and the continuity equation is enforced at the exit boundary.

In Figure (1), the streamlines are plotted for Re = 1000. The convergence histories of the finite element and staggered finite difference calculations are plotted in Figures (2) and (3). Quadratic convergence is obtained independent of the grid size and Re, provided good initial guess is available. Continuation process is used for higher Re cases. The residuals of equations (8) are always of machine accuracy since these equations are linear. Notice in figure (3), the finite difference approximation of the original system (1)-(2) is also of machine accuracy. Comparison with other numerical results and experimental data is given in figure (4). It is clear that the finite element results are less sensitive to the grid size.

The conservative and nonconservative finite difference solutions are almost the same for a smooth flow and a fine mesh. Since $[u\omega] = \langle u\rangle [w] + \langle\omega\rangle [u]$ (where the symbols $[\cdot]$ and $\langle\cdot\rangle$ represent the difference and the average quantities), the difference between the conservative and non conservative formula is proportional to $(u_x \omega_{xx} \Delta x^2 + v_y \omega_{yy} \Delta y^2)$, assuming $u_x + v_y = 0$. However, for very high Reynolds number flows, only with the conservative formulation, it is guaranteed, that the weak (inviscid) solution, with correct contact discontinuieties, is obtained in the limit of a fine mesh.

Finally, all the solutions were calculated on Cray XMP at UC San Diego. The Cpu time per Newton's iteration is only few seconds for the meshes considered. Mesh refinement is limited however by the storage requirement.

<u>References</u>

[1] Fasel, M., J.F.M., Vol 78, pp. 355-383, 1976.

[2] Kim, J. and Morn, P., J. of Comp. Phys., Vol. 59, No. 2, pp 308-323, 1985.

[3] Armaly, B. et al, J.F.M., Vol. 127, 1983.

[4] Rose, M., SIAM J. Numer. Anal. Vol.18, No.2, 1981.

[5] Gatski, T., Grosch, C. and Rose M., J. Comp. Phys., Vol. 48, No. 1, 1982.

[6] Osswald, G., Ghia, K. and Ghia, U., AIAA Paper, 87-1139.

[7] Chang, C. and Guneberger, M., to appear, 1988.

[8] Fix, G. and Rose, M., SIAM J. Numer. Anal. Vol.22, No.2, 1985.

[9] Phillips, T., IMA J. of Num. Anal. Vol. 5, 1985.

[10] Gunzeberger, M. and Peterson, J., to appear, 1988.

[11] Swartz, B. and Wendoff, B., Math. of Comp. Vol. 23, 1969.

[12] Fletcher, C., J. Comp. Phys. Vol. 51, 1983.

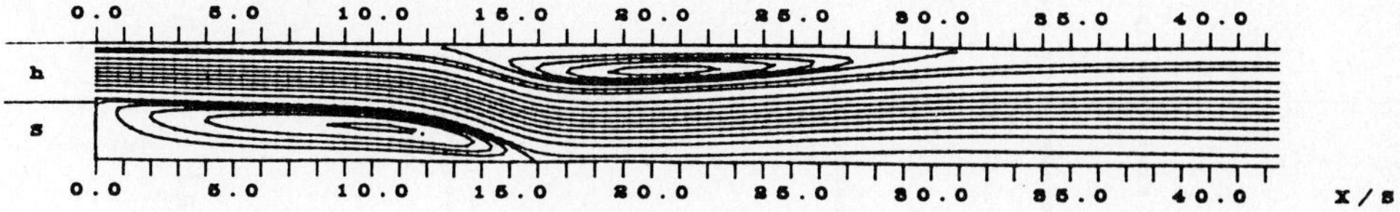

Figure (1) Streamlines of a flow over a backward facing step

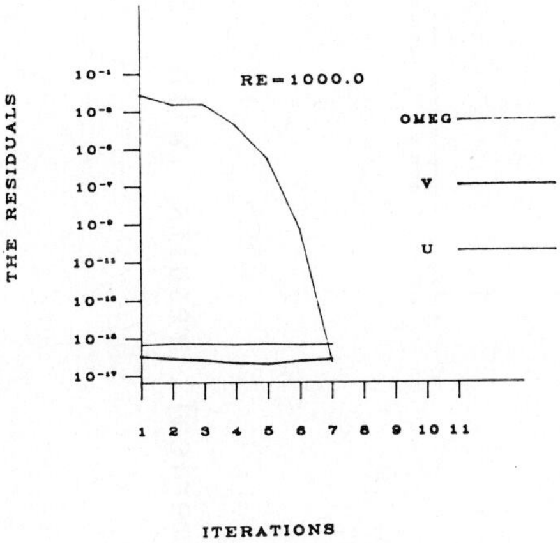

Figure (2) Convergence history of a finite element calculation

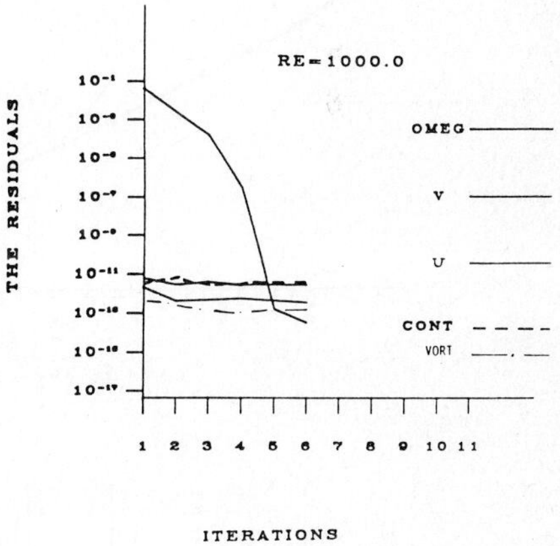

Figure (3) Convergence history of a finite difference calculation

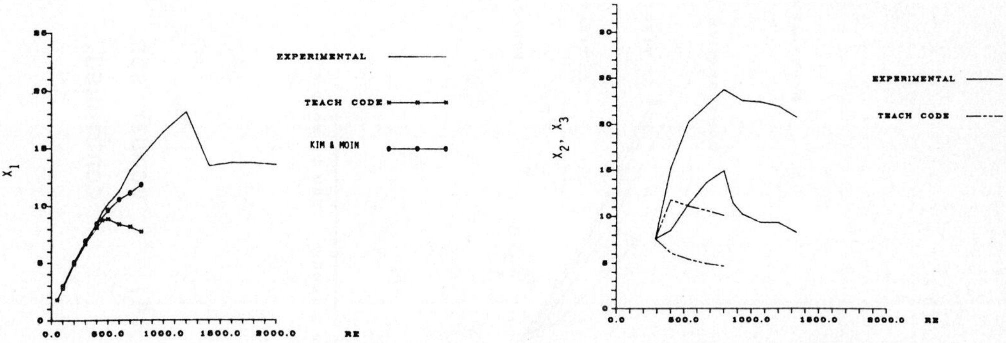

Figure (4) Comparison of Numerical results with Experimental data.

X_1 : length of separation bubble on the lower wall.

X_3-X_2 : length of separation bubble on the upper wall.

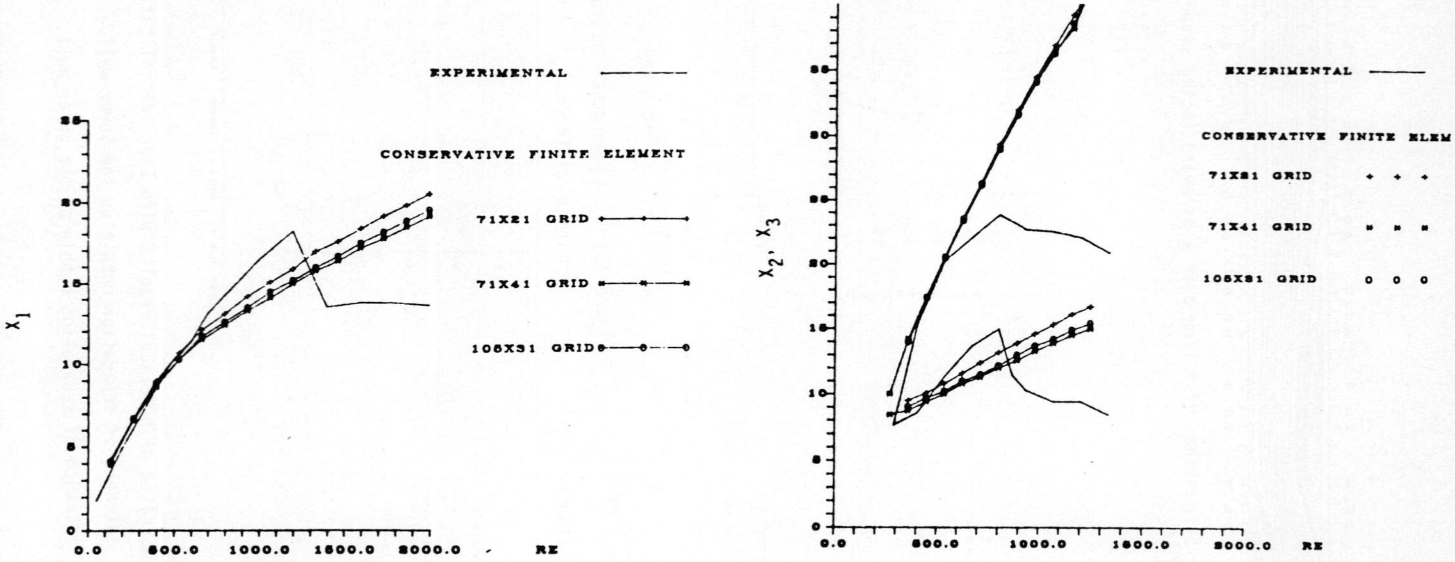

Figure (4) Comparison of Numerical results with Experimental data-continued.

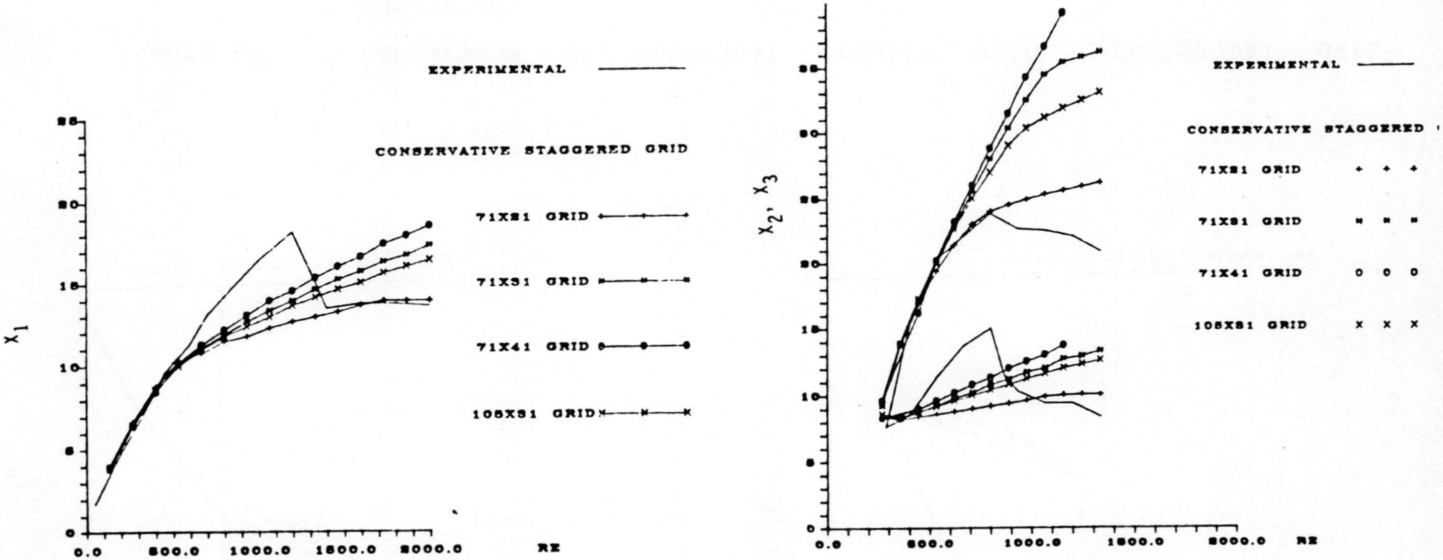

Figure (4) Comparison of Numerical results with Experimental data-continued.

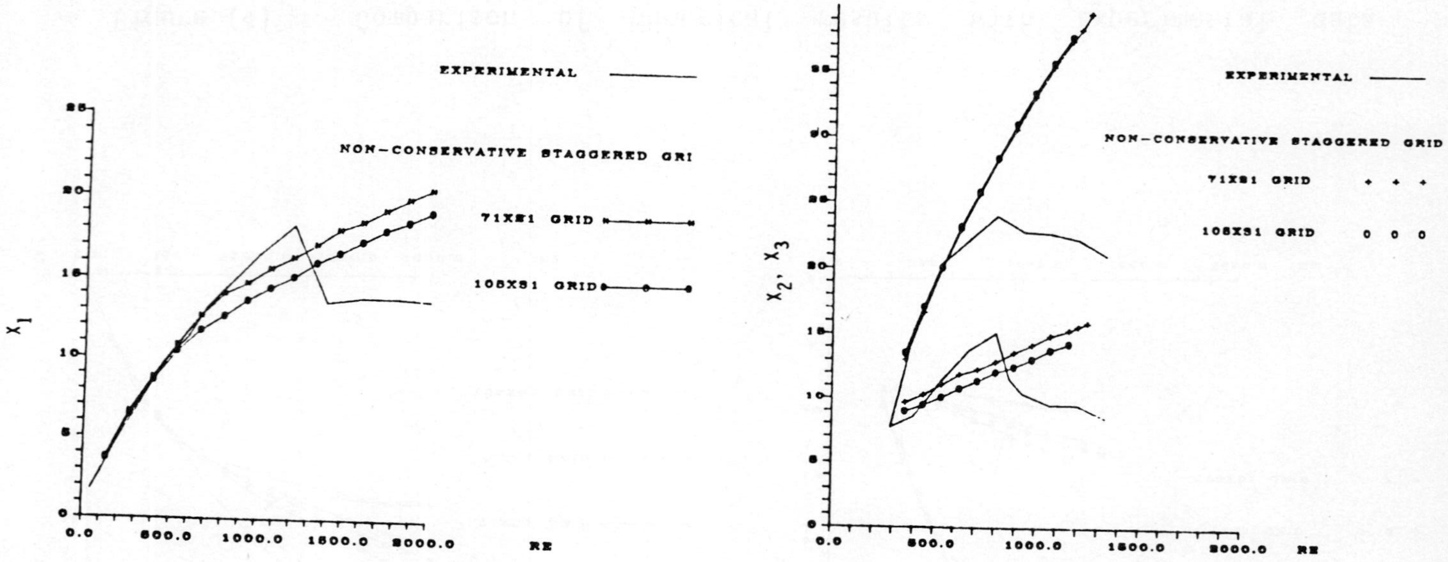

Figure (4) Comparison of Numerical results with Experimental data-continued.

PULSATILE FLOWS THROUGH CURVED PIPES

Costas C. Hamakiotes and Stanley A. Berger
Department of Mechanical Engineering
University of California, Berkeley, CA 94720

I. INTRODUCTION

Unsteady flows in curved pipes and tubes are exceedingly complex flows with important industrial, areospace, and physiological applications. Curved sections of tube are ubiquitous in piping systems and engines of all sorts. They also occur in the human vascular system, and this application is the focus of our research. In this paper we describe our procedure to obtain numerical solutions of the complete, unsteady Navier-Stokes equations for laminar, fully-developed periodic flows in curved pipes.

II. MATHEMATICAL FORMULATION

The governing equations, written in conservative form for the toroidal coordinate system shown in Figure 1 are:

$$St_m \frac{\partial u}{\partial t} + \frac{1}{rB}\left[\frac{\partial}{\partial r}\left(rBu^2\right) + \frac{\partial}{\partial \phi}\left(Buv\right) - Bv^2 - \delta r w^2 \cos\phi\right]$$
$$= -\frac{\partial P}{\partial r} + \frac{1}{Re_m}\left\{\frac{1}{rB}\left[\frac{\partial}{\partial r}\left(rB\frac{\partial u}{\partial r}\right) + \frac{\partial}{\partial \phi}\left(\frac{B}{r}\frac{\partial u}{\partial \phi}\right)\right]\right. \tag{2.1}$$
$$\left. - \frac{1}{r^2}\left(2\frac{\partial v}{\partial \phi} + u\right) + \frac{\delta v \sin\phi}{rB} + \frac{\delta^2 \cos\phi}{B^2}\left(v\sin\phi - u\cos\phi\right)\right\},$$

$$St_m \frac{\partial v}{\partial t} + \frac{1}{rB}\left[\frac{\partial}{\partial r}\left(rBuv\right) + \frac{\partial}{\partial \phi}\left(Bv^2\right) + Buv + \delta r w^2 \sin\phi\right]$$
$$= -\frac{1}{r}\frac{\partial P}{\partial \phi} + \frac{1}{Re_m}\left\{\frac{1}{rB}\left[\frac{\partial}{\partial r}\left(rB\frac{\partial v}{\partial r}\right) + \frac{\partial}{\partial \phi}\left(\frac{B}{r}\frac{\partial v}{\partial \phi}\right)\right]\right. \tag{2.2}$$
$$\left. + \frac{1}{r^2}\left(2\frac{\partial u}{\partial \phi} - v\right) - \frac{\delta u \sin\phi}{rB} - \frac{\delta^2 \sin\phi}{B^2}\left(v\sin\phi - u\cos\phi\right)\right\},$$

$$St_m \frac{\partial w}{\partial t} + \frac{1}{rB}\left[\frac{\partial}{\partial r}\left(rBuw\right) + \frac{\partial}{\partial \phi}\left(Bvw\right) + \delta r w\left(u\cos\phi - v\sin\phi\right)\right]$$
$$= -\frac{\delta}{B}\frac{\partial P}{\partial \theta} + \frac{1}{Re_m}\left\{\frac{1}{rB}\left[\frac{\partial}{\partial r}\left(rB\frac{\partial w}{\partial r}\right) + \frac{\partial}{\partial \phi}\left(\frac{B}{r}\frac{\partial w}{\partial \phi}\right)\right] - \frac{w\delta^2}{B^2}\right\}, \tag{2.3}$$

$$\frac{\partial}{\partial r}\left(rBu\right) + \frac{\partial}{\partial \phi}\left(Bv\right) = 0, \tag{2.4}$$

where $\delta = a/R$ and $B = 1 + \delta r \cos\phi$. Velocities, pressure, position vector and time have been non-dimensionalized by W_{DC}, ρW_{DC}^2, a, and ω, respectively, where a is the radius of the pipe, and ω is the frequency of the oscillation. A volumetric flow rate of the form $Q(t) = Q_{DC} + Q_{AC} \cos(t)$ is assumed, and $W_{DC} = Q_{DC}/\pi a^2$. The mean Reynolds and Strouhal numbers are defined as $Re_m = aW_{DC}/\nu$ and $St_m = a\omega/W_{DC}$, respectively. The Strouhal number can also be defined as $St_m = \alpha/Re_m^{\frac{1}{2}}$, where α is the frequency parameter, $\alpha = a(\omega/\nu)^{\frac{1}{2}}$.

The following initial and boundary conditions are assumed:

$$u(r,\phi,t=0) = v(r,\phi,t=0) = 0,\ w(r,\phi,t=0) = \frac{1}{1+\delta r\cos\phi} \qquad \text{(inviscid vortex)} \quad (2.5)$$

$$u(1,\phi,t) = v(1,\phi,t) = w(1,\phi,t) = 0 \qquad \text{(no-slip)} \quad (2.6)$$

$$\frac{\partial u}{\partial \phi} = \frac{\partial w}{\partial \phi} = 0, \text{ and } v = 0, \text{ at } \phi = 0, \pi \qquad \text{(symmetry)} \quad (2.7)$$

$$\frac{\partial u}{\partial \theta} = \frac{\partial v}{\partial \theta} = \frac{\partial w}{\partial \theta} = 0 \qquad \text{(fully-developed flow)} \quad (2.8)$$

$$Q(t) = Q_{DC} + Q_{AC}\cos(t) = \frac{2}{\pi}\int_0^{\pi}\int_0^1 w\, r\, dr\, d\phi.$$

(conservation of mass and incompressibility) (2.9)

The system of equations (2.1)-(2.4) was solved by finite differences using the Projection Method (Chorin, 1968). In the first step the momentum equations are solved without the pressure gradient to obtain an auxiliary velocity field. In the second step a Poisson equation is solved to update the pressure. Finally, in a last step, the correct velocity at the next time step is found from reduced momentum equations without convective or viscous terms.

The two-dimensional domain was discretized using a non-uniform 15x19 staggered mesh. The non-uniform mesh was characterized by a greater number of mesh points in regions where the flow gradients were expected to be large, around the entire periphery of the pipe and near the inner bend. Each time cycle of duration 2π was subdivided into 400 time steps.

III. RESULTS AND DISCUSSION

Calculations were carried out for $\delta = 1/7$, $Q_{DC} = Q_{AC}$, $\alpha = 15$, and the range $1 \le Re_m \le 1000$. If we define the mean Dean number as $\kappa_m = 2\ Re_m \cdot \delta^{\frac{1}{2}}$, this range corresponds to $0.756 \le {}_m \le 756$. For the details of the numerical calculations, and for a complete presentation of the results, the interested reader may consult Hamakiotes (1987).

In determining the solution integration of the equations must be carried out for a number of cycles sufficient for periodicity of the results to be achieved.

Typically 30-40 cycles are required before the two secondary components of the velocity converge to less than 10^{-3}. During this time the axial component of the velocity and the pressure have each converged to about 10^{-8}. The divergence is calculated to be about 10^{-8} at all times, and the numerically computed volumetric flow rate to be within $2x10^{-4}$% of the prescribed flow rate.

Figures 2a-h and 3a-h show the results for Re_m = 100 and 500, respectively. Part a of these figures shows the calculated pressure gradient (with its algebraic sign), which leads the volumetric flow rate by about 90°. Because the flow is assumed to be symmetric with respect to the centerplane, on the upper half of the pipe cross section on each of parts b-h of Figures 2 and 3 the secondary velocity vectors are shown, and on the lower half the axial isovelocity contours. The left hand side of the cross section corresponds to the inner bend and the right hand side to the outer bend.

We note that for Re_m = 100 (Figure 2) we have a Dean-like flow, one large secondary vortex on each half of the pipe cross section. In addition, when backflow (regions of negative axial flow) appears, it extends along the entire wall and occurs when the volumetric flow rate is a minimum.

The flow field for Re_m = 500 is far more complicated. At higher Reynolds numbers the Dean vortex is pushed towards the wall. Encountering the wall it mushrooms out, leading to a complex secondary flow, typified by these results for Re_m = 500, consisting of a number of smaller vortices. We note also that when backflow occurs for this case it occupies almost the entire inner half of the cross section. For both values of Reynolds numbers presented here the maximum axial velocity is located off the center near the outer bend.

ACKNOWLDEGEMENT

This work was supported by the National Science Foundation under Grant No. MEA-8116360 and ECE-8417852.

REFERENCES

Chorin, A.J. (1968), "Numerical Solution of the Navier-Stokes Equations," Math. Comp. 22, 745.

Hamakiotes, C.C. (1987), "Fully Developed Pulsatile Flow in a Curved Tube," Master of Science Thesis, University of California, Berkeley.

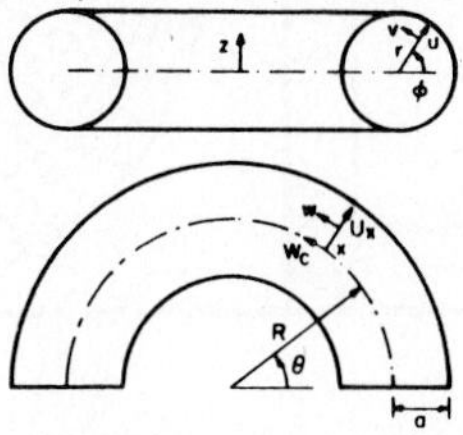

Figure 1. Totoidal Coordinate System

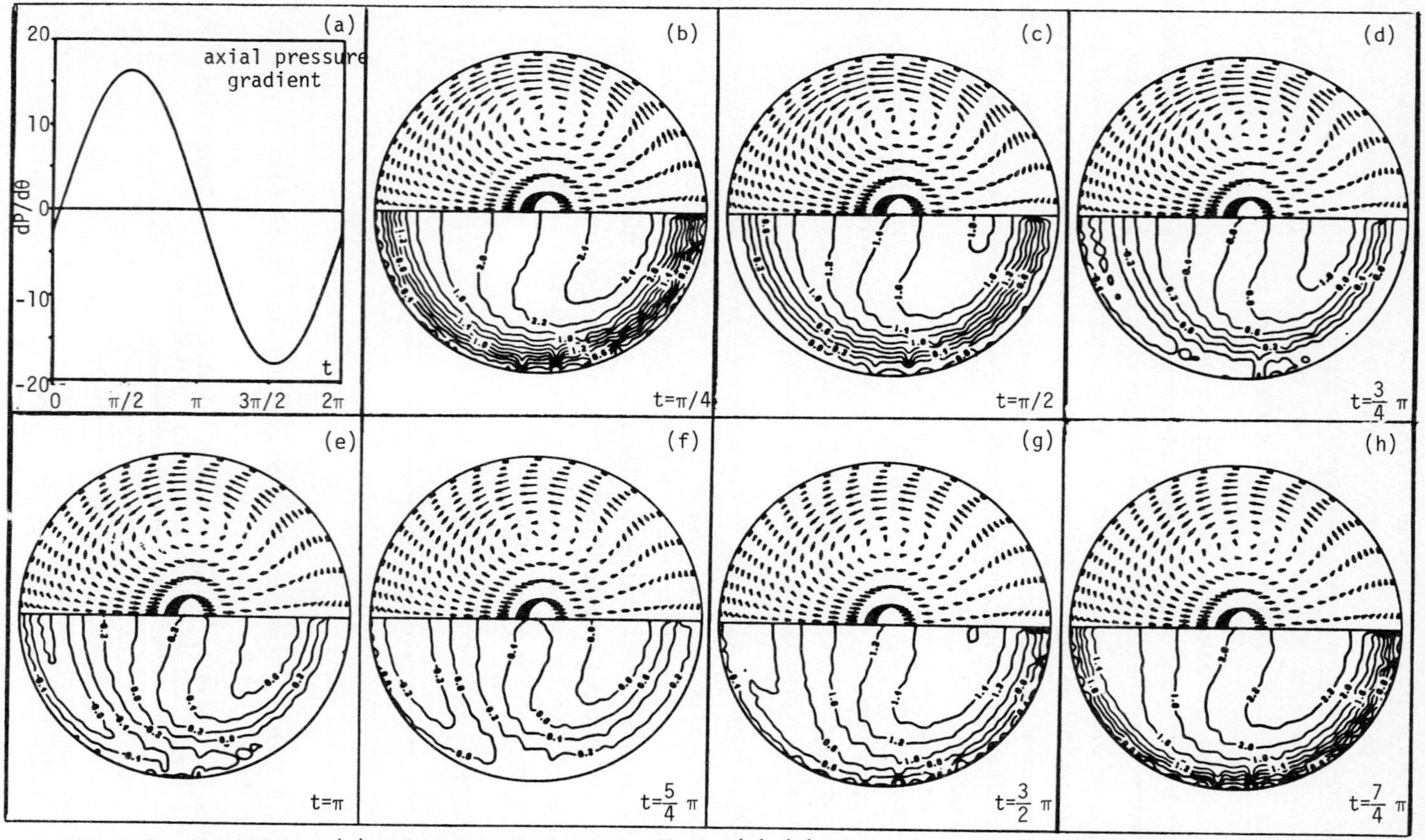

Figure 2. Re_m = 100: (a) Calculated pressure gradient; (b)-(h) Secondary velocity vectors and axial isovelocity contours for different times within a cycle.

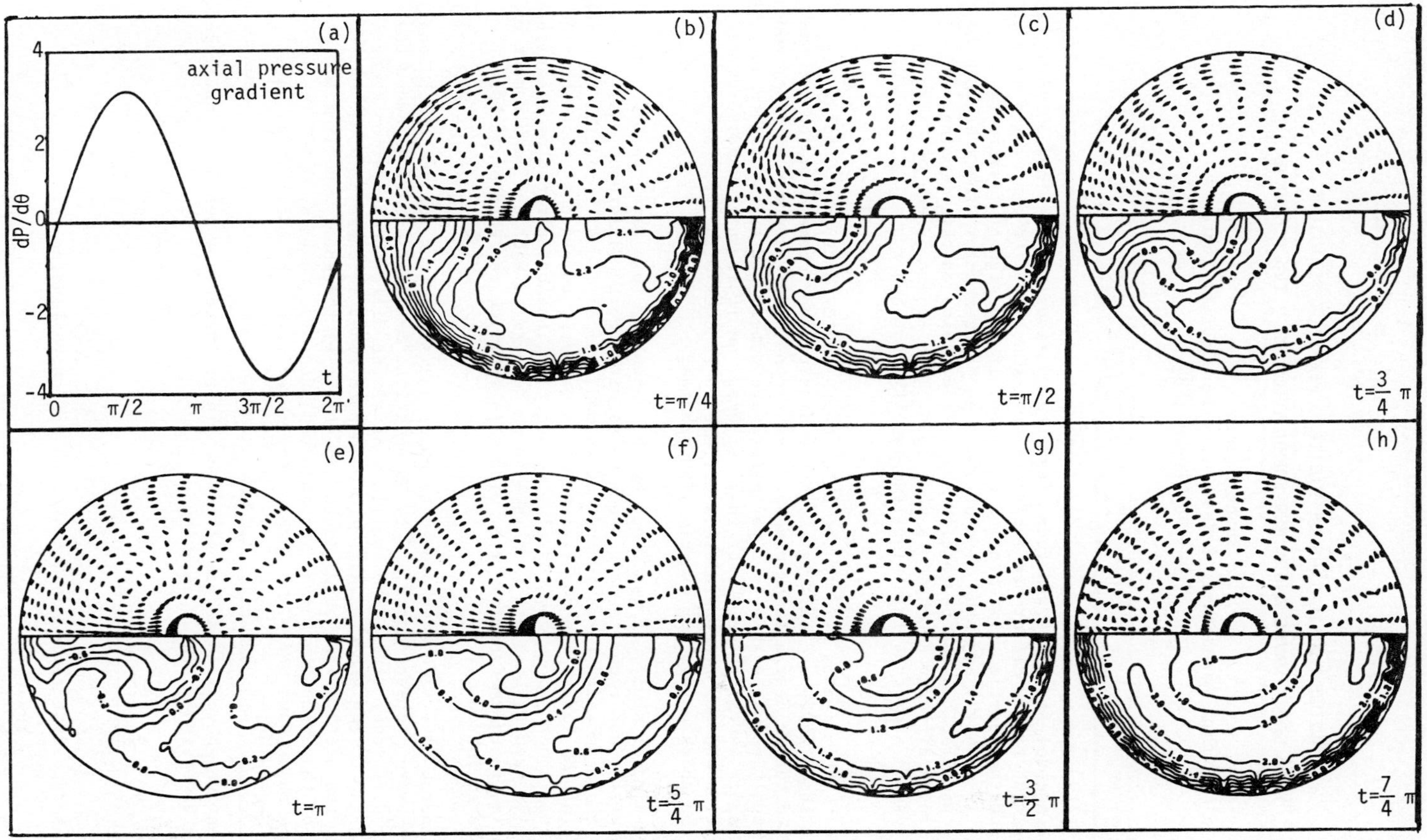

Figure 3. Re_m = 500: (a) Calculated pressure gradient; (b)-(h) Secondary velocity vectors and axial isovelocity contours for different times within a cycle.

SPURIOUS OSCILLATION OF FINITE DIFFERENCE SOLUTIONS NEAR SHOCK WAVES AND A NEW FORMULATION OF " TVD " SCHEME

Zhang Hanxin
China Aerodynamic Research and Development Center
Mianyang Sichuan PRC

Through a model study, it was found that the spurious oscillations occuring near shock waves with finite difference equations are related to the dispersion term in the corresponding modified differential equation. If the sign of dispersion coefficient is properly adjusted so that the sign changes across shock waves, the undesirable oscillations can be totally suppressed. Based on this finding, an efficient finite difference scheme is developed. This scheme is one of " TVD ". Numerical results show that this algorithm is efficient and robust.

The great attention is paid to the shock-capturing methods when complex flow fields with shock waves are calculated. Mixed dissipation schemes[1-3] which are the first order accuracy near a shock wave and the second order accuracy at other zone have been widely used. Because free parameters are determined by experience and the calculated shock waves are thick for these schemes, so it is interested to develop non-oscillation and no free-parameters difference schemes with high resolution for capturing the shock waves, for example, TVD schemes.

Through a model study, it was found by the auther of this paper in 1983 that the spurious oscillations occuring near shock waves with finite difference equations are related to the dispersion terms in the corresponding modified differential equation. If the sign of the dispersion coefficient ν_3 is properly adjusted so that the sign changes across shock waves from positive to negative, the undesirable oscillations can be totally suppressed. This fact is also in conformity with the requirement of the second law in thermodynamics, i.e. with this artifice the entropy of heat isolated system increases.

This finding have been also proved by numerical simulation for one dimensional normal shock wave problems described by NS equations or Euler equations added the third order difference terms. The oscillations of the difference solution occur in the down-stream of the shock wave if $\nu_3 > 0$ on both sides of theshock wave. On the other hand, the oscillations occur in the up-stream of the shock wave if $\nu_3 < 0$ on both sides of the shock wave. If ν_3 is adjusted properly, that is, $\nu_3 > 0$ in the front of the shock wave and $\nu_3 < 0$ behind the shock wave, the oscillations are suppressed and solution is smooth.

Based on the above finding, an efficient finite difference scheme is developed. For the one dimensional scalar equation

$$\frac{\partial u}{\partial t} + \frac{\partial f(u)}{\partial x} = 0 \qquad (1)$$

The semi-discrete difference scheme based on the above finding is

$$\left(\frac{\partial u}{\partial t}\right)_j^n = -\frac{1}{\Delta x}\left(h_{j+\frac{1}{2}}^n - h_{j-\frac{1}{2}}^n\right) \qquad (2)$$

where

$$h_{j+\frac{1}{2}} = f^+_{j+\frac{1}{2}L} + f^-_{j+\frac{1}{2}R}$$

$$f^+_{j+\frac{1}{2}L} = f^+_j + \frac{1}{2}\,\text{minmod}(\Delta f^+_{j-\frac{1}{2}}, \Delta f^+_{j+\frac{1}{2}})$$

$$f^-_{j+\frac{1}{2}R} = f^-_{j+1} - \frac{1}{2}\text{minmod}(\Delta f^-_{j+\frac{1}{2}}, \Delta f^-_{j+\frac{3}{2}}) \qquad (3)$$

$$\Delta f^{\pm}_{j+\frac{1}{2}} = f^{\pm}_{j+1} - f^{\pm}_j$$

$$\text{minmod}(\xi,\eta) = \frac{\text{sign}(\xi) + \text{sign}(\eta)}{2}\min(|\xi|,|\eta|)$$

and the superscript + or - denote the flux with positive or negative eigenvalue.

According to the concept of the total variation diminishing introduced by Harten, it can be proved that the above scheme is one of TVD. If $(\partial u/\partial t)^n_i$ is calculated with the first order or the second order difference scheme, the stability condition is $|a|\Delta t/\Delta x \leq 2/3$ or 1. This scheme is related to the Godunov's and may be considered as its extension.

This method can be extended to solving the one-dimensional Euler equations:

$$\frac{\partial U}{\partial t} + \frac{\partial F}{\partial x} = 0 \qquad (4)$$

where U is a vector function, F is a vector function of U. Its semi-discrete scheme can be given by (2),(3) if we substitute U, F for u,f.

For the one-dimensional NS equations:

$$\frac{\partial U}{\partial t} + \frac{\partial F}{\partial x} = \frac{\partial Fv}{\partial x} \qquad (5)$$

the above scheme can be used to calculate the convection term $\partial F/\partial x$, as for viscous term $\partial Fv/\partial x$, the central difference scheme can be used.

In order to solving three-dimensional flows, the time splitting method can be used, i.e. it is equivalent to solving three " one dimensional " equations.

Finally, numerical results for one dimensional unsteady flow and two or three dimensional steady flows are given with the present numerical scheme. The calculated density and pressure distributions for the shock tube flow described by Euler equations are shown in Fig. 1. The calculated density contours for the inviscid jet flow with Euler equations are shown in Fig. 2(a). The outflow Mach number(Me) is zero. The distributions of Mach number and pressure along the axis of the jet are shown in Fig. 2(b). Here Mach disk is given clearly. The hypersonic inviscid and viscous flows over two-dimensional profile and nosetip of Space Shuttle are also calculated. The computational pressure contours for two-dimensional flow with Euler equations and NS equations are given in Fig. 3(a) and 3(b). The computational pressure contours on the body surface and symmetric plane for the flows described by Euler and NS equations around the Space Shuttle are shown in Fig 4(a)(b).

The above results show that our difference algorithm is efficient and robust.

Auther thanks prof. Zhuang Fenggan and Mr. Gao Sucuen for their helps. Auther also thanks Mr. Shen Qing, Ye Youda, Zheng Min and Mao Meiliang for providing their computational results.

Reference

(1) T. H. Pulliam, AIAA Paper 85-0438, 1985.

(2) A. Jamson, et al., AIAA Paper 85-0293, 1985.

(3) H. X. Zhang, et al., Applied Math. and Mech. 4, 1, 1985.

(4) A. Harten, SIAM. J. Num. Anal, 21, 1-23, 1984.

(5) B. Van Leer, J. Comp. Phys. Vol.32, 101-136, 1979.

(6) S. R. Chakravarthy and S. Osher, AIAA Paper 85-0363. 1985.

(7) H. C. Yee, NASA TM 89464, 1987.

(8) H. X. Zhang, CARDC Report 87-3019, 1987.

(9) Tsutomu Saito, Hirogoki Nakotoyi and Koji Teshima, " Numerical Simulation and visualization of Free-jet Flow-Field ", TRans. Japan Soc. Aero. Space Sci., Vol.28, NO.82, 1986.

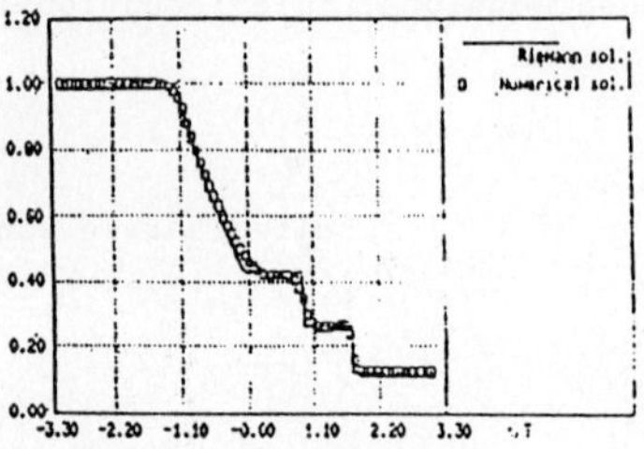

(a) Density

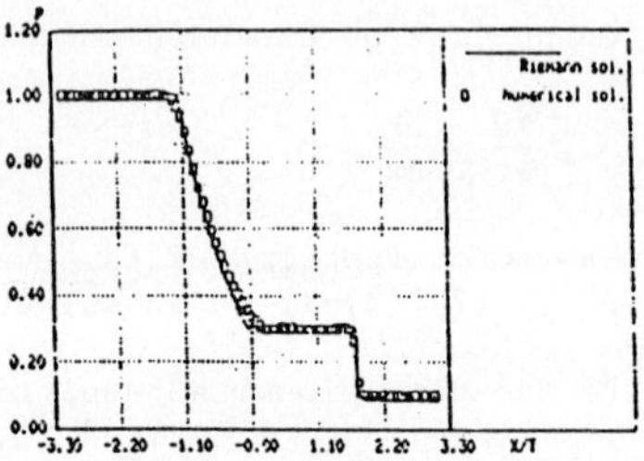

(b) Pressure

Fig. 1. Calculated density and pressure distributions for shock tube flow after 60 time steps.

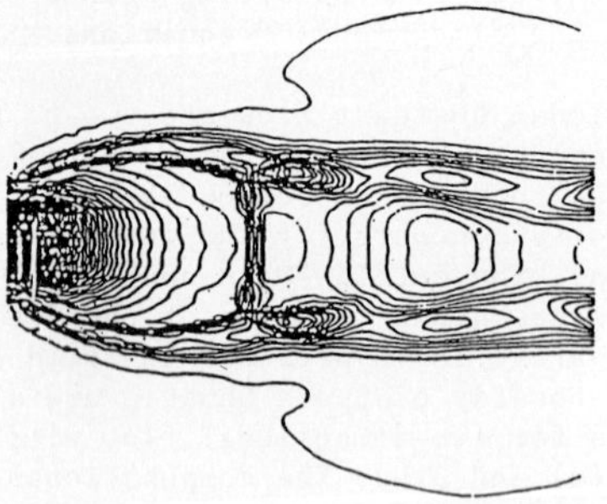

(a) Density contours

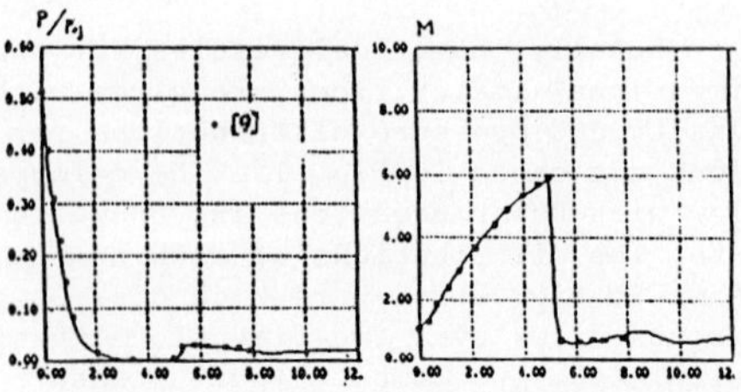

(b) Mach number and pressure distributions along the axis of jet flow

Fig. 2. Calculated results for jet flow (M_j=1.0, M_e=0., P_e/P_{0j}=1./50., H_j=H_e).

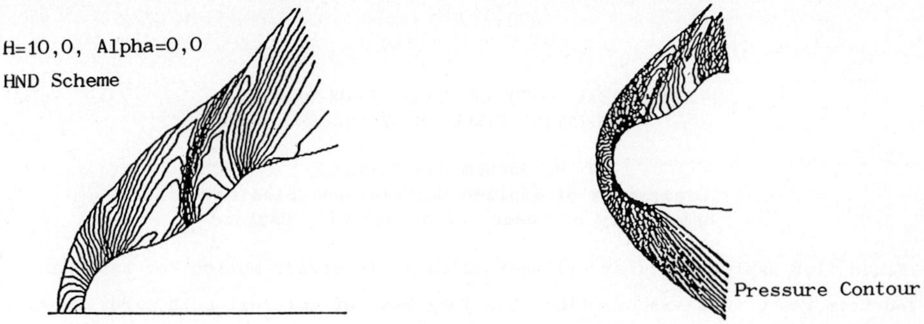

(a) Inviscid flow

(b) Viscous flow, $Re=10^6$, $T_\infty = 200^\circ K$, $Tw=1600^\circ K$.

Fig. 3. Calculated pressure contours for the flow around the profile, $\gamma = 1.4$, $M_\infty = 10$, $\alpha = 0^\circ$.

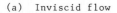

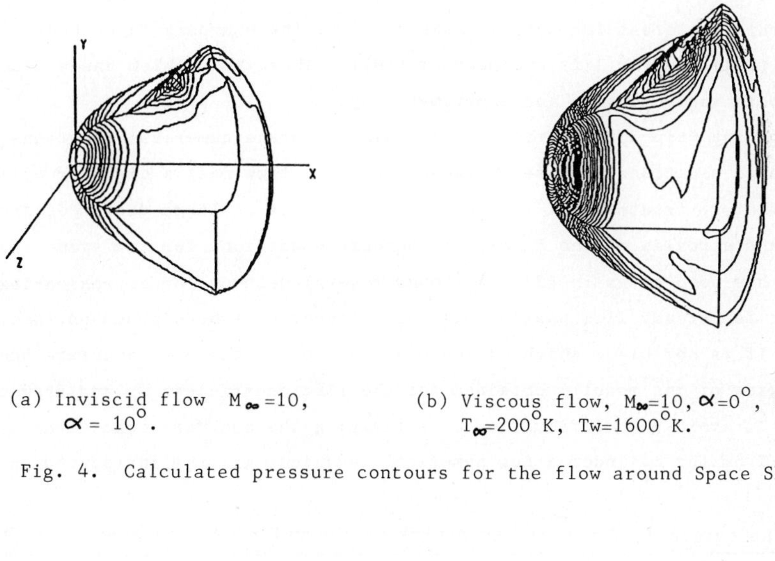

(a) Inviscid flow $M_\infty = 10$, $\alpha = 10^\circ$.

(b) Viscous flow, $M_\infty=10$, $\alpha=0^\circ$, $Re=10^6$, $T_\infty=200^\circ K$, $Tw=1600^\circ K$.

Fig. 4. Calculated pressure contours for the flow around Space Shuttle.

NUMERICAL STUDY OF STEADY FLOW PAST A ROTATING CIRCULAR CYLINDER

D. B. Ingham and T. Tang
Department of Applied Mathematical Studies
University of Leeds, Leeds LS2 9JT, England

Fluid flow past a circular cylinder which is in steady motion, or has been started from rest, in a viscous fluid has long been of interest both experimentally and theoretically (see e.g. [1-6]). In the present work we shall consider the asymmetrical flow of a viscous fluid which is generated by rotating a circular cylinder in a uniform stream of fluid. There are two basic parameters in the problem, namely, the Reynolds number, defined as $Re = 2aU/\nu$ where ν is the coefficient of kinematic viscosity of the fluid, U the unperturbated main stream speed, a the radius of the cylinder, and the rotational parameter $\alpha = a\omega_o/U$, where ω_o is the angular velocity of the rotating cylinder. When $\alpha = 0$ the motion is symmetrical about the direction of translation and this situation has previously received a considerable amount of attention, e.g. Dennis and Chang [7] and Fornberg ([3], [4]) who both provide a comprehensive reference list. The problem of the flow past a rotating cylinder is of fundamental interest for several reasons, e.g. the boundary layer control on aerofoils ([8]) and for lift enhancement ([9]). Thus the problem has attracted much interest both theoretically and experimentally.

One of the main difficulties in obtaining accurate numerical solutions of the Navier-Stokes equations for steady two-dimensional flow past a circular cylinder is the satisfactory treatment of the boundary conditions at large distances from the cylinder and a review of the numerical boundary conditions for the symmetrical flow situation has been given in [3]. Although several methods for approximating boundary conditions for steady flow past a rotating cylinder have been proposed (see, e.g. [10 - 12]), it is not clear which of these approaches is the most accurate one, especially in view of the results obtained for the lift coefficient C_L and drag coefficient C_D. To avoid the difficulties in satisfying the boundary conditions at large distances from the cylinder a new numerical technique will be introduced in this paper.

If the origin is fixed at the centre of the cylinder, the x-axis in the same direction as that of translation, then the steady flow of an incompressible fluid in a fixed two dimensional Cartesian form can be described by the non-dimensional equations

$$\frac{\partial(\Psi,\omega)}{\partial(x,y)} + \frac{2}{Re}\nabla^2\omega = 0 \qquad (1a)$$

$$\nabla^2\Psi = -\omega \qquad (1b)$$

where ω is the scalar vorticity and Ψ is the stream function. It is required to solve eqns. (1) subject to the boundary conditions

$$\Psi = 0, \quad \frac{\partial \Psi}{\partial r} = -\alpha, \qquad \text{on } r = 1, \quad 0 \leq \theta < 2\pi, \tag{2}$$

$$\frac{\partial \Psi}{\partial r} \to \sin\theta, \quad \frac{1}{r}\frac{\partial \Psi}{\partial \theta} \to \cos\theta, \quad \text{as } r \to \infty, \quad 0 \leq \theta < 2\pi, \tag{3}$$

where (r,θ) are the polar coordinates which are chosen so that $\theta = 0$ coincides with the positive x-axis.

Filon [13] showed that the asymptotic form for the stream function at large distances from the cylinder and outside the wake region is

$$\Psi \sim r \sin\theta + \frac{C_L \ln r}{2\pi} + \frac{C_D}{2\pi}(\theta - \pi), \qquad 0 < \theta < 2\pi. \tag{4}$$

Since it is very important to apply an accurate form of the boundary condition for Ψ at large distances from the cylinder we observe from equation (4) that

$$(\Psi - r \sin\theta)/r \to 0 \quad \text{as } r \to \infty . \tag{5}$$

Hence we introduce the transformations

$$\xi = 1/r, \quad \eta = 2\theta/\pi , \tag{6}$$

$$f(r,\theta) = (\Psi(r,\theta) - r \sin\theta)/r, \tag{7}$$

namely,

$$f(\xi,\eta) = \xi\, \Psi(\xi,\eta) - \sin(\pi\eta/2) . \tag{8}$$

This has the advantage that $f \equiv 0$ on $\xi = 0$ (i.e. $r = \infty$) and it requires no approximation for Ψ on the outer boundary.

With the transformation (6), the flow region ($1 \leq r < \infty$, $0 \leq \theta < 2\pi$) is transformed into a finite rectangular region of the (ξ,η) plane ($0 \leq \xi \leq 1$, $0 \leq \eta < 4$). Substituting from (6) and (8) in (1), we obtain

$$-\frac{2\xi^2}{\pi}\frac{\partial(f,\omega)}{\partial(\xi,\eta)} + \frac{2\xi}{\pi} f \frac{\partial \omega}{\partial \eta} + \xi^2 \frac{\partial \omega}{\partial \xi}\cos(\pi\eta/2) + \frac{2\xi}{\pi}\frac{\partial \omega}{\partial \eta}\sin(\pi\eta/2)$$
$$+ \frac{2}{Re}\left[\xi^4 \frac{\partial^2 \omega}{\partial \xi^2} + \frac{4\xi^2}{\pi^2}\frac{\partial^2 \omega}{\partial \eta^2} + \xi^3 \frac{\partial \omega}{\partial \xi}\right] = 0 \tag{9a}$$

$$\xi^3 \frac{\partial^2 f}{\partial \xi^2} + \frac{4\xi}{\pi^2}\frac{\partial^2 f}{\partial \eta^2} - \xi^2 \frac{\partial f}{\partial \xi} + \xi f = -\omega \tag{9b}$$

with $0 \leq \xi \leq 1$, $0 \leq \eta < 4$. The boundary conditions (2) and (3) become

$$f = -\sin(\pi\eta/2), \quad \partial f/\partial \xi = \alpha , \qquad \text{on } \xi = 1, \quad 0 \leq \eta < 4; \tag{10}$$

$$f = 0, \quad \omega = 0, \qquad \text{on } \xi = 0, \quad 0 \leq \eta < 4. \tag{11}$$

As the solution is periodic we also require that

$$f(\xi,4) = f(\xi,0), \quad \omega(\xi,4) = \omega(\xi,0), \quad 0 \leq \xi \leq 1. \tag{12}$$

In order to obtain numerical solutions of eqns. (9)-(12), the region of integration $[0,1]\times[0,4]$ is covered by a square mesh of size h and central finite difference approximations to (9) are employed. In practice, a mesh system is set up such that the mesh size in both ξ and η directions is $h = 1/N$ where N is a prescribed positive integer. In view of the periodic conditions (12), an extra line of computational grids $\eta = 4+h$ for $0 \le \xi \le 1$ is introduced. Then we have $(N+1)x(M+1)$ mesh points, where $M = 4N+1$. The mesh points (ξ_i,η_j) $(0 \le i \le N$, $0 \le j \le M)$ are (ih,jh). We now briefly outline how the boundary conditions (10)-(12) were implemented.

Boundary conditions for ω:

On $\xi = 0$, $0 \le \eta \le 4$: $i = 0$, $0 \le j < M$; $\omega_{oj} = 0$ (13)

On $\xi = 1$, $0 \le \eta \le 4$, $i = N$, $0 \le j < M$; in this case we use (10) to obtain the vorticity using the second-order accurate finite difference approximation

$$\omega_{Nj} = [f_{Nj}-f_{N-1j}-\alpha(h - \tfrac{1}{2}h^2) - \tfrac{1}{6}h^2\omega_{N-1j}]/[\tfrac{h^3}{3}(1+h)] \tag{14}$$

$$\text{On } \eta = 4+h,\quad 0 \le \xi \le 1 :\quad 0 \le i \le N,\ j = M;\quad \omega_{iM} = \omega_{i1} \tag{15}$$

$$\text{On } \eta = 0,\quad 0 \le \xi \le 1 :\ 0 \le i \le N,\quad j = 0;\quad \omega_{io} = \omega_{iM-1} \tag{16}$$

Boundary conditions for f:

$$\text{On } \xi = 0,\quad 0 \le \eta \le 4 :\quad i = 0,\quad 0 \le j < M;\quad f_{oj} = 0 \tag{17}$$

$$\text{On } \xi = 1,\quad 0 \le \eta \le 4 :\quad i = N,\quad 0 \le j < M;\quad f_{Nj} = -\sin(\pi\eta_j/2) \tag{18}$$

$$\text{On } \eta = 4+h,\ 0 \le \xi \le 1 :\ 0 \le i \le N,\ j = M;\quad f_{iM} = f_{i1} \tag{19}$$

$$\text{On } \eta = 0,\quad 0 \le \xi \le 1 :\ 0 \le i \le N,\quad j = 0;\quad f_{io} = f_{iM-1} \tag{20}$$

The resulting finite difference equations were solved iteratively in a manner similar to that as described by Ingham [11].

If L and D are the lift and drag on the cylinder, then the lift and drag coefficients are defined by

$$C_L = L/2\rho U^2 a,\quad C_D = D/2\rho U^2 a, \tag{21}$$

and each coefficient consists of components due to the friction and pressure forces. Hence

$$C_L = C_{LF} + C_{LP},\quad C_D = C_{DF} + C_{DP}, \tag{22}$$

where

$$C_{DF} = -\frac{2}{Re}\int_0^{2\pi}(\omega)_{\xi=1}\sin\theta\, d\theta,\quad C_{LF} = -\frac{2}{Re}\int_0^{2\pi}(\omega)_{\xi=1}\cos\theta\, d\theta \tag{23}$$

$$C_{DP} = -\frac{2}{Re}\int_0^{2\pi}\left(\frac{\partial\omega}{\partial\xi}\right)_{\xi=1}\sin\theta\, d\theta,\quad C_{LP} = -\frac{2}{Re}\int_0^{2\pi}\left(\frac{\partial\omega}{\partial\xi}\right)_{\xi=1}\cos\theta\, d\theta \tag{24}$$

Formulae (23) and (24) were calculated using second-order accurate finite difference approximations and Simpson's rule.

In order to compare with the limited number of solutions available for this problem, numerical results were obtained using the central difference scheme for $Re = 5$ and 20, $\alpha = 0, 0.1, 0.2, 0.4, 0.5, 1, 2$ and 3 and $h = \frac{1}{10}, \frac{1}{15}, \frac{1}{20}, \frac{1}{30}, \frac{1}{40}$. All the results presented here are as a result of using the repeated h^2-extrapolation method and these differ very little from the results which were obtained using the finest mesh sizes. In order to check the accuracy of the method the results when there is no rotation were compared with those obtained by previous investigators of this problem. The most accurate results of Fornberg [3] suggest at Re = 20 that $C_D = 2.0001$ whilst we obtain $C_D = 1.995$. However, at Re = 5 we obtain $C_D = 3.947$, $C_{DP} = 2.104$, $C_{DF} = 1.843$, whereas Dennis and Chang [7] quote values of $C_D = 4.116$, $C_{DP} = 2.119$, $C_{DF} = 1.917$ which are about 4% higher. It is perhaps relevant to point out that, in the case of intermediate Reynolds numbers, the drag coefficients estimated by [7] are also very slightly higher than those of [3].

Table 1 shows the variation of C_L and C_D with α and the accuracy of these results vary from about 0.1% at $\alpha = 0$ to 1% at $\alpha = 3$. Previous investigators have suggested that C_L is directly proportional to α for small values of α. However, comparisons of these published results show considerable discrepancies. The solution using a series expansion method with respect to α was calculated by Ingham and Tang [10] with the result that

$$C_L \sim 2.77\alpha + O(\alpha^3),\ C_D \sim 3.947 - 0.072\alpha^2; \quad \text{for Re} = 5, \tag{25}$$

$$C_L \sim 2.54\alpha + O(\alpha^3),\ C_D \sim 1.995 - 0.109\alpha^2; \quad \text{for Re} = 20, \tag{26}$$

for $\alpha \ll 1$. Although (25) and (26) are expected to be valid only for very small values of α, it can be seen from Tables 1 that they can be used over a wide range of values α. For example, at Re = 5 and $\alpha = 0.5$ the calculated C_L and C_D obtained by solving the full Navier-Stokes equations are 1.389 and 3.916 respectively, and this compares favourably with values given by (25) of 1.384 and 3.929. Similarly, the results shown in Table 1 are also in very good agreement with the series expansion results (26) for small values of α and in reasonable agreement for the large values of α considered.

Figures 1 and 2 show the variation of C_L with α for Re = 5 and 20, respectively. Also shown are the series expansion results [10] and those of [2], [11], [12] and [15]. It is observed the results of the present work and [2] are in good agreement but the results in [11], [12] and [15] are substantially lower. One of the reasons for this is that most of the other previous works have been obtained by using a coarse mesh size. Despite the many discrepancies with the previous results there is reasonable agreement between all the authors of other features of the flow, e.g., the surface vorticity and the flow patterns were found in good agreement with those obtained by Ta [12], Ingham [11], Badr and Dennis [2], all of whom considered α only in the range $0 \le \alpha \le 1$.

Re	α	0.	0.1	0.2	0.4	0.5	1	2	3
5	C_L	0.000	0.277	0.559	1.111	1.389	2.838	5.830	9.191
5	C_D	3.947	3.947	3.939	3.927	3.916	3.849	3.506	3.097
20	C_L	0.000	0.254	0.514	1.024	1.283	2.617	5.719	9.074
20	C_D	1.995	1.995	1.992	1.979	1.973	1.925	1.627	1.396

Table 1. Variation of C_L and C_D with α for Re = 5 and 20.

REFERENCES

[1] H. M. Badr & S. C. R. Dennis, J. Fluid Mech. 158, 447 (1985).

[2] ______________, Private Communication (1986).

[3] B. Fornberg, J. Fluid Mech. 97, 819 (1980).

[4] ______, J. Comput. Phys., 61, 297 (1985).

[5] F. T. Smith, J. Fluid Mech. 92, 92 (1979).

[6] ______, J. Fluid Mech. 155, 175 (1985).

[7] S. C. R. Dennis and G. Z. Chang, J. Fluid Mech. 42, 471 (1970).

[8] J. S. Tennant, W. S. Johnson & A. Krothappalli, J. Hydronautics, 10, 102 (1976).

[9] A. T. Sayers, Int. J. Mech. Elec. Engng. 7, 75 (1979).

[10] S. C. R. Dennis, Comput. Fluid Dyn., SIAM-AMS Proc. 11, 156 (1978).

[11] D. B. Ingham, Computers & Fluids, 11, 351 (1983).

[12] Ta Phouc Loc, J. Mec. 14, 109 (1975).

[13] L. N. G. Filon, Proc. Roy. Soc. A113, 7 (1926).

[14] D. B. Ingham and T. Tang, To be published.

[15] V. A. Lyul'ka, Zh. vychist. Mat. mat. Fiz, 17, 470 (1977).

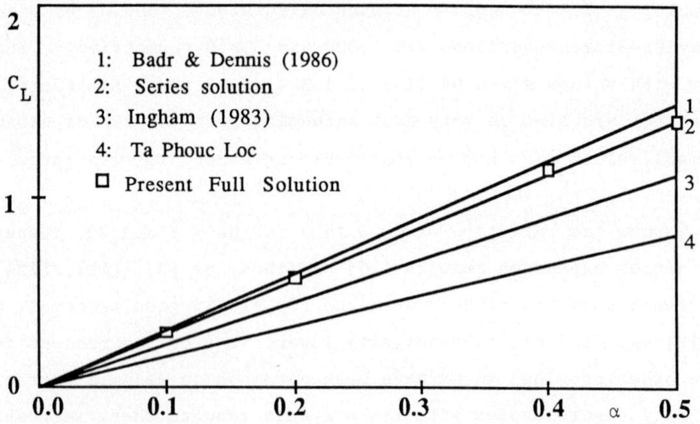

Fig. 1. The variation of C_L with α for Re = 5.

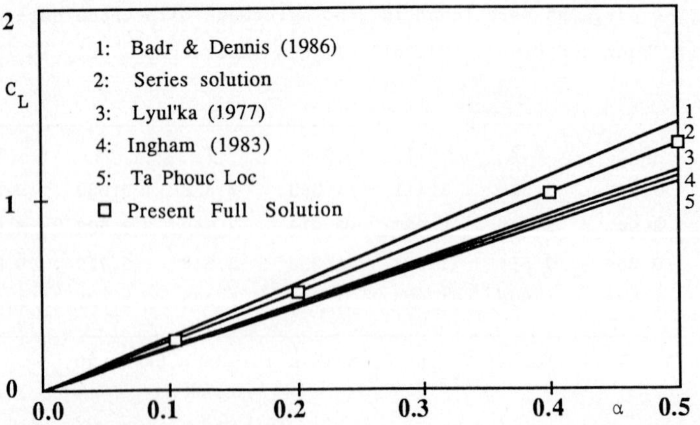

Fig. 2. The variation of C_L with α for Re = 20.

NUMERICAL SIMULATION OF THE FLOW ABOUT A WING WITH LEADING-EDGE VORTEX FLOW*)

J.M.J.W. Jacobs, H.W.M. Hoeijmakers, J.I. van den Berg and J.W. Boerstoel
National Aerospace Laboratory NLR, Amsterdam, The Netherlands

INTRODUCTION

Vortex flow associated with flow separation from wings with highly swept leading edges is of extreme importance for the high-angle-of-attack aerodynamics of various modern fighter aircraft. For wings with aerodynamically sharp leading edges the formation of the leading-edge vortex and the effect of the vortex on the flow over the upper surface of the wing is only slightly dependent on Reynolds number. This implies, that this type of flow can be simulated by inviscid methods.

For sub-critical flows potential-flow methods have met some success in modelling leading-edge vortex flow (e.g. Ref.1). In these methods the free shear layers are modelled by vortex sheets "fitted" into the potential flow field i.e. one has to decide a priori on the presence of vortex sheets and vortex cores, although their position and strength is determinated as part of the solution. This implies, that the topology of the flow must be well-defined and known in advance.

For cases with a complex vortex-flow pattern and especially for transonic vortex flow with (strong) shock waves and the mutual interaction of these nonlinear flow phenomena, the appropriate flow model to consider is based on the Euler equations. Viscous effects, which are responsible for the appearance of secondary separation and which affect the development of the primary vortex core will have to be assessed through a more elaborate flow model. The investigation in this paper concentrates on computing the compressible flow about a cropped 65-deg swept sharp-edged delta wing which is known to feature leading-edge vortex flow. Numerical results and experimental data are compared for a transonic and for a subsonic case. Attention is devoted to the influence of grid density and grid point distribution.

OUTLINE OF METHODS USED

For numerical flow simulation based on the Euler equations NLR is developing a system of computer programs. The system includes a grid generator, a flow solver and a post-processing "flow visualization" facility.

---

*) This investigation has been carried out under contract for the Netherlands Agency for Aerospace Programs (NIVR) for the Netherlands Ministry of Defence

With the grid generator multi-blocked surface fitted grids are generated using transfinite interpolation followed by tuning through elliptic methods. A description of the grid generator can be found in Ref.2. For the Euler code the Jameson, Schmidt and Turkel scheme (Ref.3) was modified in the following manner: the coefficients of the artificial dissipative terms are tuned for differences in mesh size in the three grid co-ordinate directions and the boundary condition for the fourth-order dissipative operator has been modified. For the solid-wall boundary condition the pressure is linearly extrapolated. For the far-field boundary conditions Riemann variables are used. Integration in time is carried out by a four-stage Runge-Kutta scheme. Convergence is accelerated by residual averaging. The NLR Euler code is a multi-block code with internal block boundary conditions that can handle slope-discontinuous grid lines and highly stretched cells. The flow solver runs on the NEC-SX2 computer.

FLOW COMPUTATIONS ON DIFFERENT GRIDS

Calculations have been carried out to simulate the flow about a cropped delta wing. The wing has a sharp leading edge with a sweep of 65 deg. For this wing experimental data are available obtained in the NLR high speed wind-tunnel during the International Vortex Flow Experiment (Ref.4).

Three grids of O-O type but with different dimensions and grid distributions have been used to compute the flow about the delta wing. The grid generated around the starboard half of the wing consists of six blocks with total grid dimensions 129 x 33 x 33 (131,072 cells). The grid on the wing surface is "conical" on the forward portion of the wing. The other two grids are single-block grids and were obtained from Prof. A. Rizzi of FFA Sweden. They are non-conical on the wing surface and the grid dimensions are 97 x 29 x 49 (129,024 cells) and 125 x 37 x 65 (285,696 cells) for a medium and a fine grid, respectively.

Two test cases were considered: a transonic case, $M_\infty = 0.85$ $\alpha = 10$ deg, of compressible vortex flow presumably without shocks and vortex burst and a subsonic case, $M_\infty = 0.40$ $\alpha = 20$ deg, of strong vortex flow also without vortex burst above the wing. For all calculations the solution converged well, except for the subsonic case on the NLR grid. Here the error norm of the residual remains oscillating about a constant value, after its reduction by four orders of magnitude. Inspection showed that outboard of the position of the vortex core the trailing edge Kutta condition was not fully satisfied, but the solution upstream of the trailing edge did not appear to be affected very much.

Transonic test case

Fig. 1a compares the spanwise pressure distribibition computed on the NLR grid, the one on the medium FFA-grid (referred to as FFA-M) and the measured one at the chordwise stations x/c_R = 0.3, 0.6 and 0.8. The largest differences between the results on the two grids occur on the forward portion of the wing. For the NLR grid the suction peak at x/c_R = 0.3 is compared to the FFA-M grid positioned too far outboard. This is attributed to a better grid resolution normal to the wing in the FFA-M grid. At x/c_R = 0.6 and 0.8 the position of suction peaks almost coincides for the two grids. At the wing lower surface the agreement between the two numerical results is excellent at all three sections. Both numerical results show a spike in the pressure distribution near the leading edge. It is not clear whether this is a true physical phenomenon or is caused by the numerics. The comparison between numerical results and experimental data is inconclusive because secondary separation has not been modelled in the present calculations. The secondary separation causes the primary vortex to move inboard and upward. The effect on the suction peak is that it moves inboard, decreases in height and in spanwise direction is followed by a region of about constant pressure coefficient. This means, that the computed suction peak should be more pronounced than the experimental one and should be located at a more outboard position. Clearly at x/c_R = 0.3 the suction peak is too low. In the experiment the spanwise location of the point of maximum pressure indicates a re-attachment point. For the FFA-M grid a re-attachment point appears on the upper wing surface, while for the NLR grid the flow reattaches at the plane of symmetry. On the lower wing surface the computed results agree very well with the experimental data, except at x/c_R = 0.3, where in the experiment there is a large influence from the under-wing fuselage. In Fig. 2 the isobar pattern on the wing upper surface is depicted. The left-hand side of the picture presents the result for the FFA-M grid and the right-hand side the NLR grid. It appears, that the formation of the vortex starts at the apex for the FFA-M grid and at about 20% root chord for the NLR grid. In the experiment it is known to start right at the apex.

In Fig. 1b a comparison is carried out between results computed on the FFA-F and the FFA-M grid. For reference purposes the experimental values are included in these plots. It follows from this figure, that the lateral position of the Cp-peak is about the same for the two grids. For the fine grid however, the suction peak is higher and steeper. The fine grid result also shows very clearly a maximum pressure point on the wing upper surface, indicative for flow re-attachment.

Subsonic test case

For the subsonic case the computed and measured spanwise pressure distributions are presented in Fig. 1c. Results are shown for the NLR grid and for the FFA-M grid. The lateral position of the suction peak and also its height is about the same for the two grids, although the result on the NLR grid is not fully converged. On the NLR grid the vortex formation for this case of strong vortex flow does start at the apex. However, at x/c_R = 0.3 the computed minimum Cp-value is still not low enough when compared with the experimental one. The agreement between experimental and theoretical data is excellent on the wing lower surface even for $x/c_R = 0.3$.

The results of most Euler codes include total pressure losses, caused by the numerical scheme. Although these total-pressure losses are due to the numerics of the computational method they appear to indicate regions of rotational flow (e.g. Ref.5). The total pressure losses in the vortex core at the chordwise stations $x/c_R = 0.3$, 0.6 and 0.8 are presented in Table 1. For both the transonic and the subsonic case, it is clear that there is not much difference between the value of the maximum total pressure loss and the position where it occurs. This is especially remarkable for the transonic case because on the NLR grid the vortex formation does not start at the apex. Fig. 3 depicts for the transonic case contours of constant total pressure loss in the plane $x/c_R = 0.8$. It shows, that the region of total pressure losses, indicative for the region of vortical flow, has about the same shape and position for both grids.

CONCLUDING REMARKS

The NLR Euler code has been employed to simulate the flow about a sharp-edged cropped delta wing. Solutions have been compared with experiments for two cases: a transonic case of compressible vortex flow without shocks and a subsonic case of strong vortex flow, both without vortex bursting. Solutions were obtained on three different grids and they all show a captured leading-edge vortex. However, the quality of the solution depends on the grid point density and on the grid point distribution near the leading edge, especially in the direction normal to the wing surface.

It appears that if locally a certain threshold in grid point density has been surpassed, the formation of the leading-edge vortex is simulated along the entire (sharp) leading-edge. It is remarkable, that the case with a part-span separation yields at some distance downstream of the point of vortex formation about the same amount of total pressure loss at about the same position in the flow field as a case on a denser grid with full-span separation. The differences between computed surface pressure distributions and lift coefficients on one hand and the experimental data on the other hand are primarily due to secondary separation in the experiment, which has not been modelled in the present calculations.

Further investigation is required to explain the occurrence of the spike in the pressure distribution at the leading edge and to improve the convergence of the time integration process to ensure that the Kutta condition is satisfied at edges where the grid resolution is insufficient.

ACKNOWLEDGEMENT

The authors thank Prof. A. Rizzi of FFA Sweden for providing the FFA-M and FFA-F grid.

REFERENCES

1. Hoeijmakers, H.W.M., Methods for Numerical Simulation of Leading-Edge Vortex Flow, Studies of Vortex Dominated Flows, ed. M.Y. Hussaini and M.D. Salas, Springer Verlag, pp. 223-269, 1987. Also NLR MP 85052 U.

2. Boerstoel, J.W., Preliminary Design and Analysis of Procedures for the Numerical Generation of 3D Block-Structured Grids, NLR Technical Report 86102 U, 1986.

3. Jameson, A., Schmidt, W. and Turkel, E., Numerical Solution of Euler Equations by Finite Volume Methods Using Runge-Kutta Time Stepping Scheme, AIAA Paper 81-1259, 1981.

4. Hirdes, R.H.C.M., U.S./European Vortex Flow Experiment - "Test Report of Wind-Tunnel Measurements on the 65° Wing in the NLR High Speed Tunnel HST", NLR Technical Report 85046 L, 1985.

5. Murman, E.M. and Rizzi, A., Applications of Euler Equations to Sharp Edge Delta Wings With Leading Edge Vortices, Agard CP 412, Paper 15, 1986.

Table 1
Maximum total pressure loss in the vortex core and the position of the vortex core for three grids

		Transonic			Subsonic		
x/c_R	GRID	$C_{p_{t\,max}}$	η	ζ	$C_{p_{t\,max}}$	η	ζ
0.3	NLR	0.372	0.885	0.073	0.250	0.795	0.217
	FFA-M	0.408	0.854	0.100	0.272	0.798	0.236
	FFA-F	0.405	0.846	0.116	0.291	0.806	0.224
0.6	NLR	0.422	0.857	0.081	0.241	0.791	0.220
	FFA-M	0.430	0.818	0.091	0.249	0.762	0.192
	FFA-F	0.429	0.814	0.097	0.280	0.788	0.203
0.8	NLR	0.439	0.814	0.090	0.256	0.791	0.193
	FFA-M	0.443	0.817	0.090	0.241	0.791	0.199
	FFA-F	0.422	0.811	0.089	0.260	0.786	0.200

$C_{p_{t\,max}} = 1 - p_t/p_{t\infty}$

$\eta = y/x \cot 65°$

$\zeta = z/x \cot 65°$

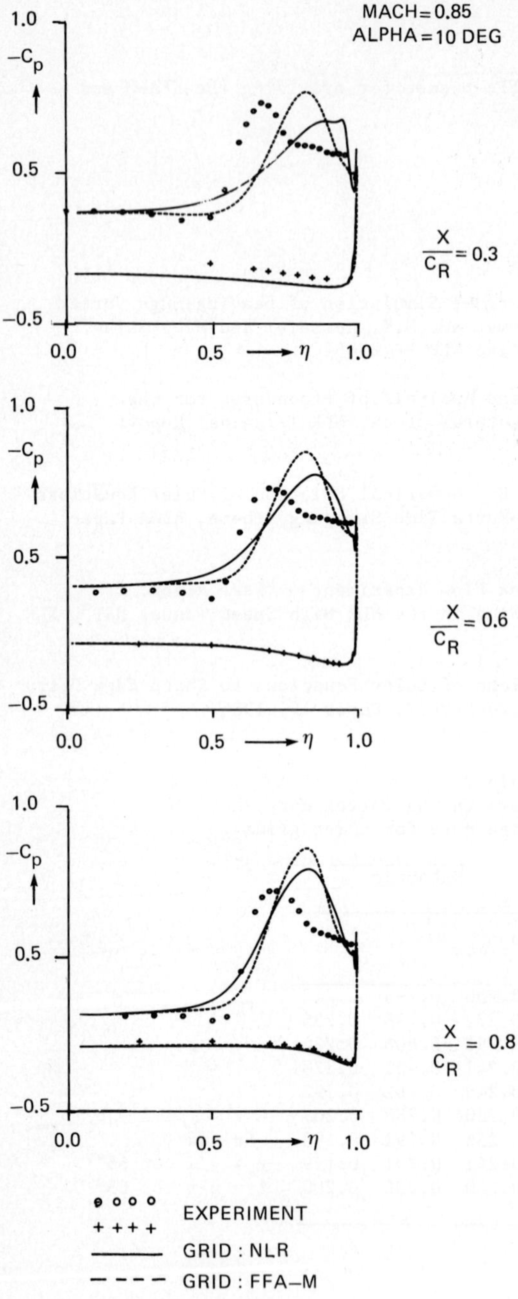

Fig. 1a Spanwise pressure distribution for two different grid point distributions

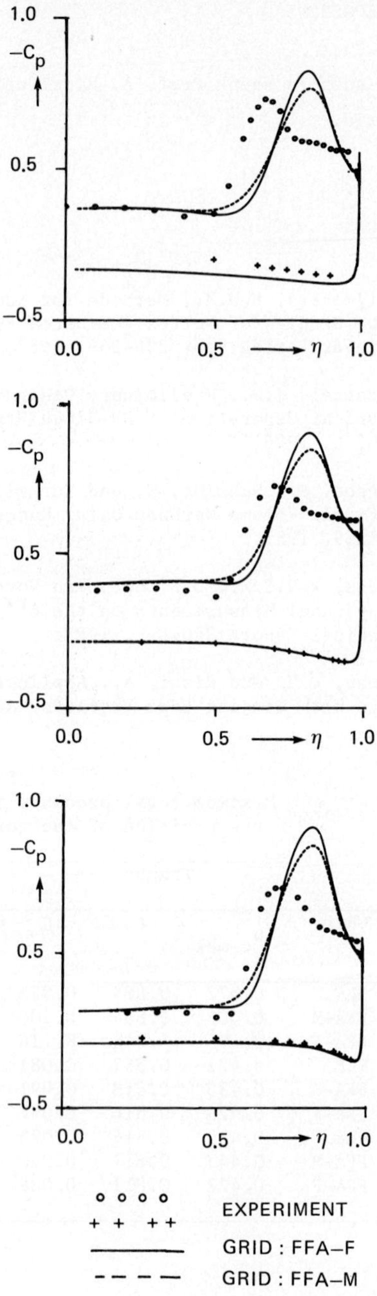

Fig. 1b Spanwise pressure distribution for two grids with different number of cells

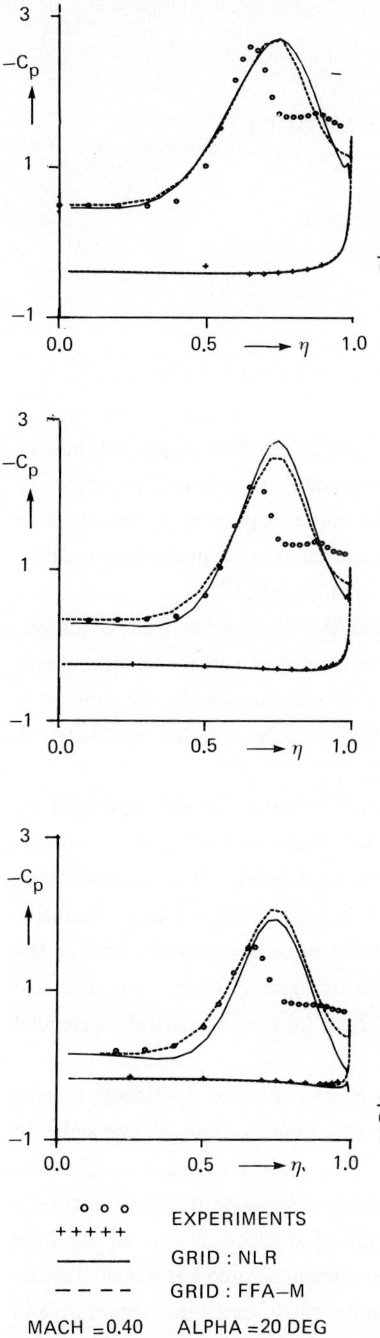

Fig. 1c Spanwise pressure distribution for two different grid point distributions

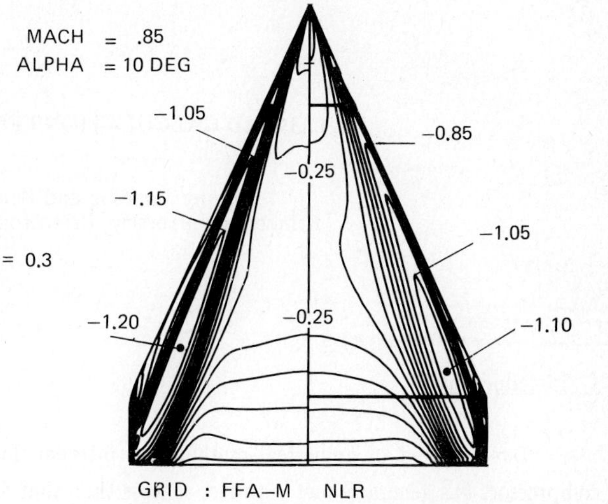

Fig. 2 Isobars on wing upper surface

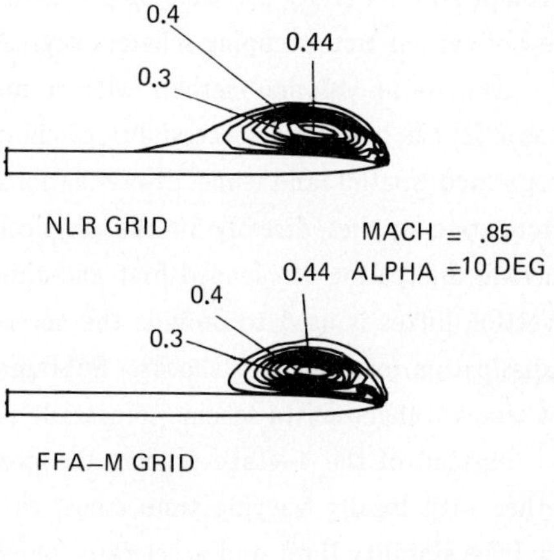

Fig. 3 Contours of constant total pressure loss at $X/C_R = 0.8$

MULTIGRID CALCULATIONS FOR CASCADES

Antony Jameson and Feng Liu
Princeton University, Princeton, NJ 08544

1. Introduction

Development of numerical methods for internal flows such as the flow in gas turbines or compressors has generally met with less success than that for external flows, due to the complexity of the flow pattern. Potential flow methods had been the major approach in cascade flow calculations until fairly recently, when solution of the Euler equations became practically feasible. The most widely used Euler method has been Denton's finite volume method [1].

Like all explicit methods, Denton's method suffers from the limit of the CFL condition. Implicit schemes have been developed to yield convergence in a smaller number of time steps. This will only pay, however, if the decrease in the number of time steps outweighs the increase in the computational effort per time step consequent upon the need to solve coupled equations. A review of various time stepping schemes is given by Jameson [2].

The finite volume method with a multiple stage time stepping scheme developed by Jameson [3] has been very successful in calculating external flows. This method has the advantage of separated spatial and time discretizations and is easy to implement. The finite–volume discretization applies directly in the physical domain and is in conservation form. Adaptive numerical dissipation of blended first and third differences in the same conservation from as the convection fluxes is used to provide the necessary higher order background dissipation, and also the dissipation for capturing shocks. Subramanian [4] and Holmes [5] have successfully extended the 4–stage Runge–Kutta scheme to cascade calculations.

Instead of the 4–stage scheme the present work uses a more flexible multistage scheme together with locally varying time steps, enthalpy damping and implicit residual averaging to extend the stability limit and accelerate convergence. Finally very rapid convergence is obtained by applying an effective multigrid method [6]. Since the necessary dissipation is added separately from the discretization of the conservation laws, the amount of dissipation can be carefully optimized to improve the accuracy. Two–dimensional subsonic, transonic and supersonic cascade flows have been calculated using this method. The results show excellent accuracy and convergence.

2. Numerical Method

With a cell centered finite volume scheme a system of ordinary differential equations is obtained by applying the integral form of the Euler equations separately to each cell

$$\frac{d}{dt}(S_{ij}W_{ij}) + Q_{ij} - D_{ij} = 0 \tag{1}$$

where S_{ij} is the cell area, W_{ij} is the flow variable (either mass, momentum or energy) defined at the center of the cell, Q_{ij} is the net flux out of the cell and D_{ij} is the artificial dissipation term. The dissipation term D_{ij} is a blending of fluxes of first and third order compared to the convection term Q_{ij}. The first order dissipation is switched off in a smooth region and on in the neighborhood of a shock by multiplying it with a pressure sensor

$$\gamma_{ij} = \frac{|P_{i+1,j} - 2P_{ij} + P_{i-1,j}|}{|P_{i+1,j} + 2P_{ij} + P_{i-1,j}|}$$

γ_{ij} is small in a smooth region and of order one across a shock. This sensor has proved to be very effective for transonic flow. For supersonic flow, however, it can produce large values in sharp expansion regions where first order dissipation is not needed. To eliminate this possibility an alternative sensor based on entropy increase can be used, since entropy increases across a shock but remains constant across expansion waves. This may be defined as

$$\gamma_{ij} = |S_{i+1/2,j} - S_{i-1/2,j}|$$

Another alternative is to try to detect compressions from the divergence of velocity. Accordingly one can set

$$\gamma_{ij} = \begin{cases} -\nabla\cdot\vec{q} & \text{if } \nabla\cdot\vec{q} < 0 \\ 0 & \text{if } \nabla\cdot\vec{q} > 0 \end{cases}$$

γ_{ij} is then proportional to $\nabla\cdot\vec{q}$ across shocks, which simulates physical viscous effects, and zero across expansion waves.

Equation (2) is integrated in time by an explicit multistage scheme. Multistage schemes are chosen because of their extended stability limit and high frequency damping properties which are appropriate for multigrid schemes. To increase the rate of convergence local time steps, enthalpy damping and implicit residual averaging techniques have been used. These techniques enable us to use Courant numbers up to almost 10. The most effective method of accelerating convergence, however, is the multigrid method. Auxiliary meshes are introduced by doubling the mesh spacing. Values of the flow variables are transferred to a coarser grid by a rule that conserves mass, momentum and energy. The multistage scheme is reformulated as described in [6] with the result that the solution on a coarse grid is driven by the residuals collected on the next finer grid. The process is repeated on successively coarser grids. Finally the corrections are passed back to the next finer grid by bilinear interpolation. Both V–cycle and W–cycle strategies have been used.

3. Boundary Conditions

The scheme has been programmed for different mesh topologies. For cascade calculations we usually encounter four types of boundaries; wall, periodic, inlet and outlet. Care must be exercised in treating these boundaries since internal flow is more sensitive to the boundary conditions than external flow. Inconsistency may cause large numerical errors and instabilities.

For wall boundaries zero normal velocity is imposed, and we use the normal momentum equation to extrapolate the pressure to the wall. Equivalent flow variables are imposed on corresponding periodic boundaries. They are updated as interior points. Based on characteristic analysis the entropy, which is equivalent to specifying total pressure, total enthalpy and the flow angle are specified on the inlet flow boundary for subsonic flow. The other independent flow variable is extrapolated from the interior by using the Riemann invariant for a one–dimensional flow normal to the boundary. Let the total entropy, enthalpy and flow angle far upstream be S_∞, H_∞ and α_∞. These values are specified for the fictitious cells upstream of the computational domain. The one–dimensional Riemann invariant normal to the boundary is

$$Re = q_{ne} + \frac{2c_e}{\gamma-1}$$

where the subscript e indicates interior values, q_n is the normal velocity to the inflow boundary and c is the speed of sound. For supersonic flow, all variables are specified on the inlet boundary. Conversely on the outlet flow boundary,only the pressure is specified for subsonic flow, while the entropy and total enthalpy plus one Riemann invariant are extrapolated. If the pressure of the initial flow field is very different from the specified back pressure, it is necessary for stability to adjust the back pressure gradually from the initial value to the specified value for the first few time steps. For supersonic flow in the axial direction all flow variables are extrapolated.

4. Test Cases

A. NACA 0012 Cascade

The first test case is a cascade with unstaggered NACA 0012 profiles and with solidity 1. An O–mesh is generated by an elliptic generator. Both a subsonic and a transonic case have been calculated. Figure 1 shows the Mach contours obtained for the transonic case with inlet Mach number 0.7. The Mach number in front of the shock is about 1.3 in contrast to the subcritical flow for the isolated airfoil at the same upstream Mach number. Figure 2 shows the convergence history of our computation with a 4 level multigrid scheme for the same cascade at inlet Mach number 0.4. The computation is started from a uniform flow. The work unit in the figure is the actual work equivalent to the number of cycles of a single grid iteration. It includes the extra work for the multigrid scheme. Comparison with the convergence history for the same case of an implicit ADI scheme by Schäfer [7] shows that the present scheme has a much faster convergence rate. Figure 3 shows the convergence history for the transonic case. We can see that the fast convergence rate remains about the same for the transonic case.

B. Hobson's Impulse Cascade

One of the most difficult test cases for cascade calculations is the Hobson shock free impulse cascade. It has very thick airfoils and was designed by the hodograph method to give a shock free supersonic pocket. Since any such shock free solution is isolated in the sense that a small perturbation in the flow conditions or airfoil shape can cause a shock wave to occur, Hobson's cascade provides a difficult test of both the accuracy and convergence for any numerical method. Ives calculated the flow of this cascade by a potential flow method [8]. The result, however, showed an obvious shock terminating the supersonic pocket in both the Cp distribution and Mach number contour pictures. Figure 4 shows the Cp distribution obtained by the present method. It is essentially shock free and in close agreement with the original hodograph design. Figure 5 shows the Mach number contours of the same solution. The smooth symmetric pattern is another indication of the accuracy. The convergence history is shown in figure 6. Again it demonstrates the fast convergence rate of the method. The results shown here, we believe, are about the best as yet published. However, it must be pointed out that the original surface definition by the hodograph method is very coarse. It has only 39 points on the upper surface and 15 points on the lower surface. A refinement of the surface definition is needed for more accurate numerical comparisons.

C. Supersonic Wedge Cascade

Denton proposed a wedge cascade as a test case for capturing oblique shocks in cascades [1]. Figure 7 shows our solution for such a cascade with inlet Mach number 2. The shock reflected from the lower blade is designed to be exactly canceled out at the corner of the upper blade, giving a uniform flow between the parallel surfaces and an expansion off the downstream corner. This design gives a good test case for numerical methods, since smearing may prevent complete cancellation of the reflected shock, and thus may produce a nonuniform region downstream. Figure 8 shows the distribution on the blade obtained by the present method. Comparison of our results with the analytic solution and that of [1] shows that the present method is of higher accuracy than Denton's. An interesting result is the prediction of two trailing edge shocks from each blade, which interact with shocks from other blades. The interaction of the shocks through the periodic boundaries was not shown in [1].

Despite the improvement over Denton's results, the reflected shock is not completely cancelled by the upper blade, and this causes oscillations in the Mach number distribution. Furthermore, the Mach number in the expansion region is lower than that of the analytic solution although the pressure has fully recovered to the analytic value as shown in Fig. 8. This is an indication of entropy production, and may have been caused by the first order dissipation in that region. Use of the alternative sensors based on entropy change or the divergence of velocity have shown improvements, but the problem remains.

D. VKI Turbine Cascade

Finally, the turbine cascade as proposed on a VKI workshop [9] is tested. Figure 10 shows the pressure contours of the solution. While the Cp distribution deviates from the experimental data near the trailing edges, the convergence shown in figure 11 is far better than any one of those reported in [9]. The deviation from experimental data, which is present in all solutions of the workshop, may be attributed to the blunt trailing edge of the blade since viscous effects can not be ignored there.

5. Conclusion

A finite volume method with a multistage time stepping scheme is used to calculate 2–dimensional cascade flow. The stability limit of the explicit scheme is extended by using implicit residual averaging. Convergence is accelerated by using locally varying time steps, enthalpy damping and most of all an effective multigrid method. Adaptive numerical dissipation is used for capturing shocks. Our results for the NACA 0012 cascade, the Hobson shock free cascade and the supersonic wedge cascade show excellent accuracy and convergence in comparison with the results by other investigators. To resolve the flow near the blunt trailing edge of a turbine cascade, however, it will be necessary to use a viscous flow model.

References

1. Denton, J.D., "A Improved Time Marching Method for Turbomachinery Flow Calculation", J. of Engineering for Gas Turbines and Power, Vol. 105, July 1983.

2. Jameson, A., "Successes and Challenges in Computational Aerodynamics", AIAA Paper 87–1184–CP, Honolulu, Hawaii, 1987.

3. Jameson, A., Schmidt, W., and Turkel, E., "Numerical Solutions of Euler Equations by Finite Volume Methods Using Runge–Kutta Time Stepping Schemes", AIAA Paper 81–1259, June 1981.

4. Subramanian, S.V. and Bozzola, R., "Computation of Transonic Flows in Turbomachines Using the Runge–Kutta Integration Scheme", AIAA J., Vol. 25, No. 1.

5. Holmes, D.G. and Tong, S.S., "A Three–Dimensional Euler Solver for Turbomachinery Blade Rows", J. of Engineering for Gas Turbines and Power, Vol. 107, April 1985.

6. Jameson, A., "Transonic Flow Calculations", MAE Report #1651, Princeton University, July 1983.

7. Schäfer, O., Frühauf, H.H., Bauer, B., and Guggolz, M., "Application of a Navier–Stokes Analysis to Flows through Plane Cascades", J. of Engineering for Gas Turbines and Power, Vol. 108, January 1986.

8. Ives, D.C. and Liutermoza, J.F., "Second–Order–Accurate Calculation of Transonic Flow Over Turbomachinery Cascades", AIAA J. Vol. 17, No. 8, 1979, p. 870.

9. "Numerical Methods for Flows in Turbomachinery Blading", Von Karman Institute for Fluid Dynamics Lecture Series 1982–05, Vol. 3, April 26–30, 1982.

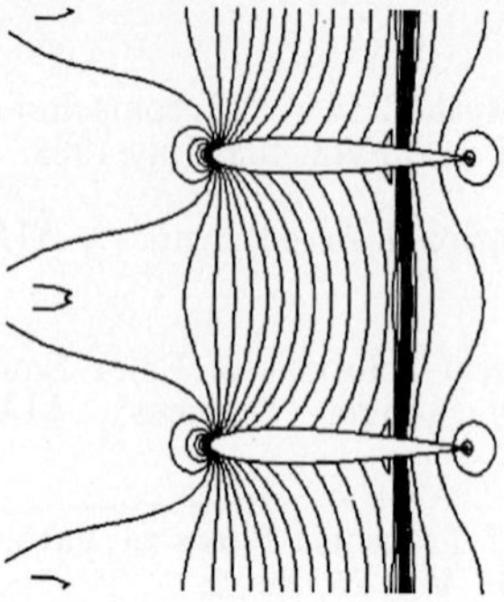

Fig. 1 Mach Number Contours for a NACA 0012
Cascade at M = 0.7, 161x33 Grid

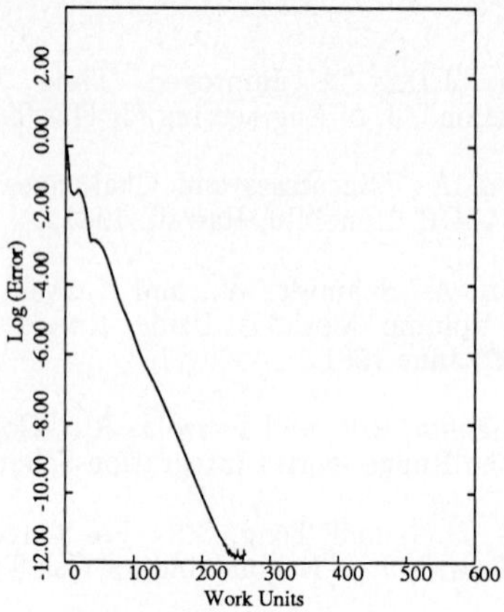

Fig. 2 Convergence History for a NACA 0012
Cascade at M = 0.4
Resid1 : 0.14e-01 Resid2 : 0.25e-13 Rate : 0.9027

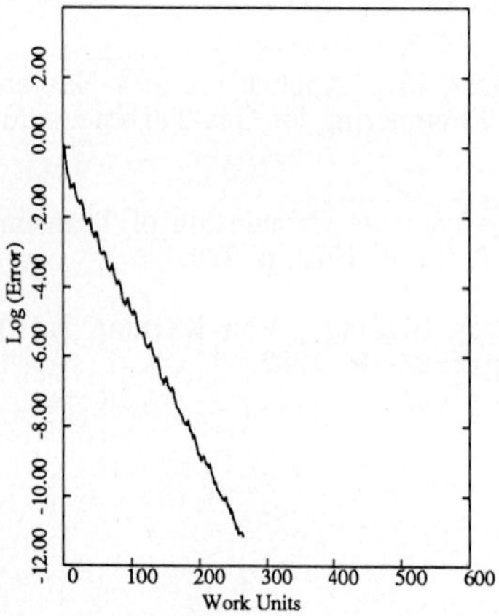

Fig. 3 Convergence History for a NACA 0012
Cascade at M = 0.7
Resid1 : 0.36e-01 Resid2 : 0.26e-12 Rate : 0.9075

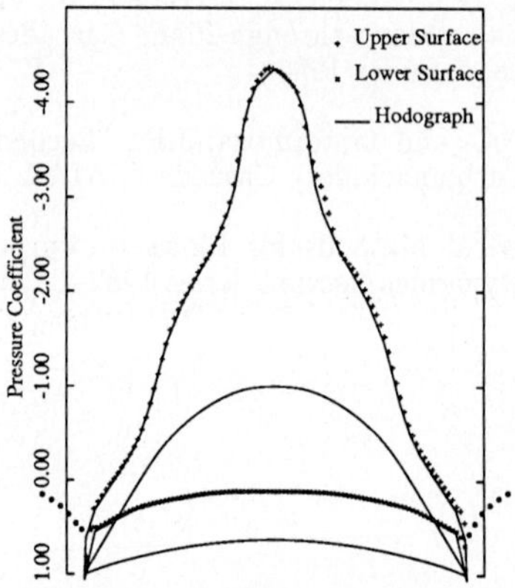

Fig. 4 Pressure Coefficient Distribution over Hobson's
Shock-Free Cascade

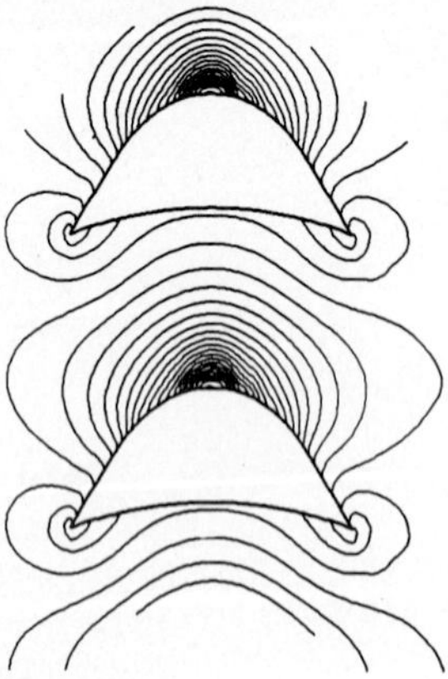

Fig. 5 Mach Contours of Hobson's Shock-Free Cascade
161x33 Grid

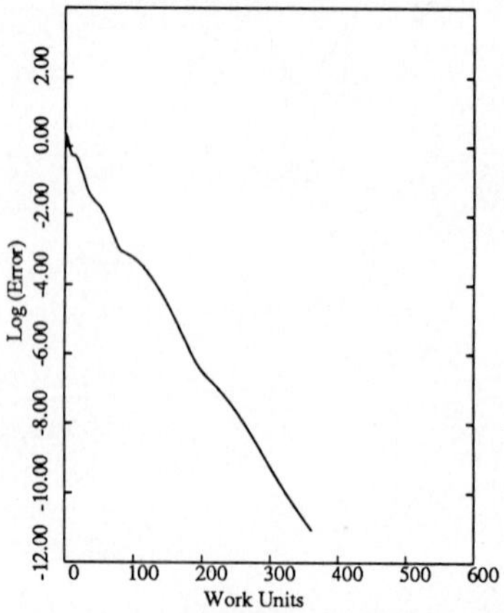

Fig. 6 Convergence History for Hobson's Cascade
Resid1 : 0.31e-01 Resid2 : 0.28e-12 Rate : 0.9319

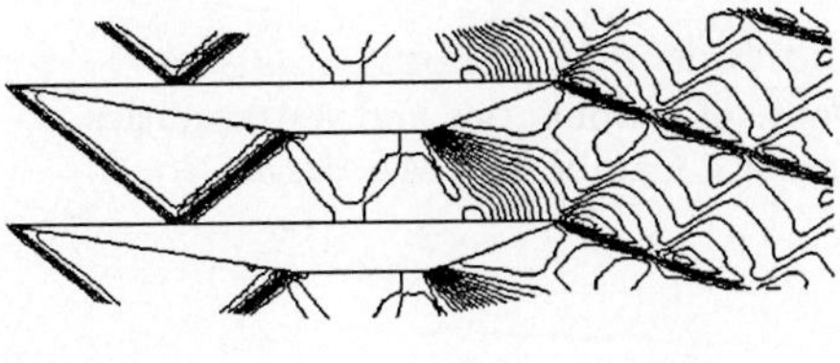

Fig. 7 Pressure Contours of a Wedge Cascade
M = 2.   161x33 Grid

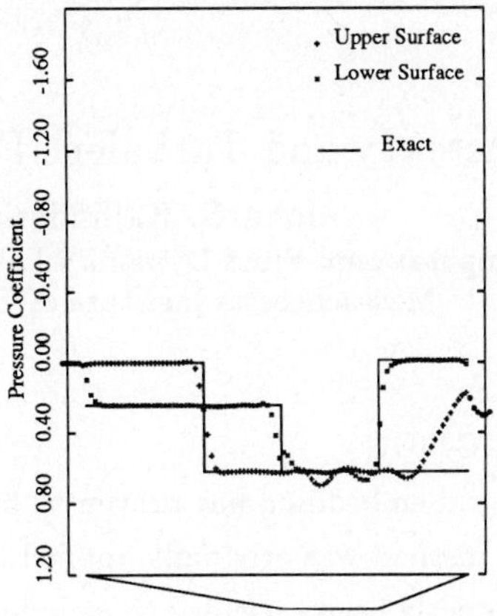

Fig. 8 Pressure Coefficient Distribution on a Wedge Cascade

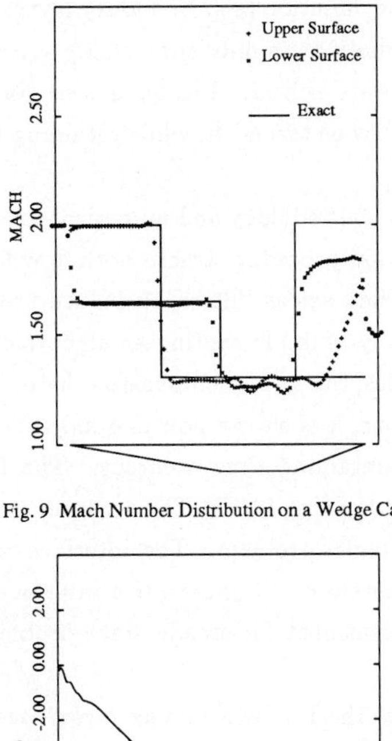

Fig. 9 Mach Number Distribution on a Wedge Cascade

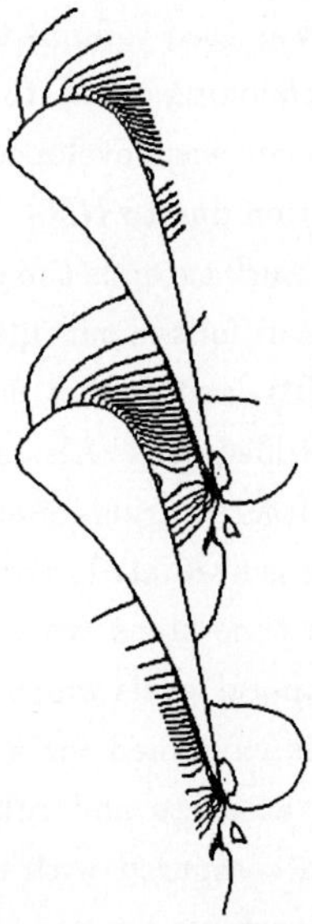

Fig. 10 Mach Number Distribution of a VKI Cascade
77x21 Grid

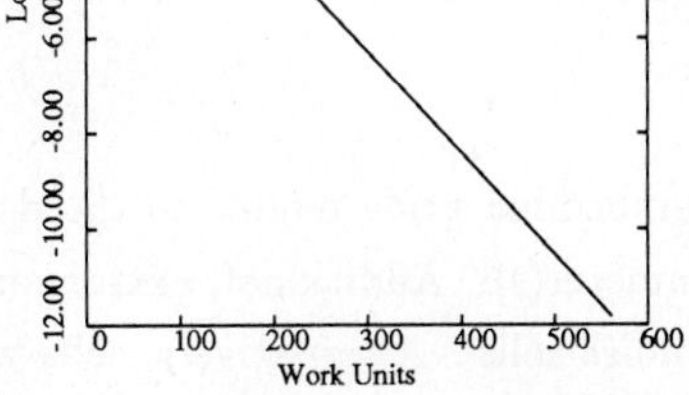

Fig. 11 Convergence History of a VKI Cascade
Resid1 : 0.66e+00   Resid2 : 0.99e-12   Rate : 0.9527

Unsteady and Turbulent Flow using Adaptation Methods

John G. Kallinderis* and Judson R. Baron†
Computational Fluid Dynamics Lab, Dept. of Aeronautics and Astronautics
Massachusetts Institute of Technology, Cambridge, MA 02139

INTRODUCTION

Adaptive grid embedding has proven to be an efficient approach for resolving important flow features. The method was originally applied to inviscid flow involving shocks as the main feature [3,2]. It has recently been extended to include viscous regions with multiple scale phenomena such as shocks and boundary layers [4]. There the ***directional embedding*** concept was introduced in regions where significant flow gradients exist only in one direction (e.g. boundary layers). ***Equation adaptation*** also was used to limit the Navier Stokes equations to only appreciably viscous regions, while for the remaining areas the Euler equations were solved. Lastly, a ***new finite volume, conservative scheme*** was developed to discretize the viscous terms [4], while retaining the inviscid terms discretization due to Ni[6].

The present work extends the previous adaptation methodology and numerical scheme to such complex flows as airfoils in turbulent flow. The detection procedure tracks both flow features and their directionality, and defines ***embedding patches*** which act as 'filters' to reduce the number of 'randomly' embedded cells. Also a new and general way of implementing an algebraic turbulence model for unstructured grids (quadrilaterals or triangles, the former emphasized here) is described and its accuracy is evaluated. For unsteady simulations, it is shown how one may allow a spatial variation of the time steps while simultaneously maintaining time accuracy. The flow field is divided into temporal levels which coincide with the embedding levels. The accuracy and efficiency of the method is examined for a forced oscillation model problem. The interface treatment is essential for an accurate and robust procedure. Therefore a ***Conservative*** interface treatment is presented and compared with a nonconservative treatment for steady state problems in both subsonic and supersonic flow.

In the following the concept of ***embedding patches***, the Baldwin-Lomax ***turbulence model implementation for unstructured meshes***, the method of ***spatially varying time steps***, and finally, a ***conservative interface treatment***, are all described and evaluated.

EMBEDDING PATCHES

An important consideration when constructing embedded grids relates to the determination of a suitable threshold of the feature detection parameter(s). Additional, extraneous cells may appear near features and are essentially embedded 'noise cells'. Alternatively, cells which should be divided often are overlooked [3]. This characteristic behavior becomes even more severe with directional embedding.

*Research Assistant
†Professor

An *embedded patch* encloses a defined fixed number of cells of the *initial* mesh. It essentially scans the domain and during the scan the included cells are examined for each patch. If the majority of its cells (e.g. 90%) are flagged for division, then *all* cells currently belonging to the patch are flagged for division. Conversely, if very few cells (e.g. 10%) are flagged, none within the patch are embedded. When repeated throughout the domain, the patch procedure can be viewed conceptually as a 'searching window', and acts somewhat like a 'noise filter' which reduces the number of 'randomly' embedded cells.

The illustrative case of a flow around an NACA 0012 airfoil shows the effectiveness of the method. Fig. 5(a) shows Mach number contours just before a *second* level of embedding. The dominant features are a normal shock on the suction side and the upper and lower surface boundary layers. Fig. 5(b) shows the resulting grid. Directional embedding is present above and below the airfoil while the leading edge region is embedded in both directions. The downstream region lying between the two shear layers is not embedded. Comparison of the grid and the resulting solution shows that both the features and their directionality are faithfully captured. The borders between the different regions are fairly free from 'noise' cells.

TURBULENCE MODELING WITH UNSTRUCTURED MESHES

The algebraic model due to Baldwin and Lomax [1] was used as a turbulent flow description. The model implicitly assumes a structured mesh, and its implementation is usually along lines normal to the surface. For an unstructured mesh (quadrilateral or triangular), such normal mesh lines generally do not exist. In the case of quadrilaterals, for example, interfaces interrupt such lines (Fig. 1).

Our approach implements the model in a 'cell-wise' manner. All necessary quantities are calculated at the center of each cell. In this way we avoid using information from outside of the cell, an approach which is common when dealing with unstructured meshes generally. For example, vorticity which is an important quantity for both the inner and outer layer formulation of the model, is calculated using Green's theorem over each cell. Specifically, $\omega = -(1/S_{cell}) \oint_{cell}(udx + vdy)$ where S_{cell} is the cell area. The distance of each cell from the wall is calculated and stored whenever the grid is updated. The only quantities that require information from outside of each cell in order to be evaluated are the Baldwin-Lomax parameters $Fmax$ and $Udiff$ which are used for the outer layer. These are evaluated by examining all cells associated with each station (cells A,B,C,D,E in Fig. 1). The assumption is that they exhibit no significant streamwise variation over cell E.

The case of a NACA 0012 airfoil in a subsonic turbulent flow tests the accuracy of the method. An initial C-mesh of 33x17 points with two levels of embedding was used for the calculation. In Fig. 4 the pressure coefficient distribution and the lift are compared favorably to experiment [8].

UNSTEADY FLOW WITH SPATIALLY VARYING TIME STEPS

The globally minimum time step required for an unsteady calculation poses a serious limitation to the application of explicit methods. We suggest a procedure which allows a spatial variation of the time steps while simultaneously maintaining time accuracy. To accomplish this the flow field is divided into temporal levels which coincide with the embedding levels (Fig. 2). The same time step is assigned at each level but between two consecutive levels the time steps vary by a factor of two. For the example in Fig. 2, the smaller cells of level 1 are integrated in time twice using a time step δt before the larger cells of level 0 are integrated using a time step $2\delta t$ [5].

Spatial adaptation (cell subdivision to resolve spatial gradients) generally results in *temporal adaptation* (reduction of the time step to resolve temporal gradients) as well [7]. Large temporal gradients may exist in regions where features are present. However, a special algorithm for detection of regions of these temporal gradients is frequently an unnecessary burden since the latter are included in the spatially embedded regions. This results in a significant simplification of the data structure required to handle adaptation.

As a test of the time accuracy of the method, we consider the case of a channel flow with a forced oscillation of the inlet Mach number (Fig. 6). A low Reynolds number ($Re = 10^3$) reduced CPU time. Using one level of embedding the time history of the U-velocity component at a point of the domain was tracked. Free stream flow was used as initial condition. The initial transient decayed after approximately 2 periods. Comparison of the solution with that for a globally embedded mesh shows the agreement between the two histories to be excellent (Fig. 6).

CONSERVATIVE INTERFACE TREATMENT

The existence of embedded regions within the interior of the domain introduces internal boundaries (interfaces) which must be treated carefully. We present a conservative interface treatment and compare it with one that is non-conservative for steady state flows in both supersonic and subsonic cases.

The inviscid terms are evaluated by performing a special line integration at cell E (Fig. 3b) which includes the hanging node b. The cell change in time for the Ni scheme is distributed to all 5 nodes of cell E as shown in Fig. 3a. The viscous terms require piecewise integration of the stresses along the dashed lines (Fig. 3b) on each cell [4]. The interface cell E is divided into five areas which are allocated to each node. The stress fluxes then cancel inside the cell and the treatment becomes conservative. Due to stretching in the cells, there is an error in the viscous terms' evaluation which is locally induced. The distributions to the five nodes due to the smoothing terms (fourth and second order) sum up to zero which makes the smoothing operator conservative. This conservative and a non-conservative treatment [4] have been compared in the following supersonic and subsonic model cases.

Interfaces near a Shock. Channel flow with a $M = 1.35$ shock was used as a test with embedding near the shock (Fig. 8). The embedded region is enclosed by the crosses. In Fig. 9 we compare mass flow across the channel for the two treatments and for a globally fine mesh without interfaces. The nonconservative treatment clearly induces a mass flow error, while the conservative one reproduces the same result as the globally fine.

Interfaces inside a Boundary Layer. The same geometry in a subsonic flow ($M = 0.5$) and with embedding at midheight inside the boundary layer was used to test the conservative and nonconservative treatments. Fig. 7 shows skin friction distributions for both the embedded and the globally fine grids. In this case the non-conservative interface treatment provides excellent agreement with the globally fine solution. An interpretation is that the stretching error in the conservative treatment is more important than the non-conservation error.

CONCLUSIONS

Concepts for adaptive procedures have been examined. Feature detection is made more accurate by the introduction of embedded patches which act as 'noise filters' in the embedding procedure. A method to implement the Baldwin-Lomax algebraic turbulence model when applied to unstructured meshes proves to be accurate. Spatially varying time steps based on embedding zones can be used for unsteady problems leading to accurate results with reduced CPU times. Lastly, a conservative interface treatment is important in regions where shocks exist, but not necessarily inside boundary layers where the stretching error in the evaluation of the viscous terms may be dominant.

References

[1] B. S. Baldwin and H. Lomax. AIAA Paper 78-257, 1978.

[2] M. Berger and A. Jameson. *AIAA Journal*, 23:561–568, Apr 1985.

[3] J. F. Dannenhoffer III and J. R. Baron. AIAA Paper 85-0484, 1985.

[4] J. G. Kallinderis and J. R. Baron. AIAA Paper 87-1167-CP, 1987.

[5] R. Lohner, K. Morgan, J. Peraire, and O. C. Zienkiewicz. AIAA Paper 85-1531, 1985.

[6] R.-H. Ni. *AIAA Journal*, 20:1565–1571, Nov 1982.

[7] M. Pervaiz and J. R. Baron. *Comm. in Applied Numerical Methods*, 4(1):97–111, Jan 1988.

[8] J. J. Thibert, M. Granjacques, and L. H. Ohman. Technical Report AGARD AR 138, 1979.

Acknowledgement *Sponsored by the Air Force Office of Scientific Research, Contract Monitor Dr. James McMichael.*

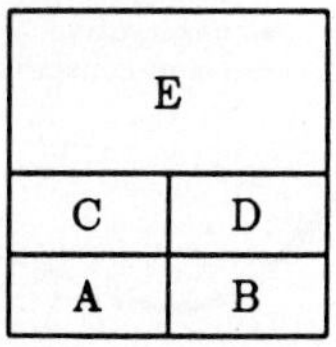

Figure 1: Cell group at a Station

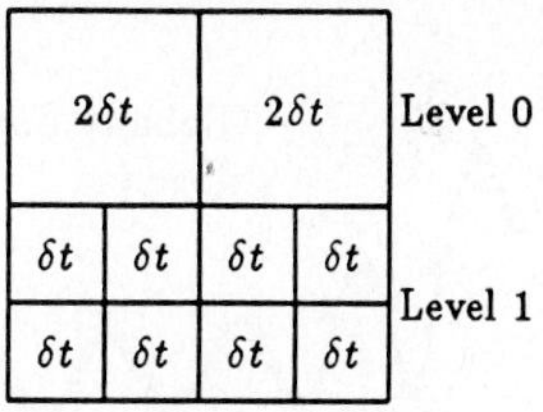

Figure 2: Spatially varying time steps

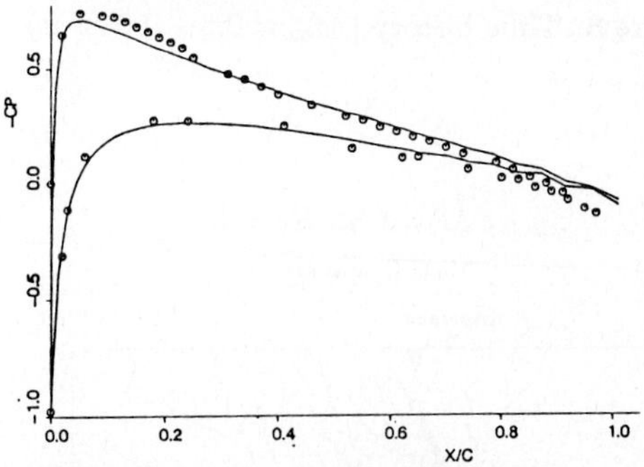

Comparison of pressure coeff. distributions
— present work ($Cl = 0.197$)
o Experiment [8] ($Cl = 0.195$)

Figure 4: NACA 0012 ($M_\infty = 0.5, Re = 2.91 \times 10^6$)

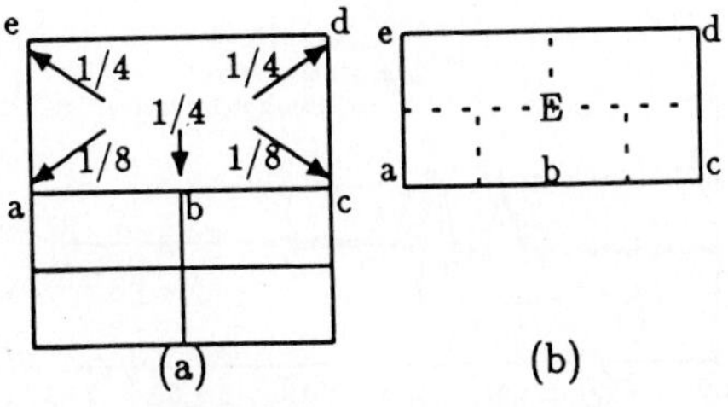

Figure 3: Conservative Interface Treatment

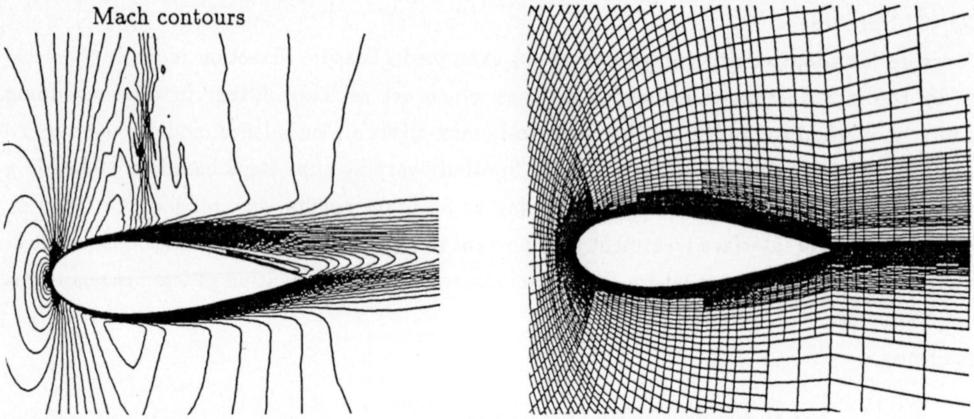

(a) solution before embedding
(vertical scale enlarged)

(b) resulting grid
(vertical scale enlarged)

Figure 5: NACA 0012 ($M_\infty = 0.8, Re = 10^5, \alpha = 2°$)

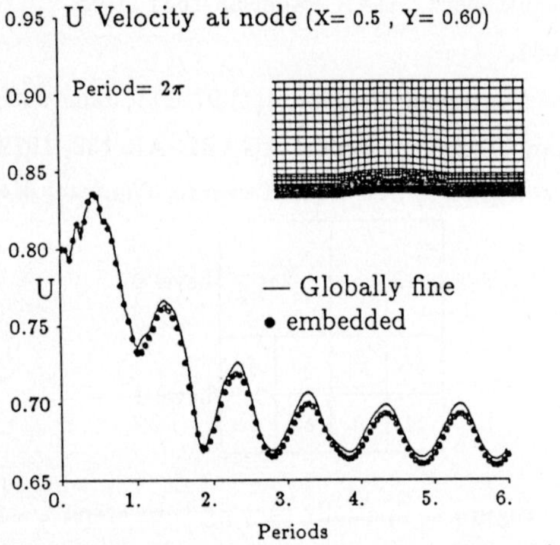

Figure 6: Time history ($M_\infty = 0.8 + 0.04 sint$)

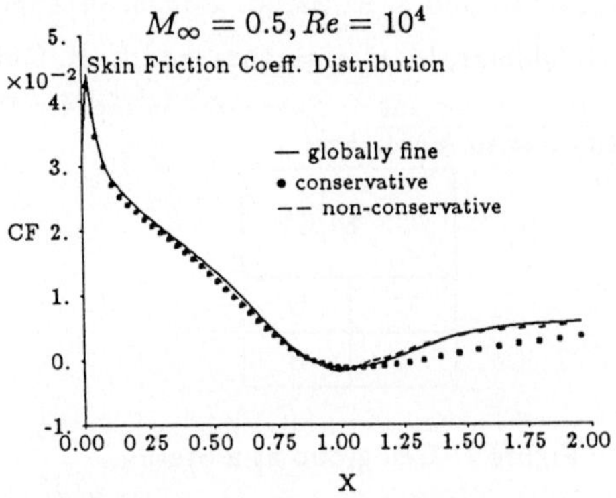

Figure 7: Comparison of interface treatments

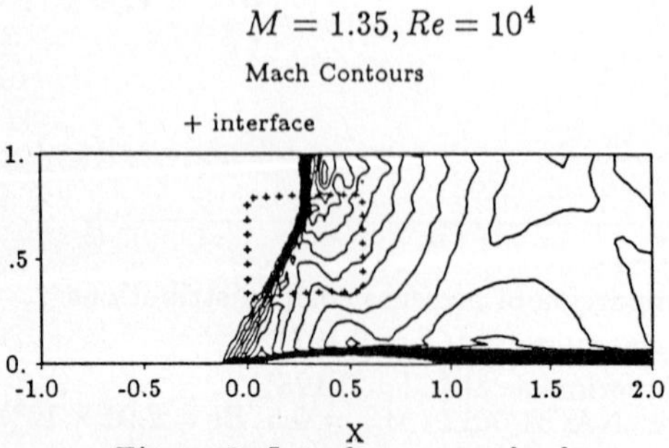

Figure 8: Interfaces near shock

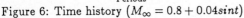

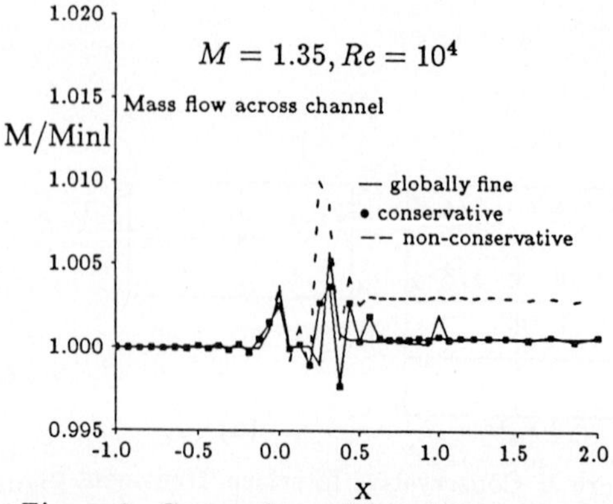

Figure 9: Comparison of interface treatments

RNS SOLUTIONS FOR THREE-DIMENSIONAL STEADY INCOMPRESSIBLE FLOWS

P.K. Khosla, S.G. Rubin and A. Himansu
University of Cincinnati, Cincinnati, OH 45221

1. Introduction

In a series of papers, the present authors have developed primitive variable and composite velocity/potential formulations for the computation of flows with viscous-invsicid interaction at large Reynolds number (Re). These procedures involve solution of the Reduced Navier-Stokes (RNS) equations, which are a large-Re asymptotic approximation to the full Navier-Stokes equations. The procedures are computationally more efficient than full NS solvers, and allow for the computation of complex flows with flow separation. The advantages of the RNS formulation and applications to two-dimensional and axisymmetric complex flow problems are discussed in [1-6].

The RNS equations retain the elliptic acoustic behavior of the full Navier-Stokes (NS) equations. This is associated with the pressure interaction, and therefore the RNS discretization requires a form of flux-split differencing. A fraction ω of the axial pressure-gradient term in the streamwise momentum equation is associated with the positive flux eigenvalues, while the remaining fraction $(1-\omega)$ corresponds to the non-positive eigenvalues. The parameter ω is a function of the local Mach number, and ranges from 0 for incompressible flow to 1 for supersonic flow. All axial convective terms are upwind differenced. The solution procedure takes advantage of this upwind differencing, and consists of streamwise marching sweeps; however, due to the non-positive values appearing in subsonic regions the pressure solution is obtained only through the relaxation process. The procedure for steady flows requires storage only for the pressure. This is a significant simplification for three-dimensional computations.

The present work builds on that of Cohen and Khosla [9], who applied the RNS global relaxation procedure in three dimensions to obtain solutions for a finite edge flat plate, the flow along an axial corner, and over a sine-wave bump. The procedure is used here to obtain laminar three-dimensional separated-flow solutions for a trough and a parabolic-arc airfoil. A sparse-matrix solver is combined with a marching or relaxation procedure. Comparisons are made with two-dimensional solutions and a finite recirculation bubble is found for the three-dimensional cases. Other three-dimensional solutions with the RNS model have also been obtained by Raven and Hoekstra [7] who have solved for the flow about a ship stern, and Rosenfeld, Israeli and Wolfshtein [8] who have used a similar relaxation procedure to calculate the flow past prolate spheroids at angle of attack.

2. Governing Equations and Solution Procedure

The three-dimensional nondimensionalized form of the RNS equations for incompressible flow in cartesian coordinates (x,y,z), are as follows :

continuity :

$$u_x + v_y + w_z = 0 \tag{2.1}$$

x- (streamwise) momentum :

$$uu_x + vu_y + wu_z + p_x = \frac{1}{Re}[u_{yy} + u_{zz}] \tag{2.2}$$

y-momentum :

$$uv_x + vv_y + wv_z + p_y = 0 \tag{2.3}$$

z-momentum :

$$uw_x + vw_y + ww_z + p_z = \frac{1}{Re}[w_{yy} + w_{zz}] \tag{2.4}$$

where x is the principal flow direction, y is normal (or nearly so) to the wall layer, z is the spanwise direction, and Re is the Reynolds number; u,v,w are the velocity components in these directions. The neglect of all viscous terms in the y-momentum equation is consistent with that of all streamwise diffusion terms. A simple shearing transformation is used to generate grids for the trough and wing geometries.

The continuity equation is discretized at the grid location (i,j-1/2,k). The streamwise (x-) and crossflow (z-) momentum equations are discretized at (i,j,k), and the normal (y-) momentum equation is discretized at (i,j+1/2,k). This reflects the boundary conditions for v at the surface and pressure at the upper boundary. All y- and z-derivatives are second-order accurate using either trapezoidal or central differencing. As discussed earlier, the elliptic or boundary-value nature of the equations for subsonic flow is reflected in the discretization of the streamwise pressure-gradient term, $\partial p/\partial x$. This leads to an initial-value problem in supersonic regions and to a boundary-value problem in subsonic regions. Inflow conditions are required for the velocities. An outflow condition is required only for the pressure; this allows the numerical outflow boundary to be placed farther upstream than is possible with full NS solvers.

During each marching sweep, at any station i the discretized equations are solved for the flow variables in that cross-plane. The discretized equations are quasilinearized to second order about the local solution iterate, and the resulting linear system is solved using a sparse matrix solver. The solver used herein is the Yale Sparse Matrix Package (YSMP), developed by Eisenstadt et al [11]. This solver is very efficient as regards memory and computation for sparse matrices. The initial value of the local iterate is taken as the solution at the previous station. In separated-flow regions, the local inversion is repeated twice to converge the solution iterate to the nonlinear discrete system.

3. Results

For each geometry considered here an initial pressure distribution is required. For steady flow solvers this choice can have a significant effect on the overall convergence. For the present calculations, an inviscid pressure distribution is first generated by solving the RNS system with slip boundary conditions. The viscous calculation is initiated after four to six global iterations and continues until the error in pressure is of the order of 0.0001. Typically, a total of about 25 global iterations is required to achieve this level of accuracy.

Carter-Wornom Trough : The geometry of the trough as shown in figure 1a is given by

$$y_b = -d(z)\ \mathrm{sech}(4(x-2.5)) \tag{3.1}$$

where $d(z) = 0.03 \times \{\sin[\pi(z/2+0.5)] + 1\}/2$ for $0 \leqq z \leqq 2$

and $d(z) = 0$ for $2 \leqq z \leqq 3$ (3.2)

The depth d, varies in the z direction, from 0.03 at the z=0 symmetry plane to zero (flat plate) for $z \geqq 2$. The outer z computational boundary is at z=3. The solution for Re=80,000 is obtained on a 101×57×31 grid. The results for c_p and c_f are depicted in figures (1b,c). The vorticity contours are shown in figure 1d. The maximum axial flow reversal is in the symmetry plane. The reversed-flow region dissipates and completely disappears as z increases. The solution near the outer z plane is similar to the flow over a flat plate. Also shown in figures (1b,c) are the two-dimensional solution for a trough of depth d=0.03 at Re=80,000. The computation was carried out on a Microvax II with a CSPI 6420 array-processor.

Wing Section (Parabolic-Arc Airfoil) : The geometry for this case is shown in figure 2a and is given by :

$$y_b = 2\,d(z)\,x(1-x) \qquad (3.3)$$

where $d(z) = 0.03 + 0.03 \times \{\sin[\pi(z/2+0.5)] + 1\}/2$ for $0 \leqq z \leqq 2$

and $d(z) = 0.03$ for $2 \leqq z \leqq 3$ (3.4)

The thickness-to-chord ratio at the symmetry plane is 0.06, and the flow Re is 80,000. A 101×50×21 grid was utilized for this case. The results are depicted in figures (2b,c,d). The flow displays axial reversal which dissipates with distance from the symmetry plane. Figure 2e shows iso-vorticity contours of the transverse or z component of vorticity in the symmetry plane. This computation was carried out on a Cray X-MP 2/4, with each cross-plane calculation requiring 19 seconds of cpu time.

Acknowledgement

This work was supported by the AFOSR under contract number F49620-85-C-0027.

References

1. Rubin, S.G., and Lin, A. : Israel Journal of Technology, 18, 1981.
2. Rubin, S.G., and Reddy, D.R. : Computers and Fluids, v.11, no.4, pp. 281-306, 1983.
3. Khosla, P.K., and Lai, H.T. : AIAA Paper No. 84-0458, 1984.
4. Ramakrishnan, S.V., and Rubin, S.G. : AIAA Journal, v.25, no.7, 1987.
5. Himansu, A., and Rubin, S.G. : AIAA Paper No. 877-1145-CP. To appear, AIAA Journal, September 1988.
6. Rubin, S.G. : Computational Methods in Viscous Flows, vol.III, W.G. Habashi, ed. Pineridge Press, Swansea, 1984.
7. Raven, H.C., and Hoekstra, M. : MARIN Report No. 50402-I-SR, 1984. Presented at the Second Intern. Symposium on Ship Viscous Resistance, Goteborg, March 1985.
8. Rosenfeld, M., Israeli, M., and Wolfshtein, M. : Preliminary version, 1987.
9. Cohen, R., and Khosla, P.K. : submitted for publication.
10. Rubin, S.G. : To appear, Computers and Fluids, 1988.
11. Eisenstadt, S.C., Gursky, M.C., Schultz, M.H., and Sherman, A.H. : 'Yale Sparse Matrix Package II. The Nonsymmetric Codes'. Yale University Department of Computer Science Research Report # 114.

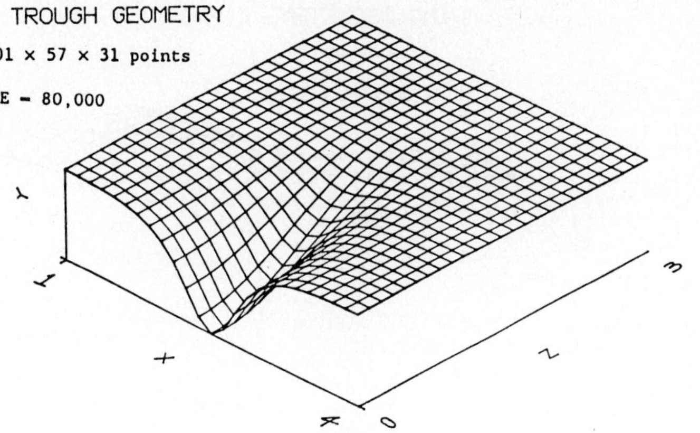

FIGURE 1A

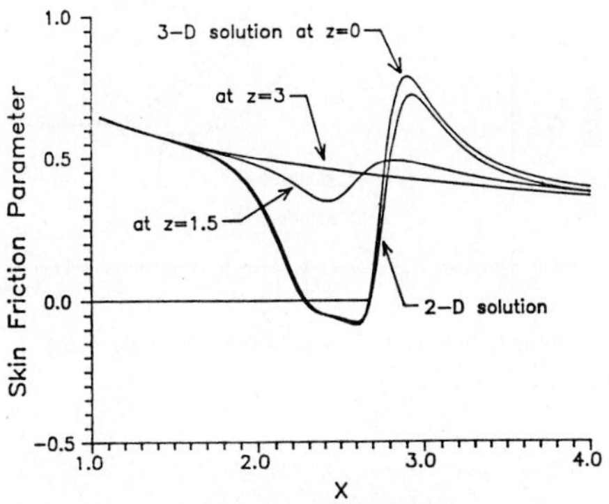

FIG. 1B TROUGH : SKIN FRICTION

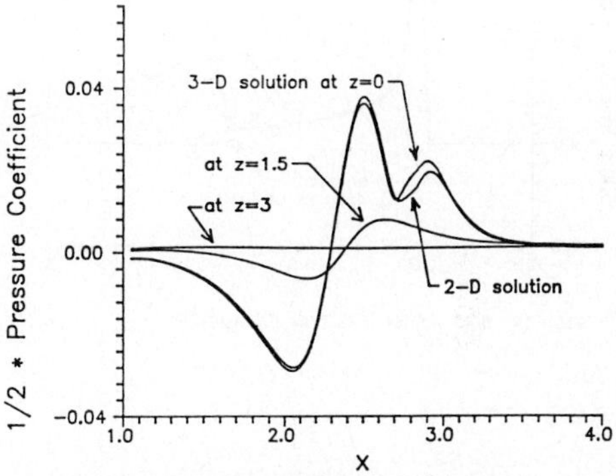

FIG. 1C TROUGH : COEFFICIENT OF PRESSURE

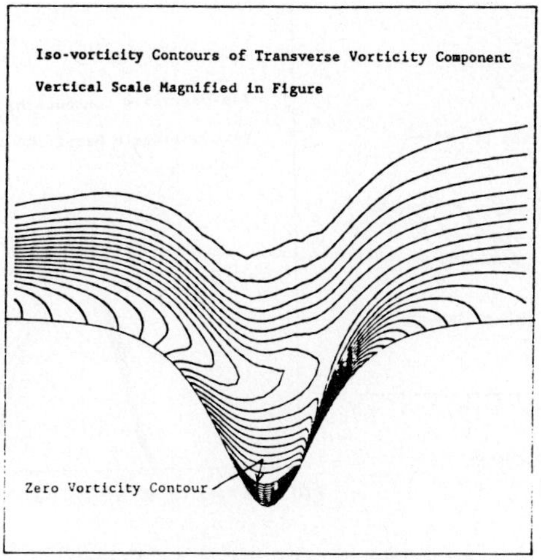

FIG. 1D TROUGH ■ VORTICITY AT SYMMETRY PLANE

WING GEOMETRY

RE - 80,000

101 × 50 × 21 points

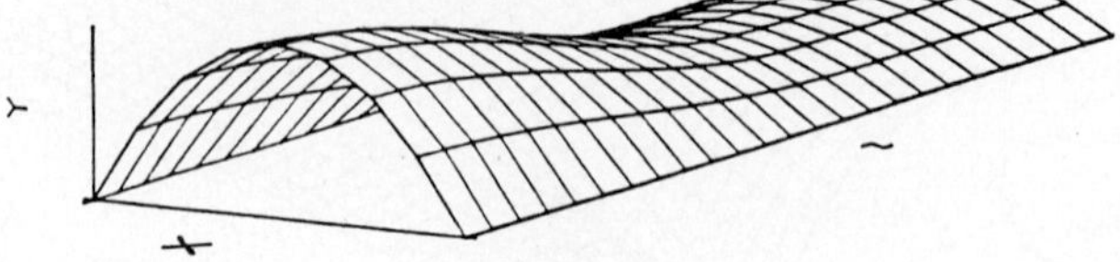

FIGURE 2A

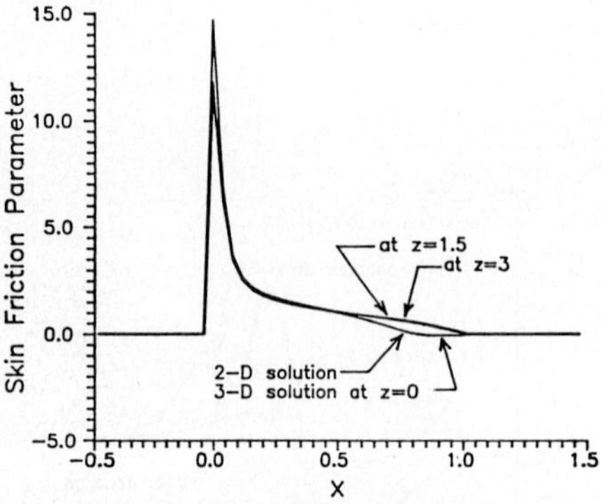

FIG. 2B WING : SKIN−FRICTION PARAMETER

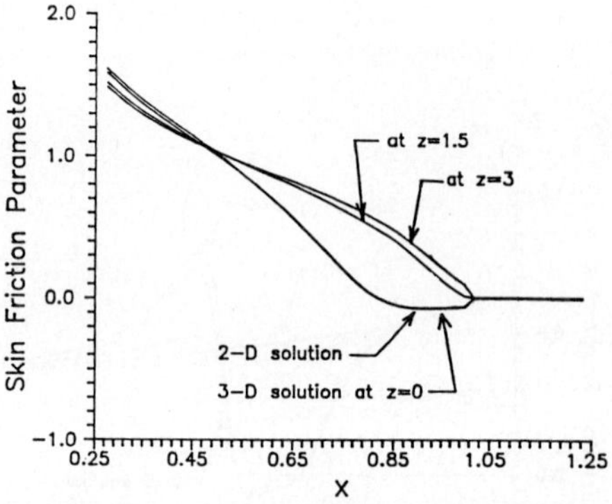

FIG. 2C WING : SKIN−FRICTION PARAMETER (ENLARGED PLOT)

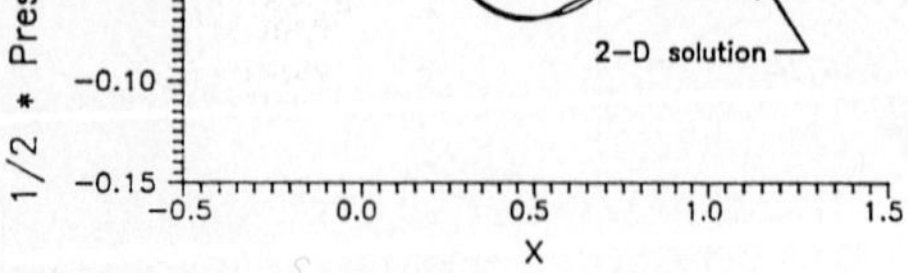

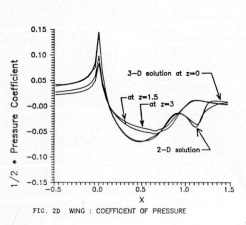

FIG. 2D WING : COEFFICIENT OF PRESSURE

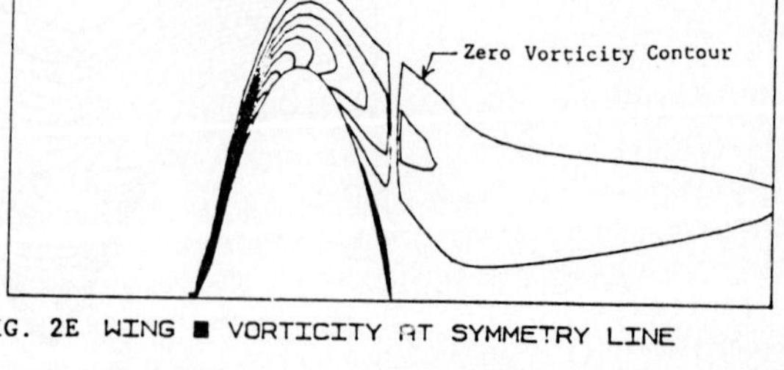

FIG. 2E WING ■ VORTICITY AT SYMMETRY LINE

NUMERICAL STUDY OF UNSTEADY VISCOUS HYPERSONIC BLUNT BODY FLOWS WITH AN IMPINGING SHOCK

G.H. Klopfer[1]
NEAR, Inc., Mountain View, CA 94043
H.C. Yee[2] and P. Kutler[3]
NASA Ames Research Center, Moffett Field, CA 94035

Abstract

A complex two-dimensional, unsteady, viscous hypersonic shock wave interaction is numerically simulated by a high-resolution, second-order fully implicit shock-capturing scheme. The physical model consists of a non-stationary oblique shock impinging on the bow shock of a blunt body. Studies indicate that the unsteady flow patterns are slightly different from their steady counterparts. However, for the sample cases investigated the peak surface pressures for the unsteady flows seem to occur at very different impingement locations than for the steady flow cases.

Introduction

The recent development of high resolution total variation diminishing (TVD) schemes [1-3] has made it possible to accurately simulate the unsteady Navier-Stokes equations for complicated shock interactions at hypersonic Mach numbers. Such flows are of current interest in hypersonic aerodynamics. In a previous study [4] the steady viscous hypersonic blunt body flows with impinging shock waves were investigated. Some numerical simulations by other numerical methods for steady blunt body flows with impinging shocks have been reported in [5-8]. Very little work has been done for viscous unsteady shock interaction flows at hypersonic speeds, in particular, numerical simulations by implicit methods. The objective of this paper is to study the unsteady viscous hypersonic blunt body flows with non-stationary impinging shocks. The rapidly changing shock waves and shear layers that occur and interact with each other make this a difficult problem to simulate by classical numerical methods. Here a two-dimensional code consisting of an implicit second-order accurate (time and space) TVD-type algorithm is used to simulate the flow field. The scheme is based on an existing TVD algorithm [9,10] for transonic and supersonic flows which has been extended recently to hypersonic and equilibrium real gas flows [2,3]. The specific objective of the study is to determine the unsteady shock interference patterns and to compare them with the corresponding patterns for steady flows at various hypersonic Mach numbers. Another objective is to determine the transient surface pressures.

The Physical Problem

The physical problem examined is the unsteady hypersonic viscous two-dimensional blunt body flow with an impinging shock. A schematic of the computational domain of the flow field is shown in figure 1. The blunt body has a thickness of D and a nose radius of R_l. The circumferential angle, θ, is measured from the

[1]Research Scientist
[2]Research Scientist, Computational Fluid Dynamics Branch
[3]Chief, Fluid Dynamics Division

symmetry line as depicted in the figure. This physical model can represent many practical applications. One example is the bow shock of a hypersonic aerodynamic vehicle interacting with the shocks of the leading edge of a wing or the cowl lip of the engine inlet. The interaction causes complicated shock-on-shock patterns in steady flows. Such steady interactions were studied in [4]. However, even under steady cruise conditions the vehicle bow shock fluctuates about some mean position, and thus the interaction is unsteady. The thickness of the cowl lip or wing is thin compared to the shock layer thickness, and even a small shock fluctuation causes a large excursion of the impinging shock in the vicinity of the inlet cowl or wing. This unsteady interaction is modeled by a non-stationary impinging shock (vehicle bow shock) moving downward with constant speed across the bow shock of the blunt body which can represent the cowl lip or wing leading edge. For steady flows, various kinds of shock interactions occur depending on the freestream conditions and impingement shock angle and location. Based on experimental data at supersonic speeds, Edney [11] has identified six different kinds of interactions labeled Type I through Type VI. For the Type IV interaction for steady flows, the peak surface pressure is many times larger than for the blunt body flow case. One motivation for the study is to see if the same type of amplification of the peak surface pressure values also occurs for the unsteady interaction.

Numerical Results

Due to space limitations, numerical results obtained for the unsteady laminar thin layer Navier-Stokes flows of a perfect gas are only summarized here. At present there are no experimental data available for the unsteady interactions to validate the numerical simulations. Therefore the unsteady results are compared to each other at different Mach numbers and to the steady state computations of reference 4 at the same Mach numbers.

Figure 2 presents the Mach contours at six instances of the diffraction process of an unsteady viscous computation. An impinging shock at an angle of 22.75° relative to the freestream moves down across the bow shock of the blunt body. The impingement shock velocity is 10% of the freestream velocity ($M_\infty = 15$). The freestream Reynolds number based on the diameter is $Re_D = 186,000$ and the freestream temperature is $T_\infty = 255.6^o K$. The Mach contours ranging from 0 to 15 in increments of 0.1 are shown in figure 2. The indicated times of each frame are normalized with the freestream velocity and cowl lip thickness. The surface pressure distributions for the last three instances compared with the steady-state case are given in figure 3. The first three surface pressure distributions are not given here and are similar to the steady case [4]. The most critical condition in terms of surface pressures no longer seems to occur for the Type IV shock interaction, as it does for the steady state case. Experimental data [11] indicates that the rates can be as much as 1000% (10 times) higher than for the non-interfering blunt body flow. Note that the shock impingement locations of the unsteady and steady computations do not coincide.

For the conditions used in this computation only the first five types of shock patterns identified by Edney are apparent in the sequence. Due to the unsteadiness, high Mach numbers, and high impingement shock angles the flow patterns are much more complex than those given by Edney. The corresponding Mach contours for the steady case at $M_\infty = 15$ are shown in figure 11 of reference 4. For both the steady and unsteady cases the same type of shock interference patterns occur although the unsteady sequence exhibits a much more complicated pattern within the shock layer. The separated boundary layer apparent for the unsteady Type I interaction does not appear for the steady Type I case with the same shock impingement angle. Also the supersonic jets flowing parallel to the blunt body surface and below the impingement point in the Type III and IV unsteady flows do not occur for the steady state cases. However for the Type V shock interference, where a new 2-D blunt body flow forms with the freestream conditions of those of a supersonic wedge flow, there are very little differences between the steady and unsteady flows.

The last set of results is shown in figures 4 and 5 for an unsteady Navier-Stokes computation at Mach 5.94. The freestream Reynolds number based on the diameter is $Re_D = 186,000$ and the freestream temperature is $T_\infty = 59.5^o K$. To demonstrate the differences at the two Mach numbers a similar sequence is used as for the Mach 15 case. For the Type I and II flows the shock patterns differ from the Mach 15 case for identical impingement shock location. However, for Type III to Type V flows, the shock patterns and locations are the same. Another notable difference at the two Mach numbers is that for the Type I flow, the transmitted shock causes the wall boundary layer to separate and thus becomes a lambda shock at the higher Mach number. The boundary layer does not separate at the lower Mach number. The boundary layer separation has not been observed for the steady flows computed so far. The surface pressure distributions compared with the steady case are given in figure 5 for the last three of the six instances shown in figure 4. The Mach contours for the steady case can be found in figure 5 of reference 4. Similar to the Mach 15 case, the unsteady peak surface pressures no longer seem to occur for the Type IV shock interaction as they do for the steady flows.

Comparison of the computed results for the surface pressures of the steady shock interaction with the experimental data of Keyes given in [12] can be found in reference 4 and is shown in figure 5. A complete description of the steady state results for various Mach numbers is also given in reference 4, where in particular the numerical computations are compared to the various experimental data in terms of the surface pressures and heat transfer rates. The comparison serves to validate the code used for the present study.

Conclusions

Numerical simulation of the unsteady Navier-Stokes equations with a high resolution time-accurate implicit TVD scheme for predicting the complicated shock-on-shock interaction on a blunt body with a non-stationary impinging shock has been carried out. It has been demonstrated that similar types of shock interactions occur for the unsteady flows as for the steady cases identified by Edney. However, for the sample cases studied in this paper the peak surface pressures do not seem to occur during the Type IV interaction as for the steady case. The significance of this fact is that the Type III interaction occurs for a much broader range of impingement shock angles and positions than for the Type IV interaction. Also the details within the shock layer differ appreciably. For example boundary layer separation and transient supersonic jets flowing parallel to the blunt body surface occur for the unsteady but not for the steady shock interactions.

References

1. Yee, H.C.; "Upwind and Symmetric Shock-Capturing Schemes," NASA TM- 89464, May 1987.

2. Montagne, J.-L., Yee, H.C., Klopfer, G.H. and Vinokur, M.; "Hypersonic Blunt Body Computations Including Real Gas Effects," Proceed. of the 2nd Int'l. Conf. on Hyperbolic Problems, March 13-18, 1988, Aachen, Germany, also NASA TM-100074, March 1988.

3. Yee, H.C., Klopfer, G.H. and Montagne, J.-L.; "High Resolution Shock Capturing Schemes for Inviscid and Viscous Hypersonic Flows," Proceed. of the BAIL V Conf., June 20-24, 1988, Shanghai, China, also NASA TM-100097, April 1988.

4. Klopfer, G.H. and Yee, H.C.; "Viscous Hypersonic Shock-on-Shock Interaction on Blunt Cowl Lips," AIAA Paper 88-0233, Jan. 1988.

5. Tannehill, J.C., Holst, T.L., and Rakich, J.V.; "Numerical Computation of Two-Dimensional Viscous Blunt Body Flows with an Impinging Shock," AIAA J., Vol. 14, No. 2, February 1976.

6. White, J.A., and Rhie, C.M.; "Numerical Analysis of Peak Heat Transfer Rates for Hypersonic Flow over a Cowl Leading Edge," AIAA Paper 87-1895, June 1987.

7. Rhie, C.M. and Stowers, S.T.; "Navier-Stokes Analysis for High Speed Flows Using a Pressure Correction Algorithm," AIAA Paper 87-1980, June 1987.

8. Thareja, R.R., Stewart, J.R., Hassan, O., Morgan, K., and Peraire, J.; "A Point Implicit Unstructured Grid Solver for the Euler and Navier-Stokes Equations," AIAA Paper 88-0036, January 1988.

9. Harten, A.; "On a Class of High Resolution Total Variation Stable Finite Difference Schemes," SIAM J. Num. Anal., Vol 21, pp.1-23, 1984.

10. Yee, H.C.; "Linearized Form of Implicit TVD Schemes for Multidimensional Euler and Navier-Stokes Equations," Comp. of Maths. with Appls., Vol. 12A, pp.413-432, 1986.

11. Edney, B.E.; "Anomalous Heat Transfer and Pressure Distributions on Blunt Bodies at Hypersonic Speeds in the Presence of an Impinging Shock," FFA Rept. 115, February 1968, The Aeronautical Research Institute of Sweden, Stockholm, Sweden.

12. Tannehill, J.C., Holst, T.L., Rakich, J.V., and Keyes, J.W.; "Comparison of a Two-Dimensional Shock Impingement Computation with Experiment," AIAA J., Vol. 14, No. 4, April 1976.

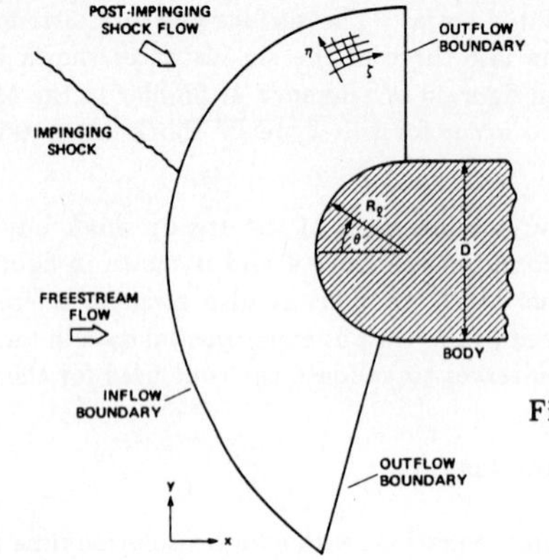

Fig. 1 Schematic of the computational domain.

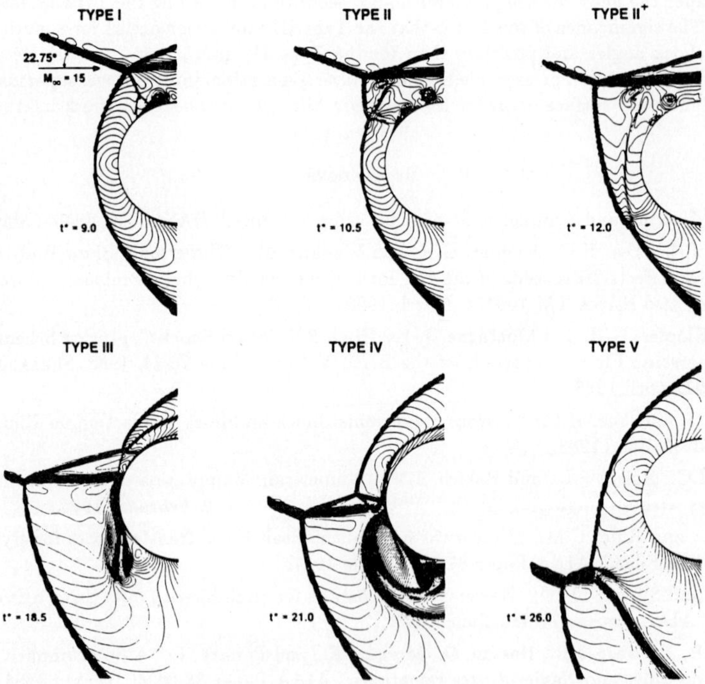

Fig. 2 Mach contours at six instances of the diffraction process with $M_\infty = 15$ and $Re_D = 186{,}000$. The indicated times (t^*) of each frame are normalized with the freestream velocity and blunt body thickness.

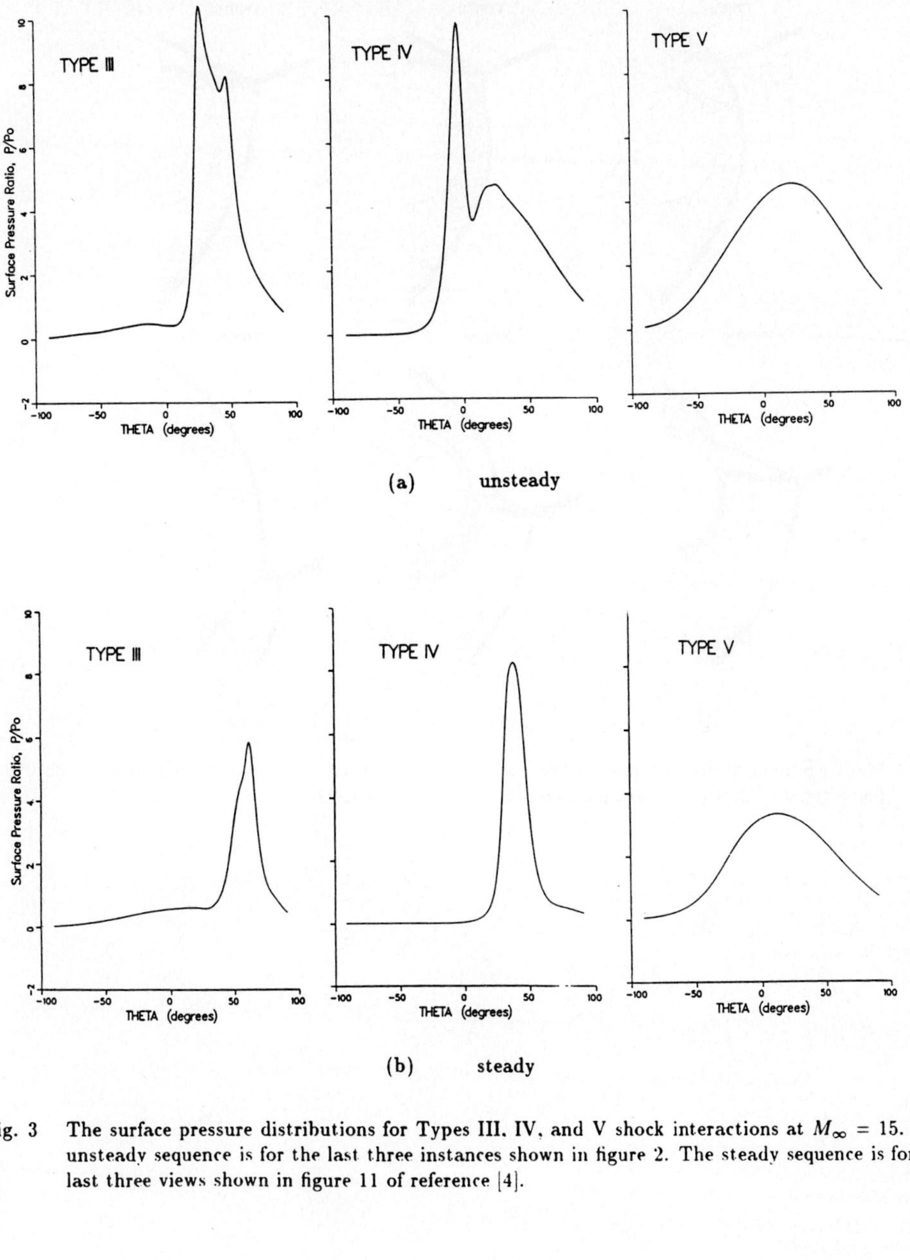

Fig. 3 The surface pressure distributions for Types III, IV, and V shock interactions at $M_\infty = 15$. The unsteady sequence is for the last three instances shown in figure 2. The steady sequence is for the last three views shown in figure 11 of reference [4].

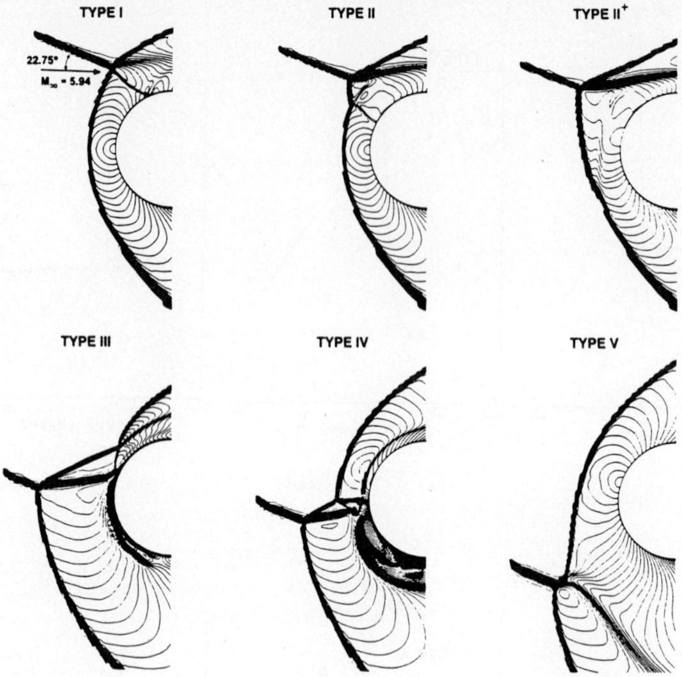

Fig. 4 Mach contours at six instances of the diffraction process with $M_\infty = 5.94$ and $Re_D = 186,000$. Same types of shock-on-shock interference as shown in figure 2.

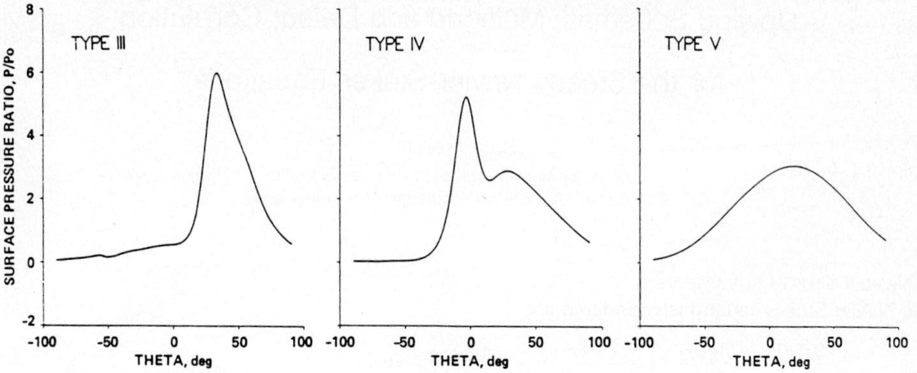

(a) unsteady

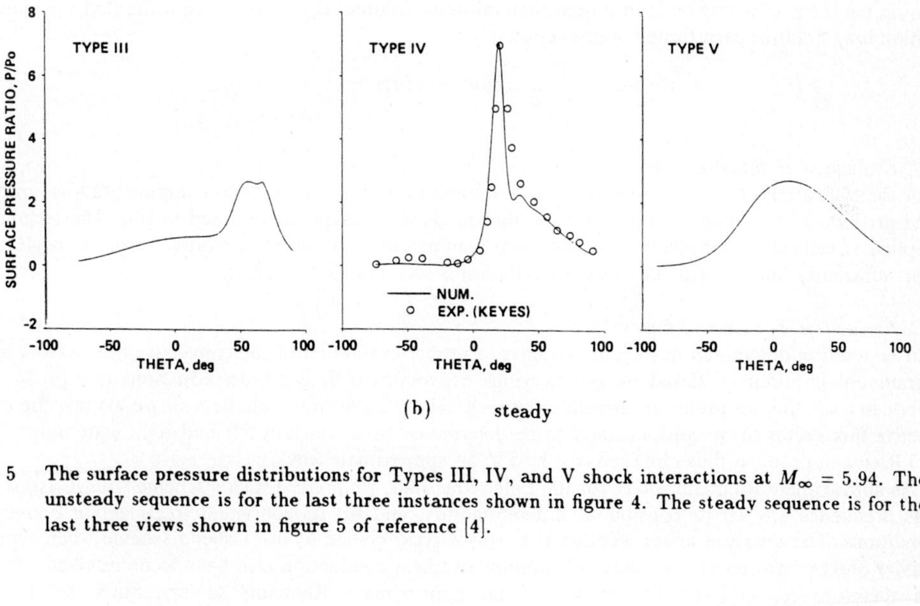

(b) steady

Fig. 5 The surface pressure distributions for Types III, IV, and V shock interactions at $M_\infty = 5.94$. The unsteady sequence is for the last three instances shown in figure 4. The steady sequence is for the last three views shown in figure 5 of reference [4].

Upwind Schemes, Multigrid and Defect Correction for the Steady Navier-Stokes Equations

Barry Koren
Centre for Mathematics and Computer Science
P.O. Box 4079, 1009 AB Amsterdam, The Netherlands

1. Navier-Stokes equations

The Navier-Stokes equations considered are

$$\frac{\partial f(q)}{\partial x} + \frac{\partial g(q)}{\partial y} - \frac{1}{Re}\left\{ \frac{\partial r(q)}{\partial x} + \frac{\partial s(q)}{\partial y} \right\} = 0, \tag{1}$$

with q the (perfect gas) conservative state vector $(\rho, \rho u, \rho v, \rho e)^T$, $f(q)$ and $g(q)$ the convective flux vectors $(\rho u, \rho u^2 + p, \rho uv, \rho u(e + p/\rho))^T$ respectively $(\rho v, \rho uv, \rho v^2 + p, \rho v(e + p/\rho))^T$, and $r(q)$ and $s(q)$ the diffusive flux vectors $(0, \tau_{xx}, \tau_{xy}, 1/((\gamma-1)Pr)\partial(c^2)/\partial x)^T$ respectively $(0, \tau_{xy}, \tau_{yy}, 1/((\gamma-1)Pr)\partial(c^2)/\partial y)^T$.

2. Discretization method

To still allow Euler flow $(1/Re = 0)$ solutions with discontinuities, the equations are discretized in the integral form. A straightforward and simple discretization of the integral form is obtained by subdividing the integration region Ω into quadrilateral finite volumes $\Omega_{i,j}$, and by requiring that the conservation laws hold for each finite volume separately:

$$\oint_{\partial\Omega_{i,j}} (f(q)n_x + g(q)n_y)ds - \frac{1}{Re}\oint_{\partial\Omega_{i,j}} (r(q)n_x + s(q)n_y)ds = 0, \quad \forall i,j. \tag{2}$$

2.1. *Evaluation of diffusive fluxes*

For the evaluation of the diffusive fluxes at a volume wall, it is necessary to compute grad(u), grad(v) and grad(c^2) at that wall. For this we use the standard technique as outlined in [11]. The technique applied is central, the directional dependence coming from the cross derivative terms is neglected. For sufficiently smooth grids this flux computation is second-order accurate.

2.2. *Evaluation of convective fluxes*

For convection dominated flows, our objective, a proper evaluation of the convective flux vectors is of paramount importance. Based on our previous experience with the Euler equations (see [3] for an overview), for this we prefer an upwind approach. Along each finite volume wall we assume the convective flux vector to be constant, and to be determined by a constant left and right state only. The 1D Riemann problem thus obtained is solved in an approximate way.

As approximate Riemann solver for the Euler equations, we prefer Osher's scheme [10]. Reasons for this preference are: (i) its continuous differentiability, and (ii) its consistent treatment of boundary conditions. The question arises whether it is still a good choice to use Osher's scheme when typical Navier-Stokes features such as shear, separation and heat conduction also have to be resolved. In [5] we therefore reconsidered the choice of an approximate Riemann solver. Since continuous differentiability is an absolute requirement for the success of our solution method, and since the only known approximate Riemann solvers with this property are Osher's [10] and van Leer's [7], our choice was confined to these two only. The requirement of accurate modelling of physical diffusion determined our choice. In [7], van Leer stated already that his flux vector splitter cannot preserve steady contact discontinuities. Since a discrete shear layer may be interpreted as a layer of contact discontinuities, doubt arose about the suitability of van Leer's scheme for Navier-Stokes codes. Recently, this doubt was confirmed in [9] where van Leer et al. made a qualitative analysis (supplemented with numerical experiments) for various upwind schemes. There, Osher's scheme turned out to be better indeed for the resolution of boundary layer flows. To shed some more light on the difference in quality between both schemes, in [5] a quantitative error analysis is presented for Osher's and van Leer's scheme. The analysis is confined to the steady, 2D, isentropic Euler equations for a perfect gas with $\gamma=1$; i.e. equation (1) with $1/Re=0$, $q=(\rho,\rho u,\rho e)^T$, $f(q)=(\rho u,\rho(u^2+c^2),\rho uv)^T$, $g(q)=(\rho v,\rho uv,\rho(v^2+c^2))^T$ and $c=$constant. For both schemes the system of modified equations is derived by considering a first-order accurate, square finite volume discretization, and a subsonic flow with u and v positive and $\rho\approx$constant. Substituting a boundary layer type solution into the error terms it appears that van Leer's scheme deteriorates compared to Osher's scheme for increasing Re.

The approximation of the left and right state in the 1D Riemann problem determines the accuracy of the convective discretization. First-order accuracy is obtained in the standard way, by taking the left and right state equal to that in the corresponding adjacent volume. Higher-order accuracy is obtained by low-degree piecewise polynomial state interpolation (MUSCL-approach) with van Leer's κ-scheme [8]. For the scalar model equation

$$\frac{\partial u}{\partial x}+\frac{\partial u}{\partial y}-\epsilon\left(\frac{\partial^2 u}{\partial x^2}+\frac{\partial^2 u}{\partial x\partial y}+\frac{\partial^2 u}{\partial y^2}\right)=0, \tag{3}$$

in [5] we derive the modified equation by considering a square finite volume discretization. From this we find as the κ that gives the highest possible (i.e. third-order) accuracy

$$\kappa=\frac{1}{3}\{1+\epsilon\left(\frac{\partial^4 u}{\partial x^4}+2\frac{\partial^4 u}{\partial x^3\partial y}+2\frac{\partial^4 u}{\partial x\partial y^3}+\frac{\partial^4 u}{\partial y^4}\right)/\left(\frac{\partial^3 u}{\partial x^3}+\frac{\partial^3 u}{\partial y^3}\right)\}. \tag{4}$$

Since convection dominated problems are our interest, we neglect the above diffusion dependence and simply take $\kappa=1/3$. To avoid spurious non-monotonicity, in [5] a new limiter is constructed for the $\kappa=1/3$ approximation. Using Sweby's notation [12], it reads

$$\varphi(r)=\frac{2r^2+r}{2r^2-r+2},\quad r\in\mathbb{R}. \tag{5}$$

3. Solution method

To efficiently solve the system of discretized equations, symmetric point Gauss-Seidel relaxation, accelerated by nonlinear multigrid (FAS), is applied. The process is started by nested iteration (FMG). Per finite volume we use a Newton iteration for the collective update of the four state vector components. With again a square finite volume discretization of (3), in [6] it is shown by local mode analysis that symmetric point Gauss-Seidel relaxation accelerated by multigrid converges fast for the first-order discretization for any value of the mesh Reynolds number Reh. However, it appears to converge very slowly for the higher-order ($\kappa=1/3$) discretization for small and moderately large values of Reh. It even appears to diverge for large values of Reh. The cause clearly is the higher-order discretization of the convection operator. No cure can be found in using some other κ. As with the Euler equations [2,3,4], the difficulty in inverting the higher-order operator is circumvented by introducing iterative defect correction (IDeC) as an outer iteration for the nonlinear multigrid cycling. With $F_h(q_h)$ the full, higher-order accurate operator and $\tilde{F}_h(q_h)$ the less accurate operator that can be easily inverted, iterative defect correction can be written as:

$$\begin{aligned}&\tilde{F}_h(q_h^1)=0,\\&\tilde{F}_h(q_h^{n+1})=\tilde{F}_h(q_h^n)-F_h(q_h^n),\ n=1,2,\cdots,N,\end{aligned} \tag{6}$$

with n referring to the nth iterand. The operator $\tilde{F}_h(q_h)$ necessarily has only first-order accurate convection, but the amount of diffusion can be chosen freely. In [6], this freedom is exploited by analyzing three approximate operators: (i) an operator without physical diffusion (first-order Euler), (ii) an operator with partial physical diffusion (zeroth-order Navier-Stokes), and (iii) an operator with full, second-order accurate physical diffusion (first-order Navier-Stokes). The last approximate operator most closely resembles the higher-order operator, and therefore was supposed to have the best convergence properties. For this operator, theory [1] predicts that already q_h^2 is second-order accurate if F_h is second-order accurate and q_h^1 sufficiently smooth. Theory does not give such a guarantee for the other approximate operators. Local mode analyses with the square finite volume discretization of (3), and experiments with the Navier-Stokes equations show the third approximate operator to have the best convergence properties indeed. Its relative complexity is taken for granted. The numerical results presented hereafter were all obtained with this operator as operator to be inverted.

4. Numerical results

To evaluate the computational method, we consider: (i) a subsonic flat plate flow at $M=0.5$, $Re=100,400$ and 1600, (ii) a supersonic flat plate flow with oblique shock wave - boundary layer interaction at $M=2$, $Re=2.96\ 10^5$, and (iii) a hypersonic blunt body flow at $M=8.15$, $Re=1.67\ 10^5$, $\alpha=30^o$.

4.1. Approximate Riemann solver

To verify the analytical results obtained for Osher's and van Leer's scheme, we consider the subsonic flat plate flow and perform an experiment with *Re*-variation. For both schemes we use the first-order accurate discretization, and identical grids and boundary conditions. The results given in Fig. 1 show the predicted deterioration of van Leer's scheme with increasing *Re*. The numerical results presented hereafter were obtained with Osher's scheme only.

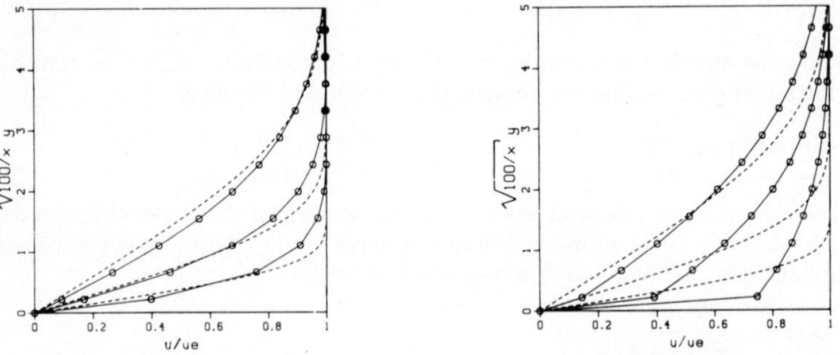

a. Osher. b. van Leer.

Fig. 1. Velocity profiles subsonic flat plate flow (-----: Blasius solution, $Re=100,400,1600$).

4.2. Monotone higher-order accuracy

To evaluate the monotone higher-order accurate discretization derived, we consider the supersonic flat plate flow. At first we compute the Euler flow solution with and without limiter. The inviscid surface pressure distributions obtained (Fig. 2a) clearly show the benefit of the limiter. To show now the benefit of higher-order accuracy we compute the Navier-Stokes flow solution with the limited $\kappa=1/3$ scheme and the first-order scheme. The Navier-Stokes solution is known to have shock-induced separation. The computed viscous surface pressure distributions (Fig. 2b) show that the higher-order solution has shock-induced separation indeed, whereas the first-order solution remains attached. (The latter lacks the plateau in the surface pressure distribution.)

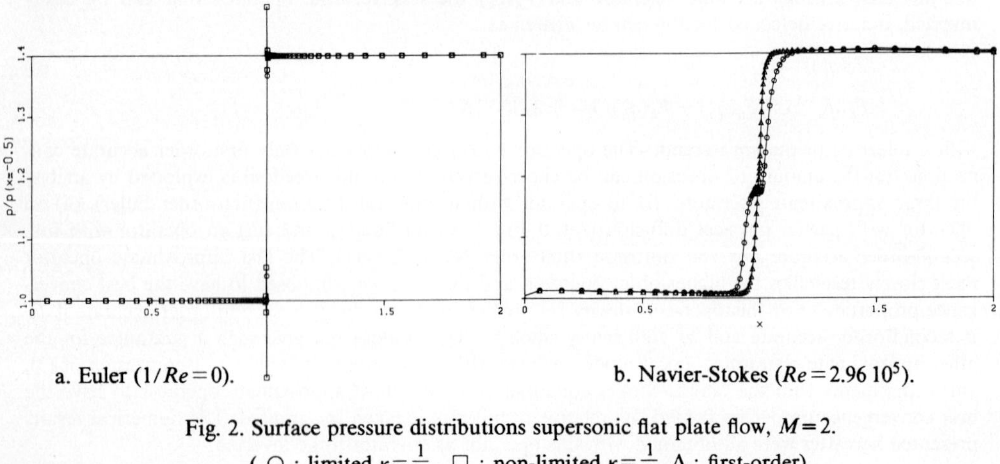

a. Euler ($1/Re=0$). b. Navier-Stokes ($Re=2.96\ 10^5$).

Fig. 2. Surface pressure distributions supersonic flat plate flow, $M=2$.
(O : limited $\kappa=\frac{1}{3}$, □ : non-limited $\kappa=\frac{1}{3}$, Δ : first-order).

4.3. *Multigrid behaviour*

Multigrid convergence results for the subsonic and supersonic flat plate flow are given in Fig. 3. The somewhat worser convergence rates for the supersonic flat plate flow are still satisfactory; for the 80×32-grid (Fig. 3b) the convergence rate with multigrid is still much faster than that without multigrid.

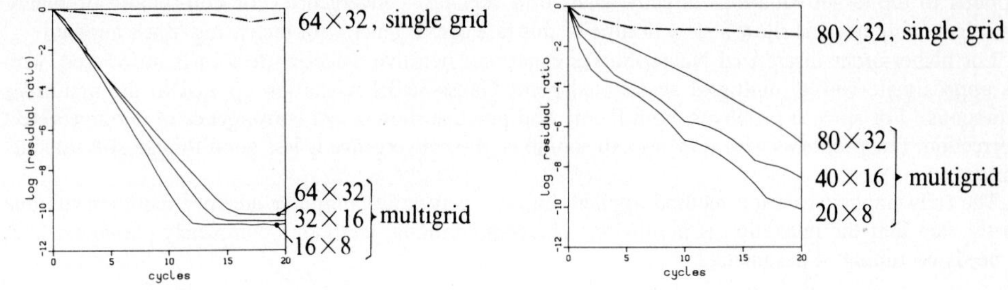

a. $M = 0.5$, $Re = 100$.

b. $M = 2$, $Re = 2.96\ 10^5$.

Fig. 3. Multigrid behaviour flat plate flows.

At present the usefulness of multigrid for flows at still higher Mach numbers is open to question. For the hypersonic blunt body flow mentioned (Fig. 4), the multigrid behaviour is shown in Fig. 5. (Notice that the finest grid, the 64×16-grid, is a locally nested grid and that the single grid comparison is done for the coarser 32×16-grid.) It appears that multigrid still pays for hypersonics, but it seems that there is a trend indeed of decreasing multigrid effectiveness from subsonic to hypersonic speeds. Our current research is devoted to ensuring a very good multigrid performance for hypersonic flow problems as well.

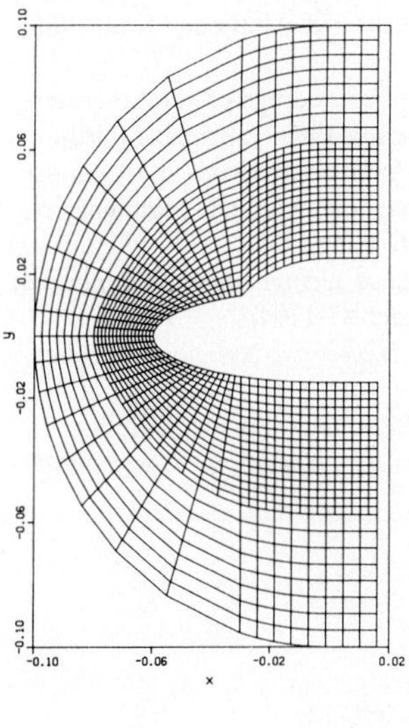

Fig. 4. Blunt body with grid.

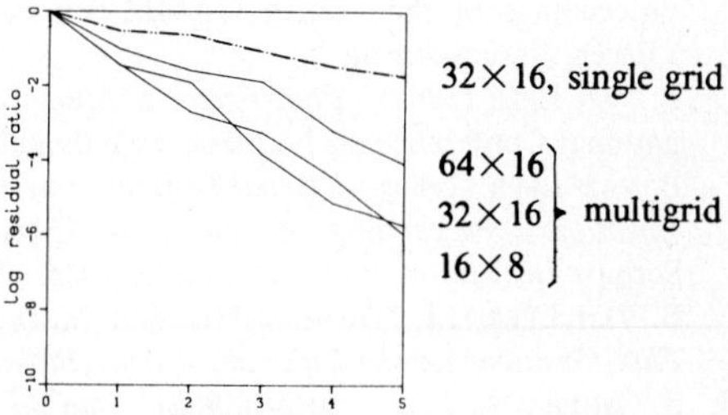

Fig. 5. Multigrid behaviour hypersonic blunt body flow ($M = 8.15$, $Re = 1.67\ 10^5$, $\alpha = 30^o$).

5. Concluding remarks

Theory and practice show that for sufficiently high Reynolds numbers, Osher's scheme leads to a more accurate resolution of boundary layer flows than van Leer's scheme. The difference in accuracy becomes larger with increasing Reynolds number.

For the first-order accurate discretized Navier-Stokes equations and flow problems with smooth solutions, theory and practice show that point Gauss-Seidel relaxation accelerated by multigrid and applied to the target equations directly, leads to a very fast convergence. For non-smooth problems, practical computations show a slow decline of this fast convergence with increasing Mach number.

For higher-order discretized Navier-Stokes equations, iterative defect correction is introduced, with as approximate solver: multigrid accelerated point Gauss-Seidel relaxation applied to the first-order equations. For smooth problems, both theory and practice show a fast convergence of iterative defect correction. For problems with non-smooth solutions, the convergence is less good though still satisfactory.

The fully implicit solution method applied, imposes very mild computer memory requirements due to the fact that the relaxation is pointwise. The computational method is completely parameter-free; it needs no tuning of parameters.

References

1. W. Hackbusch (1985). *Multi-Grid Methods and Applications.* Springer, Berlin.
2. P.W. Hemker (1986). *Defect Correction and Higher Order Schemes for the Multi Grid Solution of the Steady Euler Equations.* Proceedings of the Second European Conference on Multigrid Methods, Cologne 1985. Springer, Berlin.
3. P.W. Hemker and B. Koren (1988). *Defect Correction and Nonlinear Multigrid for the Steady Euler Equations.* Lecture Series on Computational Fluid Dynamics, von Karman Institute for Fluid Dynamics, Rhode-Saint-Genèse.
4. B. Koren (1988). *Defect Correction and Multigrid for an Efficient and Accurate Computation of Airfoil Flows.* J. Comput. Phys. (to appear).
5. B. Koren (1988). *Upwind Schemes for the Navier-Stokes Equations.* Proceedings of the Second International Conference on Hyperbolic Problems, Aachen 1988. Vieweg, Braunschweig.
6. B. Koren (1988). *Multigrid and Defect Correction for the Steady Navier-Stokes Equations.* Proceedings of the Fourth GAMM-Seminar Kiel on Robust Multi-Grid Methods, Kiel 1988. Vieweg, Braunschweig.
7. B. van Leer (1982). *Flux-Vector Splitting for the Euler Equations.* Proceedings of the 8th International Conference on Numerical Methods in Fluid Dynamics, Aachen 1982. Springer, Berlin.
8. B. van Leer (1985). *Upwind-Difference Methods for Aerodynamic Problems governed by the Euler Equations.* Proceedings of the 15th AMS-SIAM Summer Seminar on Applied Mathematics, Scripps Institution of Oceanography 1983. AMS, Providence, Rhode Island.
9. B. van Leer, J.L. Thomas, P.L. Roe, and R.W. Newsome (1987). *A Comparison of Numerical Flux Formulas for the Euler and Navier-Stokes Equations.* AIAA-paper 87-1104.
10. S. Osher and F. Solomon (1982). *Upwind-Difference Schemes for Hyperbolic Systems of Conservation Laws.* Math. Comp. 38, 339-374.
11. R. Peyret and T.D. Taylor (1983). *Computational Methods for Fluid Flow.* Springer, Berlin.
12. P.K. Sweby (1984). *High Resolution Schemes using Flux Limiters for Hyperbolic Conservation Laws.* SIAM J. Num. Anal. 21, 995-1011.

A Pseudospectral Matrix Element Method for Solution of Three-Dimensional Incompressible Flows and Its Implementation on a Parallel Computer

Hwar C. Ku, Richard S. Hirsh, Thomas D. Taylor and Allan P. Rosenberg

Johns Hopkins University Applied Physics Laboratory

Johns Hopkins Road, Laurel MD 20707

1 Introduction

In the authors' earlier paper [1] a pseudospectral matrix (PSM) method for solution of the three-dimensional incompressible Navier-Stokes equations was presented to solve 3-D driven cavity flow with Reynolds numbers up to 1000. None of these simulations exhibited the phenomenon of TGL vortices, which requires a Reynolds number of at least 1200 according to the experimental results of Koseff *et al.* [2]. Here we approach the problem by extending the techniques in our earlier paper.

In order to overcome stiff gradients in the interior of the domain as well as the strict time step size constraints, a c^0 pseudospectral matrix element (PSME) method which combines the desired features of domain decomposition and a larger time step size has been developed to solve the incompressible Navier-Stokes equations in primitive variable form.

In recent years computer designers, driven by the quest for speed and stimulated by the developing capability of single chips, have built machines of great complexity and increasing diversity. Producing efficient code for any of these machines is difficult and often involves special features, e.g., compilier directives, special hand-coded assembly language routines, etc. that make migration of the code to a different machine difficult. The algorithms used in this paper can be cleanly described using the rather minimal set of vector and parallel concepts proposed for Fortran 8x. Specifically, we use the ability to do dotproducts and matrix multiplication between subsets of arrays as well as the ability to specify that it is safe to run some loops containing subroutines concurrently.

2 Calculation of derivatives by PSME method

The spatial domain is divided into NE elements, each of which has $N+1$ collocation points selected as $x_j = (1/2)[a^e + b^e + (b^e - a^e)\cos\pi(j-1)/N]$ $(1 \le j \le N+1; 1 \le e \le NE)$. The derivatives of a function $f(x)$ in the interior of element e can be discretized as

$$f^e_x(x_i) = \frac{1}{L^e}\sum_{m=1}^{N+1} \hat{GX}^{(1)}_{i,m} f^e_m \tag{1a}$$

$$f^e_{xx}(x_i) = \frac{1}{(L^e)^2}\sum_{m=1}^{N+1} \hat{GX}^{(2)}_{i,m} f^e_m = \frac{1}{L^e}\sum_{m=1}^{N+1} \hat{GX}^{(1)}_{i,m} (f^e_x)_m. \tag{1b}$$

Here L^e is the length of the e-th element defined on the interval $[a^e, b^e]$, $f^e_m \equiv f^e(x_m)$, and $\hat{\mathbf{GX}}^{(1)}$, $\hat{\mathbf{GX}}^{(2)}$ are the invariant derivative matrices based on the domain [0,1].

The interfacial derivatives at the inter-element points approximated by weighting of the derivatives from each side; their respective fractions of the total length of two adjacent elements. C^0 continuity is explicitly assumed whenever the calculation of interface values of the first derivative is required while c^1 continuity has been implicitly assumed that $f_x^e(x_{N+1}) = f_x^{e+1}(x_1) = f_x|_{interface}$ for the second derivative calculation.

3 Governing equations and solution technique

The three-dimensional cavity flow in a rectangular box with an aspect ratio of 3 in the spanwise direction, the time-dependent Navier-Stokes equations in dimensionless form can be written as

$$\frac{\partial u_i}{\partial t} + u_j \frac{\partial u_i}{\partial x_j} = -\frac{\partial p}{\partial x_i} + \frac{1}{\mathrm{Re}} \frac{\partial^2 u_i}{\partial x_j^2} \tag{2a}$$

$$\frac{\partial u_i}{\partial x_i} = 0 \tag{2b}$$

The method applied to solve the Navier-Stokes equation is Chorin's [3] splitting technique. According to this scheme, the first step is to split the velocity into a sum of predicted and corrected values. The predicted velocity is determined by time integration of the momentum equations without the pressure term. The second step is to develop the pressure and corrected velocity fields that satisfy the continuity equation. The explicit scheme just outlined can be modified to treat the viscous term implicitly as described in [4], i.e., the Stokes problem. The current paper will be focused on the simple method and its implementation on a parallel computer.

Taking the divergence operator to the corrected velocity and incorporating the prescribed velocity boundary conditions, the pressure Poisson equations in the interior as well as the supplemental pressure equations at the boundaries are generated accordingly. Following the eigenfunction expansion technique (see detailed procedure in Ref. [4]), the original three-dimensional pressure equation is reduced to a simple one-dimensional matrix operator, which is nearly a 1-D block tri-diagonal sparse matrix. Therefore, the overall solution of pressure can be obtained through the linear superposition of each eigenvalue and its associated eigenvectors.

4 Parallel implementation of N-S Equations

It is easy to use parallelism to solve problems which can be decomposed into completely independent subproblems, but becomes progressively more difficult to do efficiently as dependencies between the subproblems increase. We assume the compilier can detect and the target machine exploit the inherent parallelism of an outer loop containing an inner loop with data dependencies which allow the outer loop to be distributed across processors. Two intrinsic functions from the proposed Fortran 8x standard - *matmul* and *dotproduct* were used to simplify the programming effort.

In our version of the time splitting approach three steps account for most of the run time. These are computing the partial derivatives involved in updating the predicted velocity, transforming back and forth from physical to eigenfunction space, and solving the sequence of reduced one-dimensional pressure equations in eigenfunction space. Each will be discussed separately below.

4.1 Predicted velocity: vector dot

Letting i, j, k be the indices of the spatial coordinate, ϕ be any of the velocity components, and $\mathbf{GX}^{(q)}, \mathbf{GY}^{(q)}, \mathbf{GZ}^{(q)}, q = 1, 2$, are the derivative matrices in x, y, and z direction, respectively. The predicted velocity hence can be written as for each $i = 2, ..., NX$

$j = 2, ..., NY$

$k = 2, ..., NZ + 1$

$$\bar{\phi}^{n+1}_{i,j,k} = \phi^n_{i,j,k} + \Delta t[\ dotproduct(\ \frac{1}{\mathrm{Re}} GX^{(2)}_{i,m_1(i):m_2(i)} - u^n_{i,j,k} GX^{(1)}_{i,m_1(i):m_2(i)},\ \phi^n_{m_1(i):m_2(i),j,k})$$
$$+\ dotproduct(\frac{1}{\mathrm{Re}} GY^{(2)}_{j,l_1(j):l_2(j)} - v^n_{i,j,k} GY^{(1)}_{j,l_1(j):l_2(j)},\ \phi^n_{i,l_1(j):l_2(j),k})$$
$$+\ dotproduct(\frac{1}{\mathrm{Re}} GZ^{(2)}_{k,n_1(k):n_2(k)} - w^n_{i,j,k} GZ^{(1)}_{k,n_1(k):n_2(k)},\ \phi^n_{i,j,n_1(k):n_2(k)})]$$

Here the three dotproduct terms correspond to the three spatial directions and an expression such as $m_1(i) : m_2(i)$ specifies the index range for the dotproduct.

4.2 Tensor product: matrix multiply

Converting the pressure and source term from eigenfunction space back to physical space require a series of matrix multiplications. This is also true for the reverse conversions from physical to eigenfunction space. To implement this we use the Fortran 8x intrinsic function *matmul.*

for each $j = 1, ..., NY + 1$

$$p_{:,j,:} = matmul[\mathbf{EX}, matmul(\hat{p}_{:,j,:},\ \mathbf{EZ}^T)\ ]$$

where":" indicates the full range including the un-necessary indices for boundary pressure terms.

4.3 Pressure equations: LU decomposition

Using Chorin's splitting scheme, our techniques lead to a separable problem for the finite dimensional analogue of the Poisson equation for the pressure. With an eigenfunction expansion in two directions, the resulting sequence of one-dimensional problems is solved by LU factorization. If the inverse is not stored, storage requirements for the inversion step are reduced to $O(N^2)$ rather than $O(N^4)$. Although forward sweep is automatically parallelizable, storage of $O(N^3)$ instead of $O(N^2)$ are still required in order to execute the backward substitution in parallel, so that the overall memory required for pressure solution is of the same order of magnitude as for the field variables.

The pressure matrix structure can be interpreted as an algebraic equation $\mathbf{A}\,\mathbf{x} = \mathbf{b}$, and the discussion of the solution is split into two steps.

4.3.1 Forward sweep

In the process of row-column reduction, rows run in a concurrent mode and columns run in a vector mode. All off-diagonal zero coefficients can be screened out without any manipulation. The scalar multiplies making a column of zeros below the diagonal are stored in the lower triangle of matrix **A** and used later for the reduction of source term **b**. The speedup is proportional to the number of processors applied to this step except for an offset caused by data addressing.

4.3.2 Backward substitution

The eigenfunction expansion is crucial to performing the backward substitution in parallel because it makes reduced one-dimensional pressure operator independent for different eigenvalues.

The following Fortran code is part of the complete solution for the pressure and illustrates the essential point. First we reduce the source term **b** by using the elements stored in the lower-triangular part of **A** during the forward sweep and next perform the backward substitution as

```
do n = 2, NUM
   b(NP,n) = b(NP,n)/A(NP,NP,n)
end do
do k = 1, NP - 1
i = NP - k
ind = iref(i)
m = NP - k + 1
   do n = 2, NUM
      b(i,n) = (b(i,n) - dotproduct(A(i,m:ind,n),b(m:ind,n)))/ A(i,i,n)
   end do
end do
```

where $iref(i)$ is the index of nonzero upper diagonal elements with respect to each row index i; NUM is the total number of points in one of the directions with the eigenvalues; and NP is the total number of points for one-dimensional pressure operator. The innermost loop performs both source term reduction and backward substitution on a given row. This is done for all the (independent) matrices labeled by eigenvalues before the row index is changed.

4.3.3 Timings

Table I shows some timings for implementation of our program for the solution of three-dimensional driven cavity flow with $NX = 36$ (single element), $NY = 42$ (7 elements) and $NZ = 48$ (8 elements with equal length), running on an Alliant FX/8-series (eight processors) computer.

Table I. Cpu time in seconds per time step

number of processors	1*	1	2	3	4	8
	132.0	80.3	42.86	30.48	24.42	14.97
speedup†		1	1.87	2.63	3.29	5.37

*: code without any optimization
†: compared to 1 processor

Note that using eight processors speeds the computation by a factor of 5.37. It is worth emphasizing that our program is written entirely in terms of the proposed Fortran 8x standard and can be run on any machine for which a compiler implementing the standard exists.

5 Results and discussion

In order to exhibit the dynamic behavior of TGL vortices at Re = 3200, the computed velocity fields are displayed in vector plots at time 15 and 20 minutes, respectively. Figs. 1a, 1b show the well-developed flow pattern visualized at plane $x = 0.765$, where between 3 $\sim$ 4 pairs of TGL vortices plus one corner pocket recirculating flow near the bottom surface were formed. These computed results demonstrate the temporal variation in corner vortex and TGL vortex strength and size due to spacial location. The presence of TGL vortices is explained by Koseff *et al.* [2] as the formation of the corner vortex inducing a rotational effect which propagates out from the side wall toward the center, and in conjunction with primary cell circulation results in the unstable interface. All the computational results are in qualitative agreement with those found by the experiment of [2]. These results, moreover, provide information on the dynamic behavior of TGL vortices.

6 Conclusions

The solution of three-dimensional Navier-Stokes equations has been solved by Chebyshev pseudospectral matrix element method employing a primitive variable formulation of splitting technique to simulate the dynamic TGL vortices in driven cavity flow.

The key feature of the work presented is that with an eigenfunction expansion in two directions, the resulting three-dimensional direct matrix inversion for the pressure Poisson equations is reduced to a set of simple one-dimensional problems for which the data dependency occurring in the backward substitution of LU decomposition is eliminated. This makes possible efficient parallel performance of both the forward and backward sweep. The other time-consuming parts of the calculation can be reduced to dotproducts and matrix multiplications - defined in the proposed Fortran 8x standard and efficiently implementable on wide classes of vector, parallel, and vector-parallel machines.

Numerical results computed by the PSME method shown dynamic lateral TGL vortices in accord with experiments.

Acknowledgement

This work was partially supported by the Office of Naval Research under the Contract Number N00039 - 87 - C - 5301.

References

[1] H. C. Ku, R. S. Hirsh and T. D. Taylor, *J. Comput. Phys.* **70**, 439 (1987).

[2] J. R. Koseff and R. L. Street, *J. Fluid Eng. ASME* **106**, 385 (1984).

[3] A. J. Chorin, *Math. Comp.* **22**, 745 (1968).

[4] H. C. Ku, R. S. Hirsh, T. D. Taylor and A. P. Rosenberg, submitted to *J. Comput. Phys.*

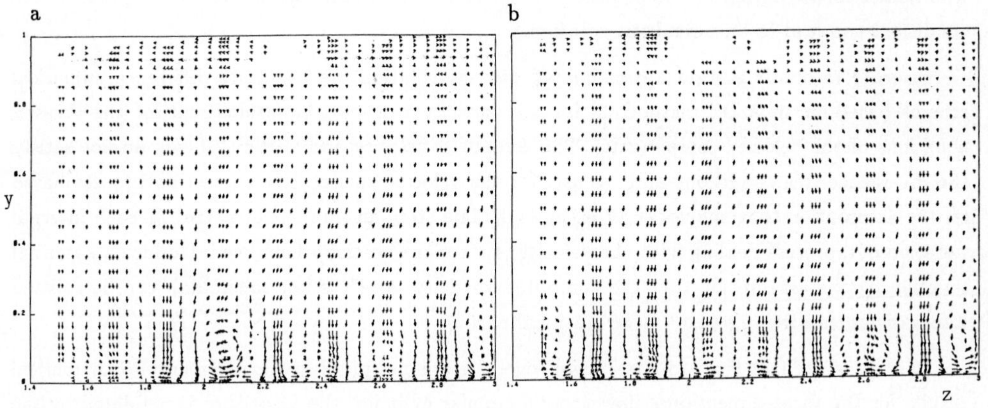

Figure 1: Flow direction vectors for Re = 3200 in the x = 0.765 plane at the time (a) 15 min. (b) 20 min.

NUMERICAL RESOLUTION OF THE THREE-DIMENSIONAL NAVIER-STOKES EQUATIONS IN VELOCITY-VORTICITY FORMULATION

W. LABIDI - L. TA PHUOC
L.I.M.S.I (C.N.R.S) - B.P.30 91406 Orsay Cedex - France

ABSTRACT

Unsteady fully three-dimensional Navier-Stokes equations of viscous incompressible flow around finite length circular cylinder placed between two circular plates is studied using finite difference numerical scheme. The algorithm uses iterative correction processes of the velocity and the vorticity fields in order to satisfy the two divergence free conditions. In order to study the three-dimensional effects, results are presented for this external flow with Re=300 for different cylinder lengths.

INTRODUCTION

The study of three-dimensional flows is the aim of several works and usually, primitive variables formulation or vorticity-potential vector formulation are used. Since few years, the use of the velocity-vorticity formulation [1], [2] appears as a good solution to avoid the difficulties encountered for the boundary conditions on the pressure for the primitive variables formulation and on the potential vector for the vorticity-potential vector formulation. In this paper, we present a method of resolution of the unsteady Navier-Stokes equations written in velocity-vorticity formulation for the viscous incompressible flow around finite length circular cylinder placed between two circular plates. For this fully three-dimensional flow, one of the difficulties, as pointed out in [1], is to satisfy the continuity equation of the velocity and the divergence free condition of the vorticity. The numerical resolution of the equations without specific treatment of the two divergence free conditions can lead to an unrealistic solution.

In order to obtain a good approximated discrete solution, we use finite difference numerical scheme based on an A.D.I method [3] for the vorticity transport equations and for the Poisson equations of the velocity components. The computed velocity and vorticity fileds do not satisfy the divergence free conditions and we use, for the velocity an iterative correction process based on the assumption that the vorticity is, at each time step, an implicit invariant at each internal iteration of a process dealing with the velocity [4]. For the vorticity field, we introduce a potential function by the assumption that the computed vorticity, which is not solenoidal, can be splitted into a divergence free part and an irrotational part [1].

In spite of the fact that, to our knownledge, there is not published experimental or numerical results for the three-dimensional flow around circular cylinder, the algorithm is validated by the comparison of results obtained in the symmetry plane with the two-dimensional ones. Results are presented for flows at Re=300 for different values of the length ratio H/D. The comparison shows that the algorithm used is able to simulate fully three-dimensional flows with good approximation in regard of the divergence free conditions.

GOVERNING EQUATIONS

For the unsteady three-dimensional flow of viscous incompressible fluid without external forces, the governing equations written in velocity-vorticity formulation are :

$$\nabla^2\vec{U} = -\nabla \times \vec{\Omega} \tag{1}$$

$$\nabla.\vec{U} = 0 \tag{2}$$

$$\frac{\partial\vec{\Omega}}{\partial t} + (\vec{U}.\nabla)\vec{\Omega} - (\vec{\Omega}.\nabla)\vec{U} = \frac{2}{Re}\nabla^2\vec{\Omega} \tag{3}$$

$$\nabla.\vec{\Omega} = 0 \tag{4}$$

$$\vec{\Omega} = \nabla \times \vec{U} \tag{5}$$

with $Re = \frac{2aU_0}{\nu}$, a the radius of the cylinder, U_0 the velocity at the infinity and ν the kinematic viscosity.

The log-polar transformation is considered :

$$r = e^{\Pi R} \quad , \quad \theta = \Pi\Theta \quad , \quad z = Z \quad .$$

In the cylindrical coordinates, (u, v, w) are the components of the velocity and $(\omega_1, \omega_2, \omega_3)$ the components of the vorticity. In order to simplify the scalar equations, the following transformed unknows are used :

$$U_1 = E.u, \quad U_2 = E.v, \quad U_3 = w$$

$$\Omega_1 = E.\omega_1, \quad \Omega_2 = E.\omega_2, \quad \Omega_3 = \omega_3 \quad with \quad E = \Pi e^{\Pi R}.$$

The conservative form of the vorticity transport equations can be written :

$$\frac{ReE^2}{2}\frac{\partial\Omega_i}{\partial t} + \Phi_{\Omega_i} = (\delta_R{}^{\Omega_i} + \delta_\Theta{}^{\Omega_i} + \delta_Z{}^{\Omega_i}).\Omega_i \tag{6}$$

where Φ_{Ω_i} is the source term and :

$$\delta_R{}^{\Omega_i} = \frac{\partial^2.}{\partial R^2} - \frac{\alpha_i Re}{2}\frac{\partial(U.)}{\partial R} \quad where \quad \alpha_i = 0 \quad if \quad i = 1 \quad and \quad \alpha_i = 1 \quad if \quad i \neq 1$$

$$\delta_\Theta{}^{\Omega_i} = \frac{\partial^2.}{\partial\Theta^2} - \frac{\beta_i Re}{2}\frac{\partial(V.)}{\partial\Theta} \quad where \quad \beta_i = 0 \quad if \quad i = 2 \quad and \quad \beta_i = 1 \quad if \quad i \neq 2$$

$$\delta_Z{}^{\Omega_i} = \frac{\partial^2.}{\partial Z^2} - \frac{\gamma_i Re}{2}\frac{\partial(W.)}{\partial Z} \quad where \quad \gamma_i = 0 \quad if \quad i = 3 \quad and \quad \gamma_i = 1 \quad if \quad i \neq 3$$

The Poisson equations of the velocity components are :

$$\frac{\partial^2 U_i}{\partial R^2} + \frac{\partial^2 U_i}{\partial\Theta^2} + E^2\frac{\partial^2 U_i}{\partial Z^2} = \Phi_{U_i} \tag{7}$$

where U_i represent one of the velocity components and Φ_{U_i} the associated source term.

The implementation of a proper set of boundary conditions is one of the sensitive point of the algorithm and sufficiently accurate approximations are needed for the vorticity components. In our case, which is an external flow problem. Different types of boundaries are encountered :

- Solid surfaces (surface of the cylinder, surfaces of the circular plates).
- external boundary.

For all the wall boundaries, the adherence condition is used for the velocity components :

$$U_1 = U_2 = U_3 = 0 \quad .$$

The walls boundary conditions of the vorticity components are obtained through Taylor expansions of relations deduced from equation (5).

At the external boundary, the normal gradient of the vorticity components is supposed to vanish. This condition can be used for the study of the first stages of little or moderate Reynolds number flows when the radial length domain is sufficiently large. For the velocity components, we explicitely use freely developed flow conditions for the U_2 and U_3 components and the continuity equation for U_1.

NUMERICAL PROCEDURE

The numerical resolution of the parabolic and elliptic equations is based on a three-dimensional A.D.I method [3]. The finite difference numerical scheme used for the resolution of the vorticity transport equations is second order accurate in space and first order in time. Combined second and fourth order finite difference scheme is developed for the Poisson equations of the velocity components. The fourth order finite difference allow us to take implicitely into account a divergence free condition on the wall surfaces [5] and it is used in the radial direction for U_1 and U_2 components and in the transverse direction for U_3. The use of these finite differences improve the results in regard of the continuity equation which remain not satisfied.

For the vorticity vector, no implicit treatment of the divergence free condition is possible and the only way to ensure it is to correct the computed field in order to make it solenoidal.

The velocity and vorticity fields are corrected with the introduction of potential functions through the assumptions that :

1) At a fixed time, the vorticity is an invariant implicit during the internal iterations of a correction process on the velocity field and we can write :

$$\vec{\Omega}^{(n\Delta t)} = \nabla \times \vec{U}^{(n\Delta t)_k} = \nabla \times \vec{U}^{(n\Delta t)_{k+1}}$$

$$\Rightarrow \quad \vec{U}^{(n\Delta t)_{k+1}} = \vec{U}^{(n\Delta t)_k} - \vec{\nabla}\vec{\Phi}^{k+1} \tag{8}$$

Assuming that $\vec{U}^{(n\Delta t)_{k+1}}$ verify the continuity equation, the divergence of equation (8) gives :

$$\nabla^2\Phi^{k+1} = \nabla.\vec{U}^{(n\Delta t)_k} = D_U{}^k \tag{9}$$

The boundary conditions on Φ^{k+1} are given on the wall surfaces by the introduction of this function (eq.(8)) and at the external boundary by the fact that the continuity equation of the velocity is not satisfied only in a three-dimensional region close to the cylinder and we can write :

$$\vec{\nabla}\Phi^{k+1}\big|_{r=r_a} = \vec{\nabla}\Phi^{k+1}\big|_{r=r_\infty} = 0$$

2) At a fixed time, the vorticity field, solution of the vorticity transport equations, does not satisfy the divergence free condition and a correction can be computed iteratively until this condition is satisfied. We can write :

$$\vec{\Omega}^{(n\Delta t)_{k+1}} = \vec{\Omega}^{(n\Delta t)_k} - \vec{\nabla}\varphi^{k+1} \tag{10}$$

where vorticity vector $\vec{\Omega}^{(n\Delta t)_{k+1}}$ satisfy the continuity equation in the whole domain. Taking the divergence of equation (10), we obtain :

$$\nabla^2 \varphi^{k+1} = \nabla . \vec{\Omega}^{(n\Delta t)_k} = D_\Omega{}^k \tag{11}$$

Using the fact that the vorticity is a rotational on the wall surfaces, the boundary values on these surfaces are not corrected during the iterative correction process. At the external boundary, as for the velocity field, the divergence free condition of the vorticity is satisfied and the field does not need to be corrected. We can write :

$$\vec{\nabla}\varphi^{k+1}|_{r=r_a} = \vec{\nabla}\varphi^{k+1}|_{r=r_\infty} = 0$$

The Poisson equations of the two potential functions are solved using combined second and fourth order finite difference numerical scheme based on the previous A.D.I method. The fourth order finite differences are used in the radial and transversal directions in order to take into account all the boundary conditions on these functions.

The correction processes use, at more, N internal iterations and they are stopped when :

$$\epsilon_D \le \epsilon_0 \quad with \quad \epsilon_D = |D_{i,j,k}|_{Max} \quad and \quad \epsilon_0 \quad is \quad a \quad fixed \quad value.$$

RESULTS

Results are obtained with this new algorithm for flows at Re=300 in two different cases : $H/D = 5; 10$, and the correction of the velocity and vorticity fields in order to satisfy the divergence free conditions appear as a necessar condition to obtain a realistic solution of the Navier-Stokes equations in velocity-vorticity formulation. For the flow with $H/D = 5$, computations are performed with $31 \times 61 \times 21$ points with $\Delta t = 5.10^{-2}$ and results can be classified as : **1)** Solution 1: the velocity and the vorticity are corrected with $N = 60$ internal iterations and $\epsilon_0 = 5.10^{-2}$. **2)** Solution 2: only the velocity is corrected using $N = 60$ internal iterations and $\epsilon_0 = 5.10^{-2}$. **3)** Solution 3: the velocity and the vorticity are corrected using $N = 160$ internal iterations and $\epsilon_0 = 5.10^{-2}$. The comparison of these solutions show that it is necessar to correct the velocity and the vorticity fields with sufficient number of iterations in the correction processes. The structure of the flow near the lateral circular plate at t=5 is given in fig.1 for the three solutions and we can see a difference on the structure of the recirculation area between the solutions 1 and 3 and the solution 2. At later time, the comparison of solution 1 with solution 3 show that the use of only 60 internal iterations is not sufficient in order to ensure the two divergence free conditions and it leads to an unrealistic solution (fig.2). The evolution with time of the three-dimensional recirculation area for solution 3 is given in fig.3 . Different views of the flow structure are given for this solution in fig.4 at different moments of the first stages of the flow.

For the flow with $H/D = 10$, the three-dimensional effects due to the lateral walls are not as important than in the case with $H/D = 5$ and it is easier to ensure the two divergence free conditions of the velocity and vorticity fields using at more 160 internal iterations in the correction processes. The structure of the flow near the lateral wall and in the symmetry plane is given for this computation performed with $31 \times 61 \times 41$ points with $\Delta t = 5.10^{-2}$ in fig.5 and the evolution with time of the three-dimensional recirculation area is given in fig.6 . The radial velocity on the symmetry axis behind the cylinder for the two three-dimensional flows ($H/D = 5; 10$) in the symmetry plane and for the two-dimensional solution is given in fig.7 and the comparison shows a good agreement.

CONCLUSION

The use of the velocity-vorticity formulation in order to simulate fully three-dimensional flows is validated and the two divergence free conditions can be assumed in a good approximation by two iterative processes. Numerical results are obtained for flow around a finite length circular cylinder and compared succesfully with two-dimensional ones. The method seems to be available for numerical simulation of external flows.

REFERENCES

[1] T.B. Gatski, C.E. Grosch, M.E. Rose, R.E. Spall, Numercial simulation of three-dimensional unsteady vortex flow using a compact vorticity-velocity formulation. Proceedings of the 7th GAMM-Conf. on Numer. Methods in Fluid Dynamics. Vol 20

[2] A. Toumi, L. Ta Phuoc, Numerical study of the viscous incompressible flow by vorticity and velocity formulation. 5th Conf. on Numer. Methods in Laminar and Turbulent flows. Montréal. July 1987

[3] K. Aziz, J.D. Hellums, Numerical solution of three-dimensional equations of motion of laminar convection. Physics of fuid, Vol 10, no 2. feb 67.

[4] N. Takamitsu, Finite difference method to solve incompressible fluid flow. J. Comp. Physics, 499-518, (1985).

[5] W. Labidi, L. Ta Phuoc, Numercial resolution of Navier-Stokes equations in velocity-vorticity formulation : application to the circular cylinder. Proceedings of the 7th GAMM-Conf. on Numer. Methods in Fluid Dynamics. Vol 20

ACKNOWLEDGEMENT

The computing means were provided by the scientific committee of the CCVR who is gratefully acknowledged.

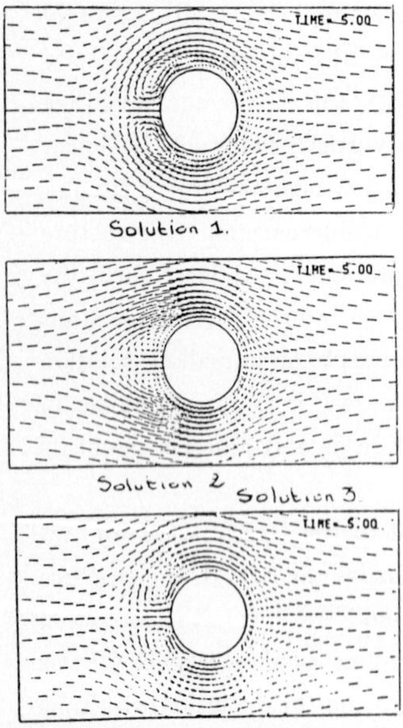

Fig. 1 - Comparison of the three solutions for the flow with $H/D = 5$.

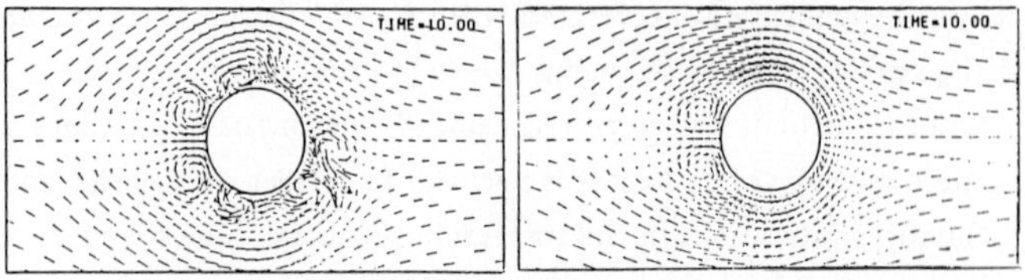

Fig. 2 - Comparison of the solutions 1 and 3 for the flow with $H/D = 5$.

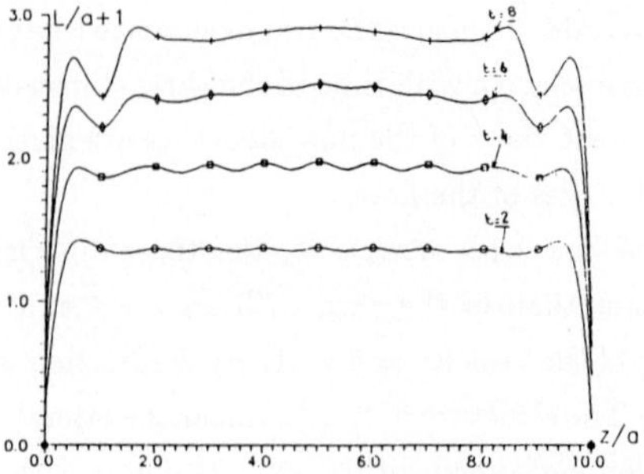

Fig.3 – Evolution with time of the three-dimensional recirculation area for Re=300 and H/D=5

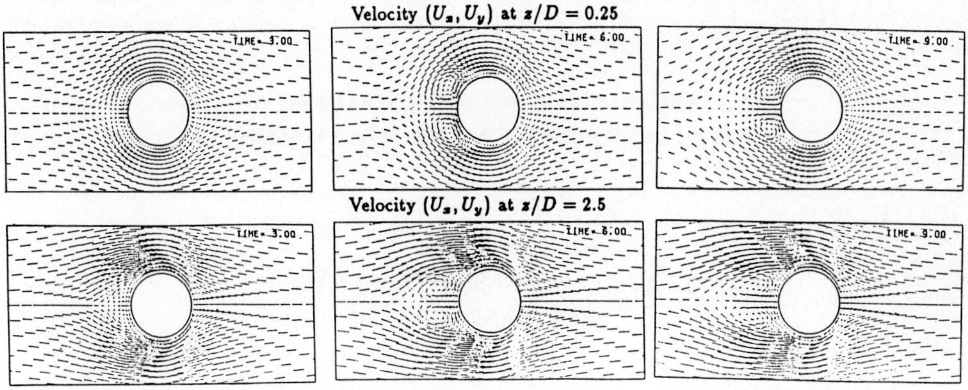

Fig. 4 - Evolution of the structure of the flow near the lateral wall and in the symmetry plane. $H/D = 5$.

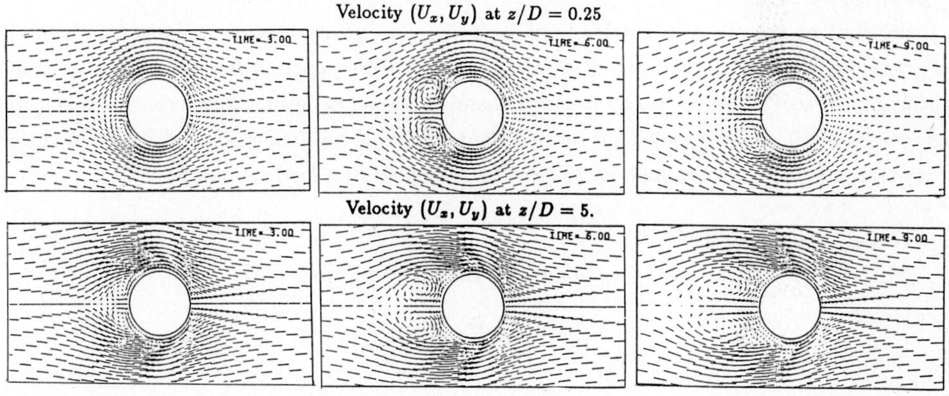

Fig. 5 - Evolution of the structure of the flow near the lateral wall and in the symmetry plane. $H/D = 10$.

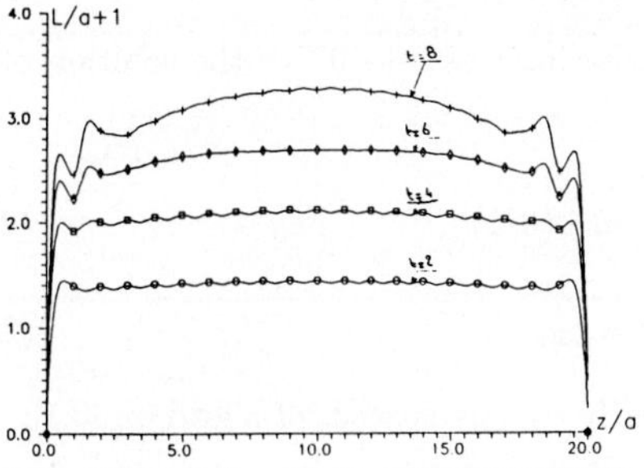

Fig. 6 – Evolution with time of the three-dimensional recirculation area for Re=300 and H/D=10

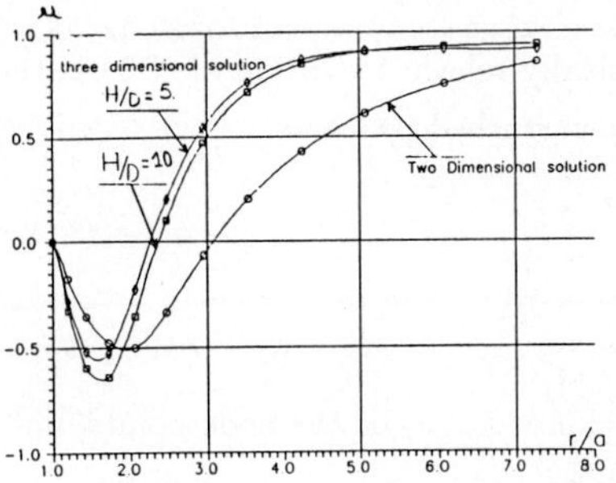

Fig. 7 – Comparison at t=5 of the 3-D solutions in the symmetry plane with the 2-D solution – Re=300

CALCULATION OF SHOCKED FLOWS BY MATHEMATICAL PROGRAMMING

John E. Lavery

NASA LEWIS RESEARCH CENTER
CLEVELAND, OH 44135

1. INTRODUCTION

Solution of the steady-state Burgers' equation on equally spaced grids by mathematical programming has been investigated in [4]. In the present paper, a framework for solving by mathematical programming Burgers' equation and the Euler equations for quasi-one-dimensional flow on grids with variable spacing is set up. Computational results for grids with abrupt changes in mesh length by factors as high as 10^4 will be presented in a forthcoming paper.

2. BURGERS' EQUATION

We consider first the steady-state inviscid Burgers' equation

$$(u^2)' = 0 \qquad \text{on} \quad (0,1) \tag{2.1a}$$

with the boundary conditions

$$u(0) = g_0, \quad u(1) = g_1 \quad (g_0 \neq g_1). \tag{2.1b}$$

The physically relevant solution of (2.1) is the pointwise limit as $\varepsilon \to 0^+$ of the solution of the singularly perturbed problem

$$-\varepsilon u'' + (u^2)' = 0 \qquad \text{on} \quad (0,1) \tag{2.2a}$$

$$u(0) = g_0, \qquad u(1) = g_1 \quad . \tag{2.2b}$$

Let x_i, $i = 0, 1, \ldots, n$, be the node points, not necessarily equally spaced, of a grid on [0,1]:

$$0 = x_0 < x_1 < x_2 < \cdots < x_{n-1} < x_n = 1. \tag{2.3}$$

Discretize Eq. (2.2) on each cell (x_i, x_{i+1}) using a four-point finite-difference scheme involving u_{i-1}, u_i, u_{i+1} and u_{i+2} for the viscous term $-\varepsilon u''$ and a two-point difference scheme involving only u_i and u_{i+1} for the inviscid term $(u^2)'$, both schemes being of second order with respect to maximum cell width at the midpoint $(x_i + x_{i+1})/2$ of the cell. The resulting finite-difference approximations of (2.2a) on the first cell (x_0, x_1), on the interior cells $(x_i, x_{i+1}), i = 1, 2, \ldots, n-2$ and on the last cell (x_{n-1}, x_n) are, respectively,

$$\varepsilon \left[\begin{array}{l} \frac{3x_0+x_1-2x_2-2x_3}{(x_0-x_1)(x_0-x_2)(x_0-x_3)} g_0 \\ +\frac{x_0+3x_1-2x_2-2x_3}{(x_1-x_0)(x_1-x_2)(x_1-x_3)} u_1 \\ +\frac{x_0+x_1-2x_3}{(x_2-x_0)(x_2-x_1)(x_2-x_3)} u_2 \\ +\frac{x_0+x_1-2x_2}{(x_3-x_0)(x_3-x_1)(x_3-x_2)} u_3 \end{array} \right] + \frac{{u_1}^2 - {g_0}^2}{x_1 - x_0} = 0, \tag{2.4a}$$

$$\varepsilon \left[\begin{array}{l} \frac{x_i+x_{i+1}-2x_{i+2}}{(x_{i-1}-x_i)(x_{i-1}-x_{i+1})(x_{i-1}-x_{i+2})} u_{i-1} \\ +\frac{-2x_{i-1}+3x_i+x_{i+1}-2x_{i+2}}{(x_i-x_{i-1})(x_i-x_{i+1})(x_i-x_{i+2})} u_i \\ +\frac{-2x_{i-1}+x_i+3x_{i+1}-2x_{i+2}}{(x_{i+1}-x_{i-1})(x_{i+1}-x_i)(x_{i+1}-x_{i+2})} u_{i+1} \\ +\frac{-2x_{i-1}+x_i+x_{i+1}}{(x_{i+2}-x_{i-1})(x_{i+2}-x_i)(x_{i+2}-x_{i+1})} u_{i+2} \end{array} \right] + \frac{{u_{i+1}}^2 - {u_i}^2}{x_{i+1} - x_i} = 0, \tag{2.4b}$$

$$\varepsilon \left[\begin{array}{l} \frac{-2x_{n-2}+x_{n-1}+x_n}{(x_{n-3}-x_{n-2})(x_{n-3}-x_{n-1})(x_{n-3}-x_n)} u_{n-3} \\ +\frac{-2x_{n-3}+x_{n-1}+x_n}{(x_{n-2}-x_{n-3})(x_{n-2}-x_{n-1})(x_{n-3}-x_n)} u_{n-2} \\ +\frac{-2x_{n-3}-2x_{n-2}+3x_{n-1}+x_n}{(x_{n-1}-x_{n-3})(x_{n-1}-x_{n-2})(x_{n-1}-x_n)} u_{n-1} \\ +\frac{-2x_{n-3}-2x_{n-2}+x_{n-1}+3x_n}{(x_n-x_{n-3})(x_n-x_{n-2})(x_n-x_{n-1})} g_1 \end{array} \right] + \frac{{g_1}^2 - {u_{n-1}}^2}{x_n - x_{n-1}} = 0. \tag{2.4c}$$

For equally spaced x_i, Eqs. (2.4) reduce to Eqs. (2.1) of [4]. In particular, Eq. (2.4b) reduces to the more familiar form

$$\varepsilon \frac{-u_{i-1} + u_i + u_{i+1} - u_{i+2}}{2h^2} + \frac{{u_{i+1}}^2 - {u_i}^2}{h} = 0 \tag{2.5}$$

$(h = x_{i+1} - x_i)$.

Eqs. (2.4) are an overdetermined system of n equations for the $n - 1$ unknowns u_i, $i = 1, 2, \ldots, n-1$. Why this system should be solved by a l_1 procedure rather than by a l_2 (least-square) procedure is discussed in [4]. The l_1 strategy for solving system (2.4) consists in finding the u_i, $i = 1, 2, \ldots, n-1$ that minimize the weighted sum

$$\sum_{i=0}^{n-1} (x_{i+1} - x_i)|r_i|, \tag{2.6}$$

of the absolute values of the residuals r_i (left sides) of Eqs. (2.4). In [4], the weights $x_{i+1} - x_i$ were all equal and were omitted. Sum (2.6) was minimized by the procedure used in [4]: each step of this procedure consisted in linearizing system (2.4) by Newton's method and solving the resulting overdetermined linear system by the Barrodale-Roberts l_1 algorithm as implemented in the IMSL (Ed. 9) subroutine RLLAV [1,2]. Convergence was deemed to have occurred when the relative l_1 error

$$\sum_{i=1}^{n-1} \frac{|u_i^{\text{cur}} - u_i^{\text{pre}}|}{|u_i^{\text{cur}}|} \tag{2.7}$$

between the current solution values u_i^{cur} and the values u_i^{pre} on the previous step was less than $0.5 * 10^{-10}$.

Computational results for Burgers' equation can be obtained using the l_1 procedure with a homotopy (continuation procedure) in ε. For $\varepsilon = 10^{-1}$, the initial guesses for the u_i were set equal to $1 - x_i$. Once convergence for $\varepsilon = 10^{-1}$ was achieved (as determined by the convergence criterion involving (2.10) discussed above), the final solution for $\varepsilon = 10^{-1}$ was used as the initial guess for the solution of the problem with $\varepsilon = 10^{-2}$. In general, the solution for $\varepsilon = 10^{-k+1}$ was used as the initial guess for the solution of the problem with $\varepsilon = 10^{-k}$, $k = 2, 3, \ldots, 15$. The results presented in [4] show that the l_1 procedure performs well on equally spaced grids. New numerical experiments on grids with abrupt changes in mesh spacing by factors as high as 10^4 show highly accurate solutions for ε as small as 10^{-15}. Two cases of abrupt changes in mesh spacing were considered: 1) fine cells in the boundary layer, coarse cells outside (goal: resolution of the boundary layer) and 2) one fine cell far from the boundary layer sandwiched between coarse cells (simulating patching of grids). Numerical results for grids with abrupt changes in mesh spacing will be presented in a forthcoming paper.

3. EULER EQUATIONS FOR QUASI-ONE-DIMENSIONAL FLOW

Steady quasi-one-dimensional flow in a nozzle of length 10 with cross-sectional area $A(x)$ can be described by

$$\frac{d}{dx}(A\rho u) = 0, \tag{3.1a}$$

$$\frac{d}{dx}(A\rho u^2) + A\frac{dp}{dx} = 0, \tag{3.1b}$$

$$\frac{d}{dx}[Au(\rho E + p)] = 0, \tag{3.1c}$$

on $(0, 10)$, where ρ is the density, u is the velocity, $E = e + (u^2/2)$ is the total energy, e is the internal energy and $p = (\gamma - 1)\rho e$ is the pressure for a perfect gas (all quantities normalized). The area function

$$A(x) = 1.398 + 0.347 \tanh(0.8x - 4) \tag{3.2}$$

and the boundary conditions

$$\begin{aligned} \rho(0) &= 0.502, & \rho(10) &= 0.776, \\ u(0) &= 1.299, & e(0) &= 1.897 \end{aligned} \tag{3.3}$$

of [6] were used. The shock in the exact solution of this problem is at $x = 4.816$.

Let x_i, $i = 0, 1, \ldots, n$, be the node points of a not necessarily equally spaced grid on [0,10] that satisfies relations (2.6) with $x_n = 1$ replaced by $x_n = 10$. Eqs. (3.1) are discretized on this grid using two-point finite-difference schemes of $O((x_{i+1} - x_i)^2)$ centered at the midpoint of the cell (x_i, x_{i+1}):

$$\frac{A_{i+1}\rho_{i+1}u_{i+1} - A_i\rho_i u_i}{x_{i+1} - x_i} = 0, \tag{3.4a}$$

$$\frac{A_{i+1}\rho_{i+1}u_{i+1} - A_i\rho_i u_i}{x_{i+1} - x_i} + \frac{A_{i+1} + A_i}{2}\,\frac{p_{i+1} - p_i}{x_{i+1} - x_i} = 0, \tag{3.4b}$$

$$\frac{A_{i+1}(\rho_{i+1}E_{i+1} + p_{i+1})u_{i+1} - A_i(\rho_i E_i + p_i)u_i}{x_{i+1} - x_i} = 0. \tag{3.4c}$$

(cf. [3,5,7]). System (3.4) is an overdetermined system of $3n$ equations for the $3n - 1$ unknowns ρ_i, $(\rho u)_i$, e_i, $i = 1, 2, \ldots, n-1$, and $(\rho u)_n$, e_n. The l_1 strategy for solving this problem is to minimize the sum

$$\sum_{i=0}^{n-1}\sum_{j=1}^{3}(x_{i+1} - x_i)|r_{ij}| \quad , \tag{3.5}$$

where r_{ij} is the residual (left side) of the jth of Eqs. (3.4) for a given i (cf. (2.9)). Sum (3.5) was minimized by the l_1 procedure used in Section 2: system (3.4) was linearized by Newton's method and the resulting overdetermined linear system was solved by the Barrodale-Roberts l_1 algorithm (IMSL Ed. 9 subroutine RLLAV). No homotopy was used since no artificial viscosity was added to Eqs. (3.4). Convergence was deemed to have occurred when the relative l_1 error

$$\sum_{i=1}^{n-1} \frac{|\rho_i^{\text{cur}} - \rho_i^{\text{pre}}|}{|\rho_i^{\text{cur}}|} + \sum_{i=1}^{n} \frac{|(\rho u)_i^{\text{cur}} - (\rho u)_i^{\text{pre}}|}{|(\rho u)_i^{\text{cur}}|} + \sum_{i=1}^{n} \frac{|e_i^{\text{cur}} - e_i^{\text{pre}}|}{|e_i^{\text{cur}}|} \tag{3.6}$$

between the solution values on the current Newton step (superscript "cur") and the solution values on the previous Newton step (superscript "pre") was less than $0.5 * 10^{-10}$. The numerical results show that the l_1 procedure produces accurate shocked solutions of (3.1) on both coarse and fine equally spaced grids as well as on grids with abrupt changes in mesh length by factors as high as $1.2 * 10^4$. Both the case of refinement near the shock (in order to find the position of the shock more accurately) and the case of refinement far from the shock (simulating patching of grids) were considered. Selected numerical results are presented in a forthcoming paper.

REFERENCES

1. I. BARRODALE AND F. D. K. ROBERTS, *SIAM J. Numer. Anal.* **10**, 839 (1973).
2. I. BARRODALE AND F. D. K. ROBERTS, *Comm. ACM* **17**, 319 (1974).
3. F. CASIER, H. DECONINCK AND C. HIRSCH, *AIAA Journal* **22**, 1556 (1984).
4. J. E. LAVERY, *J. Comp. Phys.*, to appear.
5. R. W. MacCORMACK, *AIAA Journal* **20**, 1275 (1982).
6. G. R. SHUBIN, A. B. STEPHENS AND H. M. GLAZ, *J. Comp. Phys.* **39**, 364 (1981).
7. S. F. WORNOM, *Computers & Fluids* **12**, 11 (1984).

UNIVERSAL LIMITER FOR HIGH ORDER EXPLICIT CONSERVATIVE ADVECTION SCHEMES

B. P. Leonard
Institute for Computational Mechanics in Propulsion
NASA-Lewis Research Center
and
H. S. Niknafs
Department of Mechanical Engineering
The University of Akron

For decades, computational fluid dynamicists have been searching for advection schemes which can simulate sharp changes in gradient without introducing artificial diffusion or extraneous oscillations, while handling smooth profiles without distortion. This paper describes a simple method for construction of nonoscillatory explicit conservative advection schemes of arbitrarily high accuracy. Four test profiles are considered: an isolated sine-squared wave; a unit step function; a semi-ellipse; and a narrow Gaussian. The universal limiter guarantees monotonic resolution of the step; sharpness increases uniformly with the order of the base scheme. Artificial compression is not recommended, as this may cause gross distortion of the other profiles. It is better to use a simple base method such as limited third-order upwinding, and adaptively blend into a limited higher order scheme locally, where needed. In this way, very sharp step resolution can be achieved without corrupting the other profiles. This is a cost-effective strategy since the wider computational stencil of the higher order method is required only in very isolated regions, by definition.

The universal limiter is based on a very simple idea. First, in a control-volume formulation the advected face value is estimated using high order interpolation. For locally monotonic behaviour, the face value is limited by adjacent node values in a direction normal to the face; i.e., between first-order upwinding and first-order downwinding. However, the latter is not sufficient for monotonicity maintenance. To understand the additional constraint, consider one-dimensional pure advection of a scalar ϕ at constant positive velocity and assume ϕ is locally monotonic increasing. The explicit CV update algorithm can be written

$$\phi_i^{n+1} = \phi_i^n + c\,(\phi_\ell - \phi_r) \tag{1}$$

where c is the Courant number and left and right face values are indicated. To remain monotonic, the following inequalities must be satisfied

$$\phi_{i-1}^{n+1} \le \phi_i^{n+1} \le \phi_{i+1}^{n+1} \tag{2}$$

The right-hand inequality is superceded by first-order downwinding, and assuming worst-case conditions, the left inequality becomes

$$\phi_r \leq \phi_{i-1}^n + (\phi_i^n - \phi_{i-1}^n)/c \tag{3}$$

This completes the universal limiter for monotonic behaviour. In nonmonotonic regions, first-order upwinding is adequate. Note that this does not erode the formal order of the basic scheme, which must be based on regions of low curvature. In practice, clipping of local extrema decreases with the order of the underlying method. Clipping can be entirely eliminated by an adaptive pattern-recognition strategy which uses the unlimited scheme in the vicinity of true (physical) extrema.

The paper shows that second-order shock-capturing schemes are inadequate compared with the limited QUICKEST scheme which gives excellent results using the same stencil. Arbitrarily higher order resolution can be obtained by adaptively using a higher order interpolation method locally, as needed. Narrow physical extrema can be accurately resolved by using a (locally) higher order unlimited scheme in such regions. Emphasis is now placed on the development of a robust yet sensitive indicator which can discriminate between real physical and spurious numerical extrema.

Figure 1 shows simulation of a step and a 20Δx-wide semi-ellipse using Superbee and limited QUICKEST. By design, Superbee contains artificial compression (negative viscosity), steepening all fronts, with concomitant flattening of other regions. Although not as sharp for the step, the limited QUICKEST scheme gives much lower error overall, considering several different profile shapes. Rather than using artificially compressive low-order schemes, it is better to use higher order (upwind) schemes, applying the universal limiter. This can be done globally or, more efficiently, in local regions where necessary for high resolution, using limited QUICKEST for the bulk of the flow domain. Results for limited seventh-order upwinding are shown in Figure 2. The 20Δx sine-squared profile, the step, and the semi-ellipse simulations are all excellent; only the narrow Gaussian ($\sigma = 2\Delta x$) shows significant clipping. To get good resolution of narrow peaks, one can use an unlimited higher order scheme globally, such as the eighth-order method of Figure 3(a). But, of course, unlimited schemes are highly oscillatory when modelling step-like phenomena, as seen in 3(b). Standard shock-capturing schemes are very poor in resolving narrow peaks, Figure 3(c). If physical extrema can be identified, a partially limited scheme, 3(d), performs well. This consists of limited QUICKEST in smooth regions, switching to limited seventh-order upwinding in high-curvature regions, except near the peak where an unlimited eighth-order scheme is used.

The ULTIMATE conservative difference scheme consists of applying the universal limiter (UL) to transient interpolation modelling (TIM) of the advective transport equations (ATE).

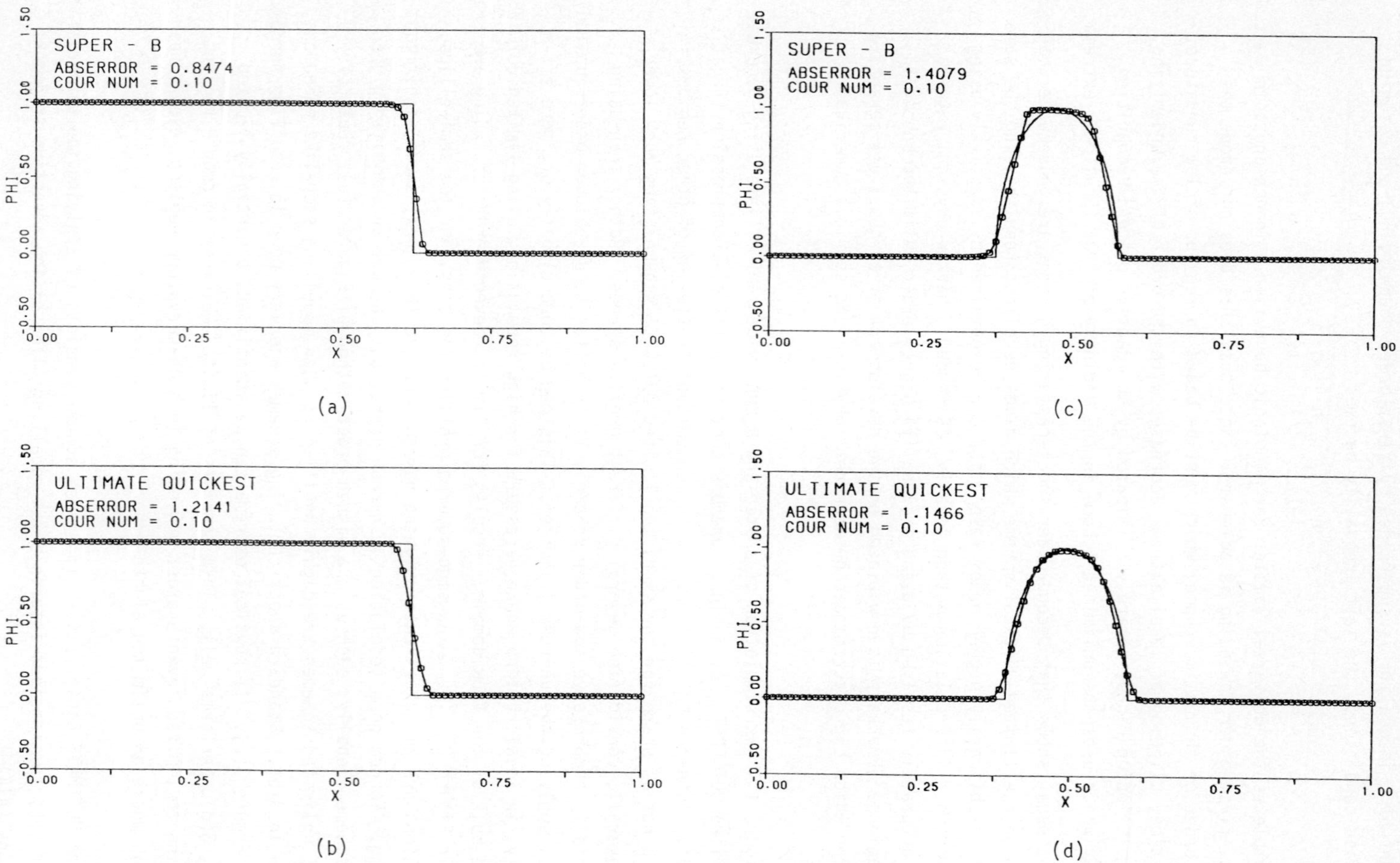

Figure 1. Step simulation: (a) Superbee, (b) ULTIMATE QUICKEST.
Semi-ellipse: (c) Superbee, (d) ULTIMATE QUICKEST.

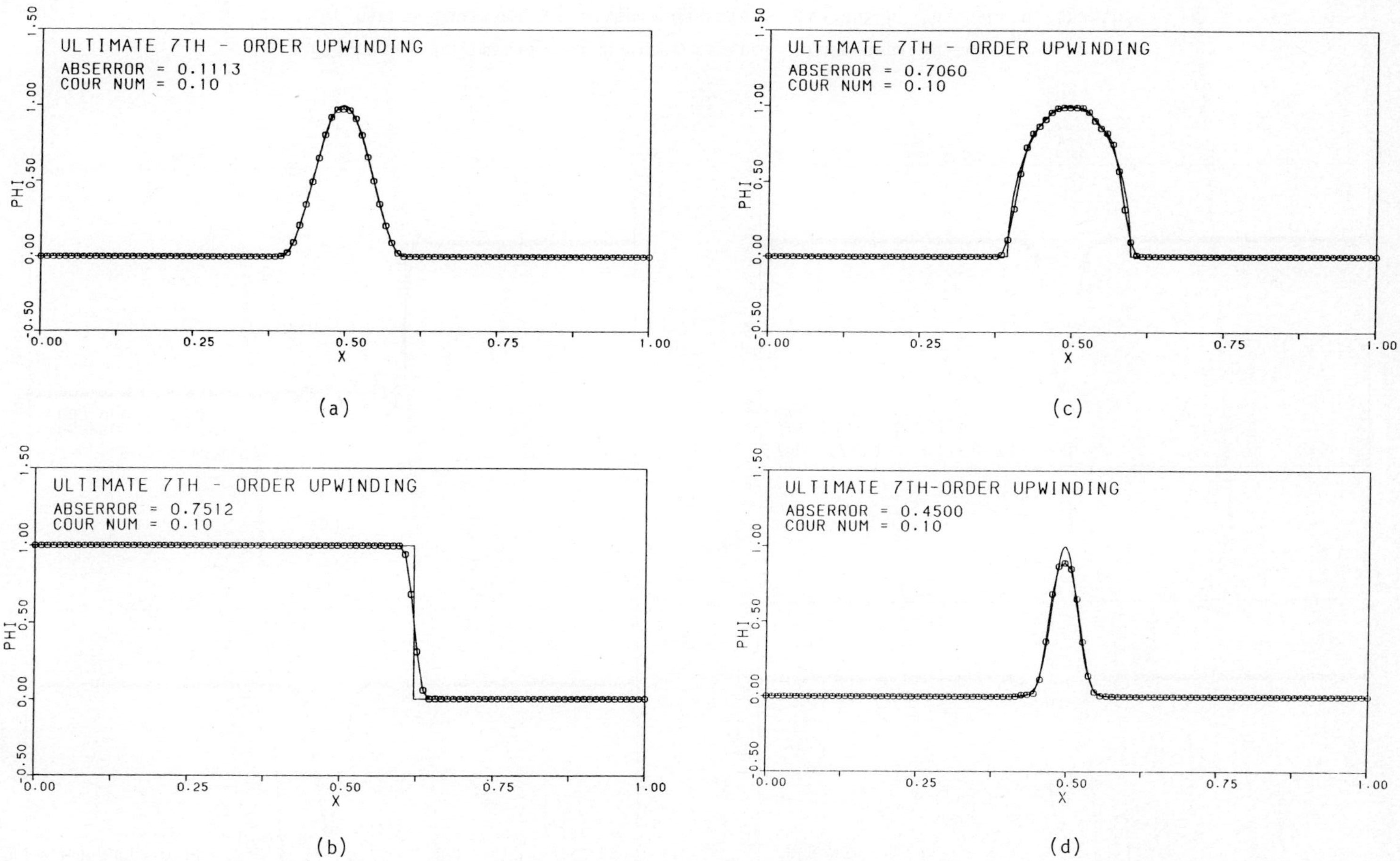

Figure 2. ULTIMATE seventh-order upwinding applied to: (a) sine-squared, (b) step, (c) semi-ellipse, (d) narrow Gaussian.

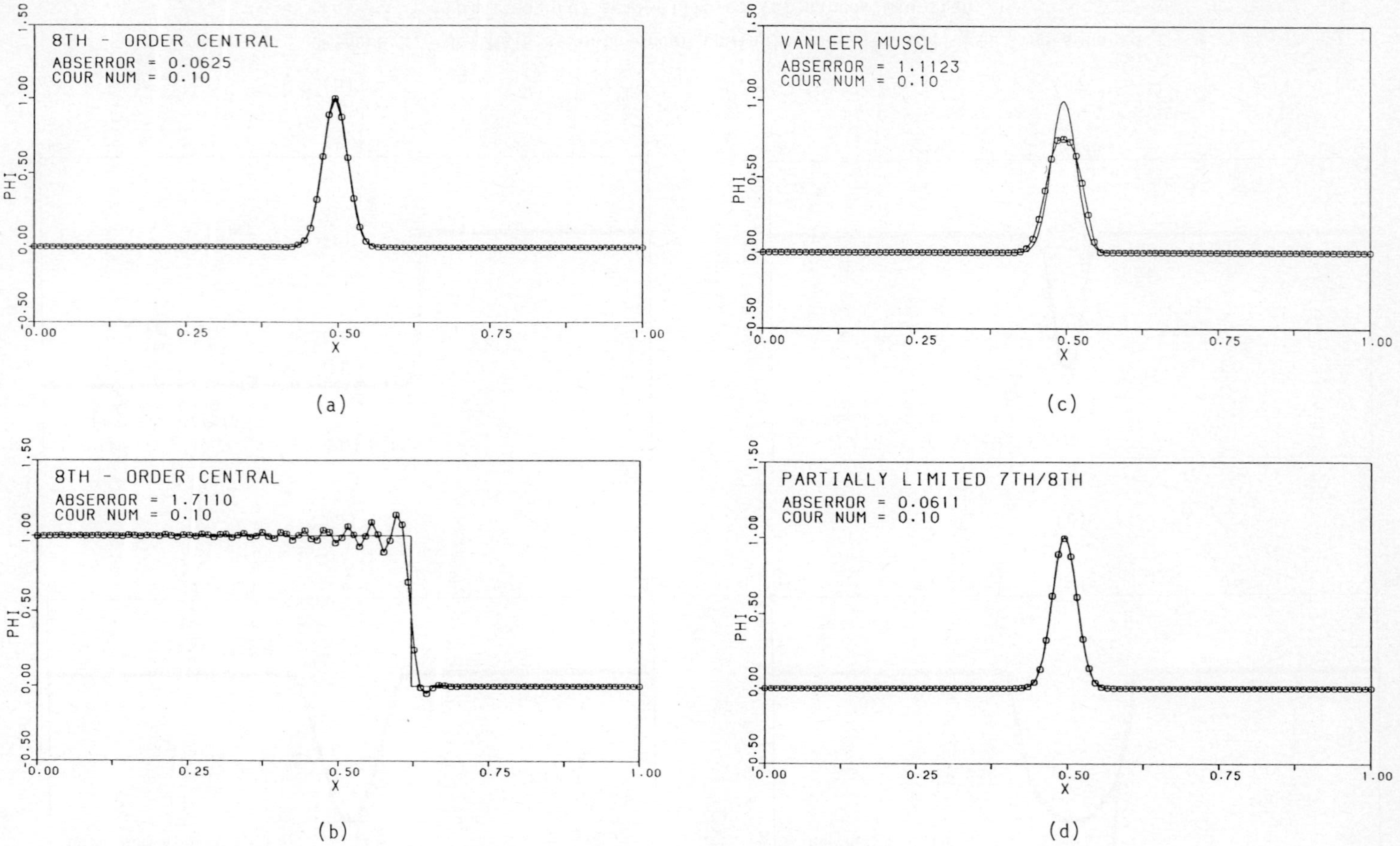

Figure 3. Unlimited eighth-order central scheme applied to:
(a) narrow Gaussian, (b) step. Gaussian: (c) MUSCL, (d) hybrid 7th/8th.

A Comparison of Numerical Schemes on Triangular and Quadrilateral Meshes

Dana R. Lindquist* and Michael B. Giles†
Department of Aeronautics and Astronautics
Massachusetts Institute of Technology
Cambridge, MA 02139

Introduction

Finite volume solution procedures for Euler equations have been extensively studied when applied to quadrilateral meshes. Many methods such as mesh adaptation and multi-grid have been developed which make these procedures efficient and highly accurate. Recently, great interest has been focussed on the development and use of unstructured triangular meshes. One of the advantages of these meshes is that they offer great flexibility in the generation of meshes around extremely complex geometries. The other advantage is that in adaptive mesh refinement the mesh always remains an unstructured triangular mesh, and so no modifications of the basic flow algorithm are required. However, a certain amount of doubt exists in the computational fluid dynamics community concerning the performance of triangular mesh solution schemes; in particular a recent paper by Roe [1] proved that the local truncation error is only first order on very irregular meshes. Giles [2] argues, however, that these triangular schemes can still be globally second order accurate.

The goal of this study is to address this question of how well a given computational method can perform on a triangular mesh as compared to the more commonly used quadrilateral meshes. In particular, two node-based finite volume schemes will be examined: the node-based quadrilateral cell Jameson scheme [3] which has been modified for triangular meshes by Mavriplis and Jameson [4], and the quadrilateral cell Lax-Wendroff method developed by Ni [5] which has been modified by Lindquist for use on triangular meshes [6]. Care has been taken to keep the triangular and quadrilateral versions of a scheme similar to provide a fair basis for comparison.

Development of Schemes

Both the quadrilateral and triangular meshes used here are described by an unstructured pointer system. In this system the most important element in the calculation, which in this case is the cell, points to surrounding elements which are needed in the calculation, such as the nodes which make up the cell. By using an unstructured pointer system, a slightly longer computation time and more storage space is required, but greater ease in programming and more flexibility in mesh type is gained.

Both the Jameson and Ni schemes which are examined here are node-based four equation models of the Euler equations. It was found that the numerical smoothing used had a larger effect on the solutions than the base solvers, therefore the reader is referred to the references for more details about the base solvers.

*Research Assistant

†Assistant Professor

Numerical Smoothing

Numerical smoothing is a dissipative operator which is added to numerical schemes to damp out oscillations in the solution and provide stability. A fourth difference operator is used for numerical smoothing which is applied throughout the flow field. Another operator is required for shock capturing and will not be discussed here.

To compute the fourth difference smoothing operator, a second difference of a second difference is computed. For both quadrilaterals and triangles two second difference operators are examined. The first is a relatively simple operator which gives a non-zero second difference for a linear function on an irregular mesh. The second operator is more complex, but results in a zero second difference for a linear function. By examining the effect of the second difference operator on a linear function the accuracy of the operator is tested, since for second order or higher accuracy the contribution must be zero.

Typical triangular and quadrilateral cells which will be used to describe the operators are shown in Figure 1 with their corresponding nodes.

The low-accuracy second difference operator is not dependent on the location of the nodes surrounding the node for which the second difference is computed, but merely on the function values at these nodes. For a triangular mesh the contribution from cell A to the second difference at node 1 is

$$(D^2S)_{1A} = (S_3 + S_2 - 2S_1) \tag{1}$$

where S is the variable for which the second difference is computed. For a quadrilateral mesh the contribution from cell A to the second difference at node 1 is

$$(D^2S)_{1A} = (S_4 + S_3 + S_2 - 3S_1) \tag{2}$$

Similar contributions are found for the other nodes of the cell. This second difference is conservative for both triangular and quadrilateral meshes since the total contribution of each cell is zero.

The high-accuracy second difference operator consists of finding the first derivative for each cell and then combining the derivatives on the cells surrounding a node to form a second difference. Unlike the low-accuracy second difference operator, this operator is dependent on the mesh geometry. The operator for a triangular cell mesh is examined first. Referring to Figure 1 the first derivative with respect to x is found for cell A

$$\begin{aligned}(S_x)_A &= \frac{1}{A_A}\int\!\!\int_{cell A}\frac{\partial S}{\partial x}dx\,dy\\ &= \frac{1}{A_A}\int_{1-2-3} S\,dy\\ &= \frac{1}{2\,A_A}[(S_2+S_1)(y_2-y_1)+(S_3+S_2)(y_3-y_2)+(S_1+S_3)(y_1-y_3)]\end{aligned} \tag{3}$$

and similarly for the derivative with respect to y. A similar process is performed to create a second difference. The integration is taken around all the triangles which surround the node for which the second difference is computed, using the derivative values calculated at the cells. To get a second difference instead of a second derivative, there is no division by the area of the integrated region. The contribution to the second difference at node 1 from cell A is

$$(D^2S)_{1A} = \int_{2-3} [(S_x)_A dy - (S_y)_A dx] = (S_x)_A(y_3 - y_2) - (S_y)_A(x_3 - x_2) \tag{4}$$

This second difference operator is conservative since again the total contribution of each cell is zero.

On all boundaries, solid wall or farfield, boundary conditions must be implemented for the second difference operators. For the low-accuracy operator the contribution to a node on the boundary is simply the contribution from the cells surrounding that node which are inside the domain as given in Equations (1) and (2). The high-accuracy operator involves a line integral in Equation (4) which must be closed when considering a node on the boundary. To do this the integral is continued along the boundary faces on either side of the node in question using the value of (S_x) and (S_y) from the cell directly inside the boundary.

To formulate a second difference operator for a quadrilateral mesh a similar process is employed. It turns out that if the first derivative is found by integrating around the complete quadrilateral, the oscillatory modes are not damped out, and the primary purpose of the operator is not fulfilled. To prevent this problem, the quadrilateral is broken into triangles consisting of three of the four nodes of the quadrilateral, and the triangular operator is applied [6].

The first method of creating a fourth difference is to use the low-accuracy second difference twice by operating first on the state vector and then operating on this second difference. This fourth difference is conservative, but is second order accurate only on a uniform mesh since the second difference operator used is only second order accurate on a uniform mesh. The second method is to compute a second difference of the state vector using the high-accuracy method and operate on this second difference with the low-accuracy second difference. This operator is second order accurate since the first operator is second order accurate and conservative since the second operator is conservative. The second method is more expensive than the first, but the effect per iteration is an increase of only 5-10% which is a small increase for the gain in accuracy. The fourth difference is multiplied by a coefficient, between 0.0001 and 0.01, to control the amount of smoothing which is added to the scheme.

Numerical Accuracy

The accuracy of triangular schemes is of great concern to many people. Triangular schemes by their nature have no sense of directionality, unlike quadrilateral schemes where there are two discrete directions. Different analytical arguments have been made which provide conflicting results concerning the accuracy of triangular schemes. Analytical accuracy studies show that both the Jameson and Ni schemes are second-order accurate for smooth solutions on smooth quadrilateral meshes. The question is whether the accuracy degenerates to first order on triangular meshes and irregular quadrilateral meshes.

A numerical study was performed to determine the accuracy of the schemes and the effect of the numerical smoothing to try to answer this question. For inviscid flow, there should be no total pressure loss for smooth, subsonic flow. Thus, for such flows any total pressure loss is purely numerical in nature. With this in mind, the total pressure loss is defined as the error in subsonic flow with a 0.5 inlet Mach number through a duct with a $\sin^2 x$ bump on the lower surface. This geometry is shown in Figure 2 with Mach number and % total pressure loss contours for the Ni scheme on an irregular triangular mesh. The global error is defined as the root mean square of the total pressure loss. For various meshes on which a typical mesh length h was varied the error was calculated and the order of accuracy for the scheme was found by plotting $\log(h)$ verses $\log(error)$ and finding the slope of the resulting line. An example of this curve fitting is shown in Figure 3 with the mesh used for the computations in Figure 2 indicated. This irregular triangular mesh is shown in Figure 4.

In Table 1 the results of the numerical accuracy study are summarized. It is interesting to note that the absolute value of the error is on the same order of magnitude for the cases shown in Table 1. For the high-accuracy numerical smoothing, both the Ni and Jameson schemes for both quadrilateral and triangular meshes are second order accurate. This applies for both regular and irregular meshes. The low-accuracy numerical smoothing does not produce second order accuracy on either a regular or irregular, quadrilateral or triangular mesh. The decrease in accuracy for a regular mesh is due to dissipative errors caused by enforcing $\frac{\partial u}{\partial n} = 0$ in the second difference operator, which is an improper boundary condition for the base solver. This causes a numerical boundary layer to be formed which is at least one cell wide causing a decrease in accuracy. This effect is further discussed by Lindquist in [6].

Conclusions

The accuracy of node-based Ni and Jameson algorithms on both quadrilateral and triangular meshes are evaluated numerically for a subsonic test case. The results show that all four algorithms can indeed be second order accurate, but that extreme care must be taken when implementing numerical smoothing so the accuracy of the scheme is not reduced by smoothing which is less accurate than the flux calculation. The triangular schemes perform as well as the quadrilateral schemes, and offer flexibility in mesh generation and adaptation.

References

[1] Roe, P., *Error Estimates for Cell-Vertex Solutions of the Compressible Euler Equations*, ICASE Report No. 87-6, 1987.

[2] Giles, M. B., *Accuracy of Node-Based Solutions on Irregular Meshes.* 11th International Conference on Numerical Methods in Fluid Dynamics, June 1988.

[3] Jameson, A., *Current Status and Future Directions of Computational Transonics*, Computational Mechanics-Advances and Trends, ASME Publication AMD 75, 1986.

[4] Mavriplis, D. and Jameson, A., *Multigrid Solution of the Two-Dimensional Euler Equations on Unstructured Triangular Meshes*, AIAA-87-0353, January 1987.

[5] Ni, R.-H., *A Multiple-Grid Scheme for Solving the Euler Equations*, AIAA 181-025R, June 1981.

[6] Lindquist, D. R. *A Comparison of Numerical Schemes on Triangular and Quadrilateral Meshes.* SM thesis, Massachusetts Institute of Technology, May 1988.

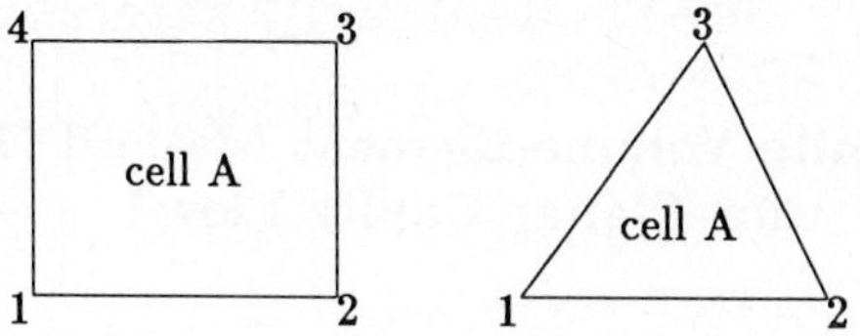

Figure 1: Typical quadrilateral and triangular cells

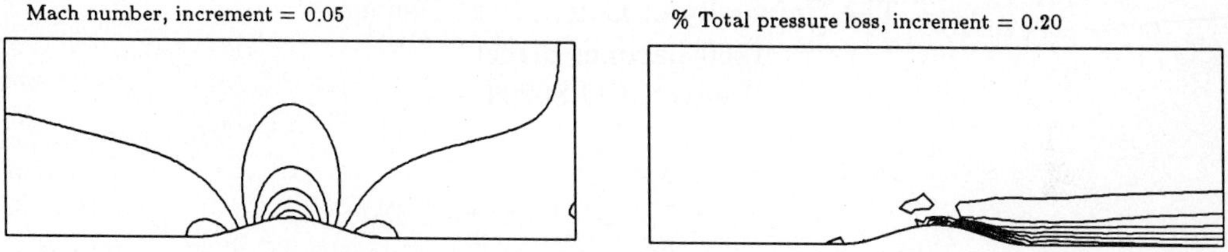

Figure 2: Mach number and % total pressure loss contours for $\sin^2 x$ duct

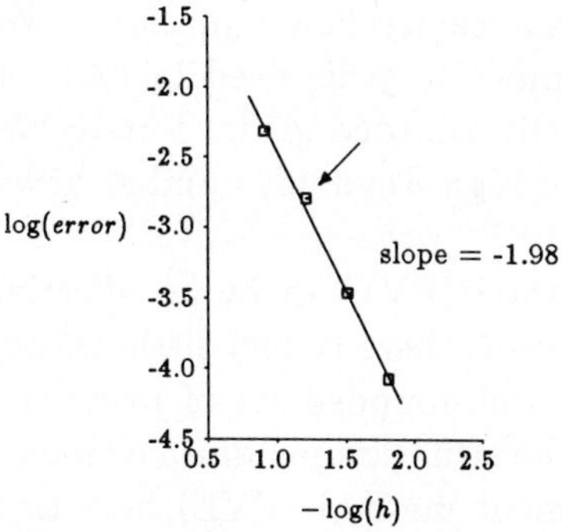

Figure 3: Accuracy data for triangular Ni scheme with high-accuracy smoothing on an irregular mesh

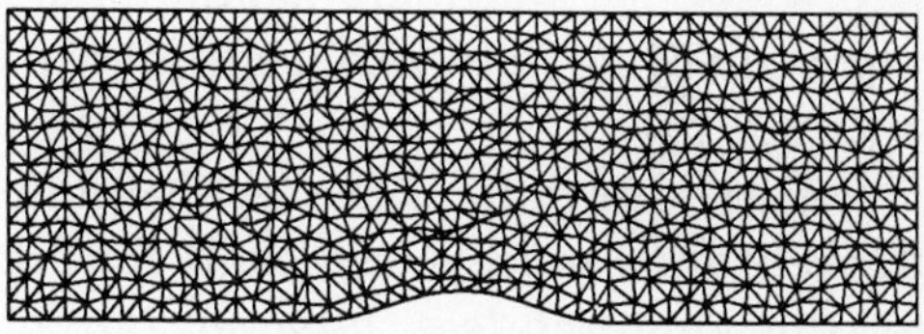

Figure 4: Irregular triangular mesh with 16 faces per unit length for $\sin^2 x$ duxt

Case	Order of Accuracy Regular Mesh	Order of Accuracy Irregular Mesh
Triangular Ni, high-accuracy smoothing	2.33	1.98
Quadrilateral Ni, high-accuracy smoothing	2.04	2.17
Triangular Jameson, high-accuracy smoothing	1.95	2.01
Quadrilateral Jameson, high-accuracy smoothing	2.01	2.05
Triangular Jameson, low-accuracy smoothing	1.55	1.69
Quadrilateral Jameson, low-accuracy smoothing	1.52	1.64

Table 1: Numerical accuracy results

The Finite Volume-Element Method (FVE) for Planar Cavity Flow*

C. LIU AND S. MCCORMICK
Computational Mathematics Group, Campus Box 170
The University of Colorado at Denver
1200 Larimer Street
Denver, CO 80204

Abstract. In this paper, we introduce a discretization technique, called the finite volume-element method (FVE), which combines finite volumes (cf. [2]) and finite elements (cf. [1]) into a general approach for problems posed in conservative form. We illustrate its use and analyze its accuracy by combining FVE with multigrid to solve incompressible potential flow and planar cavity flow equations. We then use this approach for discretization on the composite grids used in FAC [4] and AFAC [5], which are multilevel methods for locally adapted grids. Finally, we report on some tests with a special version of FVE for high Reynolds number flows.

Introduction. The classical finite volume method (FV) can be an effective discretization process, especially for fluid flows. However, there is very little theory for this approach (see, however, [2]) and there is no well-founded set of principles for guiding FV implementation. Contrast this with the situation for finite elements (FE). (See [1].) We introduce the finite volume-element method (FVE) here as a first step in overcoming these deficiencies.

The basic idea behind FVE is to use certain aspects of both the FV and FE discretization principles: FE is used to discretize the function space by providing an element partition of the computational domain, $\overline{\Omega}$, and restricting the functions to be piecewise polynomials with respect to their elements; FV is used to discretize the equations by specifying a fixed volume partition of $\overline{\Omega}$; and these approaches are combined by replacing the surface fluxes of the continuous solution by the surface fluxes of the discrete element functions.

This paper is a shortened version of an earlier report available at the University of Colorado at Denver. Subsequent to the earlier version, we discovered the method 'control volume finite elements' (cf. [3]) which takes the same approach on which

*This work was supported by the Air Force Office of Scientific Research under grant number AFOSR-86-0126 and the National Science Foundation under grant NSF DMS-8704169.

we base FVE. However, in our use of FVE we more closely adhere to the principles that found it (e.g., we treat systems of equations in a unified way). This ensures greater accuracy, which is especially evident from our results at high Reynolds (Re) numbers.

II. Driven Cavity Flow Discretization. The governing equations for a driven cavity may be written in the following vorticity-stream function form:

$$(\psi_y \xi)_x - (\psi_x \xi)_y - \frac{1}{Re}(\xi_{xx} + \xi_{yy}) = 0 \qquad (1)$$
$$\psi_{xx} + \psi_{yy} + \xi = 0,$$

where ψ is the stream function and ξ is the vorticity. We take these equations to be defined on the unit square Ω with the following boundary conditions:

$$\begin{array}{lll} \psi = \psi_x = 0 & \text{at} & x = 0 \text{ or } 1 \\ \psi = \psi_y = 0 & \text{at} & y = 1 \\ \psi = 0, \psi_y = 1 & \text{at} & y = 0. \end{array} \qquad (2)$$

The basic idea behind FVE starts with the integral form of (1) over a given control volume, V, which for the first equation is

$$\int_S \tilde{n} \cdot \left[\begin{pmatrix} \psi_y \xi \\ -\psi_x \xi \end{pmatrix} - \frac{1}{Re} \nabla \xi \right] dS = 0, \qquad (3)$$

where S is the surface of V and $\tilde{n}$ is the outward unit normal to S. Using the triangulation depicted in Fig. 1, we then simply replace the unknowns in (3) by piecewise linear functions on this triangulation. We do this for the second equation in (1) as well.

III. Multigrid. For the driven cavity equations discretized by FVE, we thus have two algebraic equations at every grid point. To avoid the poor smoothing rate that point relaxation exhibits, we use collective relaxation. This technique solves both equations simultaneously to update the approximation at each stage of a full sweep through the grid points. Special care is needed at the boundary with this scheme. Finally, we use the nonlinear (FAS) version of multigrid.

IV. Computational Results for Uniform Grids at Moderate Re. We computed four low Re number flows (Re = 0, 10, 50, and 100) using 17×17 and 33×33 meshes. The convergence histories are illustrated in Table 1, showing that good efficiency is obtained in every case. In fact, the residual reduction factor for each V-cycle averages less than 0.1. The maximum stream function values and their positions are listed in Table 2, showing that the computational results are quite dependable (cf. [6]).

V. Computational Results for Composite Grids. The FVE method provides an effective scheme for discretization on the adapted grids used by FAC [4] and AFAC [5]. To illustrate this, we again use a basic 17×17 or 33×33 grid on $\overline{\Omega}$, then add a local grid with the same number of points in the corner $\overline{\Omega}_f = [\frac{1}{2}, 1] \times [\frac{1}{2}, 1]$. This resolves the secondary vortex in the flow field. Tables 3 and 4 depict the convergence bahavior of these methods for $Re = 0$ and 100, respectively. Here, 'resid' refers to the Euclidean norm of the residual errors on the composite grid, 'factor' to the ratio of successive values of 'resid,' and 'multiplication average' to the seventh root of the ratio of the initial and final values of 'resid.'

VI. High Reynolds Number Flow. For high Re, in place of piecewise linear basis functions, we use the following basis functions on each of the triangles depicted in Fig. 1:

$$\psi = ax + by + c,$$
$$\xi = e^{-(2Ax+2By)}(\alpha x + \beta y + \gamma).$$

A and B are determined by

$$A = \frac{Re \cdot u}{2}, \quad B = \frac{Re \cdot v}{2}$$

where u and v are local values of ψ_y and ψ_x, respectively. Table 5 shows the results for a 65×65 grid using $Re = 400$ and 1000. These results compare favorably to the finer grid results of [7].

References

[1] I. Babuska, J. Chandra and J. Flaherty, *Adaptive Computational Methods for Partial Differential Equations*, SIAM, Philadelphia, 1983.

[2] A. Samarskii, R. Lazaroff and L. Makarov, *Finite difference schemes for differential equations with weak solutions*, Moscow Vissaya Skola, 1987.

[3] B. R. Baliga and S. V. Patankar, *A new finite element formulation for convection-diffusion problems*, Num. Heat Transfer 3 (1980), 393–409.

[4] S. McCormick, *Fast adaptive composite grid (FAC) methods: theory for the variational case*, in: Defect Correction Methods: Theory and Applications (K. Bohmer and H. J. Stetter, eds.), Comp. Suppl. 5 (1984), 115–122.

[5] L. Hart and S. McCormick, *Asynchronous multilevel adaptive methods for solving partial differential equations on multiprocessors*, Lecture Notes in Pure and Applied Mathematics/110, Marcel-Dekker, May, 1988.

[6] R. D. Mills, J. R. Aeronaut. Soc. 69 (1965), 116–126.

[7] V. Ghia, K. N. Ghia and C. T. Shin, *High-Re solutions for incompressible flow using the Navier-Stokes equations and a multigrid method*, S. Comp. Phys. 48 (1982), 387–411.

Table I. Convergence History.

Cycle	Work	Mesh 17 x 17				Mesh 33 x 33			
		Re=0	Re=10	Re=50	Re=100	Re=0	Re=10	Re=50	Re=100
1	6.875	0.16EO	0.15EO	0.13EO	0.22EO	0.32EO	0.30EO	0.24EO	0.26EO
2	13.750	0.12E-1	0.11E-1	0.11E-1	0.18E-1	0.32E-1	0.29E-1	0.27E-1	0.31E-1
3	20.625	0.48E-3	0.44E-3	0.71E-3	0.27E-2	0.14E-2	0.13E-2	0.14E-2	0.23E-2
4	27.500	0.44E-4	0.36E-4	0.45E-4	0.64E-3	0.20E-3	0.19E-3	0.12E-3	0.12E-3
5	34.375	0.30E-5	0.24E-5	0.39E-5	0.13E-4	0.24E-4	0.24E-4	0.17E-4	0.10E-4
6	41.250	0.16E-6	0.13E-6	0.45E-6	0.27E-5	0.12E-5	0.12E-5	0.85E-6	0.13E-5

Table II. Computational Results (Maximum stream function values and their positions)

Mesh	17 x 17				33 x 33			
Re	0.0	10.0	50.0	100.0	0.0	10.0	50.0	100.0
ψ_{max}	0.0989	0.0987	0.0983	0.0977	0.0995	0.0996	0.100	0.102
ξ_c	2.97	2.99	3.00	3.12	3.02	3.02	3.15	3.19
X_c	0.500	0.500	0.563	0.625	0.500	0.516	0.578	0.610
Y_c	0.250	0.250	0.251	0.255	0.250	0.250	0.251	0.265

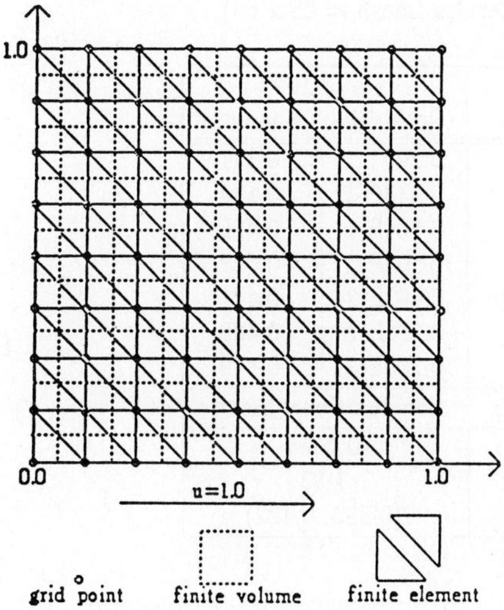

Fig. 1. Driven Cavity Flow

Table III. FAC and AFAC for Re = 0.0.

FAC Cycle	Work Unit	17x17 Residual	17x17 Factor	33x33 Residual	33x33 Factor
1	5.562	0.142E-2	0.151	0.578E-3	0.156
2	11.125	0.227E-3	0.160	0.976E-4	0.169
3	16.688	0.349E-4	0.154	0.135E-4	0.138
4	22.250	0.532E-5	0.152	0.200E-5	0.149
5	27.812	0.804E-6	0.151	0.311E-6	0.155
6	33.375	0.121E-6	0.151	0.482E-7	0.155
7	38.937	0.182E-7	0.150	0.690E-8	0.143
Multiplication Average			0.153		0.152
AFAC Cycle	**Work Unit**	**17x17 Residual**	**17x17 Factor**	**33x33 Residual**	**33x33 Factor**
1	5.562	0.268E-3	0.0518	0.244E-3	0.0953
2	11.125	0.596E-3	2.224	0.283E-3	1.158
3	16.688	0.295E-4	0.0495	0.250E-4	0.0884
4	22.250	0.663E-4	2.250	0.327E-4	1.310
5	27.812	0.311E-5	0.0468	0.271E-5	0.0828
6	33.375	0.730E-5	2.351	0.365E-5	1.344
7	38.937	0.324E-6	0.0444	0.286E-6	0.0785
Multiplication Average			0.251		0.273

Table IV. FAC and AFAC for Re = 100.0.

FAC Cycle	Work Unit	17x17 Residual	17x17 Factor	33x33 Residual	33x33 Factor
1	5.562	0.138E-1	0.177	0.605E-2	0.199
2	11.125	0.452E-2	0.327	0.116E-2	0.192
3	16.688	0.769E-3	0.170	0.206E-3	0.178
4	22.250	0.179E-3	0.232	0.319E-4	0.155
5	27.812	0.350E-4	0.196	0.528E-5	0.165
6	33.375	0.765E-5	0.218	0.764E-6	0.145
7	38.937	0.151E-5	0.197	0.122E-6	0.160
Multiplication Average			0.212		0.169
AFAC Cycle	Work Unit	17x17 Residual	17x17 Factor	33x33 Residual	33x33 Factor
1	5.562	0.101E-1	0.129	0.513E-2	0.169
2	11.125	0.270E-2	0.267	0.852E-3	0.166
3	16.688	0.121E-2	0.447	0.342E-3	0.402
4	22.250	0.192E-3	0.159	0.507E-4	0.148
5	27.812	0.153E-3	0.796	0.313E-4	0.618
6	33.375	0.145E-4	0.0952	0.383E-5	0.122
7	38.937	0.309E-4	2.128	0.339E-5	0.885
Multiplication Average			0.326		0.273

Table V. Convergence History and Results (mesh = 65 x 65).

FAC Cycle	Work Unit	Re = 400 Residual	Re = 400 Factor	Re = 1000 Residual	Re = 1000 Factor
1	5.64	0.622E-1	0.132	0.297	0.321
2	11.28	0.141E-1	0.226	0.338E-1	0.114
3	16.92	0.244E-2	0.173	0.164E-1	0.485
4	22.56	0.500E-3	0.214	0.104E-1	0.637
5	28.20	0.127E-3	0.244	0.493E-2	0.471
6	33.84	0.443E-4	0.349	0.236E-2	0.480
7	39.48	0.149E-4	0.336	0.106E-2	0.449
		Computational results		Computational results	
ψ_{max}		0.108		0.0944	
ξ_c		2.10		1.83	
(x_c, y_c)		(0.563, 0.391)		(0.538, 0.422)	

ADAPTIVE REMESHING FOR TRANSIENT PROBLEMS WITH MOVING BODIES

Rainald Löhner

Berkeley Research Associates, Springfield, VA 22150, USA
and
Laboratory for Computational Physics and Fluid Dynamics
Naval Research Laboratory, Washington, D.C. 20375, USA

The accurate simulation of time-dependent compressible flows with moving bodies has been an outstanding goal for a long time [1-3]. Several major capabilities are required to simulate efficiently this class of problems: a) Flow solvers that can handle moving frames of reference, b) high-order, monotonicity preserving schemes, c) a consistent integration scheme for the rigid body motion, and d) fast regridding capabilities. It is also convenient to incorporate an adaptive refinement capability for the flow domain in order to reduce the overall CPU-cost of the algorithm.

The Equations of Motion for the Fluid

As portions of the mesh are allowed to move in time, we need to recast the PDEs describing the fluid. This is most easily accomplished by the Arbitrary Lagrangian-Eulerian (ALE) formulation. Given the velocity field $\mathbf{w}$ for the elements

$$\mathbf{w} = (w^x, w^y) \quad , \tag{1}$$

the Euler equations that describe an inviscid, compressible fluid may be written [4] as

$$\begin{Bmatrix} \rho \\ \rho u \\ \rho v \\ \rho e \end{Bmatrix}_{,t} + \begin{Bmatrix} (u-w^x)\rho \\ (u-w^x)\rho u + p \\ (u-w^x)\rho v \\ (u-w^x)\rho e + up \end{Bmatrix}_{,x} + \begin{Bmatrix} (v-w^y)\rho \\ (v-w^y)\rho u \\ (v-w^y)\rho v + p \\ (v-w^y)\rho e + vp \end{Bmatrix}_{,y} = -\nabla\cdot\mathbf{w}\begin{Bmatrix} \rho \\ \rho u \\ \rho v \\ \rho e \end{Bmatrix} \quad . \tag{2}$$

The Flow Solver (FEM-FCT)

For the compressible flows described by Eqn.(2), discontinuities in the variables may arise (e.g., shocks or contact discontinuities). As the flow solver is explicit, the allowable timestep is directly proportional to the element size. Therefore, for this class of transient problems, it pays off immensely to devise flow solvers that can capture shocks over a minimum number of elements. In the present case this is accomplished by combining, in a conservative manner, a high-order scheme from the Lax-Wedroff class with a low-order scheme [5]. The temporal discretization of Eqn.(2) yields

$$U^{n+1} = U^n + \Delta U \quad , \tag{3}$$

where ΔU is the increment of the unknowns obtained for a given scheme at time $t = t^n$. Our aim is to obtain a ΔU of as high an order as possible without introducing overshoots. To this end, we rewrite Eqn.(7) as:

$$U^{n+1} = U^n + \Delta U^l + lim(\Delta U^h - \Delta U^l). \tag{4}$$

Here ΔU^h and ΔU^l denote the increments obtained by the high- and low-order scheme respectively. Further details on the limiting procedure, its algorithmic implementation, and the high- and low-order schemes employed may be found in [5].

The Equations of Motion for the Rigid Bodies

The movement of rigid bodies can be found in standard textbooks on classical mechanics. See, for example [6]. The situation under consideration is shown in Figure 1. Given the position vector of any point of the body

$$\mathbf{r} = \mathbf{r}_c + \mathbf{r}_0 \ , \tag{5}$$

the velocity and acceleration of this point will be

$$\dot{\mathbf{r}} = \dot{\mathbf{r}_c} + \dot{\mathbf{r}_0} = \mathbf{v}_c + \omega \times \mathbf{r}_0 \ , \quad \ddot{\mathbf{r}} = \dot{\mathbf{v}_c} + \dot{\omega} \times \mathbf{r}_0 + \omega \times (\omega \times \mathbf{r}_0) \ . \tag{6}$$

With the following abbreviations

$$m = \int_\Omega dm = \int_\Omega \rho d\Omega \ , \quad I_{ij} = \int_\Omega r_0^i r_0^j \rho d\Omega \ , \quad \Theta = \begin{Bmatrix} I_{yy} + I_{zz} \ , & -I_{xy} \ , & -I_{xz} \\ -I_{xy} \ , & I_{xx} + I_{zz} \ , & -I_{yz} \\ -I_{xz} \ , & -I_{yz} \ , & I_{xx} + I_{yy} \end{Bmatrix} \ ,$$

we then have the following equations describing balance of forces and moments:

$$m\dot{\mathbf{v}}_c = \sum \mathbf{F} = m\mathbf{g} - \int_\Gamma p\mathbf{n} d\Gamma \ , \tag{7}$$

$$\Theta\dot{\omega} + \int_\Omega (\omega \cdot \mathbf{r}_0) \cdot (\mathbf{r}_0 \times \omega) d\Omega = \sum \mathbf{r}_0 \times \mathbf{F} = -\int_\Gamma p\mathbf{r}_0 \times \mathbf{n} d\Gamma \ . \tag{8}$$

In order to update the velocities and positions of the bodies in time, we employ an explicit time-marching scheme. This seems reasonable, as in practical calculations the time-scales of the body-movement are much larger than those associated with the fluid flow.

Adaptive Remeshing

As the bodies interacting with the flowfield may undergo arbitrary movement, a fixed mesh structure will lead to badly distorted elements. This means that at least a partial regeneration of the computational domain will be required. The idea is to regenerate the whole computational domain adaptively, taking into consideration the current flowfield solution. In particular, we employ the advancing-front technique [7,8] in conjuction with appropriate error-indicators [9]. It was found that a significant speed-up could be achieved by first generating a coarser, but stretched mesh, and then refining globally this mesh with classic h-refinement.

The Overall Algorithm

The for the advancement of the solution in time looks as follows:

O.1 Advance the solution one timestep

- A.1 Compute the body forces and moments from the pressure field and any exterior forces.
- A.2 Taking into consideration the kinematic contraints for the body movements, update the velocities of the bodies at $t = t^{n+1}$: $\mathbf{v}_c^{n+1}, \omega^{n+1}$. At the same time, obtain the average velocities $\mathbf{v}_c^{av}, \omega^{av}$ for the time-interval $[t^n, t^{n+1}]$.
- A.3 With the average velocities $\mathbf{v}_c^{av}, \omega^{av}$, obtain the velocities $\mathbf{w}_\Gamma$ on the surface of each body for the time-interval $[t^n, t^{n+1}]$.

A.4 Given the surface velocities $\mathbf{w}_\Gamma$ on the boundaries of the global domain, obtain the global velocity field $\mathbf{w}_\Omega$ for the element movement.

A.5 Advance the solution by one timestep using the ALE-FEM-FCT solver.

A.6 Given the actual timestep Δt^n and the velocity field for the element movement $\mathbf{w}_\Omega$, update the coordinates of the points.

A.7 Update the geometric parameters for the elements that have been moved.

A.8 Update the centers of mass $\mathbf{r}_c$ for the bodies, as well as the coordinates of the points defining the body geometry.

O.2 If the grid has become too distorted close to the moving bodies or if the desired number of timesteps between refinements has elapsed: adaptively remesh the computational domain.

R.1 Obtain the error indicator matrix for the gridpoints of the present grid.

R.2 Given the error indicator matrix, get the element size, element stretching and stretching direction for the new grid.

R.3 Using the old grid as the 'background grid', remesh the computational domain using the advancing front technique.

R.4 If further levels of global h-refinement are desired: refine the new grid globally.

R.5 Interpolate the solution from the old grid to the new one.

O.3 If the desired time-interval has elapsed: Stop. Otherwise, advance the solution further (go to O.1)

Numerical Examples

Object Falling into Supersonic Free Stream:
The problem statement is as follows: an object is placed in a cavity surrounded by a free stream at $M_\infty = 1.5$. After the steady-state solution is reached (time T=0.0), a body motion is prescribed, and the resulting flowfield disturbance is computed. Adaptive remeshing was performed every 100 timesteps initially, while at later times the grid was modified every 50 timesteps. No subsequent h-refinement was used in this case, as the mesh adaptation process is not done very often. The maximum stretching ratio specified was $S = 5.0$. Figures 1a,b show two stages during the computation at times T=20 and T=175. One can see how the location and strength of the shocks changes due to the motion of the object. Notice how the directionality of the flow features is reflected in the mesh.

Shock-Box Interaction:
The problem statement is shown schematically in Figure 2a. A very strong shock ($M_s = 10.0$) collides with a box. The ensuing sharp pressure rise on the surface of the box generates a strong horizontal force. The body tilts on account of it. The kinematic conditions for the body movement were set as follows: until the y-velocity at point A (see Figure 2a) reaches a positive value, the position of this point is assumed fixed. As a result of this constraint, the box rotates around point A. After reaching a positive y-velocity at point A, the position of point A is no longer considered fixed. Thus, the box is 'released'. Adaptive remeshing was performed every 10 timesteps throughout the computation. The maximum stretching ratio specified was $S = 5.0$, and one level of global h-refinement was used. Figures 2b,c show two stages of the simulation at times T=0.4 and T=3.2. Observe again the change in strength and position of the shocks due to body motion. Figure 2d shows the trajectory of the box in time.

Acknowledgements
This work was funded by the Office of Naval Reseach through the Naval Research Laboratory, the Applied Computational and Mathemetics Program of DARPA, and the NASA Johnson Space Center.

References

[1] J.A. Benek, J.L. Steger and F.C. Dougherty; AIAA-83-1944-CP (1983).

[2] F.C. Dougherty, J.A. Benek and J.L. Steger; NASA Ames TM 88193 (1985).

[3] L. Formaggia, J. Peraire and K. Morgan - *Appl. Math. Mod.* 12, 175-181 (1988).

[4] J. Donea - *Comp. Meth. Appl. Mech. Eng.* 33, 689-723 (1982).

[5] R. Löhner, K. Morgan, J. Peraire and M. Vahdati - *Int. J. Num. Meth. Fluids* 7, 1093-1109 (1987).

[6] A. Sommerfeld - *Vorlesungen über Theoretische Mechanik*; Verlag Harri Deutsch, Frankfurt (1976).

[7] J. Peraire, M. Vahdati, K. Morgan and O.C. Zienkiewicz - *J. Comp. Phys.* 72, 449-466 (1987).

[8] R. Löhner - *Comm. Appl. Num. Meth.* 4, 123-135 (1988).

[9] R. Löhner - *Comp. Meth. Appl. Mech. Eng.* 61, 323-338 (1987).

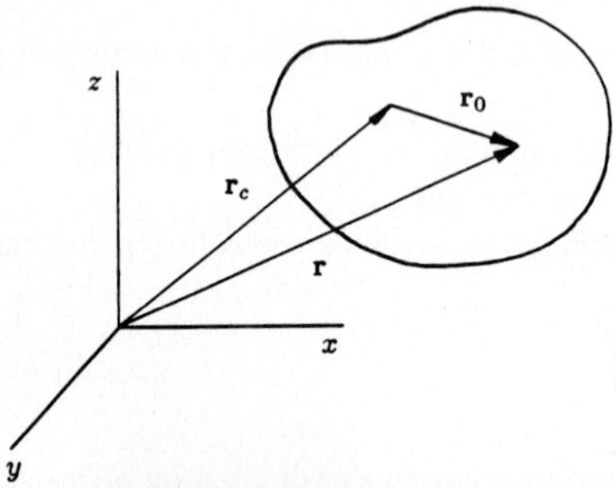

Figure 1: Kinematics of Rigid Body Motion

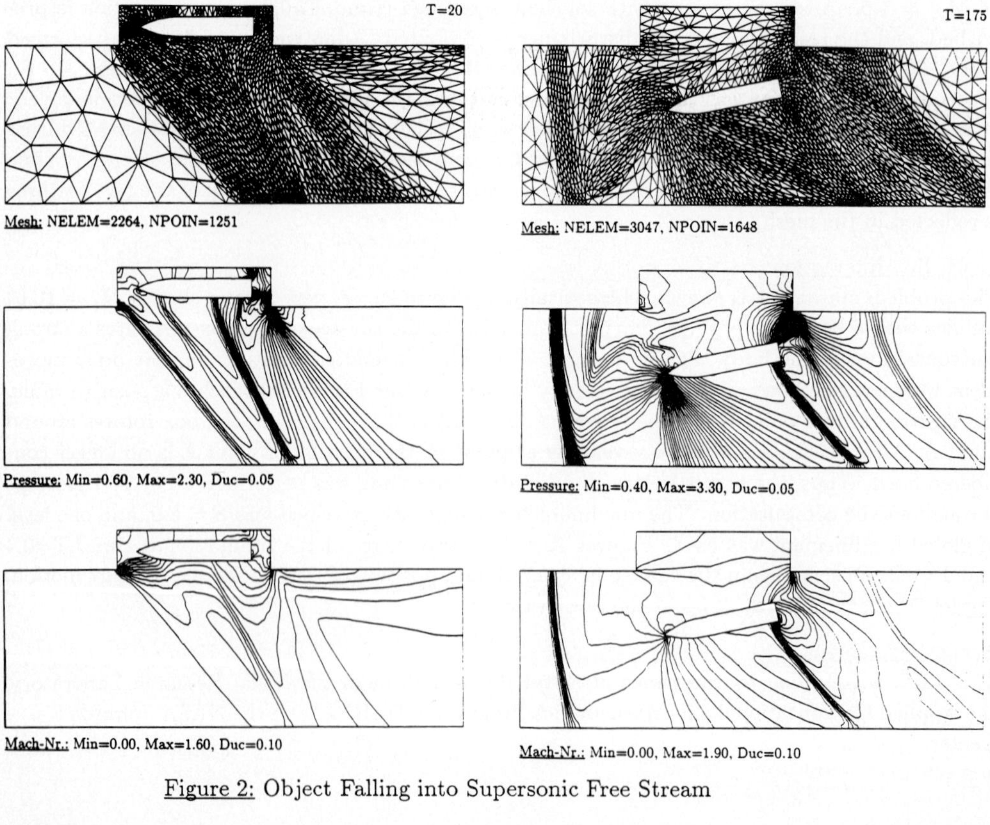

Figure 2: Object Falling into Supersonic Free Stream

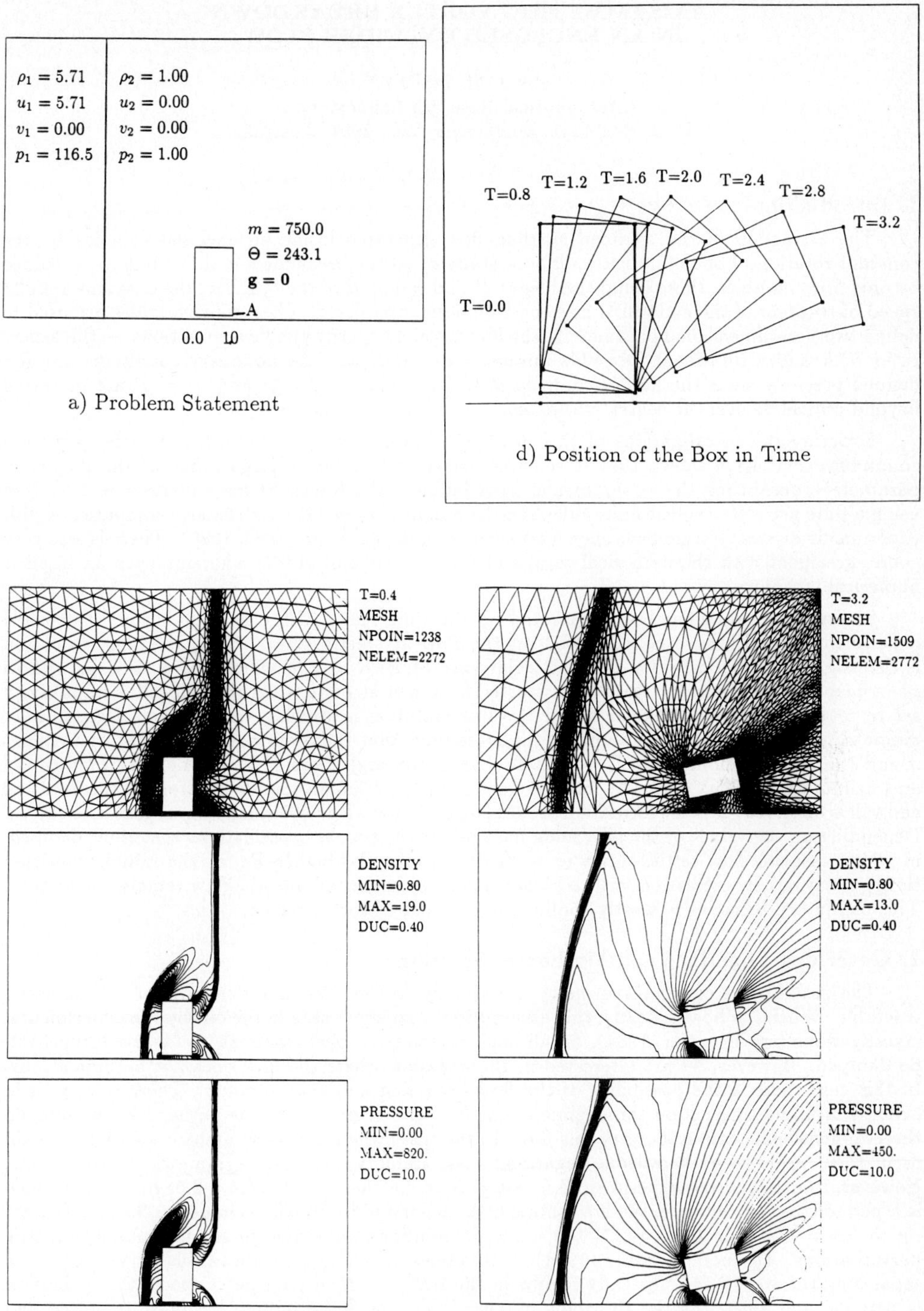

Figure 3: Shock-Box Interaction

AXISYMMETRIC VORTEX BREAKDOWN IN AN ENCLOSED CYLINDER FLOW.

by *J. M. LOPEZ**
Aeronautical Research Laboratory,
P. O. Box 4331, Melbourne, Vic., 3001, Australia.

1. Introduction

The attraction of the confined swirling flow generated inside an enclosed cylinder by the constant rotation of one of its endwalls to a study of vortex breakdown is that the flow is defined by only four variables, these being the height H and radius R of the cylinder, the constant angular speed of rotation of the endwall Ω and the kinematic viscosity ν. These four variables are used to define two non-dimensional parameters which completely specify the flow conditions — the aspect ratio H/R and a rotational Reynolds number $Re = \Omega R^2/\nu$. The boundary conditions are also defined precisely since the flow is confined in a fixed volume. As Re and/or H/R are increased beyond critical values, the central vortex core undergoes breakdown.

Experimental investigations of this flow (eg. Vogel 1968, Escudier 1984, Roesner - private communication) have shown that it remains axisymmetric over a large range of the governing parameters, permitting the axisymmetric formulation of the Navier-Stokes equations to be solved using a time-accurate explicit finite-difference technique (Lopez 1987). Extensive comparisons with experiments at steady state have shown excellent agreement (Lopez 1987, 1988). There is also very good agreement with the numerical results of Lugt & Abboud (1987) who employed an implicit numerical technique.

So far, little work has been presented on the time scales of the flow evolution. Published experimental results have concentrated on the flow at steady state. Here, the time evolution is investigated numerically for the particular case of $H/R = 2.5$, which has a richness of flow phenomena as Re is varied. The flow is started from rest at $t = 0$ when the endwall is impulsively set to rotate at constant angular speed Ω. The evolution is studied with the aid of graphically animated sequences of the contours of the streamfunction ψ, azimuthal component of vorticity η and the angular momentum $\Gamma = rv$. The advection and diffusion of the angular momentum and azimuthal vorticity into the interior flow from the Ekman boundary layer on the rotating endwall is observed, as is the formation of the central vortex which eventually undergoes breakdown. Depending on the values of the governing parameters the transient oscillations are either damped, in which case the flow settles down to a steady state, or for higher Re, as the initial transients slowly decay, a new oscillation with a higher frequency is excited and the flow remains oscillatory. This oscillatory behavior is mostly confined to the central vortex region.

2. Governing Equations and Method of Solution

The assumption of axial symmetry is imposed on the flow in order to keep the numerics tractable. When the flow is steady, this assumption is satisfied, as is borne out by the experimental investigations (eg. Escudier, 1984). Small non-axisymmetric perturbations on these solutions will be damped. However, we are interested in the situation where the flow does not become steady. In this case there is the possibility of the flow becoming non-axisymmetric. There is very little experimental information on this regime of the flow and there are, to the author's knowledge, no three-dimensional computations of this flow. In the cases where the axisymmetric solutions remain oscillatory, we have no knowledge regarding their stability to non-axisymmetric perturbations. However, from Escudier's experimental investigations, for the case of $H/R = 2.5$, $Re = 2765$ which is a periodic oscillatory case, the flow shows little departure from axial symmetry. So that at least up to $Re = 2765$, for $H/R = 2.5$, the unsteady solutions are stable to small non-axisymmetric perturbations. To quote Escudier (1984): "For values of $\Omega R^2/\nu$ just above this curve (the curve separating the steady and unsteady regime in the $\Omega R^2/\nu$ - H/R parameter space) the oscillation is periodic and axial. As the Reynolds number is increased the motion becomes disturbed and ultimately turbulent. For $H/R > 3.1$, the first sign of non-steady motion is a precession of the

* Present address:— NASA Ames Research Center, Moffett Field, CA. 94035, U.S.A.

lower breakdown structure." Hence, for the case of $H/R = 2.5$, there exists a range of Re for which the flow is oscillatory and axisymmetric. Unfortunately, there is no estimate of the Reynolds number at which the flow loses stability to non-axisymmetric perturbations. Fortunately, we can proceed to investigate the transition to unsteadiness for the case of $H/R = 2.5$ using the assumption of axial symmetry, knowing that at least for $Re \leq 2765$, the axisymmetric oscillatory solutions are stable.

The governing equations used are the unsteady axisymmetric Navier-Stokes equations, written in conservative streamfunction-vorticity form, with time and length scaled by Ω and $1/R$ respectively. The method of solution and details of the boundary conditions are given in Lopez (1988), where the accuracy of the method is also discussed.

3. Results and Discussion

When the cylinder aspect ratio is 2.5, for $Re < 1000$, the steady flow consists of an Ekman boundary layer on the rotating endwall pumping fluid radially outward in a spiral path. At the same time, fluid is drawn into the boundary layer from the interior to satisfy continuity. This centrifugal pumping, together with the presence of the cylinder sidewall, establishes a secondary overturning meridional flow which advects fluid from the Ekman layer, where it has acquired angular momentum and total head, into the interior of the flow. The fluid loses a proportion of its acquired angular momentum and total head through the action of viscous stresses in the boundary layers on the cylinder sidewall and on the stationary endwall. The boundary layer on the stationary endwall separates on the axis and forms a central vortex which returns the fluid to the Ekman boundary layer. The structure of this central vortex is primarily determined by the structure of the stationary endwall boundary layer, which for $Re < 1000$ is broad and diffuse. At these low Reynolds numbers, the central vortex is essentially cylindrical and its associated vorticity vector is primarily directed in the axial direction, with some azimuthal component due to the secondary overturning motion.

When Re is increased to 1600, a new feature is present in the steady flow field. The streamlines on the central vortex now show the features of a weak standing centrifugal wave, as can be seen in figure 1. The features and mechanics responsible for this behavior have been discussed at length in Lopez (1988) and Brown & Lopez (1988). In summary, at the higher Reynolds numbers, the structure of the stationary endwall boundary layer has changed such that its depth is reduced and its flow has larger gradients of angular momentum Γ, and total head H. The distribution of Γ and H in the central vortex, as it emerges from the stationary endwall boundary layer, is such that a region of azimuthal vorticity is generated by the tilt and stretch mechanism described in Brown & Lopez (1988), which induces a reversed axial velocity and an undulation in the central streamtubes. As Re is increased, this effect is enhanced and the induced reversed axial velocity is large enough to stagnate the axial flow. At $Re = 1918$, a small spherical recirculation bubble is evident on the axis. Figure 1 illustrates the changes in flow structure at steady state with Re. The direction of the meridional flow in figure 1 is downward. Of particular interest is the trend that the wavelength of the undulations is decreasing with increasing Re whilst their amplitudes increase and the stagnation point on the axis occurs further upstream. Also, the time taken to reach steady state increases with Re. Figure 2 illustrates the time history of the streamfunction at a particular point in the flow ($r = R/6$, $z = 2H/3$). For $Re = 2200$, the flow settles down to a steady state after approximately 700 rotations of the bottom endwall, whereas for $Re = 2400$ it took about 1200 rotations and about 2000 rotations for $Re = 2500$. For $Re = 2600$, the system was integrated out to 3000 rotations and there was still evidence of oscillations, however these were slowly damped. Up to this value of Re, Escudier (1984) reports that the experimental flow also settles down to a steady state and, provided $\Omega R^2/\nu$ is well below the value at which the flow becomes oscillatory (which for $H/R = 2.5$ is approximately 2600), the flow reached steady state when started from rest in ten's of seconds. In Escudier's experiment, Ω was typically of the order of ten revolutions per second. As the Reynolds number for unsteadiness was approached, Escudier reports that the time taken to reach steady state became very long. This is precisely the behavior indicated in figure 2.

From figures 1 and 2, we note that persistent oscillations, both damped and undamped, are associated with a bifurcation in the flow streamlines. For $Re < 2400$, the flow field at steady state consists of two individual recirculation bubbles located on the axis of symmetry. The vortical flow between the two bubbles is similar in nature to that upstream of the leading bubble, however it has a larger vortex core and is less intense. This reconverged vortical flow posses a distribution of Γ and H which is responsible for the presence of the second bubble via the mechanism described earlier. However, when Re is increased beyond 2400, the two bubbles coalesce and the vortical flow past the leading bubble does not have a chance to reconverge fully. For $2400 < Re < 2600$, the coalescing of the two bubbles is not very great. There is still a considerable amount of reconvergence past the first bubble and the flow is able to find a steady equilibrium configuration after some time. For higher Re (eg. $Re = 2765$), no such steady equilibrium is found. In this case, the boundary layer flow on the endwalls and sidewall reaches steady state relatively quickly and the emergent central vortex undergoes breakdown further upstream than it does at lower Re, following the trend shown in figure 1. The resultant recirculation bubble is larger, as is the second bubble which forms in closer proximity to the leading bubble. This larger second bubble causes the vortical flow to diverge at a much larger r than it would have in the absence of the second bubble. The result is a very broad, near stagnant central vortex flow past the leading bubble. The subsequent distribution of Γ and H is not conducive to breakdown and hence the second recirculation bubble cannot be maintained and is seen to be washed downstream. As it is washed downstream, the vortical flow past the leading bubble is able to re-converge in the absence of the second bubble. The leading bubble is now closed at the downstream end by a stagnation point on the axis and the central vortex has a similar structure to that upstream of the leading bubble. The flow topology resembles that of lower Re cases where two individual bubbles are present. However, the reconverged central vortex is prone to a more pronounced breakdown and it does so in a manner which results in a re-coalescence of the two bubbles and the whole process repeats itself. The frequency of this oscillatory behavior is found from the power spectra in figure 3 and is f1. The oscillation, due to the interaction between the two bubbles, occurs approximately 27 times every 1000 rotations of the bottom endwall. It appears that the non-dimensional frequency f1, is independent of Re for the range considered (i.e. $2600 \leq Re \leq 4000$).

From the power spectra shown on the left hand side of figure 3, the transient oscillations due to the impulsive start have a frequency f0 which is also independent of Re. Only the rate at which it is damped is Re dependent. At $Re = 4000$, there is still a weak signal due to f0 in the spectra taken for $2000 \leq t \leq 3000$. For $Re < 3500$, the f1 oscillation is the predominant mechanism responsible for the unsteadiness. However, at $Re = 3000$, there is a faint signal f2 evident at late times and at $Re = 3500$ this f2 signal is present at early times as well. At late times for $Re = 3500$, the power spectra clearly shows the non-linear interaction between the two incommensurate frequencies f1 and f2 and the corresponding time-series from figure 2 shows this situation as an almost chaotic signal. At $Re = 4000$, this new f2 frequency dominates the time dependence of the flow solution, although there are still faint signals at f0 and f1 which have not completely died out at those times.

An examination of the streamfunction at various times allows an identification of the mechanism responsible for this new f2 frequency which emerges at higher Re. At these higher Re, the central vortex undergoes vortex breakdown even closer to the stationary endwall, following the trend illustrated in figure 1. Also, the recirculation bubble is larger, as are the recirculating velocities inside the bubble. The effect of having this active recirculating bubble in such close proximity to the endwall is to increase the thickness of the endwall boundary layer. A particle of fluid which would have spiraled inward to a very small r in the absence of the bubble, now is diverted at a larger r and the result is a broader and less intense central vortex than would have been expected at the high Re values. Now, with a weakened central vortex, the active recirculation bubble cannot be maintained and it is washed downstream. In the absence of the bubble, the endwall boundary layer 'heals' and once again separates at $r = 0$ to form a tight intense central vortex which undergoes a violent breakdown and results in a large recirculation bubble. The whole process repeats itself with a frequency f2.

References

Brown, G. L. & Lopez, J. M. 1988 *A.R.L. Aero. Report 174, AR-004-573, also submitted to J. Fluid Mech.*

Escudier, M. P. 1984 *Experiments in Fluids* **2**, 189–196.

Lopez, J. M. 1987 *in press: Proc. Int. Sym. Comp. Fluid Dynamics (North Holland).*

Lopez, J. M. 1988 *A.R.L. Aero. Report 173, AR-004-572, also submitted to J. Fluid Mech.*

Lugt, H. J. & Abboud, M. 1987 *J. Fluid Mech.* **179**, 179–200.

Vogel, H. U. 1968 *Max-Planck-Inst. Bericht 6.*

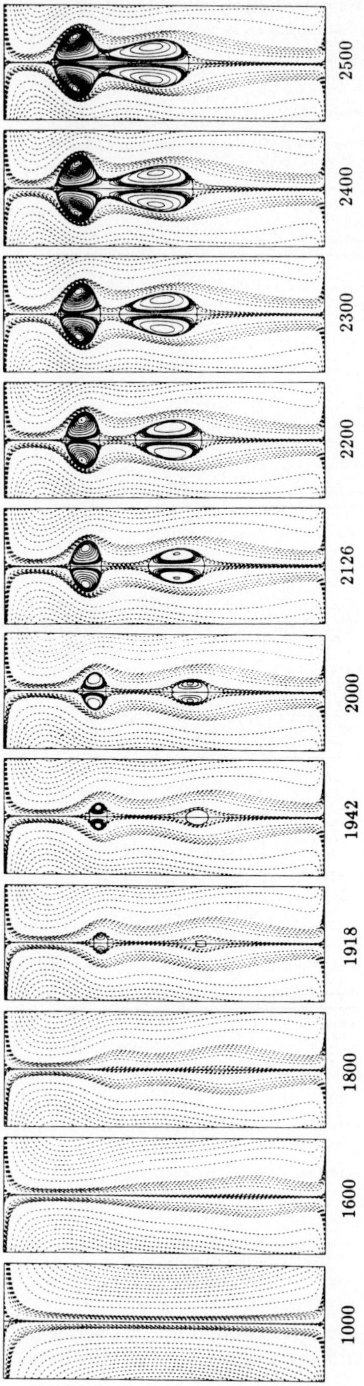

Fig. 1: Details of the streamfunction at steady state ($0 \leq r \leq 5R/12$) for $H/R = 2.5$ and Re as indicated.

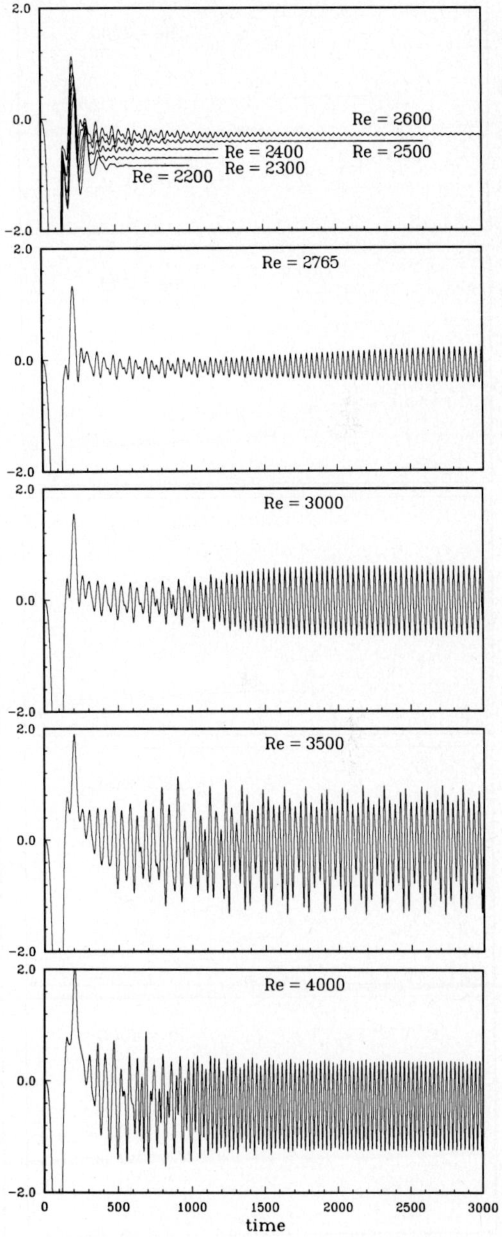

Fig. 2: Time history of the streamfunction at the point $r = R/6$, $z = 2H/3$ for $H/R = 2.5$ and Re as indicated. Time is scaled with Ω.

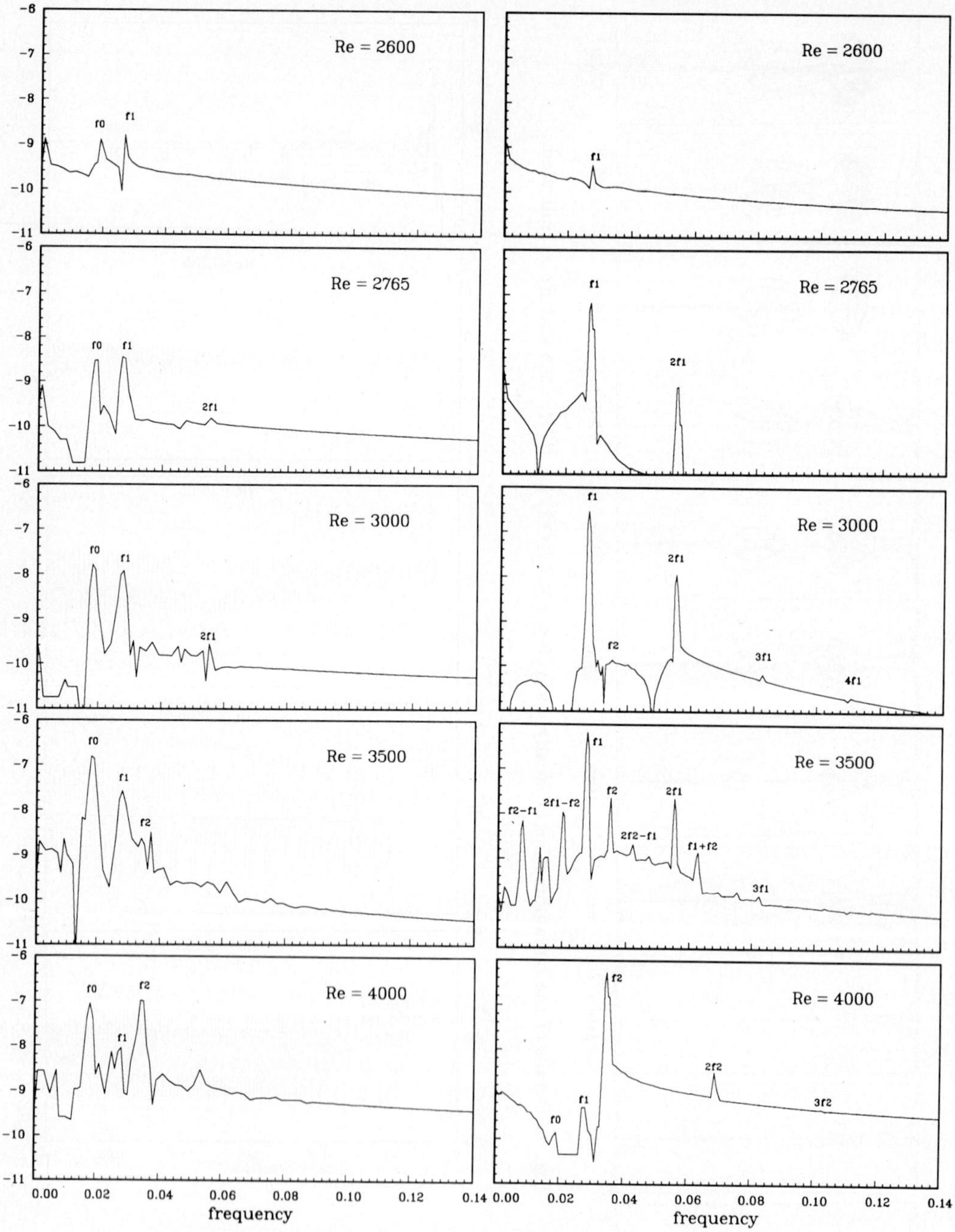

Fig. 3: Log_{10} of the power spectrum corresponding to the streamfunction time-series in Fig. 2. The spectra on the left hand side are for $500 \leq t \leq 1000$ and the spectra on the right hand side are for $2000 \leq t \leq 3000$. The frequency is scaled with $1/\Omega$.

Numerical Analysis of a Multigrid Method for Spectral Approximations

Yvon MADAY
Université Paris XII, Laboratoire d'Analyse Numérique, Université Pierre et Marie Curie, Paris, 75252, Fr.
and Brown University, Providence RI, USA.

Rafael MUÑOZ
I.N.R.I.A. Domaine de Voluceau - Rocquencourt, Fr.

Introduction

Spectral methods use polynomials of high degree to approximate the solution of some partial differential equations. The discretization parameter is the degree of the polynomial and the convergence toward the exact solution is obtained when the degree of the polynomial tends to infinity. The main contribution is due to D. Gottlieb and S. A. Orszag [1] and their initial monograph is well worth looking at. We refer also to the book of C. Canuto, M. Y. Hussaini, A. Quarteroni and T. A. Zang [2] for a new review on the method.

In these methods, the numerical solution consists of functions, generally globally continuous, that are piecewise polynomial of degree less than or equal to some given integer with respect to each variable. As noted by S. A. Orszag [1], one important factor for the competitivety of the method is the ability to use tensorial bases for the discrete spaces. This results in some drawback for the geometry of the domain that has to be equal to a parallelotope, or can be decomposed into bricks with an exact or a deformed parallelotope shape. This explains the great number of contributions on the use of domain decomposition techniques coupled with spectral type methods. The flexibility of the new algorithms allows now to minimize the previous drawback on the geometry of the domain (see e.g. [4, 5, 6, 7]).

Most of the works, both from the numerical and the theoretical points of view, deal with the approximation of first or second-order problems and mainly in the computational fluid mechanics area. The current conclusion reveals that the method is competitive with respect to more conventional approaches and provides even better results when the solution is regular in some sense. The main reason for this competitivity is related to the high (infinite) order of the method that uses all the regularity of the solution. For instance if the exact solution is analytic, the resulting error is exponentially small with respect to the discretization parameter.

Resulting algebraic systems

As regard the implementation of the methods, of importance is the care that is taken to derive simple algebraic systems that are easily solvable. The main drawback of the spectral techniques, relies on the fact that the resulting systems are **full and ill conditioned.** A condition number of N^4 for second order problems is standard. Direct methods are thus generally not used to solve the problems and **iterative techniques** are needed. It is well known that two major kinds of spectral methods are encountered in the literature, they are based either on Chebysheff approximations or Legendre approximations (see [2] for more details). One of the interest of the Legendre approach relies on the fact that the resulting system is **symmetric**. The standard iterative techniques are demonstrated to work better in such cases.

When domain decomposition is used, it is often that **parallel computation** is involved. For spectral type techniques, the domain decomposition is more a question of viability of the method since complex geometries could not be handled otherwise. Parallel algorithms are then naturally combined with spectral methods.

If the value of the solution is known on the interfaces of the domain decomposition, then the approximate problem can be set in as many independent subproblems as many subelements, then the parallel computation is highly efficient. This is the case when the global matrix is "condensed" so as to provide a smaller matrix that works only on the interface values. Once this small problem is solved, the interface values are known and the full system can then be solved in parallel. The resulting influence matrix technique has been used, for instance [7, 8, 9]. The main problem with this approach is the important cost required for solving the interface problem that is smaller only in the 2-D case. Besides, the non rectilinear geometries so as the moving boundaries do not fit well to this approach.When a great number of domains has to be used, or a 3-D calculation is involved, an iterative technique to guess the values at the interface has to be preferred. Two level of iterations are then to be distinguished :

The first one deals with the unknown values of the solution on the interfaces.
The second one is required by the bad conditioning of the matrices.

These two level can be either nested or treated simultaneously. As regard the second level, we have already noticed the poor condition number of the resulting matrix. A standard iterative technique will thus require a great number of iterations, in particular not independent of the degree N of the approximation. A good preconditioner of the stiffness matrix has to be used in order to diminish the condition number of the preconditioned matrix and thus results in a number of iterations, independent of N.

A lot of work exist in the literature for the study of preconditioner. These are generally related either to the matrix by itself or at a higher level to another discretization of the initial P.D.E. this discretization is either of Finite difference or Finite element type (see [10, 11]).

Various preconditioners.

The simpler preconditioners are diagonal ones.They are very simple to invert and also portable to more than one spatial dimension. Since we are interested also in global approaches, we shall indicate, when it is possible, the dependance on both N (the polynomial degree) and K (the number of subdomains). In this term, the condition number of the stiffness matrix resulting from the discretization of a second order problem scales like K^2N^3. A very sensitive improvement and simple to obtain is provided by the use of the **diagonal part of the stiffness matrix**. Indeed, we are able to prove that the resulting condition number scales now like K^2N^2. This results for instance in a conjugate gradient algorithm that will converge at a rate of 1-c/KN.

A more impressive result is provided by the use of more complex preconditioners based on

lower order methods for discretizing the P.D.E.. This idea was suggested by S. A. Orszag and he mainly used it in the framework of the finite difference methods. The result is now a preconditioned matrix that has a precondition number independent of N (and certainly of K also). What could be dreamed of better!!

The problem with this last approach relies on the fact that the preconditioner, if simple to invert in one spatial dimension, is still difficult to invert in more spacial dimensions. To cope with this problem, various **incomplete LU decompositions** or **approximate factorizations** can be proposed. This is discussed interalia in [12,13]. Their conclusions reveal that a loss of independence with respect to the discretization parameters is observed when such simplifications are done in the original preconditioners. Besides, if nonstandard boundary conditions are used in the formulation of the problem, the resulting condition number is still worse.

The condition number of the preconditioned matrix is the only factor of importance when a Richardson or a conjugate gradient method is used. Besides, if a multigrid approach is considered, this is no more the case. Indeed, the multigrid approach is based on two distinct steps that have to combine efficiently. One is the smoother and is very close to a precondition procedure, the second one is a coarse grid correction that results from a resolution of the problem that is studied with a similar method that uses fewer parameters.

The two previous kind of preconditioners have been used as smoothers. We refer to [14] for the finite difference approach and to [15] for the Jacobi approach. The first method seems to work mainly due to the good properties of the preconditioner. The second method, very simple to generalize to complex geometries and 3-D calculation appears as providing parameter independent rate of convergence in 1-D and a very low dependency on N only in the multidimensional case. The robustness of the first method to problems with oscillatory coefficients has been numerically proved, this is not yet the case for the second one which is only at its beginning. The aim of this paper is to justify the good properties of the second method in order to foresee its generality.

Multigrid algorithm, basic concepts.

The multigrid algorithm is based on two different concepts. The first one consists in the definition of the nested spaces and the way one goes from one space to the other one. The second one consists of the definition of the smoother, i.e. the operation that is done on the finest grid

The spectral context is very simple to handle nested spaces since the degree of the polynomials that are used in the discretization is a good way to generate fine and coarse spaces. The spaces will consist of piecewise polynomials, the degree of which varies from one space to the other but the domain decomposition will remain exactly the same. For instance, in the case where two domains are used, and if $\mathbf{P}^n$ denotes the set of all functions that are polynomial of degree $\leq$ n over each domain, then a possible choice of nested spaces is the following :

$$\mathbf{P}^{16} \supset \mathbf{P}^8 \supset \mathbf{P}^4 \supset \mathbf{P}^2 \supset \mathbf{P}^1 .$$

The way to pass the information from one space to the other one is then trivial since from a coarse

space to a finer space, we only use the **identity**. This is the **prolongement operator** $\mathcal{P}_{n/2\to n}$. On the other way , the **restriction operator** $\mathcal{R}_{n\to n/2}$ is, as in most multigrid techniques, equal to the transpose of the restriction operator.

Let us now present the smoother **B** that we have analyzed. We shall first deal with the case of the simple problem

$$- u_{xx} = f \quad \text{over } I,$$

where f is a given force and I is an interval]a,b[. We suppose that the interval I is broken into K subintervals $]a_k, a_{k+1}[$, $k = 1,..,K$.

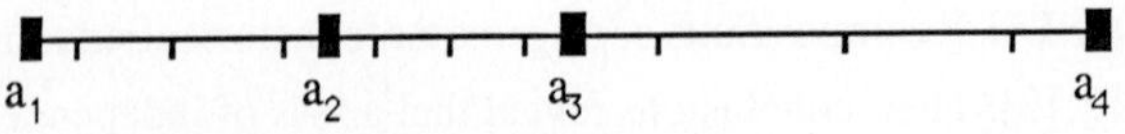

The space of approximation $\mathbf{X}^N$ will then be the subspace of all continuous functions in $\mathbf{P}^N$. The numerical solution u_N will be characterized in $\mathbf{X}^N$ so that

$$\Sigma_{GL}\, u_N v_N = \Sigma_{GL}\, f v_N \quad \text{for any } v_N \text{ in } \mathbf{X}^N .$$

The resulting system of algebraic equations is obtained by choosing a well suited base for the space $\mathbf{X}^N$. The base of Lagrangian interpolant provides a good candidate. This results in a set of unknowns $\mathbf{U}_N$ to define u_N equal to the values at the set of collocation points. As is well known, the interpretation of the method is a **pure collocation method** at any Gauss Lobatto point interior to some subinterval.

The resulting system can be written as follows

$$\mathbf{A}_N \mathbf{U}_N = \mathbf{F}_N \quad .$$

Multigrid algorithm, the method.

Let now $\mathbf{D}_N$ denote the diagonal of the stiffness matrix $\mathbf{A}_N$, the 2 level multigrid V-cycle is based on m smoothing $\mathcal{S}$ where we solve m times a problem defined as follows

$$\mathbf{D}_N (\mathcal{S}\mathbf{V}_N - \mathbf{V}_N) = \mathbf{F}_N - \mathbf{A}_N \mathbf{V}_N \ ,$$

and one coarse grid correction where we solve once a problem defined as follows

$$\mathbf{A}_{N/2} (\tilde{\mathcal{C}}\mathbf{V}_N - \mathbf{V}_N) = \mathcal{R}_{N\to N/2}(\mathbf{F}_N - \mathbf{A}_N \mathbf{V}_N) \ ,$$

then define the operator of correction by

$$\mathcal{C}\mathbf{V}_N = \mathcal{P}_{N/2\to N}(\tilde{\mathcal{C}}\mathbf{V}_N)$$

We are now in position to define the V-Cycle, it is based on the iteration of procedure that can be written as follows

$\mathbf{U}_N^0$ is given a an initial guess, then suppose that $\mathbf{U}_N^n$ has been obtained after n iterations, then $\mathbf{U}_N^{n+1}$ is defined as follows

$$\mathbf{U}_N^{n+1} = \mathcal{S}^m\ \mathcal{C}\ \mathcal{S}^m\ \mathbf{U}_N^n \quad .$$

This approach has been used by [15] and in this situation (1-D, constant coefficients), good numerical conclusions could be derived for the quality of this multigrid procedure. Since this very beginning, theoretical proofs of this behavior and numerical evidences both adapted to more complex problems have been provided in [16].

Note that the previous procedure was explained on a two level method but the generalization of the method to more levels is strateforward since it only requires to solve the problem

$A_{N/2}\,(\tilde{\mathcal{C}}V_N - V_N) = \mathcal{R}_{N\to N/2}(F_N - A_N\,V_N)$,

using a V-cycle between the grids of level N/2 and N/4.

Multigrid algorithm, theoretical justification, 1-D. (see [16]])

The numerical analysis that we have done in the one dimensional case is rather complete since in the case of any elliptic second order problem, including the nonconstant coefficients ones, the convergence rate is here proved to be independent of both N and K. The main theoretical tool is the use of the general theory of [17] based on the analysis of the eigenvalue problem defined as follows :

Find an element Ψ_N in X_N and a real number λ such that

(*) $\quad A_N\,\Psi_N = \lambda\,D_N\,\Psi_N$,

We can prove that the highest eigenvalue to this problem is of order 1, besides, the smallest eigenvalue scales like $(KN)^{-2}$. This property is straightforward when K = 1 and the interval is I =]-1, 1[with the

Proposition 1 : For any polynomial ϕ and ψ of P_N that vanish at the boundary

$$\Phi_N\,D_N\,\Psi_N = (N)(N+1)/3 \int_I \phi(x)\,\psi(x)/(1-x^2)\,dx \ ,$$

where Φ_N (resp Ψ_N) is the vector of the values of ϕ (resp. ψ) at the Gauss-Lobatto points.

Proposition 2 : There exists a constant C independent of N such that, for any polynomial ϕ of P_N that vanishes at the boundary we have

$$\Phi_N\,A_N\,\Phi_N \le C\,\Phi_N\,D_N\,\Phi_N \ .$$

The last proposition relates the size of the similar eigenvalue problem but set now on the subspace X_N of all elements of X_N that are orthogonal to $X_{N/2}$, more precisely we have

Proposition 3 : There exists a constant C' independent of N such that, for any polynomial ϕ of P_N that vanishes at the boundary and that verifies

$\mathcal{P}_{N/2\to N}\,(\Psi_{N/2})\,A_N\,\Phi_N = 0$, for any ψ of $P_{N/2}$ that vanishes at the boundary,

we have

$$\Phi_N\,D_N\,\Phi_N \le C\,\Phi_N\,A_N\,\Phi_N \ .$$

These three lemmas prove that the diagonal part of the stiffness matrix is very similar to the stiffness matrix for the high modes, ie the polynomials of high degree. This shows that the smoothing procedure is very efficient. The coarse grid correction takes care of the low degree polynomials. The theory can also prove justify the numerical results proved in [15] as regard the independence of the convergence rate of the method with respect to the parameters of the discretization.

Multigrid algorithm, theoretical justification, 2-D.

Let us now turn to the multidimensional case. The theoretical justification is right now less complete however, the results yet obtain help in designing an efficient numerical algorithm.

As an easy consequence of the one dimensional result, the equivalent of proposition 2 is valid also in this situation. The result is now, both in 2-D and 3-D

Proposition 3' : There exists a constant C' independent of N such that, for any polynomial ϕ of $\mathbf{P}_N$ that vanishes at the boundary and that verifies

$$\mathcal{P}_{N/2\to N}(\Psi_{N/2})\,\mathbf{A}_N\,\Phi_N = 0 \quad \text{, for any } \psi \text{ of } \mathbf{P}_{N/2} \text{ that vanishes at the boundary,}$$

we have

$$\Phi_N\,\mathbf{D}_N\,\Phi_N \leq C\,N\,\Phi_N\,\mathbf{A}_N\,\Phi_N \;.$$

This result is optimal and proves that the convergence rate of the procedure based on the same strategy as in the one dimensional case is no more independent of N but scales like 1-c/N. This analysis of the spectrum of the eigenvalue problem (*) enables however to combine a Chebyshev acceleration to the multigrid algorithm and arrive at the conclusion that the convergence rate is then in a low way dependent of N since it scales now like $\mathbf{1 - c/N^{1/2}}$ as will be proved in a future paper. This is also observed in the numerical simulations and since the current polynomial degree is less than 20, the convergence rate is still spectacular and better than the preconditioned conjugate gradient algorithm. The numerical experiments show that even in this case, the convergence rate is independent of K (the number of elements). The theoretical proof of this fact is under consideration.

Remark: The analysis reported here has been extended by the second author to treat the same algorithm in the Chebyshev context. The independence of the convergence rate with respect to the discretization parameters is also obtained in this situation.

References

[1] D. GOTTLIEB & S.A. ORSZAG - *Numerical Analysis of Spectral Methods : Theory and Applications*, SIAM (1977).

[2] C. CANUTO, M.Y. HUSSAINI, A. QUARTERONI & T.A. ZANG - *Spectral Methods in Fluid Dynamics*, Springer-Verlag (1987).

[3] S.A. ORSZAG - Comparison of Pseudospectral and Spectral Approximations, Stud. Appl. Math. **51**, pp253-259 1972.

[4] C. BERNARDI, Y. MADAY & A.T.PATERA - Nonconforming spectral element methods, Analysis of the method - in preparation.

[5] Y. MADAY, C. MAVRIPLIS & A.T.PATERA - Nonconforming spectral element methods, Numerical results - to appear in the proceedings of the second conference on Domain Decomposition Methods, (UCLA).

[6] D. FUNARO, A. QUARTERONI & P. ZANOLLI - An iterative procedure with interface relaxation for domain decomposition methods, to appear in SINUM.

[7] M.MACARAEG & C. L. STREET - Improvement in spectral collocation through multiple domain techniques, J. Appl. Numer. Math., **2** pp 95,108, 1986.

[8] Y. MARION & B. G. GAY - Résolution des equations de Navier-Stokes par une méthode spectrale via une technique de coordination, in Proc. 6th Int. Symp. Finite Element Methods in Flow Problems, 1986.

[9] R. PEYRET - Introduction to spectral methods, preprint **98**, Université de Nice, lecture series **1986-04**, V.K.I.

[10] C. CANUTO & A. QUARTERONI - Preconditioned minimal residual methods for Chebyshev spectral calculations, JCP, **60**, pp 315-337, 1985.

[11] M. DEVILLE & E.MUND - Chebyshev pseudospectral solution of second order elliptic equations with finite element preconditioning JCP, **60**, pp 517-533, 1985.

[12] Y. MORCHOISNE - Résolution des équations de Navier-Stokes par une méthode spectrale de sous-domaines, Comptes-Rendus du 3ème Congrès Intern. sur les Méthodes Numériques de l'Ingénieur, publiés par P. Lascaux, Paris (1983).

[13] C. L. STEET, T. A. ZANG & M. Y. HUSSAINI - Spectral multigrid methods with applications to transonic potential flows, JCP **57**, pp 43-76, 1985.

[14] T.A. ZANG, Y. S. WONG & M.Y. HUSSAINI - Spectral multigrid methods for elliptic equations, J.C.P. **48**, pp485-501, 1982 and J.C.P. **52**,pp489-507, 1984.

[15] E. M. RONQUIST & A.T. PATERA - Spectral element multigrid, part 1 Formulation and numerical results, J. of Scient. Comp. 2-2 pp389-406, 1987.

[16] Y. MADAY & R. MUNOZ - Spectral element multigrid, part 2 Theoretical Justification, submitted to J. of Scient. Comp.

[17] R. E. BANK & C. C. DOUGLAS - Sharp estimates for multigrid rates of convergence with general smoothing and acceleration, SINUM **22**, pp 617,633, 1985.

ASYMMETRIC SEPARATED FLOWS ABOUT SHARP CONES IN A SUPERSONIC STREAM

F. Marconi
Grumman Corporate Research Center
Bethpage, New York 11714-3580

INTRODUCTION

In the late 1940's, the phenomenon of lateral instability due to asymmetric vortex formation was discovered. On slender-nosed aircraft flying at a high angle of attack ($\alpha \cong 3$ x nose half angle), the flow which usually is symmetric about the plane containing the free stream velocity vector can be very asymmetric due to flow separation. While the separation points are only slightly asymmetric, the resulting vortices are very asymmetric, causing a large side force and lateral instability. The cause of this asymmetry has been debated since it was first discovered. The basic issue of whether it is a viscous or inviscid phenomenon is still uncertain. The inviscid flow separation model used here is an ideal tool for investigating this issue.

One of the first indications that this phenomenon was inviscid in nature came from the work of Fiddes.[1] He showed small disturbance calculations with vortex sheets modeled by vortex filaments that exhibit asymmetric vortices. The geometry was a slender cone and the flow was incompressible. Fiddes showed that two solutions existed for a symmetrically separated flow. One solution has symmetrically located vortices and the other has asymmetrically located vortices. The second, asymmetric, solution exists only in the very high angle-of-attack regime, $\alpha > 2$ x cone half angle. At these conditions, the asymmetric solution compares much better with experimental data than does the symmetric one. Of course, in the wind tunnel experiment, Fiddes[1] found that in asymmetric flow the separation points cannot be symmetric. Very small asymmetries in the separation locations resulted in very large side forces, indicating very large asymmetries in the vortex locations. Computationally, Fiddes found that for slightly asymmetric separation point locations there existed two solutions, one which was slightly asymmetric and the other very asymmetric. Experimental data indicate that the onset of the rise in side force is insensitive to Reynolds number, indicating again that this phenomenon is inherently inviscid.

The question of whether the small disturbance approximation could be the cause of the asymmetry in Fiddes computation can be answered quite simply by the Euler computation presented here. The work presented here is confined to the supersonic regime, where experimentalists find reduced side forces.[2] The Euler code used here in conjunction with a flow-separation model was able to predict asymmetric flows, indicating again that the phenomenon is inviscid in nature. The issue of interest will be investigated fully by considering simple geometries; slender right circular cones will be considered here.

NUMERICAL METHOD - SEPARATION MODEL

Although employing a number of new and important features, the overall numerical procedure used in this study is essentially standard.[3] The fully three-dimensional Euler equations are solved with an explicit marching technique. The marching direction, z (Fig. 1), is an iterative coordinate for the conical flows considered here. The finite difference scheme used is Moretti's characteristic-based λ-scheme.[4] In each case shown, the maximum residual (i.e., the derivative of pressure in the marching direction) was reduced at least six orders of magnitude; the grid had 89 (radial) x 178 (circumferential, right and left) points. The basic scheme is presented in more detail in Ref. 5.

The model used here to force separation at a specified location is founded in the work of J.H.B. Smith.[6] Smith assumed irrotational flow outside a vortex sheet in order to analyze the local flow at separation. The present work uses only the basic concept, which does not depend on the irrotational assumption. The basic concept acquired from Ref. 6 is that, at a specified separation point, there is a vortex sheet or contact surface which has a jump in velocity direction. The condition at the intersection of the sheet and a smooth surface (i.e., the separation point) is that the crossflow velocity (Fig. 1) stagnates on the lee side of the sheet (Fig. 2a). The concept of crossflow applies only to conical flow. The extension of this model to fully 3-D flow has been discussed in Ref. 7. The

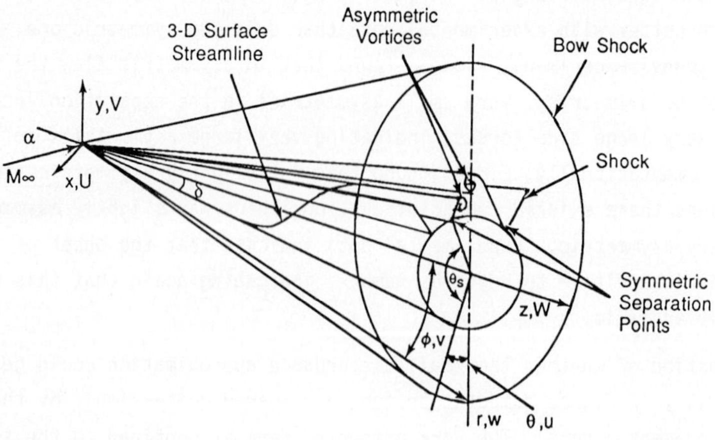

Fig. 1 Three-dimensional Sketch of Flow Field & Coordinate System

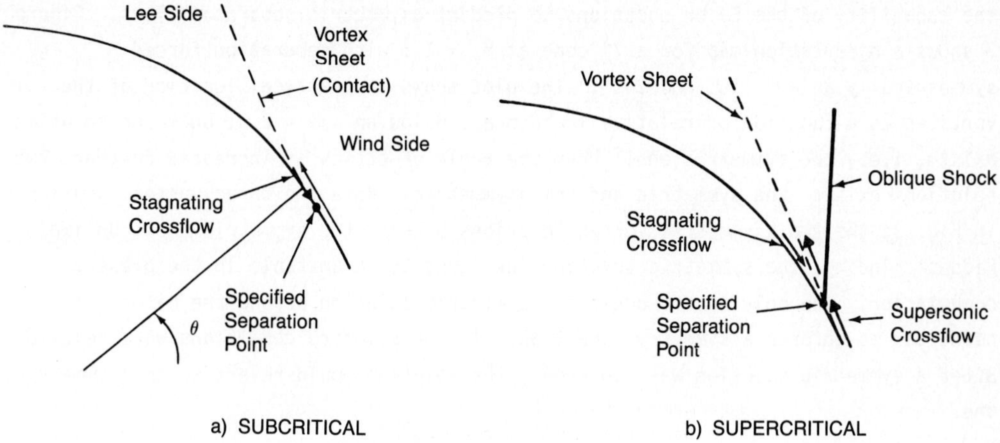

Fig. 2 Sketch of Forced Separation Model

crossflow velocity on the wind side of the sheet is determined by the global solution. The vortex sheet is a stream surface and the pressure is continuous across it. In the case of isentropic irrotational flow considered in Ref. 6, these conditions imply that the modulus of velocity is continuous across the sheet. If the crossflow stagnates on both sides of the sheet ($u = v = 0$) at separation, the isentropic condition implies that the radial component of velocity (w, Fig. 1) is continuous and the sheet does not exist, i.e., no vorticity is shed into the flow field. These arguments conclude that, in isentropic flow, the separating sheet must leave the cone surface tangentially if the wind side crossflow is subsonic. If the surface crossflow is subsonic just upstream of separation and the sheet leaves the surface at any angle relative to the surface, then the flow on the windside of the sheet must stagnate and no vorticity will be shed into the flow. On the other hand, if the crossflow is supersonic, the sheet can leave the surface at an angle, an oblique crossflow shock will occur, and the flow on the wind side of sheet will not stagnate so that vorticity is shed. The supersonic crossflow model is sketched in Fig. 2b. This model does not violate the concepts of vorticity shedding proposed by Smith,[6] since the flow on the wind side of the separation point need not stagnate, and thus a jump in velocity exists across the sheet. In order to capture the shock at separation in one mesh interval, no pressure derivatives were taken across the separation point on the cone surface. In addition, a small pressure plateau is imposed just after separation, the level being taken from just downstream of separation.

NUMERICAL RESULTS

This section is intended to demonstrate the computational capability of the techniques discussed in the previous section. More importantly, it will demonstrate the capability of the Euler equations to predict asymmetric separated flows. Figure 3 shows a bifurcation map for a 7° cone at $M_\infty = 1.6$ with separation forced symmetrically at $\theta = 115°$ and $245°$. The plot shows the vertical location of the vortices as a function of relative incidence. Below an $\alpha/\delta = 2.3$, only one solution exists, i.e., the symmetric one. When the angle of attack is increased further, two solutions exist: the symmetric and the asymmetric. Note the three vortex locations in Fig. 3; the two asymmetric vortex locations bracket the symmetric one. Unlike Fiddes' finding, the symmetric solution was found to be unstable in the present computation. The only way to achieve a symmetric solution beyond the bifurcation point was to enforce a symmetry condition. If the symmetry conditions were relaxed after a symmetric solution was converged, the solution would revert to an asymmetric one.

Figures 4 to 7 show results for this 7° cone at $M_\infty = 1.6$ and $\alpha = 20°$. Figure 4 shows the crossflow streamlines in the vicinity of the lee plane; the asymmetry of the vortices is obvious. The separation points are placed symmetrically ($\theta = 115°$ and $245°$), as is the case in all the calculations to follow. Figure 5 shows the surface pressure distribution; it should be pointed out that θ is measured counterclockwise. The crossflow is locally supersonic so that shocks precede the separation points (see the isobars of Fig. 6). The vortex to the left is much closer to the surface (Fig. 4) than that on the right; this can be seen in its influence on the surface pressure. The minimum surface pressure at $\theta = 210°$ is a result of this vortex interaction. The isobars of Fig. 6 show the shocks before the separation points and the closed isobars representing the local minimum in pressure

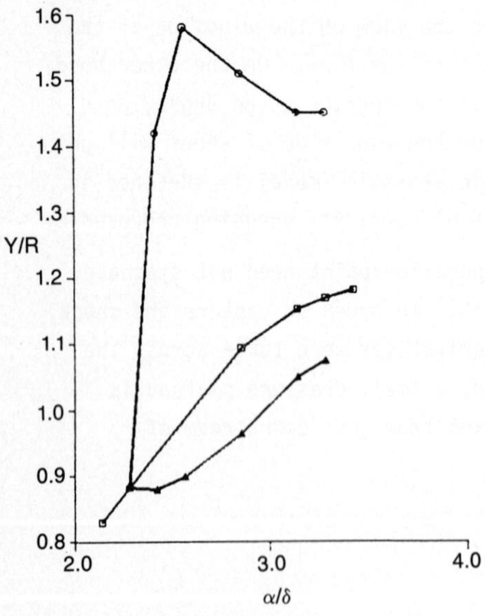

Fig. 3 Bifurcation Map, 7° Cone at $M_\infty = 1.6$, Symmetric Separation at $\theta_1 = 115°$, $\theta_2 = 245°$

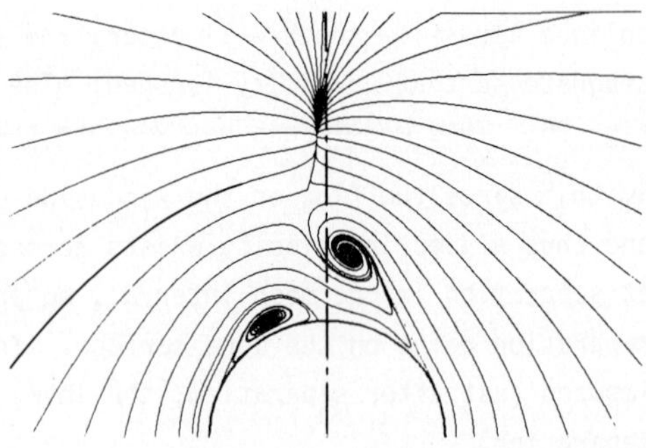

Fig. 4 Crossflow Streamlines, 7° Cone at $M_\infty = 1.6$, $\alpha = 20°$

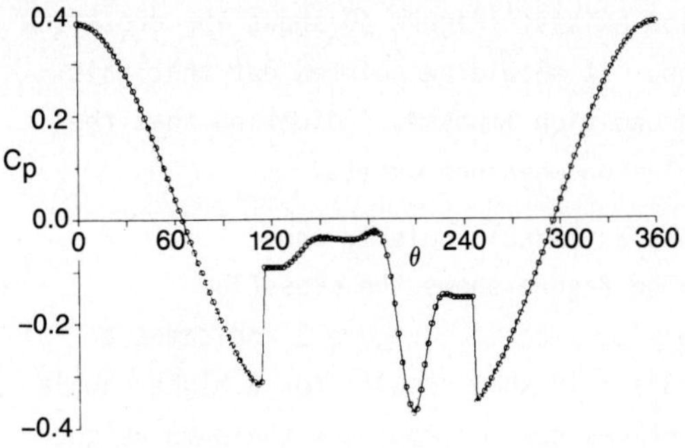

Fig. 5 Surface Pressure Distribution, 7° Cone at $M_\infty = 1.6$, $\alpha = 20°$

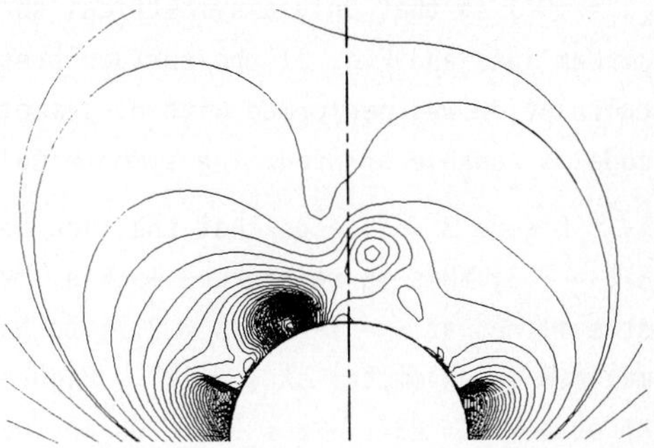

Fig. 6 Cross-sectional Isobars, 7° Cone at $M_\infty = 1.6$, $\alpha = 20°$

at the vortex centers. While the asymmetry is significant, it is localized in supersonic flow because of the limited domain of influence. Figure 5 shows that the surface pressure is symmetric, i.e., unaffected, before the separation points. This is the reason side forces due to asymmetries decrease with increasing Mach number. The asymmetry may be significant, but because its domain of influence is limited, the side forces may be small. Figure 7 shows the entire crossflow plane streamline pattern; note that the flow seems symmetric except in the vicinity of the vortices.

Figures 8 and 9 show the same 20° angle-of-attack case with symmetry enforced. Figure 8 shows the crossflow streamlines and Fig. 9 the surface pressure distribution. Note that the asymmetric vortices of Fig. 4 bracket those of Fig. 8. Figures 10 and 11 show results of a computation at small enough angles of attack

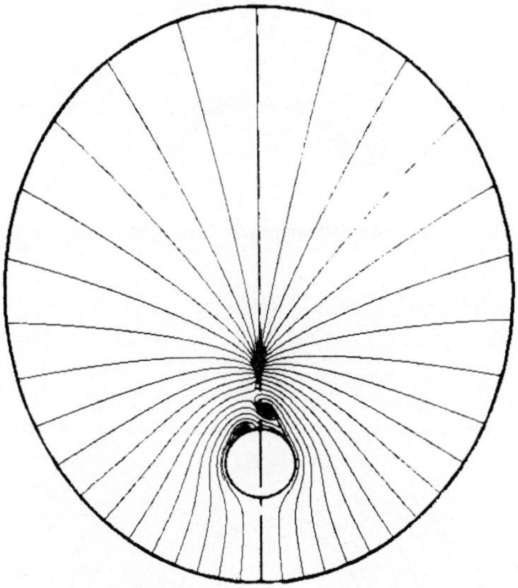

Fig. 7 Crossflow Streamlines (Whole Plane), 7° Cone, $M_\infty = 1.6$, $\alpha = 20°$

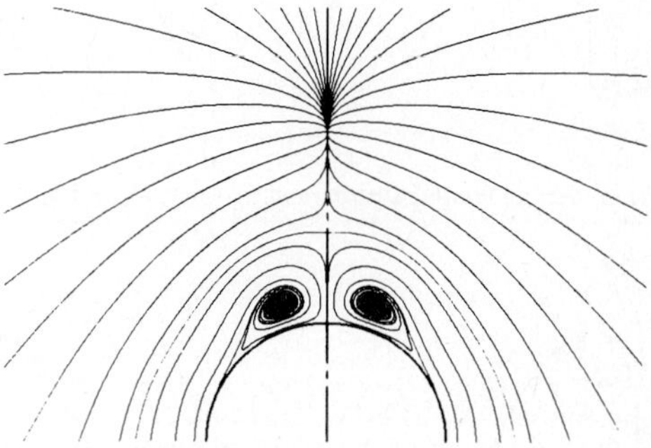

Fig. 8 Crossflow Streamlines, 7° Cone, $M_\infty = 1.6$, $\alpha = 20°$

(α = 15°) so that only a symmetrical solution exists. Figure 10 shows the crossflow streamlines and Fig. 11 the surface pressure. It should be pointed out that this calculation was performed with no symmetry condition imposed, indicating that the code is capable of producing symmetrical solution when one exists.

Figure 3 indicates that the flow becomes asymmetric quickly above α/δ = 2.3; this is reinforced in Fig. 12. The figure shows the crossflow streamlines at α = 19° (α/δ = 2.7) to be very asymmetric. Figure 3 indicates a maximum in asymmetry at α = 19°. Figures 13 and 14 show results for a higher angle of attack α = 23°, α/δ = 3.3. While the vortices seem to approach the same height above the cone (see Fig. 13), the flow is by no means approaching symmetry. This is reinforced by the surface pressure plot of Fig. 14. It seems that the vortex height may not be the best indication of asymmetry. Figure 15 shows another indication of asymmetry, the side force. It can be seen that there is no side force until α/δ > 2.3. The rapid rise in side force is indicated, with a maximum value at α/δ = 2.7 (α = 19°). There is a drop in side force to α = 20°, but this is followed by a leveling off of the side force.

Figure 16 shows the onset of asymmetry for the 5° cone tested by Peake et al[2]. Two computations are compared with the data of Peake. The computational

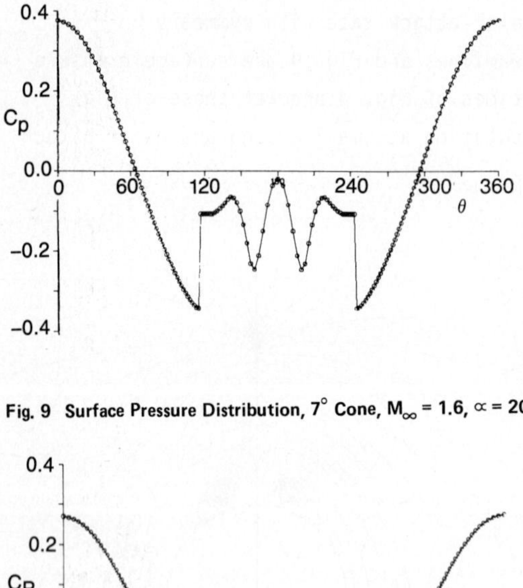

Fig. 9 Surface Pressure Distribution, 7° Cone, M_∞ = 1.6, α = 20°

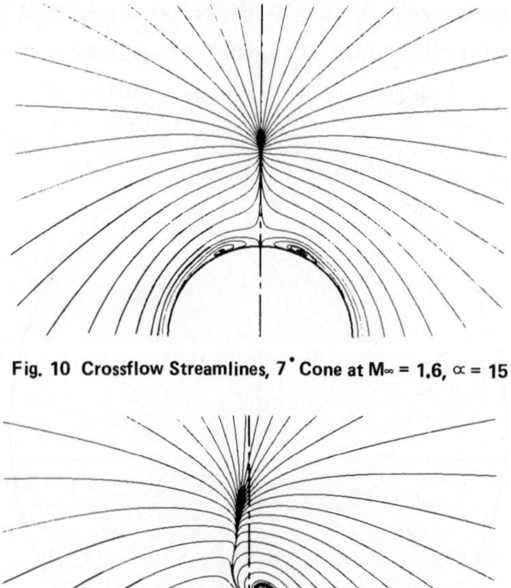

Fig. 10 Crossflow Streamlines, 7° Cone at M_∞ = 1.6, α = 15°

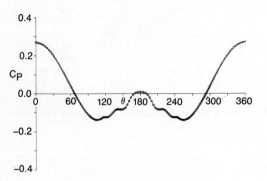

Fig. 11 Surface Pressure Distribution, 7° Cone, M_∞ = 1.6, α = 15°

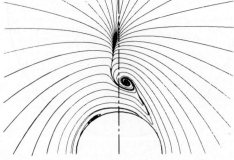

Fig. 12 Crossflow Streamlines, 7° Cone, M_∞ = 1.6, α = 19°

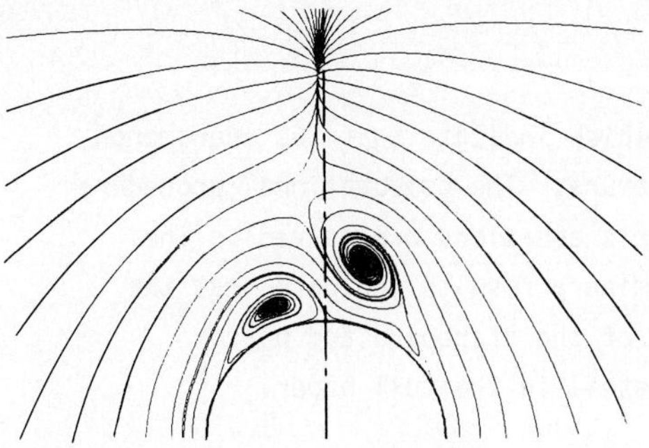

Fig. 13 Crossflow Streamlines, 7° Cone, M_∞ = 1.6, α = 23°

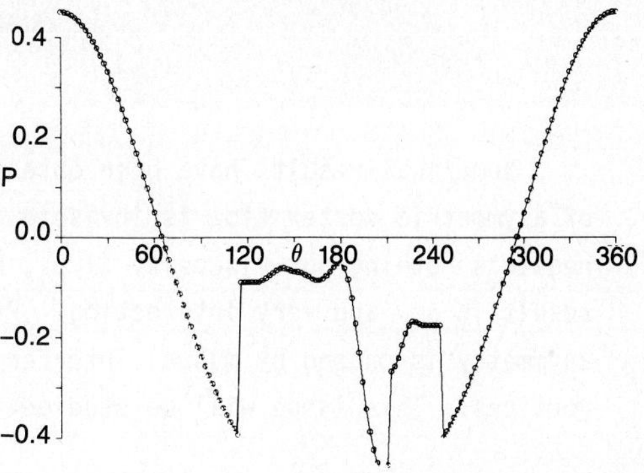

Fig. 14 Surface Pressure Distribution, 7° Cone, M_∞ = 1.6, α = 23°

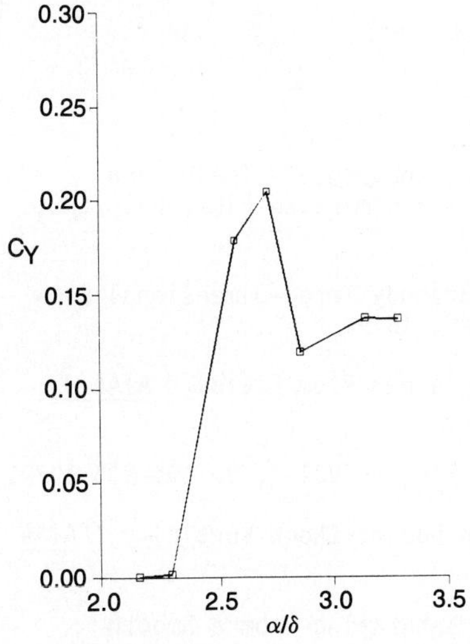

Fig. 15 Onset of Asymmetry, Side Force vs Angle of Attack (7° Cone, M_∞ = 1.6)

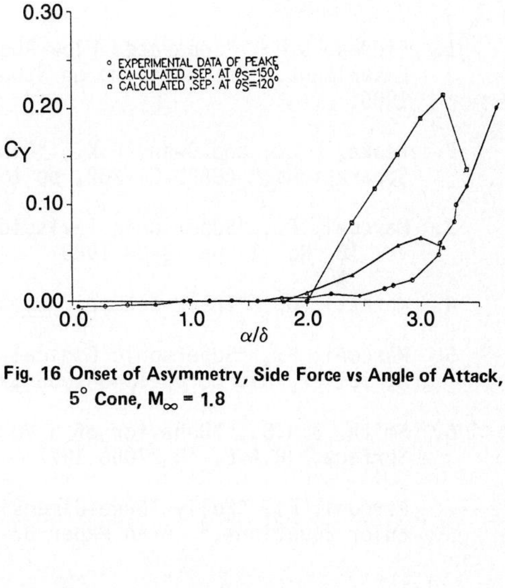

Fig. 16 Onset of Asymmetry, Side Force vs Angle of Attack, 5° Cone, M_∞ = 1.8

results differ in separation point location; these locations are fixed over the angle of attack range considered. It should also be noted that the separation locations are symmetric in both computations. Moving the separation locations 30° closer to the lee plane seems to make the side force result compare better with the experimental data. It is obvious that once the critical incidence is passed, viscous effects become important. For example, once the vortices become asymmetric, the separation points will move asymmetrically due to viscid/inviscid interaction. Also the separation points' locations are a function of angle of attack. This function is viscosity dependent and so has not been included there. These issues will be addressed in future work.

SUMMARY

Numerical results have been obtained which indicate that the phenomenon of asymmetric vortex flow is inviscid in nature. The computational procedure neglects nothing but viscosity (i.e., Euler's equations are solved). The result is new and very interesting. Preliminary results indicate that the asymmetry is caused by mutual interference of the starboard and port vortices. This issue will be studied in detail in the full paper.

ACKNOWLEDGEMENT

The work was supported in part by the Air Force Office of Scientific Research under contract No. F49620-84-C-0036 with Dr. J. Wilson as contract monitor.

REFERENCES

1. Fiddes, S.P., "Separated Flow About Cones at Incidence - Theory and Experiment," Proc of Symp on Studies of Vortex Dominated Flow, NASA/LRC, 1985.

2. Peake, D.J., and Owen, F.K., "Control of Forebody Three-Dimensional Flow Separations," AGARD-CP-262, pp 15-1 to 15-46, 1979.

3. Marconi, F., "Supersonic Inviscid, Conical Corner Flow Fields," AIAA J, Vol 18, No. 1, pp 78-84 1980.

4. Moretti, G., "The λ-Scheme," Computers and Fields, Vol 7, pp 191-205 1979.

5. Marconi, F., "Supersonic Conical Separation Due to Shock Vorticity," AIAA J, Vol 22, No. 8, pp 1048-1055 1984.

6. Smith, J.H.B., "Behavior of a Vortex Sheet Separating from a Smooth Surface," R.A.E. TR 77058 1977.

7. Marconi, F., "Fully Three-Dimensional Separated Flows Computed with the Euler Equations," AIAA Paper 87-0451, 1987.

A FLUX SPLIT ALGORITHM FOR UNSTEADY INCOMPRESSIBLE FLOW

Charles L. Merkle and Mahesh M. Athavale
Department of Mechanical Engineering
The Pennsylvania State University
University Park, PA 16802

INTRODUCTION

The primary difficulty in computing incompressible flows is associated with finding a satisfactory way to link changes in the velocity field to changes in the pressure field. Of the several generic choices that are available, including stream function-vorticity, velocity-vorticity and primitive variables formulations, we here consider one of the latter. Primitive variables techniques can broadly be divided into methods based upon a pressure Poisson equation[1-4], and methods based upon artificial compressibility[5-10]. In the present paper, we consider a direct marching procedure that is philosophically related to artificial compressibility methods, although it does not introduce a fictitious time derivative into the continuity equation. The particular method chosen is completely analogous to compressible flow schemes based upon flux difference splitting[11,12].

The flux splitting procedures we use for direct time-marching solutions of the unsteady incompressible equations are developed on the basis of the Euler equations. This simplifies the notation somewhat and allows us to focus attention on those (inviscid) terms that cause the difficulties in linking pressure and velocity. The viscous terms are then added later by a simple, straightforward procedure. The incompressible equations are written in the general conservation law form,

$$\Gamma \frac{\partial Q}{\partial t} + \frac{\partial E}{\partial \xi} + \frac{\partial F}{\partial \eta} = 0 \qquad (1)$$

where $Q = (p, u, v)^T/J$ is a complete vector with all non-zero entries and Γ = diag (0,1,1) is a diagonal matrix that preserves the null entry in the time derivative of the continuity equation by means of the zero on the top diagonal. The flux vectors E and F are given in their standard form as:

$$Q = \frac{1}{J}\begin{bmatrix} p \\ u \\ v \end{bmatrix} \qquad E = \frac{1}{J}\begin{bmatrix} U \\ uU + \xi_x p \\ vU + \xi_y p \end{bmatrix} \qquad F = \frac{1}{J}\begin{bmatrix} V \\ uV + \eta_x p \\ vV + \eta_y p \end{bmatrix} \qquad (2)$$

For convenience, we have used the Euler equations in two-dimensional generalized coordinates. The three-dimensional formulation follows directly. We have also absorbed the density in the pressure, p. The symbols u and v represent the velocity components in the original Cartesian coordinates, while U and V represent the contravariant velocities. The remaining terms are the metric derivatives.

The choice of Q in Eqn. 1 as a complete vector allows us to define the flux Jacobians $A = \partial E/\partial Q$ and $B = \partial F/\partial Q$ in exactly the same fashion as for the artificial compressibility formulation. We do, however, note that the inability to invert the matrix, Γ, implies that the equations are not totally hyperbolic as they would be if Γ^{-1} existed. (Note that if we choose $\Gamma = \text{diag}(1/\beta,1,1)$ where β is a scaling component, Γ^{-1} exists, and we have the totally hyperbolic, artificial compressibility formulation.)

UPWIND DIFFERENCING PROCEDURE

Because the incompressible equations are not totally hyperbolic, flux split systems cannot be interpreted in a physical sense as a Riemann solver, but they may still be interpreted as a finite difference representation having the indicated order of accuracy. In our analysis we use flux differencing splitting and a third order upwind biased system. Following Rai[11] and Barth[12], the spatial derivative of a flux vector is differenced in the general form:

$$\frac{\partial E}{\partial x} = \frac{\tilde{E}_{i+1/2} - \tilde{E}_{i-1/2}}{\Delta x} \tag{3}$$

where $\tilde{E}$ is a numerical flux,

$$\tilde{E}_{i+1/2} = \frac{1}{2}[E_{i+1} + E_i] - k_1[\Delta E^+_{i+1/2} - \Delta E^-_{i+1/2}] - k_2[\Delta E^+_{i-1/2} - \Delta E^-_{i+3/2}] \tag{4}$$

Specific values of k_1 and k_2 provide first, second or third order accuracy. For example, $k_1=k_2=1/6$ leads to third order accurate upwind-biased differencing, while $k_1=k_2=0$ leads to second order central differencing and $k_1=1/2, k_2=0$ provides a first order upwind scheme.

The quantity, $\Delta E^{\pm}_{i+1/2}$, represents the flux difference in the respective directions, and is given by:

$$\Delta E^{\pm}_{i+1/2} = A^{\pm}_{i+1/2}(\bar{Q})\,[Q_{i+1} - Q_i] \tag{5}$$

where A is the Jacobian matrix and is evaluated using an appropriate averaging method such as that of Roe[13]. A previous application of flux difference splitting for the incompressible equations was presented by Hartwich and Hsu[14], but their analysis was for the artificial compressibility formulation and was limited to steady flows.

In the present formulation, boundary conditions are enforced by a method analogous to the MOC procedure used earlier for the artificial compressibility equations[7,8]. The multidimensional equations are approximately factored by an LU procedure[15] which uses third order accuracy on the right-hand side and first order on the left-hand side. This method is somewhat different from the compressible procedure of Rai[11] and Barth[12], but still requires from four to eight iterations at each time step to maintain third order accuracy.

The viscous terms are added to the right-hand side of the above equations by using central differences. These second order terms are also added on the left-hand side and factored in LU fashion.

RESULTS

As demonstrations of the capabilities of the direct marching algorithm, we present results for both inviscid and viscous flows. The inviscid flow problem is for the convection of periodic vorticity through a circular arc cascade. Vorticity was introduced at the upstream end by imposing an inflow containing a series of wake-like regions as shown on the left of Fig. 1. The location of the periodic wakes translated upward at constant velocity to simulate a wake-rotor interaction. The results shown in Fig. 1 are for a non-dimensional cycle time, $T=0.35$ where T is the time required for the wake to traverse one blade passage normalized by the time for a particle at the maximum freestream speed to traverse the chord length. These results show the vorticity contours associated with this flow and demonstrate the manner in which the incoming vorticity is "cut" by the blades and rolls up as it propagates through the cascade.

The viscous flow cases are for the flow in a square cavity with an oscillating top lid. The results shown in Fig. 2 are for a Reynolds number of 400 (based on the maximum lid velocity). The lid oscillation is given by a sinusoidal function and the present results are for a non-dimensional frequency of unity (based on the maximum lid velocity and the cavity width). The results in Fig. 2 show streamline contours at various instants in a half cycle. The arrows above each picture show the magnitude and direction of the lid velocity for each instant. Results for the second half cycle are mirror images of these results. The dashed lines are results obtained for the same problem by Soh and Goodrich[10] and show good agreement with the present results.

Additional calculations are in progress, but overall, the present procedure appears to be an excellent method for computing either viscous or inviscid unsteady flows. The LU approximate factorization promises equally good performance in three dimensions as well. The fact that the method is exactly analogous to unsteady compressible flow procedures[11,12] also suggests the versatility of the algorithm. The incompressible formulation, of course, requires the solution of one less equation which means it will be somewhat faster than its compressible counterpart.

ACKNOWLEDGEMENT

This work was sponsored by the Office of Naval Research under Contract N00014-87-K-0123.

REFERENCES

1) Harlow, F. H. and Welch, J. E., "Numerical Calculation of Time-Dependent Viscous Incompressible Flow with Free Surface", Physics of Fluids, Vol. 8, Dec. 1965, pp. 2182-2189.
2) Patankar, S. V., Numerical Heat Transfer and Fluid Flow, Hemisphere Publishing Co., New York, 1980.
3) Raithby, G. D. and Schneider, G. E., "Numerical Solution of Problems in Incompressible Fluid Flow: Treatment of the Velocity-Pressure Coupling", Numerical Heat Transfer, Vol. 2, 1979, pp. 417-440.
4) Dennis, S. C. R., Ingham, D. B., and Cook, R. N., J. Comp. Physics, 33 (1979), p. 325.
5) Chorin, A. J., "A Numerical Method for Solving Incompressible Viscous Flow Problems", Journal of Computational Physics, Vol. 2, 1967, pp. 12-26.
6) Steger, J. L. and Kutler, P., "Implicit Finite Difference Procedures for the Computation of Vortex Wakes", AIAA Journal, Vol. 15, No. 4, April 1977, pp. 581-590.
7) Choi, D. and Merkle, C. L., "Application of Time-Iterative Schemes to Incompressible Flow", AIAA Journal, Vol. 23, No. 10, October 1985, pp. 1518-1524.
8) Merkle, C. L. and Tsai, Y.-L. Peter, "Application of Runge-Kutta Schemes to Incompressible Flows", AIAA Paper 86-0553, AIAA 26th Aerospace Sciences Meeting, January 6-8, 1986, Reno, NV.
9) Merkle, C. L. and Athavale, M. M., "Time-Accurate Unsteady Incompressible Flow Algorithms Based on Artificial Compressibility", AIAA Paper 87-1137, AIAA 8th Computational Fluid Dynamics Conference, June 9-11, 1987, Honolulu, HA.
10) Soh, W. Y. and Goodrich, J. W., "Unsteady Solution of Incompressible Navier-Stokes Equations", to appear in Journal of Computational Physics.
11) Rai, M. M., "Navier-Stokes Simulations of Blade-Vortex Interaction Using High-Order Accurate Upwind Schemes", AIAA Paper 87-0543, Reno, NV, 1987.
12) Barth, T. J., "Analysis of Implicit Local Linearization Techniques for Upwind and TVD Algorithms", AIAA Paper 87-0595, Reno, NV, 1987.
13) Roe, P. L., "Approximate Reimann Solvers, Parameter Vectors and Difference Schemes", Journal of Computational Physics, Vol. 43, 1981, pp. 357-372.
14) Hartwich, P. M. and Hsu, C. H., "Incompressible Navier-Stokes Solutions for a Sharp-Edged Double-Delta Wing", AIAA Paper 87-0206, Reno, NV, 1987.
15) Athavale, M. M. and Merkle, C. L., "An Upwind Differencing Scheme for Time-Accurate Solutions of Unsteady Incompressible Flow", to be presented at the First National Fluid Dynamics Congress, July 24-28, 1988, Cincinnati, OH.

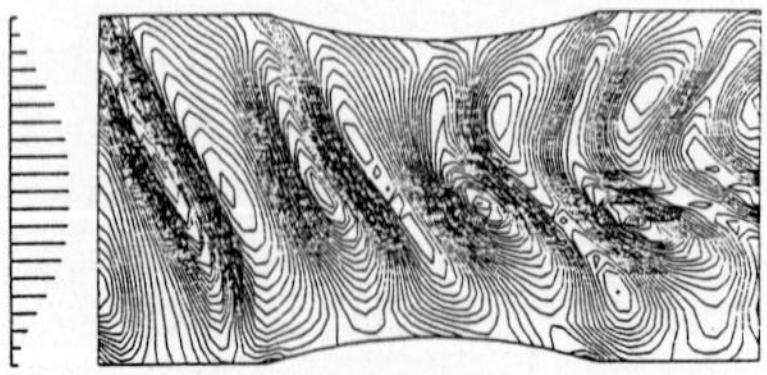

t=nT

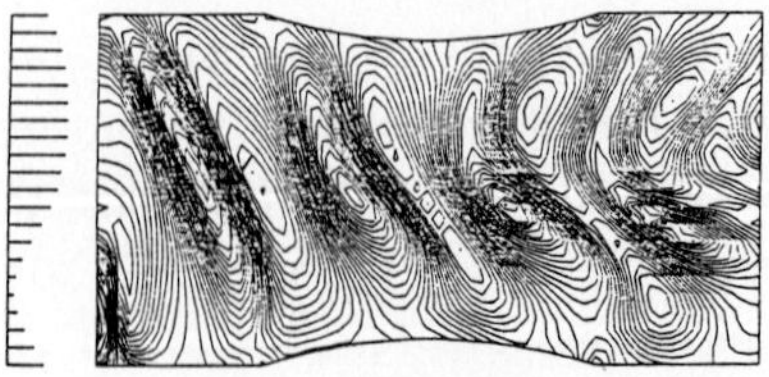

t=nT+0.2T

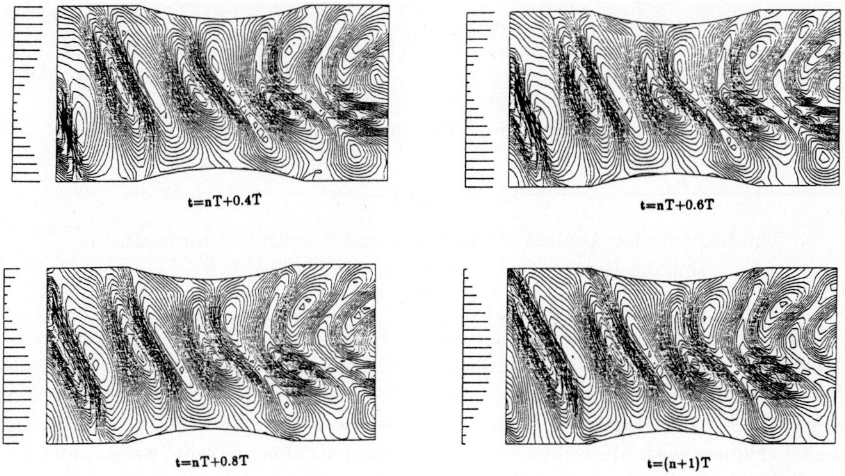

Fig. 1 Vorticity contour plots for the unsteady inviscid flow test case at various time instants. Cycle time T = 0.35 L/u_n. Inlet u velocity profiles are shown on the left.

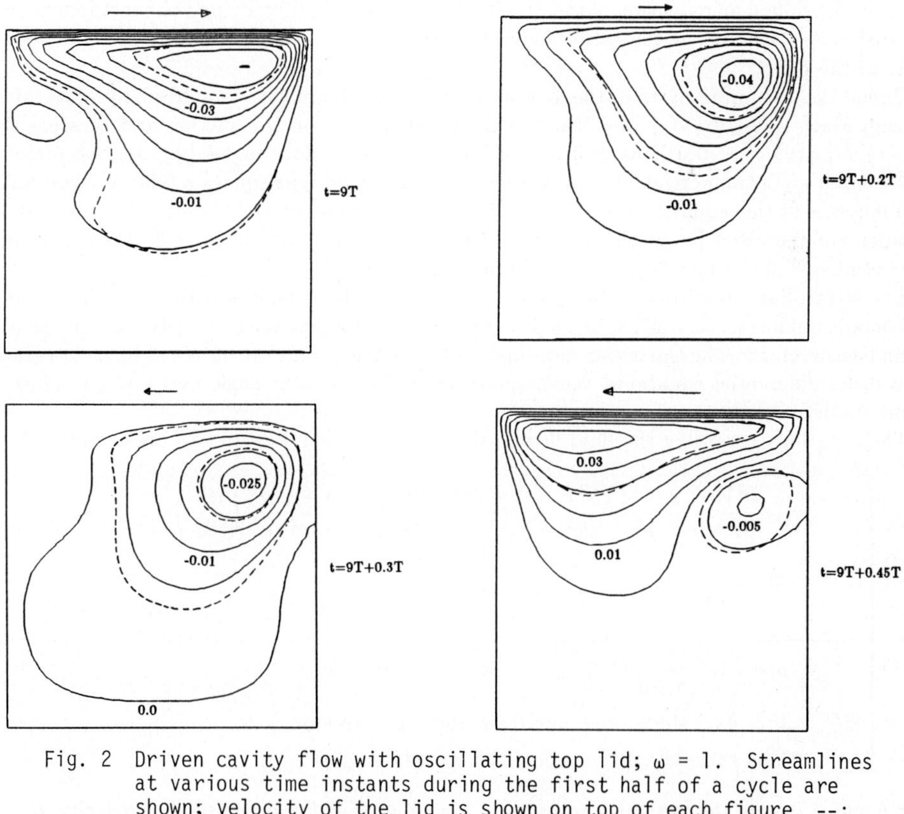

Fig. 2 Driven cavity flow with oscillating top lid; ω = 1. Streamlines at various time instants during the first half of a cycle are shown; velocity of the lid is shown on top of each figure. --: representative streamlines from Soh and Goodrich[10].

Inverse Method for the Determination of Transonic Blade Profiles of Turbomachineries

B. MICHAUX

The Institute for Applied Mathematics and Scientific Computing
Indiana University, Bloomington, IN 47405 U.S.A

We present a new numerical method allowing the computation of transonic blade profile geometry for turbomachineries as well as an investigation about an equivalence of the constraints of Lighthill [1] which allows us to obtain the existence of a closed profile. This type of method was originally developed by J.D Stanitz [3] for the computation of the inverse problem for designing two-dimensional channels with prescribed velocity distributions along channel walls. Later he proposed with L.T Sheldrake [4] to use it to define the blade profiles as follows: they replace a blade row by a succession of channels unwedged by the blade cascade spacing. Hence, the profiles are defined by the part between two channels. Then, they transform the physical domain to a domain of computation defined by the potential lines and the streamlines of fluid. From this original idea which gives us a rectangular computational domain (see fig. 1) whose boundaries are fixed and known, we developed an extension of the Stanitz method for a fluid verifying the exact isentropicity law: $p/\rho^\gamma = cst$ (where γ is the ratio of specific heats (≈ 1.4), p and ρ are the pressure and the density of the fluid).

Under the assumptions that the flow is perfect and isentropic, and from the data of the upstream Mach number (M_1), the Mach number distribution on the suction and pressure sides ($M_{suc}(\tau), M_{pre}(\tau), \tau \in [0,1]$), as well as the inlet and outlet flow angles (ϕ_1, ϕ_2), we present a method based on the numerical integration of a nonlinear mixed type equation (elliptic-hyperbolic). The unknowns of this equation are the velocity, the Mach number and the density - the last two quantities are expressed by an algebraic function of the velocity by virtue of St-Venant relations for the isentropic fluids (see [2] and the references therein).

The streamline curvatures in the physical domain can be determined by a function of these aerodynamic unknowns, as well as the angle between the streamline tangent vector and the physical domain basis vector $\vec{\imath}$. The Cartesian coordinates of the blade profile are obtained by an integration of first order differential equations; these equations are functions of angle and velocity, along the streamlines defining the profile.

The equations governing the fluid flow in the computational domain ($\mathcal{R}$) are as follows:

$$\left[\begin{array}{ll} & -\dfrac{\partial^2 V}{\partial \psi^2} - \dfrac{1-M^2}{\rho^2}\dfrac{\partial^2 V}{\partial \xi^2} + \dfrac{1+M^2}{V}\left(\dfrac{\partial V}{\partial \psi}\right)^2 + \dfrac{1+\gamma M^4}{\rho^2 V}\left(\dfrac{\partial V}{\partial \xi}\right)^2 = 0 \quad \text{in } \mathcal{R} \\ & M = V\sqrt{\dfrac{2}{\gamma+1}\dfrac{1}{1-\frac{\gamma-1}{\gamma+1}V^2}} \quad \text{in } \overline{\mathcal{R}} \\ & \rho = \left(1 - \dfrac{\gamma-1}{\gamma+1}V^2\right)^{\frac{1}{\gamma-1}} \quad \text{in } \overline{\mathcal{R}} \\ BC: & V = g \text{ on the profile, upstream and downstream} \\ & V \text{ is periodic on the rest of the boundary} \end{array}\right. \tag{1}$$

where ξ and ψ denote the potential and stream functions, V is the dimensionless velocity, ρ is the dimensionless density and g is obtained from the Mach number distributions prescribed on the suction and pressure side of the profile and the upstream Mach number.

In addition, we have the following expression for the streamline curvatures:

$$\mathcal{X} = \rho \frac{\partial V}{\partial \psi} \quad \text{in } \overline{\mathcal{R}} \tag{2}$$

Finally, the deviation that generates the blade profiles, as well as their coordinates, are obtained by integrations of the following equations in $\overline{\mathcal{R}}$:

$$\left[\begin{array}{l} \dfrac{\partial \phi}{\partial \xi} = \dfrac{\mathcal{X}}{V} \\ \dfrac{\partial x}{\partial \xi} = \dfrac{\cos\phi}{V} \\ \dfrac{\partial y}{\partial \xi} = \dfrac{\sin\phi}{V} \\ BC: \quad \phi(0,\psi) = \phi_1, \quad x(0,\psi) = x_1(\psi), \quad y(0,\psi) = y_1(\psi) \end{array}\right. \tag{3}$$

where $\phi_1, x_1(\psi)$ and $y_1(\psi)$ are physical data.

1] Existence of solution.

- The three Lighthill constraints (1945) (see [1]).

For an incompressible flow around an isolated profile, the solution of the inverse problem for the previous data does not exist, except if the prescribed velocity distribution $V_{profile}$ on the profile verifies the following constraints:

$$\int_{Profile} \log\left(\frac{V_{profile}(\omega)}{V_\infty}\right) \begin{Bmatrix} 1 \\ \cos\omega \\ \sin\omega \end{Bmatrix} d\omega = 0.$$

Where ω is the polar angle in the plane defined by conform transformation of the profile in a circle.

2] Case of a blade row.

Role of the spacing g: this quantity is unknown. Writing the circulation of the velocity around the physical domain we obtain:

$$L_{suc}A_{suc} - L_{pre}A_{pre} = g(V_1 \sin\phi_1 - V_2 \sin\phi_2)$$

where $A_{suc} = \int_0^1 V_{suc}(\tau)\,d\tau$, $A_{pre} = \int_0^1 V_{pre}(\tau)\,d\tau$ and L_{suc} is the length of the suction side, L_{pre} is the length of the pressure side.

The spacing g can be fixed and the length L_{suc} and L_{pre} let free but linked by the precedent relation. Thus, we get the first necessary degree of freedom, which is equivalent to the first Lighthill constraints. We need two other freedom degrees to get a closed profile.

3] Computational domain and boundary conditions.

We compute the potential difference $\Delta\xi'_{suc}$ on the suction side, on which the length is equal to 1.

$$\Delta\xi'_{suc} = \frac{1}{deb}\int_0^1 V_{suc}(\tau)\,d\tau$$

with $deb = \rho_1 V_1 \cos\phi_1 = \rho_2 V_2 \cos\phi_2$ and we multiply it by a constant L_{suc} to get: $\Delta\xi_{suc} = L_{suc}\Delta\xi'_{suc}$. On the other hand we have: $\Delta\xi_{pre} = \Delta\xi_{suc} - (\Delta\xi_{AB} - \Delta\xi_{GF})$ where $\Delta\xi_{AB} = \frac{\tan\phi_1}{\rho_1}$, $\Delta\xi_{GF} = \frac{\tan\phi_2}{\rho_2}$. In the same way, we have

$$\Delta\xi'_{pre} = \frac{1}{deb}\int_0^1 V_{pre}(\tau)\,d\tau$$

which gives us a second constant L_{pre}: $L_{pre} = \frac{\Delta\xi_{pre}}{\Delta\xi'_{pre}}$

If we assume that we know a determination of L_{suc} (near the solution), we can get L_{pre} from the precedent relations and we can build the numerical domain. We also get the boundary conditions for the velocity on the suction and the pressure side. On the other hand, from the relation of conservation of debit and the S[t]-Venant relations, we get the outlet boundary conditions.

4] Computation of the profile.

The existence of the stagnation points on the profile generates a discontinuity of the streamline curvatures at these points. Thus, there exist two angles η_{suc} et η_{pre} such that:

$$\left[\begin{array}{l}\phi(\xi_{I-},0)=\phi(\xi_{I+},0)-\eta_{suc}\,,\ \phi(\xi_{C-},1)=\phi(\xi_{C+},1)-\eta_{pre}\\ \phi(\xi_{H+},0)=\phi(\xi_{H-},0)+\eta_{suc}\,,\ \phi(\xi_{D+},1)=\phi(\xi_{D-},1)+\eta_{pre}\end{array}\right.$$

with $\phi(\xi_{I\pm},.)=\lim_{h\mapsto\pm0}\phi(\xi_I+h,.)$. We also have the periodic conditions: $\phi(\xi_{I-},0)=\phi(\xi_{C-},1)$ and $\phi(\xi_{H+},0)=\phi(\xi_{D+},1)$ and from the relation between velocity and angle we obtain:

$$\left[\begin{array}{l}\phi(\xi_{I-},0)=\phi_1+\displaystyle\int_{\xi_B}^{\xi_C}\frac{\rho(\xi,0)}{V(\xi,0)}\frac{\partial V}{\partial\psi}(\xi,\psi)_{|\psi=0}\,d\xi\\ \phi(\xi_{H+},0)=\phi_2-\displaystyle\int_{\xi_H}^{\xi_G}\frac{\rho(\xi,0)}{V(\xi,0)}\frac{\partial V}{\partial\psi}(\xi,\psi)_{|\psi=0}\,d\xi\end{array}\right.$$

Let φ_{cam} be the camber of the profile: $\varphi_{cam}=\phi(\xi_{I-},0)-\phi(\xi_{H+},0)=\phi(\xi_{C-},1)-\phi(\xi_{D+},1)$ and let $\delta\phi_{suc}$ and $\delta\phi_{pre}$ be the variation of angle on each point of the suction and pressure side. Then we have:

$$\left[\begin{array}{ll}\delta\phi_{suc}(s_{suc})=\displaystyle\int_{s_{suc,I}}^{s_{suc}}\rho(s,0)\frac{\partial V}{\partial\psi}(s,\psi)_{|\psi=0}\,ds & s_{suc}\in[s_{suc,I},s_{suc,H}]\\ \delta\phi_{pre}(s_{pre})=\displaystyle\int_{s_{pre,C}}^{s_{pre}}\rho(s,1)\frac{\partial V}{\partial\psi}(s,\psi)_{|\psi=1}\,ds & s_{pre}\in[s_{pre,C},s_{pre,D}]\end{array}\right.$$

Finally, we get: $\eta_{suc}=-\frac{1}{2}\left(\varphi_{cam}+\delta\phi_{suc}(s_{suc,H})\right)$ and $\eta_{pre}=-\frac{1}{2}\left(\varphi_{cam}+\delta\phi_{pre}(s_{pre,D})\right)$.

REMARK: The two degrees of freedom on η_{suc} and η_{pre} correspond to the equivalent of the last two Lighthill constraints for the incompressible flow. Finally, we have

- suction side: $\phi(s_{suc},0)=\phi(\xi_{I+},0)+\delta\phi_{suc}(s_{suc})$ $s_{suc}\in[s_{suc,I},s_{suc,H}]$
- pressure side: $\phi(s_{pre},1)=\phi(\xi_{C+},0)+\delta\phi_{pre}(s_{pre})$ $s_{pre}\in[s_{pre,C},s_{pre,D}]$

5] Coordinates of the profile.

- suction side:

$$\left\{\begin{array}{l}s_{suc}\in[s_{suc,I},s_{suc,H}]\\ x(s_{suc},0)=x_{LE}+\displaystyle\int_{s_{suc,I}}^{s_{suc}}\cos\phi(s,0)ds\\ y(s_{suc},0)=y_{LE}+\displaystyle\int_{s_{suc,I}}^{s_{suc}}\sin\phi(s,0)ds\end{array}\right.$$

- pressure side

$$\left\{\begin{array}{l}s_{pre}\in[s_{pre,C},s_{pre,D}]\\ x(s_{pre},1)=x_{LE}+\displaystyle\int_{s_{pre,C}}^{s_{pre}}\cos\phi(s,1)ds\\ y(s_{pre},1)=y_{LE}+\displaystyle\int_{s_{pre,C}}^{s_{pre}}\sin\phi(s,1)ds\end{array}\right.$$

where (x_{LE},y_{LE}) denote the Cartesian coordinates of the leading edge.

To get a closed profile we need to have: $x(s_{suc,H},0)=x(s_{pre,D},1)$ and $y(s_{suc,H},0)=y(s_{pre,D},1)$ which is equivalent to the following condition: $\frac{L_{suc}^4}{L_{pre}^4}\frac{S_1^2+S_2^2}{S_3^2+S_4^2}=1$, with:

$$\left[\begin{array}{ll}S_1=\displaystyle\int_0^1\cos\delta\phi_{suc}(\tau)\,d\tau, & S_2=\displaystyle\int_0^1\sin\delta\phi_{suc}(\tau)\,d\tau\\ S_3=\displaystyle\int_0^1\cos\delta\phi_{pre}(\tau)\,d\tau, & S_4=\displaystyle\int_0^1\sin\delta\phi_{pre}(\tau)\,d\tau\end{array}\right.$$

REMARK: By using a bisection method on L_{suc}, we get the right length of the suction side (and also of the pressure side), which gives us a closed profile. But, if we want to get some round leading and trailing edges, we need to change the initial distributions of Mach number on the profile, to obtain:

$$\eta_{suc} = \frac{\pi}{2} \quad \text{and} \quad \eta_{pre} = -\frac{\pi}{2}$$

6] Numerical Methods.

Due to the mixed type of the equation (1) (elliptic-hyperbolic), we approximated the partial derivatives by the classical scheme of center difference with three points for the mesh points where the flow is subsonic and by the upwind scheme with three points for the mesh points where the flow is supersonic. For the computation of the aerodynamic unknowns, we used a fixed point method for the determination of the Mach number and the density, and a Newton method at each iteration of the fixed point method to compute the velocity. The periodic boundary conditions were also treated during the application of the Newton method (see [2] for more details). Finally, we integrated the first order equations (2) and (3) by the trapezoidal numerical integration rules.

7] Numerical Results and Conclusion.

Example 1 presents the geometry of a blade profile from data corresponding to a transonic flow. For this case, the inlet and outlet angles are respectively $45°26'$ and $11°28'$. The upstream Mach number is 0.85. We present in figure 2 the distributions of Mach number on the profile, the closed profile corresponding to these data and the lines of isomach through the blade row. The maximal value of the Mach number on the profile is 1.15.

Example 2 presents a profile obtained from the same data as example 1, except for the Mach number distributions on the profile. These last distributions correspond to a more realistic case. The maximal value of the Mach number is then 1.08.

In both cases, the Mach number distributions were adapted to get some round leading and trailing edges.

Finally this method, which can be integrated in an interactive process, seems to correspond well to industrial requirements.

Acknowledgement. This research was supported in part by SNECMA, *Societe Nationale d'Etude et de Construction de Moteurs d'Aviation*, under contract # 265843 KY with the *Laboratoire d'Analyse Numerique* of *Université Paris-Sud*, France, by AFOSR, under contract # 88-103 and by the Research Fund of Indiana University, U.S.A.

References.

[1] **M.J.Lighthill** A new method of two-dimensional aerodynamic design. *Aeronautical research council reports and memoranda N° 2112, April 1945*

[2] **B.Michaux** Méthodes inverses pour la détermination des profils d'aubes transsoniques des turbomachines. *Thèse, Univ. de Paris-Sud, Janvier 1988*

[3] **J.D.Stanitz** Design of two-dimensional channels with prescribed velocity distributions along channels walls. *NACA 2593/2595*

[4] **J.D.Stanitz & L.T.Sheldrake** Application of design method to high-solidity cascades and tests of an impulse cascade with 90° of turning. *NACA TN 2652*

Physical and computational domains.

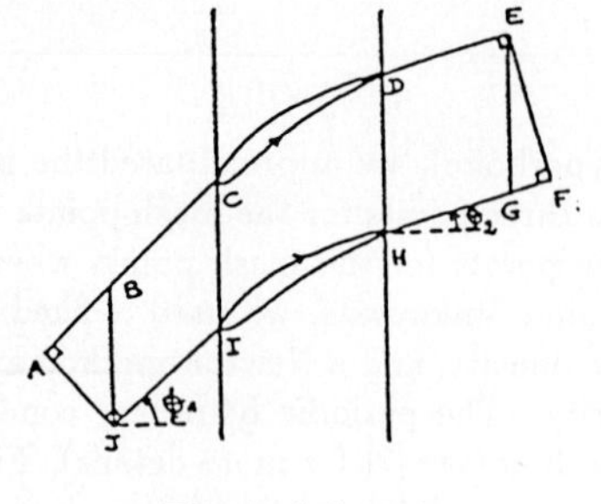

Physical domain.

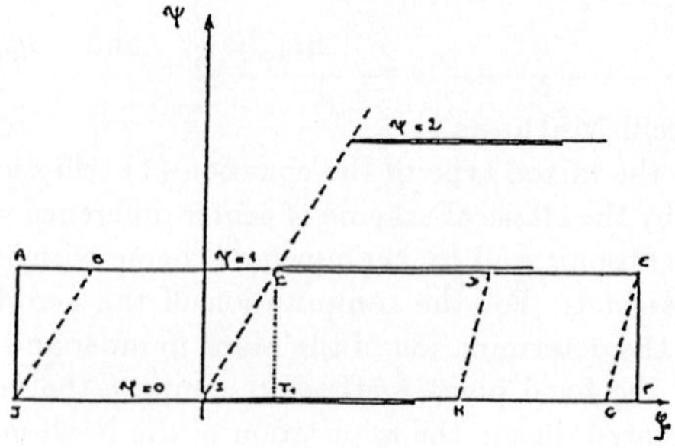

Computational domain.

Figure 1.

Example 1.

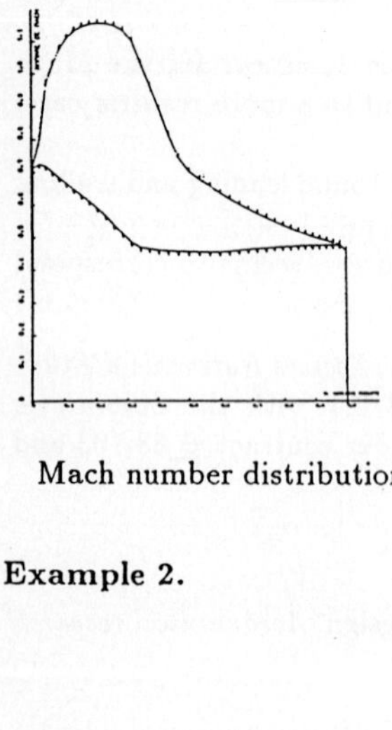

Mach number distributions.

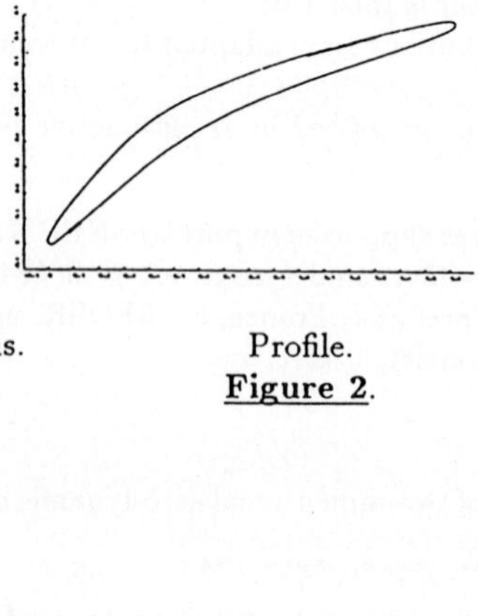

Profile.

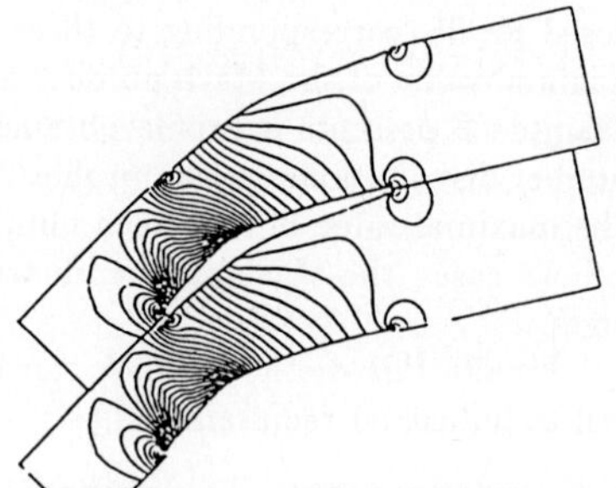

Lines of isomach.

Figure 2.

Example 2.

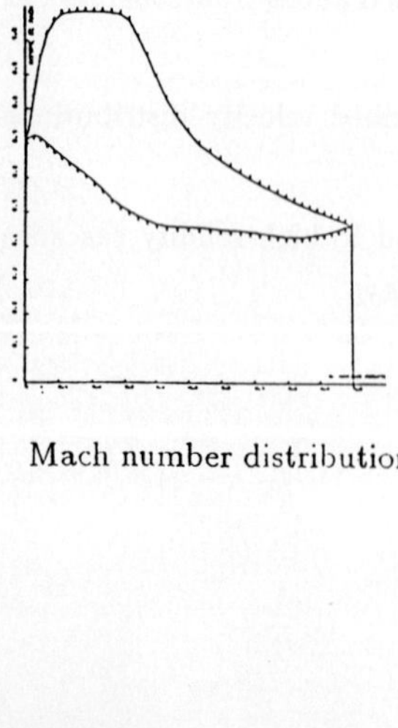

Mach number distributions.

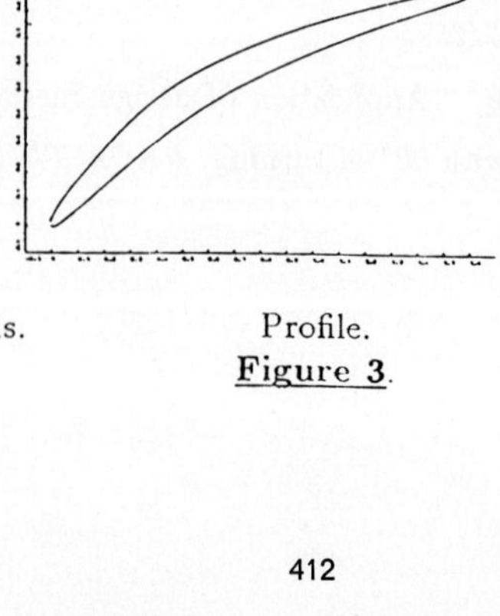

Profile.

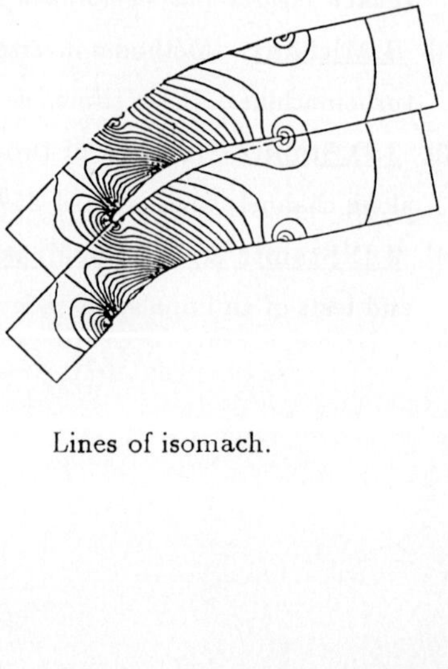

Lines of isomach.

Figure 3.

Large Eddy Simulation of the Turbulent Flow in a Curved Channel

Su Mingde

Tsinghua University
Beijing, PR China

Large eddy simulation (LES) is an efficient method, which is developed for the last ten years and more. In 1986 [1-3] we developed an algebraic modelling of the LES and used it in the numerical simulation of the turbulent flow in a straight channel. In the present paper this modelling is extended into the numerical simulation of the turbulent flow in a curved channel and some useful results are recieved.

1. Principle equations of LES

In Fig.1 the geometric shape of the flow field and the coordination system are shown, where s is longitudinal coordinate along the inner wall, n is normal distance from the inner wall, z is width direction. Here it is assumed that the fluid is incompressible and Newtonian. The principle equations in vector form are:

Equation of continuum:

$$\nabla \cdot \vec{V} \quad 0 \tag{1}$$

Equation of momentum:

$$\frac{d\vec{V}}{dt} = -\frac{1}{\rho}\nabla p + \nu \Delta \vec{V} \tag{2}$$

Filtering the N-S equations (1),(2) with the white-noise filter, we obtain the filtered principle equations:

$$\nabla \cdot \overline{\vec{V}} = 0 \tag{3}$$

$$\overline{\frac{\partial \vec{V}}{\partial t}} + (\overline{\vec{V}} \cdot \nabla)\overline{\vec{V}} = -\frac{1}{\rho}\nabla \bar{p} + \mathrm{Div}(2\nu \overline{\overline{S}} + \overline{\overline{\tau}}) \tag{4}$$

where

$$\overline{\overline{\tau}} = \{-\overline{v_i' v_j'}\} \qquad (i,j = s, n, z) \tag{5}$$

2. Algebraic modelling

To closure the eqs. (3),(4), the modelling for terms $-\overline{v_i' v_j'}$ ... must be introduced. Here the algebraic modelling is used, where are some assumptions:

The quasi-Reynolds stresses are diveded into two parts: isotropic and inhomogeneous part. The usual modelling for the isotropic part is Bossinesq assumption:

$$-(\overline{V_i' V_j'})_{iso} = 2\nu^* \bar{S}_{ij} - \frac{1}{3}\delta_{ij}\overline{V_\ell' V_\ell'} ;$$

$$\nu^* = f(c\Delta)^2 S ; \quad S = (\tfrac{1}{2}\bar{S}_{\ell k}\bar{S}_{\ell k})^{1/2} ; \quad f = \frac{c}{s}\frac{\overline{E'}^{1/2}}{\Delta} ; \quad \Delta = (h\Delta s \Delta n \Delta z)^{1/3} \tag{6}$$

The inhomogeneous part is

$$-(\overline{V_i' V_j'}) = \nu^{ij} < \bar{S}_{ij} > \tag{7}$$

where $\langle . \rangle$ denote the time-average value.

In addition we assume that the Re number of the flow is high, the tuebulent flow is 2-d, steady in the statistical means. The 3rd correlations of velocities are ignored, and the grids are enough small. After the above mentioned basic assumptions, we can establish the algebraic eqs. about the value $\sigma = \overline{E'}^{1/2}/\Delta$:

$$|\mathring{A}_{ss}| S_{ss} + 2|\mathring{A}_{sn}| S_{sn} + |\mathring{A}_{nn}| S_{nn} + C_E \frac{\overline{E'}^{1/2}}{\Delta} |A| = 0 \tag{8}$$

where $\mathring{A}_{ss}$, ... are functions of σ . This is a 4-power equation. The minimum non-negativ root of eq. (8) is the value of σ. Because the value of $\overline{E'}^{1/2}$ is obtained with solution of an algebraic equation, so this modelling is named as algebraic modelling.

3. Conditions to determinate solution

Boundary conditions:

In z-direction, $\bar{V}_s$, $\bar{V}_n$, $\bar{V}_z$, $\bar{p}$ are periodic with period 2H, where H is width of the channel.

There are adhesive conditions at the well:

$$\bar{V}_s = \bar{V}_n = \bar{V}_z = 0 \tag{9}$$

The condition of pressure p at wall is

$$\partial\bar{p}/\partial n = 0 \tag{10}$$

THe distribution of velocities and pressure at entry are known, which are obtained after the computational results at the exit of a straight channel.

The condition at exit is

$$\partial f/\partial s = 0 \tag{11}$$

where f can be $\bar{V}_s$, $\bar{V}_n$, $\bar{V}_z$.

Initial conditions:

For simplification, we take the computational velocities and pressure in a straight channel as initial field.

4. Numerical results and conclusion

Because of the violent variation of the flow near the wall, so we take denser grids near wall than the grids far away the wall. To provide the stability of the computation the semi-explicit and semi-implicit difference scheme is used, in addition the spectral method is used.

The values of the constant coefficients in the modelling are shown in the following table:

c_ε	c_p	c_1	c_2	c
1.10	.3	2.5	.75	.17

In Fig.1, L_1=6H, L_2=7.5H, H=1, radius of the curvature R is equal to 9.25H. There are two nets: sparse net 34X16X16 and fine net 68X32X32. The CPU time for every time step is 125.0 seconds. At first the calculation is made at sparse net, then at fine net. Figs. 2-11 shows the results.

These results show that the velocities of the fluid in the inner side is greater than in the outside. On the wall with concave curvature the intensity of turbulence is increased, and on the wall with convex curvature it is reduced. The second vortex is not so clear. In general, the computational results coincide with the experiment. So the algebraic modelling is available.

References

1.Su, M.D. ACTA MECHANICA SINICA, implement (1987)
2.Su, M.D. Lecture Notes in Phys. No.264(1986)
3.Su, M.D. Inner report (1987)

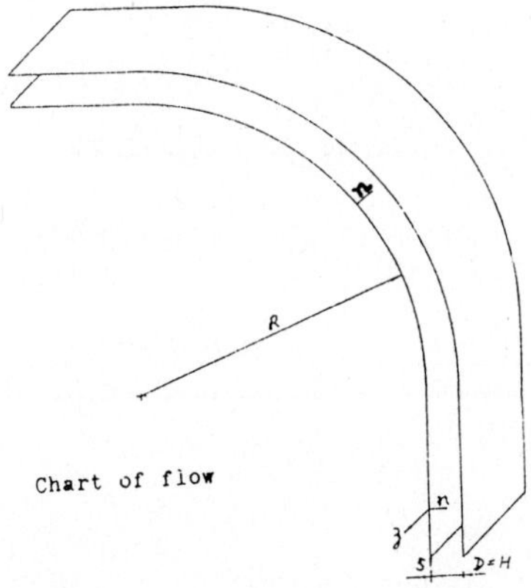

Fig.1 Chart of flow

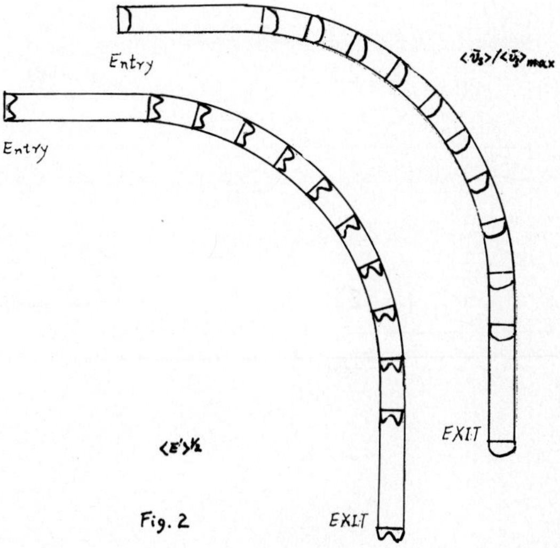

Fig. 2

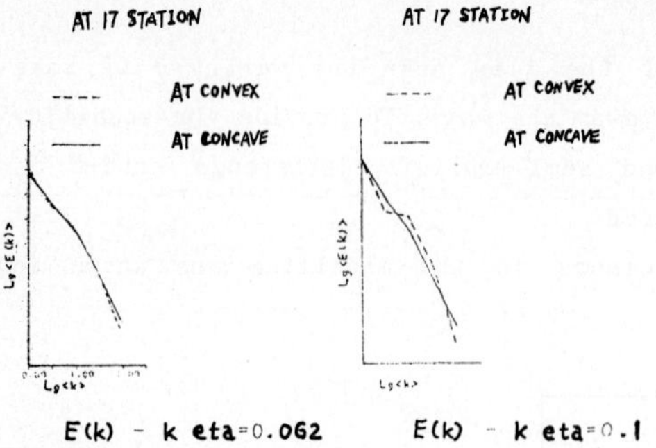

Fig. 3

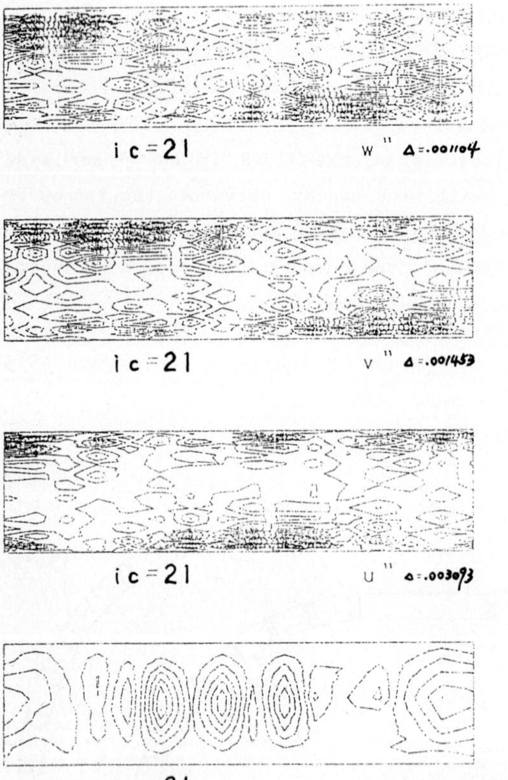

Fig. 4

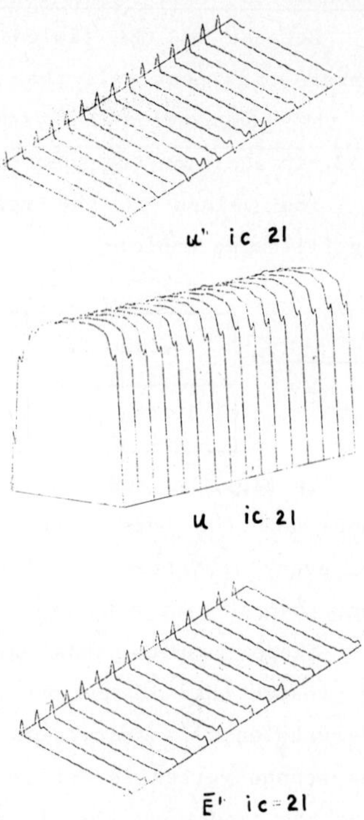

Fig. 5

omegn ic 17

Fig.8

omegz ic=17

Fig.9

$\Delta V_s'=.09129$

$\Delta V_n'=.001256$

$kc=12$

$\Delta V_z'=.00099$

V_s', V_n', V_z' c o u n t o r

Fig.6

omegz" ic 17

Fig.10

omegs ic=17

$\omega_s, \omega_n, \omega_z''$ - c o u n t o r

Fig.7

$\Delta \omega_z=.00256$

$kc=12$

$\Delta \omega_z''=.06925$

$\Delta \omega_n=.007845$

Fig. 11

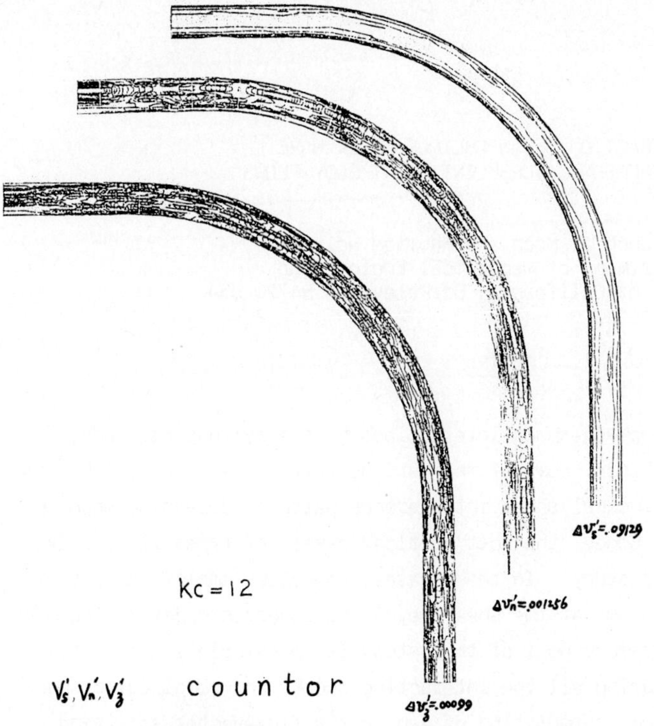

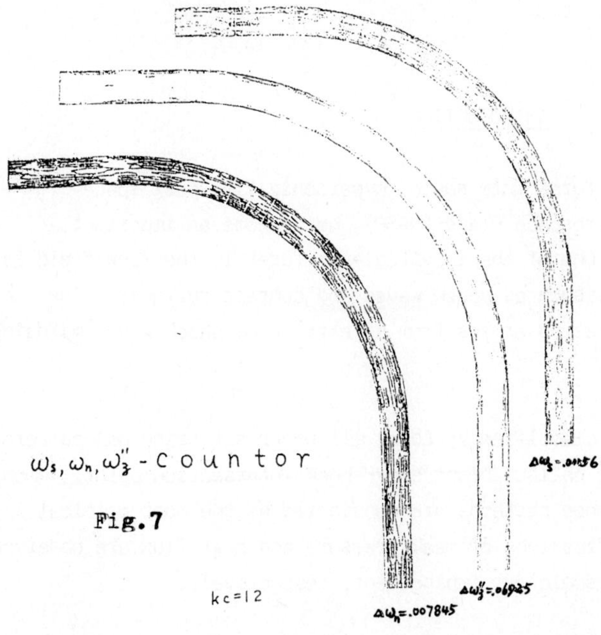

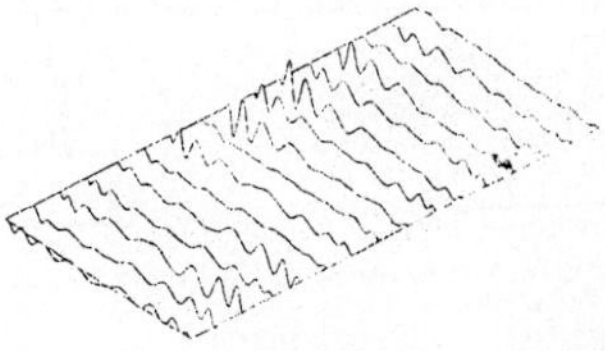

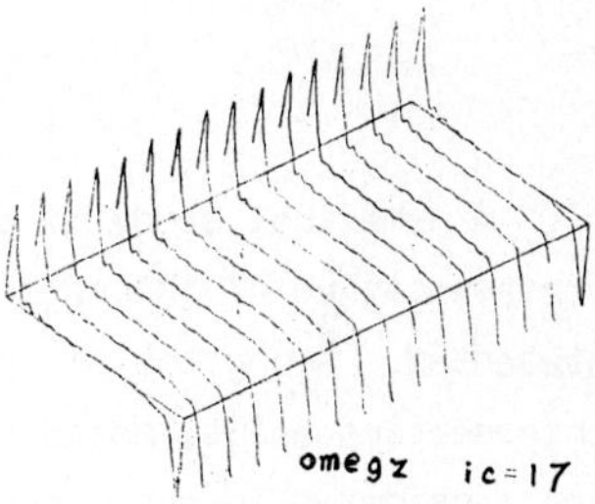

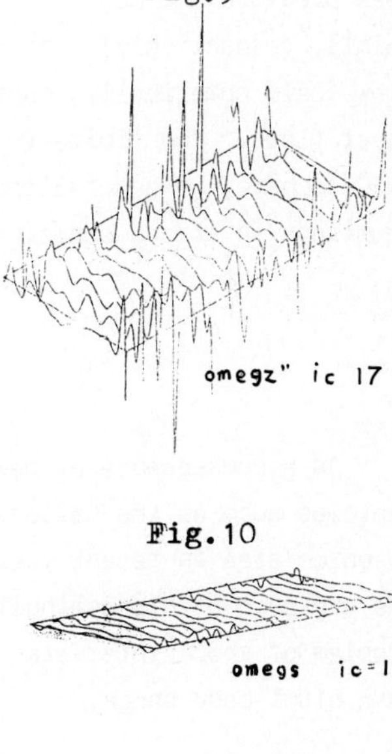

INTERACTION OF AN OBLIQUE SHOCK WAVE WITH SUPERSONIC TURBULENT BLUNT BODY FLOWS

Young J. Moon and Maurice Holt
Department of Mechanical Engineering
University of California, Berkeley, CA 94720 USA

ABSTRACT

A numerical study of shock-on-shock interactions near a cylindrical body, representative of the engine inlet cowl of the National Aerospace Plane (NASP), is presented. Among the six principal shock interference patterns depending upon the intersection point, noted by Edney, the most critical cases, of types III and IV, are considered in the present study. In these cases, anomalous amplifications of peak pressure and heat flux occur at the shear layer and supersonic jet impingement points, respectively. The primary goal of this study is to calculate the entire flow field numerically, capturing all the interacting shocks and complicated shock-layer flows. The finite volume formulation of Van Leer's flux-vector splitting MUSCL scheme, in generalized coordinates, is used to solve the full Navier-Stokes equations in strong conservative form.

INTRODUCTION

As a consequence of new ventures into space, hypersonic flow past space vehicles such as the National Aerospace Plane (NASP), has become an important research area in recent years. One of the inevitable features in the flow field is the presence of discontinuities, such as shock waves and contact surfaces. The problem of shock interference heating arises from an extraneous shock wave impinging on a blunt body shock.

The interactions have been classified by Edney [1] under six principal patterns depending on the location of the oblique-blunt body shock intersection point. Among them, types III and IV interference patterns are considered as the most critical cases, in which anomalous amplifications of peak pressure and heat flux are observed due to the shear layer and supersonic jet impingement, respectively.

A type III interference pattern occurs when an oblique shock impinges on a blunt bow shock just inside the upper sonic line. Downstream of two triple points generated by the shock interactions free shear layers emanate, one of which impinges on the body. Key parameters in this type III pattern are the angle the shear layer makes with the tangent to the body surface (shear layer angle) and the status of the shear layer. The shear layer angle will determine the deflection of the shear layer and its attachment to the body. It also has been shown [1-3] that the heat transfer rate is strongly dependent on whether the shear layer and boundary layers are laminar, transitional, or turbulent. Correlations of the experimental data and Schlieren photographs indicate that the free shear layer and boundary layers are turbulent.

The type IV interference pattern occurs when the impinging shock strikes the near-normal part of the blunt bow shock. In consequence, the supersonic jet embedded in the subsonic portion of the blunt body flow impinges nearly normal to the surface of the body. This interference pattern produces the largest heat transfer rate and pressure peaks due to the jet impingement on the body. The flow downstream of the embedded oblique shock still remains supersonic and therefore terminates by forming a jet bow shock near the body. In the region between the first embedded oblique shock and terminating bow shock, the flow passes through expansion waves and embedded oblique shocks in sequence. This process continues as long as the distance is sufficient and the flow is not exhausted to subsonic. Correlations [1,2] of the experimental data for the nearly perpendicular impingement case indicated that the peak heat transfer rate varies proportionally to the square root of the peak pressure, which supports the presence of laminar interaction.

NUMERICAL ALGORITHMS

The finite volume formulation of the Van Leer's flux-vector splitting MUSCL scheme [4] in generalized coordinates is used to solve the time-dependent compressible full Navier-Stokes equations in strong conservation law form. The scheme is fully upwinded, second order accurate, or upwind-biased, third order accurate, for the convective and pressure terms, and central-differenced, second order accurate, for the diffusion terms. The min-mod type limiter is also employed to limit the higher order terms whenever they are responsible for generating oscillatory solutions near discontinuities. The backward Euler implicit scheme is used for time integration, since only the steady-state solution is of interest. The first order approximation is made in the left hand side, which results in a block tridiagonal matrix structure. The left hand side matrix is then inverted with the approximate factorization method and LU decomposition technique. The free stream condition is used as an initial condition with the no-slip and constant wall temperature conditions at the wall surface and with the second order exptrapolation at the supersonic outflow boundaries. A Baldwin-Lomax algebraic turbulence model [5] is also used with the present method for the boundary layer in the type III+ case, defined as a transitional stage between the types III and IV.

NUMERICAL RESULTS

The present method is applied to the cases of type III+ and IV interference patterns which correspond to the cases of Runs 29 and 21, respectively, in the experiment [2].

The free stream flow conditions chosen for typeIII+ case are $M_\infty = 8.03$, $Re_D = 3.84\times10^5$, $Pr = 0.72$, $T_\infty = 222.6^\circ R$, $\rho_\infty = 1.546\times10^{-3}$ lbm/ft^3, $P_\infty = 0.128$ psia, $Q_{stag} = 62.4$ Btu/ft^{2}sec, and $\gamma = 1.4$, with a cylinder diameter $D = 3$ in. The wall temperature of $T_w = 530^\circ R$ is also assumed constant along the cylinder surface for both cases. The grid, consisting of 151 and 81 points in the ξ and η directions, respectively, is shown in Fig. 1. The Mach number contours plotted in Fig. 2 with $\Delta M = 0.25$ indicate that the interference pattern of this case is type III+ which is characterized by three flow features: (i) infant stage of supersonic jet, (ii) terminating jet bow shock, and (iii) boundary layer separation. Since the shear layer angle exceeds the maximum turning angle, the boundary layers are separated. A very mild separation is generated in the upward moving boundary layer, meanwhile the downward moving boundary layer possesses two separation bubbles. These are also indicated in Figs. 3 and 4. The pressure distribution along the body is compared with experiment in Fig. 3. It agrees reasonably well with experiment, matching the stagnation point correctly. The overprediction of peak pressure in the present calculation may result partially from grid resolution for the shear layer region but also from the assumption of laminar flow in the impinging shear layer. The laminar and turbulent heat flux distributions along the body are compared with experiment in Fig. 4. The global trend agrees reasonably with experiment, but heat fluxes of the present laminar solution are underpredicted. However the turbulent solution agrees favorably with experiment with the same magnitudes of heat fluxes in the region near the impingement point. The present turbulence model does not correctly predict heat fluxes at the local shear layer impinging region and the two separation regions. Regionally proper turbulence models will be further investigated and included in ref. [6].

The free stream flow conditions chosen for the type IV case are $M_\infty = 8.03$, $Re_D = 3.88\times10^5$, $T_\infty = 219.8^\circ R$, $\rho_\infty = 1.554\times10^{-3}$ lbm/ft^3, $P_\infty = 0.126$ psia, and $Q_{stag} = 61.7$ Btu/ft^{2}sec. The grid, consisting of 121 and 81 points in the ξ and η directions respectively, is shown in Fig. 5. The Mach number contours plotted in

Fig. 6 with $\Delta M = 0.25$ clearly show type IV interference pattern. Also, due to the excess of maximum turning angle, some downward moving streams, beyond jet impingement, depart from the wall, carrying high values of momentum. As they move downstream, they are pushed back to the wall by the flow downstream of the lower displaced bow shock. Therefore another departure occurs at a certain point in the rear region. This behavior of the downward moving wall jet results in a very low pressure regime along the downstream section of the body. In the region near the second departure, the wall jet reaches a Mach number close to 5 and the resulting pressure reaches a value close to the vacuum state. The pressure distribution along the body is compared with experiment in Fig. 7 and agrees quite favorably with it. However the peak pressure is overpredicted by approximately 35%. Probably the grid resolution in the region along the stagnating streamline is mainly responsible for this inconsistency. The heat flux distribution along the body is also compared with experiment in Fig. 8. In that plot, two sets of experimental data are displayed. The white circles correspond to the original data [2], while the corrected version of the measurements is represented by the black circles. The present calculation substantially underpredicts the heat fluxes along the body, especially for the impinging region, even though the global trend resembles the experiment. The numerical diffusions inherently generated from the present upwind scheme result in severe underprediction of heat fluxes for this extremely thin thermal boundary layer.

CONCLUSIONS

The present method captures complicated interacting shocks with monotonic properties, and compares anomalous surface pressures and heat fluxes reasonably well with experiment. However the peak pressure is overpredicted but the peak heat flux is underpredicted. The grid resolution in the overall interaction region seems to be responsible for the disagreement with experiment. In addition the heat flux calculation is quite sensitive to the grid resolution in the boundary layer, especially for the type IV case. The present turbulence model used for the boundary layer gives some encouraging results for surface heat flux calculations. Regionally proper turbulence models will be further investigated.

REFERENCES

[1] Edney, B.E., "Anomalous Heat Transfer and Pressure Distributions on Blunt Bodies at Hypersonic Speeds in the Presence of an Impinging Shock," FFA Rept. 115, Feb. 1968, The Aeronautical Research Inst. of Sweden, Stockholm, Sweden.

[2] Wieting, A.R., "Experimental Study of Shock Wave Interference Heating on a Cylindrical Leading Edge," NASA TM 100484, May 1987.

[3] Keyes, J.W., "Correlation of Turbulent Shear Layer Attachment Peak Heating Near Mach 6," AIAA Journal, Vol. 15, Dec. 1977, pp. 1821-1823.

[4] Van Leer, B., "Flux-Vector Splitting for the Euler Equations," Lecture Notes in Physics, Vol. 170, 1982, pp. 507-512.

[5] Baldwin, B.S., and Lomax, H., "Thin Layer Approximation and Algebraic Model for Separated Turbulent Flows," AIAA Paper 78-257, presented at AIAA 16th Aerospace Science Meeting, Huntsville, Alabama, Jan. 1978.

[6] Moon, Y.J., "Turbulence Modelling on Interaction of an Oblique Shock with Viscous Hypersonic Blunt Body Flows," abstract submitted to AIAA 27th Aerospace Science Meeting to be held in Reno, Nevada, Jan. 1989.

[7] Moon, Y.J., "Interaction of an Oblique Shock Wave with Supersonic Flow Past a Blunt Body," Ph.D. Dissertation, University of California at Berkeley, June 1988.

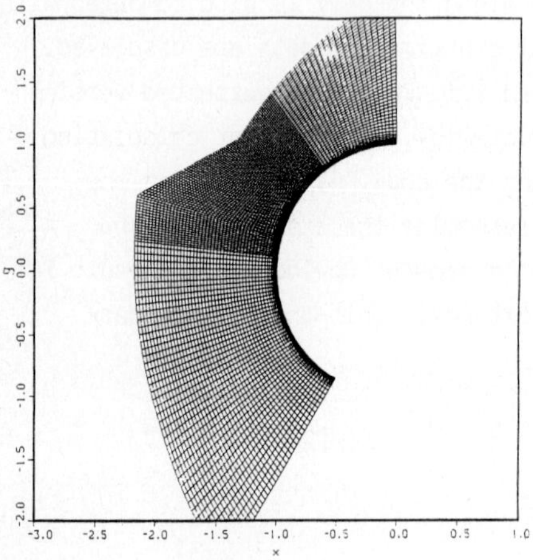

Fig. 1. Grid (151x81), $M_\infty = 8.03$, type III$^+$

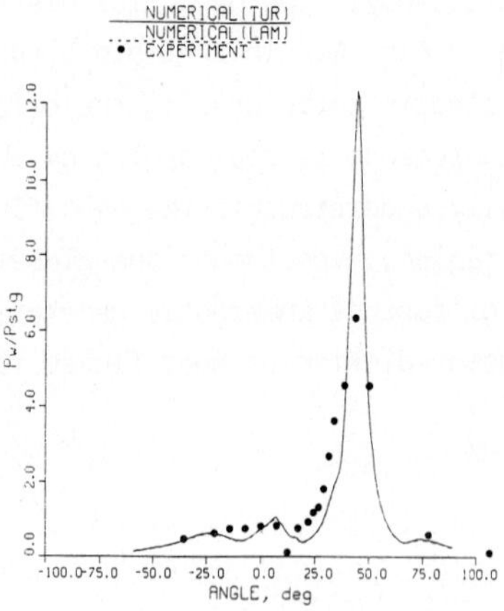

Fig. 3. Pressure distribution along the cylinder, $M_\infty = 8.03$, type III$^+$

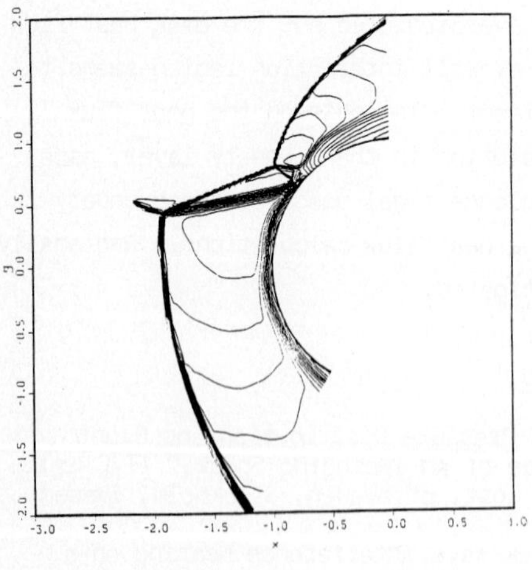

Fig. 2. Mach number contours, $M_\infty = 8.03$, $\Delta M = 0.25$, type III$^+$

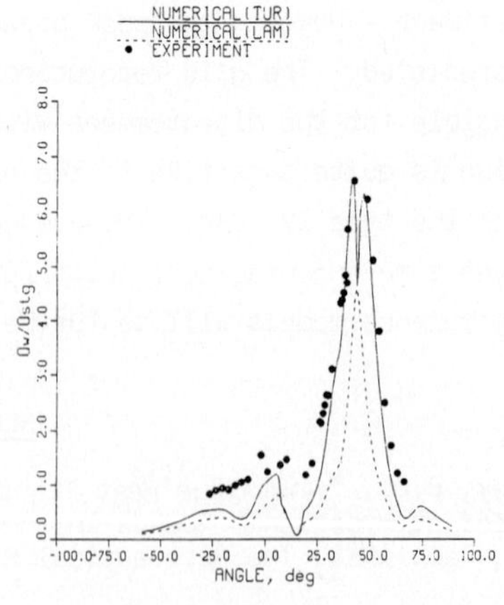

Fig. 4. Heat flux distribution along the cylinder, $M_\infty = 8.03$, type III$^+$, Turbulent

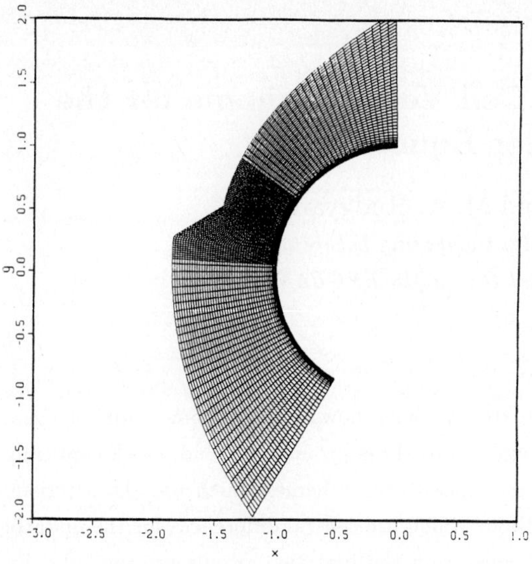

Fig. 5. Grid (121x81), $M_\infty = 8.03$, type IV

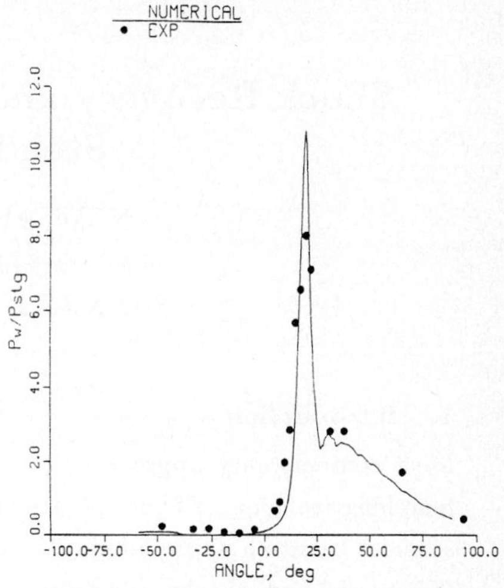

Fig. 7. Pressure distribution along the cylinder, $M_\infty = 8.03$, type IV

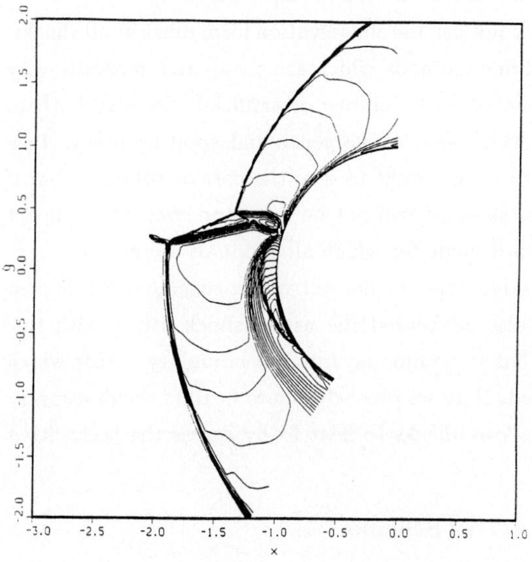

Fig. 6. Mach number contours, $M_\infty = 8.03$, $\Delta M = 0.25$, type IV

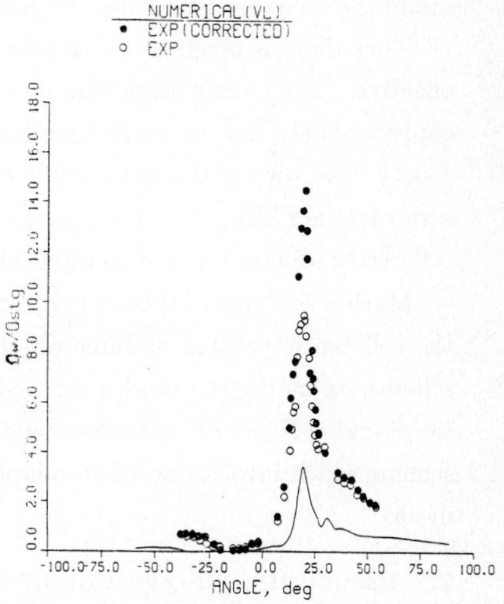

Fig. 8. Heat flux distribution along the cylinder, $M_\infty = 8.03$, type IV

Shock Recovery and the Cell Vertex Scheme for the Steady Euler Equations

K. W. Morton and M. A. Rudgyard
ICFD, Oxford University Computing Laboratory,
8-11 Keble Road, Oxford OX1 3QD, ENGLAND.

1. Introduction

Most contemporary approaches to calculating steady Euler flows employ some form of time-marching technique. They can be broadly divided into three groups: upwind shock-capturing schemes based on approximate Riemann solvers; shock-fitting schemes which use characteristic forms away from shocks; and finite volume methods which use central differences for the updates and artificial viscosity to capture shocks. Schemes from the first two groups are undoubtedly needed for truly unsteady flows. However, they are often unnecessarily expensive for steady flows and in any case far from perfect: the upwind schemes essentially superimpose one dimensional calculations and cannot adequately propagate waves travelling at oblique angles to a body-fitted mesh; and characteristic-based schemes which do not use the conservation form must fit all shocks.

Thus there is much to be said for finite volume methods which are cheap and generally very effective. Their main disadvantages are associated with the use of artificial viscosity both to capture shocks and to damp spurious modes which slow convergence and spoil accuracy. In a steady flow, setting the finite volume residuals to zero ought to give the correct solution: but if a shock is not aligned with the mesh a residual across it will not be zero; and even the compact cell vertex scheme has the spurious chequer-board mode for which all residuals are zero.

Morton & Paisley [4] have pointed out the advantages of the cell vertex scheme of Ni [5] over the cell centre scheme of Jameson [1]. They also advocated the use of shock fitting with this scheme, adjusting the mesh near a shock so that it became an internal boundary across which the Rankine-Hugoniot relations could be applied. Here we present a more flexible shock recovery scheme which involves no mesh adaption and allows shocks to float freely across the body-fitted mesh.

2. Pseudo-time Integration of the Cell Vertex Scheme

In the cell vertex scheme the steady integral law,

$$\oint_{\partial\Omega} \mathbf{f}dy - \mathbf{g}dx = \mathbf{0}, \tag{2.1}$$

is replaced by the compact approximation,

$$\begin{aligned}&(\mathbf{f}_1 - \mathbf{f}_3)(y_2 - y_4) + (\mathbf{f}_2 - \mathbf{f}_4)(y_3 - y_1) - \\ &(\mathbf{g}_1 - \mathbf{g}_3)(x_2 - x_4) - (\mathbf{g}_2 - \mathbf{g}_4)(x_3 - x_1) = \mathbf{0},\end{aligned} \tag{2.2}$$

where Ω is taken to be the quadrilateral C of Fig.1. For error analysis, it is convenient to normalise the left-hand side of (2.2) by twice the cell area V_Ω and denote the result as the cell *residual*, $\mathbf{R}_\Omega$.

To drive the cell residuals to zero, we use the time stepping algorithm of Morton [2], which was derived as a mass-lumped Taylor-Galerkin technique. Like the Lax-Wendroff formulation of Ni [5], this has the useful property that a nodal update uses a linear combination (with matrix coefficients) of neighbouring cell residuals. Such an update scheme is easily programmed as a distribution of each cell residual to its four nodes. For example, referring once again to Fig.1 for the integration of $div(\mathbf{AR}, \mathbf{BR})$ round the dashed line, the contribution from cell C to node 1 is

$$\delta \mathbf{w}_1^C = -\frac{\Delta t_1}{\Sigma V_\Omega}\{V_C \mathbf{I} - \Delta t_C[(y_4 - y_2)\mathbf{A}_C - (x_4 - x_2)\mathbf{B}_C]\}\mathbf{R}_C, \tag{2.3}$$

where $\mathbf{w}$ is the vector of conserved variables, Δt_C is a local time step associated with cell C and $\mathbf{A}_C$, $\mathbf{B}_C$ are the jacobian matrices $\partial \mathbf{f}/\partial \mathbf{w}$, $\partial \mathbf{g}/\partial \mathbf{w}$ approximated at the cell-centre: ΣV_Ω denotes the sum of the areas of the four neighbouring cells and Δt_1 is the minimum of the time steps associated with them.

During the shock capturing phase used to roughly locate shocks, a simple artificial viscosity as in [4] is used. This can also be written as distributing a cell contribution to its vertices. For example, the contribution from cell C to node 1 is

$$\delta \mathbf{w}_1^C \,|_{diss} = (\epsilon_1 + \epsilon_2)(\mathbf{w}_C - \mathbf{w}_1), \tag{2.4}$$

where $\mathbf{w}_C = (\mathbf{w}_1 + \mathbf{w}_2 + \mathbf{w}_3 + \mathbf{w}_4)/4$. The first coefficient ϵ_1 is scaled by a norm of the cell residual so that it introduces heavy damping near an oblique captured shock: the second coefficient ϵ_2 is numerically much smaller and is used to control the chequer-board mode.

3. The Shock Detection and Recovery Algorithms

The first stage of any shock recovery scheme is to recognise the position of shocks relative to the underlying mesh. This needs to be relatively cheap, since for generality most of the mesh should be scanned, as well as reliable and local enough to detect interacting shocks. We have chosen for the present study a test which is applied to every cell edge to detect whether a shock crosses it. From the state vectors at the two nodes defining the edge, one can also obtain approximations to the shock orientation and upstream and downstream states. First, we define a possible shock normal,

$$\hat{n} = -\text{sgn}(\Delta p)\frac{\Delta \vec{u}}{|\Delta \vec{u}|}, \tag{3.1}$$

where Δp and $\Delta \vec{u}$ denote the differences in the pressure and the velocity vector for the pair of nodes. (Note that the pressure difference is used to distinguish between the upstream and downstream sides of the compressive shock.)

Criteria for detecting steady shocks can be much simpler than those for unsteady shocks, for the shock is fully developed and should be stationary. We need the normal components of the velocities, $U = \vec{u}.\hat{n}$, and by neglecting the shock speed we would detect a shock if

$$
\begin{aligned}
(U-a)_u &\geq 0, \\
(U-a)_d &\leq 0,
\end{aligned}
\tag{3.2}
$$

where the subscripts u and d denote the upstream (low pressure) and downstream (high pressure) sides of the shock respectively, and a is the speed of sound, $\sqrt{(\partial p/\partial \rho)}$. Finally, we check to see if the shock is physically compressive and is stronger than some positive pressure tolerence, ϵ_p, i.e.

$$
(\Delta p)\mathrm{sgn}(\Delta U) \leq -\epsilon_p. \tag{3.3}
$$

Once a shock is detected crossing an edge, further information about it can be recovered: for example, one may refine the approximation to the states either side by using other than nearest-node values, although this has not so far been found necessary; one can also estimate the shock position and, for a set of edge-crossings defining a shock, gaps can be filled in and positions smoothed—little of this has been done because shock positions are not much used in the update algorithm. Shock information can be held in lists, ordered according to the mesh structure rather than explicitly identifying individual shocks, so that a mesh-wide detection scan is not necessary at each update.

4. The Update Algorithm with Shock Recovery

The update algorithm consists of adjusting nodal values to resolve discrepancies from the steady state, i.e. nonzero cell residuals and unsatisfied Rankine-Hugoniot conditions at shock edge-crossings. The latter are dealt with first by adjusting nodal fluxes before calculating residuals, through imposing the unsteady jump condition,

$$
\mathbf{F}_d = \mathbf{F}_u - \sigma(\mathbf{w}_u - \mathbf{w}_d), \tag{4.1}
$$

where $\mathbf{F}$ denotes the resolved normal flux $(\mathbf{f}, \mathbf{g}).\hat{n}$: it has been found sufficient to do this by calculating the shock speed σ from the mass conservation condition, whilst resetting all other downstream fluxes with (4.1) and the unchanged tangential component of the fluxes.

The cell residuals can then be calculated in the usual way. However, residuals for shocked cells (i.e. those with at least one edge crossed by a shock) will not be used in the nodal update: these are therefore simply set to zero so that the update procedure of §2 can be applied directly.

Note that the simple, very local procedure so far described makes no use of the shock position within an edge. However, it would not easily allow a shock to adjust its position to a neighbouring cell edge. The appropriateness of such a move can be deduced from the shock position in the edge and the shock speed given by (4.1): after the move, the values at the node crossed by the shock are reset by extrapolation.

5. Numerical Results

The algorithm has been tested on a variety of transonic and supersonic model problems and compared with the adapted mesh algorithm of [4] and the non-adapted mesh algorithm of Morton, Childs and Rudgyard [3] which used a more global representation of the shock curve. We show results in Figs.2-4 for supersonic flow at $M_\infty = 8.15$ past a double ellipse at angle of attack $\alpha = 30°$: Fig.2 shows a partially detected bow shock and a canopy shock with normals at each edge-crossing; Fig.3 gives contour plots of the pressure coefficient (Note that the contouring routine's use of bilinear interpolation between the known vertex values gives the shock a jagged appearance with a thickness of one mesh interval); and Fig.4 gives the pressure coefficient along $y = 0$ through the shock and on the upper and lower surfaces.

References

[1] A. Jameson. *Proc. IMA Conference on Numerical Methods in Aeronautical Fluid Dynamics*, (ed. P. L. Roe) pages 289–308, Academic Press, London, 1982.

[2] K. W. Morton. *The Mathematics of Finite Elements and Applications.*, (ed. J. R. Whiteman) pages 353–378, Academic Press, London, 1988.

[3] K. W. Morton, P. N. Childs, and M. A. Rudgyard. Oxford Univ. Comp. Lab. Rep. NA88/5. (to appear in *Proc. IMA Conference on Numerical Methods for Fluid Dynamics*, Oxford, March 1988).

[4] K. W. Morton and M. F. Paisley. Oxford Univ. Comp. Lab. Rep. NA87/6. (to appear in J. Comp. Phys.).

[5] R. H. Ni. AIAA J., 20, No.11, pages 1565–1571, 1981.

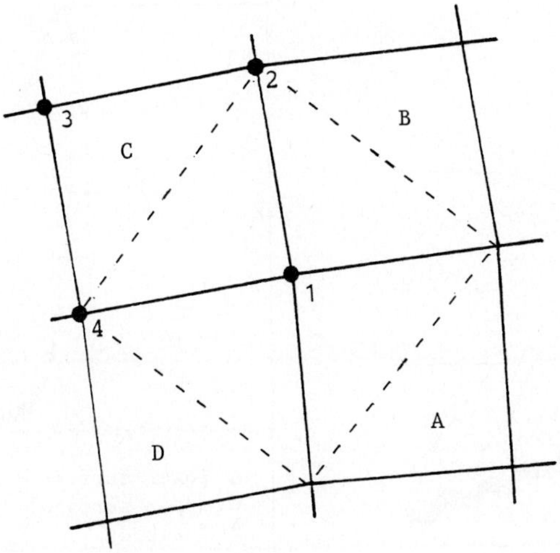

Fig.1 Geometry for flow variables at cell vertices; and the cells used to update vertex 1.

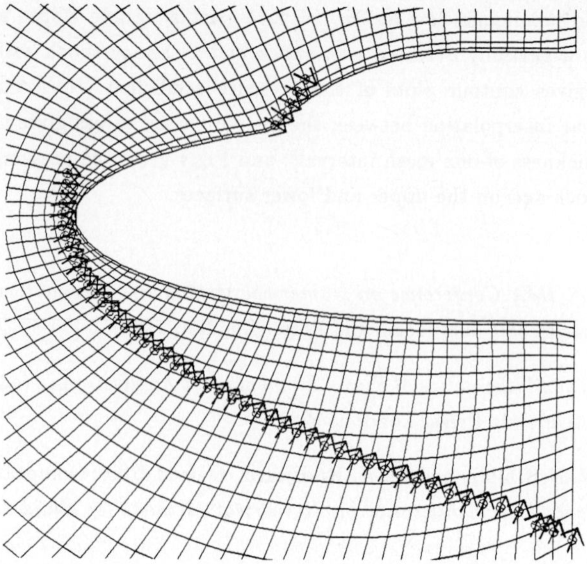

Fig.2 Detected bow and canopy shocks, with shock normals, for the double-ellipse problem.

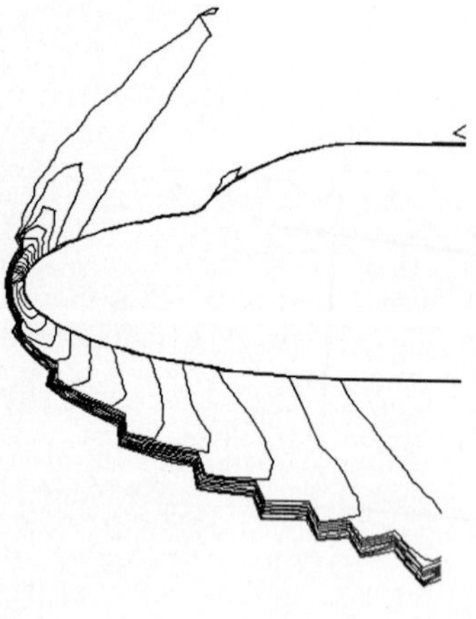

Fig.3 Contours of C_p as for Fig.2.

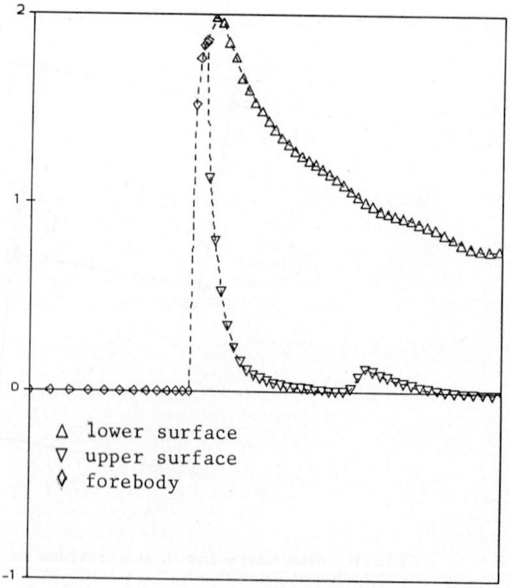

Fig.4 Plot along centre-line and upper and lower body surfaces of C_p, as for Fig.2.

A new multigrid approach to convection problems*

Wim A. Mulder
Department of Mathematics
University of California at Los Angeles
Los Angeles, CA 90024-1555

1. Introduction

One of the bottlenecks in the application of the multigrid method to the computation of steady-state flows is alignment, the flow being aligned with the grid [1,4]. Consider, as an example, the hyperbolic equation $u_t + au_x + bu_y = 0$. The steady-state problem becomes one-dimensional for $b \to 0$, $a \neq 0$, namely, $au_x = 0$. Thus the solution will be independent of y, unless some structure in the y-direction is imposed by the boundary conditions. A proper discretization will reflect this by being independent from neighboring values in the y-direction. Suppose now that the iteration error (the deviation from the steady state) has an arbitrary structure in the x-direction and is highly oscillatory in the y-direction. Restriction to a coarser grid will involve some sort of averaging in both co-ordinate directions. The oscillatory y-component will cause cancellation, and as a result, this error will not show up on the coarser grid and can not be removed by the multigrid method. It also can not be removed by smoothing on the finest grid, as would normally be done, because there is no coupling in the y-component. This component actually *must* remain unaffected if the discrete operator reflects the character of the differential equation. The only way to remove the error is by some relaxation scheme acting along the x-direction. If the x-component of the error is smooth, this will be an inefficient process. For a relaxation scheme that uses only local data, the communication of boundary data to the interior will take $O(N^{\frac{1}{2}})$ iterations, where N is the total number of points in the computational domain. Thus, grid-independent convergence rates can not be obtained.

Line relaxation is one way to overcome the problem of alignment. Damped alternating-direction Line-Jacobi relaxation provides uniformly good convergence rates for the Euler equations of gas dynamics in two dimensions [3]. Line relaxation requires a global linearisation of the residual, and can be avoided by using semi-coarsening. In this paper we present a new method that employs semi-coarsening in d co-ordinate directions simultaneously, where d is the number of space dimensions. We will mainly concentrate on the two-dimensional case, the generalization to three dimensions being obvious.

2. Method

Semi-coarsening can be implemented in several ways. The most obvious approach involves coarsening in alternating directions, but this does not solve the problem of alignment. It is necessary to apply semi-coarsening in two directions simultaneously. A straightforward approach starts out with, say, a 8×8 grid, and coarsens to a 4×8 and an 8×4 grid. The 4×8 grid is coarsened to 2×8 and 4×4, etc. At each coarser level, the number of grids doubles. This leads to an $O(N \log N)$ method in 2 dimensions. It becomes useless in 3 dimensions because the number of points increases too rapidly on progressively coarser levels.

The method proposed here has $O(N)$ complexity. Figure 1 shows the various grids employed if the finest grid has 8×8 points and the coarsest 1×1. In d dimension the cost of a V-cycle is proportional to $2^d N$, and of an F-cycle $(d+1)2^d N$. For a W-cycle, the $O(N)$ complexity is lost.

The method requires a modification of the usual restriction and prolongation operators, because data from more than one coarser or finer grid have to be combined. This modification is the crucial part of the present method. A convenient and simple scheme is proposed, which allows the use of parallel smoothing (parallel to the computation of the coarse-grid corrections).

Let a grid be denoted by (m_1, m_2). In the finite-volume case, it would have 2^{m_1} cells in the x-direction and 2^{m_2} in the y-direction. In the finite-difference case these numbers have to be increased by 1. On the finest grid, $m_1 = M_1$ and $m_2 = M_2$, whereas the coarsest has $m_1 = m_2 = 0$. The current solution on a grid is denoted by $u^{(m_1,m_2)}$, and the corresponding residual by $r^{(m_1,m_2)} = f^{(m_1,m_2)} - \mathcal{L}^{(m_1,m_2)}(u^{(m_1,m_2)})$, where $\mathcal{L}(u) = f$ is the nonlinear problem to be solved. Let the restriction operator be denoted by R for the residual and by $\tilde{R}$ for the solution. The initial solution on a coarser grid (m_1, m_2) on level $l = m_1 + m_2$ is found as the restriction of one or two solutions on level $l+1$, depending on the number of links to finer grids (cf. Fig.

* *This work has been supported in part by the Center for Large Scale Scientific Computing (CLaSSiC) Project at Stanford under ONR grant N00014-82-K-0335, and in part by the NSF grant DMS85-03294 and the ONR grant N00014-86-K-0691 at UCLA.*

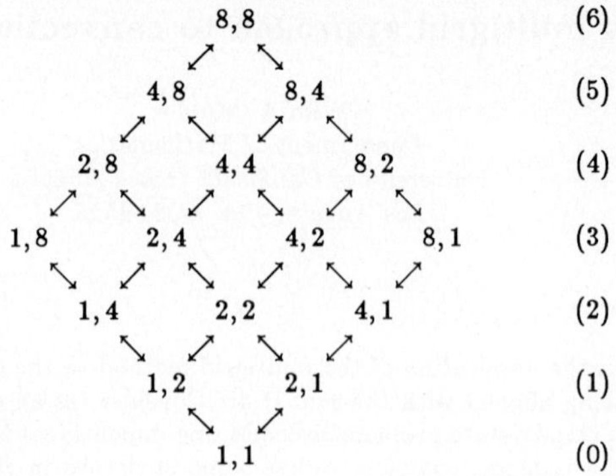

Fig. 1. *Arrangement of finest* (8×8) *and coarser grids, that leads to an* $O(N)$ *multigrid method for problems with alignment. The level numbers are given on the right. The arrows indicate how the grids are linked by restriction (downward) and prolongation (upward).*

1). Thus, the restriction operators R and $\tilde{R}$ correspond to semi-coarsening in the x-direction (R_x and $\tilde{R}_x$), or semi-coarsening in the y-direction (R_y and $\tilde{R}_y$), or a combination of both. The last is necessary if the coarser grid is linked to two finer grids. In that case, we give equal weights to the appropriately restricted data from each grid. An algorithmic description of the operations involved in restriction from level $l+1$ to l is:

$$
\begin{aligned}
&\text{for all } (m_1, m_2) \text{ with } m_1 + m_2 = l,\ 0 \le m_1 < M_1,\ 0 \le m_2 < M_2 \text{ do}\\
&\quad \text{if there are links to } (m_1+1, m_2) \text{ and } (m_1, m_2+1) \text{ then}\\
&\qquad u^{(m_1,m_2)} := \tfrac{1}{2}[\tilde{R}_x u^{(m_1+1,m_2)} + \tilde{R}_y u^{(m_1,m_2+1)}]\\
&\qquad \tau^{(m_1,m_2)} := \tfrac{1}{2}[R_x r^{(m_1+1,m_2)} + R_y r^{(m_1,m_2+1)}] - r^{(m_1,m_2)}(u^{(m_1,m_2)})\\
&\quad \text{else if there is only one link to } (m_1+1, m_2) \text{ then}\\
&\qquad u^{(m_1,m_2)} := \tilde{R}_x u^{(m_1+1,m_2)}\\
&\qquad \tau^{(m_1,m_2)} := R_x r^{(m_1+1,m_2)} - r^{(m_1,m_2)}(u^{(m_1,m_2)})\\
&\quad \text{else if there is only one link to } (m_1, m_2+1) \text{ then}\\
&\qquad u^{(m_1,m_2)} := \tilde{R}_y u^{(m_1,m_2+1)}\\
&\qquad \tau^{(m_1,m_2)} := R_y r^{(m_1,m_2+1)} - r^{(m_1,m_2)}(u^{(m_1,m_2)})\\
&\quad \text{end if}\\
&\text{end do}
\end{aligned}
\tag{2.1}
$$

The resulting coarse-grid problem becomes

$$\mathcal{L}^{(m_1,m_2)}(\tilde{u}^{(m_1,m_2)}) = f^{(m_1,m_2)} + \tau^{(m_1,m_2)}.$$

Here the solution $\tilde{u}^{(m_1,m_2)}$ can be determined approximately by a multigrid cycle with respect to the grid (m_1, m_2), or by a single-grid solver if this grid is chosen to be the coarsest.

The basic difference with the standard multigrid approach is the occurrence of two links to finer grids. We have chosen equal weighting of data from the two grids. If $u^{(m_1+1,m_2)}$ and $u^{(m_1,m_2+1)}$ have been obtained directly by restriction from the finest grid (M_1, M_2), then

$$
\begin{aligned}
u^{(M_1+1,M_2)} &= \tilde{R}_y u^{(M_1,M_2)}, \qquad u^{(M_1,M_2+1)} = \tilde{R}_x u^{(M_1,M_2)},\\
u^{(M_1+1,M_2+1)} &= \tfrac{1}{2}[\tilde{R}_x \tilde{R}_y + \tilde{R}_y \tilde{R}_x] u^{(M_1,M_2)}.
\end{aligned}
\tag{2.2}
$$

A similar expression is obtained for the restriction of the residual. For common choices of the restriction operator, such as nine-point restriction or volume-weighted restriction, we have $\tilde{R}_{xy} = \tilde{R}_x \tilde{R}_y = \tilde{R}_y \tilde{R}_x$, implying that (2.2) results in the same expression as when $\tilde{R}_{xy}$ is applied directly. The same is true for all coarser grids, that is, (2.1) produces the same result when going from the finest grid (M_1, M_2) to any arbitrary coarser grid (m_1, m_2), as one obtains by selecting an arbitrary path between these two grids and applying only $\tilde{R}_x$ or $\tilde{R}_y$. In practice, the solution will be modified on each grid before it is restricted to still coarser grids, and this equivalence disappears.

The prolongation operator brings corrections from one or more coarser grids to the current grid. We choose the following approach:

$$
\begin{aligned}
&\text{for all } (m_1, m_2) \text{ with } m_1 + m_2 = l,\ 0 < m_1 \le M_1,\ 0 < m_2 \le M_2 \text{ do}\\
&\quad u^{(m_1,m_2)} := \mathcal{S}^{\nu_p}\left(u^{(m_1,m_2)}, f^{(m_1,m_2)}, \mathcal{L}^{(m_1,m_2)}(\cdot)\right)\\
&\quad \text{if there is a link to } (m_1, m_2 - 1) \text{ then}\\
&\qquad u^{(m_1,m_2)} := u^{(m_1,m_2)} + P_y[u^{(m_1,m_2-1)} - \tilde{R}_y u^{(m_1,m_2)}]\\
&\quad \text{end if}\\
&\quad \text{if there is a link to } (m_1 - 1, m_2) \text{ then}\\
&\qquad u^{(m_1,m_2)} := u^{(m_1,m_2)} + P_x[u^{(m_1-1,m_2)} - \tilde{R}_x u^{(m_1,m_2)}]\\
&\quad \text{end if}\\
&\text{end do}
\end{aligned}
\tag{2.3}
$$

Here we have included ν_p parallel smoothing steps with some relaxation scheme $\mathcal{S}$. Next the one or more coarse-grid corrections are added to the solution on the current grid (m_1, m_2). The fundamental difference with standard multigrid is that the corrections are always computed with respect to the restriction of the most recent solution rather than the one at the begin of the coarse-grid correction cycle. Choosing the y-direction first in (2.3) results in an asymmetry of the algorithm with respect to the co-ordinate directions. In the actual code used in Sect. 3, this asymmetry is reduced by alternating the order of prolongation in x and y each time (2.3) is carried out.

The restriction operator (2.1) uses equal weights when more than one finer grid exists. The prolongation operator (2.3) can also be expressed in terms of weights:

$$
\begin{aligned}
u^{(m_1,m_2)} := &\left[P_x u^{(m_1-1,m_2)}\right] + (I - P_x\tilde{R}_x)\left[P_y u^{(m_1,m_2-1)}\right]\\
&+ (I - P_x\tilde{R}_x)(I - P_y\tilde{R}_y)\left[\mathcal{S}^{\nu_p}\left(u^{(m_1,m_2)}, f^{(m_1,m_2)}, \mathcal{L}^{(m_1,m_2)}(\cdot)\right)\right]
\end{aligned}
\tag{2.4}
$$

This shows that the weight for the first term on the right-hand side is I, for the second $(I - P_x\tilde{R}_x)$, and for the third $(I - P_x\tilde{R}_x)(I - P_y\tilde{R}_y)$. Because $P_y\tilde{R}_y$ and $P_x\tilde{R}_x$ approach I for the low frequencies, the low-frequency component of the new solution is mainly determined by $P_x u^{(m_1,m_2-1)}$. This preference with respect to one co-ordinate direction can be reduced by the alternating approach mentioned above. Then the next prolongation would result in an expression similar to (2.4), but with x and y interchanged.

In addition to the above parallel smoothing, one can perform ν_1 pre- and ν_2 post-smoothing steps, just as in the standard multigrid method. The method can in principle be used for any problem with strong anisotropy. However, it can not handle alignment at 45°. This is not important for the example discussed next.

3. Application to the Euler equations of gas dynamics

The method has been applied to the nonlinear Euler equations that describe the flow of an inviscid ideal compressible gas. The system of 4 equations is discretized by upwind differencing, providing solutions with first-order accuracy. The restriction operator is based on volume-averaging, the prolongation operator employs piecewise constant interpolation (cf.[4]).

Convergence rates are estimated by two-level analysis for the linear constant-coefficient case. With damped Point-Jacobi as smoother, a worst-case convergence factor of 0.730 per multigrid cycle is obtained if one smoothing step parallel to the coarse-grid corrections is performed. If only one pre- or post-smoothing step is carried out, the worst-case convergence factor becomes 0.500. These convergence factors are the worst-case values for any grid-size and aspect ratio of the grid.

Figure 2 shows the two-level convergence factor $\overline{\lambda}_r(u/c, v/c)$ for a grid with aspect ratio $h_y/h_x = 1$ and size 64×64, based on (2.4). Here h_x and h_y are the cell-sizes in both co-ordinate directions. The u and v are the velocities in the x- and y-direction, respectively, and c is the sound speed.

For the nonlinear equations, it is recommended to use post-smoothing because of shocks. The coarse-grid correction may want to move the shock a bit, and this results in large local errors after prolongation to a finer grid. These have to be removed by smoothing, otherwise convergence may be lost. Here we use one post-smoothing step with damped Point-Jacobi.

The nonlinear equations are discretized by either van Leer's Flux-Vector Splitting [8] or Osher's scheme in the natural ordering [6,7]. The matrix used for Point-Jacobi smoothing can be obtained directly from the Jacobians of the fluxes if Flux-Vector Splitting is used. With Osher's scheme one has to choose an inconsistent linearization to avoid eigenvalues close to zero.

Both schemes have been applied to flow through a channel with a circular arc at one of the walls. Convergence rates between 0.3 and 0.4 are found for subsonic, transonic, and supersonic flow, independent of grid-size, up to a finest grid of 128×64. The coarsest grid used during the F-cycle has 4×2 cells. In a Full Multigrid [1] code, where refinement doubles the number of cells in *both* co-ordinate directions, a stationary solution with an iteration error smaller than the discretization error is found in only 3 to 6 multigrid cycles. A coarser version of the grid used in the computations is displayed in Fig. 3a, whereas Figs. 3b-d show the converged solutions for the three cases.

The results shown are obtained with a first-order scheme. Higher-order accuracy is a requirement for practical computations, and can be obtained with the current method by using defect correction [2:(14.3.2)]. This will be discussed elsewhere [5]. A longer version of the current paper has been submitted to *J. Comp. Phys.*

References

[1] A. Brandt, *Guide to Multigrid Development*, Lecture Notes in Mathematics 960 (1981), 220-312.

[2] W. Hackbusch, *Multi-Grid Methods and Applications*, Springer Series in Computational Mathematics 4 (1985), Springer Verlag, Berlin/Heidelberg.

[3] W. A. Mulder, *A note on the use of Symmetric Line Gauss-Seidel for the upwind differenced Euler equations*, Manuscript CLaSSiC-87-20 (1987), Stanford University.

[4] W. A. Mulder, *Analysis of a multigrid method for the Euler equations of gas dynamics in two dimensions*, in *Multigrid Methods: Theory, Applications, and Supercomputing*, ed. S. McCormick, Marcel Dekker, New York (1988).

[5] W. A. Mulder, *A high-resolution Euler solver*, in prep.

[6] S. Osher and F. Solomon, *Upwind difference schemes for hyperbolic systems of conservation laws*, Math. Comp. 38 (1982), pp. 339-374.

[7] M. M. Rai and S. R. Chakravarty, *An Implicit Form for the Osher Upwind Scheme*, AIAA J. 24 (1986), 735-743.

[8] B. van Leer, *Flux-vector splitting for the Euler equations*, Lecture Notes in Physics 170 (1982), pp. 507-512.

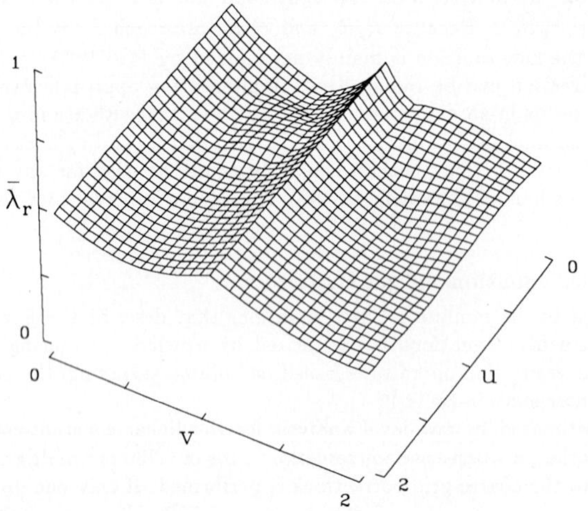

Fig. 2. *Two-level convergence rate* $\overline{\lambda}_r$ *as a function of* u *and* v $(c = 1)$, *for a* 64×64 *grid with* $h_x = h_y$, *using damped Point-Jacobi smoothing with* $\nu_1 + \nu_2 = 1$, $\nu_p = 0$. *The worst convergence factor is* $\frac{1}{2}$, *for* $v = 0$ *and* $|v| = c$.

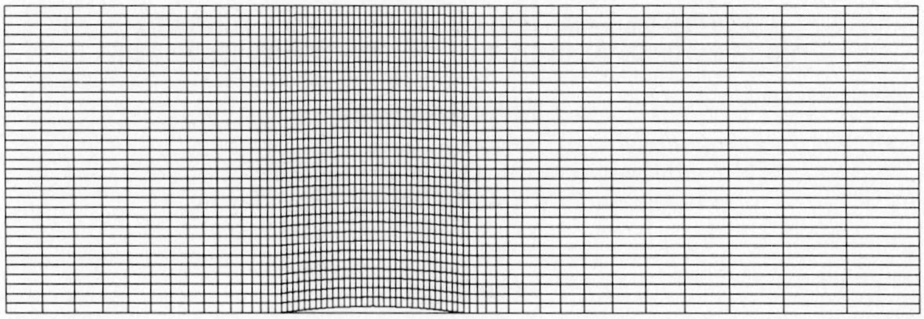

Fig. 3a. *64 × 32 grid used in the computations. The channel has dimensions 5 by 2, with the circular arc between 1.5 and 2.5 if the left lower corner is at $x = 0$. The thickness of the arc is 4.2% of the chord. Inflow is from the left, outflow to the right. The results in the following figures are computed on a 128 × 64 grid.*

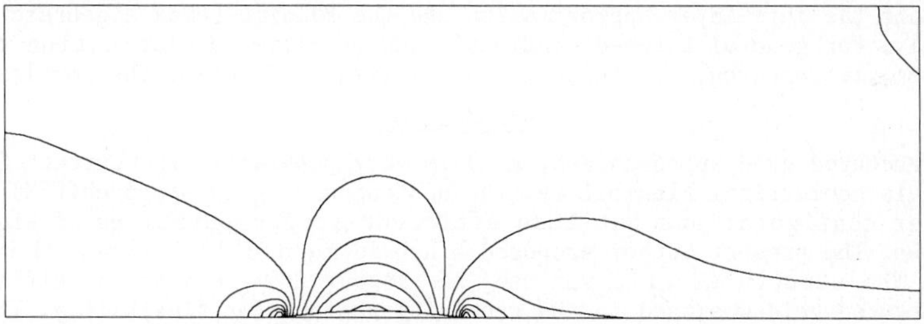

Fig. 3b. *Iso-Mach lines for mach 0.5 inflow. Contours are 0.01 apart (first-order Osher scheme). The asymmetry gives an indication of the low accuracy of the first-order scheme.*

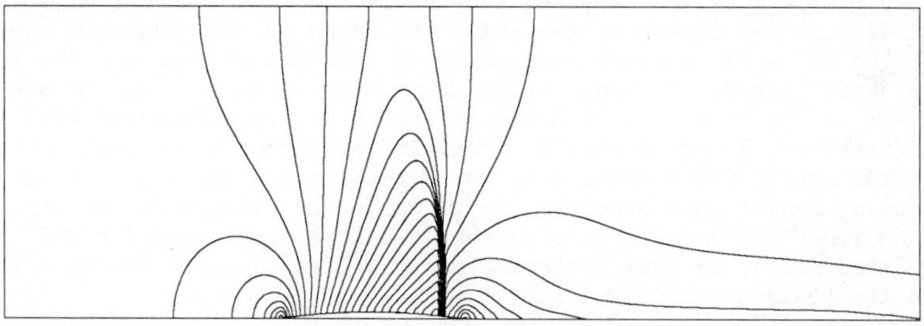

Fig. 3c. *Iso-Mach lines for mach 0.85 inflow. Contours are 0.025 apart (first-order Osher scheme).*

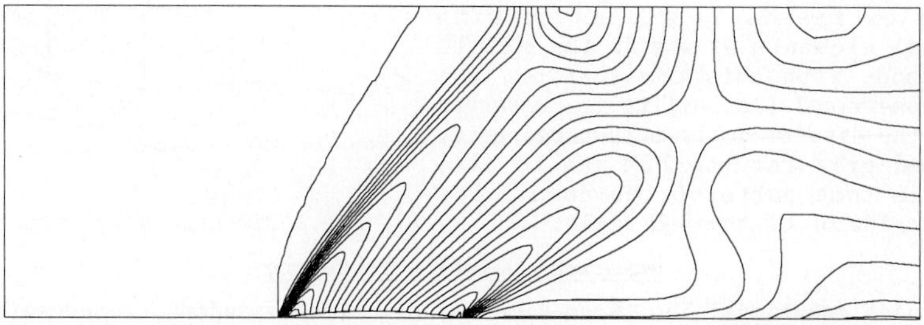

Fig. 3d. *Iso-Mach lines for mach 1.4 inflow. Contours are 0.025 apart (first-order Osher scheme).*

A FINITE-ELEMENT METHOD ON PRISMATIC ELEMENTS FOR THE THREE-DIMENSIONAL NAVIER-STOKES EQUATIONS

Kazuhiro Nakahashi
Department of Aeronautical Engineering
University of Osaka Prefecture
Mozu-umemachi, Sakai, Osaka 591, Japan

Abstract

A numerical scheme using a prismatic element is developed for the three-dimensional, compressible Navier-Stokes equations. With the prismatic elements, a body surface is covered by an unstructured triangular mesh which gives the geometrical flexibility. In the direction normal to the surface, the grid is structured so that an efficient computational scheme using the implicit residual averaging and the multigrid technique can be developed. This directionally structured grid also enables to implement the Thin-Layer approximation and the Baldwin-Lomax algebraic turbulence model easily. For general three-dimensional flow problems, a combination of several grids; prism, tetrahedron, and hexagon, is utilized to describe the complex field.

Introduction

The structured grid approach(FDM, FVM) is very popular in the current CFD, but it is lacking in geometrical flexibility. The unstructured grid approach(FEM) is suited for complex configurations but less efficient for computations of high-Reynolds number flows. The present author proposed a hybrid method[1] between these two approaches. The structured grid was used in viscous flow regions for efficiency, and the unstructured grid was used in the remaining regions for flexibility. It has been shown that the method can efficiently predict the complex viscous flow fields[1,2]. In three-dimensional problems, however, this method does not have the complete flexibility, because the body surface is covered by the structured grid for the FDM.

The objective of this paper is to propose an FEM-like scheme which has both efficiency and flexibility in computing the three-dimensional Navier-Stokes equations. A prismatic element was chosen as the grid. The bases of the element are triangles which cover the three-dimensional surface in an unstructured manner. The sides of the element are quadrilaterals so that the grid system is structured in the direction away from the surface(Fig.1). To use this directionally-structured grid has several important advantages; **1)** geometrically flexible for three-dimensional body, **2)** efficient since the severe CFL condition in the direction outward from the surface can be relaxed by using convergence acceleration techniques developed in the structured-grid approach, **3)** easy implementations of the Thin-Layer approximation and the Baldwin-Lomax turbulence model, **4)** easy vectorization and less computer storage requirement compared to the fully-unstructured FEM.

For general three-dimensional problems, the prismatic elements are used to cover the viscous flow regions near the body surface, and the remaining region outside of the viscous flow fields is filled up with tetrahedral elements as was in the FDM-FEM hybrid method. With this combination, the overall geometrical flexibility is preserved even with the directionally-structured grid. Structured grid for the FDM can be also combined in some parts of the domain for easier description of the 3-D field.

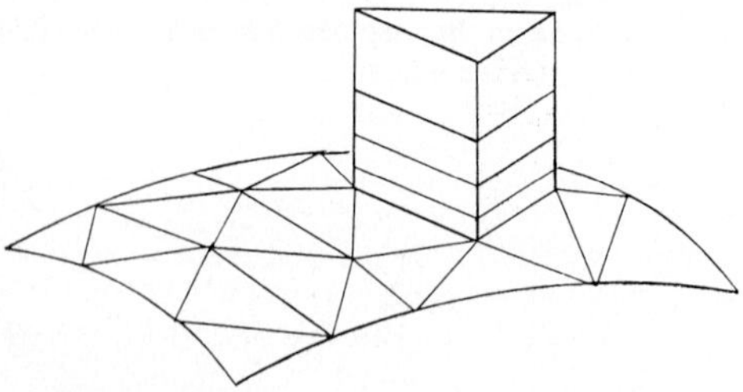

Fig.1 Directionally-structured grid.

Numerical Solution Procedure

3.1 Governing equations: The three-dimensional, time-dependent, compressible Navier-Stokes equations in Cartesian coordinates can be written as

$$\frac{\partial Q}{\partial t} + \frac{\partial F_i}{\partial x_i} = R_e^{-1} \frac{\partial H_i}{\partial x_i} \tag{1}$$

where $Q=[\rho, \rho v_1, \rho v_2, \rho v_3, e]^T$ is the vector of conserved dependent variables, and F_i and H_i are the convective and viscous flux vectors respectively. The pressure is related to the total energy e by the equation of state; $p = (\gamma-1)(e-\frac{1}{2}\rho v_i v_i)$.

3.2 Spatial discretization: The computational domain is divided into prismatic elements whose triangular bases are parallel to the body surface. A concept similar to the Jameson's finite-volume approach[3] is applied to this directionally structured grid system. With an integral form, Eq.(1) becomes

$$\frac{\partial}{\partial t}\iiint Q\,dV = -\iint \mathbf{F}\,d\mathbf{S} + R_e^{-1}\iint \mathbf{H}\,d\mathbf{S} \tag{2}$$

where the left-hand side is a volume integral and the right-hand side is a surface integral of a control volume.

Assuming that flow variables are stored at mesh nodes, we consider a control volume as a union of prisms surrounding each node as shown in Fig.2. In this figure, numbers 0, 1, 2, ,, is the nodal number in the plane parallel to the body surface. The number j is the plane number counted in the direction away from the body surface. In Fig.2, the control volume is a union of twelve prisms who have a common node at (0,j), but the number of elements to make a control volume is arbitrary.

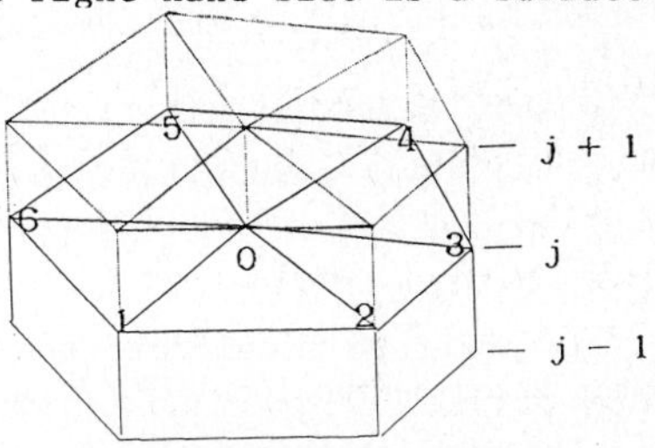

Fig.2 Control volume

By considering the flux balance for the control volume, we obtain a following discretized form of equations.

$$\frac{\partial}{\partial t}(\sum_{el} V_e Q) = -(\sum_{face-i} \mathbf{F}\mathbf{S} + \sum_{face-j} \mathbf{F}\mathbf{S}) + R_e^{-1}(\sum_{face-i} \mathbf{H}\mathbf{S} + \sum_{face-j} \mathbf{H}\mathbf{S}) \tag{3}$$

In this equations, V_e denotes the volume of each prismatic element, $\sum_{el}$ is a sum of all elements in the control volume, $\sum_{face-i}$ is a sum of all side faces, and $\sum_{face-j}$ is of all upper and lower faces of the control volume.

Assuming Q in the left-hand side be constant in the control volume, we obtain a ordinary differential equations as

$$\frac{dQ}{dt} = -f_i(F) - f_j(F) + R_e^{-1}[f_i(H) + f_j(H)] + D_i + D_j \tag{4}$$

where f_i and f_j are flux operators and are defined as

$$f_i(F) = (\sum_{face-i} \mathbf{F}\mathbf{S})/\sum_{el} V_e\,, \quad f_j(F) = (\sum_{face-j} \mathbf{F}\mathbf{S})/\sum_{el} V_e \tag{5}$$

In the Eq.(4), D_i and D_j are the artificial dissipative terms in the direction parallel and normal to the surface, respectively. The dissipative term D_i has a form

$$D_i = \sum_k \varepsilon^{(2)}_{i,k}\phi_{i,k}(Q_k - Q_0) - \sum_k \varepsilon^{(4)}_{i,k}\phi_{i,k}(E_k - E_0)\,, \quad E_0 = \sum_k (Q_k - Q_0)$$

$$\varepsilon^{(2)} = k^{(2)}\left|\sum_k \frac{p_k - p_0}{p_k + p_0}\right|, \quad \varepsilon^{(4)} = \mathrm{MAX}\,[0, k^{(4)} - \varepsilon^{(2)}] \tag{6}$$

The scaling factor ϕ is defined as; $\phi_{i,k} = \{|\hat{v}| + c\}\overline{S}$, where $\hat{v}$ is the velocity component parallel to the edge, c is the sonic velocity and $\overline{S}$ is the sectional area of the control volume normal to the edge. The dissipative term normal to the surface, Dj, has the similar form to Di but is multiplied by the local Mach number so as to decrease the artificial viscosity in the boundary layers. The constants in Eq.(6), $k^{(2)}$ and $k^{(4)}$, must be determined empirically.

3.3 Viscous terms: To evaluate the viscous term, we need to compute the first derivatives of velocity components and sonic velocity at each node. This is done using the Green theorem as

$$\left.\frac{\partial v_i}{\partial x_j}\right|_{node} = (\sum_{face} S_{xj} v_{i-face})/\sum_{el} V_e \tag{7}$$

The Thin-Layer approximation[4] has been commonly used in the structured-grid approach, and shown to be effective to improve the computational efficiency without decreasing the accuracy. In the present formulation, this approximation can be easily implemented owing to the directionally-structured grid. This is achieved by just neglecting Hi in Eq.(4) which is the viscous flux parallel to the body surface.

The Baldwin-Lomax turbulence model[4] also prefers the structured grid in the direction normal to the surface, thus the present formulation can easily implement the model.

3.4 Time integration: A three-stage time stepping scheme is used for integrating Eq.(4) in time. For economy, the viscous and dissipative terms are evaluated only at the first stage. The solution is therefore advanced from time level n to n+1 as;

$$
\begin{aligned}
Q^{(1)} &= Q^{n} \\
Q^{(1)} &= Q^{(0)} - 0.6\Delta t\,[f_i(F^{(0)}) + f_j(F^{(0)}) - R_e^{-1} f_j(H^{(0)}) + D_i^{(0)} + D_j^{(0)}] \\
Q^{(2)} &= Q^{(0)} - 0.6\Delta t\,[f_i(F^{(1)}) + f_j(F^{(1)}) - R_e^{-1} f_j(H^{(0)}) + D_i^{(0)} + D_j^{(0)}] \qquad (8) \\
Q^{(3)} &= Q^{(0)} - 1.0\Delta t\,[f_i(F^{(2)}) + f_j(F^{(2)}) - R_e^{-1} f_j(H^{(0)}) + D_i^{(0)} + D_j^{(0)}] \\
Q^{n+1} &= Q^{(3)}
\end{aligned}
$$

In order to accelerate the convergence to steady state, spatially variable time step based on the local CFL condition is used.

The implicit residual averaging[5] is also applied to the equation. The residual, R(Q), at a mesh point is replaces by a weighted average of neighboring residuals. This average can be efficiently calculated implicitly in the structured direction of the present grid by the following tridiagonal equation;

$$-\varepsilon_j \overline{R}_{j-1} + (1+2\varepsilon_j)\overline{R}_j - \varepsilon_j \overline{R}_{j+1} = R_j \qquad (9)$$

A constant ε_j is determined locally by a condition; $\varepsilon_j \geq .25\,\{(\Delta t_j/\Delta t_j^{*})^2 - 1\}$ where Δt_j^{*} is the maximum stable time step of the scheme, and Δt_j is the actual time step.

The prismatic grid also enables to use the multigrid strategy[6] in a straightforward manner. Since the mesh parallel to the body surface is unstructured, this technique is applied only to the direction normal to the wall. Auxiliary meshes are introduced by doubling the mesh spacing in j-direction. Here a simple saw-tooth cycle is used. Values of the flow variables are transferred to a coarser grid. Then the multistage scheme is reformulated with the result that the solution on a coarse grid is driven by the residuals collected on the next finer grid. The process is repeated in successively coarser grids. Finally the corrections are interpolated back from each grid to the next finer grid without any intermediate Navier-Stokes computations.

Computational Results

Shock-Boundary Layer Interaction on a Flat Plate: At first, the method was applied to a shock-boundary layer interaction problem on a flat plate[7]. The flow is two-dimensional but the computation was performed in a three-dimensional field to check the ability of the present three-dimensional Navier-Stokes code. Shown in Fig.3 are the grid system and the computed pressure contours. The bottom surface is covered by 1188 triangles with 700 points and the number of points in the direction normal to the surface is 49. The velocity vectors near the shock impinging point are shown in Fig.4 by comparing with the result obtained by a well-checked two-dimensional implicit finite-difference solver[8]. The velocity profiles and separation length of both results show good agreements with each other. The computations were performed by using both fully viscous and thin-layer model, and the results did not show any critical differences.

Hemisphere-cylinder: In Fig.5, grid and computed density contours on a hemisphere-cylinder are shown. The computations were performed without meridian symmetry assumption. The triangular prismatic grid enables us to discretize the three-dimensional surface appropriately without futile grid points.

Turbine cascade with hub wall: A combination of the present method with a FDM is shown in Fig.6-8 for a 3-D turbine cascade with the hub wall. As shown in Fig.6, the turbine blade is covered by the FDM grid(hexagonal mesh) and the remaining region is filled up by the prismatic elements. This combination gives easy grid generation, and is capable to treat more complex 3-D internal configurations. Shown in Fig.7 is the computed density contours on the blade, hub and midspan plane. A comparison of the midspan isentropic surface Mach number with the experiment[9] is shown in Fig.8.

Supersonic Interacting Flow along a Corner: In Fig.9, a grid system with a combination of prismatic and tetrahedra elements is shown for an interactive flow fields along a symmetric corner. Prismatic elements covers viscous flow region near the corner surface, and the tetrahedra elements covers the remaining region. Although this flow configuration is simple enough for the structured-grid approach as was computed by Shang[10], it is of basic configuration for numerous three-dimensional flow problems encountered in the engineering applications, such as in fuselage wing junctions and in rectangular inlet. Thus this flow field is chosen for examining the capability of the present method. Computed pressure contours shown in Fig.10 reproduces the interacting shock wave structure.

Conclusion

The results shown here demonstrated the feasibility of the method for computations of the three-dimensional Navier-Stokes equations. Even for a complex three-dimensional body, to cover the surface by triangular elements is relatively easy. Therefore the present method using the prismatic elements has quite attractive features for three-dimensional high Reynolds number flow computations. For the successful extension of the method to more complex flow problems, however, it is indispensable to develop grid generation techniques for the prismatic and tetrahedral elements.

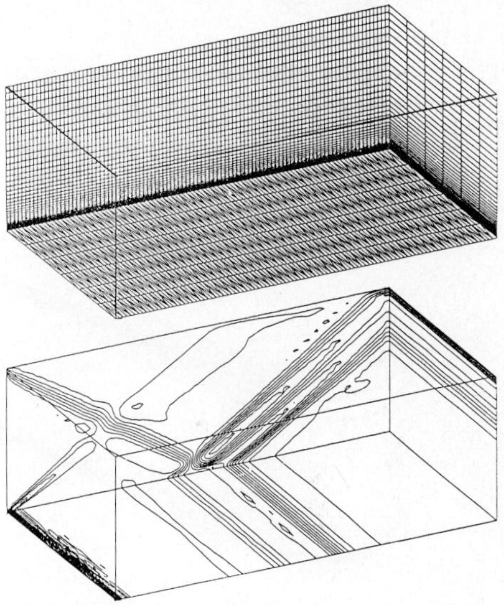

Fig.3 Prismatic grid and computed pressure contours of shock-boundary layer interaction flow; M_∞=2.0, Re =2.96x10^5.

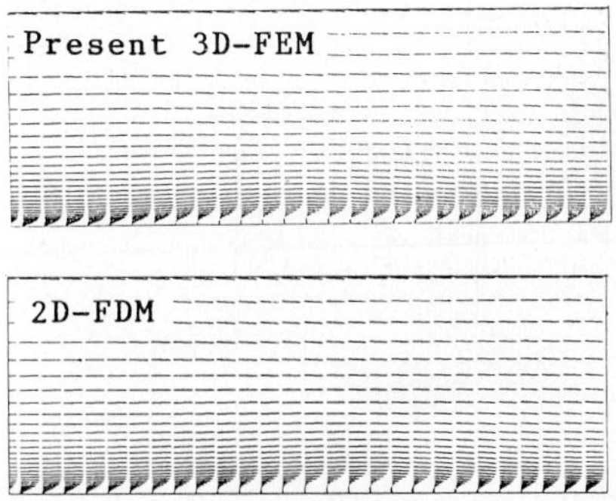

Fig.4 Computed velocity profiles near the shock impinging point.

References

[1] Nakahashi, K., "FDM-FEM Zonal Approach for Computations of Compressible Viscous Flows," Proc. of 10th ICNMFD, Lecture Notes in Physics, Vol.264, pp.494-497, Springer-Verlag, 1986.

[2] Nakahashi, K. and Obayashi, S., "Viscous Flow Computations Using a Composite Grid," AIAA Paper 87-1128, 1987.

[3] Jameson, A. and Baker,T.J.,"Improvements to the Aircraft Euler Method," AIAA Paper 87-0452, 1987.

[4] Baldwin, B.S, and Lomax, H., "Thin Layer Approximation and Algebraic Model for Separated Turbulent Flows," AIAA Paper 78-257, 1978.

[5] Jameson, A. and Baker, T.J., "Solution of the Euler Equations for Complex Configurations" AIAA Paper 83-1929, 1983.

[6] Jameson, A. and Baker, T.J., "Multigrid Solution of the Euler Equations for Aircraft Configurations," AIAA Paper 84-0093, 1984.

[7] MacCormack, R.W., "Numerical Solution of the Interaction of a Shock Wave with a Laminar Boundary Layer," Proc. 2nd ICNMFD, Lecture Notes in Physics, Vol.8, p.151, Springer-Verlag, 1971.

[8] Pulliam, T.H. and Steger, J.L., "Recent Improvements in Efficiency, Accuracy, and Convergence for Implicit Approximate Factorization Algorithms," AIAA Paper 85-0360, 1985.

[9] Hodson, H.P. and Dominy, R.G.,"Three-Dimensional Flow in a Low-Pressure Turbine Cascade at Its Design Condition," ASME Paper No.86-GT-106, 1986.

[10] Shang, J.S. and Hankey, W.L., "Numerical Solution of the Navier-Stokes Equations for a Three-Dimensional Corner," AIAA J., Vol.15, No.11, pp.1575-1582, 1977.

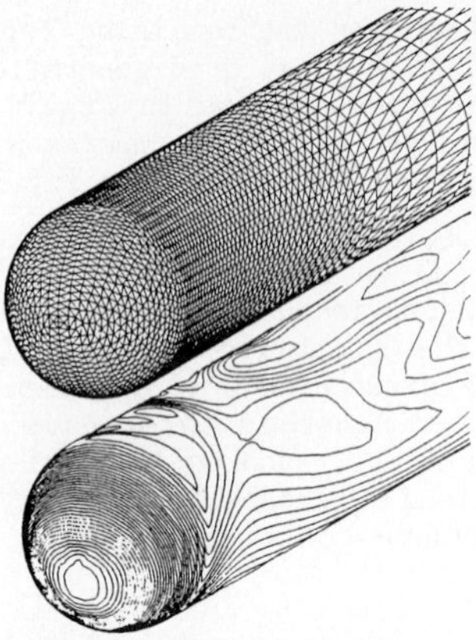

Fig.5 Surface grid and computed density contours of hemisphere-cylinder flow; $M_\infty=1.2$, $\alpha=19^0$, $Re_\infty=2.225\times10^5$.

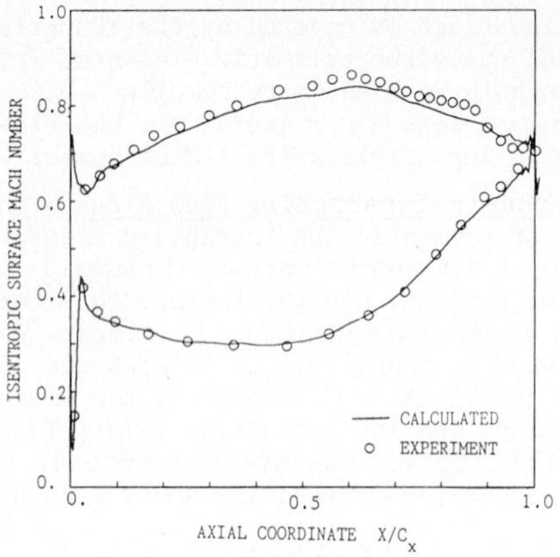

Fig.8 Midspan surface isentropic Mach numbers of turbine cascade; $M_{2th}=0.702$, $\alpha_1=38.8^0$, and $Re_2=2.9\times10^5$.

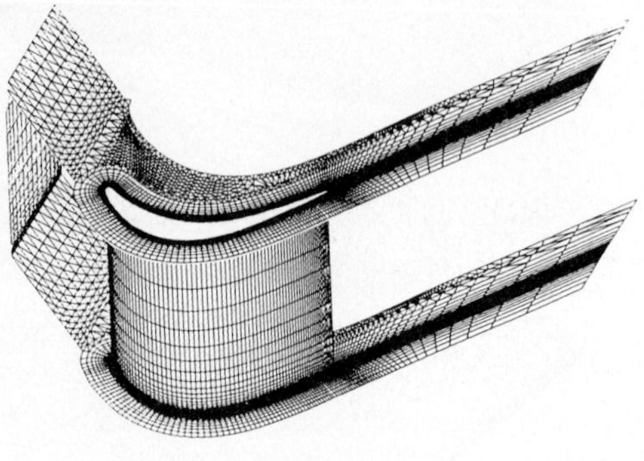

Fig.6 Prismatic and hexagonal composite grid for computation of turbine-hub junction flow field.

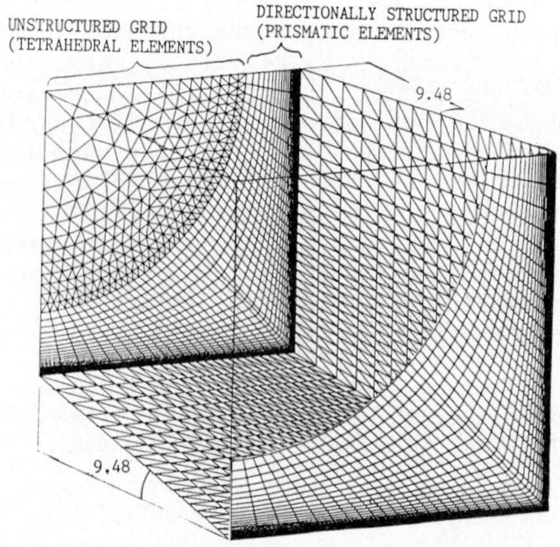

Fig.9 Prismatic and tetrahedral composite grid for computation of corner flow.

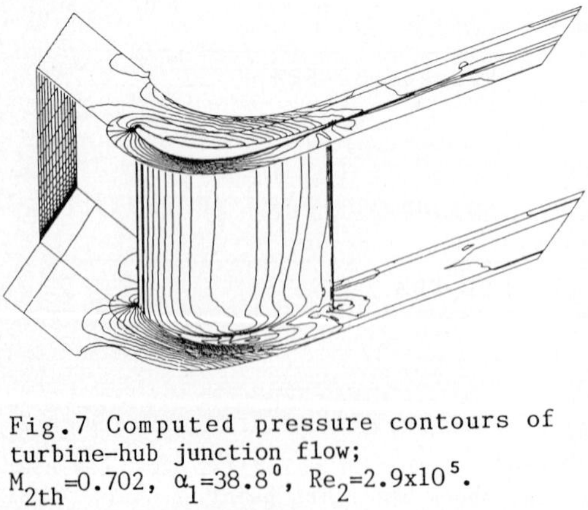

Fig.7 Computed pressure contours of turbine-hub junction flow; $M_{2th}=0.702$, $\alpha_1=38.8^0$, $Re_2=2.9\times10^5$.

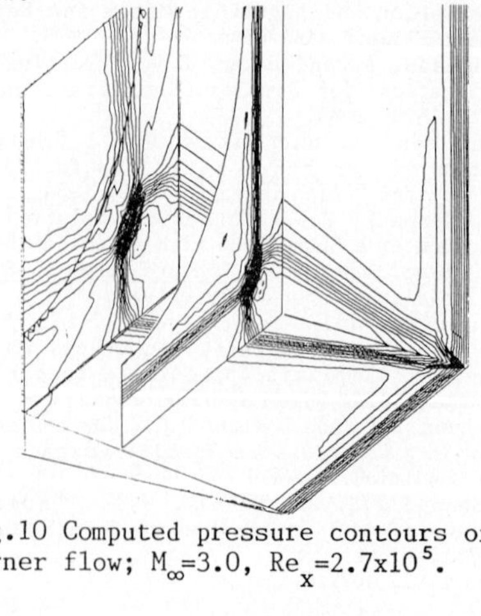

Fig.10 Computed pressure contours of the corner flow; $M_\infty=3.0$, $Re_x=2.7\times10^5$.

VORTICES AROUND CYLINDER IN CONFINED FLOWS

N. Nishikawa, S. Akiyama, and O. Akune

Faculty of Engineering, Chiba University
Yayoi 1-33 Chiba, Japan

INTRODUCTION

In the present study the three dimensional full Navier-Stokes equations are solved for the unsteady internal flows around the cylinder across the rectangular channel for Re<5000. A scheme is developed with 3rd order differencing, general curvilinear coordinates where HSMAC code is accerelated by local implicit expressions for a velocity corrections at a cell.

By the development of environment for numerical computation, incompressible 3-D flows at relatively simple configurations have been solved for three dimensional Navier-Stokes equation. For steady external flow around a circular cylinder perpendicular to the flat plate the INS3D code for incompressible N. S. equation has been applied by Kwak et al. [1] Their code utilizes the Approximate Factorized form and requires to apply a block tridiagonal inversion. The experimental study for the similar configuration as the calculation by Kwak et al has been reported by Baker[2] which shows visualization picture of laminar horse shoe vortex and so on for Reynolds number up to 10000.

While the flow around a cylinder across the channel has been clarified with the results for Re<500. One of the preceding example for this configuration is the calculation by Mitra et al[3] which shows slight symptoms of unsteady flow behavior, and where the Reynolds number is not larger than 440. They use the overlapping coordinates of rectilinear and polar coordinates which need the 'global' iteration between solutions in each coordinate system, however the vortices flowing along wake are not counted accurately by a few existing numerical results.

BASIC EQUATIONS

$$div\ \mathbf{v} = 0$$

$$\frac{\partial \mathbf{v}}{\partial t} + (\mathbf{v} \cdot grad)\,\mathbf{v} = -grad\ p + Re^{-1}\Delta\mathbf{v}$$

HSMAC type velocity correction 、

$$\delta v_{x_i} = -\Delta t\, J \left[\sum_{i=1,3} (J^{-1}\xi_{x_i}\delta p)_{\xi} + (J^{-1}\eta_{x_i}\delta p)_{\eta} + (J^{-1}\zeta_{x_i}\delta p)_{\zeta} \right]$$

Boundary conditions are imposed as follows. The adhesion conditions are applied at the cylinder surface, bottom and side walls of the channel. Inlet velocity profiles are parabolics along both horizontal and perpendicular direction.

NUMERICAL SCHEME

Among incompressible solvers the relatively new Navier Stokes solvers, all of which uses 3rd order upwind differences, have been compared as for the efficiency in solving two dimensional cavity flow[4]. This shows that the Quick scheme is best in stability and accuracy, on comparing with Agarwal scheme and Kawamura scheme, which is applied to a very high Reynolds number 3-D flows[5]. Thus the 3rd order upwind experssion is recommended to be introduced here to attain comparable level as these schemes.

In the present study we developed a scheme with the following points

1. 3rd order upwind differeces for convection terms
2. General curvilinear coordinates
3. Accerelated HSMAC code

Among the standard style 3-D codes such as SIMPLER, HSMAC, or INS3D, we improved Mitra-Kiehm's HSMAC type code [3] to the one with general coordinates. However large iteration cycles, which is one of disadvantage of HSMAC scheme, is attempted to reduce. The velocity corrections in original relaxation do not explicitly proportional to the magnitude of velocity component along grid lines. This is caused by too simplified estimation of the correction to velocity components, because the correction in each component depends upon the aspect ratio of a grid cell not upon magnitude of flux-component umbalance.

The 3-D application of the accerelation method by the first author for 2-D flow[6] is introduced as follows. That is, The improved value of velocity through iteration are obtaind from simultaneous equations for velocity corrections as for chosen two components of velocity. This system of equations is deduced from elimination of pressure term in a iteration procedure of original HSMAC scheme[3], which has an implicit form for δu

$$\delta u = 1.8\Omega [D + \beta \Sigma \delta u / \Delta x] / \Sigma \Lambda x^{-2} \quad \text{(D: residual of divV.)}$$

Thus, in contrast to the usual HSMAC scheme, where hundreds iterations per time step are repeated for several hundreds time steps, the iteration number is expected to be reduced here considerably. Another strategy to avoid such situation is the use ofthe integrated and discretized continuity equations at each cells, like as the line relaxation in 2-D flow [7]. However, the residual of continuity equation has bad effect for this initial stage of relaxation when the residuals are relatively large.

The solutions are obtained for Re=1000, and 5000 referred to the cylinder diameter-D with grid distribution 23x51x24, respectively for vertical half height, circumferential direction along cylinder surface, and radial direction. The half height of channel H/D is more or less 5, and both the width of channel and the distance between the cylinder-center and inlet are fixed as 5D. The distance between the cylinder center and downstream end of channel is 10*D. At the downstream end the extrapolation condition is applied. The two dimensional O-type grids generated by elliptic p.d.e are stacked up vertically to the bottom to form a three-dimensional grid as shown in Fig. 1. Dense mesh distributions are used near the cylinder surface and the channel walls. The CPU time per 5000 time steps(T=19.1) is 2.5 hour on Fujitsu VP-400.

RESULTS

In Fig. 2 the surface streamlines on the cylinder surface is illustrated at Re=1000, which can be compared with the experimental visualization photograph in Fig. 3 in water towing tank. In Fig. 2 we can observe the separation lines as converging streamlines and some flow singularities, especially the focus at rear upper part of the cylinder surface.

The separation-point angle measured from rear stagnation at the half height are 74°, 69°, 66°, respectively at Re=1000, 300, and 100, which behavior agree with 2-D cylinder results.

In Fig. 4. the crosssection of flow across stagnation points is shown which shows a saddle point on the bottom of channel, together with the foot of horseshoe vortex around the cylinder at T=40.

The solutions for the elliptic cylinder of 1/2 thickness ratio and with 4 degree slant angle from the oncoming flow are also obtained for Re=400 with grid distribution 12x23x25 respectively for vertical direction and crosswise, lateral direction. The contours of vorticity is shown in Fig. 5 where the contours at horizontal levels are shown in each Figures 7(a, b, c).

In Fig. 6 the iteration cycles spent in the initial time steps are shown for the two cases. That is, the technique with varing Ω over-relaxation parameter, and another with varying the weight β to implicit increment of δu, δv. The latter is better and the small improvement to usual HSMAC (β=0, Ω=1) is realized. More improvement is possible in the near future.

AKNOWLEDGEMENT

The numerical computation was performed partly under the support of Grant in Aid of Ministry of Education:Grant No. 62613009 through Prof. Honma, No. 61550042 and mainly by VP-400 by the courtesy of Dr. R. Mendez :the Director of Institute for Supercomputer Research of Recruit Inc.

REFERENCES

1. Kwak, D. et al. AIAA J. Vol. 24 p390-396(1986)
2. Baker, C. J. J. Fluid Mech 95, p347-367(1979)
3. Mitra, N., Kiehm, P., Fiebig, M., Proc. 10th ICNMFD Springer ver. p481-487(1986)
4. Kawai, M. & Ando, Y. J. Jap. Mech. Eng Vol. 52 p2067(1986)
5. Shirayama, S. Doctoral Dissertation, Tokyo Univ. (1987).
6. Nishikawa. N., Suzuki, A., and Suzuki, T., Proc. 10th ICNMFD Springer ver. p499-504(1986)
7. Nishikawa, N. et al Intn. Sympo. Compu. Fluid. Dyna. Sydney(198 (1987), or Computational Fluid Dynamics , Ed. G. V. Davis, Elsevier, pp569-578(1988).

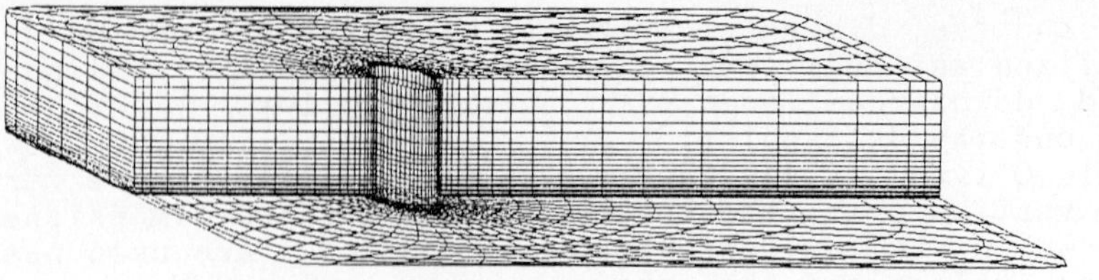

Fig. 1 GRID DISTRIBUTION 23x51x24

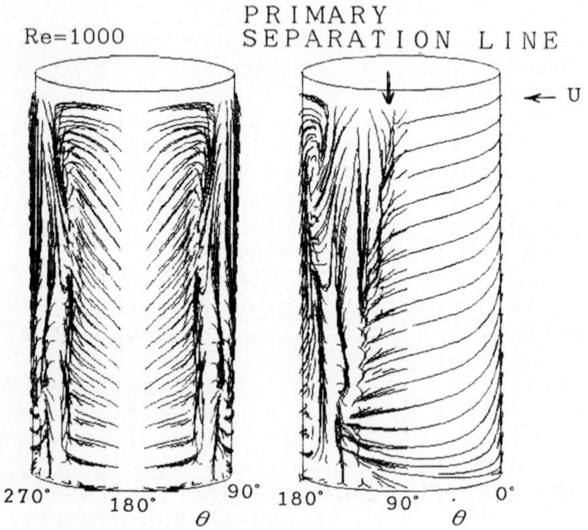

(a) REAR VIEW (b) SIDE VIEW

Fig. 2

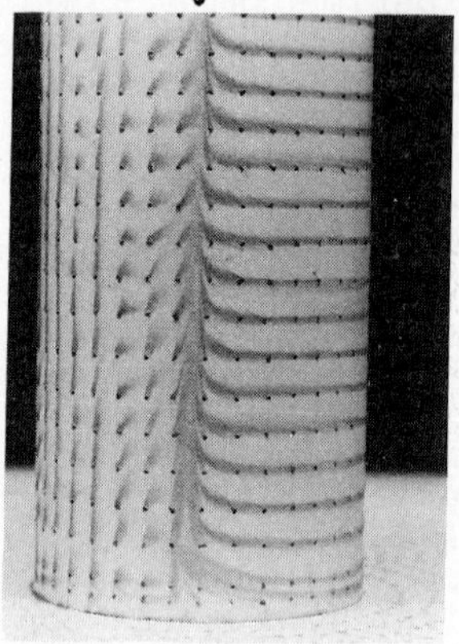

EXPERIMENTAL FLOW-VISUALIZATION Re=1800
ELECTLORYSIS METHOD

Fig. 3

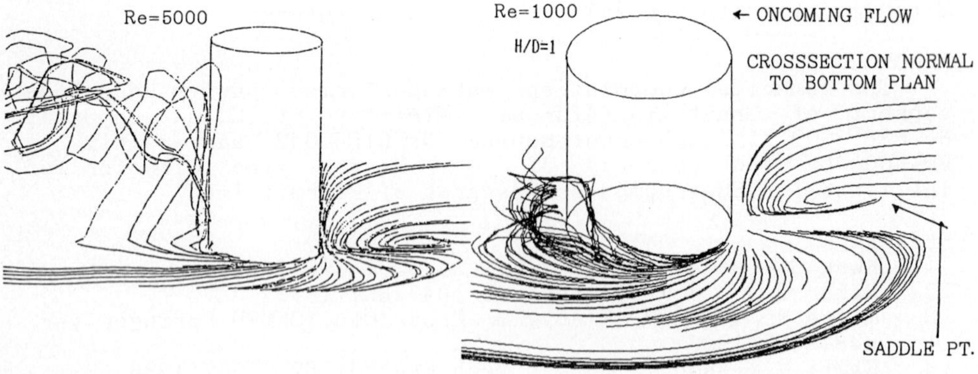

Fig. 4

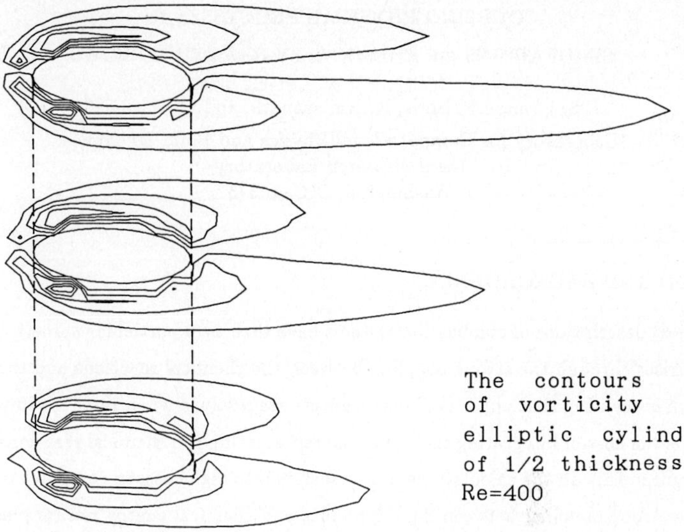
The contours
of vorticity
elliptic cylinder
of 1/2 thickness
Re=400

Fig. 5

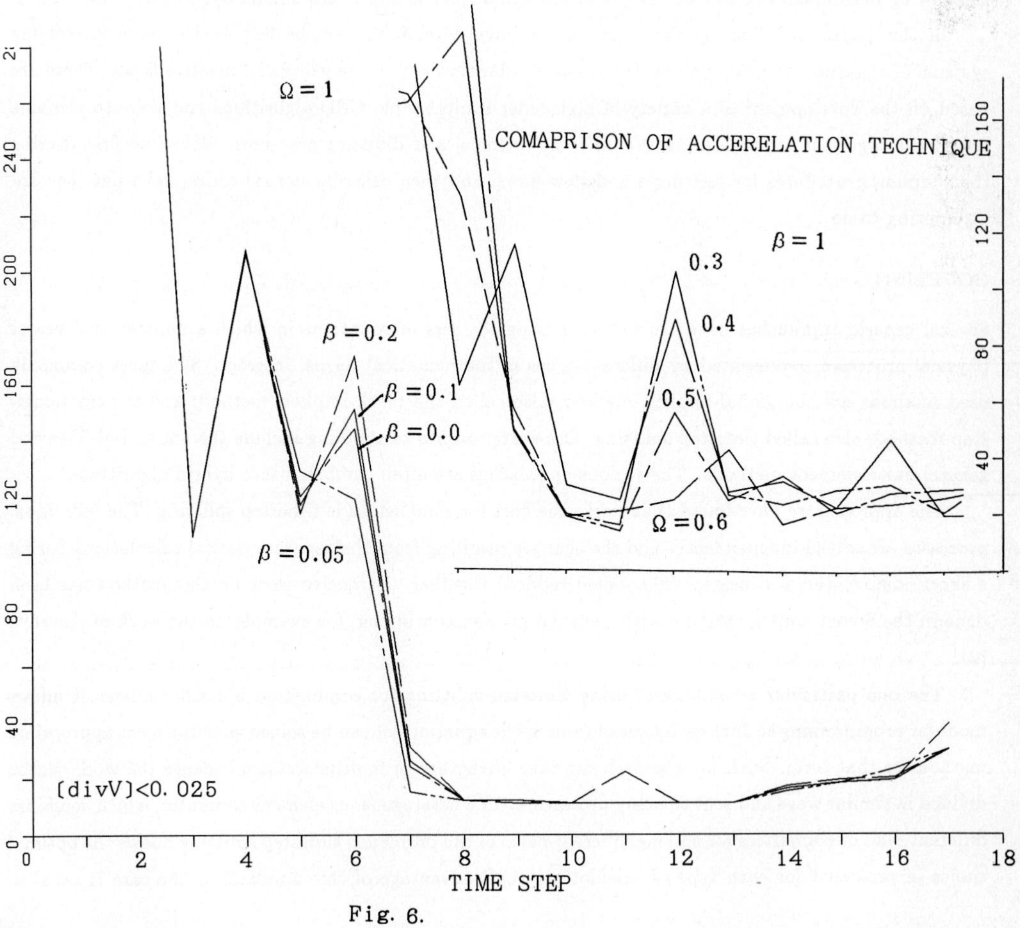
COMAPRISON OF ACCERELATION TECHNIQUE
NUMBER OF ITERATIONS CYCLES
TIME STEP
Ω = 1
Ω = 0.6
β = 1
β = 0.2
β = 0.1
β = 0.0
β = 0.05
0.3
0.4
0.5
[divV]<0. 025

Fig. 6.

COUPLING PHYSICAL PROCESSES IN SIMULATIONS OF CHEMICALLY REACTIVE FLOWS

E.S. Oran, J.P. Boris, K. Kailasanath, and G. Patnaik*
Laboratory for Computational Physics and Fluid Dynamics
Naval Research Laboratory
Washington, D.C. 20375

INTRODUCTION AND BACKGROUND

Accurate numerical descriptions of combustion systems have their own particular numerical problems. One major problem arises through the strong coupling between the chemical reactions and the fluid dynamics. A second problem is ensuring that physical diffusion effects are resolved when these are important. Another typical problem is the cost of integrating the equations representing the chemical reactions. Such problems put accuracy requirements on the calculations that translate into high costs in computer time and memory. For example, the strong coupling between fluid dynamics and chemical energy release places requirements on the accuracy of the fluid dynamics variables, particularly on the temperature. Further, accurately calculating the amount of diffusion of light species can be just as important as fluid dynamics and chemistry.

In this paper, we describe the approach we have developed over the last twelve years for solving systems of equations typical of those describing combustion and, more generally, reactive flows. These are based on the development of a variety of high-order compressible CFD algorithms and ways to combine these with algorithms representing the different chemical and diffusive processes. Below we first decribe the coupling procedures for fast flows and slow flows, and then describe several codes and what they are attempting to do.

COUPLING

Several generic approaches have evolved for solving the sets of equations in which a number of different physical processes, represented by different types of mathematical terms, interact. The most commonly used methods are the global-implicit method, also called the block-implicit method, and the fractional-step method, also called timestep splitting. Other approaches to coupling include the method-of-lines and general finite-element methods. The various approaches are often combined into hybrid algorithms.

The approach we have found that works the best for combustion is timestep splitting. The individual processes are solved independently and the changes resulting from the separate partial calculations during a short computational timestep are coupled (added) together. Extensive work on this method has been done in the Soviet Union, starting with Yanenko (1) and continuing, for example, in the work of Kovenya (2).

The one particular advantage of using timestep splitting for combustion is its flexibility: it allows modular programming so that each type of term in the equation set can be solved with the most appropriate method for that term. Such an approach can take advantage of multiprocessors because the work can be divided in similar ways and sent to many processors. In a heterogeneous element computer, which combines different kind of computers for solving different parts of the problems, timestep splitting allows the optimal choice of processor for each type of calculation. A disadvantage of this approach is the care it takes to get it to work correctly. Stability is not guaranteed; it depends on the separate criteria of the individual algorithms and the properties of their coupling. Convergence is usually tested by increasing the spatial and temporal resolution. When some aspect of the program uses asymptotic methods, however, situations arise for which reducing the timestep makes the answers less accurate, and ways have to be developed to deal with conflicting requirements. In a reactive flow calculation, the timestep must always be monitored to ensure that the extent of changes from one process do not cause the algorithm representing the next process to become unstable.

We describe several combustion models and use them as vehicles to illustrate several approches to constructing numerical models for reactive flows. The first is a multidimensional detonation model, in which the major physical processes are convection and chemical energy release (3). This is based on an explicit Eulerian convection algorithm and therefore the coupling is straightforward. The other models are both implicit, a one-dimensional Lagrangian flame model (4) and a two-dimensional Eulerian flame model (5). In addition to convection and chemical energy release, the effects of thermal conduction, molecular diffusion, thermal diffusion, physical viscosity, and radiation transport can be important. Here a more complicated coupling is used that occurs through the implicit pressure equation.

HIGH-SPEED DETONATION CALCULATIONS

In a propagating multidimensional gas-phase detonation, the detonation front, either in free space or confined in a chamber, is not smooth and displays a complicated pattern of intersecting shock waves. Triple points at the intersections of the incident shocks, transverse shocks, and Mach stems trace out repeating, rhomboid-like patterns called detonation cells. The size as well as the degree of regularity of the cell structure is a property of the reacting material as well as the bounding conditions.

The computational difficulties involved in simulating such a system consist of resolving the complicated shock interactions at the shock front and in representing and solving the energy-release processes. We have taken two approches to this, depending on the level at which we want to resolve the chemical processes. First, when a detailed set of chemical reactions is known well enough, the solution of a set of nonlinear ordinary differential equations describing this set can be coupled to the solution of the compressible continuity equations for density, momentum, and energy. However, if the reaction mechanism is not known, or the solution of the set of equations at every computational cell is prohibitively expensive, models of the chemical properties can be substituted for the full reaction set.

Figure 1 is taken from a two-dimensional calculation of a propagating detonation in which the convective transport is calculated by a high-order monotone Eulerian FCT method (6) and the chemical energy release is represented by a phenomenological model calibrated using a detailed set of chemical reactions. Coupling the processes here is straightforward, as long as not too much energy is released in one fluid dynamic timestep. The fast acoustic phenomena must be resolved, and diffusive transport phenomena can generally be neglected over such short timescales.

A surprising feature of these calculations is that integration of the coupled continuity equations for the primary variables must be more accurate than would be required if they were integrated alone. The convection terms provide the background temperature bath in which the chemical reactions occur. These reactions depend exponentially on the temperature. Overshoots in the solution can be catastrophic to the fluid dynamic temperature. We have generally found that a fourth-order monotone scheme is adequate. This required accuracy extends to calculations of condensed-phase sytems where the equation of state can be very sensitive to fluid variables.

However, we find that the order of accuracy of the solution of the ordinary differential equations representing the chemical reactions can be considerably less than would be required if they were integrated alone. This is an unexpected gain because such chemical integrations are by far the most costly processes in the calculation. The reason for this windfall is that one of the most expensive parts of the solution of the ODEs is when the system reaches equilibrium and the equations become very stiff. At this point, the chemical production and loss terms nearly balance, and the required timestep to resolve them can be very small. However, when the chemistry is coupled to fluid dynamics, there is always some change in the system due to fluctuations of the fluid variables amd the system never really quite reaches equilibrium.

LOW-SPEED FLAME CALCULATIONS

Numerical simulations of flames are complicated because of the number of different physical processes that are all important to a quantitative description of their behavior. In addition to convective transport and chemical reactions, it is now necessary to consider the effects of thermal conduction, molecular diffusion, radiation transport, and thermal diffusion. Because it may be important to resolve these physical diffusion terms and not allow them to be overshadowed by numerical diffusion, a Lagrangian method could be the optimal approach.

Thus we used a Lagrangian approach for the fluid dynamics in the one-dimensional flame model we initially developed (4), and to this couple models for other physical processe. The coupling here, however, was not nearly as straightforward as for the high-speed flow. Characteristic timescales for important physical changes in the sytem are now considerably longer and resolving the full spectrum of acoustic waves is no longer necessary. Thus we use an implicit method which solves a pressure equation instead of an energy equation. For such a system, the direct coupling approach results in large oscillations of the system variables, even when the timesteps are substantially reduced, and we must therefore take another approach. At the end of a timestep, we evaluate those changes that occur in the pressure due to all the processes, and use these in an elliptic pressure eqution to produce the final variables at the end of a timestep. Note that the changes in species variables are updated after they are calculated, but temperature is help constant during each chemical timestep.

An interesting issue here has to do with adaptive gridding in a subsonic reactive flow problems. In the high-speed supersonic flows, if the shock or detonation moves into a quiescent region, it can always move into a finely gridded region to preserve accuracy. Regridding is more difficult for subsonic flows in regions where there is anything interesting happening, and it can introduce errors as variables are interpolated onto finer grids. These errors look like numerical diffusion in the solution or they can even add unwanted oscillations to the problem. How to regrid in an arbitrary subsonic reactive flow is an important issue to be resolved.

There are inherent difficulties in extending Lagrangian method to multidimensions. These are based on the problems of separating the solution of the continuity equations from the geometrical aspects of the algorithm. Such a separation is essentially impossible in a Lagrangian algorithm and has been discussed extensively in the literature. Whereas we hoped to be able to extend the ideas of the Lagrangian dynamically restructuring triange-based grids (7) to flame simulations, we found that this approach is extremely complex to implement. Thus we were faced with the problem of how we could use an Eulerian method to resolve the flame structure adequately.

Our solution to this is BIC-FCT (5), a one-step correction to FCT that makes the solution implicit in the sound waves, but maintains the high-order and monotonicity of the FCT algorithms. In the new step, we add up the changes in energy due to processes other than convection, and use these as source terms in an elliptic equation. When this correction is made, the solution is stable for timesteps greater than allowed by the CFL condition. This algorithm has been tested extensively, was coupled by the pressure-coupling techniques described above to representations of other physical processes, and is now being used in a number of different flame simulations. An example is shown in Figure 2, taken from a calculation of the instability at the front of a hydrogen-oxygen-nitrogen flame. This model included a detailed nonequilibrium chemical reaction mechanism, as well as models for thermal conduction, physical viscosity, and molecular diffusion.

SOME CONCLUSIONS

The exact way the processes are coupled depends on the individual properties of the different algorithms used and the regimes in which the competing physical processes interact. In particular, we have found that the best form of the coupling to use varies according to whether the convection is treated by an implicit or explicit approach. Other processes may be done explicitly or implicitly, and still not alter the basic procedure. In an explicit computation, we believe that the order in which the processes are integrated and their effects applied should not affect the final answer. However, implicit acoustic coupling involves using a pressure-like equation either instead of or in addition to an energy equation. In this case, we add up the pressure differences caused by each process and then at the end of a timestep use the pressure equation to evaluate the fluid variables for the new timestep.

ACKNOWLEDGMENTS

This work was sponsored in part by the Naval Research Laboratory through the Office of Naval Research and in part by NASA through the Microgravity Sciences Program.

REFERENCES

1. N.N. Yanenko, 1971, *The Method of Fractional Steps*, Springer-Verlag, New York.
2. V.M. Kovenya and N.N. Yanenko, *Splitting Methods in Gas-Dynamics Problems*, Nauka, Novosibirsk, 1981.
3. E.S. Oran, T.R. Young, J.P. Boris, J.M. Picone, and D.H. Edwards, *Proceedings of the 19th Symposium (International) on Combustion*, The Combustion Institute, Pittsburgh, PA, 573–582, 1982; also K. Kailasanath, E.S. Oran, and J.P. Boris, *Comb. Flame* 61, 199–209, 1985.
4. K. Kailasanath, E.S. Oran, and J.P. Boris, in *Proceedings of the GAMM-Workshop on Numerical Methods in Laminar Flame Propagation*, Vieweg, Wiesbaden, 152–166, 1982; also K. Kailasanath, E.S. Oran, and J.P. Boris, *A One-Dimensional Time-Dependent Model for Flame Initiation, Propagation and Quenching*, NRL Memorandum Report 4910, Naval Research Laboratory, Washington, DC, 1982.
5. G. Patnaik, J.P. Boris, R.H. Guirguis, and E.S. Oran, *J. Computational Physics*, 71, 1–20, 1987; also, G. Patnaik, K. Kailasanath, K.J. Laskey, and E.S. Oran, to appear in the *Proceedings of the 22nd Symposium (International) on Combustion*, The Combustion Institute, Pittsburgh, PA, 1989.

6. J.P. Boris, and D.L. Book, 1976, Solution of the Continuity Equation by the Method of Flux-Corrected Transport, *Methods in Computational Physics*, 16: 85–129; also, J.H. Gardner, Boris, J.P., E.S. Oran, R.H. Guirguis, and G. Patnaik, 1987, *LCPFCT — Flux-Corrected Transport for Generalized Continuity Equations*, NRL Memorandum Report, Naval Research Laboratory, Washington, DC, to appear.
7. M.J. Fritts, and J.P. Boris, *J. Comp. Phys.* 31: 173–215, 1979; also D.E. Fyfe, E.S. Oran, and M.J. Fritts, 1988, Surface Tension and Viscosity with Lagrangian Hydrodynamics on a Triangular Mesh, *J. Comp. Phys.*, to appear, 1988.

* Berkeley Research Associates, Springfield, VA 22150

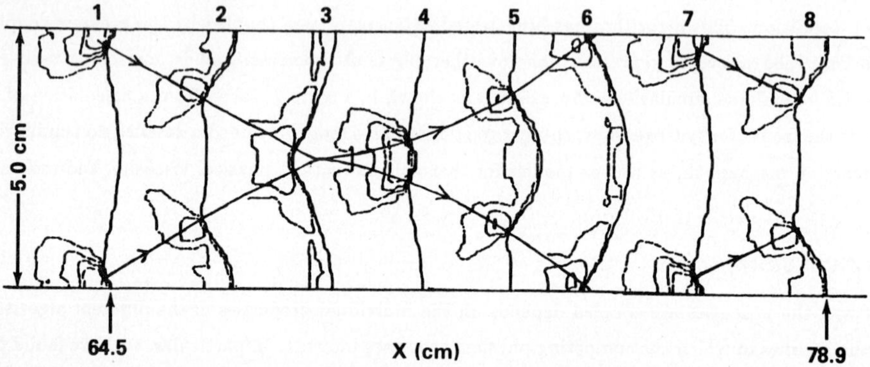

Figure 1. A composite of pressure contours at 10 μs intervals showing the structure of a detonation propagating in a room temperature stoichiometric mixture of hydrogen and oxygen diluted with 60 % argon. The figure shows the evolution of the incident shocks, Mach stems, and triple points. The lines with arrows show the movement of the triple points (3).

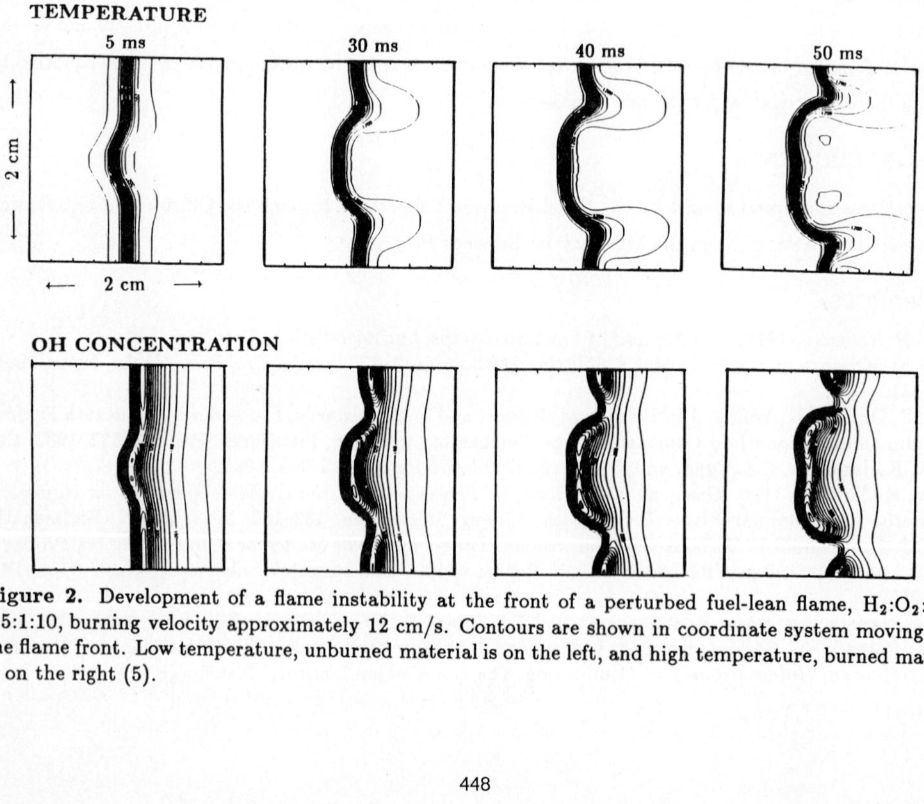

Figure 2. Development of a flame instability at the front of a perturbed fuel-lean flame, H_2:O_2:N_2 / 1.5:1:10, burning velocity approximately 12 cm/s. Contours are shown in coordinate system moving with the flame front. Low temperature, unburned material is on the left, and high temperature, burned material is on the right (5).

EXPLICIT EVALUATION OF DISCONTINUITIES IN 2-D UNSTEADY FLOWS SOLVED BY THE METHOD OF CHARACTERISTICS

C. Osnaghi
Dipartimento di Energetica, Politecnico di Milano
Piazza L. da Vinci 32, 20133 Milano, Italy

INTRODUCTION

When shock waves appear in the numerical solution of flows, a choice is necessary between *shock capturing* techniques, possible when equations are written in conservative form, and *shock fitting* techniques. If the second one is preferred, e.g. in order to obtain better definition and more physical description of the shock evolution in time, the method of characteristics is advantageous in the vicinity of the shock and it seems natural to use this method everywhere. This choice requires to improve the efficiency of the numerical scheme in order to produce competitive codes, preserving accuracy and flexibility, which are intrinsic features of the method: this is the goal of the present work.

WORKING EQUATIONS

The equations of inviscid flows (Euler) have been written in an axisymmetric coordinate system (m,n,θ), rotating with angular velocity and reduced to a quasi 2-D form on a generic surface of revolution. Locally this surface is developed and the axes rotated with respect to the meridional direction, in order to obtain a convenient orthogonal system (x,y). Energy equation is not considered. This set of equations is hyperbolic in time and it is possible to find a family of characteristic surfaces (characteristic conoids) with special propagation properties. The system can be reduced to a set of compatibility equations, which contain only derivatives with respect to surface directions. It is convenient to write the compatibility equations along the generatrices MQ of the conoid, named bicharacteristics (see Fig.1).

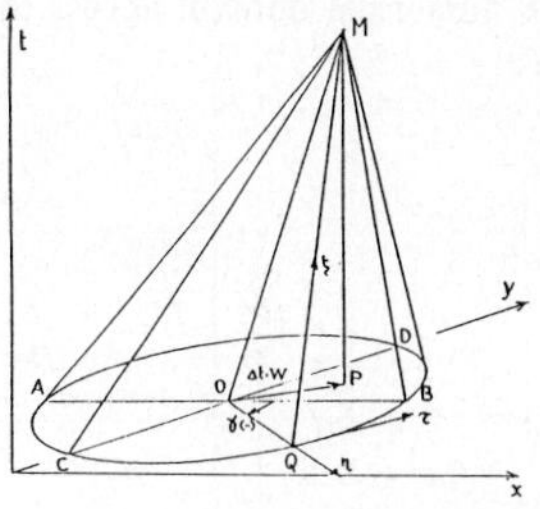

Fig.1 Characteristic cone

$$\epsilon\left(\frac{dc}{dt}\right)_\xi - \cos\gamma\left(\frac{du}{dt}\right)_\xi - \sin\gamma\left(\frac{dv}{dt}\right)_\xi - \frac{c}{k}\left(\frac{ds}{dt}\right)_\xi = D - c\left(\frac{\partial W_\tau}{\partial \tau}\right) \qquad (1)$$

where

$$\left(\frac{d}{dt}\right)_\xi = \frac{\partial}{\partial t} + (u - c\cos\gamma)\frac{\partial}{\partial x} + (v - c\sin\gamma)\frac{\partial}{\partial y}$$

ξ and τ are defined on the surface and derivatives in τ direction are considered as perturbative terms, k is the isentropic exponent, assuming the fluid locally as a perfect gas and ϵ=2/(k-1). The angle γ defines the point Q.

A special characteristic direction is defined by the path line OM, along wich entropy propagates. The correspondig compatibility equation is:

$$\epsilon\left(\frac{dc}{dt}\right)_{OM} = C_1 - c\left(\frac{\partial u}{\partial x} + \frac{\partial v}{\partial y}\right) \qquad \text{where} \qquad \left(\frac{d}{dt}\right)_{OM} = \frac{\partial}{\partial t} + u\frac{\partial}{\partial x} + v\frac{\partial}{\partial y} \quad (2)$$

The terms D in (1) and C_1 in (2) depend on angular velocity, radius and channel height. In the present plane applications they are zero.

To obtain the solution at a point M, at time t, the speed of sound c and the components of the relative velocity W, u and v, are calculated from a set of compatibility equations on the path line OM and on AM, BM, CM, DM, where cosγ and sinγ are 0 or ±1. Along these directions the Riemann constants: $R_A=\epsilon c+u$, $R_B=\epsilon c-u$, $R_C=\epsilon c+v$, $R_D=\epsilon c-v$ propagate. Note that for A, B, C, D the direction τ coincides with x or y and W_τ coincides with u and v. Equations are solved by a mean square method in order to minimize the influence of the partial derivatives

NUMERICAL SCHEME

The solution is explicit in time, starting from the flow properties at points A, B, C, D and O (see Figg.1 and 2), which are obtained by bilinear or biquadratic interpolation using the adjacent grid points. For example the point A and D are interpolated from points 0, 3, 2, 6 to 1st order or 0, 3, 11, 2, 10, 6 to 2nd order. The grid is not cartesian, nor orthogonal (x',y' in Fig.2) and it is necessary to introduce the metric coefficients to perform interpolation in curvilinear coordinates. This must be done with an efficient alghoritm, to preserve computer time. Note that the point M to be solved is not necessarily a mesh point or, if so, the new mesh may change with time.

For stability it is requested that the numerical domain contains the physical one; due to the presence of the points 5, 6, 7, 8 this condition leads to a CFL number larger than one (< 1.25). Many reasons suggest the choice of a streamvise H-mesh, in order to get a better description of the slip-lines and shock waves. This choice leads to a singular mesh in front of a blunt body, which must be removed by using a special wall mesh, shown in Fig.3. Another problem near the leading edge arises from the severe variation of the mesh size. To overcome this difficulty it was decided to implement a rigorous methodology for mesh refinement on 2 or 3 levels with any number of points per level. The method of characteristics is particularly well suited for matching different domains. If the flow is supersonic and strongly accelerated (e.g. expansion over corners), a considerable improvement is possible by choosing the numerical domain across the Mach lines; i.e., points 2, 6 and 4, 7 in Fig.4.

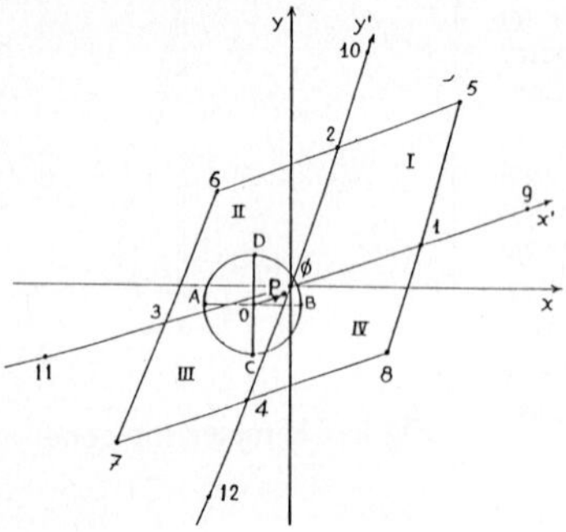

Fig.2 Numerical scheme

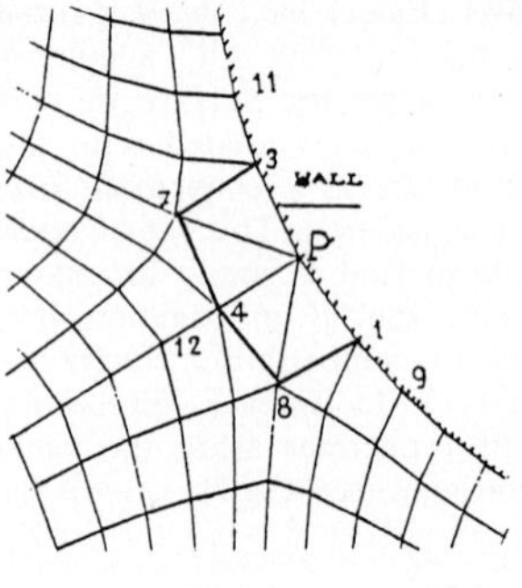

Fig.3 L.E. Mesh

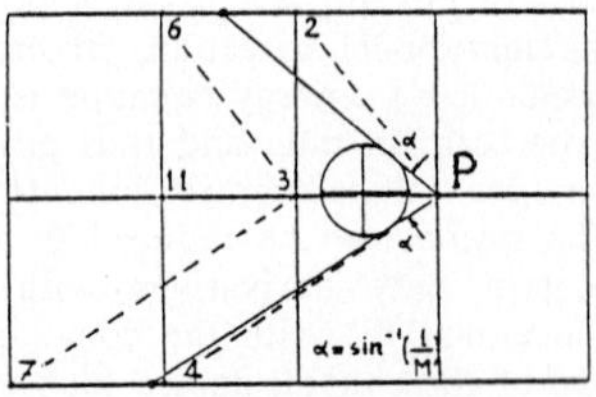

Fig.4 Supersonic point

BOUNDARY CONDITION AND DISCONTINUITIES

Several conditions are possible upstream and downstream, depending on the Mach number normal to the boundary: entropy and total pressure are given upstream, permeability condition, pressure and angle may be done, if bicharacteristics come from inside. On right and left boundary or internal bodies permeability, wall, jet or periodicity conditions can be imposed.

Slip-lines are considered as grid lines, wich are recalculated at each time step. Slip points are double points and get informations from separate regions: they are calculated as points with the same direction, by repeating the calculation until convergence is achieved in terms of pressure. Velocity discontinuities are possible due to unsteadiness or entropy jumps across the slip line. Special consideration is requested for separation points near the trailing edge of a profile: if the flow is subsonic velocity and pressure are calculated by considering them as slip line points, but angle is imposed by the wall direction. This continuity condition for pressure is a good model of the classical Kutta condition.

If the flow is supersonic in the trailing edge region a separation-reattachment criterion is used to evaluate the dead water and the shock or Prandtl-Meyer expansion in the corners.

The method of characteristics captures the shock wave in a single cell, due to the correct choice of the dependence domain. No spurious oscillations are present upstream and downstream, but the shock position depends on the initial solution and does not change if downstream conditions change. An explicit shock treatment is needed to replace correctly the shock, or to move it; the shock is considered as a boundary between two regular domains. The point M in Fig.6 is regular, because the flow relative to the shock is supersonic; downstream a compatibility equation can be used, e.g. for the bicharacteristic NB', which exists both for strong and weak shocks. If the slope of the shock is defined by the geometrical congruence, the Rankine-Ugoniot relations define uniquely a local shock speed, which matches upstream and downstream compatibility equations. It is very important to respect the correct domain of dependence in defining the slope of the shock; in other words, the choice of the branch of the shock to be used is critical.

The steps of the shocks detection, tracking and calculation are as follows:

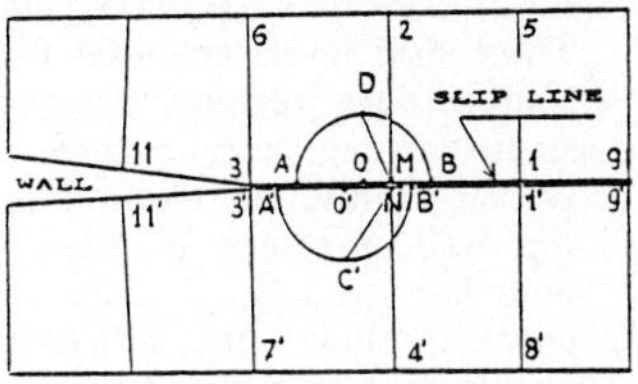

Fig.5 Slip line points

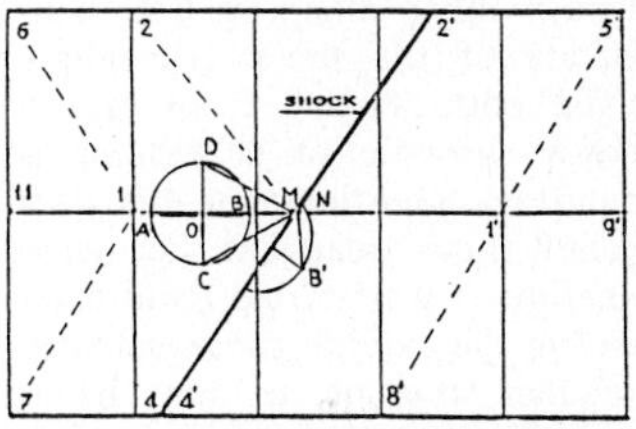

Fig.6 Shock wave points

1) Shock points are detected and flagged by testing the existence of flexes in the pressure distribution along the streamlines.
2) When 3 or more neighbouring points may be connected with a smooth line, a shock is defined and possibly connected to other existing shocks.
3) Shock points are calculated as previously explained and moved according to the upstream and downstream flow.
4) The first and the last point of the shock need special considerations, because they are wall or focalisation points, strong or weak points. Each situation need a special algorithm.
5) A test is necessary, to eliminate shock points when the discontinuity is too weak or, in steady problems, if the flow is subsonic.

It is impossible here to describe all difficulties to overcome ambiguous situations expecially where very weak shock points are present or situations involving interaction of two or more shocks, which are not considered in the present work.

RESULTS AND DISCUSSION

A great number of calculation was performed to test different geometries, boundary conditions and shock typologies. Only cases with one shock are considered here; more complex situations are at present under investigation.

Three results are presented:

1) CHANNEL FLOW (Fig.7)

A convergent-divergent channel with A_{throat}/A_{∞} =.74 is assumed. For isentropic, one-dimensional, chocked flow: p_{∞}/p_0 =.849.

Two pressure values are given downstream: the first value generates a supersonic bubble with shock, the second one generates a passage shock. In all cases the solution with explicit shock calculation is compared with the *shock capturing* one. The explicit calculation was started both from the implicit one and from the isentropic flow, by lowering continuously the downstream pressure; identical results are obtained. Remark that the position of the shock "captured" is not correct.

2) SYMMETRIC AIRFOIL (Fig.8)

Starting from an uniform flow at infinity at Mach=1.325, a bow wave is generated in front of an isolated profile. The grid is structured on 3 levels, two of which are crossed by the shock. In figures 8c and 8e the shock is captured in one mesh; the change in width is due to the interpolation of the iso-Mach lines for meshes of different size. The explicit calculation of the shock (Fig. 8d and 8f) moves upstream the discontinuity and defines accurately the leading edge region.

3) PROFILES IN CASCADE (Fig.9)
Three grid levels are used for channel, blade and leading edge region, to avoid errors due to the stagnation region approximation, wich can seriously compromise the explicit shock calculation near the wall. Pressure is given on the downstream boundary (p_{out}/p_0 = .70), in order to obtain a chocked flow with a passage shock.The Kutta condition is imposed at the trailing edge and a slip line is considered downstream.

All *shock fitting* calculations are stopped when Mach number change is less then 10^{-5} per step and the velocity of the shock is negligible. Many cases were solved both starting from arbitrary solutions or from known subsonic or supersonic solutions by changing boundary conditions; in the first case sometimes the explicit shock treatment was started from the beginning, sometime it was started from a *shock capturing* solution. Starting the explicit shock calculation from an arbitrary, irregular situation it may be impossible to connect automatically shock points; in this case the calculus aborts. Typically 3 or 4000 time steps are necessary to reach convergence in steady problems.

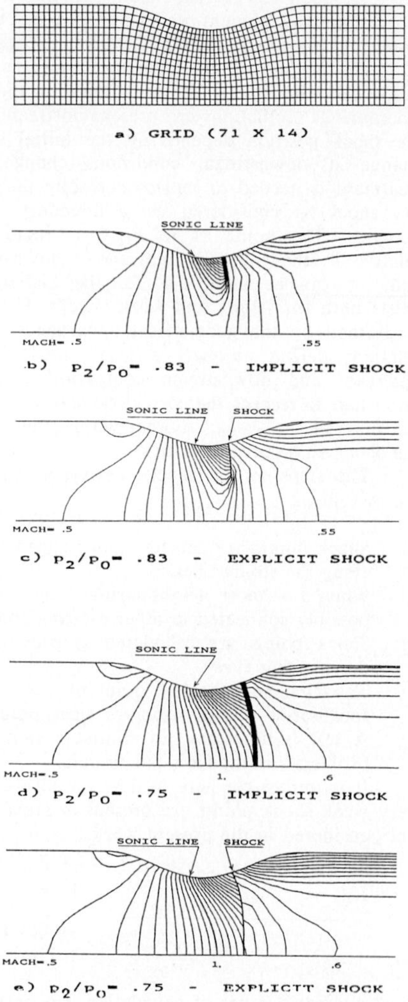

Fig.7 Channel flow - IsoMach contours

CONCLUSIONS

Shock fitting is possible and reliable even in complex flow situations and the method of characteristics is suitable to this technique, because of its capability of matching different domains in a simple way and its intrinsic ability to respect the propagation direction of flow disturbancies. It seems more appropriate to apply this technique to describe the evolution of flow with time, than to calculate steady situations starting from unphysical initial solutions. Many efforts occur to solve flows involving complex shock interaction.

REFERENCES

[1] Osnaghi, C.,Macchi, E. "On the Application of a Time Dependent Technique in Transonic Double Flow Nozzle Solutions", AGARD Conference Proceedings,No.91, Oslo (Norway), September, 1971.

[2] Osnaghi, C.,Perdichizzi, A., Bassi, F."Transonic Flow Calculation in Centrifugal Compressors by a Time Dependent Method of Characteristics", AGARD-CPP-282, Brussels, 7-9 May 1980.

[3] Moretti, G.,Zanetti, L., "A New and Improved Computational Technique for Two-Dimensional, Unsteady, Compressible Flows", AIAA Journal, Vol.22,No 6, pp.758-765,June 1984.

[4] Marcum, D.L.,Hoffman, J.D., "Calculation of Three-Dimensional Flowfields by the Unsteady Method of Characteristics", AIAA Journal, Vol.23, No 10, pp. 1497-1505, October 1985.

[5] Moretti, G., "A Technique for Integrating Two-Dimensional Euler Equations", Computers and Fluids, Vol. 15, No 1, pp.59-76, 1987.

[6] Henshaw, W.D., "A Scheme for the Numerical Solution of Hyperbolic Systems of Conservation Laws", Journal of Computational Physics, No 68, pp. 25-47, 1987.

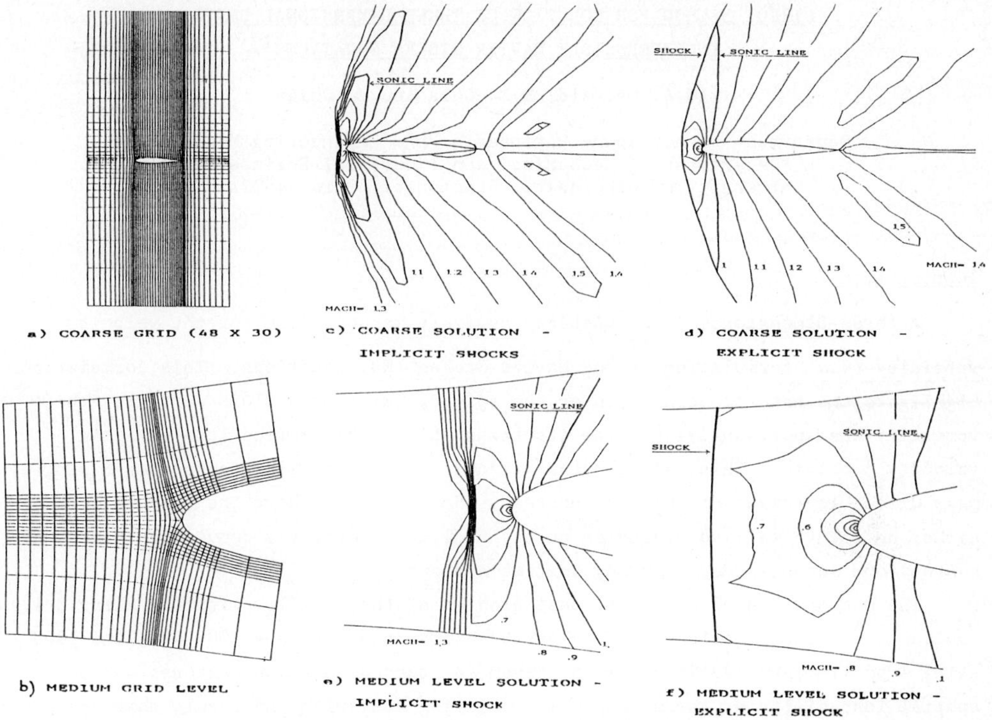

Fig.8 Symmetric Airfoil -IsoMach contours

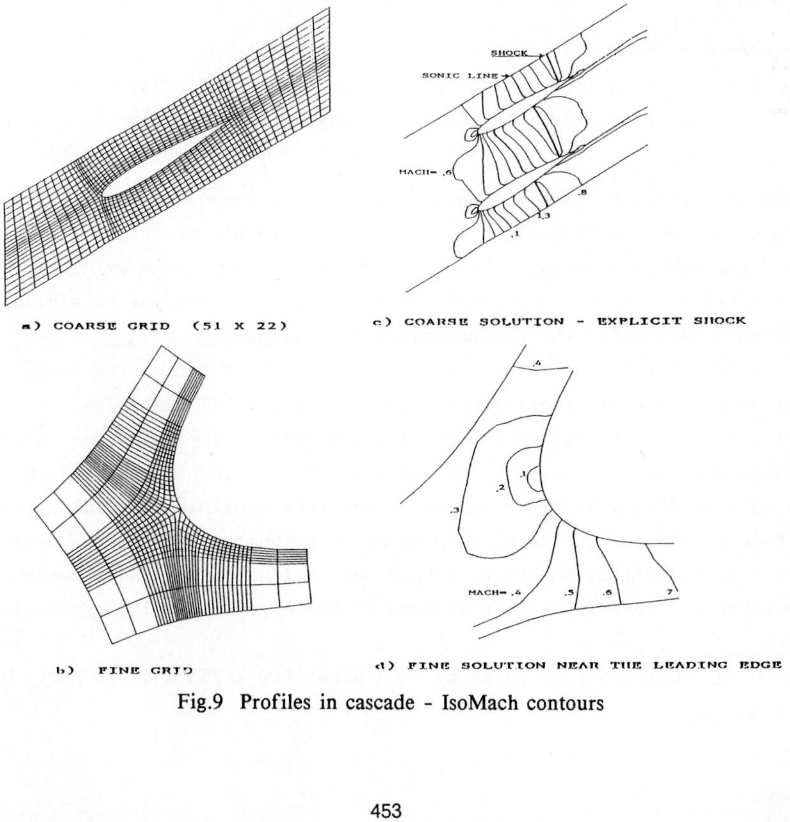

Fig.9 Profiles in cascade - IsoMach contours

DIRECT METHOD FOR SOLUTION OF THREE-DIMENSIONAL UNSTEADY INCOMPRESSIBLE NAVIER-STOKES EQUATIONS[†]

G.A. Osswald, K.N. Ghia and U. Ghia*

Department of Aerospace Engineering and Engineering Mechanics
*Department of Mechanical and Industrial Engineering
University of Cincinnati, Cincinnati, Ohio 45221 USA

Introduction

A three-dimensional (3-D) unsteady analysis is described for the velocity-vorticity $(\bar{V},\bar{\omega})$ formulation of the Navier-Stokes (NS) equations. This formulation highlights the natural separation of the spin dynamics of a fluid particle (the vorticity transport analysis) from its translational kinematics (the elliptic velocity problem). This natural separation is exploited below to produce a **single pass direct inversion** of the full unsteady NS operator without the need to iterate either on future wall vorticity or on the nonlinear vorticity convection terms, and has uniform second-order accuracy in time and space.

The method is applied to the test problem of the 3-D flow within a shear driven cubical box. Illustrative results have been obtained for Re = 100, 400, and 1000 using a preliminary (25,25,25) mesh. Even so, the calculation captures a 3-D spatial instability of the primary recirculating eddy with the result that, even for the Re = 100 case, the steady state flow within the cubic box is nowhere 2-D.

The Numerical Method

The governing equations are discretized using central differences on a staggered computational grid (Fig. 1). The unknowns are the covariant components (V_1,V_2,V_3) and $(\omega_1,\omega_2,\omega_3)$ of velocity and vorticity, respectively. The discretization maintains complete consistency, i.e., integral vorticity and velocity constraints, solenoidal velocity condition, etc., are algebraically guaranteed throughout the entire flow evolution. The analysis permits the exact removal of the over-determined nature of the elliptic velocity problem, leading to a nonsingular matrix-vector problem for the divergence-curl operator.

The vorticity equations are solved using a modification of the Douglas-Gunn (1964) version of the alternating-direction implicit (ADI) technique. A part of this modification is the use of a three-level "forward" implicit time integration scheme which provides second-order temporal accuracy without requiring any iteration upon the wall vorticity boundary values or upon the nonlinear vorticity convection terms. Even more important is the manner of distributing the partial derivatives in the convection and diffusion terms amongst the various directional sweeps. The analysis permits direct implicit treatment of the mixed partial derivative terms

† This work was supported, in part, by NASA Grant NAG 1-753 and, in part, by AFOSR Grant 87-0074.

which occur in the generalized-coordinate diffusion operator. Also, the entire 3-D vorticity transport inversion is reduced to a sequence of 1-D scalar tridiagonal inversions swept through the 3-D domain. No block-tridiagonal inversions are required. Finally, the implicit solution for the three unknown vorticity components is achieved using only six, rather than the expected nine, directional sweeps. One directional sweep per unknown vorticity component reduces to the identity transformation exactly. Thus, the computational effort of one implicit unknown can be effectively removed from the vorticity transport inversion prior to coding, as shown next.

The Vorticity Transport Algorithm OVORT3D

The vorticity transport equation, written in conservation law form, is

$$\frac{\partial \omega}{\partial t} + \nabla \times (\bar{\omega} \times \bar{V}) + \frac{1}{Re} \nabla \times (\nabla \times \bar{\omega}) = 0 \ . \tag{1}$$

Here, Re is a Reynolds number defined as $Re = (U_R L_R)/\nu$, with U_R a reference speed, L_R a reference length, and ν the constant kinematic viscosity of the incompressible fluid. Use is made of generalized orthogonal coordinates to facilitate the treatment of complex flow geometries and permit the resolution of the often disparate length scales of unsteady viscous flows. The ω_1 vorticity transport equation can be written as

$$\left(\frac{\partial \omega_1}{\partial t}\right) + \frac{g_{11}}{\sqrt{g}} \left[\frac{\partial}{\partial \xi^2} \left\{\frac{g_{33}}{\sqrt{g}} (\omega_1 V_2 - \omega_2 V_1) - \frac{1}{Re} \frac{g_{33}}{\sqrt{g}} \left(\frac{\partial \omega_1}{\partial \xi^2} - \frac{\partial \omega_2}{\partial \xi^1}\right)\right\} + \frac{\partial}{\partial \xi^3} \left\{\frac{g_{22}}{\sqrt{g}} (\omega_1 V_3 - \omega_3 V_1) - \frac{1}{Re} \frac{g_{22}}{\sqrt{g}} \left(\frac{\partial \omega_1}{\partial \xi^3} - \frac{\partial \omega_3}{\partial \xi^1}\right)\right\}\right] = 0 \tag{2}$$

Here, g_{ij} are the elements of the covariant metric tensor and g is its determinant. Similar equations result for the ω_2 and ω_3 vorticity components.

The only mixed partial derivative terms occurring in the ω_1-transport equation are associated with the ω_2 and ω_3 components. This permits the implicit treatment of these terms within a Douglas-Gunn (1964) ADI scheme without disrupting the scalar tridiagonal structure of any directional sweep. This contrasts the structure of the 3-D linear momentum equations for which mixed partials of V_1 occur within the V_1-transport equation.

Equation (2) is discretized using the two-level central-time central-space Crank-Nicolson technique with the velocity field extrapolated in time to the (n+1/2) time level from known velocities at the (n) and (n-1) time levels, thus forming a three-level forward implicit technique. The delta form is introduced where

$$(\delta W1)^{n+1} \triangleq W1^{n+1} - W1^{n} \quad \text{and} \quad W1^{n} \triangleq (\omega_1)^{n}_{ijk} \ , \tag{3}$$

with similar definitions for $(\delta W2)^{n+1}$ and $(\delta W3)^{n+1}$. The discrete analogue of Eq. (2) is approximately factored and, in finite-difference operator notation, the result can be written as

$$[I][I + (\frac{\Delta t}{2}) A_{12}][I + (\frac{\Delta t}{2}) A_{13}](\delta W1)^{n+1} - (\frac{\Delta t}{2})[B_{12}](\delta W2)^{*2*} - (\frac{\Delta t}{2})[B_{13}](\delta W3)^{n+1}$$

$$+ (\Delta t)\{[A_{12}+A_{13}](W1)^n - [B_{12}](W2)^n - [B_{13}](W3)^n\} = 0 \qquad (4a)$$

where (Δt) is the discrete time step,

$$[I](\delta W1)^{n+1} \triangleq (\delta W1)^{n+1}_{ijk} ,$$

$$[A_{12}](W1)^n \triangleq GW1_{ijk}[G33_{ijk}\{\frac{(V2^{n+1/2}_{ij+1k} + V2^{n+1/2}_{ij+1k-1})}{2} - \frac{1}{Re}\}(W1^n_{ij+1k})$$

$$+ [G33_{ijk}\{\frac{(V2^{n+1/2}_{ijk}+V2^{n+1/2}_{ijk-1})}{2} + \frac{1}{Re}\} - G33_{ij-1k}\{\frac{(V2^{n+1/2}_{ijk}+V2^{n+1/2}_{ijk-1})}{2} - \frac{1}{Re}\}](W1^n_{ijk})$$

$$- G22_{ij-1k}\{\frac{(V2^{n+1/2}_{ij-1k} + V2^{n+1/2}_{ij-1k-1})}{2} + \frac{1}{Re}\} (W1^n_{ij-1k})] \quad . \qquad (4b)$$

Similar forms can be obtained for the operators A_{13}, B_{12}, and B_{13}. Also, $(\delta W2)^{*2*}$ represents the second intermediate estimate of $(\delta W2)^{n+1}$ that occurs as a part of the ADI solution methodology. Similarly, the ω_2 and ω_3 vorticity transport equations are obtained as:

$$[I + (\frac{\Delta t}{2})A_{21}] [I] [I+(\frac{\Delta t}{2})A_{23}] (\delta W2)^{n+1} - (\frac{\Delta t}{2})[B_{21}](\delta W1)^{*1*}- (\frac{\Delta t}{2})[B_{23}](\delta W3)^{n+1}$$

$$+ (\Delta t) \{[A_{21}+A_{23}](W2)^n - [B_{21}](W1)^n - [B_{23}](W3)^n\} = 0 \qquad (5)$$

and

$$[I + (\frac{\Delta t}{2})A_{31}][I + (\frac{\Delta t}{2})A_{32}] [I] (\delta W3)^{n+1}- (\frac{\Delta t}{2})[B_{31}](\delta W1)^{*1*}- (\frac{\Delta t}{2})[B_{32}](\delta W2)^{*2*}$$

$$+ (\Delta t) \{[A_{31}+A_{32}](W3)^n - [B_{31}](W1)^n - [B_{32}](W2)^n\} = 0 \quad . \qquad (6)$$

The direct implicit solution of Eqs. (4)-(6) proceeds as follows

(1-Sweep of ω_1-Equation)

$$[I](\delta W1)^{*1*} = (\Delta t) \{[B_{12}](W2)^n + [B_{13}[(W3)^n - [A_{12}+A_{13}](W1)^n\} \qquad (7a)$$

(1-Sweep of ω_2-Equation)

$$[I + (\frac{\Delta t}{2})A_{21}](\delta W2)^{*1*} = (\Delta t) \{[B_{21}](W1)^n+[B_{23}](W3)^n - [A_{21}+A_{23}](W2)^n\}$$

$$+ (\frac{\Delta t}{2}) [B_{21}](\delta W1)^{*1*} \qquad (7b)$$

(1-Sweep of ω_3-Equation)

$$[I + (\frac{\Delta t}{2})A_{31}](\delta W3)^{*1*} = (\Delta t) \{[B_{31}](W1)^n+[B_{32}](W2)^n - [A_{31}+A_{32}](W3)^n\}$$

$$+ (\frac{\Delta t}{2}) [B_{31}](\delta W1)^{*1*} \qquad (7c)$$

(2-Sweep of ω_2-Equation): $\quad [I](\delta W2)^{*2*} = (\delta W2)^{*1*} \qquad (8a)$

(2-Sweep of ω_1-Equation)

$$[I + (\frac{\Delta t}{2}) A_{12}] (\delta W1)^{*2*} = (\delta W1)^{*1*} + (\frac{\Delta t}{2}) [B_{12}](\delta W2)^{*2*} \qquad (8b)$$

(2-Sweep of ω_3-Equation)

$$[I + (\frac{\Delta t}{2}) A_{32}] (\delta W3)^{*2*} = (\delta W3)^{*1*} + (\frac{\Delta t}{2})[B_{32}](\delta W2)^{*2*} \quad (8c)$$

(3-Sweep of ω_3-Equation): $[I](\delta W3)^{n+1} = (\delta W3)^{*2*}$ (9a)

(3-Sweep of ω_1-Equation)

$$[I + (\frac{\Delta t}{2}) A_{13}](\delta W1)^{n+1} = (\delta W1)^{*2*} + (\frac{\Delta t}{2})[B_{13}](\Delta W3)^{n+1} \quad (9b)$$

(3-Sweep of ω_2-Equation)

$$[I + (\frac{\Delta t}{2}) A_{23}](\delta W2)^{n+1} = (\delta W2)^{*2*} + (\frac{\Delta t}{2})[B_{23}](\delta W3)^{n+1} . \quad (9c)$$

Finally, $(W1)^{n+1}$, $(W2)^{n+1}$, $(W3)^{n+1}$ are obtained using the defining equations similar to Eq. (3).

Three of the nine sweeps described above do NOT require any matrix inversions, and the remaining six sweeps require the inversion of only scalar (as opposed to block) tridiagonal matrices. Consequently, OVORT3D is a very efficient 3-D transport procedure. Indeed, the computational intensity of OVORT3D more closely reflects the cost of solving for two, rather than three, unknowns.

The Elliptic Velocity Inversion DIVCRL

The elliptic velocity problem consists of specifying both the divergence, $d = \nabla\cdot\bar{V}$, and curl, $\bar{\omega} = \nabla\times\bar{V}$, of the velocity vector $\bar{V}$ throughout the interior of the flow domain, together with consistent velocity conditions on the boundary of the flow domain. The direct inversion algorithm DIVCRL for the elliptic velocity problem was developed earlier by the authors in Ref. (3). Proper choice of discretization led to sparse matrices with repetitive block structure, which was exploited to produce efficient direct inversion techniques. Vector speed ups of 15 are achieved for the DIVCRL algorithm on (17,17,17) test calculations. During the (25,25,25) flow calculations, the vectorized version of the DIVCRL algorithm achieved a computational effort index of 55 CPU μ seconds/mesh point/time step. The computational index of the scalar version of OVORT3D was 41 μ sec. for the (25,25,25) calculations. Vectorization of OVORT3D is currently in progress.

Results and Discussion

The viscous incompressible flow within a shear driven cubical box has been solved using the (25,25,25) uniformly distributed grid shown in Fig. 2. The velocity of the top wall is assumed to be aligned with the x^1-coordinate axis. For discussion purposes, the front of the box is defined as the x^3=1 plane and the side of the box is defined as the x^1=1 plane. Thus, Fig. 2 presents the top, front and side views of the grid distribution used in this study.

An impulsive start from rest was employed as the initial condition. A time step Δt = 0.025 characteristic time units was employed for all cases, and the 3-D unsteady, flow simulation was run until steady-state was reached using 32-bit machine accuracy. The mid-plane tangential velocity vectors corresponding to steady state for Re=100 are plotted in Fig. 3 where the top, front and side views provide the component of velocity tangent to the x^2=1/2, x^3=1/2, and x^1=1/2 planes, respectively. The corresponding contours of the normal vorticity component in Fig. 4 show the strong cellular nature of the vorticity field depicted in the top and side views of Fig. 3. This is the signature of a pair of Taylor-Görtler vortices that have formed within the primary recirculation eddy on opposite sides of the mid-span x^3=1/2 plane. The energy provided by the moving wall feeds this 3-D spatial instability with the result that the flow within the mid-span plane is no longer 2-D.

Figure 5 shows the tangential velocity field within the x^2=0.688 plane, which is parallel to the moving top and just below the primary-eddy center. It also provides direct visual evidence that $(\partial^2u/\partial z^2)$ is not zero at x^3=1/2 plane and thus the mid-span flow is influenced by nonuniformly distributed viscous forcing terms. Due to the presence of the off mid-span Taylor-Görtler vortices, the 3-D cubic box flow is nowhere 2-D even at Re=100. Indeed, the highest kinematic energy fluid is located near the cores of these off mid-span vortices and energy transfers away from the mid-span to drive this 3-D spatial instability.

Figures 6 and 7 present steady-state results for Re=400. The 3-D effect is clearly visible in the top view of Fig. 6; the strength of the Taylor-Görtler vortex pair clearly increases as seen in the top and side views of Fig. 7. The vertical extent L_1 of the secondary downstream corner eddy within the mid-span plane from the present (25,25,25) 3-D computation is compared in Fig. 8 with the 2-D computations using the procedure developed by the authors in Ref. (4) using a (129,129) mesh as well as with the experiment of Pan and Acrivos (1967). The three-dimensional effects are clearly seen. Also, the predicted results show slight discrepancy with the data and this may be due to the use of a rotating cylinder as the moving top wall in the experimental study, while the calculations assume a flat upper surface. The energy transfer mechanism between the moving wall, the primary eddy and primary 3-D instabilities are critical in establishing the actual L_1 location.

Conclusion

A uniformly second-order accurate 3-D unsteady incompressible NS analysis is developed in generalized orthogonal coordinates. The flow simulation is carried out on the NASA-Langley CYBER 205 supercomputer. The direct inversion methodology developed for the nonlinear operator of the governing equations is very robust and fully vectorizable, can exploit parallelism, and has been efficiently implemented. The present strategy is to enhance CPU efficiency at the expense of additional storage, which therefore limits permissible grid size. With the large high-speed memories (2 Giga words +) of the next generation supercomputers, this strategy will permit fine-grid 3-D computations to be performed efficiently.

References

1. Douglas, J. and Gunn, J.E., (1964), Numerische Mathematik, Vol. 6, pp. 428-453.
2. Koseff, J.R. and Street, R.L., (1983), Three Dimensional Turbulent Shear Flows, ASME Publication, pp. 23-31.
3. Osswald, G.A., Ghia, K.N. and Ghia, U., (1987), AIAA CP 874, pp. 408-421.
4. Osswald, G.A., Ghia, K.N. and Ghia, U., (1988), Computational Mechanics '88 - Theory and Applications, Editors: S.N. Atluri and G. Yagawa, Springer-Verlag, New York, Vol. 2, pp. 5151-5154.
5. Pan, F. and Acrivos, A., (1967), Journal of Fluid Mechanics, Vol. 28, pp. 643-655.

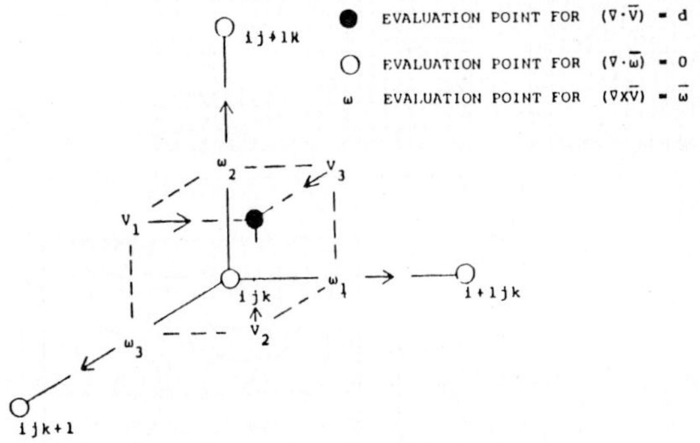

Fig. 1. Typical Computational Cell Showing Staggered Evaluation of Covariant Velocity and Vorticity Components.

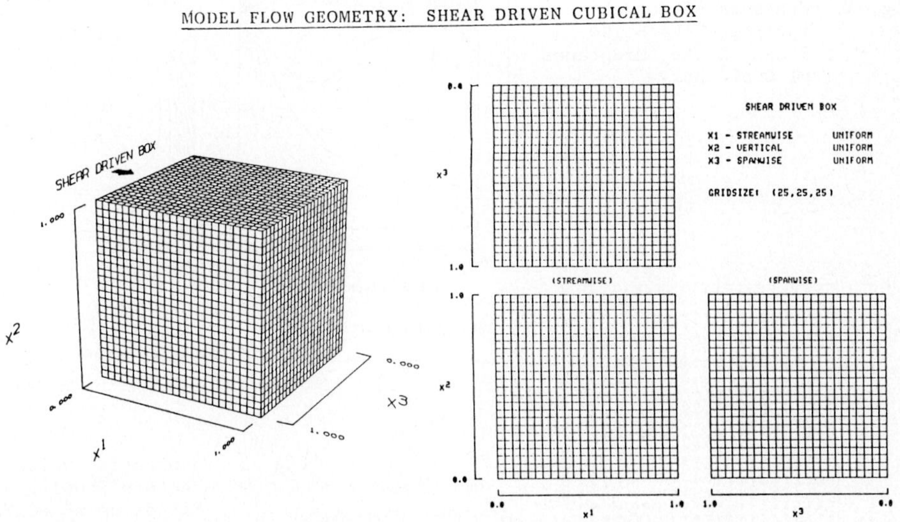

Fig. 2. Uniform Grid Distribution for Shear Driven Box Flow Simulation. Impulsive Start From Rest.

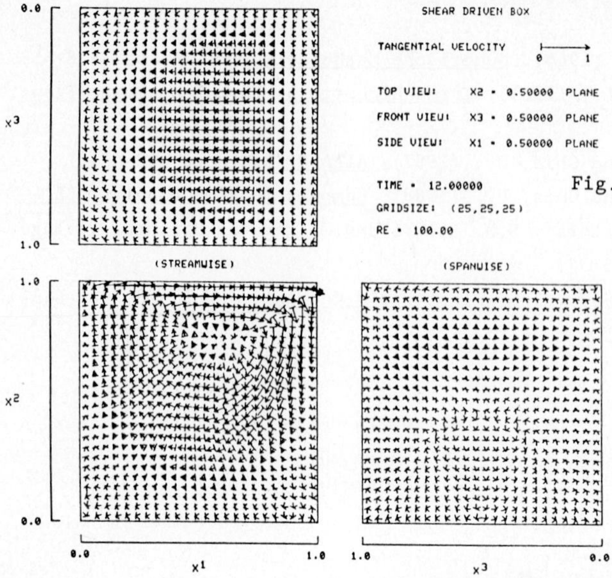

Fig. 3. Tangential Velocity Vectors, Re = 100, Steady State. Midplanes of Cubic Box.

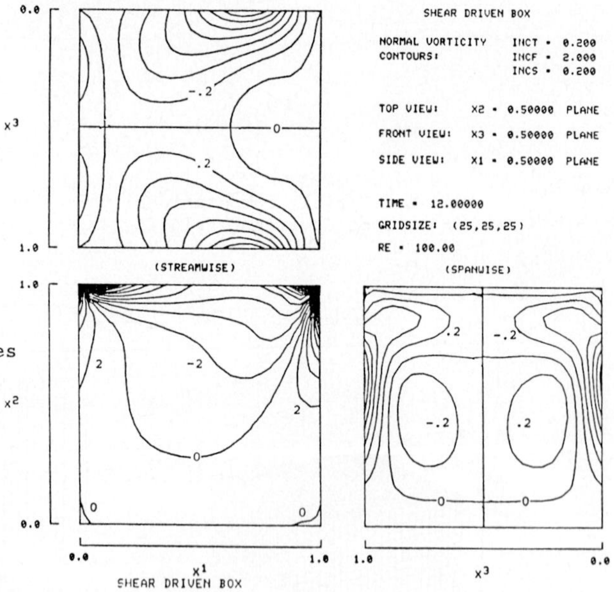

Fig. 4. Contours of Normal Vorticity, Re = 100, Steady State, Midplanes of Cubic Box.

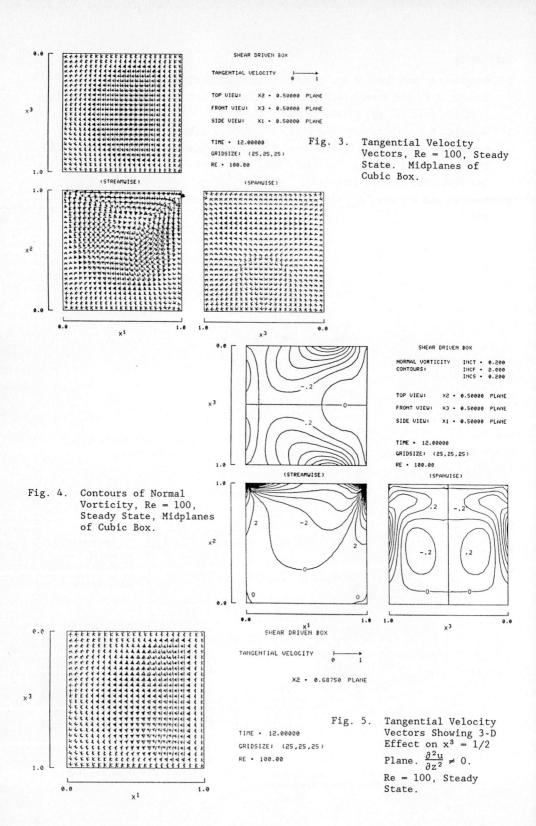

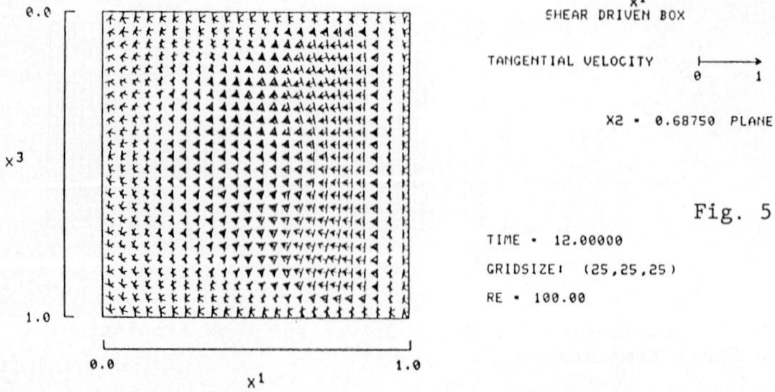

Fig. 5. Tangential Velocity Vectors Showing 3-D Effect on $x^3 = 1/2$ Plane. $\frac{\partial^2 u}{\partial z^2} \neq 0$. Re = 100, Steady State.

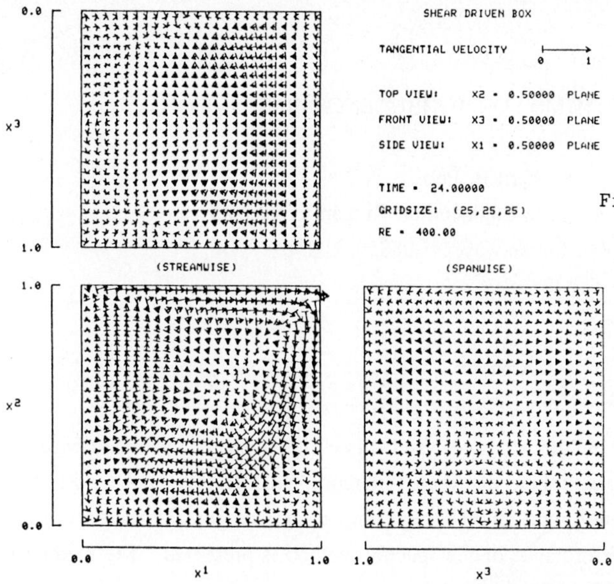

Fig. 6. Tangential Velocity Vectors, Re = 400, Steady State. Midplanes of Cubic Box.

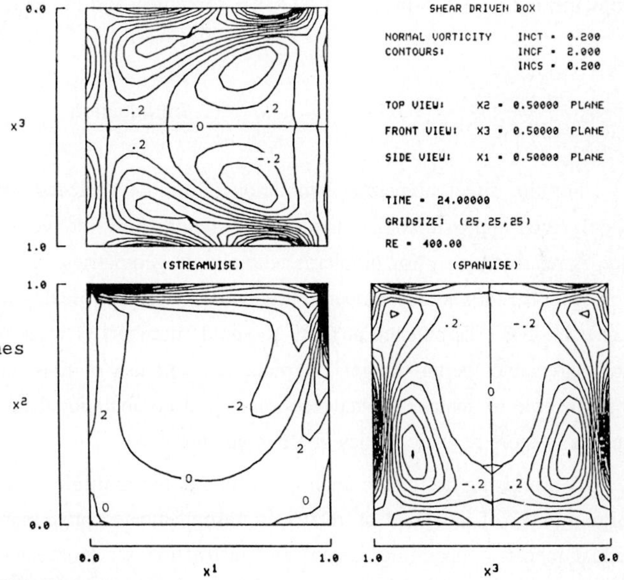

Fig. 7. Contours of Normal Vorticity, Re = 400, Steady State, Midplanes of Cubic Box.

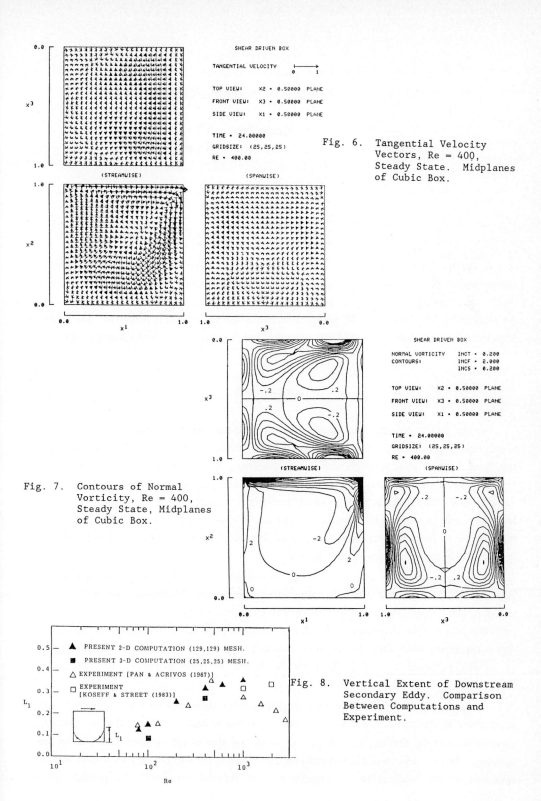

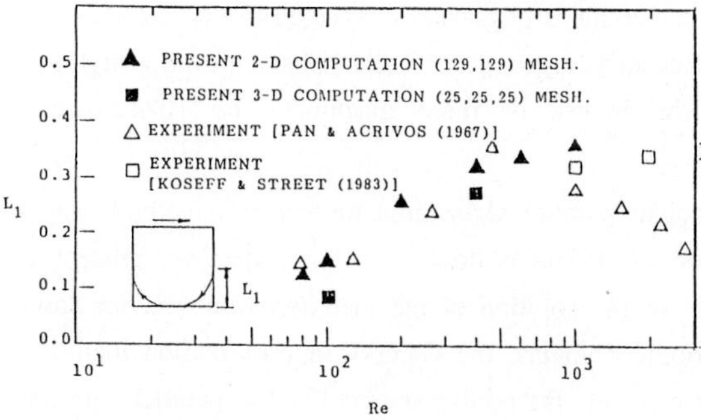

Fig. 8. Vertical Extent of Downstream Secondary Eddy. Comparison Between Computations and Experiment.

HYPERCUBE ALGORITHMS FOR TURBULENCE SIMULATION

Richard B. Pelz
Mechanical and Aerospace Engineering, Rutgers University
P.O. Box 909, Piscataway, NJ 08855, U.S.A.

Abstract

A spectral method for the simulation of two- and three-dimensional, unsteady, viscous, incompressible fluid flow is developed for a distributed-memory, MIMD hypercube. Periodic domains are decomposed in one direction, and an FFT algorithm is derived for distributed vectors. An analytic formula for the parallel performance of the spectral method is presented. The method is implemented on a 128-processor NCUBE computer, and the experimental performance is reported.

1. Introduction

For the direct numerical simulation of fluid turbulence, spectral methods have been the most widely used approximation scheme. For smooth flows convergence is exponential, and implementation is relatively easy for problems with simple geometries. Even with the most powerful computers, however, the Reynolds number (based on Taylor microscale) for well-resolved simulations has not exceeded 100. Turbulence at these Reynolds numbers is weak, having a limited range of scales and no appreciable inertial range. Through efficient use of massively-parallel processors, however, we may be able to achieve higher resolutions and computational rates, and thus extend turbulence simulations to more realistic values of Reynolds number.

A promising design for such a machine is the multiple instruction-multiple data computer with a large number of independent, relatively unsophisticated processors. Each processor possesses an individual memory, operating system and instruction set (computer code). The processors are loosely connected in a hypercube network, in which a processor is physically connected to that set of processors whose binary labels differ in one place. Recently, a group at Sandia Labs achieved a high level of parallel performance solving PDEs on the largest of these machines, the 1024-processor NCUBE/ten.[1]

The goal of our work is to develop efficient parallel algorithms for spectral methods and to apply them to the direct numerical simulation of turbulent flows. In this paper we present the parallelization of the simplest spectral method for the solution of the incompressible Navier-Stokes equations. In two- and three-dimensional periodic domains, the Galerkin approximation method is employed, and the transform method is used to calculate the nonlinear term.[2] The parallel algorithm is described in Section 2; an analytic model for parallel performance is derived in Section 3; and numerical experiments on the 128-processor NCUBE/seven are presented in Section 4.

A question that will be answered in this analysis is how does the parallel performance of spectral methods compare with the performance of finite difference methods. Since finite difference methods have local support, wouldn't they be more suitable for domain decomposition techniques? We shall show that in many cases spectral methods are competitive with finite difference methods.

2. The Parallel Algorithm

In this section, we first review the spectral method and time-integration scheme. The parallelization of the method is then discussed. An FFT algorithm for distributed vectors is developed, and the whole algorithm is presented.

The problem to be solved is the time-integration of the Navier-Stokes equations for an incompressible flow. A second-order accurate, time-split scheme using leap-frog and Crank-Nicholson methods is employed for the integration. The spatial domains are the 2-D rectangle and 3-D box with periodic boundary conditions.

Under the Galerkin approximation, the flow variables are expanded in a finite Fourier series, and the problem is recast in terms of the Fourier coefficients. The evaluation of the nonlinear term requires the calculation of a convolution involving the Fourier coefficients. The number of operations for this calculation is reduced by using the transform method. [2] Here, fast Fourier transforms are employed to calculate the nonlinear term in "physical space."

In the parallel spectral method, the domain is divided in one direction producing strips in two-dimensions and slabs in three-dimensions. The subdomains are mapped one-to-one and onto the processors of the hypercube. The derivative, Laplacian and inverse Laplacian operations and the evaluation of the nonlinear term are all point-wise calculations. No data must be communicated between processors to perform these operations, and the speedup is linear in the number of processors. The Fourier transform of vectors that are complete within each processor also requires no communication.

The Fourier transform of the vectors that are distributed across processors does require data communication. In Figure 1 we show the FFT algorithm for a complex vector of length M=8.[3] Suppose that the vector is distributed across P=8 processors (one element per processor). During each of the $\log_2 P$ stages, data must be sent and received on each processor. If each processor index is identified with the corresponding element index, all communication is to and from processors that are nearest neighbors on the hypercube.

When the calculation shown in Figure 1 is complete, the input vector has been overwritten in memory by its Fourier transform in bit-reversed order. A reordering of the vector requires (non-nearest neighbor) communication which can be time-consuming. The reordering, however, is not necessary, because the order of the elements is not important for the point-wise evaluation of the nonlinear term. The inverse FFT of the bit-reverse-ordered vector is similar to the original FFT; only nearest neighbor communication is required.

Figure 2 shows the speedup versus processor dimension, $\log_2 P$, for the complex FFT of a vector with length M=128. Speedup is defined as the ratio of computation times of one processor and P processors. The number of transforms is N, and the two-dimensional array is arranged as (N,M/P). Low parallel performance is obtained when only one transform is performed, especially if a large number of processors is used. Performance increases markedly as the number of transforms increases; speedup reaches 60 for 800 transforms on 128 processors. This can be explained by examining the formula for the time spent in data communication, $T = \log_2 P\,(T_{st} + T_{wd} 2N \cdot M/P)$, where T_{st} and T_{wd} are the start-up time for a write (473 μs) and the communication rate (12.6 μs per 32-bit word), respectively. The numerical values are for an NCUBE computer.[4] At each of the $\log_2 P$ stages, $2N \cdot M/P$ real numbers are transferred. It is the relative magnitudes of the two terms in parenthesis that determine the performance scaling. If the first term dominates, the communication time increases with processor dimension. If the second term dominates, the communication time decreases as $\log_2 P/P$. The latter case can occur when N is large. More discussion of communication times is in Section 3.

Another method of performing the Fourier transform of a distributed vector is to rearrange the array so that the vector that was distributed, becomes complete within a processor. The transform can then be performed locally. When doing multiple transforms this is achieved by rearranging the array so that an index that was complete becomes distributed. This operation is the transpose. The block transpose algorithm [5] requires approximately the same amount of communication between processors as the FFT. Chu [6] has recently shown that for two-dimensional Fourier transform problems, the parallel performance of the transpose and distributed FFT methods is about the same. See Pelz [7] for more detail on implementation and results of the transpose for parallel spectral methods.

The complete parallel spectral scheme can be written. For the 3-D problem, velocities are given in Fourier representation and arranged such that a processor p has arrays of the form (1:nx,p·nz+1:(p+1)·nz,1:ny), where p=0,1, · · · ,P−1. The vorticity field is calculated, the y Fourier transform of the velocity and vorticity is performed, and the 2nd and 3rd indices are switched. Six distributed transforms in z are then performed. After conjugate symmetric-to-real transforms are performed, the arrays are in physical representation. The nonlinear term is calculated, and the inverses of the last three steps are performed on the velocity and nonlinear terms. The Fourier coefficients of velocity are then updated to the next time-level using the leap-frog/Crank-Nicholson method. The update operation can be performed with linear speedup.

3. Model for Parallel Performance of the Spectral Method

In this section, an analytic formula is presented for the parallel performance of the algorithm proposed in the previous section. We shall denote T_0 the total time required by one processor to complete one time-step and shall assume that there is no inherent serial time in T_0. The calculation time using an idealized parallel processor with no communication time will be T_0/P. Using the equation for communication time of an FFT that was given in Section 2, the speedup for a parallel spectral method is given by

$$\text{Speedup} = P\left[1 + \frac{N_{FFT}\, P \log_2 P}{T_0}\,(T_{st} + T_{wd}\, N_{tot}/P)\right]^{-1} \tag{1}$$

where N_{FFT} is the number of transforms required per time-step and N_{tot} is the total number of grid points.

In order to maximize the speedup for a given number of processors, the second term in brackets should be minimized. The parameters T_{st} and T_{wd} (given in Section 2) are determined by the type of computer being used. The parameter T_0 is related to the computational rate of the individual processors. Parameters over which the user has direct control are N_{tot} and N_{FFT}. In some cases the number of transforms can be decreased at the expense of larger memory requirements, e.g., for the 3-D calculation, the number of FFTs can be reduced from 12 to 9 by adding 3 more arrays. Since the start-up time T_{st} is multiplied by $P\log_2 P$, there will be serious degradation if this term dominates. Thus, N_{tot}/P should be set so that $T_{st} \ll T_{wd}\, N_{tot}/P$.

The expression for speedup of a residual calculation in a finite difference scheme with nearest-grid-point support and with similar domain decomposition is

$$\text{Speedup} = P\left[1 + \frac{2\,P}{T_0}\,(T_{st} + T_{wd}\, N_{tot}^{(d-1)/d})\right]^{-1} \tag{2}$$

where d is the dimension of the physical problem. The main difference between Eqns. (1) and (2) is the length of the transferred vectors. If $P = O(N_{tot}^{1/d})$, then these vector lengths are of the same order, and communication times are comparable.

To complete the implicit time-stepping scheme for the finite difference method would require solving a number of sparse matrix equations. A recursive-type scheme, such as the LU factorization, does not parallelize well. It is concluded that spectral methods will have competitive parallel performance with methods having local support.

4. Numerical Experiments

In this section, we present timing results of numerical experiments performed on a hypercube. Both two- and three-dimensional problems are solved. The computer employed is the NCUBE/seven with 128 supermini processors each having 1/2 MByte RAM. Because of the memory limitations and the fact that in order to compute speedups, the whole problem must be performed on one processor, N_{tot} was limited to 2^{13} 32-bit numbers. Much larger problems were performed (64x64x128 and 1024x512), but could not be compared to timings from one processor.

The initial condition was the Taylor-Green vortex.[8] No special symmetries were assumed and no removal of aliased modes was performed. Energy and enstrophy were monitored at each time-step, a calculation which also requires inter-processor communication. They compared exactly with results obtained on other computers.

Figure 3 shows a graph of $\log_2$(speedup) versus processor dimension for a grid of 8x32x32. Experimental results (circles) were obtained using the distributed FFT method. The speedup predicted by the analytic model, c.f. Eq (1), is plotted with triangles. The optimal speedup, linear with the number of processors, is given by the solid line. The table on the figure shows the speedup values. The experimental speedup does not deviate much from the optimal; efficiency is monotonically decreasing, however, reaching 82% for 32 processors. The analytic formula predicts accurately the speedup. Calculations done using the transpose method show a slightly lower speedup.[5]

In Figure 4 we plot seconds per time-step versus processor dimension for the following cases: 8x8x16 with 16 processors, 16x16x32 with 32 processors, 32x32x64 with 64 processors and 64x64x128 with 128 processors. On a serial computer, the increase in CPU time from a grid (N/2,N/2,N) to (N,N,2N) is from 8 to 8(d+1/3)/(d-2/3) where $d=\log_2 N$. The extent to which parallel processing "two-dimensionalizes" the problem is shown in the figure. The dotted and dashed lines represent the linear and log scalings for a two-dimensional serial problem, which are 4 and 4/(1-1/d), respectively. The parallel performance of the 3-D problem is within the bounds of the 2-D serial problem. The time necessary to complete one time-step with the grid 64x64x128 is 45 seconds. This is about 7 times slower than the CPU time for a Q8-optimized code on the CYBER 205. It has been reported that the computational rate on a single NCUBE processor is from 4 to 7 times slower than the optimal rate of 0.5 MFLOPS.[1] We have experienced the slowest rates. More optimal rates will be obtained when the compiler is more efficient. All code is presently written in FORTRAN 77. If the processors were to run at the optimal rate, the overall computation rate for a problem with a 64x64x128 grid and 128 processors would be comparable to that of the CYBER 205.

In Figure 5, we plot $\log_2$(speedup) versus processor dimension for the two-dimensional Navier-Stokes problem with a grid of 128x128. Shown are results of numerical experiment (circles), analytic model (triangles) and optimal speedup (solid line). The table in the figure contains the values for speedup. The efficiency for 128 processors is 68%. The model predicts accurately the performance of the experiment except at large dimension. The reason for the discrepancy is as follows. As the ratio N_{tot}/P decreases, an increasingly important factor in the communication time is the synchronization time. If a pair of processors that are out of synchronization must exchange data, one processor will be idle until the other "catches up." This effect was not accounted for in the model.

De-synchronization will occur in an exchange of data. Node B writes to node A, issues a read and waits for the data to be sent from node A. Node A reads the data sent from node B and writes its data to node B. Node A can then start its calculation again, while node B lags behind because it must read the data sent to it by node A. The concurrent write-and-read exchange is not used because of the limited buffer space. The extent to which nodes become de-synchronized is a function of the data length and the number of processors involved in the FFT.

Conclusions

A parallel-spectral method for turbulence simulation has been developed and implemented on a hypercube computer. The parallel performance of the distributed-vector FFT algorithm increases with the number of transforms, which means that large-scale problems will be efficient on massively-parallel computers. The analytic formula for parallel performance of the algorithm closely models the performance of numerical experiments. The communication time for the spectral method is competitive with finite difference methods. The scaling of computation time is one dimension less than that of the physical problem, provided the number of processors scales with a linear dimension. De-synchronization and communication start-up time degrade the parallel performance.

I would like to thank R. Peskin and E. Siegel for helpful comments. The computations reported here were performed at the Parallel Computing Laboratory of the Center for Computer Aids for Industrial Productivity (CAIP), an Advanced Technology Center of the New Jersey Commission of Science and Technology, at Rutgers University, Piscataway, NJ. This work was supported by the National Science Foundation under grant EE 88-08780 and CAIP.

References

[1] J. L. GUSTAFSON, G. R. MONTRY, and R. E. BENNER, Development of Parallel Methods for a 1024-Processor Hypercube, SIAM J. Sci. Stat. Comp. , 9, No. 4, (1988).

[2] S. A. ORSZAG, Numerical Simulation of Incompressible Flows within Simple Boundaries. I: Galerkin (Spectral) Representation, Stud. in Appl. Math., 50, (1971) pp. 293-327.

[3] J. W. COOLEY and J. W. TUKEY, An Algorithm for the Machine Calculation of Complex Fourier Series, Maths Comput., 19, (1965) pp. 297-301.

[4] A. BEGUELIN and D. J. VASICEK, Communication between Nodes on a Hypercube, in Hypercube 1987, SIAM, Philadelphia, (1987) pp. 353-381.

[5] O. A. MCBRYAN and E. F. VAN DE VELDE, Hypercube Algorithms and Implementations, SIAM J. Sci. Stat. Comp., 8, No. 2, (1986).

[6] C. Y. CHU, Comparison of Two-Dimensional FFT Methods on the Hypercube: The Choice Between Strips and Patches, TR–87–850, Dept. Comp. Sci., Cornell U., (1987).

[7] R. B. PELZ, Large-Scale Spectral Simulation of the Navier-Stokes Equations on a Hypercube Computer, AIAA Paper 88–3643, 1st Natl. Fluid Dyn. Congress., Cincinatti, OH, (1988).

[8] M. E. BRACHET, D. I. MEIRON, S. A. ORSZAG, B. G. NICKEL, R. H. MORF and U. FRISCH, Small-scale Structure of the Taylor-Green Vortex, J. Fluid Mech., 130, (1983), pp. 411-452.

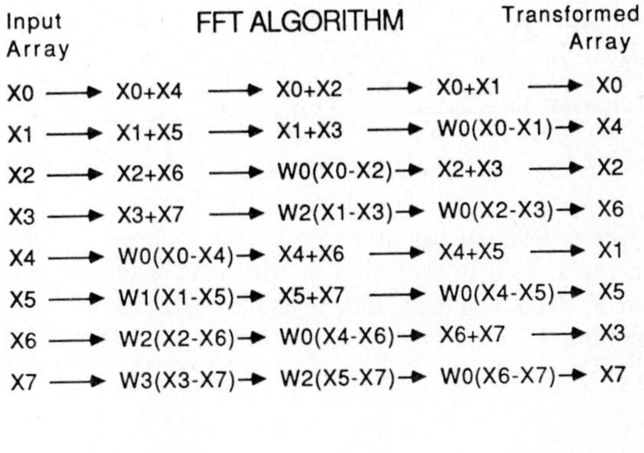

Figure 1: A base-2 FFT algorithm of a complex 8-tuple. Arrows denote "is replaced in memory by"; Wn is the nth root of unity.

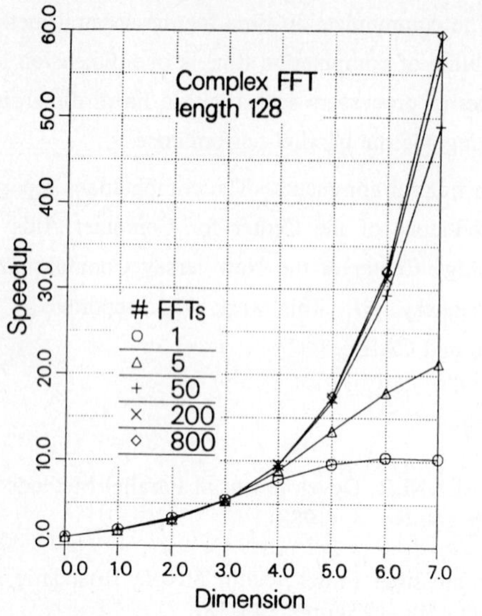

Figure 2: Speedup versus processor dimension for an FFT of complex vectors with length 128. The curves are for different numbers of transforms.

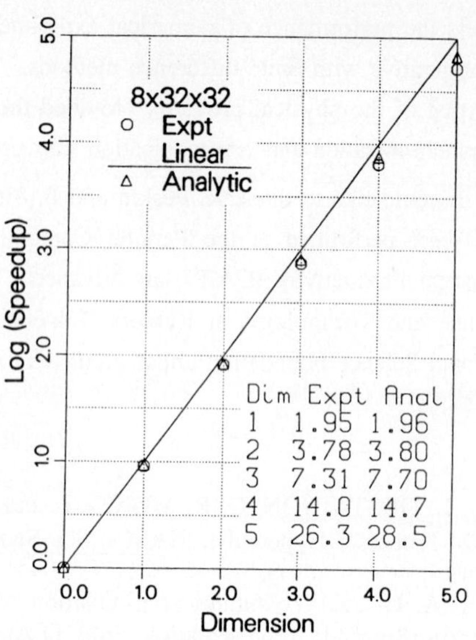

Figure 3: $\log_2$(speedup) versus processor dimension for 3-D case.

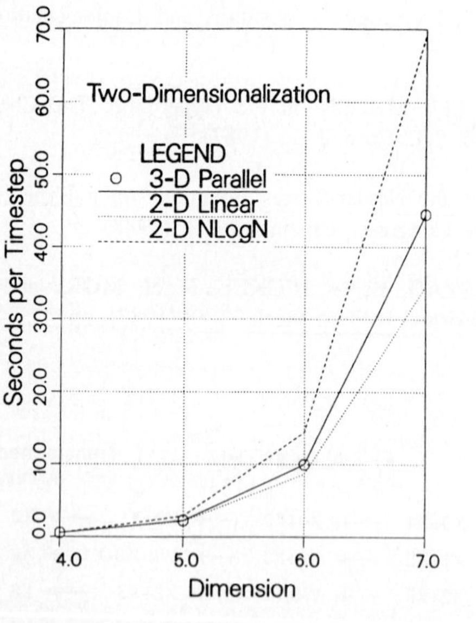

Figure 4: Seconds per time-step for the following cases: 8x8x16 with dim 4, 16x16x32 with dim 5, 32x32x64 with dim 6 and 64x64x128 with dim 7.

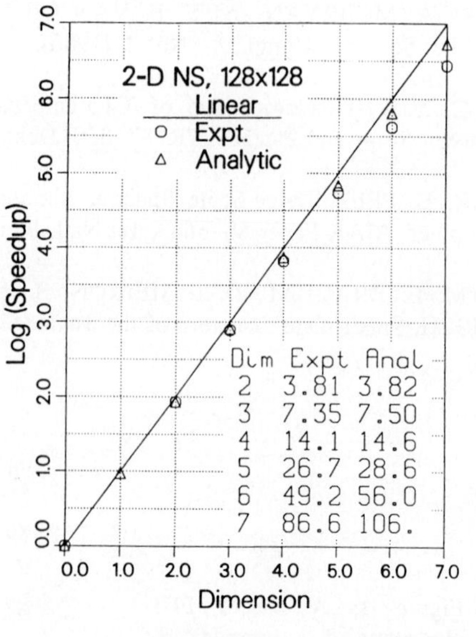

Figure 5: $\log_2$(speedup) versus processor dimension for 2-D case.

ADAPTIVE NUMERICAL SOLUTIONS OF THE
EULER EQUATIONS IN 3D USING FINITE ELEMENTS

J.Peraire, J.Peiro, L.Formaggia and K.Morgan
Institute for Numerical Methods in Engineering
University College
SWANSEA SA2 8PP
United Kingdom

INTRODUCTION

Recent years have seen an upsurge in interest in the application of unstructured grid based solution techniques for the solution of compressible flows. These methods appear to offer certain advantages for flows involving complex features and/or geometries but, to date, the practical demonstrations of these advantages in 3D have been limited. In this paper we report on our experiences in producing an adaptive mesh solution for a flow involving shock interaction on a swept cylinder and an initial solution for a flow past a complex fighter configuration.

FINITE ELEMENT SOLUTION ALGORITHM

The 3D compressible Euler equations are expressed in the conservative form

$$\frac{\partial U}{\partial t} + \frac{\partial F^i}{\partial x_i} = 0 \qquad i=1,2,3 \qquad (1)$$

and the computational spatial domain, Ω, is discretised with an unstructured assembly of 4-noded linear tetrahedral finite elements. The application of the Galerkin weighted residual process leads to the semi-discrete system

$$M_{IK}\frac{dU_K^*}{dt} = \int_\Omega F^{i*}\frac{\partial N_I}{\partial x_i}\,d\Omega - \int_\Gamma F^{i*}\,n_i\,N_I\,d\Gamma \qquad (2)$$

where N_I is the piecewise linear shape function associated with node I of the mesh, M_{IK} is the consistent mass matrix and (n_1,n_2,n_3) denotes the unit outward normal vector to the boundary surface Γ of Ω. The superscript * denotes the finite element approximations and the fluxes F^{i*} are interpolated directly in terms of their nodal values. Equation (2) is discretised in time using a two-step explicit method, incorporating steady state acceleration procedures, which has been fully described previously [1]. The resulting equation forms a 'high order' solution algorithm and the same equation with the addition of artificial viscosity forms a 'low order' scheme. The artificial viscosity is controlled by incorporating an element pressure switch and is constructed, in a finite element fashion, using a combination of the consistent and lumped mass matrices [2]. The flux corrected transport method is used to combine the high and low order schemes in an attempt to produce a monotone high order solution [3]. The solution code incorporating a colouring algorithm to avoid vector dependency in scatter-add operations, runs at an average rate of 55 Mflops on a Cray XMP using a single processor. The average CPU time/node/timestep is $9*10^{-5}$ seconds and requires 68 storage locations of memory per node. A recent version of the code utilises the microtasking facility to make use of the multiprocessing capability of the Cray XMP-48.

MESH GENERATION AND ADAPTIVITY

The generation of tetrahedral elements is accomplished by an advancing front method in which nodal points and elements are simultaneously created [4]. This algorithm is capable of automatically generating meshes over domains of arbitrary geometry, starting from a triangulation of the domain boundary surface. The generation proceeds at a rate of $3.1*10^4$ tetrahedra per hour on a VAX 8700. The meshes are generated according to an externally specified distribution of mesh size. The method can account for directionality by generating elements which are stretched in prescribed directions. With these features, this approach is especially well suited for the generation of meshes which are adapted to a computed flow solution. The determination of the mesh geometrical characteristics in the adaptive process is achieved via a directional error estimating procedure based upon the results of interpolation theory [4]. In one dimension the approach is to select a certain 'key' scalar variable θ for the Euler equation system and, assuming exact nodal values, the root mean square of the local error is estimated as

$$E_e^{RMS} = \frac{1}{11} h_e^2 \left| \frac{d^2\theta}{dx^2} \right|_e \qquad (3)$$

The requirement for equidistribution of the error then means that the new mesh will be generated with a spacing δ which is computed according to

$$\delta^2 \left| \frac{d^2\theta}{dx^2} \right|_e = \text{constant} \qquad (4)$$

For the multi-dimensional case this 1D criterion is applied to each of the principal directions of the matrix of second derivatives separately. This leads to the requirement

$$\delta_\beta^2 \left| \frac{\partial^2\theta}{\partial x_i \partial x_j} \beta_i \beta_j \right| = \text{constant} \qquad (5)$$

where δ_β is the spacing to be used in the β_i direction.

RESULTS

Two examples are included to illustrate the numerical performance of the algorithms described above.

(i) Flow past an F-18 configuration at Mach 0.9, $\alpha = 3°$. An initial solution for this problem has been computed that includes simulation of engine effects. Engine inlet conditions have been given in the form of a specified Mach number of 0.4 and a jet pressure rate of 3 was assumed to determine outlet conditions. The surface discretisation is shown in figure 1 and the computed surface pressure contour distribution is given in figure 2.

(ii) Shock interaction on a swept cylinder at Mach 8.04 This is a problem of current practical interest because of its implications to the design of the hypersonic vehicles [5]. The experimental configuration is indicated in figure 3 and the initial mesh employed is displayed in figure 4, together with a sequence of two adaptively regenerated meshes, with corresponding computed pressure contour distributions.

ACKNOWLEDGEMENTS

The authors would like to thank the Aerothermal Loads Branch of the NASA Langley Research Center for partial support of this work under grant NAGW-478 and also the UK Science and Engineering Research Council for grant GR/E/6176.2 which enabled the flow analysis code to be run on a Cray XMP-48. Jaime Peraire thanks The Research Corporation Trust for their support of his work.

REFERENCES

[1] J. Peraire, J. Peiro, L. Formaggia, K. Morgan and O.C.Zienkiewicz. AIAA paper 88-0032.

[2] K. Morgan and J. Peraire. 'An artificial viscosity model for Euler schemes using unstructured grids', in preparation.

[3] R. Löhner, K.Morgan, M. Vahdati, J. Boris and D.L. Book, 'FEM-FCT : Combining unstructured meshes with high resolution', Comm. Appl. Num. Methods, to appear (1988).

[4] J. Peraire, M. Vahdati, K. Morgan and O.C.Zienkiewicz. J. Comp. Physics. 72, 449-460,1987.

[5] A.R.Wieting, Ph.D. Dissertation, Old Dominion University, Norfolk, Virginia, 1987.

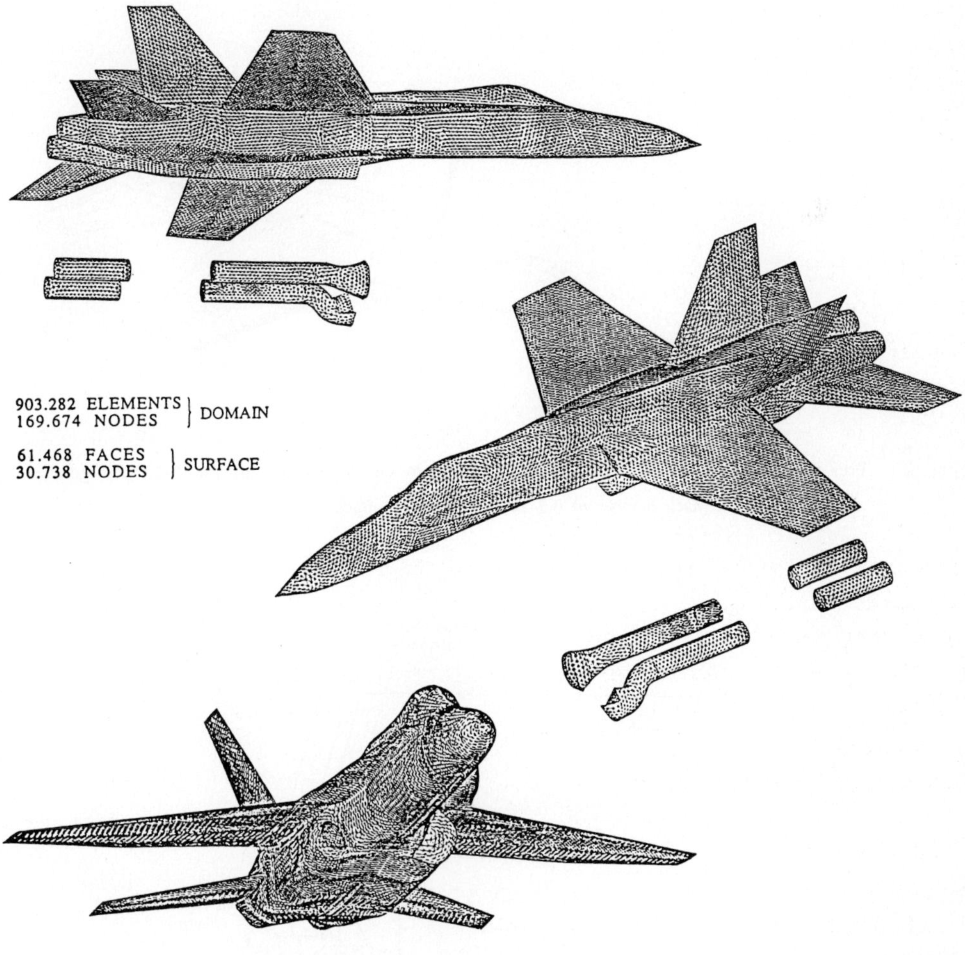

Figure 1 Surface discretisation for an F-18 configuration, including detail of the engine duct discretisation (shown detached)

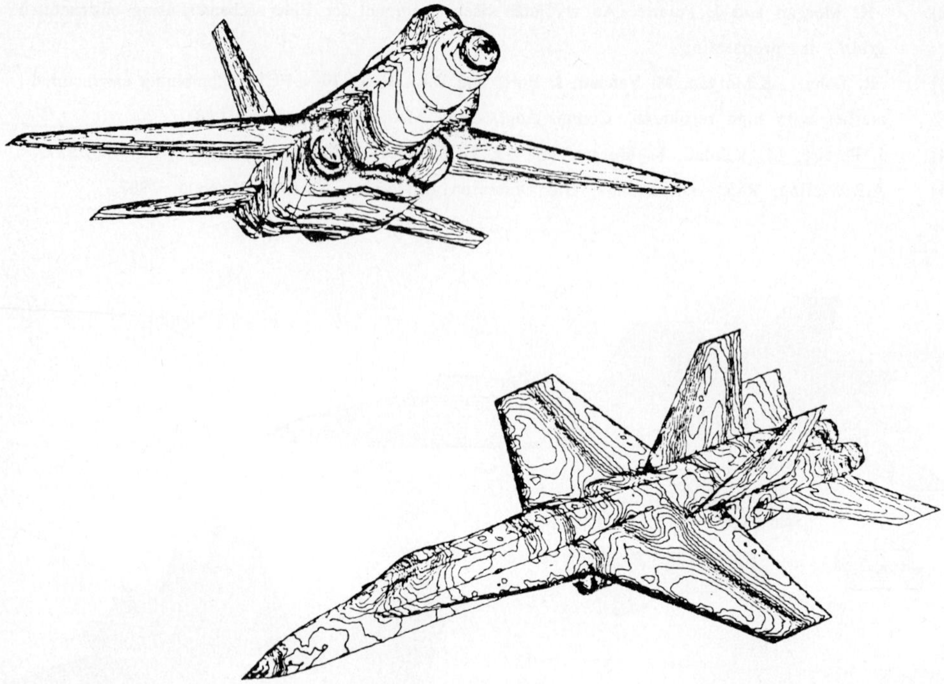

Figure 2 Preliminary results for the distribution of pressure on the surface of the F-18 configuration for a free stream Mach number of 0.9 and 3° angle of attack

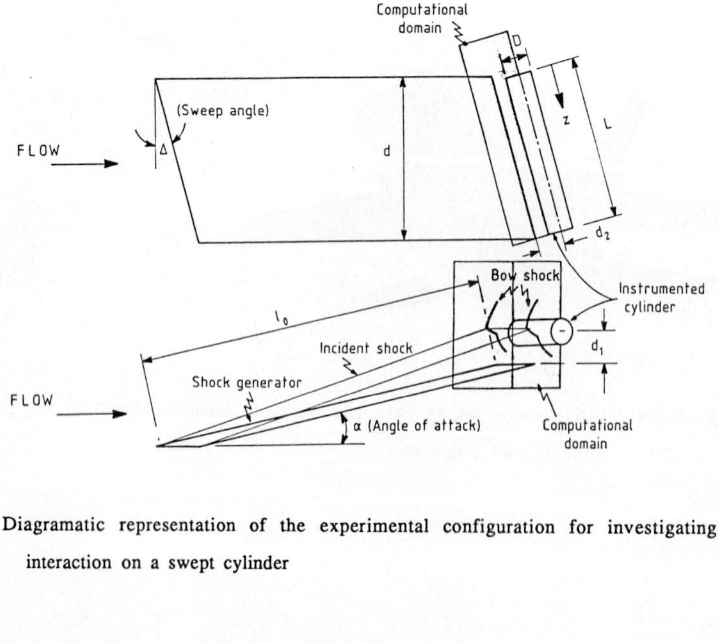

Figure 3 Diagramatic representation of the experimental configuration for investigating shock interaction on a swept cylinder

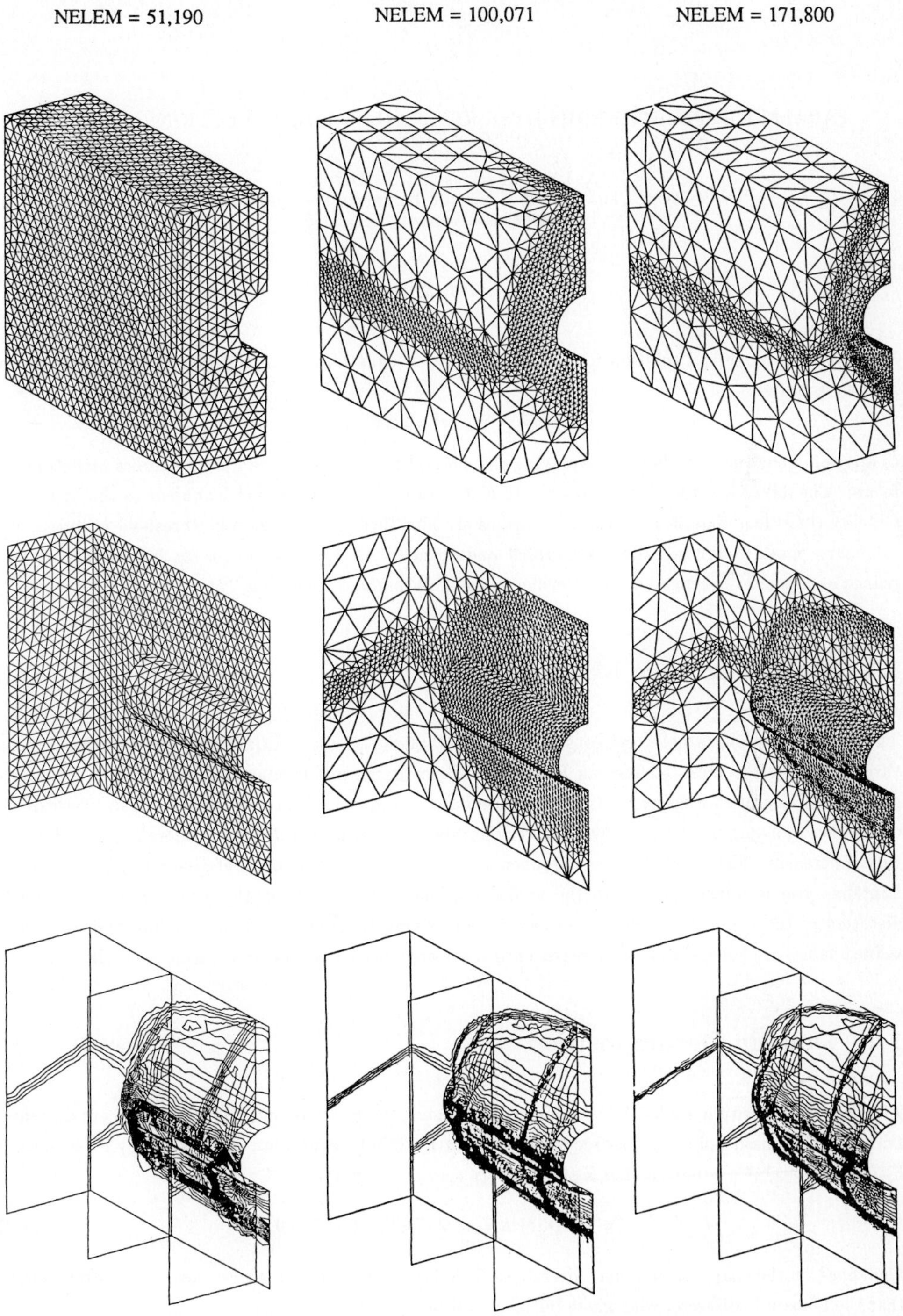

Figure 4 Shock interaction on a swept cylinder at Mach number 8.04 : meshes employed and corresponding pressure contours

PARALLEL HETEROGENEOUS MESH REFINEMENT FOR ADVECTION-DIFFUSION EQUATIONS

A. Louise Perkins and G. Rodrigue
University of California at Davis and Lawrence Livermore Laboratory
PO Box 808, Livermore CA, 94550

Abstract

We develop a numerical solution for advection-diffusion equations; here we solve the two dimensional Burgers equation,

$$\vec{v}_t + \vec{v} \cdot \nabla \vec{v} = \epsilon \Delta \vec{v} \ , \tag{1.1}$$

as a proof of concept with hopes of applying the method to more complete fluid dynamics models in the future. The advection-diffusion equations are first advanced on a coarse mesh subject to the hyperbolic stability condition. Significant parabolic regions are identified with a parabolic threshold criterion, and the coarse mesh is dynamically decomposed and refined. An Euler-Lagrange method advances these refined subdomains independently, allowing coarse grain parallel scheduling.

1.1 Euler-Lagrange Method

The solution method is heterogeneous in the sense that the problem formulation and solution method varies from mesh to mesh. It uses an Eulerian reference frame to formulate the equations on the coarse grid, and a Lagrangian reference frame to formulate the equations on the refined grids. The explicit coarse mesh advancement uses upwind advection differences, centered diffusion differences, and forward time differences. The implicit Lagrange scheme is an extension of the method given in [Rodr 82]. The Lagrange grid is interpolated onto the underlying Eulerian mesh when the grid becomes excessively distorted or the Lagrangian grid moves past its underlying Eulerian coarse mesh refinement area. The refined values are conservatively averaged onto the coarse mesh at every coarse time step [Berg].

1.2 Domain Decomposition

The grid decomposition is a local uniform mesh refinement [Berg 82] about regions of significant diffusion. Local uniform mesh refinement is frequently used for hyperbolic equations. Here we apply it to a mixed hyperbolic parabolic equation. For $\vec{x}$ the Eulerian spatial coordinates, let

$$F_E(\vec{x}, \vec{v}, \epsilon, t) \equiv \vec{v}_t + \vec{v} \cdot \nabla_{\vec{x}} \vec{v} - \epsilon \Delta_{\vec{x}} \vec{v} = 0 \tag{1.2}$$

be defined on the coarse rectangular discretized Eulerian mesh $\Omega_d \subset R_d^2$, where the subscript reminds us that the domain is discrete, with given initial conditions

$$\vec{v} = \begin{bmatrix} u \\ \\ v \end{bmatrix} = \begin{bmatrix} \left\{ \begin{array}{ll} 1 & \text{if } x \leq 1.0 \\ 0 & \text{if } x > 1.0 \end{array} \right\} \\ 0 \end{bmatrix} , \tag{1.3}$$

and Neumann boundary conditions. For $\vec{x}'$ the Lagrangian spatial coordinates, with coordinate transformation

$$\nabla_{\vec{x}} = J^{-1}\nabla_{\vec{x}'} , \tag{1.4}$$

the associated Lagrangian operator is

$$F_L(\vec{x}', \vec{v}, \epsilon, t) \equiv \vec{v}_t - \epsilon(J^{-1}\nabla_{\vec{x}'} \cdot J^{-1}\nabla_{\vec{x}'}) = 0 \tag{1.5}$$

The initial conditions for the refined Lagrangian grids are given by Eq. (1.3) at time zero, and are interpolated from the coarse mesh at advanced time steps. Neumann boundary conditions are again applied. The coarse mesh provides flux values for the advection interpolation scheme from Lagrange to Eulerian reference frames insuring conservation across subdomain interfaces.

Domain decomposition adds modeling precision without significantly increasing computational costs in part because it is parallel. It lends itself to heterogeneity, and can be made conservative. Here it is used as a tool to combine the explicit and implicit methods for the varying time scale requirements.

Grid points with curvature greater than the parabolic threshold are marked as above-threshold points. We cluster these points into subdomains. Our clustering algorithm consists of two parts, cluster analysis and overlap detection. The cluster analysis is based on a similarity matrix threshold. We modify the usual isotropic threshold [Meis 72] to a threshold that generates refined grids aligned with the coarse mesh. For each fixed above-threshold point (i, j), we cluster it with another above-threshold point (k, l) when $s_{ij}^{kl} = 1$ for

$$s_{ij}^{kl} = \left\{ \begin{array}{ll} 1 & |i-k| \leq T_x \quad \text{and} \quad |j-l| \leq T_y \\ 0 & \text{otherwise} \end{array} \right\} . \tag{1.6}$$

This allows us to construct rectangular clusters on one pass through the above-threshold points. Thresholds T_x and T_y are used to create rectangular zones customed to the application and may vary throughout the domain as well. The clustering algorithm is used to create nonoverlapping meshes; this is accomplished by resetting all above-threshold points once they are clustered. The overlap detection algorithm decides whether the clustered points are a new grid or an extension of an old grid, whether an existing grid is separating, or whether two existing grids are merging. Once the new grids are created we append domain of dependence buffer zones. These buffer zones are stripped off before merging back onto the coarse mesh.

After the grid points are clustered into subdomains, a check is made to compare the number of available processors with the number of refined subdomains. At this point primitive load balancing is used to distribute the work load among the available processors equitably.

1.3 Parallel Implementation

On parallel computers it is important to have well-defined numerical tasks that can be executed simultaneously. Domain decomposition is used as described above to introduce coarse grain parallelism. Each available processor is given a subproblem $F_L^j = 0$ defined on the subdomain $\Omega_d^j \subset R_d^2$. The refined meshes are merged onto the coarse mesh at well defined coarse time step barriers [Rodr 87]. The mesh configuration is static between these barriers. Consider this static configuration as a graph with both directed and undirected edges, $G = (V, E_D, E_U)$, where V is the set of all meshes to advance, E_D are directed edges that express the refined-coarse relationship, and E_U are undirected edges that represent overlapping refined meshes. The directed graph G_D is used to insure that the underlying coarser meshes have advanced and filled in the flux boundary information for any refined meshes that require a Lagrangian to Eulerian interpolation. The undirected graph G_U is used to schedule parallel mesh advancement.

Let t_i^m be the CPU time to advance mesh i one time step to time step m, let n be the total number of meshes, let K be the number of coarse time steps between coarse barriers, and let the temporal refinement ratio be l. Number the coarse mesh 1, and assume there is only one level of refinement to simplify the presentation. Then the parallel advancement time between barriers, letting the time step m begin anew at each barrier, is given by

$$t_p \doteq \sum_{m=1}^{K} t_1^m + \max_{2 \leq i \leq n} \left(\sum_{m=1}^{K \cdot l} t_i^m \right) . \tag{1.7}$$

The serial time required is given by

$$t_s \doteq \sum_{m=1}^{K} t_1^m + \sum_{i=2}^{n} \sum_{m=1}^{K \cdot \ell} t_i^m . \tag{1.8}$$

We can now approximate speedup as the ratio:

$$S \doteq \frac{t_s}{t_p} . \tag{1.9}$$

Computers have been compared based on many metrics including throughput. We measure the throughput of our algorithm by utilizing the following theorem.

Theorem 1.1 *Throughput (Eq. (1.11)) is directly proportional to speedup (Eq. (1.10)).*

Let

- Os - set of all serial operations
- Mx - set of all operations on the maximal thread (path) through a program
- ct - clock tick (unit of time)
- ω_j - number of clock cycles for the jth operation (we count pipelined instructions as a function of pipe usage and vector instructions as one operation)
- PU - set of and number of processors
- Opu - set of all operations on processor PU
- t_s - time in ct units to execute the code on one processor
- t_p - time in ct units to execute the code on PU processors
- t_j - computer time for j^{th} operation, $t_j = \omega_j \cdot ct$

Substitute the total operation count for time in the speedup ratio (Eq. (1.9)),

$$S = \frac{t_s}{t_p} = \frac{\sum_{Os} t_j}{\sum_{Mx} t_j} = \frac{\sum_{Os} \omega_j}{\sum_{Mx} \omega_j} . \tag{1.10}$$

Let throughput be the number of operations per clock tick,

$$T_p = \frac{\sum_{PU} Opu}{\sum_{Mx} \omega_j} , \quad T_s = \frac{Os}{\sum_{Os} \omega_j} . \tag{1.11}$$

Then the ratio of serial to parallel throughput can be written as

$$\frac{T_s}{T_p} = \left(\frac{Os}{\sum_{PU} Opu} \right) \left(\frac{\sum_{Mx} \omega_j}{\sum_{Os} \omega_j} \right) = \left(\frac{Os}{\sum_{PU} Opu} \right) \left(\frac{1}{S} \right) , \tag{1.12}$$

simply relating speedup to throughput by a parallelization efficiency factor, PE,

$$S = \left(\frac{T_p}{T_s} \right) \left(\frac{Os}{\sum_{PU} Opu} \right) = \left(\frac{T_p}{T_s} \right) PE , \tag{1.13}$$

proving the theorem by construction.

1.4 Results

For diffusion coefficient $\epsilon = 5 \times 10^{-3}$, we use a coarse grid spatial step of 0.1 and temporal step of 0.04, and advance the coarse mesh to time 0.16. The spatial and temporal refinement ratio is 4. The solution on the coarse grid is shown at time 0.0 and 0.16 in Fig. 1.1 and Fig. 1.2 respectively. The domain decomposition is shown in Fig. 1.3, and a refined mesh solution at time 0.16 is shown in Fig. 1.4. The timing measurements are summarized in Table 1.1. The parallel efficiency is driven by the overlap of the refined solvers. This overlap is due to the domain of dependence addendum. Total parallel efficiency is 0.643. The speedup is 3.55 for the four processors available on the Cray 2, but the parallel throughput is less than that at 2.284. That is, the cost of separating the refined meshes into independent tasks adds additional work that degrades overall performance. The method yields good quality results with significant improvement over serial throughput.

[Berg 82] Marsha J. Berger, *Adaptive Mesh Refinement For Hyperbolic Partial Differential equations*, PhD Dissertation, Stanford University, 1982.

[Berg] Marsha J. Berger, and Phillip Colella, *Local Adaptive Mesh Refinement for Shock Hydrodynamics*, to appear, J. Comput. Phys.

[Meis 72] William S. Meisel, *Computer-Oriented Approaches to Pattern Recognition*, Academic Press, 1972.

[Rodr 82] Garry Rodrigue, Chris Hendrickson, and Mike Pratt, *An Implicit Numerical Solution of the Two Dimensional Diffusion Equation and Vectorization Experiments*, in Parallel Computations, edited by Garry Rodrigue, Academic Press, NY, 1982. (0 12 592101 2).

[Rodr 87] Garry Rodrigue, and A. Louise Perkins, *Locating Parallel Numerical Tasks in the Solution of Viscous Fluid Flow*, Lecture Notes in Computer Science, Parallel Computing in Science and Engineering 4th International DFVLR Seminar on Foundations of Engineering Sciences Proceedings, Bonn, Federal Republic of Germany, Springer-Verlag, June 1987.

This work performed under the auspices of the U.S. Department of Energy by the Lawrence Livermore Laboratory under Contract W-7405-Eng-48.

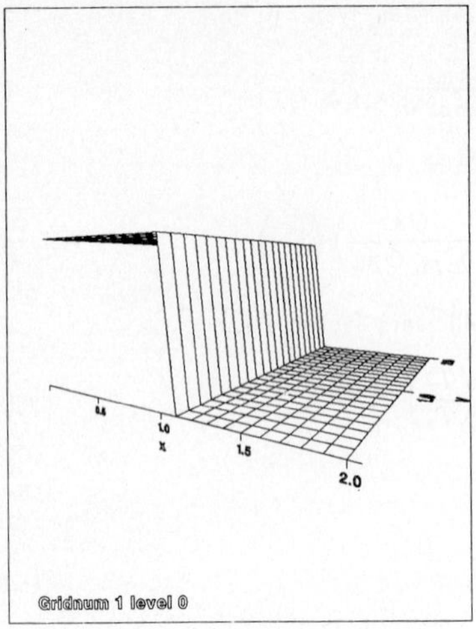

Figure 1.1: Coarse Grid at t=0

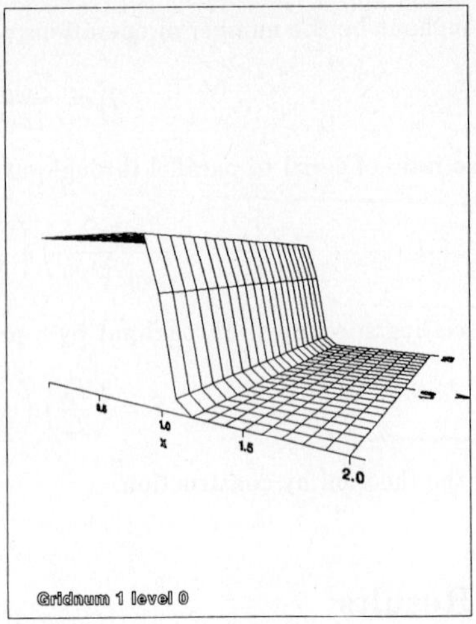

Figure 1.2: Coarse Grid at t=.16

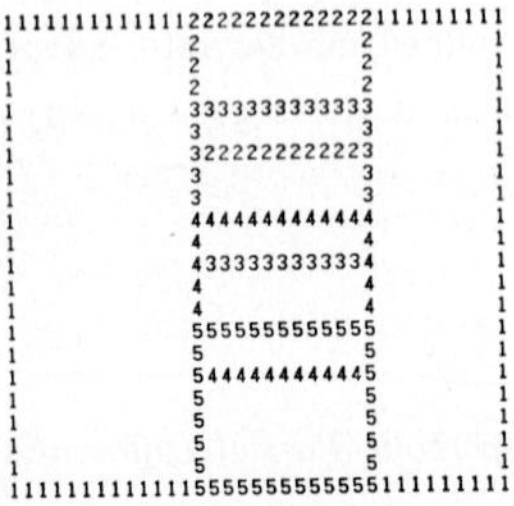

Figure 1.3: Domain Decomposition

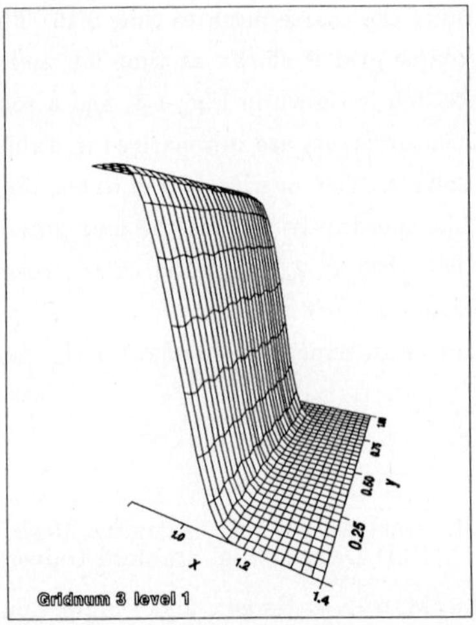

Figure 1.4: A Refined Grid at t=.16

Task	Serial	Parallel	PE	Nonredundant Work
Multitasking	4.6	4.6	0	0
Scheduling	5.6	3.2	1	3.2
Coarse Solver	4.8	4.8	1	4.8
Refined Solver	1848	496	.625	310
Cluster analysis	16.3	13.1	1	13.1
Grid Management	10.2	10.2	1	10.2
Totals	1889.5	531.9	.643	341.3

Table 1.1: Timing Measurements in microseconds

SIMULATION OF INVISCID HYPERSONIC REAL GAS FLOWS

M. Pfitzner, C. Weiland, G. Hartmann

Messerschmitt-Boelkow-Blohm GmbH
Ottobrunn , FRG

The development of new space transport systems and of hypersonic aircraft requires the simulation of high speed air flows about realistic configurations. New concepts of propulsion systems have to be studied and here the numerical simulation of the flowfield can lead to new insights. The high temperatures which appear in these flows require real gas effects to be taken into account.

We present 2-D and 3-D calculations of inviscid supersonic and hypersonic flows about a variety of configurations including orbiters, missiles and through nozzles. Equilibrium real gas and ideal gas results are compared and the differences are discussed. The free stream Mach numbers are in the range $4<=M_\infty<=30$. The supersonic part of the flow field is calculated with a pseudo space marching method, which uses the algorithm for the unsteady Euler equations as relaxation procedure. Shock fitting as well as shock capturing capabilities for the ideal and real gas cases are demonstrated.

We integrate the unsteady Euler equations in quasi-conservative form in generalized coordinates:

$$Q_\tau + A^+Q_\xi{}^+ + A^-Q_\xi{}^- + B^+Q_\eta{}^+ + B^-Q_\eta{}^- + C^+Q_\zeta{}^+ + C^-Q_\zeta{}^- = 0 \qquad (1)$$

where $Q = (\rho,\rho v_x,\rho v_y,\rho v_z,e)^T$ are the conservative variables.

The matrices $A^\pm,B^\pm,C^\pm$ are split according to the sign of their eigenvalues /1-3/. The space derivatives of Q are calculated with a third order accurate upwind-biased formula /4/:

$$Q^+{}_{\xi}\Big|_m = \frac{1}{6\Delta\xi}(Q_{m-2} - 6Q_{m-1} + 3Q_m + 2Q_{m+1})$$

$$Q^-{}_{\xi}\Big|_m = -\frac{1}{6\Delta\xi}(Q_{m+2} - 6Q_{m+1} + 3Q_m + 2Q_{m-1}) \qquad (2)$$

and a MacCormack-type artificial diffusion term prevents wiggles near shocks.

The time-stepping from time level n to level (n+1) towards the steady state is done with an explicit three-step Runge-Kutta method /5/:

$$Q^{(1)} = Q^n - a_1\,\Delta t\,P(Q^n)$$
$$Q^{(2)} = Q^n - a_2\,\Delta t\,P(Q^{(1)}) \qquad (3)$$
$$Q^{n+1} = Q^n - a_3\,\Delta t\,P(Q^{(2)})$$

where the coefficients a_1, a_2 and a_3 are optimized for the third order upwind biased space discretization scheme /6/. Second order accuracy in time requires $a_3 = 1$, $a_2 = 0.5$. The coefficient a_1 is used to maximize the region of (linear) stability of the algorithm. A von-Neumann stability analysis of the 1-d advection equation yields a maximum CFL number of 1.75 for a_1 ranging between 1/4 and 1/3. The Runge-Kutta scheme was used in the present calculations with $a_1 = 1/4$ and CFL = 1.25.

The previously used first order one step timestepping scheme has a very weak linear instability region at intermediate wave numbers which vanishes quickly as the CFL number approaches zero. We have used this scheme sucessfully on relatively coarse meshes with CFL numbers between 0.2 and 0.45 in 2-D and 3-D calculations /2,4/.

Hypersonic (blunt) reentry bodies are normally of a sufficiently simple geometry permitting a (bow) shock fitting procedure. The bow shock fitting option is preferred whenever possible because of its superior accuracy leading to reduced demand for number of grid points and computer memory. Imbedded shocks (which are mostly relatively weak) are always captured.

The equilibrium real gas equation of state has been incorporated in the algorithm using the real gas equivalents of the flux matrices A,B,C and the corresponding eigenvalues and eigenvectors in eq. (1) /7/. The equilibrium pressure with the corresponding derivatives with respect to internal energy and to density are calculated using a curve fit routine. For air flows we used the routines from refs. /8,9/. For flows of general gas mixtures we calculate the pressure surface of state pointwise as a function of the logarithm of density and internal energy with the aid of the Gordon-McBride code /10/. A complete representation of the surface of state from these discrete data points is constructed by interpolation with standard bicubic splines. A comparison with the equilibrium air curve fit routines shows this approach to yield a slightly less accurate, but smoother approximation to the equation of state.

Supersonic and hypersonic flows about pointed nose configurations are calculated using a pure space-marching shock-fitting code formulated for the quasi-conservative Euler equations /11/. The code has been modified to include equilibrium real gas effects /12/.

For hypersonic flows the velocity component in the coordinate direction of the main stream is mainly supersonic. The Euler equations then are hyperbolic in that coordinate. In recompression regions, however, subsonic pockets my appear. For the simulation of flows of that type we developed a marching strategy using the unsteady Runge-Kutta code as basis /6/:

A zone of coordinate planes is defined in which the unsteady equations are iterated toward convergence. The inflow boundary is supersonic and the variables there are prescribed. At the outflow boundary the flow are extrapolated. The zone is marched downstream over the body.

If local subsonic pockets appear in the flowfield, the width of the zone is adapted such that the subsonic part of the field is completely covered by the zone. After convergence has been achieved in the adapted zone, the width of the zone is reset to its original value and the marching continues from the supersonic downstream boundary of the adapted zone.

The blunt body flow may be simulated using the same code with a fixed zone width. The marching strategy saves much CPU time since it generates excellent starting values for the unsteady relaxation scheme. It does not break down if the flow becomes locally subsonic like pure space-marching schemes do. The pseudo space marching method is, however, slightly less efficient than pure space marching. A similar idea has been proposed in /13/.

Fig. 1 demonstrates the influence of 3-D effects on ideal gas flow about a HERMES-like configuration at $M_\infty = 10$, $\alpha = 30°$. The Mach number isolines in the planes of symmetry in both cases clearly reveal the canopy shock. In the 3-D case (fig. 1b) the canopy shock is weaker and the bow shock standoff distance is smaller than in the 2-D case (fig. 1a). The entropy layer on the windward side of the body is much thinner in the 3-D case. Fig. 2 contains a comparison of the ideal gas and real gas static temperature distributions on the windward side of a HERMES-like body for $M_\infty = 10$, $\alpha = 30°$ at an altitude of H = 100 km. Equal temperature increments of $\Delta T = 100$ are used for the isolines. As expected, the real gas calculation yields a much lower maximum temperature. Fig. 3 shows isolines of static temperature for ideal gas flow at $M_\infty = 25$, $\alpha = 35°$. Note the temperature increment of $\Delta T = 1000$ used for this case. The qualitative features of the isolines are similar to that in fig. 2 in spite of the higher angle of attack and the much higher freestream Mach number. Fig. 4 contains the static temperature isolines in a cross-section halfway down the body of the real gas flowfield. The imbedded crossflow shocks are clearly seen. Shown in fig. 5 is a comparison of the ideal gas and real gas temperature distribution on the body in the plane of symmetry. Also shown is the temperature distribution for the $M_\infty = 25$ case. Every second grid point is marked by a symbol on the curves. The leeside canopy shock is captured within three grids without oscillations. The real gas temperature shows less variation and is (on the leeward side) partly above the ideal gas one. Apart of the much higher level the $M_\infty = 25$ case temperature curves are similar to the $M_\infty = 10$ case. The capabilities of the space marching code are demonstrated in fig. 6. Fig. 6a gives a view of real gas static temperature isolines in flow about a missile at $M_\infty = 8$, $\alpha = 10°$, H = 0 km. A comparison of real gas and ideal gas temperatures on the body on the leeward side is shown in fig. 6b. Because of the slenderness of this body the real gas effects are not very pronounced in spite of the high Mach number. Fig. 7 shows a comparison of lines of constant Mach numbers for ideal gas (γ=1.4) and equilibrium real gas air flow through a rocket nozzle. An imbedded captured crossflow shock can be seen, which moves toward the axis of symmetry and gets weaker if real gas effects are included. The shock is captured in 2-3 grids crosses cleanly from one grid line to the next.

In conclusion, we have presented 2-D and 3-D simulations of ideal and equilibrium real gas hypersonic flows using a quasi-conservative scheme with 3rd order upwind-biased space discretisation and optimized three step Runge-Kutta time stepping. Shock fitting and shock capturing capabilities are demonstrated.

REFERENCES

/1/: Chakravarthy S.R., Anderson D.A., Salas M.D.: AIAA paper 80-0268
/2/: Weiland C.: Notes on Numerical Fluid Mechanics, Vol.13, pp. 383-390, Vieweg 1986
/3/: Whitfield D.L. and Janus J.M.: AIAA paper 84-1552 (1984)
/4/: Weiland C. and Pfitzner M.: Lectures Notes in Physics, Vol.264, pp. 654-659, Springer Verlag (1986)
/5/: Turkel E. and Van Leer B.: Lecture Notes in Physics, Vol.218, pp. 566-570, Springer Verlag (1985)
/6/: Pfitzner M.: 2. International Conference on Hyperbolic Problems, Aachen 1988, Notes on Numerical Fluid Dynamics, Vieweg (1988)
/7/: Pfitzner M. and Weiland C.: AGARD-CP 428, Paper 22, Bristol (1987)
/8/: Tannehill J.C. and Mugge P.H.: NASA CR-2470 (1974)
/9/: Srinivasan S., Tannehill J.C. and Weilmunster K.J.: ISU-ERI-Ames 86401; ERI Project 1626; CFD15,1986
/10/: Gordon S., McBride B.J.: NASA SP-273 (1971)
/11/: Weiland C.: J.Comp.Phys. Vol.29, pp. 173-198 (1978)
le Neuve 1987, Notes on Numerical Fluid Mechanics, Vieweg (1988)
/13/: Chakravarthy S.R., Szema K.Y.: J.Aircraft 24, pp. 73-83 (1987)

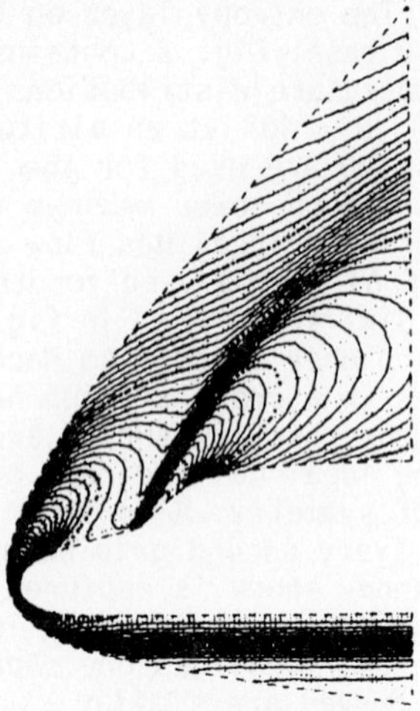

Fig.1a : Mach number isolines in symmetry plane, 2-D-flow
$\Delta M = 0.1$

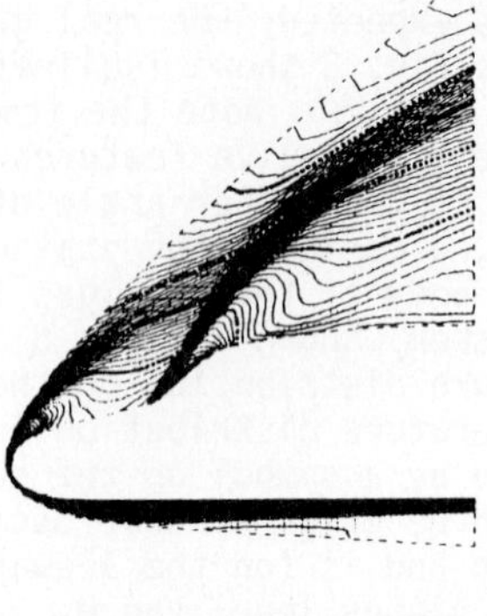

Fig.1b : Mach number isolines in symmetry plane, 3-D-flow
$\Delta M = 0.1$

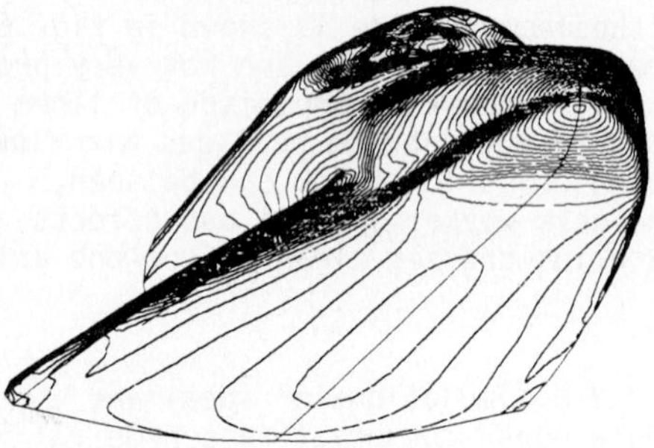

Fig.2a : Ideal gas static temperature isolines
($\Delta T = 100$)

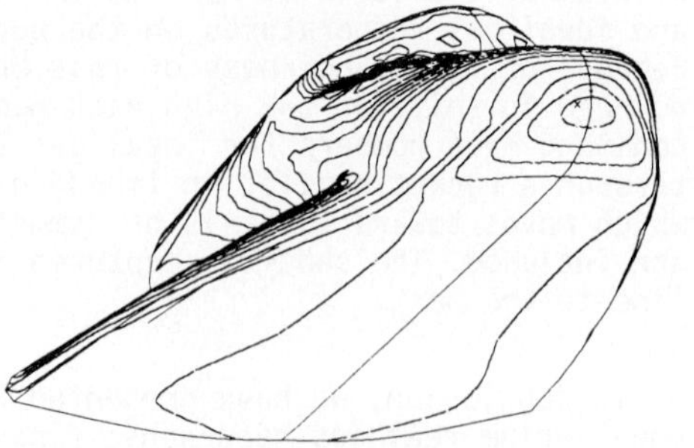

Fig.2b : Real gas static temperature isolines
($\Delta T = 100$)

Fig.2 : HERMES-like forebody, $M_\infty = 10$, $\alpha = 30°$, H = 50 km
Comparison of ideal and real gas
Grid: 17*121*61 for half body

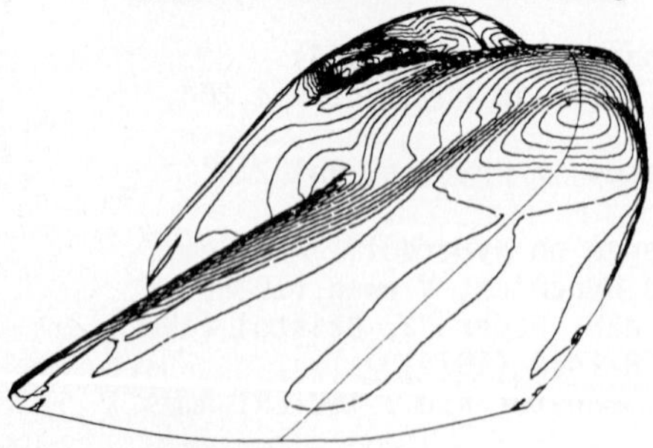

Fig.3 : HERMES-like forebody, $M_\infty = 25$, $\alpha = 35°$,
Ideal gas static temperature isolines
($\Delta T = 1000$)

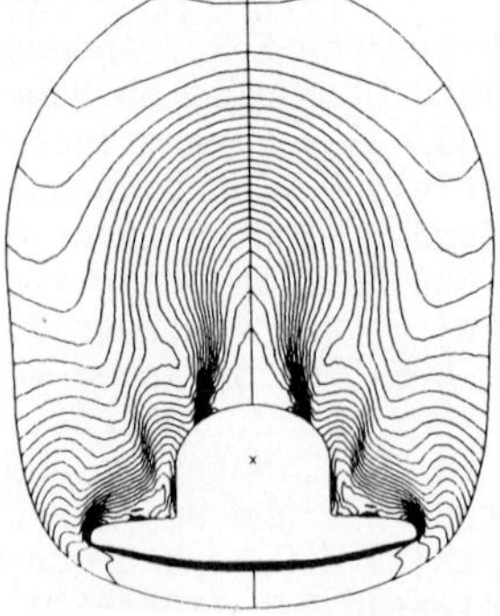

Fig.4 : HERMES-like forebody, $M_\infty = 10$, $\alpha = 30°$, H = 50 km
Real gas static temperature cross sectional plane
($\Delta T = 50$)

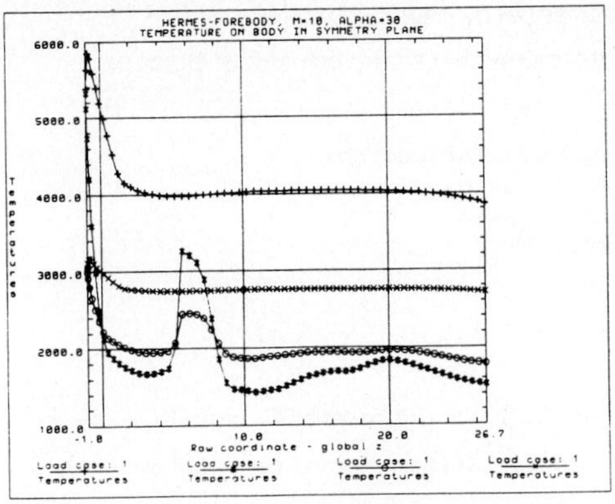

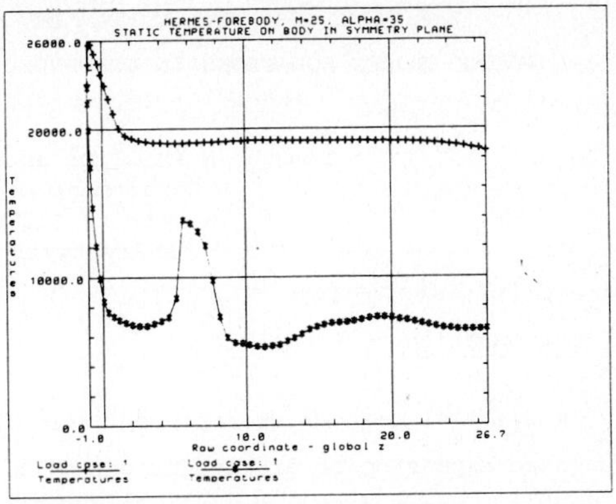

Fig.5 : HERMES-like forebody, M_∞ = 10, α = 30, H = 50 km
Ideal gas and real gas static temperature on body in symmetry plane

HERMES-like forebody, M_∞ = 25, α = 35
Ideal gas static temperature on body in symmetry plane

+ : ideal gas luv side, * : ideal gas lee side
x : real gas luv side, o : real gas lee side

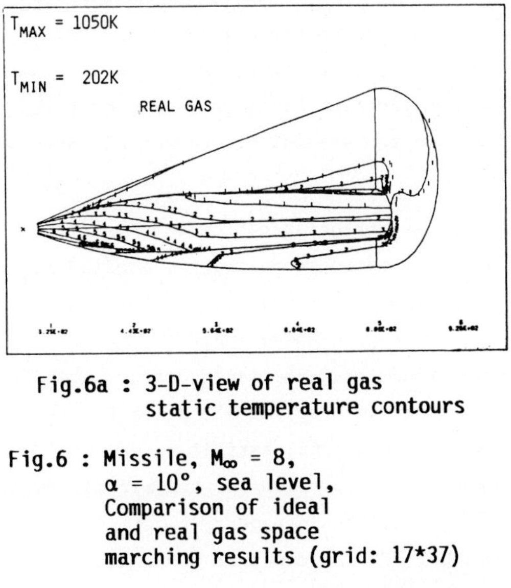

Fig.6a : 3-D-view of real gas static temperature contours

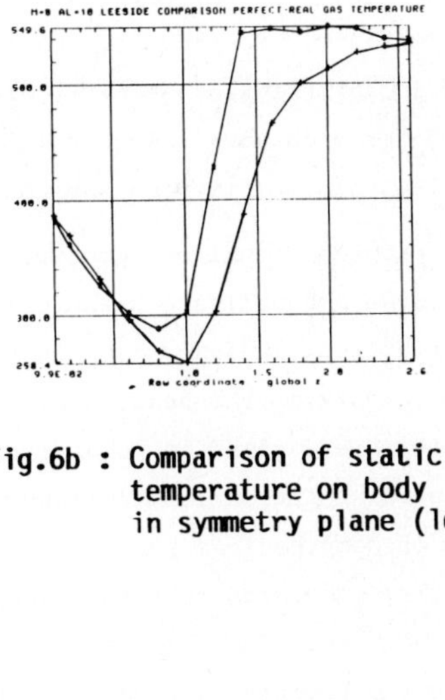

Fig.6b : Comparison of static temperature on body in symmetry plane (lee)

Fig.6 : Missile, M_∞ = 8, α = 10°, sea level, Comparison of ideal and real gas space marching results (grid: 17*37)

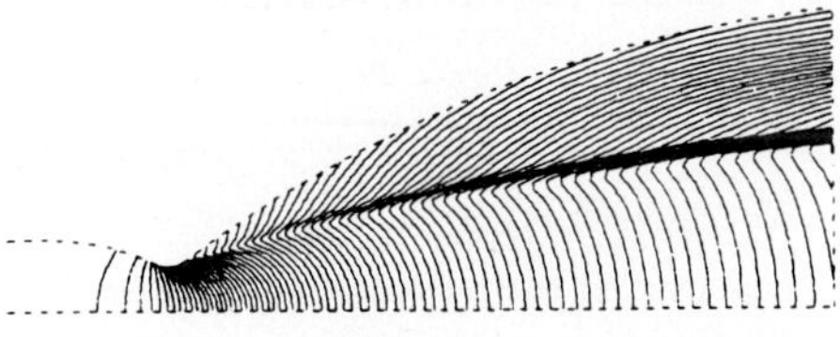

Fig.7a : Mach number contours, ideal gas, γ = 1.4

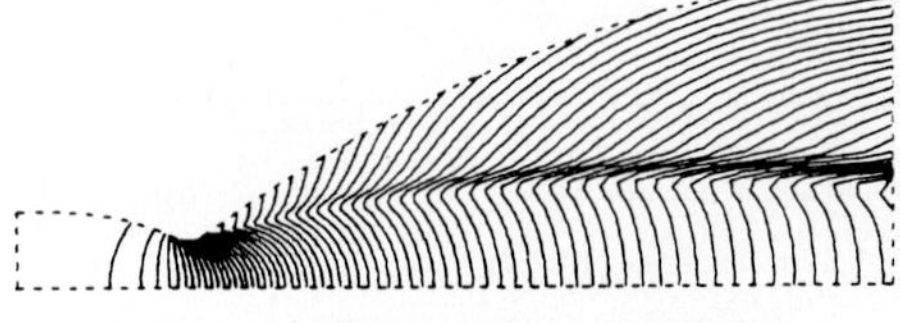

Fig.7b : Mach number contours, real gas air flow

Fig.7 : Nozzle flow simulations (rotationally symmetric, grid : 101*40)

EFFICIENT SPECTRAL ALGORITHMS FOR SOLVING THE INCOMPRESSIBLE NAVIER-STOKES EQUATIONS IN UNBOUNDED RECTANGULARLY DECOMPOSABLE DOMAINS

Timothy N Phillips and Andreas Karageorghis
Department of Mathematics
University College of Wales
Aberystwyth SY23 3BZ

1. Introduction

In spectral methods we represent the solution to a differential equation by a truncated expansion of eigenfunctions of a singular Sturm-Liouville problem. This choice of trial functions is responsible for the superior approximation properties of spectral methods when compared with finite difference and finite element methods. This is, of course, most evident for problems with smooth solutions in which case the series expansion converges faster than any finite power of $1/n$ as $n\to\infty$. This phenomenon is known as exponential convergence.

Spectral methods are most easily applied to problems defined in rectangular or circular regions in which case Chebyshev or Fourier series expansions, respectively, are appropriate. However, the natural choices of expansion functions for a problem defined in an irregular geometry are unwieldy and inefficient to use and need to be computed for each new irregular region. There are two ways of circumventing this difficulty, namely, mapping and matching.

The mapping technique involves transforming the irregular region into a simpler one by using a coordinate transformation. Spectral methods can then be applied in the simpler region using standard expansion techniques. Orszag [1] illustrates this process in solving the heat equation in an annular region whose inner and outer boundaries are defined in polar coordinates and finds that the rapid convergence of the spectral solution is maintained. Streett, Zang and Hussaini [2] solve the problem of compressible potential flow past a two-dimensional airfoil by using a numerically generated conformal mapping of Jameson [3] to transform the airfoil onto the unit circle. The appearance of shock discontinuities in supercritical flows degrades the spectral accuracy of the solution but it is found that fewer degrees of freedom are required than for finite difference methods to obtain equivalent solutions.

The second technique, matching, divides the region into a number of simpler sub-regions or elements. A spectral approximation to the solution of the differential equation is sought within each element. The representations are matched by imposing continuity conditions across element interfaces. This procedure results in a coupling of the expansion coefficients in two contiguous elements. These methods combine the flexibility of finite element methods to treat problems in irregular regions with the superior approximation properties of spectral methods for problems possessing smooth solutions. The global element of Delves and Hall [4] and the spectral element methods of Patera [5] and Phillips and Davies [6] are methods which have been developed with this philosophy in mind. These methods differ in the choice of trial functions and the continuity conditions satisfied at element interfaces. In this paper we consider representations in the form of a double series of Chebyshev polynomials.

2. The Governing Equations

We consider the flow through a $1:\alpha$ contracting channel. The flow is assumed to be symmetric about the centre line $y=0$ so that only the upper half of the channel need be considered. The upper half of the channel is shown in Fig 1 together with some of the boundary conditions. At entry and exit we impose Poiseuille flow.

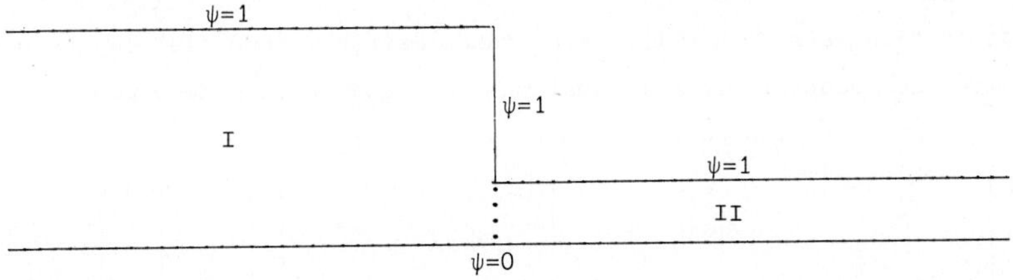

Figure 1. Upper half of contraction geometry.

The incompressible Navier-Stokes equations are

$$(\underline{v}.\nabla)\underline{v} = -\nabla p + (Re)^{-1}\nabla^2\underline{v}, \tag{1}$$

$$\nabla.\underline{v} = 0. \tag{2}$$

The introduction of a stream function defined by

$$u = -\frac{\partial\psi}{\partial y}, \quad v = \frac{\partial\psi}{\partial x},$$

where $\underline{v} = (u,v)$ means that the continuity equation (2) is automatically satisfied and (1) becomes

$$\nabla^4\psi - Re\left[\frac{\partial\psi}{\partial y}\frac{\partial}{\partial x}(\nabla^2\psi) - \frac{\partial\psi}{\partial x}\frac{\partial}{\partial y}(\nabla^2\psi)\right] = 0. \tag{3}$$

Equation (3) is nonlinear in the stream function and is linearized using a Newton-type technique [7].

The linearized equation is solved in the collocation framework at each stage. This means that the equation is satisfied at a chosen set of points in each element. The representations in each of the two elements are matched by imposing continuity conditions in a collocation sense across the element interface. The use of high order trial functions permits high order continuity to be imposed, in this case C^3.

3. Method of Solution

A disadvantage of spectral methods is that they produce algebraic systems for the expansion coefficients which do not exhibit the same sparsity features that one associates with the finite difference or finite element methods. However, the spectral element method produces systems of equations which are block tridiagonal in which the blocks are full. This fact was not exploited by Karageorghis and Phillips [8] who considered Stokes flow i.e. Re = 0, and solved the full system using a Crout factorization technique from the NAG Library [9].

In this paper we take advantage of the structure of the spectral element matrix in constructing direct solution techniques. We consider an application of the capacitance matrix technique [10]. One should note that the interface conditions in the spectral element method are much more complex than in the finite difference method. Typically they involve all the expansion coefficients in adjoining elements. The rows of the spectral element matrix corresponding to these conditions are full.

In this section we show how the capacitance matrix method [10] may be generalized to solve the spectral element equations in rectangularly decomposable domains. These may be written in the partitioned form:

$$\left(\begin{array}{c|c|c} E & F & O \\ \hline G & H & R \\ \hline O & P & Q \end{array}\right) \left(\begin{array}{c} \underline{x} \\ \hline \underline{y} \\ \hline \underline{z} \end{array}\right) = \left(\begin{array}{c} \underline{r} \\ \hline \underline{s} \\ \hline \underline{t} \end{array}\right) . \qquad (4)$$

In order to apply the capacitance matrix approach we write system (4) in the natural component form suggested by the partitioning

$$E\underline{x} + F\underline{y} = \underline{r} \,, \qquad (5)$$

$$G\underline{x} + H\underline{y} + R\underline{z} = \underline{s} \,, \qquad (6)$$

$$P\underline{y} + Q\underline{z} = \underline{t} \,. \qquad (7)$$

We write $\underline{x}$ and $\underline{z}$ in terms of $\underline{y}$ by premultiplying (5) and (7) by the inverses of E and Q, respectively, i.e.

$$\underline{x} = E^{-1}\underline{r} - E^{-1}F\underline{y}, \quad \underline{z} = Q^{-1}\underline{t} - Q^{-1}P\underline{y}. \qquad (8)$$

Eliminating $\underline{x}$ and $\underline{z}$ from (6) we obtain the following system for $\underline{y}$:

$$(H - GE^{-1}F - RQ^{-1}P)\underline{y} = \underline{s} - GE^{-1}\underline{r} - RQ^{-1}\underline{t} \qquad (9)$$

where the coefficient matrix on the left-hand side of (9) is known as the capacitance matrix. This system which is much smaller than the full system can be solved efficiently using the Crout factorization technique.

4. Numerical Results

The use of the capacitance matrix technique produced significant savings in computational cost and time over the original method. The efficiency of the method is greater when the system of equations becomes larger. For Reynolds numbers less than 120 no continuation is needed when a zero initial guess for the coefficients is made. Thereafter continuation in Reynolds number is used in steps of ten. In Figures 2, 3, 4 and 5 we give contour plots of the stream function when $\alpha=1/2$ for Reynolds numbers of 0, 50, 100 and 150, respectively. The results were produced with a total of 1144 degrees of freedom. This means that in regions I and II we have double Chebyshev representations of order 25×25. Qualitative and quantitative agreement with the work of Dennis and Smith [11] is reached. Further details can be found in Karageorghis and Phillips [12].

References

[1] S A Orszag, J Comput Phys 37, 70-92 (1980).
[2] C L Streett, T A Zang and M Y Hussaini, J Comput Phys 57, 43-76 (1985).
[3] A Jameson, AIAA Paper 79-1458 (1979).
[4] L M Delves and C A Hall, J Inst Maths Applics 23, 223-234 (1979).
[5] A Patera, J Comput Phys 54, 468-488 (1984).
[6] T N Phillips and A R Davies, J Comp Appl Math 21, 173-188 (1988).
[7] T N Phillips, J Comput Phys 54, 365-381 (1984).
[8] A Karageorghis and T N Phillips, J Comput Phys, to appear.
[9] N A G Library Mark 11, NAG (UK) Ltd, Banbury Road, Oxford, UK.
[10] B L Buzbee, F W Dorr, J A George and G H Golub, SIAM J Numer Anal 8, 722-736 (1971).
[11] S C R Dennis and F T Smith, Proc Roy Soc Lond A 372, 393-414, (1980).
[12] A Karageorghis and T N Phillips, Technical Report ABERNA 22, Department of Mathematics, University College of Wales (1988).

Acknowledgement

The authors are grateful to the UK Science and Engineering Research Council for financial support which enabled this work to be performed.

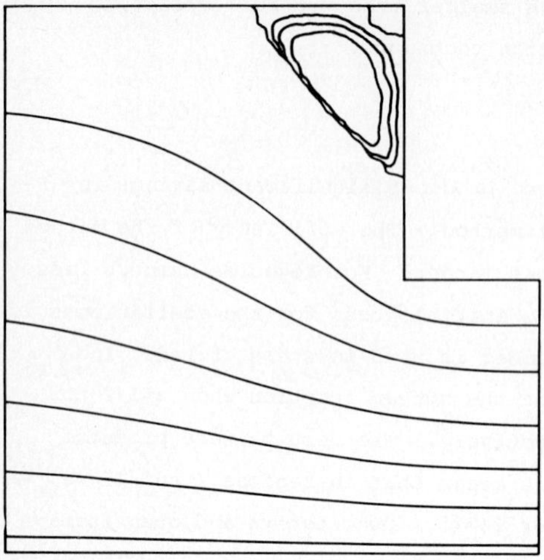

Fig 2. Re = 0.

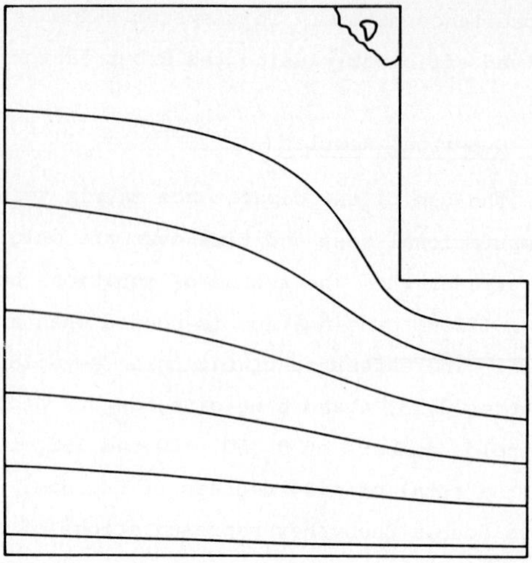

Fig 3. Re = 50.

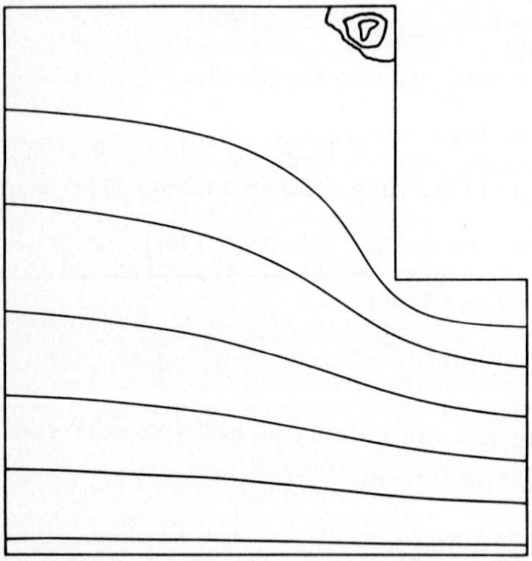

Fig 4. Re = 100.

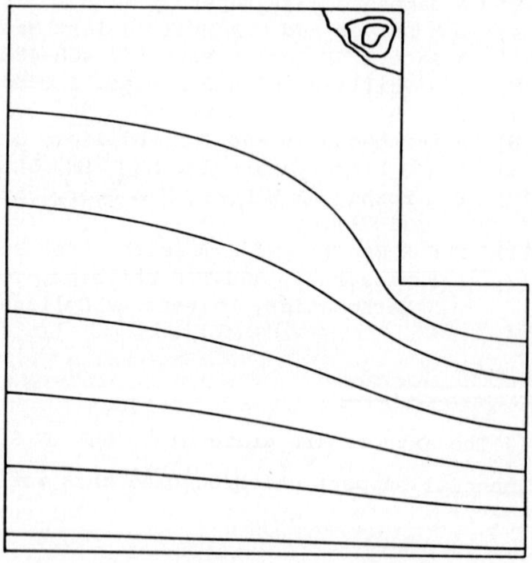

Fig 5. Re = 150.

Contour plots of the stream function obtained with 1144 degrees of freedom.

COMPUTATION OF THE THREEDIMENSIONAL WAKE OF A SHIPLIKE BODY

J. Piquet, P. Queutey & M. Visonneau
CFD Group ; UA1217 ; ENSM
1 Rue de la Noe, 44072 Nantes Cedex, France

1. Introduction

Both computations and experiments indicate that the thin boundary layer assumptions are no more valid in the stern region of an afterbody because of the dramatic thickening of the viscous zone across which important variations of pressure occur. Other significant features of this problem include strong viscous inviscid interaction, significant hull curvature effects, abnormally low levels of turbulence in the greatest (outer) part of the viscous layer and occurrence of a large longitudinal vorticity component which occasionally leads to a free vortex separation without noticeable flow reversal in the direction of the ship motion. Solutions of this problem have been developped by several authors, e.g. Abdel Meguid, Muraoka, Tzabiras & Loukakis, Chen & Patel, Hoekstra & Raven among others.

The developped numerical method follows the framework of [2]. It uses the most common partial transformation approach where only the physical cartesian coordinates (x,y,z) are transformed into a body-fitted coordinate system (λ,η,ζ) while the unknowns remain, besides the pressure, the cartesian velocity components U, V, W. The body-fitted coordinate system is generated by means of the classical elliptic approach resting on Poisson equations. In order to allow systematic comparisons with experiments, a particular solution is retained for the mesh generator in the stern region where λ = const. planes coincide with x = const. planes [2]. The used topology is clearly evidenced in Figs. 1 where views of the mesh for the SSPA ship liner are presented.

The mean momentum equations are retained under a partially parabolic form, elliptic terms being treated as source terms. The space discretisation is built with the finite analytic method of [1] : a combination of exponential and linear boundary conditions allows, on each cell volume, the analytic solution of the equations to be obtained using the method of separation of variables. From this analytic solution, influence coefficients of a nine point scheme result in each λ=const. plane.

Node placements of velocity unknowns rest on a MAC staggered mesh in order to avoid the classical chequerboard uncoupling. The SIMPLER procedure [5] is used in order to satisfy the continuity constraint but it has to be modified to account for the curvilinear and the non orthogonal character of the coordinate system and for the fact that the velocity components are neither orthogonal to the control surfaces, nor parallel to the direction of the coordinate lines. In these last particular aspects, the developped numerical method departs significantly from [2] and [6].

2. The iterative method for the pressure

The continuity equation is written

$$(1)\ \left[b_1^1 U + b_2^1 V + b_3^1 W \right]_\lambda + \left[b_1^2 U + b_2^2 U + b_3^2 W \right]_\eta + \left[b_1^3 U + b_2^3 V + b_3^3 W \right]_\zeta = 0$$

where the chain rule derivative has been used. The area measures b_i^j are defined from the covariant base vectors $\mathbf{a}_i = \partial r/\partial\lambda^i$ by $b_\ell^i = (\mathbf{a}_j x \mathbf{a}_k)_\ell$, i, j, k in cyclic order. J is the jacobian of the transformation $\lambda^i \to x^i$: $J = D(x,y,z)/D(\lambda,\eta,\zeta)$.

Once momentum equations are solved for each velocity node point, continuity is enforced using the projection phase (2):

$$U_d = U_d^* - C_u \left[b_1^1 P_\lambda + b_1^2 P_\eta + b_1^3 P_\zeta \right]_d$$

$$(2)\ V_n = V_n^* - C_v \left[b_2^1 P_\lambda + b_2^2 P_\eta + b_2^3 P_\zeta \right]_n$$

$$W_e = W_e^* - C_w \left[b_3^1 P_\lambda + b_3^2 P_\eta + b_3^3 P_\zeta \right]_e$$

where the values of C_u, C_v, C_w are not necessary for the argument. Substituting (2) in (1) gives the full pressure equation to be solved

$$(3)\ \frac{\partial}{\partial\lambda^i}\left[a_{ij} \frac{\partial P}{\partial\lambda^j} \right] = -\operatorname{div} V^* \ ;\ a_{ij} = C_u b_1^i b_1^j + C_v b_2^i b_2^j + C_w b_3^i b_3^j$$

Performing the substitution on the discrete forms of (2) and (3) and introducing suitable interpolations leads to a pressure matrix that is the discrete analogue of (3) and remains to be solved. If velocity components are specified over all the boundaries, consistency requires the exact satisfaction of the global mass conservation in the discrete sense. This is accomplished by means of an additive plane-by-plane correction on the pressure equation [7] and on the far field normal component of the velocity.

The pressure matrix A to invert is neither positive definite, nor symmetric because of the interpolation procedure. Therefore a preconditionned biconjugate gradient (PBCG) is used. If M is the preconditionning matrix for A, the PBCG method is defined by the following sequence (4)

(4) PBCG algorithm

$$x^0 \in R^N$$

$$r^0 = b - A.x^0 \ ;\ p^0 = z^0 = M^{-1}.r^0$$

$$r^{*0} = b - A^T.x^0 \ ;\ p^{*0} = z^{*0} = M^{-T}.r^{*0}$$

Iterate on :

$$x^{n+1} = x^n + \alpha_n.p^n$$

$$r^{n+1} = r^n - \alpha_n.A.p^n$$

$$
\left|
\begin{array}{l}
r^{*n+1} = r^{*n} - \alpha_n . A^T . p^{*n} \\
p^{n+1} = z^{n+1} + \beta_{n-1} . p^n \quad ; \quad z^n = M^{-1} . r^n \\
p^{*n+1} = z^{*n+1} + \beta_{n-1} . p^{*n} \quad ; \quad z^{*n} = M^{-T} . r^{*n} \\
\alpha_n = \dfrac{\langle z^n , r^{*n} \rangle}{\langle A . p^n , p^{*n} \rangle} \quad ; \quad \beta_n = \dfrac{\langle z^{n+1} , r^{*n+1} \rangle}{\langle z^n , r^{*n} \rangle}
\end{array}
\right.
$$

The whole sequence is easily vectorized except the computation of z^n and of z^{*n} : M is taken as an incomplete LU decomposition without fill-in as extradiagonals have been found to worsen the convergence, probably because the lack of diagonal dominance is increased. The solution of the triangular L and U systems is therefore highly recursive. Fortunately, L and U can be written $L = D_L[I-\ell]$, $U = D_U[I-u]$ where D_L and D_U are the diagonal parts of L and U. ℓ and u are in fact convergent matrices so that they can be approximated by their truncated Neumann series, e.g. $[I-\ell]^{-1} \cong I+\ell+\ell^2+...+\ell^m$, which are computed using the Horner scheme for $y = L^{-1}r \cong [I+\ell+\ell^2+...+\ell^m]D_L^{-1}r$.

The cost is 8N for the computation of the inner products and of the coefficients, 54N for the computation of Ax and A^Tx from x, 65N for the inversions of M and of M^T so that the computational cost per iteration is 127N. The storage cost is 27N for the pressure coefficients that define A, N for M and M^T (only the diagonal is necessary) and 10N for temporary vectors. The storage cost is therefore less than 40N instead of N^2 for a full system.

The convergence to machine zero needs five times less than with a direct solver using the block-tridiagonal structure of the matrix as well as optimized routines of the linpack library. The vectorization speedup is about 10. Retaining four terms in the Neumann series gives a convergence rate similar to the one which would result from the SSOR preconditionning : $M = \omega[\omega^{-1}D-\ell]D^{-1}[\omega^{-1}D-u]/(2-\omega)$ while twelve terms give roughly the same convergence characteristics as the incomplete LU factorization.

3. Numerical Results

The SSPA 720 has been used extensively as a test case for boundary layer calculation methods [3], all of which break down some distance ahead of the stern. Meanflow data are found in [4]. For this hull, the domain solution extends from 2x/L = 0. (midship station) to 2x/L = 6. so that the fictive rudder starts at 2x/L=.95 and the hull disappears at 2x/L=1.025 . In the cross sectional planes, the domain extends from the hull surface and wake centreline to a cylindrical boundary one ship length away from the ship axis where the application of uniform flow conditions U=1 : P=0 is valid. The grid used to test the pressure solver consists in N = 60x30x13 = 23400 points.

Fig. 2 compares the convergence histories for the SLOR solution of the full 27 points pressure matrix at the first sweep with the solution using the BCG method without preconditionning (M = I) and with the incomplete LU PBCG method. The considered pressure matrix is such that 20600 rows are not diagonally dominant, the sum of the absolute values of the offdiagonal entries being 30% in excess with respect to the diagonal entry. This result shows that roughly 30 BCG iterations lower enough the residual of the pressure gradient while the SLOR method is ineffective.

Girthwise pressure distributions are given in fig. 3, using the experimental values [4] corrected for blockage [2]. Velocity profiles along the keel and the waterline are compared with the experiments in fig. 4 which demonstrates that the computation is successful in the thin as well as in the thick boundary layer region. U (—) and V (--) are interpolated back into the boundary layer coordinate system and plotted with respect to the normal distance to the wall. The method can therefore cope with highly distorted and anisotropic grids: the first point away from the wall is about $y^+ = 1$ while aspect ratios vary between 10^5 and 10^{-1}. Endly the cpu computational effort (per point, per iteration) is less than 2. 10^{-4} s on VP200 ; 35 global iterations are needed.

Acknowledgments: Authors gratefully acknowledge Pr. Patel who provided them with an earlier source listing of his partially parabolic code -described in [8]- with wall functions (polar case afterbody 3:1). Computations have been performed on the VP200 (CIRCE) and with a dotation of 15 hours on the Cray 2 (CCVR) by the scientific Committee of CCVR. Financial support of DRET through contract 86.34.104 is also gratefully acknowledged.

References

[1] Chen, J. & Chen H.C., (1984) J. Comp. Phys., Vol. 53 ; N°2, pp. 209-226

[2] Chen, H.C. & Patel, V.C. (1985) Proc. 4th. Num. Conf. on Numerical Ship Hydrodynamics (Washington D.C.) pp. 492-511.

[3] Larsson, L (1980) "Proc. SSPA-ITTC Workshop on Ship Boundary Layers, SSPA Rept N°90 (1981).

[4] Larsson, L. (1075) "Boundary Layers on Ships", PHD Thesis, Chalmers Univ., Goteborg.

[5] Patankar, S.V. (1980) "Numerical Heat Transfer and Fluid Flow" ; Hemisphere Publishing Corp. New-York.

[6] Piquet, J. & Visonneau, M. (1985) Proc. 7th. GAMM Conf. Num. Meth. in Fluid Mech. Vieweg Verlag, Deville, M. Ed.

[7] Settari, A. & Aziz, K. (1973) SIAM J. Numer. Anal. ; Vol. 8, pp. 99-113.

[8] Chen, H.C. & Patel, V.C. (1985) IIHR Report No. 285.

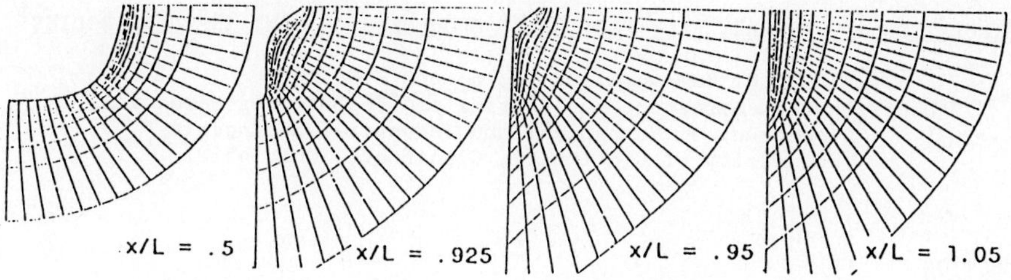

Fig. 1 Grid for the SSPA 720 : transverse crossections

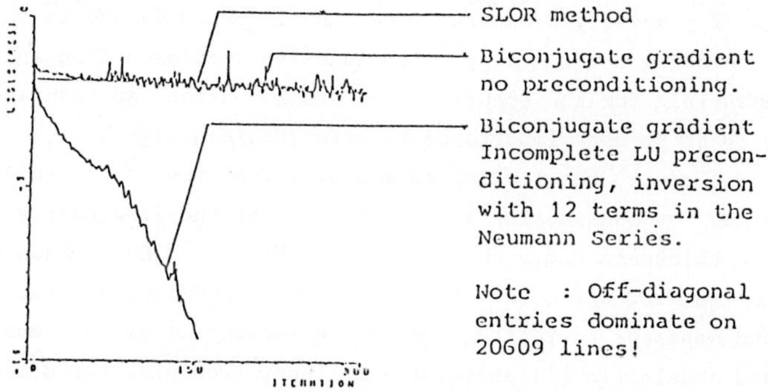

Fig. 2 History of Convergence on the Pressure field within one given sweep.

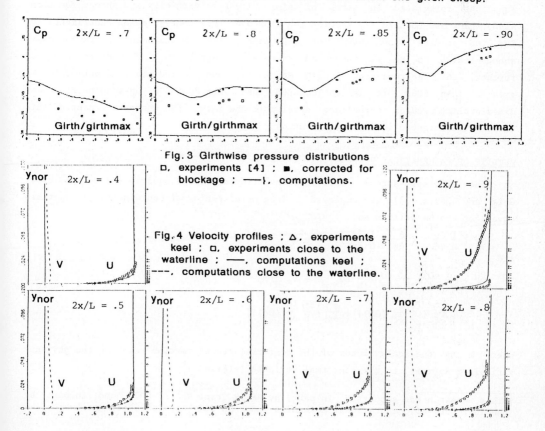

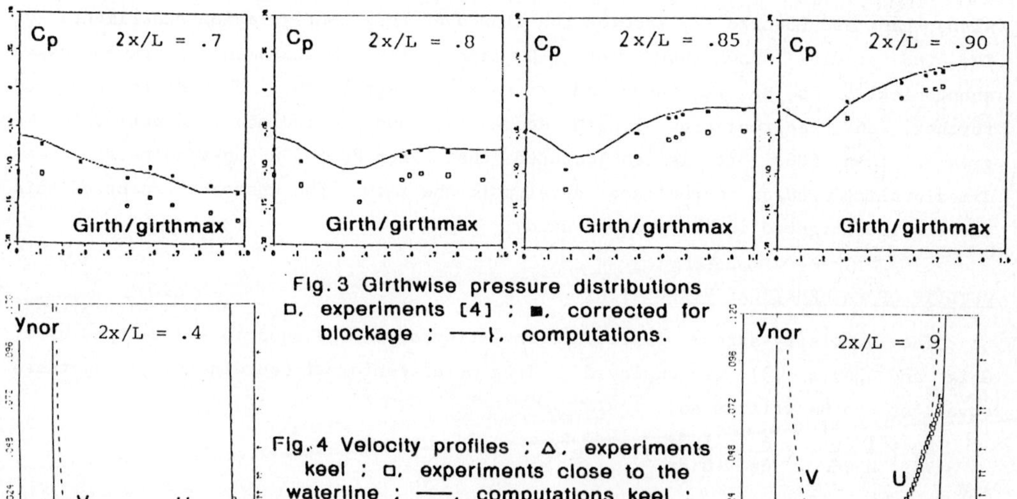

Fig. 3 Girthwise pressure distributions □, experiments [4] ; ■, corrected for blockage : ——, computations.

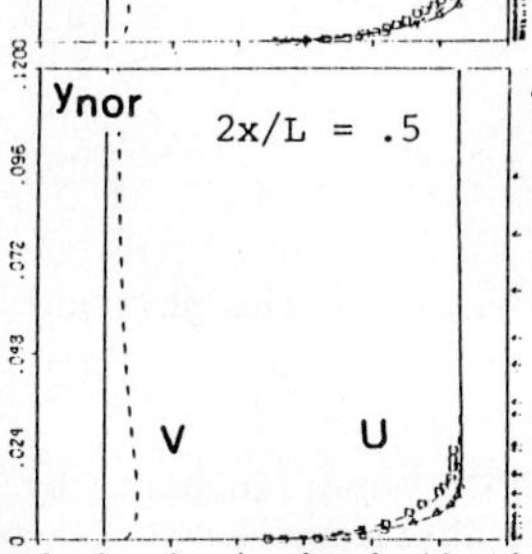

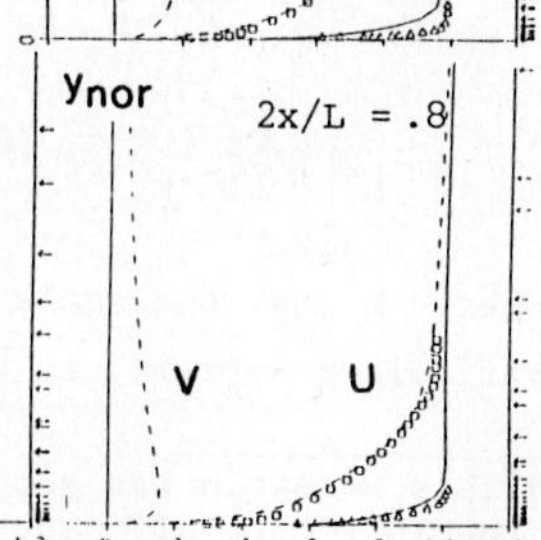

Fig. 4 Velocity profiles : △, experiments keel ; □, experiments close to the waterline : ——, computations keel ; ---, computations close to the waterline.

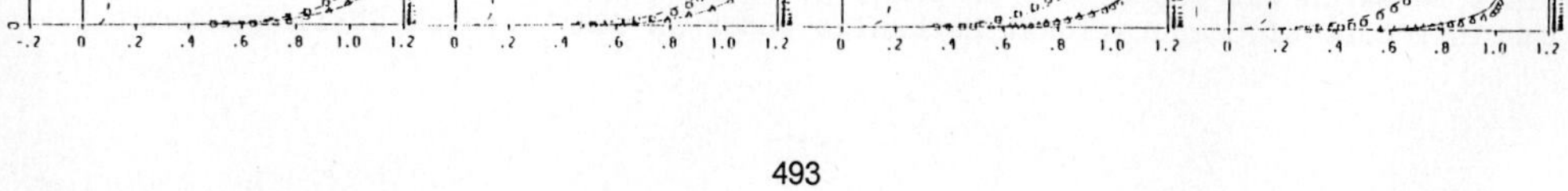

SIMULATION OF UNSTEADY FLOW PAST SHARP SHOULDERS ON SEMI-INFINITE BODIES†

R. Ramamurti, U. Ghia* and K.N. Ghia
Department of Aerospace Engineering and Engineering Mechanics
*Department of Mechanical and Industrial Engineering
University of Cincinnati, Cincinnati, Ohio 45221

INTRODUCTION

The flow over airfoils at high Reynolds number (Re) or high angle of attack exhibits regions of massively separated flow, and is unsteady due to repeated vortex formation and shedding. Flow separation leads to increased drag and reduced lift and is also responsible for the occurrence of stall. Hence, an accurate analysis of separated flows is of primary importance to aerodynamic design.

In the present study, the leading edge separation encountered in high-incidence flows is examined by considering a model problem of the flow past a semi-infinite body with finite thickness and a sharp shoulder (Fig. 1). Both these flows undergo a large change in flow direction, around the sharp shoulder and the leading edge, respectively, and separate thereafter; hence, the selection of this model problem.

U. Ghia and Abdelhalim [1] analyzed the steady flow past two-dimensional (2-D) as well as axisymmetric semi-infinite bodies. They employed the Navier-Stokes (NS) equations formulated in terms of similarity-type vorticity and stream function. Also, Lane and Loehrke [2] studied the flow over this configuration experimentally. The two studies show that, for 2-D cases, the reattachment length increases monotonically as Re is increased up to approximately 300. When Re is increased further, the recirculation length decreases, and remains nearly constant for Re greater than 600. It is conjectured that, for Re > 300, unsteadiness, three-dimesionality and/or turbulence develop in the flow. The unsteady aspect of this flow is investigated in the present study.

OUTLINE OF MATHEMATICAL FORMULATION

The vorticity-stream function formulation of the NS equations developed by U. Ghia and Davis [3] is employed. In general conformal coordinates (ξ,η), these equations can be written as

$$\frac{\partial}{\partial\xi}\left(\frac{1}{H}\frac{\partial\psi}{\partial\xi}\right)+\frac{\partial}{\partial\eta}\left(\frac{1}{H}\frac{\partial\psi}{\partial\eta}\right)=-h^2\Omega \tag{1}$$

and

$$\frac{\partial\Omega}{\partial t}+\frac{v_1}{h}\frac{\partial\Omega}{\partial\xi}+\frac{v_2}{h}\frac{\partial\Omega}{\partial\eta}-\frac{\Omega}{h}\left(v_1\frac{H_\xi}{H}+v_2\frac{H_\eta}{H}\right)$$

$$=\frac{1}{h^2}\left[\frac{\partial}{\partial\xi}\left\{\frac{1}{H}\frac{\partial}{\partial\xi}(\Omega H)\right\}+\frac{\partial}{\partial\eta}\left\{\frac{1}{H}\frac{\partial}{\partial\eta}(\Omega H)\right\}\right] \tag{2}$$

where h is the scale factor of the conformal transformation relating the physical variable $z=(x+iy)$ to the conformal variable $\zeta=(\xi+i\eta)$,

† This research was supported, in part, by AFOSR Grant No. 87-0074, and, in part, by the Institute of Computational Mechanics under an OBR Research Challenge Award.

$$z = \frac{1}{2} [\zeta(Re-\zeta^2)^{1/2} + Re \sin^{-1}(\zeta/Re)] \quad , \tag{3}$$

and

$$H = r^j, \quad v_1 = \frac{1}{hH}\frac{\partial\psi}{\partial\eta} \quad \text{and} \quad v_2 = -\frac{1}{hH}\frac{\partial\psi}{\partial\xi} \quad . \tag{4}$$

Here, the index j is equal to zero for 2-D flow and unity for axisymmetric flow. The Reynolds number is defined as

$$Re = \frac{2}{\pi}\frac{U_\infty d}{\nu} \tag{5}$$

where d is the thickness of the plate.

To enhance the accuracy and efficiency of the ensuing numerical calculations, Eqs. (1)-(2) are further transformed as follows. In terms of the (ξ,η) coordinates, the potential flow past the bodies considered is the stagnation point flow, with $\psi_{inv} \propto \xi^{j+1}$. This suggests a similarity-type stream-function variable for the present viscous flow; subsequently, Eq. (1) leads to a similarity variable for vorticity; hence,

$$\psi(\xi,\eta) = \xi^{j+1} f(\xi,\eta) \quad \text{and} \quad \Omega(\xi,\eta) = -\frac{\xi^{j+1}}{Hh^2} g(\xi,\eta) \quad . \tag{6}$$

In order to facilitate the implementation of the asymptotic boundary conditions far downstream as well as in the outer far-field, further transformations are employed. The final transformed variables are given as

$$G = g/(Re+\eta_w)^{j/2} \quad , \quad F = f-(\eta^{j+1}-\eta_w^{j+1})/(j+1), \quad N = (\eta-\eta_w)/(C+\eta-\eta_w) \quad \text{and}$$

$$S = 1 - a(A_t/\xi)\,\ell n(1+\xi/A_t) + \sum_{i=1}^{NCL} b_i \tan^{-1}(\xi-\xi_i) + c \quad . \tag{7a-d}$$

Here, η_w is the value of η along the body and is zero for plates with sharp shoulders and nonzero for blunt bodies; C, a, A_t, b_i and c are grid-distribution control parameters. For example, the transformation relating the conformal coordinate ξ to the computational coordinate S, given by Eq. (7d), provides coordinate clustering at various streamwise locations ξ_i; one of these corresponds to the shoulder, while the remaining are placed in the region of flow separation. Also, these clustering locations are important to maintain a slowly increasing mesh interval, which would otherwise be overwhelmed by the logarithmic function.

In terms of the final variables, Eqs. (1)-(2) can be written as

$$Q_S F + Q_N F = G\,(Re+\eta_w)^{j/2} + P_F \tag{8}$$

and

$$O_S G + O_N G = G_t + P_G \quad , \tag{9}$$

where Q and O are differential operators and P may be viewed as a source term. The details of the form of Q, O and P are not relevant to the present study; these can be found in Ref. 1.

<u>Boundary Conditions</u>

The no-slip condition <u>at the wall</u> gives

$$F = 0 \quad \text{and} \quad F_N = -C\eta_w^j \quad . \tag{10}$$

As $\eta \to \infty$, the flow approaches the inviscid flow, and hence,

$$F_N = 0 \text{ and } G = 0 \quad . \tag{11}$$

Far downstream, i.e., as $\xi \to \infty$, all derivatives with respect to ξ vanish and the governing equations reduce to a self-similar form. The solution of the resulting ordinary differential equations provides the downstream boundary condition. Along the stagnation line, symmetry conditions are employed. The boundary condition for the vorticity at the wall is obtained from the stream-function field, by writing the stream-function equation along the wall.

NUMERICAL PROCEDURE

An essential feature of the method of solution appropriate for simulation of unsteady flow is that, for each temporal-step advance of the vorticity field, the stream-function field must constitute a converged solution of Eq. (8). Hence, the governing equations, Eqs. (8) and (9), are solved in an uncoupled manner. The stream-function equation is solved using the multigrid (MG) scheme, coupled with the strongly implicit (SI) procedure as the basic relaxation operator; being an iterative procedure, the SI scheme requires the inclusion of a fictitious time derivative term in the equation. The vorticity equation is solved using the factored-ADI scheme.

MG-SI Scheme for the Stream-function Equation

The stream-function equation is a Poisson equation. U. Ghia et al. [4] have employed the MG-SI procedure successfully for solving the Poisson equation for pressure and found that the SI relaxation operator was the least sensitive to the problem parameters and the most efficient among the procedures they considered. Also, comparison of the computational effort for a standard problem shows that the MG-SI scheme requires only twice the computer time as that needed by direct Gaussian elimination; the computer memory requirement for the MG-SI scheme is many times less. This latter requirement becomes a major concern in direct elimination techniques, especially as the grid is refined and when nearly equal number of grid points are used in both directions.

The MG procedure, described in detail in Ref. 4, is employed to solve Eq. (8).

Factored ADI Scheme for the Vorticity-Function Equation

In order to capture the boundary-layer-like features of the flow, the streamwise convective term is upwind differenced and included in the normal sweep of the ADI procedure; all other spatial derivatives are discretized using second-order accurate central differences. The discretized vorticity equation is then factored and solved using the Thomas algorithm for each sweep. When only steady-state solutions are sought, spatially varying time steps are employed to enhance convergence. The variation of the time step is of the form

$$\Delta t = \Delta t_G \, /(1\text{-}N)h^2 \quad , \tag{12}$$

where Δt_G is a specified constant.

RESULTS AND DISCUSSION

Results were first obtained for steady flow. For blunt-shouldered configurations, the flow is quite benign, with a thin elongated separation bubble aft of the shoulder region. Therefore, only sharp shouldered configurations are discussed here. Figure 2 shows the streamline pattern for Re=200 as well as the corresponding surface distributions of vorticity function and pressure. Reattachment occurs at $x = 7.91$ or $x/d = L_R = 5.03$. This compares well with the available results of Refs. 1 and 2. For Re ≥ 300, the steady-flow calculations lead to some anomalous flow features as indicated in Fig. 3 by a very sharp and localized dip in the surface vorticity function at the shoulder, followed by a second dip prior to reattachment. Hence, only true time-dependent calculations were pursued for Re ≥ 300.

Figures 4 and 5 show the unsteady breakdown of the separated flow, in terms of contours of ψ and G, respectively, for Re=300 and $42.98 \leq t/Re \leq 52.98$, with an (81x81) nonuniform grid. The separation bubble breaks up into two, with the downstream one dissipating as it convects along the plate. The process nearly repeats itself over the 10 characteristic time units shown. The lack of smoothness in some of the G contours corresponds to small oscillations (at the level of ±0.001) around $G \simeq 0$ and are attributed to the streamwise grid-point distribution. This is confirmed by a grid refinement study using a (161x81) grid (Fig. 6) leading to the results shown in Figs. 7 and 8. The vorticity-function contours are now much smoother. The recirculation region is larger, but the vortex breakdown and shedding process remain essentially unchanged, occupying approximately 11 characteristic time units.

The refined (161x81) grid was used to analyze the case with Re=400. The early-time results shown in Figs. 9 and 10 indicate a larger width of the separation region than for Re=300. The calculation needs to be pursued further in time. This is suggested by the concavity in the zero streamline just upstream of reattachment in the time-averaged results shown in Fig. 11. Also, the time-averaged reattachment length L_R is yet considerably smaller than that observed experimentally (Ref. 2).

CONCLUSION

A Navier-Stokes analysis has been developed for simulating unsteady flows using general clustered conformal coordinates and a 3-level multigrid procedure for the stream-function equation. A fixed algorithm for the multigrid cycle is employed, wherein one iteration is performed on all but the coarsest grid, on which three iterations are performed. Convergence is said to be achieved when the L_2-norm of the residuals becomes less than 10^{-6}. This level of convergence is usually achieved in 2 or 3 cycles of the multigrid scheme. The calculations for Re=300 and 400 need to be continued for longer time, to definitively determine the time-asymptotic behaviour of these flows. Also, finer grids are to be developed for use with higher-Re cases (Re≃1000) for which high-quality water-tunnel experimental measurements are available from Hourigan [5].

REFERENCES

1. Ghia, U. and Abdelhalim, A., AIAA Paper No. 82-0024, 1982; also, AIAA Journal, Vol. 25, May 1987.

2. Lane, J.C. and Loehrke, R.I., ASME HTD, Vol. 13, 1980.

3. Ghia, U. and Davis, R.T., AIAA Journal, Vol. 12, No. 12, 1974.

4. Ghia, U., Ghia, K.N. and Ramamurti, R., AIAA Paper No. 83-0551, 1983; also, AIAA Journal, August 1988.

5. Hourigan, K., Private Communication, CSIRO, Melbourne, 1987.

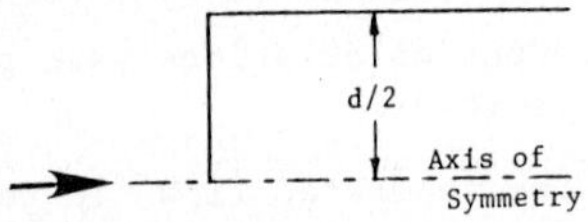

Fig. 1. Schematic of Flow Configuration

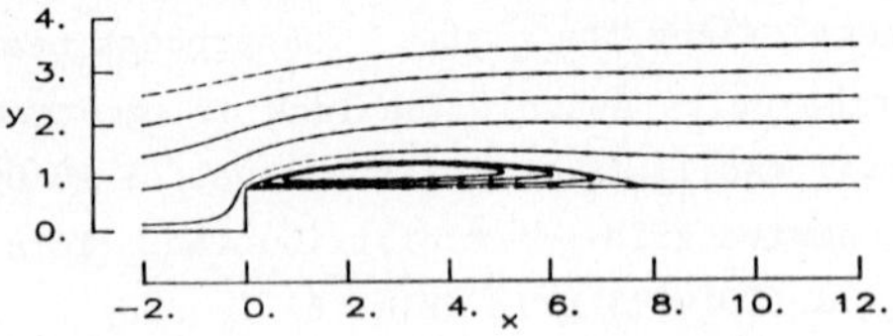

Fig. 2a. Steady-State Streamlines; Re = 200.

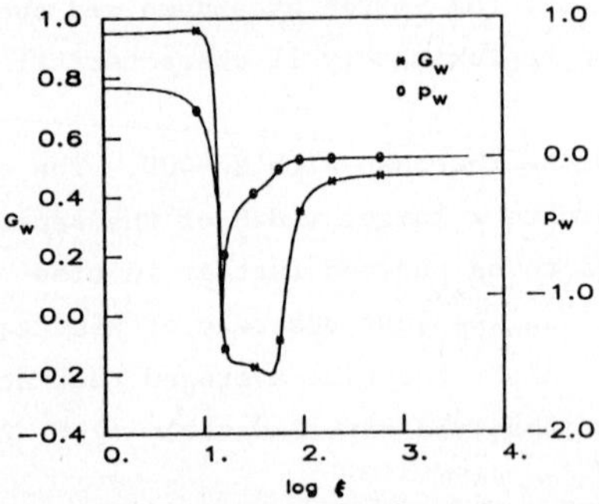

Fig. 2b. Steady-State Surface Pressure and Skin Friction Function; Re = 200, Shoulder at ξ_c = Re = 14.14.

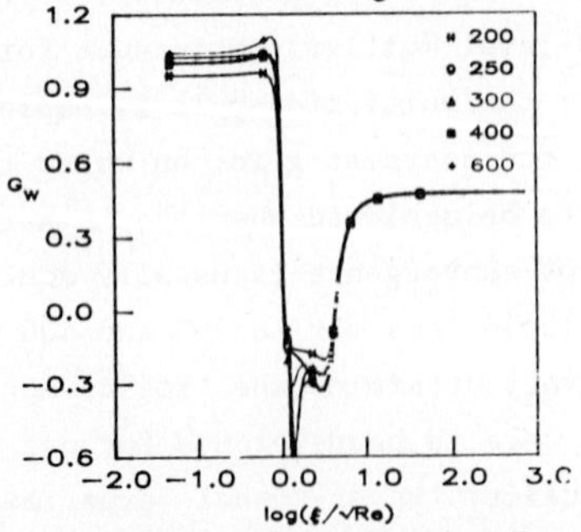

Fig. 3. Effect of Increasing Re on Skin-Friction Function; Steady-State Results via Pseudo-Time Approach.

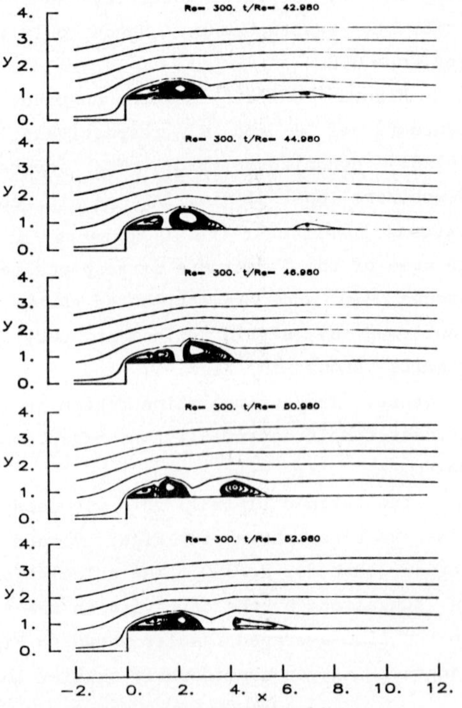

Fig. 4. Instantaneous Streamlines via True Transient Approach; Re = 300, (81x81) Non-Uniform Grid.

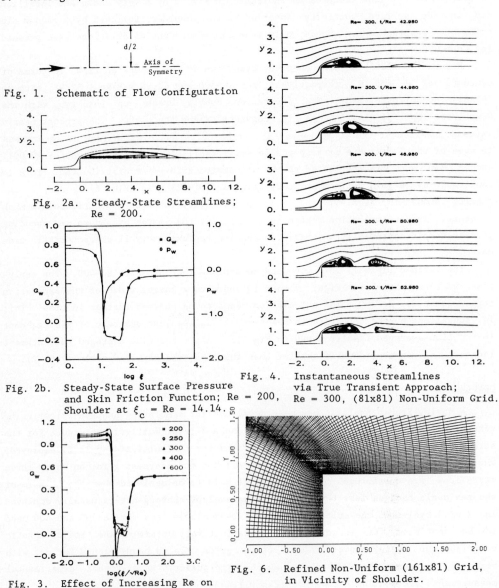

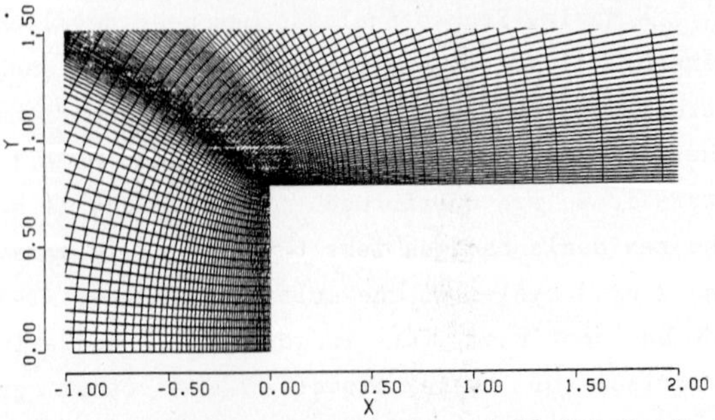

Fig. 6. Refined Non-Uniform (161x81) Grid, in Vicinity of Shoulder.

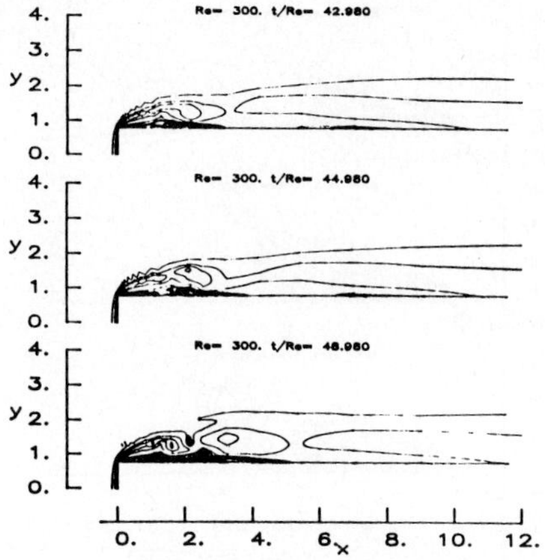

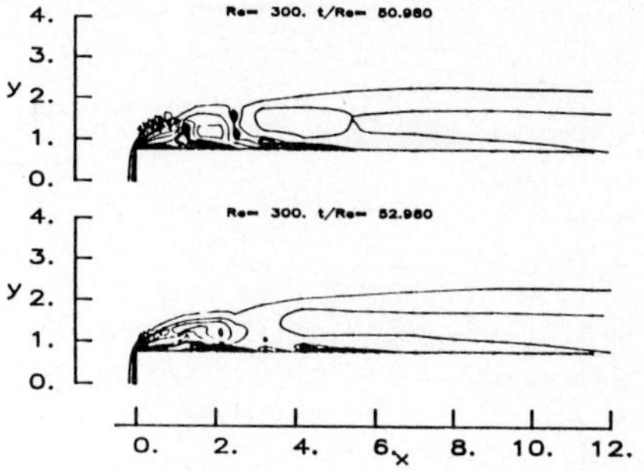

Fig. 5. Vorticity Contours for True Transient Solution; Re = 300, (81x81) Non-Uniform Grid.

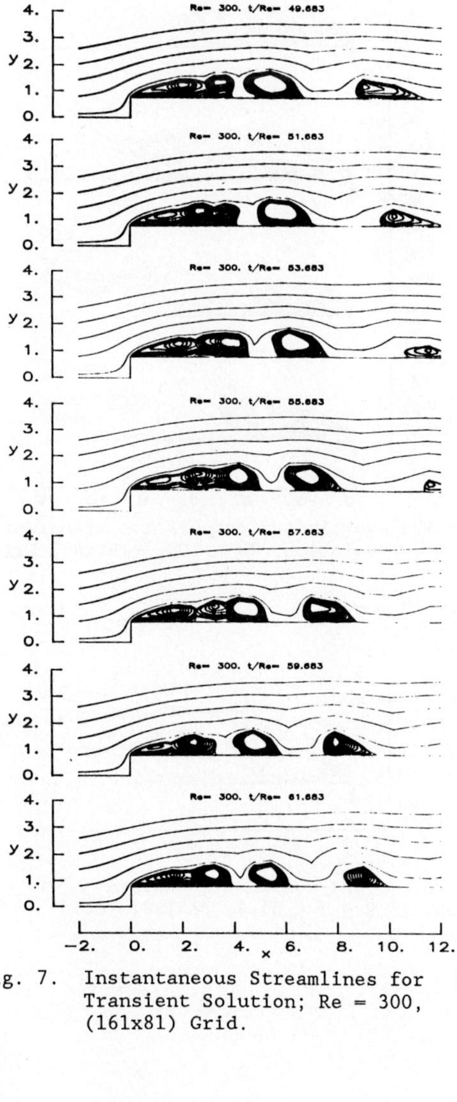

Fig. 7. Instantaneous Streamlines for Transient Solution; Re = 300, (161x81) Grid.

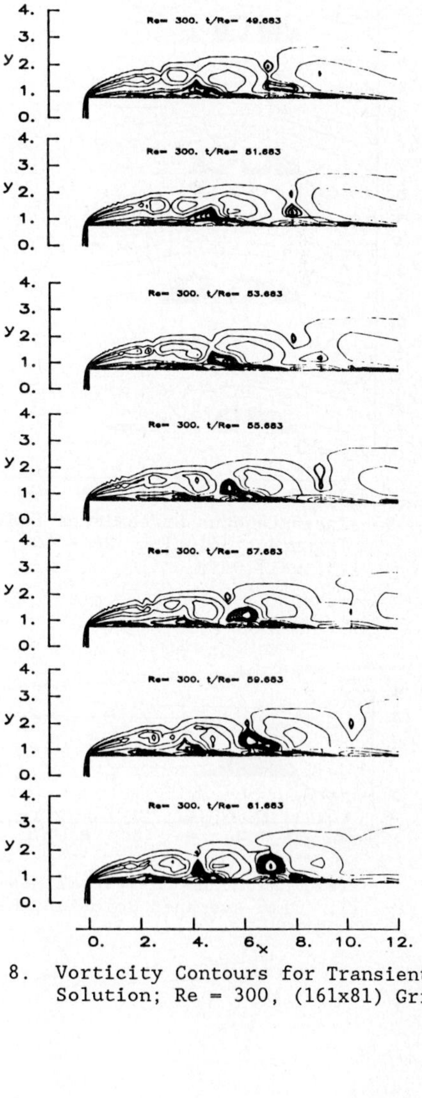

Fig. 8. Vorticity Contours for Transient Solution; Re = 300, (161x81) Grid.

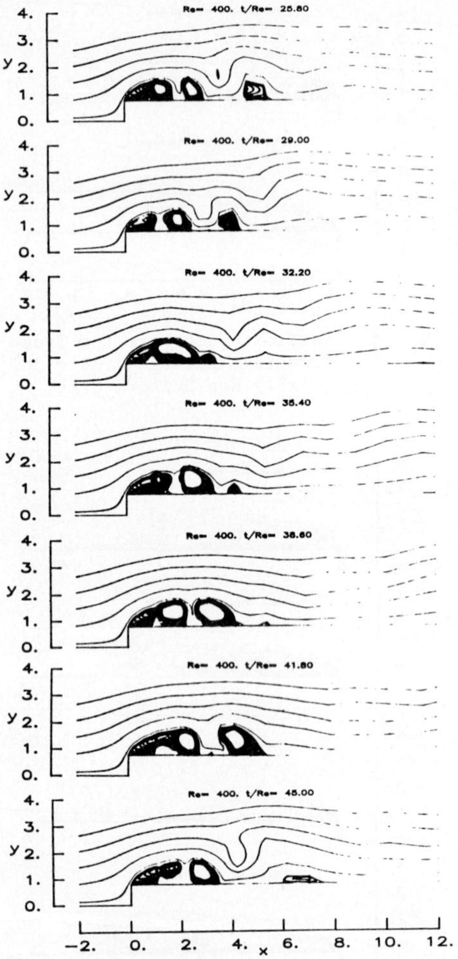

Fig. 9. Instantaneous Streamlines for Transient Solution; Re = 400, (161x81) Grid.

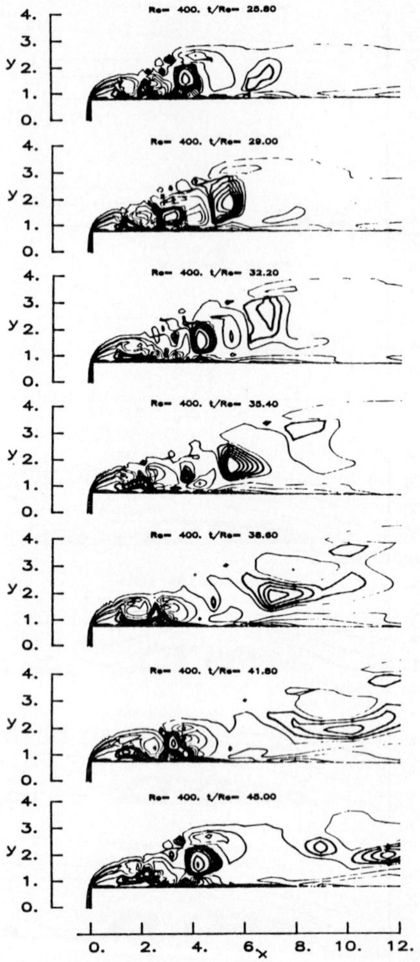

Fig. 10. Vorticity Contours for Transient Solution; Re = 400, (161x81) Grid.

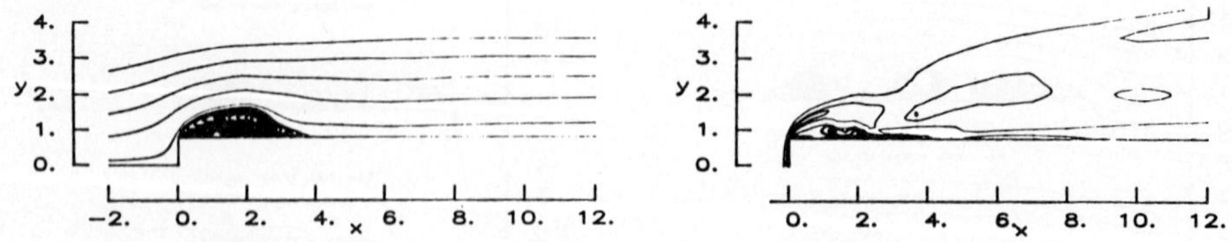

(a) Instantaneous Streamlines (b) Vorticity Contours

Fig. 11. Time-Averaged Solution; Re = 400, $25.0 \leq T \leq 51.4$, (161x81) Grid.

SEMI-IMPLICIT FINITE-DIFFERENCE SIMULATION OF LAMINAR HYPERSONIC FLOW OVER BLUNT BODIES

S. Riedelbauch, B. Müller, W. Kordulla
DFVLR Institute for Theoretical Fluid Mechanics,
Bunsenstr. 10, D-3400 Göttingen, FRG

Summary

Laminar hypersonic flows over the nose of a spacecraft and an ellipsoid at angle of attack are simulated by a semi-implicit finite-difference method. The central discretization of the thin-layer Navier-Stokes equations is explicit in the surface tangential directions and implicit in the normal direction. The scheme operates as a checkerboard line relaxation method, and requires only the conservative variables and the Cartesian coordinates to be permanently stored. The code is verified by comparison with axisymmetric computational results and with experiments. Attached flow is predicted for the spacecraft nose, and separated flow is observed for the ellipsoid.

1. Introduction

The development of new spacecrafts has renewed the interest in hypersonics. Today the aerodynamic design can take advantage of improved numerical methods and more powerful computers to analyse hypersonic flow around spacecrafts.

The present paper outlines the development, validation, and application of a Navier-Stokes code to simulate three-dimensional laminar supersonic and 'cold' hypersonic flow over blunt bodies using the perfect gas assumption.

2. Governing Equations

The Newtonian fluid considered is assumed to be a perfect gas with constant specific heats (ratio $\gamma = 1.4$) and without external forces and heat sources. The independent variables of the Navier-Stokes equations in dimensionless conservation law form are transformed to boundary fitted moving coordinates.

For the high Reynolds number flows investigated, the thin-layer approximation [1] is employed by neglecting all streamwise ξ- and circumferential η- derivatives of the viscous terms and retaining only the corresponding pure near-normal ζ-derivatives [2]:

$$\frac{\partial \hat{q}}{\partial \tau} + \frac{\partial \hat{E}}{\partial \xi} + \frac{\partial \hat{F}}{\partial \eta} + \frac{\partial \hat{G}}{\partial \zeta} = \frac{1}{Re_{\infty, L}} \frac{\partial \hat{S}}{\partial \zeta} \qquad (1)$$

The boundary conditions on the body contour are no-slip condition, isothermal or adiabatic wall, and $\partial p/\partial \zeta = 0$. The pressure behind the bow shock is calculated from the inviscid energy equation in non-conservative form [3]. The shock velocity and the remaining flow variables behind the bow shock are evaluated from the Rankine-Hugoniot relations. Bilateral symmetry conditions are used. At the outflow boundary, the conservative variables are obtained by linear extrapolation.

3. Numerical Method

The time derivative in the thin-layer Navier-Stokes equations (1) is approximated by the first-order Euler implicit formula. The fluxes $\hat{E}^{n+1}$ and $\hat{F}^{n+1}$ are approximated by $\hat{E}^n$ and $\hat{F}^n$ leading to an explicit CFL condition, whereas the fluxes $\hat{G}^{n+1}$ and $\hat{S}^{n+1}$ are linearized to remove the stiffness in the near normal direction [4]. Here, a linearization with respect to q^n [5] instead of $\hat{q}^n$ [4] is performed to facilitate the numerical treatment of the singularity of the grid at the nose. The spatial derivatives are approximated by second-order central differences. Thus, the following semi-implicit finite-difference scheme is obtained [6]:

$$\left[J^{-1}I + \Delta\tau\left(\mu_\zeta\delta_\zeta \frac{\partial\hat{G}^n}{\partial q} - Re_{\infty,L}^{-1}\,\delta_\zeta \frac{\partial\hat{S}^n}{\partial q}\right) - \varepsilon_I J^{-1}\delta_\zeta^2\right]\Delta q^n =$$

$$= -q^n\Delta(J^{-1})^n - \Delta\tau\left[\mu_\xi\,\delta_\xi\,\hat{E}^m + \mu_\eta\,\delta_\eta\,\hat{F}^m + \mu_\zeta\,\delta_\zeta\,\hat{G}^n - Re_{\infty,L}^{-1}\,\delta_\zeta\,\hat{S}^n\right] \tag{2a}$$

$$-\varepsilon_E J^{-1}\left[\delta_\xi^4\,\bar{q} + \delta_\eta^4\,\bar{q} + \delta_\zeta^4\,q^n\right]$$

where $m = \begin{Bmatrix} n \\ n+1 \end{Bmatrix}$ $\quad for\, n + i + j \begin{Bmatrix} odd \\ even \end{Bmatrix}$,

$$q^{n+1} = q^n + \Delta q^n \tag{2b}$$

and $\bar{q}$ is taken at the latest time level. I denotes the 5 x 5 identity matrix. The numerical damping coefficients have been chosen according to [2]: $\varepsilon_E = \Delta\tau$ and $\varepsilon_I = 2\varepsilon_E$. The classical finite-difference operators are defined by

$$\delta_\xi\,a_{i,j,k} = a_{i+1/2,j,k} - a_{i-1/2,j,k}\ , \qquad \mu_\xi\,a_{i,j,k} = (a_{i+1/2,j,k} + a_{i-1/2,j,k})/2$$

etc. For convenience, $\Delta\xi = \Delta\eta = \Delta\zeta = 1$ is assumed.

In the surface-tangential directions, a checkerboard relaxation scheme is employed, which changes black and white after each time step. In the normal direction, the block-tridiagonal linear system is solved by the Richtmyer algorithm.

The major advantages in using the semi-implicit algorithm (2) instead of the implicit Beam and Warming scheme [7] are: a) the elimination of the block-matrix inversions in the two surface-tangential directions, and b) the reduction of storage requirements, because the 5 delta variables Δq^n need not to be stored. A disadvantage of (2) occurs, if the time step admissible with [7] has to be reduced due to the CFL condition because of small spacings in the surface-tangential directions.

The blunt body code is implemented on the CRAY-XMP 216 of the DFVLR and operates at about 41 microseconds per time step and per grid point. The scheme (2) is vectorized, except for the recursive Richtmyer algorithm.

4. Results

Validation

The code based on the semi-implicit algorithm (2) is validated for laminar supersonic and hypersonic flow past three simple blunt bodies: an adiabatic hemisphere-cylinder at $M_\infty = 2.94$, $\alpha = 0°$, $Re_{\infty,R} = 2.2\ 10^5$, an isothermal blunt cone with 15° half angle at $M_\infty = 10.6$, $\alpha = 0°, 15°, 20°$, $Re_{\infty,R} = 1.1\ 10^5$, $T_w = 300K$, $T_{0\infty} = 1111K$, and an adiabatic hemisphere at $M_\infty = 4.15$, $\alpha = 0°$, $Re_{\infty,R} = 1.5\ 10^6$. The results show that the pressure and the temperature or heat flux on the bodies are correctly predicted compared with numerical solutions of the axisymmetric Navier-Stokes equations and with experimental data [8].

Hypersonic flow past a blunt spacecraft nose

Laminar flow past an adiabatic blunt spacecraft nose is simulated at $M_\infty = 12.2$, $\alpha = 10°$, $Re_{\infty,1m} = 141200$ and $T_\infty = 266.6K$. Although real gas effects have to be taken into account at these freestream conditions, the perfect gas assumption is used to investigate the hypersonic flow field over this non-axisymmetric body. The 79 x 79 x 41 (streamwise x circumferential x near normal) mesh of the converged solution is shown in Fig. 1a. The farfield boundary corresponds to the location of the bow shock. The temperature on the windward side of the body decreases downstream of the stagnation point up to the juncture between the blunt nose and the flat part of the windward side, and then increases again (Fig. 1b). On the leeward side, however, the temperature decreases continuously in downstream direction. The pressure contours of the exit surface show a local maximum, located at the lower 'corner' (Fig. 2a). The flow expands over the lower corner, the pressure increases to a smaller local maximum about in the middle of the right hand side, and decreases slowly behind the upper corner. The skin friction lines indicate divergent flow along the lower 'corner' but no separation (Fig. 2b).

Hypersonic flow past an ellipsoid

The perfect gas assumption is valid for laminar flow past a 12:5:3 ellipsoid at $M_\infty = 8.15$, $\alpha = 25°$, $Re_{\infty,L} = 9.6\ 10^5$ (L = large half axis), $T_w = 296K$ and $T_{0\infty} = 800K$. The 122 x 63 x 41 (only every other point in ξ-direction is plotted) mesh shows the ellipsoid elongated until x/L = 3.5, and indicates the bow shock (Fig. 3a). The pressure and heat flux have their maxima at the stagnation point, and local maxima in the leeward symmetry plane (Figs. 3b,c). Note that the increments on the windward sides are higher than on the leeward sides. The Mach number contours and the velocity vectors show a separation region on the leeward side with a primary and secondary separation line (Fig. 4). Under the primary vortex core, the pressure increases in the clockwise direction (Fig. 3b) and causes the secondary separation (Fig. 4). The separated flow is essentially subsonic. The foot prints of both the primary and secondary streamwise vortices are clearly visible in the skin-friction lines (Fig. 5). Those lines show a similar behaviour as oil flow visualizations [9] (Fig. 6). The skin-friction lines (Fig. 5) also indicate, that a region of reverse flow relative to the main flow exists between both separation lines. The location is approximately between x/L $\approx$ 1.1 and x/L $\approx$ 2.0. The experimental model [9], however, ends at x/L $\approx$ 1.6 with the consequence that the base flow may interact with the leeside vortex structure and influence the separation lines.

5. Conclusions

The present paper outlines a semi-implicit finite-difference algorithm to solve the thin-layer Navier-Stokes equations. The method is explicit in the two near wall-tangential directions and implicit in the near wall-normal direction. The code is validated, and is applied to simulate laminar hypersonic flow over blunt bodies. The flow field past a spacecraft nose shows divergent flow along the lower 'corner' at a moderate angle of attack. Primary and secondary crossflow vortices are found on the leeward side of an ellipsoid at $\alpha = 25°$. The skin-friction lines are qualitatively similar to oil flow visualizations.

References

[1] Baldwin, B.S., Lomax, H.: Thin-Layer Approximation and Algebraic Model for Separated Turbulent Flows. AIAA Paper 78-257 (1978).

[2] Pulliam, T.H., Steger, J.L.: Implicit Finite-Difference Simulations of Three-Dimensional Compressible Flow. AIAA J. 18 (1980), pp. 159-167.

[3] Kutler, P., Pedelty, J.A., Pulliam, T.H.: Supersonic Flow Over Three-Dimensional Nosetips Using an Unsteady Implicit Numerical Procedure. AIAA Paper 80-0063 (1980).

[4] Rizk, Y.M., Chaussee, D.S.: Three-Dimensional Viscous-Flow Computations Using a Directionally Hybrid Implicit-Explicit Procedure. AIAA Paper 83-1910 (1983).

[5] Thomas, P.D., Lombard, C.K.: Geometric Conservation Law and Its Application to Flow Computations on Moving Grids. AIAA J. 17 (1979), pp. 1030 - 1037.

[6] Müller, B., Rues, D.: Finite-Difference Method for the Simulation of Three-Dimensional Laminar Supersonic Flow over Blunt Bodies. DFVLR - Interner Bericht IB 221-86 A 03 (1986).

[7] Beam, R.M., Warming, R.F.: An Implicit Factored Scheme for the Compressible Navier-Stokes Equations. AIAA J. 16 (1978), pp. 393-402.

[8] Riedelbauch, S., Müller, B.: The Simulation of Three-Dimensional Viscous Supersonic Flow Past Blunt Bodies with a Hybrid Implicit/Explicit Finite-Difference Method. DFVLR-FB 87-32, Göttingen, 1987.

[9] Da Costa, J.L., Aymer de la Chevalerie, D., Alziary de Roquefort, T.: Ecoulement Tridimensionnel Hypersonique sur une Combinaison d'Ellipsoides. CEAT, Rapport No. 4 RDMF 86, Université de Poitiers, Juillet 1987.

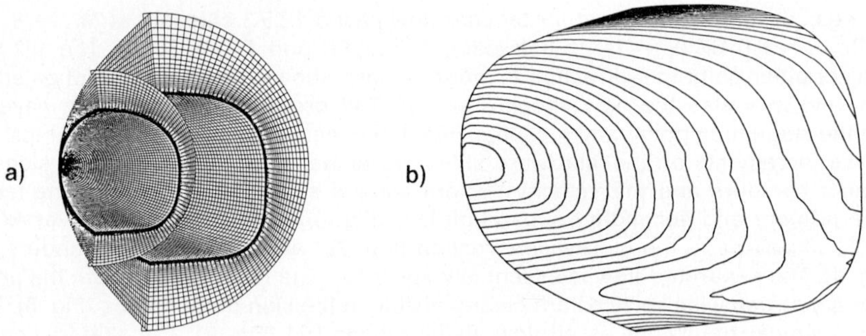

Fig. 1: a) Mesh and b) temperature contours of a blunt spacecraft nose for $M_\infty = 12.2$, $\alpha = 10°$, $Re_{\infty, 1m} = 141\,200$.

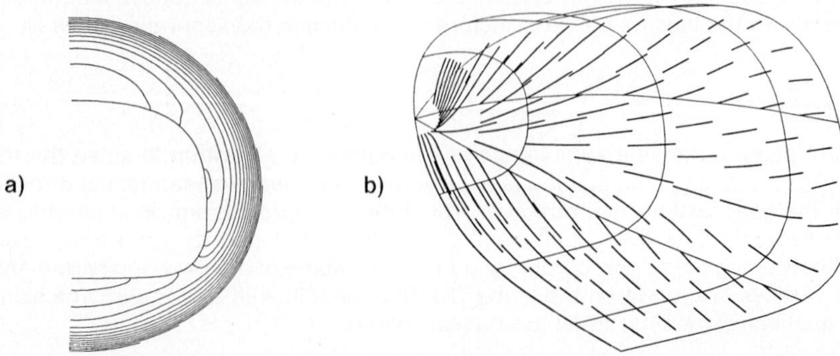

Fig. 2: a) Pressure contours and b) skin-friction lines at the outflow plane of a blunt spacecraft nose for $M_\infty = 12.2$, $\alpha = 10°$, $Re_{\infty, 1m} = 141\,200$.

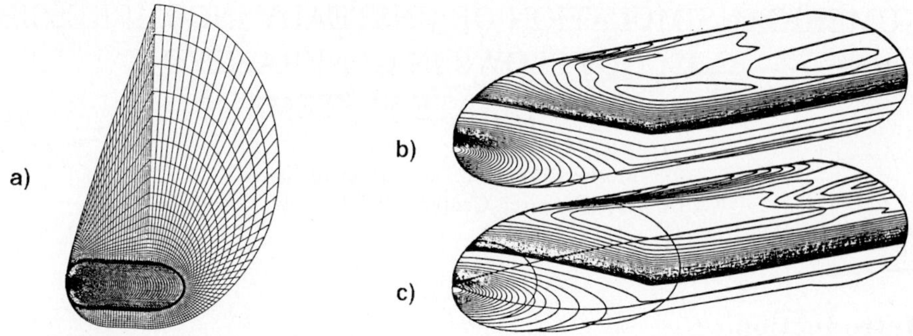

Fig. 3: a) Mesh, b) pressure and c) heat flux contours of 12:5:3 ellipsoid for $M_\infty = 8.15$, $\alpha = 25^\circ$, $Re_{\infty,L} = 960\ 000$.

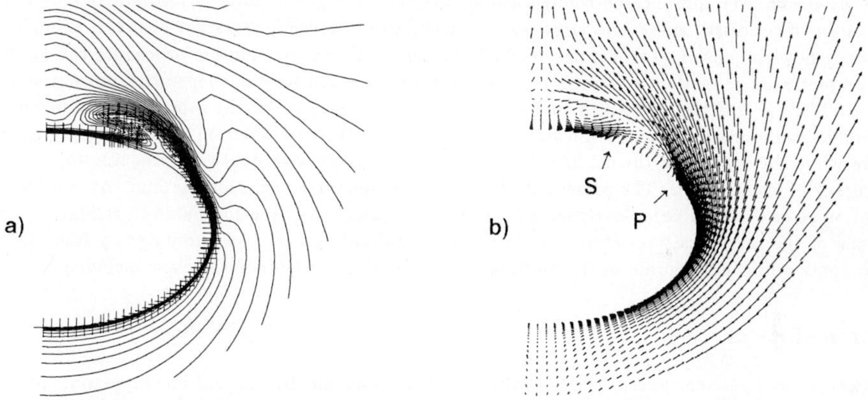

Fig. 4: a) Mach number contours and b) velocity vectors at x/L ≈ 1.3 of ellipsoid for $M_\infty = 8.15$, $\alpha = 25^\circ$, $Re_{\infty,L} = 960\ 000$, (+ + + : sonic line, P : primary, S: secondary separation).

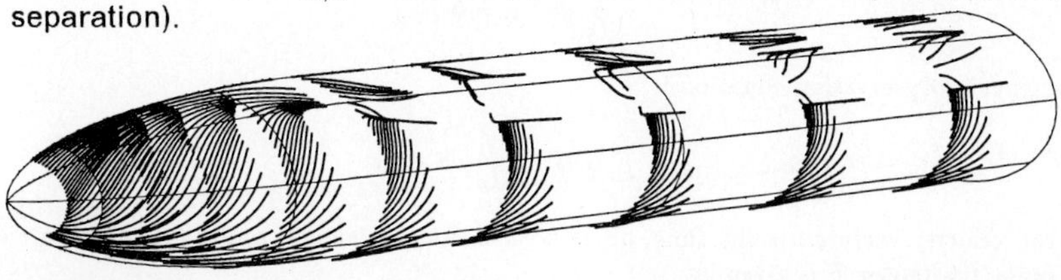

Fig. 5: Predicted skin-friction lines on ellipsoid for $M_\infty = 8.15$, $\alpha = 25^\circ$, $Re_{\infty,L} = 960\ 000$.

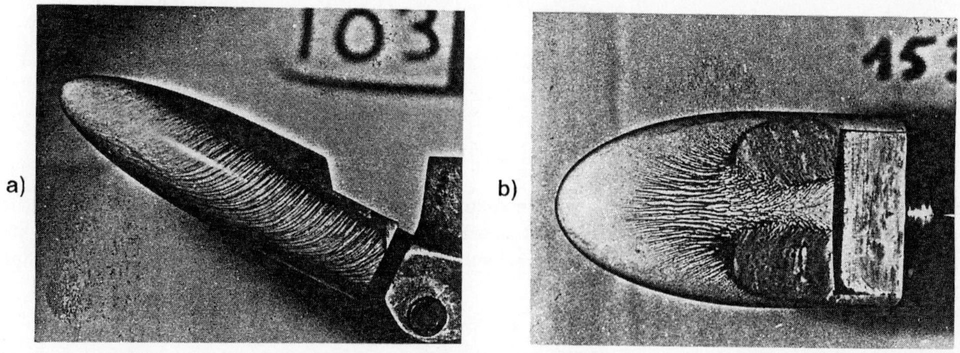

Fig. 6: a) Side view and b) leeward view of oil flow visualizations [9] on ellipsoid for $M_\infty = 8.15$, $\alpha = 25^\circ$, $Re_{\infty,L} = 960\ 000$.

NUMERICAL SIMULATION OF UNSTEADY INCOMPRESSIBLE VISCOUS FLOWS IN GENERALIZED COORDINATE SYSTEMS

Moshe Rosenfeld* and Dochan Kwak†
NASA Ames Research Center, Moffett Field, CA 94035

1 Introduction

Numerous fluid flow phenomena are essentially time-dependent. Until recently, the numerical simulation of unsteady flowfields over complicated configurations was very difficult because of limited computer resources both in terms of processor speed and storage. The new large supercomputers are powerful enough to make the unsteady flow simulation possible, provided an efficient and accurate solution procedure can be devised.

The present work presents a numerical simulation method for viscous time-dependent incompressible flowfields in arbitrary geometries. Two principal approaches have been used for the solution of the incompressible Navier-Stokes equations in primitive formulation [1]. One method is based on the fractional-step solution procedure while the other uses the artificial-compressibility concept with dual-time stepping for restoring the time-accuracy of the solution. The present study uses a fractional-step solution method. An efficient and robust solution procedure has been developed and several test cases have been computed to validate the method. The details of the numerical procedure and some of the validation cases are presented in Refs. [2] and [3]. Therefore, only a brief description of the method is given here and a few solutions are included.

2 Formulation

The Navier-Stokes equations govern the flow of isothermal, constant-density and incompressible fluids. For a control volume V with surface S the mass conservation equation is

$$\oint_S d\mathbf{S} \cdot \mathbf{u} = 0 \,, \tag{1}$$

and the momentum conservation equation is

$$\frac{d}{dt} \int_V \mathbf{u}\, dV = \oint_S d\mathbf{S} \cdot \bar{T} \,, \tag{2}$$

where $\mathbf{u}$ is the velocity vector, t is the time, $d\mathbf{S}$ is a surface-area element, and dV is a volume-element. For Newtonian fluids the tensor $\bar{T}$ is given by

$$\bar{T} = -\mathbf{u}\mathbf{u} - P\,\bar{I} + \nu\left(\nabla\mathbf{u} + (\nabla\mathbf{u})^T\right) \,, \tag{3}$$

where P is the pressure, ν is the effective kinematic viscosity, $\bar{I}$ is the identity tensor, $\nabla\mathbf{u}$ is the gradient of $\mathbf{u}$, and $(\cdot)^T$ is the transpose operator.

3 Discretization and Numerical Solution

Equations (1-3) are directly discretized over finite volumes to yield a fully conservative second-order scheme. Finite-volume discretization is preferred since it usually results in more accurate and stable solutions for generalized coordinate systems, especially for meshes with heavily clustered points and large curvature. Special

*NRC Research Associate
†Research Scientist

care is given to satisfy of the "geometrical conservation laws" [2] to minimize the errors resulting from the spatial discretization.

The numerical problems associated with the absence of a pressure time-derivative in the mass conservation equation are handled by a fractional-step procedure. In each time step, the momentum equations are solved for an approximate velocity field which does not satisfy the continuity equation. In the second stage, the velocity and pressure fields are corrected such that the mass conservation equation is satisfied. This step leads to a Poisson equation with Neumann-type boundary conditions which may exhibit very poor convergence properties, especially in generalized coordinate systems. The CPU time consumed by the Poisson solver may be as high as 80% of the total computational time even in a Cartesian case [4]. In the present work an attempt is made to minimize these difficulties.

This attempt has led to the choice of the pressure (defined at the center of the computational cells) and the *volume fluxes* across each computational cell face as the dependent variables instead of the familiar Cartesian components of the velocity. (The volume fluxes are equivalent to the scaled contravariant velocity components in a staggered grid.) This choice ensures the satisfaction of the *discrete* mass conservation equation to round-off errors in any coordinate system and has favorable effect on the convergence properties of the Poisson solver (see details in Refs. [2] and [3]).

A consistent Poisson solver is obtained by deriving the Laplacian operator from the discrete equivalent of the operator $\nabla \cdot \nabla$ (where $\nabla\cdot$ and ∇ are the divergence and the gradient operators, respectively), rather than discretizing the Laplacian operator directly. A novel and efficient ZEBRA scheme with four-color ordering is devised for the efficient solution of the nonorthogonal Poisson equation on vector computers. The Poisson solver could be converged to any specified small error in all the cases solved so far.

4 Results

Various test cases were solved to validate the present solution method. The three-dimensional flow in a cubic cavity as well as the symmetric flow over a circular cylinder at $Re = 40$ and at $Re = 200$ with vortex shedding are described in Ref. [2]. The present paper reports additional cases to assess further the capabilities of the method. All the cases were solved using the three-dimensional code. Two-dimensional solutions were obtained by specifying along the third direction two intervals and periodic boundary conditions.

4.1 Lid-driven Cavity, Re = 10^4

In this case the time-dependent flow in a lid-driven square cavity was solved for a Reynolds number of 10^4. A Cartesian nonuniform grid of $81 \times 81 \times 3$ mesh points was used with clustering near the walls. The minimal interval is $0.28 \cdot 10^{-2}$ and the maximal interval at the center is $0.23 \cdot 10^{-1}$. The upper lid started moving impulsively at $T = 0$ and the time-accurate flowfield was computed until the nondimensional time $T = 120$ was reached. Previously, steady solutions of this case were reported in the literature, see for example Ghia *et al.* [5]. However, these solutions were not obtained by a time-accurate method. For $Re = 10^3$, the steady solution obtained by the present solution procedure compares favorably with other computations (see Ref. [3]).

For the case $Re = 10^4$ the evolution of the main vortex and of the secondary vortices at the corners of the cavity exhibit complicated patterns that are repeated almost periodically without approaching a steady state. Figure 1 shows the instantaneous streamlines for $102 > T > 92$ which corresponds approximately to one "period." During this time interval, the main vortex rotates around a fictitious center and the flowfield at the lower two corners goes through a complicated series of states which include the generation of secondary and tertiary vortices and their merging with one another. Recently, Bruneau and Jouron [6] have found that for this case the solution has not converged to a steady state if a fine grid is used. However, no experimental results are available for comparison at this time.

4.2 Symmetrical Flow over a Circular Cylinder

The symmetrical flowfield over a circular cylinder at $Re = 40$ was solved employing a nonorthogonal grid with $45 \times 73 \times 3$ mesh points in the radial, circumferential, and axial directions, respectively. Figure 2 compares the time-evolution of the separation length with other numerical and experimental results. Good agreement is obtained, especially for $T < 6$. Slight disagreement is found for $T > 6$ due to the "wall effect," which is the result of the finite distance of the outer boundary and the Dirichlet-type boundary conditions given in the present solution. This wall effect was studied earlier by varying the outer boundary distance (see Ref. [2]).

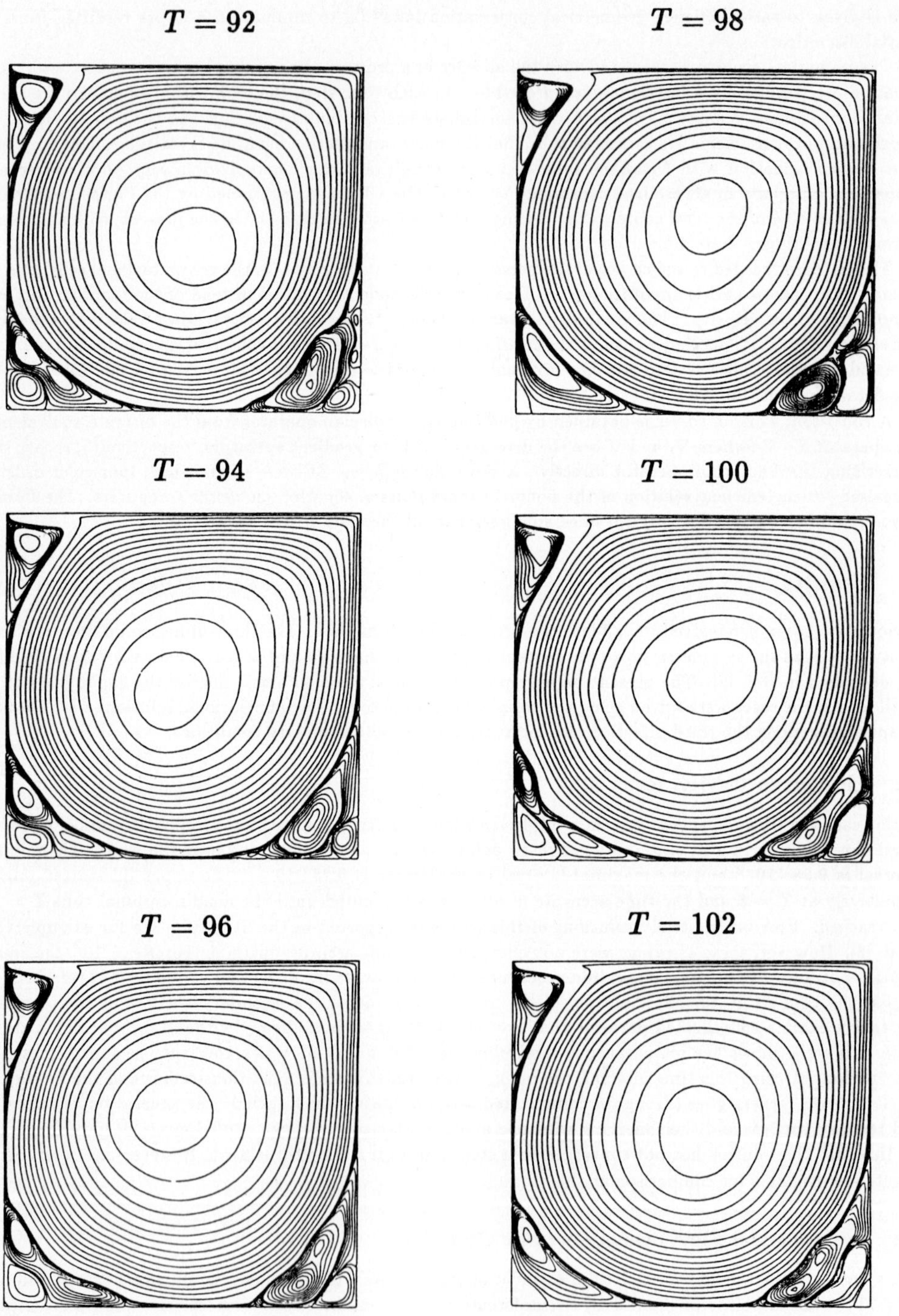

Figure 1: Instantaneous streamlines, $Re = 10^4$

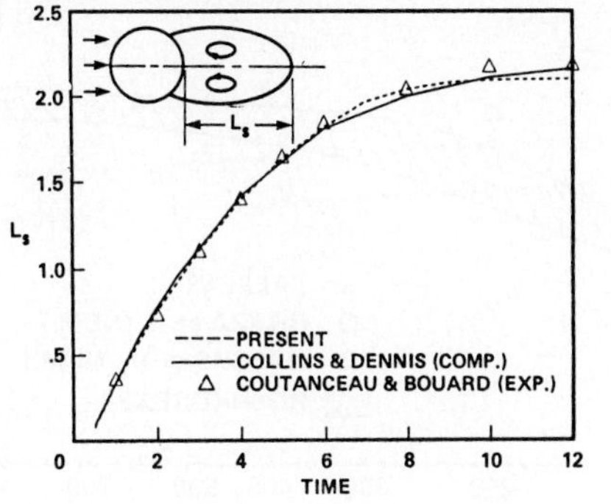

Figure 2: Time-evolution of the separation-length ($Re = 40$)

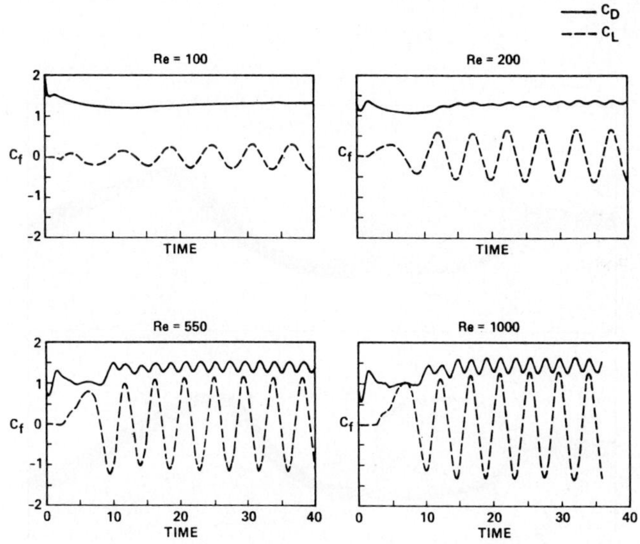

Figure 3: Time-evolution of the force coefficients

4.3 Vortex Shedding from a Circular Cylinder

This case considers the shedding of vortices behind a circular cylinder at Reynolds numbers 100-1000. A cylindrical grid of $81 \times 84 \times 3$ mesh points was employed with points clustered near the cylinder and in the wake region. The vortex shedding was triggered by rotating the cylinder for a short period of time in both clockwise and counterclockwise directions. Figure 3 shows the time evolution of the lift coefficient (dashed line) and the drag coefficient (full line) for several Reynolds numbers. In each case, the onset of the vortex-shedding and the periodic flowfield can be observed after a relatively short transient flow. Figure 4 compares the Strouhal number with other experimental and numerical results given in Refs. [9]-[11]. The present solution yields a somewhat high Strouhal number for all the range of the Reynolds numbers. This behavior was also found by Gresho *et al.* [10]. Figure 5 shows the instantaneous streamlines for Reynolds numbers 100, 200, 550, and 1000 at the instant which corresponds to the maximal lift coefficient.

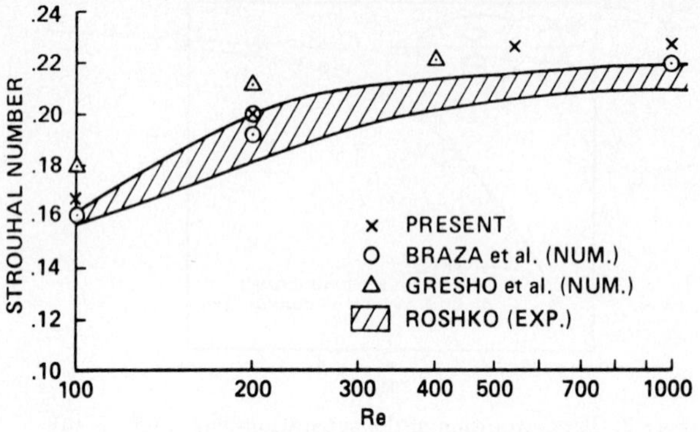

Figure 4: Dependence of the Strouhal number on the Reynolds number (circular cylinder)

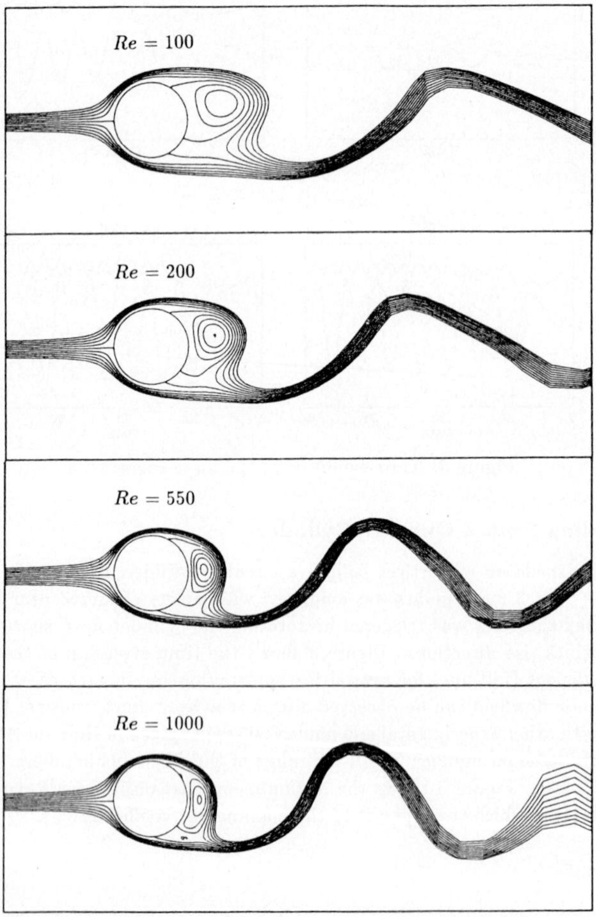

Figure 5: Effect of the Reynolds number on the instantaneous streamlines

5 Concluding Remarks

A time-accurate method for the solution of the incompressible Navier-Stokes equations is developed based on a fractional-step method. The dependent variables are the pressure and the volume-fluxes across the computational cells. The governing equations are discretized by the finite-volume method with special care to satisfy the geometric conservation laws. The method is second-order-accurate both spatially and temporarily.

The computed results compare favorably with other numerical and experimental studies. In the present paper some of the numerical results for the flow in a lid-driven square cavity and over a circular cylinder are presented.

References

[1] Kwak, D., Chang, J. L. C. , Rogers, S. E. and Rosenfeld, M., "Three-Dimensional Incompressible Navier-Stokes Computations of Internal Flows," Proc. of the 12th IMACS World Congress on Scientific Computation, Paris, France, July 18-22, 1988 (also NASA TM 100076, 1988).

[2] Rosenfeld, M., Kwak, D. and Vinokur, M., "A Solution Method for the Unsteady and Incompressible Navier-Stokes Equations in Generalized Coordinate Systems," AIAA-88-0718, 1988.

[3] Rosenfeld, M., Kwak, D. and Vinokur, M., "Development of an Accurate Solution Method for the Unsteady Incompressible Navier-Stokes Equations in Generalized Coordinate Systems," NASA TM, in process, 1988.

[4] Van Doormaal, J. P. and Raithby, G. D., "Enhancements of the SIMPLE Method for Predicting Incompressible Fluid Flows," *Num. Heat Transfer*, **7**, 1984, pp. 147-163.

[5] Ghia, U., Ghia, K. N. and Shin, C. T., "High-Re Solutions for Incompressible Flow Using the Navier-Stokes Equations and a Multigrid Method," *J. Comp. Phys.*, **48**, 1982, pp. 387-411.

[6] Bruneau, C. H. and Jouron, C., "An Efficient Scheme for Solving Steady Incompressible Navier-Stokes Equations," Université de PARIS-SUD MATHÉMATIQUES Prépublications 88-22, 1988.

[7] Collins, W. M. and Dennis, S. C. R., "Flow Past an Impulsively Started Circular Cylinder," *J. Fluid Mech.*, **60**, 1973, pp. 105-127.

[8] Coutanceau, M. and Bouard, R., "Experimental Determination of the Main Features of the Viscous Flow in the Wake of a Circular Cylinder in Uniform Translation, Part 2. Unsteady Flow," *J. Fluid Mech.*, **79**, 1977, pp. 257-272.

[9] Braza, M., Chassaing, P. and Ha Minh, H., "Numerical Study and Physical Analysis of the Pressure and Velocity Fields in the Near Wake of a Circular Cylinder," *J. Fluid Mech.*, **165**, 1986, pp. 79-130.

[10] Gresho, P. M., Chan, S. T., Lee, L. and Upson, C. D., "A Modified Finite Element Method for Solving the Time-Dependent, Incompressible Navier-Stokes Equations. Part 2: Applications," *Int. J. Num. Meth. Fluids*, **4**, 1984, pp. 619-640.

[11] Roshko, A., "On the Development of Turbulent Wakes from Vortex Streets," NACA TN 2913, 1953.

ACCURACY OF THE MARCHING METHOD FOR PARABOLIZED NAVIER-STOKES EQUATIONS

V.V. Rusanov, O.N. Belova, V.A. Karlin
Keldish Institute of Applied Mathematics, Moscow, USSR

Space-marching method for solving the parabolized Navier-Stokes equations (PNS) is widely used in the steady viscous gas flow problems [1]-[6]. Its popularity has grown despite the incorrectness of the resulting mixed problem (with initial and boundary conditions) for steady PNS. Difficulties due to incorrectness of the mixed problem appear to be paid off by computer time savings, and accuracy proves quite plausible for specific applications. In order to expand the field of applications of the space-marching method a detailed analysis of solution accuracies is required. The accuracy dependence on the regularization methods used should be also studied. In this paper the results of investigating the accuracy of viscous flow computations in the mixed steady PNS problem are presented.

The mixed steady PNS problem corresponding to a supersonic 2D flow problem is formulated as follows:

$$G_x \frac{\partial U}{\partial x} + G_y \frac{\partial U}{\partial y} - G_{yy} \frac{\partial^2 U}{\partial y^2} = F(U, \frac{\partial U}{\partial y}), \quad x \in [X_o, X_\infty], \quad y \in [0,1], \tag{1}$$

$$U(X_o, y) = U^{(o)}(y), \qquad U(x,1) = U_\infty, \qquad aU(x,0) = aU_B, \tag{2}$$

where dependent variables $U = (\rho, v_x, v_y, T)$ are the density, the velocity components and the temperature; $a = diag(0,1,1,1)$; U_∞, U_B and $U^{(o)}$ are values of U at $y = 1$, $y = 0$ and $x = X_o$, respectively. Matrices of the system (1) are:

$$G_x = \begin{bmatrix} v_x & \rho & 0 & 0 \\ \frac{T}{\gamma \rho M^2} & v_x & 0 & \frac{1}{\gamma M^2} \\ 0 & 0 & v_x & 0 \\ 0 & (\gamma-1)T & 0 & v_x \end{bmatrix}, \qquad G_y = \begin{bmatrix} v_y & 0 & 0 & 0 \\ 0 & v_y & 0 & 0 \\ \frac{T}{\gamma \rho M^2} & 0 & v_y & \frac{1}{\gamma M^2} \\ 0 & 0 & (\gamma-1)T & v_y \end{bmatrix},$$

$G_{yy} = (\rho R)^{-1} diag(0, \mu, 4\mu/3, \mu\gamma/P)$, μ - viscosity coefficient, γ - the ratio of the specific heats and M, R, P - Mach, Reynolds, Prandtl numbers respectively.

The statement of the problem (1)-(2) does not take into account the upstream propagation of perturbations, and its regularization distorts the downstream propagation of perturbations. Computational experience shows that this distortions of viscous flow physical properties may deteriorate the quality of numerical solutions of (1)-(2) in the vicinity of initial data, especially if these are obtained by method non-consistent with space-marching. To estimate the influence of these distortions on the accuracy of the numerical solution the linearized version of (1)-(2) was thoroughly investigated.

Let $\mathcal{K}$ be a set of initial conditions $U^{(o)}(y)$ for which the solution of problem (1)-(2) exists. It is assumed that for every $U^{(o)}(y) \in \mathcal{K}$ and $\bar{U}^{(o)}(y) \in \mathcal{K}$ the corresponding solutions of (1)-(2) $U_o(x,y)$ and $\bar{U}_o(x,y)$ are tending to each other when $\bar{U}^{(o)}(y) \to U^{(o)}(y)$. If $\bar{U}^{(o)}(y)$ is sufficiently close to $U^{(o)}(y)$ then the $u_o(x,y) = \bar{U}_o(x,y) - U_o(x,y)$ with good accuracy is the solution of the problem obtained by linearization of (1)-(2) for $U(x,y) = U_o(x,y)$:

$$A_x \frac{\partial u}{\partial x} + A_y \frac{\partial u}{\partial y} - A_{yy} \frac{\partial^2 u}{\partial y^2} = B_y \frac{\partial u}{\partial y} + Bu, \tag{3}$$

$$u(X_o,y) = u^{(o)}(y), \tag{4}$$

$$u(x,1) = 0, \; au(x,0) = 0, \tag{5}$$

where $A_\alpha = G_\alpha(U_o(x,y))$, $\alpha \in \{x,y\}$ and $A_{yy} = G_{yy}(U_o(x,y))$.

If $U_o(x,y)$ changes insignificantly when $x \in [X_o, X_\infty]$, we can ignore the x dependence of the coefficients in (3) and search for the solution in the form $u(x,y) = c(x)w(y)$. Then $w(y)$ must be the solution of the eigenvalue problem:

$$\sigma\mathcal{A}w = \mathcal{B}w, \quad aw(0) = 0, \quad w(1) = 0, \tag{6}$$

and $c(x)$ the solution of the initial value problem:

$$\frac{dc}{dx} + \sigma c = 0, \; x \in [X_o, X_\infty], \; c(X_o) = c_\sigma^{(o)}. \tag{7}$$

Here $\mathcal{A} = A_x$, $\mathcal{B} = (A_y - B_y)\frac{d}{dy} - A_{yy}\frac{d^2}{dy^2} - B$, and $c_\sigma^{(o)}$ is the coefficient at $w_\sigma(y)$ in the $u^{(o)}(y)$ eigenfunction expansion.

The solutions of (6) depend on the $U_o(x,y)$ function. We investigated (6) for $U_o(x,y)$ which is in qualitative agreement with Blasius's solution for the 2D plan flow about a semi-infinite thin plate: $U_\infty=(1,1,1,1)$, $U_B=(0,0,0,1)$. In fact, we choose for $U_o(x,y) = (\rho_o, v_{x,o}, v_{y,o}, T_o)$:

$$v_{x,o}(y) = (1 - exp(-y/\varepsilon))/(1 - exp(-1/\varepsilon)), \; \varepsilon = C_\varepsilon/\sqrt{R},$$
$$\rho_o = T_o = 1, \quad v_{y,o}(y) = \varepsilon v_{x,o}(y). \tag{8}$$

Taking into account a small deviation of flow along the normal to the plate we exclude the y-momentum equation from (3)-(5).

Under conditions (8) the problem (6) was approximated with symmetric finite differences on the 81 point grid and solved by the QR - iteration method. The typical localization of its spectrum is showen in Fig.1 at M = 3 and R = 1000. There are four grid points, in subsonic layer of the flow.

Investigation of problems of the type (6) and their discrete analogs has shown that the spectrum is situated both in the right and left half-planes of the complex variable. Incorrectness of the problem (3)-(5) can be seen in a nonphysical growth of harmonics corresponding to σ with $\mathcal{Re}(\sigma) < 0$, which follows from (7). The incorrectness can be eliminated by transferring some conditions (4) from the boundary $x = X_0$ to $x = X_\infty$. However, it deprives us of the possibility to use the space-marching method of solution for the flow problem.

The other method consists in constructing the finite difference scheme for (3)-(5) or (1)-(2) such that for growing x it provides damping of the harmonics corresponding to $\mathcal{Re}(\sigma) < 0$.

Let us consider the four-point in x difference scheme for (3)-(5) (its extension on (1)-(2) is obvious):

$$\begin{aligned}
&[\beta_1 (A_x)_{m+1,l} + \beta_2 (A_x)_{m,l} + \beta_3 (A_x)_{m-1,l} + (1 - \beta_1 - \beta_2 - \beta_3)(A_x)_{m-2,l}] \cdot \\
&\cdot [\alpha_1 \delta_x u_{m,l} + \alpha_2 \delta_x u_{m-1,l} + (1 - \alpha_1 - \alpha_2)\delta_x u_{m-2,l}] = \\
&= \beta_1 [(B_y - A_y)\delta_y u + A_{yy}\delta_{yy} u + Bu]_{m+1,l} + \\
&+ \beta_2 [(B_y - A_y)\delta_y u + A_{yy}\delta_{yy} u + Bu]_{m,l} + \\
&+ \beta_3 [(B_y - A_y)\delta_y u + A_{yy}\delta_{yy} u + Bu]_{m-1,l} + \\
&+ (1-\beta_1-\beta_2-\beta_3)[(B_y - A_y)\delta_y u + A_{yy}\delta_{yy} u + Bu]_{m-2,l} ,
\end{aligned} \tag{9}$$

where

$$\delta_x u_{m,l} = h_x^{-1}(u_{m+1,l} - u_{m,l}), \qquad \delta_y u_{m,l} = (2h_y)^{-1}(u_{m,l+1} - u_{m,l-1}),$$

$$\delta_{yy} u_{m,l} = h_y^{-2}(u_{m,l+1} - 2u_{m,l} + u_{m,l-1}).$$

If in (9): $\alpha_1 = 1$, $\alpha_2 = 0$, $\beta_2 = 1-\beta_1$, $\beta_3 = 0$, then the scheme is two-point in x and has the first order of approximation. Let $\{\sigma_{n,h}\}$ be a spectrum of finite difference analog of (6) due to (9), and $\{\sigma^-_{n,h}\}$ be a subset of $\{\sigma_{n,h}\}$ in the left complex half-plane. Then necessary stability conditions for two-point case of (9) are:

$$\beta_1 > 1/2 \quad \text{and} \quad h_x > 2/[(2\beta_1 - 1)\min_n\{|\sigma^-_{n,h}|\}], \tag{10}$$

In our investigations $\sigma^-_{n,h}$ with nonzero imaginary part was not found. So, we consider real $\sigma^-_{n,h}$ only.

Under limitations $\alpha_2 = 1-\alpha_1$, $\beta_2 = 1/2 + \alpha_1 - 2\beta_1$, $\beta_3 = 1-\beta_1-\beta_2$ the scheme (9) is three-point in x and has the second order of approximation at $x = x_m + \xi h_x$, where $\xi = \alpha_1 - 1/2$. Its necessary stability conditions are:

$$\begin{aligned}
&\alpha_1 > 1/2, \quad \beta_1 > \alpha_1/2, \\
&h_x > (2\alpha_1 - 1)/[(2\beta_1 - \alpha_1)\min_n\{|\sigma^-_{n,h}|\}] \quad \text{for } \tfrac{1}{2}\alpha_1 < \beta_1 < \tfrac{1}{4}(4\alpha_1^2 - 2\alpha_1 + 1) \\
&h_x > 2/[(2\alpha_1 - 1)\min_n\{|\sigma^-_{n,h}|\}] \quad \text{for } \beta_1 > \tfrac{1}{4}(4\alpha_1^2 - 2\alpha_1 + 1),
\end{aligned} \tag{11}$$

For each α_1 in the (σ_x, β_1) plane, where $\sigma_x = \sigma h_x$, they define a region. An example of such a region at $\alpha_1 = 3/2$ is shown in Fig. 2.

Finally, if

$$\alpha_2 = 1/2 - 2\alpha_1 + \sqrt{2\alpha_1 + 1/3}\,, \qquad \beta_3 = 2/3 - 2\alpha_1 + 3\beta_1\,, \tag{12}$$
$$\beta_2 = 1/6 + \alpha_1 - 3\beta_1 + (1/2)\sqrt{2\alpha_1 + 1/3}\,,$$

then the scheme (9) is effectively four-point and at $x_m + \xi h_x$, where $\xi = \sqrt{2\alpha_1 + 1/3} - 1$, has the third order of approximation in x. Stability regions in the (σ_x, β_1) plane for (9),(12) are shown at $\alpha_1 = 3$ and 5 in Figs. 3 and 4 respectively.

Stability conditions for the schemes mentioned above may be expressed as

$$h_x > \Omega(\alpha_1, \beta_1)/\min_n\{|\sigma^-_{n,h}|\}, \qquad (\alpha_1, \beta_1) \in \mathbb{D}, \tag{13}$$

where the region $\mathbb{D}$ and function $\Omega(\alpha_1, \beta_1)$ on $\mathbb{D}$ depend on the order of approximation. Most significant part of $\mathbb{D}$ is shown in Fig.5 for scheme (9),(12). For other cases these regions and functions are obvious due to (10) and (11). Although (9) has the second order of approximation in y, the step size h_y, as opposed to h_x, may be chosen as small as desired. Thus, accuracy of (9) is defined by $\sigma_{o,h} = \min_n\{|\sigma^-_{n,h}|\}$ only.

For parametric investigation we simplified the model (8) once again. Specifically, interval [0,1] was divided onto two subintervals $[0,\delta)$ and $(\delta,1]$, where $\delta = -\varepsilon\, \ln\{1-[1-exp(-1/\varepsilon)]/M\}$ is the value of y at which the velocity of main flow U_o is equal to the sound speed. Values of $v_{x,o}, v'_{x,o}, v''_{x,o}$ and $v_{y,o}$, $v'_{y,o}$ were averaged on these subintervals:

$$v_\delta = \langle v_{x,o}\rangle_{[o,\delta)} = \delta^{-1}\int_o^\delta v_{x,o}(y)dy\,, \qquad v_\Delta = \langle v_{x,o}\rangle_{(\delta,1]} \tag{14}$$

and were used instead of $v_{x,o}$, $v_{y,o}$ and their derivatives depending on the fact whether $y \in [0,\delta)$ or $(\delta,1]$.

The eigenvalue problem (6) for (8),(14) has a charakteristic equation in elementary functions. The roots of this equation may be localized. Appropriate graphics of $\sigma_o = \min_n\{|\sigma^-_n|\}$ versus the Reynolds number R are shown in Fig.6 at Mach number 2 and 3 and in Fig.7 at M=6. Also are shown values of $\sigma_{o,h}$ computed with the QR -algorithm for finite difference approximation of (6),(8). Graphics show that at sufficiently large Reynolds numbers values of σ_o are independent of R. Besides, the results for models (8),(14) and (8) are in agreement with other ones not only qualitatively but quantitatively too at moderate Mach numbers.

In Figs.6 and 7 we may see an enormous growth of $\sigma_{o,h}$ as R increases, which is due to the reduction of the subsonic layer thikness. Consequently, the grid points exit from the subsonic layer. The exit of each grid point from the subsonic layer (with increase of

R) is associated with exit of maximum module eigenvalue from the left complex halfplane through the $-\infty$. The exit of the last inside grid point from $[0,\delta)$ results in an enormous growth of $\sigma_{o,h}$ as R increases.

The calculations were carried out at $\gamma = 1.4$, $P = 0.72$ and $C_\varepsilon = 3.354$. This value of C_ε was obtained as the closest approach of $v_{x,o}(y)$ from formulae (8) to the velocity profile, which was calculated from (1)-(2) at $M = 3$, $R = 1000$ and $x = 1$.

Test calculations with (1)-(2) and (3)-(5) confirmed estimates of σ_o and stability conditions (10), (11), (13). The last is not trivial as (6) seems to have an incomplete system of eigenfunctions. Calculations were carried out on the 101 point grid without any condensation. According to [7], a violation of stability condition caused a primary increase of eigenfunction corresponding to σ_o and not having an oscillating form.

REFERENCES

1. Lin T.C., Rubin S.G. - Computers and Fluids, 1973, v.1, pp.37-57.
2. Lubard S.G., Helliwell W.S. - AIAA J., 1974, v.12, No.7, pp.965-974.
3. Schiff L.B., Steger J.L. - AIAA Paper 79-0130, January 1979.
4. Vigneron Y.C., Rakich J.V., Tannehill J.C. - AIAA Paper 78-1137, July 1978.
5. Belova O.N., Kokoshinskaya N.S., Paskonov V.M. In: Library of routines in aerohydrodynamics (methods and algorithms). - Moscow States University, 1984, pp.24-37 (in Russian).
6. Karlin V.A. - Preprint, Keldysh Inst. Appl. Math., USSR Ac. Sci., 1986, N. 201 (in Russian).
7. Belova O.N., Karlin V.A. - Preprint, Keldysh Inst. Appl. Math., USSR Ac. Sci., 1988, N. 24 (in Russian).

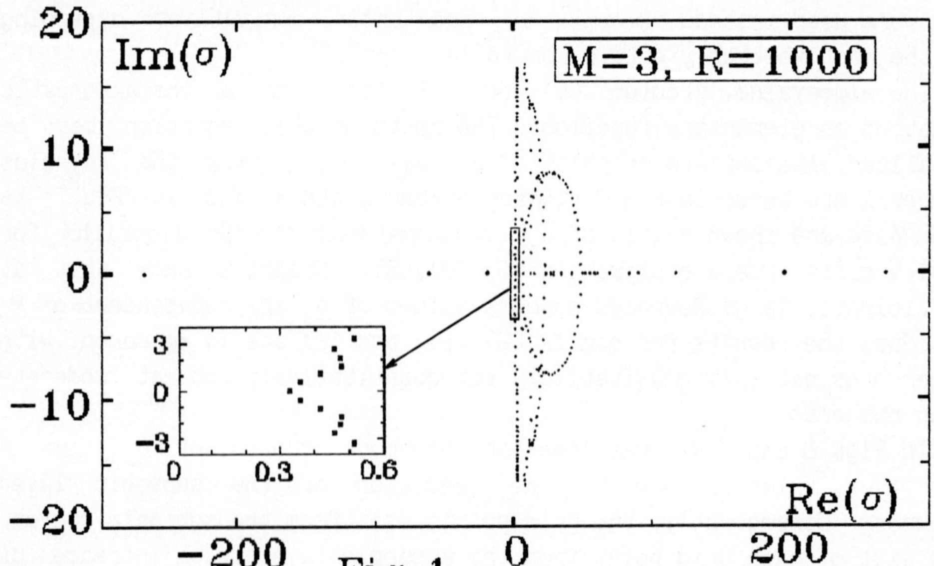

Fig.1

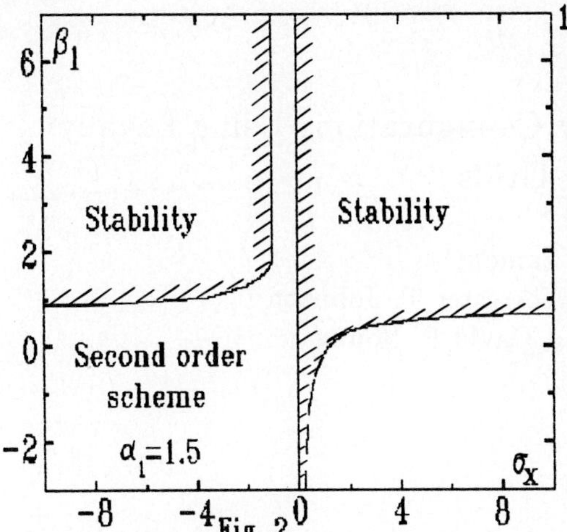

Fig. 2

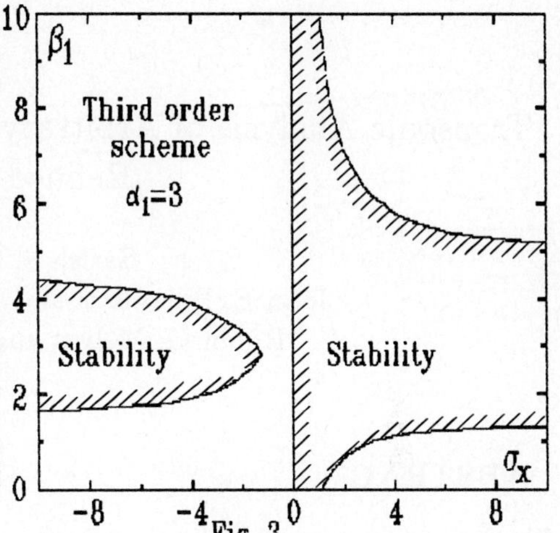

Fig. 3

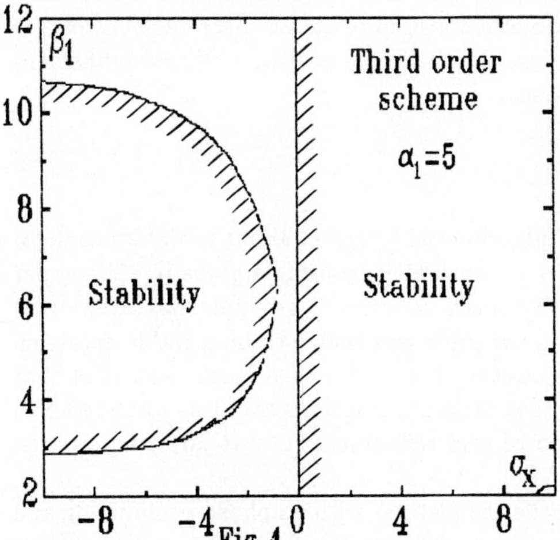

Fig. 4

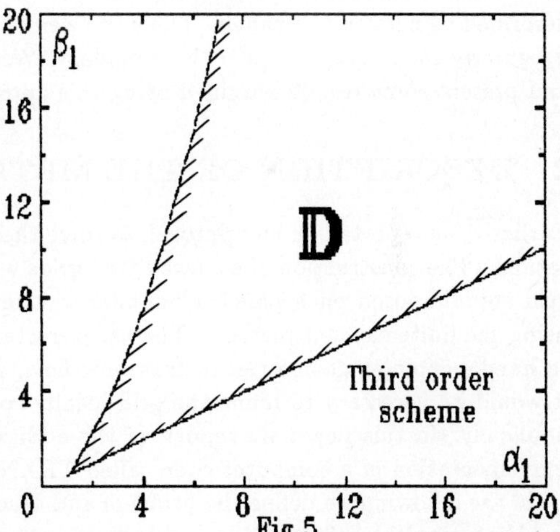

Fig. 5

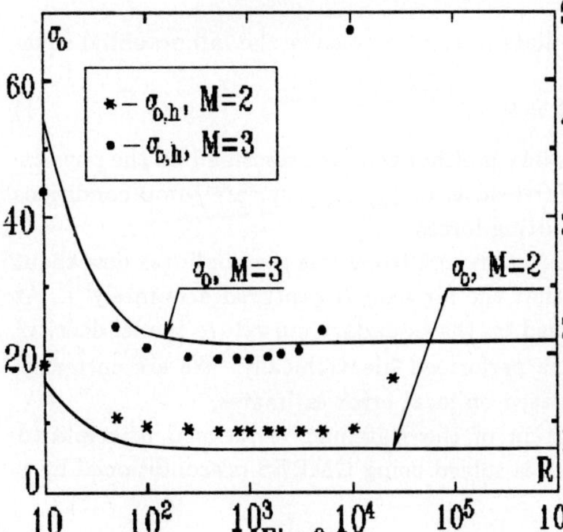

Fig. 6

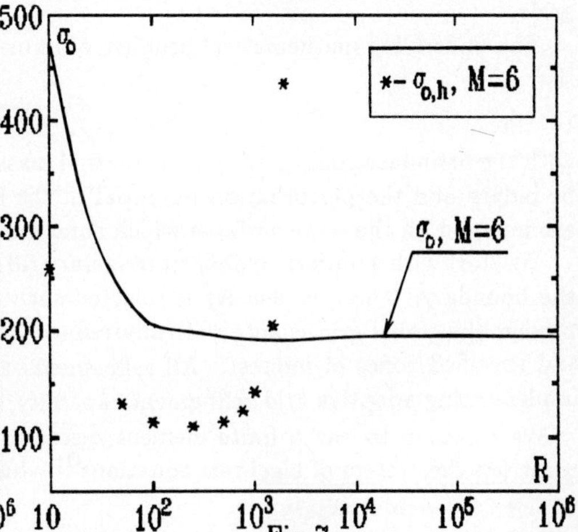

Fig. 7

Transonic Analysis of Arbitrary Configurations Using Locally Refined Grids

Satish S. Samant*
John E. Bussoletti and Forrester T. Johnson †
Robin G. Melvin and David P. Young ‡

1 ABSTRACT

We present an approach to solve the full potential equation about arbitrary configurations. A hierarchical refinement of a globally uniform rectangular grid is superimposed over a boundary described by networks of panels. The finite element method is used to obtain discrete operators for irregularly shaped regions near the boundary. We describe the implementation of grid refinement and present some results obtained using this approach.

2 DESCRIPTION OF THE METHOD

Earlier[1] we reported on an approach to solve the full potential equation about arbitrary configurations. The construction of surface fitted grids was circumvented by using a globally rectangular grid superimposed on a paneled boundary. The Bateman variational principle was discretized using the finite element method. The use of rectangular grids was shown to be a viable approach to handle complex geometries in transonic flow. However, it was also clear from the outset that it would be necessary to refine the grid locally so that accurate solutions could be obtained economically. In this paper we report on the addition of grid refinement to that approach and its implementation in a computer code called TRANAIR.

In the following we define the problem and describe the method with emphasis on building and implementing the refined grids and in Section 3 we show some results obtained by using refined grids.

The underlying mathematical problem of immediate interest is to solve the full potential equation

$$\vec{\nabla} \cdot \rho \vec{\nabla} \Phi = 0 \tag{1}$$

with the boundary condition that the normal mass flux is either zero or prescribed on the physical boundary and the perturbation potential in the far field is $\mathcal{O}(\frac{1}{r})$. Appropriate jump conditions are imposed on the wake surfaces which simulate lifting forces.

We start with a uniform global rectangular grid large enough to enclose the nonlinear flow about the boundary. The grid density is selected such that the far field is captured accurately[1]. At present the global grid is automatically refined based on the boundary curvature (panel density) and specified zones of interest. All refinements are performed hierarchically. We are currently implementing adaptive grid refinement capabilty based on local error estimates.

We continue to use a finite element discretization of the Bateman variational principle to construct the system of algebraic equations[1] which is solved using GMRES preconditioned by a

*Boeing Commercial Airplanes
†Boeing Advanced Systems
‡Boeing Computer Services

direct sparse matrix solver[2] and an exterior Poisson solver[1].

We have developed a compact data structure which contains essentially all the information regarding the refined grid. Assymtotically this data structure requires storage equal to 10/7 the number of unrefined boxes; however, the factor is closer to two in practice. It allows efficient extraction of a variety of information, such as the location of nodes and element centroids, box size, box level, node indices, box adjacency and identity of boundary boxes.

We use a modification of an oct-tree type data structure described by Samet[3]. At the beginning of the oct-tree data structure we store certain overhead information which describes the size of the grid. This is followed by two blocks of data, each as long as the size of the global grid, describing its refinement. Then follow the branch data elements of the tree. These describe the hierarchical refinement and form the bulk of the oct-tree.

A typical branch data element in an oct-tree is illustrated in Figure 1. The first word in a branch data element points to the father box, the next eight words point to the refinement branches of the sons if any, and the last word contains an accumulation index specifying the number of nodes encountered up to that point in the oct-tree.

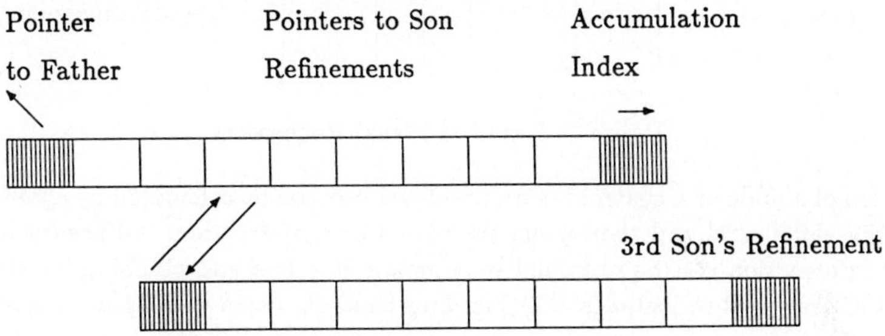

Figure 1: Oct-tree Data Structure Element

A null pointer in the son entry in any branch data element represents an unrefined box. Any unrefined box cut by a boundary is called a T-box. A negative number is placed in the son entry in a branch data element to give the index of that T-box in a T-box list. This is a convenient and compact way of accounting for the presence of the boundary.

We have extended the oct-tree data structure further to accomodate nodal information. We assign the index of a box to the node at its lower-left-near corner. In order to account for all nodes at refinement interfaces we perform psuedo refinement (see Figure 2). This allows us to keep track of the nodes as well as the boxes using the same oct-tree at only a modest increase in storage. Boxes added by the psuedo refinement are used only to identify nodes and do not contribute to the finite element formulas.

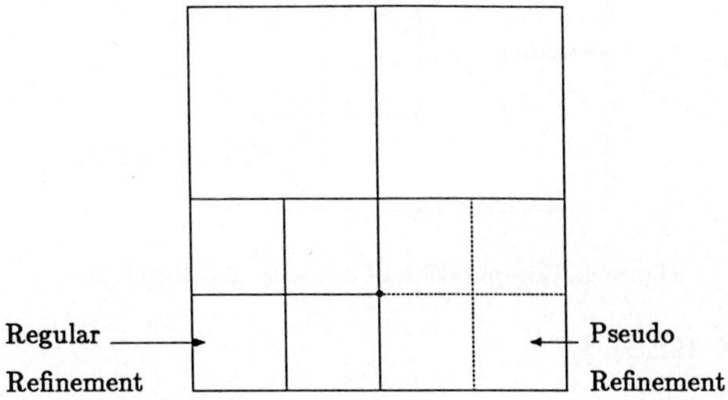

Figure 2: Some Details of the Hierarchical Refinement

Even though the oct-tree data structure described above is able to reflect an arbitrary collection of hierarchically refined grid cells, we have found it convenient to restrict the refinement pattern. We require that no two face or edge neighbours in a "legal" refined grid differ by more than one level (see Figure 3). This rule prevents pathologically large stencils under certain circumstances, but allows refinement down to an arbitrary level within one adjacent coarse grid box.

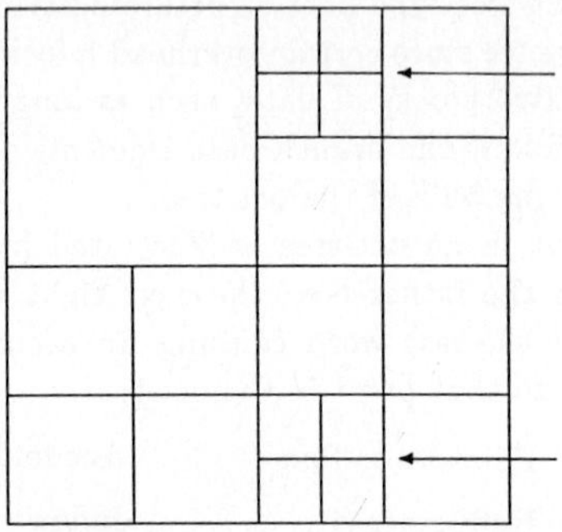

Figure 3: Legal Vs Illegal Refinement

The location of a node or a centroid of an unrefined box can be calculated by climbing up to the ancestor in the global grid and then using the global box information. Adjacency or box-to-box connectivity information can be obtained by starting at a box and climbing up the oct-tree to the root of the branch that includes that box and climbing down a complementary path to the neighbouring box.

We also use another type of data structure which takes advantage of the rectangular nature of the boxes in the refined grid. In calculating the velocity at the centroids of the boxes or the residuals at the nodes we take advantage of the fact that all boxes have eight nodes and only eight boxes contribute to the residual at each node (see Figure 4). The operations suggested here, throwing contributions from a box to the nodes (residual evaluation) or collecting contributions from the nodes to the box (velocity computation), is used rather extensively in the code.

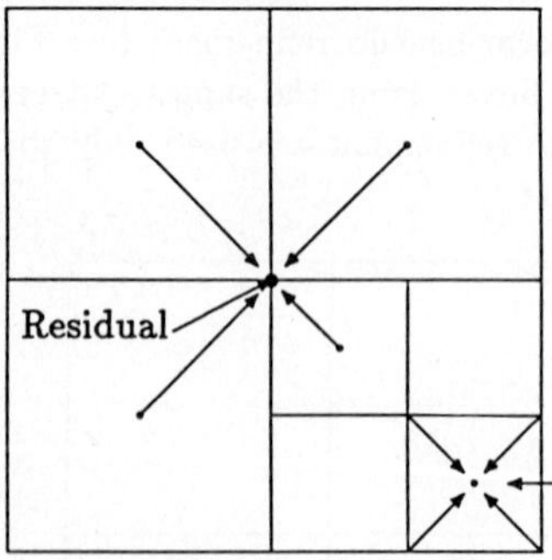

Figure 4: Computation of Velocity and Residual

3 SAMPLE RESULTS

Our first example illustrates results for the ONERA wing in incompressible flow. Here we solve the same problem with two different grids. The first grid is a global grid with about 35,000 boxes.

The second is a locally refined grid with approximately the same number of boxes. The results of the calculations are shown in Figure 5. The locally refined grids produce much more accurate results.

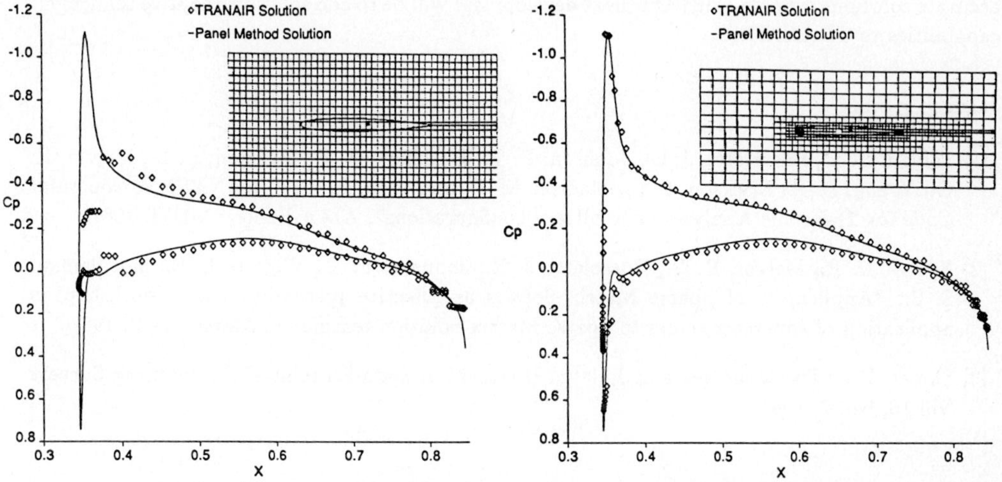

Figure 5: Subcritical ONERA M6 Wing Results Using Refined Grids

Next we show results for an F16 at Mach 0.9 at an angle of attack of 4.0 degrees. This example illustrates the ability of TRANAIR to handle complex geometries. The TRANAIR results are obtained on a grid with about 125,000 boxes. The results of the calculations are shown in Figure 6. The comparison between the computation and test data is quite good.

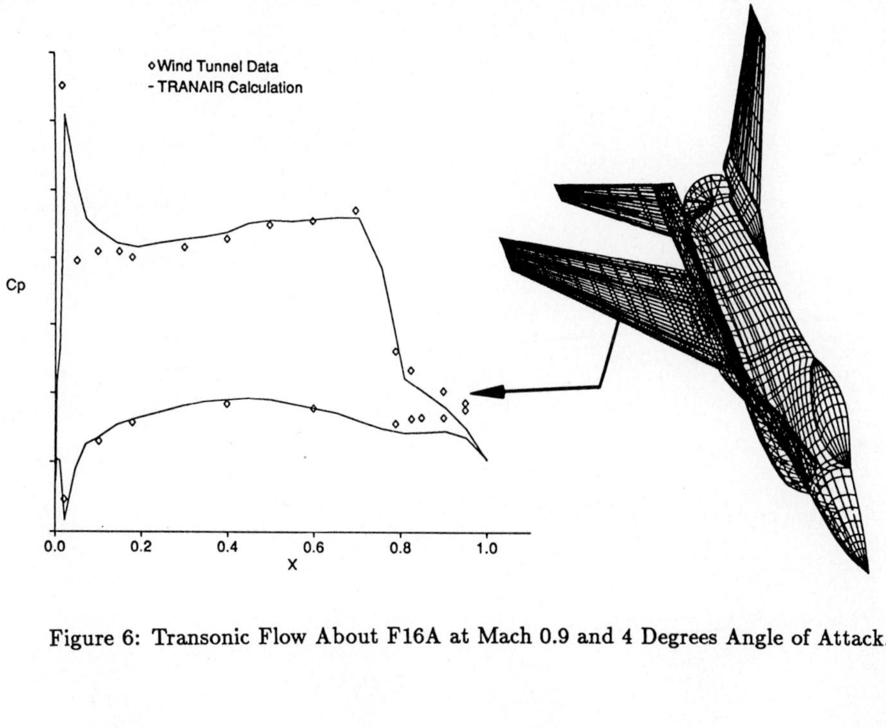

Figure 6: Transonic Flow About F16A at Mach 0.9 and 4 Degrees Angle of Attack.

4 SUMMARY

In this paper we have described a data structure to handle hierarchically refined grids. The data structure is compact and efficient to interrogate. The local refinement allows calculation of accurate solutions economically. Our next development will be to add solution adaptive refinement capabilities to TRANAIR.

References

[1] Samant S. S., Bussoletti, J. E., Johnson, F. T., Burkhart, R. H., Everson, B. L., Melvin, R. G., Young, D. P., Erickson, L. L., Madson, M. D. and Woo, A. C., "TRANAIR: A Computer Code for Transonic Analyses of Arbitrary configurations", AIAA Paper 87-0034, 1987.

[2] Young, D. P., Melvin, R. G., Bussoletti, J. E., Johnson, F. T., Wigton, L. B., and Samant S. S., "Application of Sparse Matrix Solvers as Effective preconditioners", Workshop on application of supercomputers to Sparse Matrix Solution techniques, March 27-30, 1988.

[3] Samet, H., "The Quadtree and Related Hierarchical Data-Structures", Computing Surveys Vol 16, No. 2, 1984.

GROUP EXPLICIT METHODS FOR SOLVING COMPRESSIBLE FLOW EQUATIONS ON VECTOR AND PARALLEL COMPUTERS

Nobuyuki Satofuka, Koji Morinishi and Ruriko Hashino
Department of Mechanical Engineering
Kyoto Institute of Technology
Matsugasaki, Sakyo-ku, Kyoto 606, Japan

1 Introduction

During the past decade installation of vector supercomputers such as Cray 1 has brought about a dramatic increase in computing power. Further, it is anticipated that all future supercomputers will be a combination of vector and parallel architectures. In order to exploit fully the potential of these machines, algorithms should adapt to these aichitectural features. Those which cannot be efficiently used on these architectures will be of little utility. Most straight forward technique of achieving vectorization and parallelization is by use of explicit methods. Unfortunately the classical explicit methods suffer from severe stability restriction and poor convergence characteristics, and therefore require unacceptable computing time.

Recently, Evans has proposed new explicit methods for the finite difference solution of parabolic equations [1][2]. The central idea of the new methods is the use of asymmetric finite difference to groups of 2 adjacent points (4(8) points for 2(3) dimensions) successively on the grid which results in a set of implicit equations which can be easily converted to explicit form. The methods called Group Explicit (G.E.) method have many advantages especially for use on parallel computers and have been applied to a class of scalar parabolic equations.

In this paper, we try to extend the concept of G.E. method to the solution of fluid dynamic equations and implement the method on VP-200 (vector) and NCUBE-4 (parallel) computers.

2 Concept of Group Explicit Methods

The basic concept of the present G.E. methods can be easily understood when they are applied to the following one-dimensional model equation,

$$\frac{\partial u}{\partial t} = -\sigma\frac{\partial u}{\partial x} + \nu\frac{\partial^2 u}{\partial x^2} \tag{1}$$

In the Group Explicit method [1] asymmetric finite difference approximations to the Eq.(1) at two adjacent points are written as

$$(1+\gamma)u_i^{n+1} + (\lambda-\gamma)u_{i+1}^{n+1} = (\lambda+\gamma)u_{i-1}^n + (1-\gamma)u_i^n \tag{2.a}$$

$$-(\lambda+\gamma)u_i^{n+1} + (1+\gamma)u_{i+1}^{n+1} = (1-\gamma)u_{i+1}^n - (\lambda-\gamma)u_{i+2}^n \tag{2.b}$$

where $\lambda = \sigma\frac{\Delta t}{\Delta x}$, $\gamma = \nu\frac{\Delta t}{\Delta x^2}$

Rearranging the Eq.(2), we obtain the following simultaneous equations in matrix form,

$$\begin{bmatrix} 1+\gamma & \lambda-\gamma \\ -(\lambda+\gamma) & 1+\gamma \end{bmatrix} \begin{bmatrix} u_i^{n+1} \\ u_{i+1}^{n+1} \end{bmatrix} = \begin{bmatrix} (\lambda+\gamma)u_{i-1}^n + (1-\gamma)u_i^n \\ (1-\gamma)u_{i+1}^n - (\lambda-\gamma)u_{i+2}^n \end{bmatrix} \tag{3}$$

The system of equations (3) can be easily converted into explicit form as:

$$\begin{bmatrix} u_i^{n+1} \\ u_{i+1}^{n+1} \end{bmatrix} = \frac{1}{|A|} \begin{bmatrix} 1+\gamma & -(\lambda-\gamma) \\ \lambda+\gamma & 1+\gamma \end{bmatrix} \begin{bmatrix} (\lambda+\gamma)u_{i-1}^n + (1-\gamma)u_i^n \\ (1-\gamma)u_{i+1}^n - (\lambda-\gamma)u_{i+2}^n \end{bmatrix} \tag{4}$$

where $|A| = 1+2\gamma+\lambda^2$

If we use an odd number of grid points (an even number of intervals) in computational domain, an ungrouped (single) point appears at adjacent to either right or left boundary (Fig.1). For those ungrouped points, we can use Eq.(2.a) or (2,b) with known values of u at boundaries. We adopt the strategy that the ungrouped point is alternately placed on the right or left boundary at each time step as shown in Fig.2.

Extension to multi-dimensional as well as system of equations is straightforward. In the case of two-dimensional Navier-Stokes equations, the resulting system of equations yields 4×4 block matrix, which can be inverted directly by LU decomposition or by using approximate factorization. For steady flows a matrix diagonalization technique can be incorporated for saving the computational work.

3 Numerical Results

The first example of the application to the Navier-Stokes equations, deals with a quasi-one-dimensional nozzle flow problem. In Fig.3, the convergence histories of the L_2 -norm of residuals are compared with that of the Rational-Runge-Kutta (RRK) method [3]. The calculation with a uniform 65 grid points was carried out for the Reynolds number Re=500. The computed pressure distribution is plotted in Fig.4 and compared with the ideal inviscid solution.

The next example for the Navier-Stokes equations shows the results of the Couette flow problem with $u(w)$=0.2 and Re=50. A 9×9 uniform grid with $dx=dy$=0.125 is used. At inlet boundary, three out of four flow variables are fixed whereas one is fixed at the exit. The rest of variables are exterapolated from those of the interior points. The convergence histories of L_2– norm of residuals for the present, RRK and Beam-Warming [4] methods are plotted in Fig.5. The convergence rate of the present methods is comparable to that of Beam-Warming method and much faster than that of RRK.

As an example of implementation of the present method on vector computer, Fig.6 shows the pressure distribution for the RAE2822 airfoil at M_∞=0.73, α=2.8° and Re=6.5×10^6. The corresponding experimental data of Cook et. al. [5] as well as those obtained with RRK scheme [6] are also plotted on the figure. The distribution obtained with the present method is in excellent agreement with that of RRK scheme. The computing time required for one time step is 28.5 milliseconds with 193×33 grid points on FACOM VP-200. The ratio of scalar to vector execution time is 54.5.

The final example shows the results for a flow past a 20% thick symmetric circular arc airfoil in a channel computed on a parallel computer, NCUBE-4. Numerical calculations were carried out with 1, 2, and 4 nodes of NCUBE. In the case with 4 nodes, the computational domain of 65×33 grid points is split into 4 subdomains as shown in Fig.7. Figure 8 shows the iso-Mach lines for M_∞=0.5 and Re=100. Speedup factors for using 2 and 4 nodes are 1.91 and 3.51, respectively.

References

[1] Evans, D.J. and Abdullah, A.R.B., Intern. J. Computer Math., Vol.14 (1983) pp.73-105.
[2] Evans, D.J., New Parallel Algorithms for Partial Differential Equations, in: Feilmeier, M., Joubert, G. and Schendel, U.,(eds.), Parallel Computing 83 (North-Holland, 1984) pp.3-56.
[3] Satofuka, N., Proceedings of the Fifth GAMM-Conference on Numerical Methods in Fluid Mechanics, Rome (1983).
[4] Beam, R.M. and Warming, R.F., AIAA J., Vol.16, No.4 (1978) pp.393-402.
[5] Cook, P.H., McDonald, M.A., and Firmin, M.C.P., AGARD-AR-138.
[6] Morinishi, K. and Satofuka, N., AIAA Paper 87-0426.

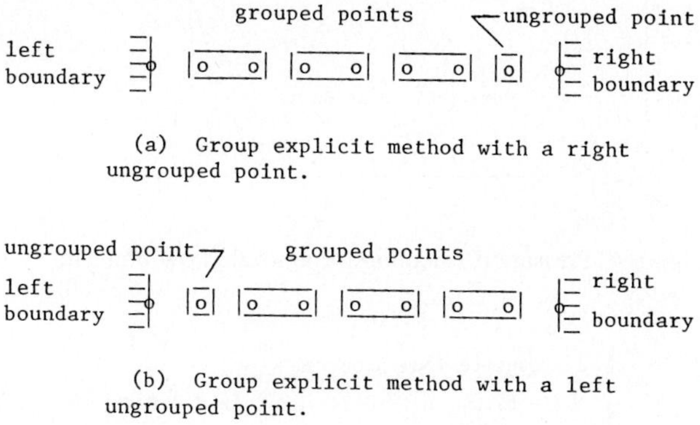

Figure 1 Group explicit methods with a right ungrouped point and a left ungrouped points.

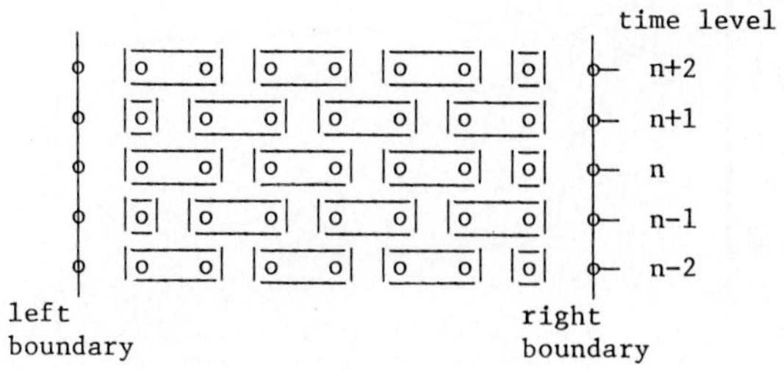

Figure 2 Single alternating group explicit strategy.

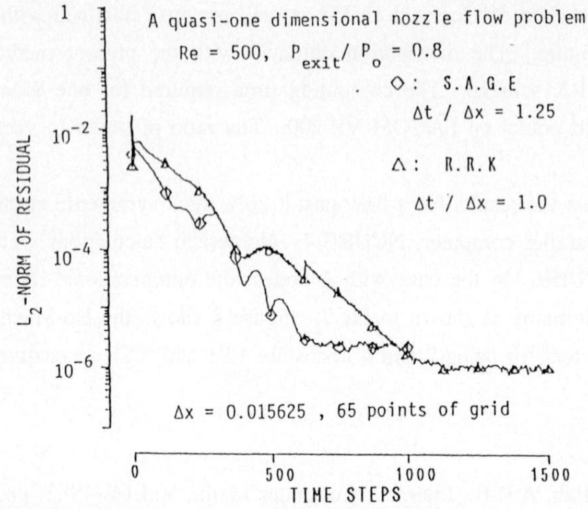

Figure 3 Convergence history for a nozzle flow problem.

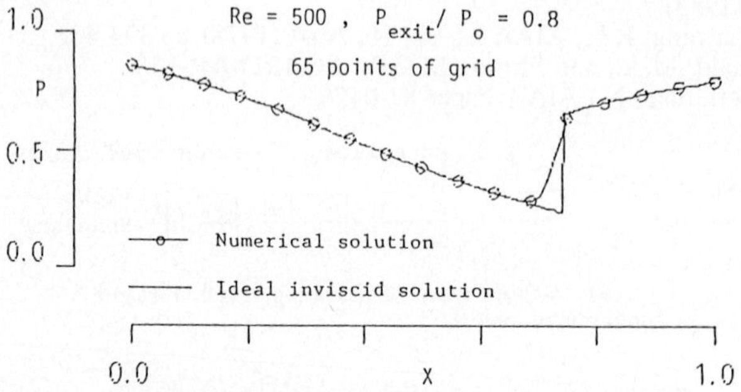

Figure 4 Pressure distribution for a nozzle flow problem.

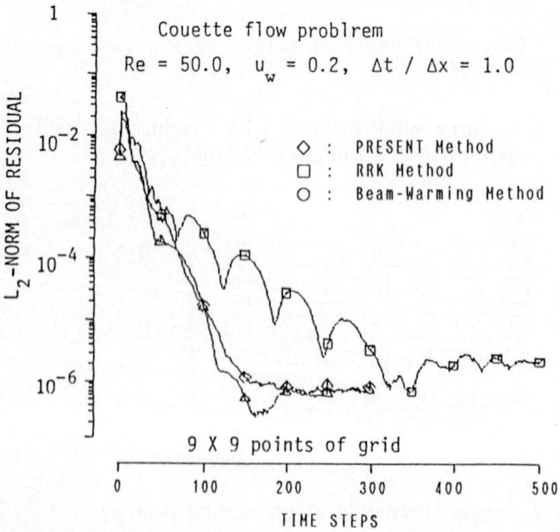

Figure 5 Convergence history for the Couette flow problem.

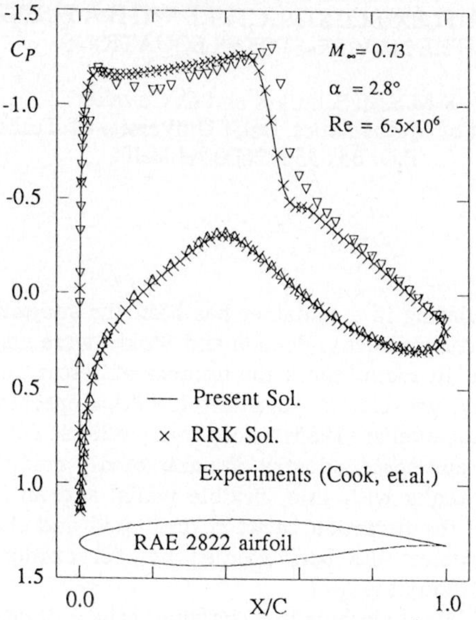

Figure 6 Pressure distribution for RAE2822 airfoil.

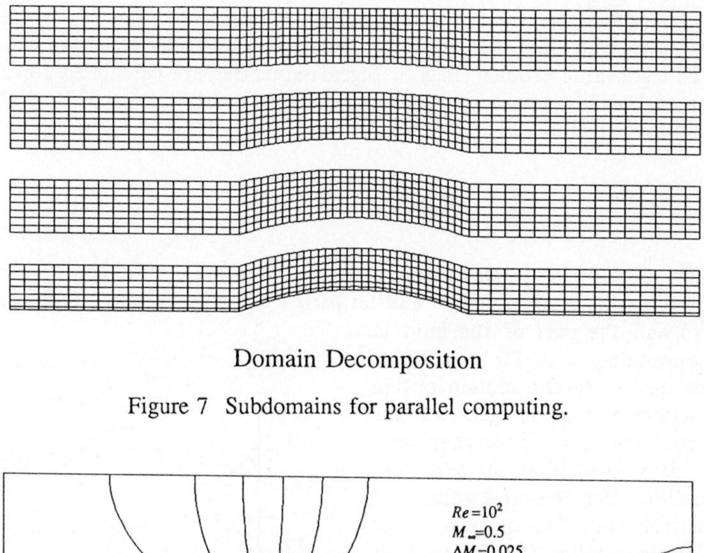

Domain Decomposition

Figure 7 Subdomains for parallel computing.

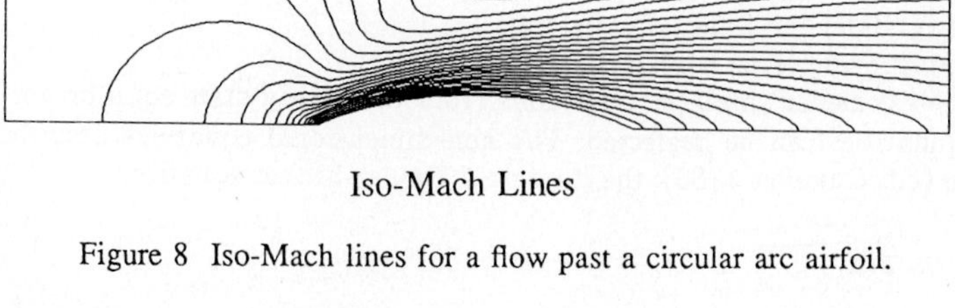

Iso-Mach Lines

Figure 8 Iso-Mach lines for a flow past a circular arc airfoil.

INTERACTIONS OF A FLEXIBLE STRUCTURE WITH A FLUID GOVERNED BY THE NAVIER-STOKES EQUATIONS

R.M.S.M. Schulkes and C. Cuvelier
Department of Mathematics, Delft University of Technology
P.O. Box 356, 2600 AJ Delft

1. Introduction

The problem of a liquid oscillating in a container has been the subject of research over a long period of time. In the last century Cauchy, Poisson and Stokes were among those who's interest was aroused by this problem. In recent times the number of researchers studying liquid slosh problems has increased greatly, we refer to Abramson (1966), Kopachevskii & Myshkis (1967), Veldman & Vogels (1984) and Cuvelier (1985) among many others. The importance of the liquid slosh problem in the aviation and space industry, is undoubtedly credit to this increase. Liquid propellants, sloshing in fuel tanks with thin, flexible walls, add an interesting aspect to the liquid slosh problem namely, the iteraction between the liquid and the flexible container. The fluid-structure interaction problem has been studied by, for example, Berger *et al* (1975), Morand & Ohayon (1979) and Boujot (1987).

The general problem of liquid sloshing in a container can be divided into two main areas - the linearized problem of small oscillations about an equilibrium configuration and the non-linear problem of large amplitude oscillations. In this paper we will only deal with the linearized problem. In § 2 the problem to be dealt with is formulated. We show how the linearized Navier-Stokes equations in variational form, together with a finite element discretization, can be written as a quadratic eigenvalue problem. A method based on an inverse iteration procedure is used to solve the eigenvalue problem in § 3. Eigenfrequencies are calculated for a number of representative cases. Flexible container walls are introduced in § 4 and the effect of this on the eigenfrequencies is investigated.

2. Problem formulation

Consider a viscous, incompressible, Newtonian fluid, partly filling a container C. Let $\Omega(t)$ denote the region occupied by the fluid at time t and let $\partial\Omega(t)$ be its boundary, which consists of the free surface $S(t)$ and the part of the fluid in contact with the container wall $\Gamma(t)$, see figure 1. The equations describing the motion of the fluid in the container are, in general, the Navier-Stokes equations in $\Omega(t)$ together with appropriate boundary conditions on $S(t)$ and $\Gamma(t)$, and the condition that the total volume of the fluid remains constant. The container C is placed in a micro-gravitational field, so that capillary forces on $S(t)$ have to be taken into account.

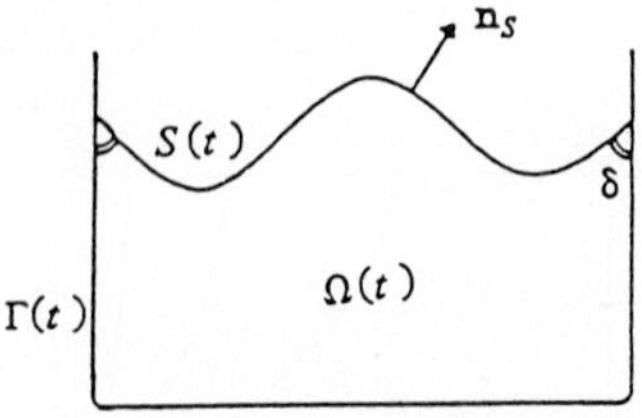

Figure 1.

In the steady state, the fluid occupies the region Ω_0 with equilibrium free surface S_0 and $\Gamma_0=\partial\Omega_0 \bigcap C$. Let us consider small perturbations from the steady state equilibrium so that products of small quantities can be neglected. The non-dimensional equations describing the fluid motion $\mathbf{u}$ are then (c.f. Cuvelier 1985): the linearized Navier-Stokes equations

$$\frac{\partial \mathbf{u}}{\partial t} + \nabla p - Oh\,\nabla^2\mathbf{u} = 0, \quad \text{in } \Omega_0 \tag{2.1}$$

the incompressibility condition

$$\nabla \cdot \mathbf{u} = 0, \qquad \text{in } \Omega_0 \tag{2.2}$$

together with the boundary conditions on the container wall

$$\mathbf{u} = 0 \quad \text{or} \tag{2.3}$$

$$u_n = 0 \, , \sigma_\tau = 0 \quad \text{on } \Gamma_0,$$

boundary conditions on the free surface

$$\left.\begin{aligned} \sigma_\tau &= 0 \\ \sigma_n &= \left(\frac{1}{R_{S_0}^2} - Bo\,\hat{\mathbf{k}}\cdot\mathbf{n}_S\right)\eta + \frac{\partial^2\eta}{\partial s_0^2} \\ u_n &= \frac{\partial\eta}{\partial t} \end{aligned}\right| \text{ on } S_0, \tag{2.4}$$

and the contact angle conditions

$$\frac{\partial\eta}{\partial s_0} = \eta\frac{\cot\delta}{R_{S_0}} \text{ at } s_0(x_1{=}0) \quad , \quad \frac{\partial\eta}{\partial s_0} = -\eta\frac{\cot\delta}{R_{S_0}} \text{ at } s_0(x_1{=}1). \tag{2.5}$$

In the above equations p denotes the pressure, η the small normal deviation from S_0, $\hat{\mathbf{k}}$ a unit vector directed along the positive z-axis, $\mathbf{n}_S$ a unit vector normal to S_0, R_{S_0} the radius of curvature of S_0, δ is the contact angle and s_0 a curvilinear coordinate parallel to S_0. The Ohnesorge number Oh is a gross measure of how the viscous forces compare to capillary forces and the Bond number Bo is a measure of how gravitational forces compare to capillary forces. The normal and tangential stresses and velocities are denoted by σ_n, σ_τ and u_n, u_τ respectively; σ_{ij} is the usual Cauchy stress tensor.

Let us assume that $\mathbf{u}$, p and η exhibit a temporal behaviour of the form $e^{\lambda t}$, where λ is in general a complex quantity. The weak formulation of (2.1)-(2.5) can, on eliminating η from the equations and using Green's theorem, be written as follows (c.f. Cuvelier 1985);

Find $\mathbf{u}$ and p such that

$$\int_{\Omega_0} (\lambda \mathbf{u}\cdot\mathbf{v}^* - p\,\nabla\cdot\mathbf{v}^*)dx + Oh\,a(\mathbf{u},\mathbf{v}) + \frac{1}{\lambda}b(\mathbf{u},\mathbf{v}) = 0, \tag{2.6}$$

$$\int_{\Omega_0} q\,\nabla\cdot\mathbf{u}dx = 0, \tag{2.7}$$

for all $\mathbf{v}$ and q,

where $\mathbf{v}$ and q are test functions and the * in (2.6) denotes the complex conjugate. The bilinear form $a(\mathbf{u},\mathbf{v})$ is related to the rate of dissipation of energy due to viscous forces and $b(\mathbf{u},\mathbf{v})$ is a functional depending of the free surface boundary conditions. A finite element discretization of (2.6) and (2.7), together with a penalty function approach to uncouple the continuity and momentum equation, yields the following eigenvalue problem

$$(\lambda^2\mathbf{M} + \lambda\mathbf{S} + \mathbf{B})\mathbf{u} = 0, \tag{2.8}$$

where $\mathbf{M}$ is the mass matrix, $\mathbf{S}$ the stiffness matrix and $\mathbf{B}$ the boundary matrix related to the boundary conditions on S_0.

3. Solution of the eigenvalue problem

Let $\mathbf{w}=\lambda\mathbf{u}$ then (2.8) can be written in the standard eigenvalue problem form, viz.

$$\begin{bmatrix} 0 & \mathbf{I} \\ -\mathbf{M}^{-1}\mathbf{B} & -\mathbf{M}^{-1}\mathbf{S} \end{bmatrix} \begin{bmatrix} \mathbf{u} \\ \mathbf{w} \end{bmatrix} = \lambda \begin{bmatrix} \mathbf{u} \\ \mathbf{w} \end{bmatrix}. \tag{3.1}$$

Solving (3.1) using an inverse iteration technique, requires an estimate of some simple eigenvalue λ_p of (3.1) with corresponding eigenvector $(\mathbf{u}_p, \mathbf{w}_p)^T$. Define the following inverse iteration procedure

$$(\mu^2\mathbf{M}+\mu\mathbf{S}+\mathbf{B})\hat{\mathbf{w}}^{(n+1)} = \mathbf{B}\mathbf{u}^{(n)} - \mu\mathbf{M}\mathbf{w}^{(n)},$$

$$\hat{\mathbf{u}}^{(n+1)} = \frac{1}{\mu}(\hat{\mathbf{w}}^{(n+1)} - \mathbf{u}^{(n)}),$$

$$\mathbf{u}^{(n+1)} = \frac{\hat{\mathbf{u}}^{(n+1)}}{N}, \quad \mathbf{w}^{(n+1)} = \frac{\hat{\mathbf{w}}^{(n+1)}}{N}, \quad \text{with } N = \max(\|\hat{\mathbf{u}}^{(n+1)}\|, \|\hat{\mathbf{w}}^{(n+1)}\|).$$

then it can be shown readily that $\mathbf{u}^{(n)} \rightarrow \mathbf{u}_p$, $\mathbf{w}^{(n)} \rightarrow \mathbf{w}_p$ as $n \rightarrow \infty$ if $|\mu-\lambda_p| < |\mu-\lambda_i|$ for all $i \neq p$. The above iterative procedure converges fast when μ is a good estimate of λ_p, but not so when μ is situated in between two eigenvalues.

In the first experiment we compare theoretical results with those obtained numerically. The motion of an inviscid, incompressible fluid can be described in terms of a velocity potential which statisfies Laplace's equation. In the case of a rectangular container of length 1 and height 1, half filled with a capillary fluid with contact angle $\delta=\pi/2$, the eigenfrequencies of the inviscid fluid motion are given by

$$\omega_k^2 = (Bok\pi+(k\pi)^3)\tanh(k\pi/2)\ , \ k=1,2,... \tag{3.2}$$

We expect the imaginary part of the eigenfrequencies of a low-viscous fluid to be approximated well by (3.2). In table 1 we show the eigenvalues obtained numerically for the case $Oh=10^{-3}$, $Bo=0{,}1000$ using a regular mesh of 200 quadratic elements and free-slip boundary conditions on the container wall. Observe that $\text{Im}(\lambda_k)$ agrees well with (3.2).

eigenmode	$Bo=0$		$Bo=1000$	
	λ_k	ω_k (3.2)	λ_k	ω_k (3.2)
1	-0.019+5.333i	5.330	-0.021+53.94i	53.94
2	-0.080+15.72i	15.72	-0.084+80.68i	80.67
3	-0.19+28.94i	28.89	-0.20+101.3i	101.3
4	-0.35+44.59i	44.88	-0.36+120.6i	120.6

TABLE 1

We next investigate the effect of the contact angle on the eigenfrequencies. We take $Oh=10^{-3}$ and $Bo=0$ so that the free surface is considerably curved when $\delta\neq\pi/2$. In table 2 we show how the first three eigenfrequencies depend on δ. Vector plots of a number of representative cases are shown in figure 2.

δ	λ_1	λ_2	λ_3
$\pi/2$	-0.019+5.33i	-0.080+15.7i	-0.19+28.9i
$\pi/4$	-0.020+4.57i	-0.079+13.0i	-0.16+24.2i
$\pi/8$	-0.021+4.38i	-0.27+11.1i	-0.35+20.0i

TABLE 2

For additional numerical results we refer to Schulkes & Cuvelier (1988).

4. flexible container walls

Consider a rectangular container of length 1, height ½, completely filled with a viscous, incompressible liquid. The bottom of the container is assumed to be rigid, but the side walls consist of thin, flexible shells which are fixed at the bottom and top points. The boundary conditions at the rigid bottom are as in (2.3), but on the side walls a new set of conditions is prescribed, namely

$$\sigma_n = \frac{1}{Bo_W}\frac{\partial^2\xi}{\partial z^2} - r_\rho\frac{\partial^2\xi}{\partial t^2}, \quad \sigma_\tau = 0, \quad \frac{\partial\xi}{\partial t} = u_n . \tag{4.1}$$

In the above equations ξ denotes the displacement of the container wall from the steady state, r_ρ is a measure of the ratio of the shell and fluid densities and Bo_W is a measure of the ratio of gravitational forces and tension forces in the shell. The above equations are valid for $Bo_W \leqslant Bo^{-1}$ in which case the shell is only slightly bent in the steady state. Introduction of the boundary conditions (4.1) does not alter the structure of the eigenvalue problem (2.8) but terms are added to the mass and boundary matrices. In table 3 we show the eigenfrequencies for various values of Bo_W and r_ρ; the remaining parameter values are $Oh=10^{-3}$, $\delta=\pi/2$ and $Bo=1000$.

r_ρ	$Bo_W=6\times10^{-6}$		$Bo_W=6\times10^{-4}$	
	λ_1	λ_2	λ_1	λ_2
10^2	-0.110+53.84i	-0.10+80.67i	-0.10+57.76i	-0.15+85.66i
10^1	-0.093+53.85i	-0.10+80.73i	-0.12+59.57i	-0.16+86.44i
10^0	-0.090+53.86i	-0.10+80.73i	-0.44+74.34i	-0.37+93.28i
10^{-1}	-0.090+53.86i	-0.10+80.73i	-0.38+89.90i	-0.54+107.5i
10^{-2}	-0.090+53.86i	-0.10+80.73i	-0.34+91.91i	-0.52+110.1i

TABLE 3

Note that when $Bo_W \ll Bo^{-1}$, the eigenfrequencies approach those in the case of rigid walls (c.f. table 1) and are virtually independent of the density ratio r_ρ. When the tension in the container wall decreases ($Bo_W=6\times10^{-4}$), it is found that the eigenfrequencies do depend on the parameter r_ρ. When r_ρ is large, which corresponds to a heavy shell, the fluid motion corresponds to that in a rigid container. For smaller values of r_ρ, the motion of the shell alters the fluid flow significantly as is shown in figure 3.

References

Abramson, N.H. 1966 The dynamic behavior of liquids in moving containers. *NASA SP*–106.

Berger, H., Boujot, J. & Ohayon, R. 1975 On a spectral problem in vibration mechanics: computation of elastic tanks partially filled with liquids. *J. Math. Anal. Appl.* **51**, 272-298.

Boujot, J. 1987 Mathematical formulation of fluid-structure interaction problems. *Math. Mod. Num. Anal.* **21**, 239-260.

Cuvelier, C. 1985 A capillary free boundary problem governed by the Navier-Stokes equations. *Comp. Meths. Appl. Mech. Engng.* **48**, 45-80.

Kopachevskii, N.D. & Myshkis, A.D. 1967 Hydrodynamics in weak fields of force. *J. Phys. Math.* and *Math. Phys.* **6**, 1054-1063.

Morand, H. & Ohayon, R. 1979 Substructure variational analysis of the vibrations of coupled fluid-structure systems, finite element results. *Int. J. Num. Meths. Engng.* **14**, 741-755.

Schulkes, R.M.S.M. & Cuvelier, C. 1988 To appear.

Veldman, A.E.P. & Vogels, M.E.S. 1984 Axisymmetric liquid sloshing under low-gravity conditions. *Rept. NLR MP 82037U*.

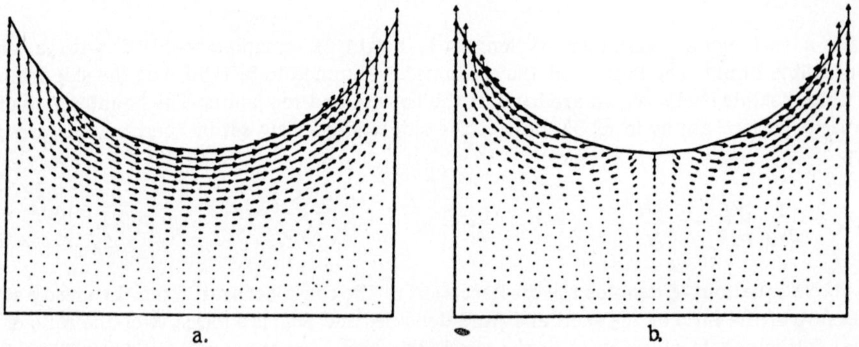

Figure 2.
Velocity vectors corresponding to the first (a) and the second (b) eigenmode, with parameter values $\delta=\pi/8$, $Bo=0$ and $Oh=10^{-3}$.

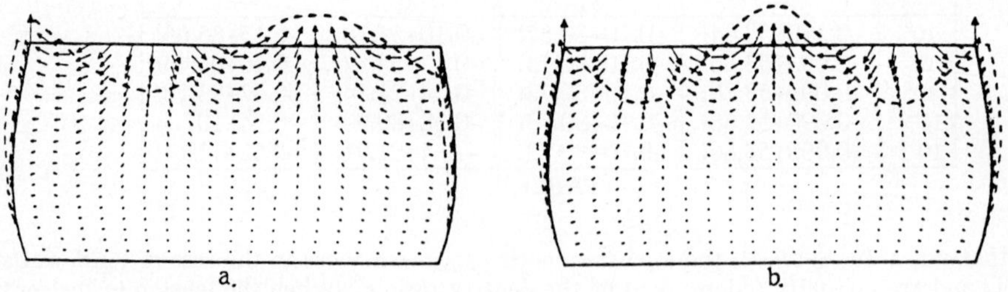

Figure 3.
Velocity vectors and container wall deflections corresponding to the first (a) and second (b) eigenmodes, with parameter values $\delta=\pi/2$, $Bo=1000$, $Oh=10^{-3}$, $Bo_W=6\times10^{-4}$ and and $r_\rho=0.01$.

Navier-Stokes Simulation of Transonic Flow About Wings Using a Block Structured Approach

D. Schwamborn
DFVLR Institute for Theoretical Fluid Mechanics,
Bunsenstr. 10, D-3400 Göttingen, FRG

Summary

An explicit finite volume method for the integration of the Navier-Stokes equations is presented. The method is cast into a very flexible block-structured approach. Results are presented for the simulation of the DFVLR-F5 experiment which has been designed to serve as a well defined boundary value problem for transonic viscous flow computations.

Introduction

The recent development of supercomputers with respect to speed and memory enabled a substantial improvement in the numerical simulation of transonic flows in two and three dimensions. Two dimensional solutions of the Euler and Navier-Stokes equations have become more and more commonplace. Calculations for wings and wing-body combinations or even for complete aircrafts are no longer curiosities, at least regarding Euler solutions. Three dimensional solutions, however, for turbulent flows with high resolution near walls or for highly complex geometries cannot be handled yet in the main memory of most of today's supercomputers. Hence codes for such configurations have to work with input/output (I/O) efficiently in order not to downgrade the performance or turn-around time on vector computers. This efficiency can be achieved by dividing the computational domain into a number of subdomains, called "blocks" in the following, which can be handled in the central memory one after the other. Although there exist other I/O concepts, e.g. to hold successive "slices" of the computational domain in core memory as they are needed for the calculation, the block structure provides, in addition, an excellent means to handle very complex geometries that cannot easily be dealt with in the usual approach, where the computational domain has to be a topological box in the computational space. (This argument does not apply to the unstructured finite element approach.) For an explicit method such a block structure has no serious drawbacks whereas it might lead to some I/O problems for implicit methods when the implicitness has to be maintained across block boundaries. Therefore a block structured solver was developed for an existing explicit code for three dimensional flows.

Numerical approach

The integration of the full Reynolds-averaged Navier-Stokes equations is carried out with a cell-centered finite volume Runge Kutta method [1]. Since the approach corresponds to central differencing the use of artificial damping terms is necessary. For these damping terms the usual blend of second and fourth order differences is employed. In contrast to the use in the Euler equations we scale the numerical diffusion by a weighting function. In this function the modulus of the velocity and the reciprocal of $\sqrt{1+\varepsilon}$ is used, where ε is the eddy viscosity. The idea is to diminish the numerical diffusion not only very close to walls but also in the main portion of the boundary layer and the wake. Without such a scaling the artificial diffusion would become much larger than the physical one in viscous near-wall regions.

We use a linearized four-stage Runge Kutta scheme to save memory because no intermediate results have to be stored. Since we are, here, only interested in steady state solutions we accelerate the convergence with the help of local time stepping, i.e. the largest possible time step for each cell - as it is derived from stability analysis - is used. We don't use further acceleration techniques to date. In order to save computation time the artificial and physical diffusion terms are only updated once per time step resulting in a reduction of execution time of more than 50 percent.

We tried to develop a block structure concept as flexible as possible. Therefore we allow not only for arbitrary blocksizes but also for almost arbitrary interface relations between blocks. This is achieved by dividing each block face into different segments where each segment may have different boundary conditions such as wall, far-field or symmetry conditions or connections to other segments. The block faces are rectangular in the computational domain, and their boundaries are defined by corresponding integer coordinates.

With this block structure very complex geometries even with multiply-connected domains and steps can be handled, where the complete computational domain does no longer exhibit a simple box shape. Each block has a layer of dummy cells around it which are used for information transport between blocks or to fulfill the boundary conditions. In the former case the dummy cells correspond to the innermost cells of the neighboring block - which may also be identical with a different part of the block under consideration. This situation occurs for example if the C-type mesh around an airfoil segment is cast into a single block. One face of this block has three segments then: the first is the lower part of the wake cut, the second is the airfoil wall and the third the upper side of the wake cut. If there where no segmentation of the block faces such a simple situation would already result in three blocks.

Those dummy cells transfering the information between neighboring blocks are updated after the block has been stored in memory and their contents are kept fixed during the calculations for this block. For the other dummy cells the boundary conditions are applied after each stage of a time step. The algorithm for all cells within the block is thus maintained the same which is important for vector computers. The number of time steps computed in a block before switching to the next block can be prescribed. Since the dummy cell information from other blocks is kept fixed during those time steps, its influence spreads the more into the interior of the block the more time steps are made. Thus the solution cannot be independent of the block structure in the transient phase. If only one time step is made for each block at a time the difference with respect to a solution with just one block is very small, but it may become larger and even slow down the convergence if many time steps are made for each block at a time, although this is desirable to minimize I/O overhead. The solution for the flow field has, of course, to be independent of the block structure once the steady state is reached.

Discussion of results

The method has been applied to simulate the transonic flow about the symmetric DFVLR-F5 wing [2], which has been investigated experimentally [3]. Figure 1 shows the experimental set-up. The wing root consists of an extremely smooth fairing between a splitter plate and the main portion of the 20° swept wing. This realistic fairing avoids the generation of a horseshoe vortex. The experiment was designed to provide a well defined boundary value problem for transonic flow. Therefore the wind tunnel walls were closed, and measurements were made not only at the wing but also at all the walls, and in the entrance and exit planes. The experiments were carried out for a Mach number of 0.82 and a Reynolds number of $10^7/m$ and two angles of attack (α = 0° and 2°). The boundary layer was fully turbulent along the wind tunnel walls, and along the splitter plate [3]. Transition on the wing was natural, and the transition line was determined in the experiments with some uncertainties to which we will refer later. All results presented here for the wing in free air flow and in the wind tunnel are obtained using C-O type meshes which were divided into four blocks for the computation.

Figure 2 shows the view of a such a mesh for the wind tunnel case, when the windtunnel walls and splitterplate are assumed inviscid. The total number of cells is 200 x 40 x 34 (chordwise x spanwise x normal to the surface direction), and the first cell-center away from the wing is at about $y^+ = 2$, where y^+ is the law-of-the-wall coordinate. Boundary conditions for this case are: no-slip, adiabatic wall and zero normal pressure gradient at the wing surface, symmetry condition in the plane of the splitter plate and along all other walls. In the inlet plane mach number, temperature and flow direction are specified using one dimensional Riemann invariants, thus allowing for some flexibility of the solution, there. At the outlet a non-reflecting condition is used for the pressure without prescribing the experimental values. In the free air case symmetry is also enforced at the splitter plate, and at the far-field boundary a formulation via Riemann invariants is used.

The pressure distribution for the non-lifting wing is presented in figure 3 at three representive sections for both the free air and the wind tunnel results. At the root of the wing the pressure distributions compare quite well to the experimental data despite of the neglected splitter plate boundary layer. Over the main part of the span this is no longer true; discrepancies are found regarding the position of the weak shock (and its strength in the wind tunnel simulation). At the wing tip the agreement becames better again. Figure 4 shows the skin friction line pattern on the wing surface for the free air and the wind tunnel case, respectively. In both cases we find a shock induced laminar separation with turbulent reattachment of the flow and a small separated region with a focus near the fairing close to the trailing edge. The appearence of the shock induced separation depends strongly on how the turbulence is switched on at the position of the transition line taken from the experiment. If the switching is done in such a way that the flow is already fully turbulent at the transition line the separation vanishes but this has practically no influence on the pressure distribution. Unfortunately experimental information from oil-flow visualization does not yet exist providing us with some information about the structure of the separation.

The pressure distribution for the lifting cases in the wind tunnel and in free air are both shown in figure 5. Again the agreement is quite good near the root, especially in the free air case. Away from the root section the shock is stronger now and its position agrees better with the experiment than for $\alpha = 0°$. But there is an overexpansion in front of the shock, where the experimental pressure distribution indicates a laminar separation which extends over the main spanwise portion of the wing. At the lower surface the sharpness of the recompression is underpredicted in the numerical results. At the tip the discrepancies between calculation and experiment are smaller again at least for the upper surface.

Figure 6 shows the skin friction lines on the upper surface for the lifting wing. The shock induced separation is now fully turbulent and for the free air case it has become larger than in the nonlifting case at the outboard half of the span. The same is true for the second separated area. In the wind tunnel case the shock induced separation extends now almost up to the trailing edge and due to this the second separation has become smaller.

Concluding remarks

Although the boundary layer on the splitter plate has been neglected in all the computations the results are astonishingly close to the experimental pressure distribution near the root especially for the free air case. This seems to be due to the very thin turbulent boundary layer on the plate and the extremly smooth fairing of the wing.

Unfortunately there exist some uncertainties regarding the exact position of the transition line - actually a transition zone - in the experiment which can be very important because this line is immediately in front of the shock position for $\alpha = 0°$ and is between laminar separation and shock position for $\alpha = 2°$. Thus a variation of the transition line of a few percent may result in larger changes of the numerical solution. The results are also influenced by the way the turbulence model is switched on, and certainly the turbulence model itself is of crucial importance. Although the inviscid splitter plate assumption seems to be justified from the results the influence of viscous tunnel walls is not yet clear

References

[1] Schwamborn, D.: Simulation of the DFVLR-F5 Wing Experiment Using a Block Structured Explicit Navier-Stokes Method,in "Numerical Simulation of the Transonic DFVLR-F5 Wing Experiment". Notes on Numerical Fluid Mechanics, Vol.22, Vieweg Verlag, Wiesbaden, 1988.

[2] Sobieczky, H.: Geometry Generation for Transonic Design, in Recent Advances in Numerical Methods in Fluids, Vol. 4. Swansea, Pineridge Press 1985, pp. 163-182.

[3] Sobieczky, H. Hefer, G. Tusche, S.: DFVLR-F5 Test Wing Experiment For Computational Aerodynamics. AIAA Paper 87-2485 CP, 1987. (see also Proceedings of [1])

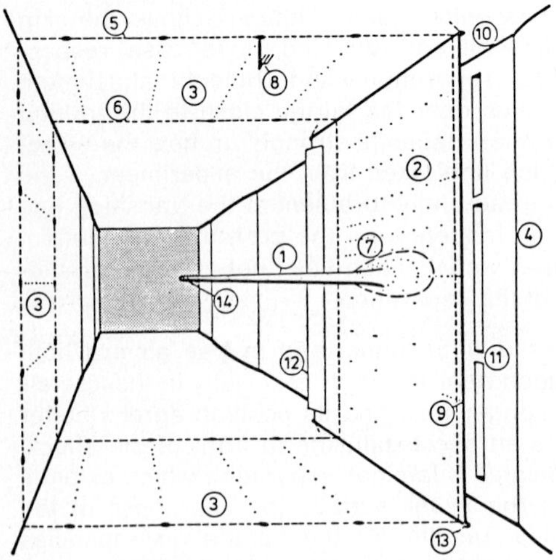

Fig. 1: Schematic wind tunnel arrangement (wing (1) on splitter plate (2), inlet (5) and exit plane (6) of the control volume)

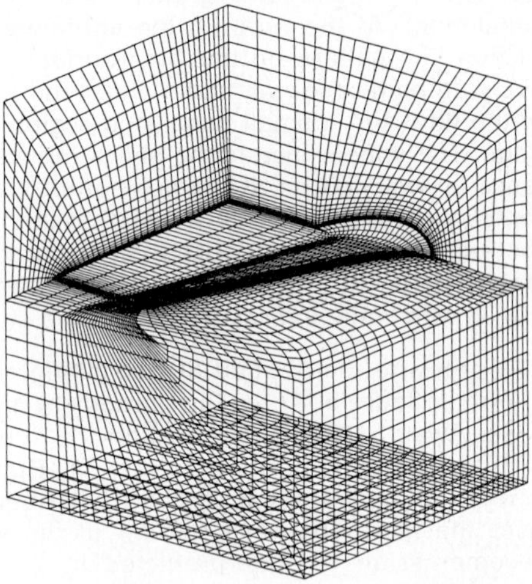

Fig. 2: C-O mesh for wind tunnel case (only every third line chordwise and every second line normal to wing surface shown)

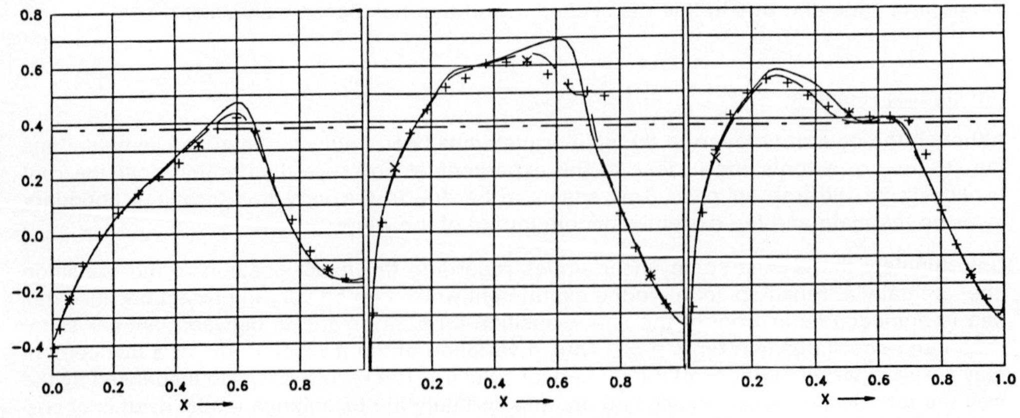

Fig. 3: Cp-distribution ($\alpha = 0°$) at 0.1, 49.2, 95.4 % span (solid line: wind tunnel, broken line: free air, Symbols: Experiment [3])

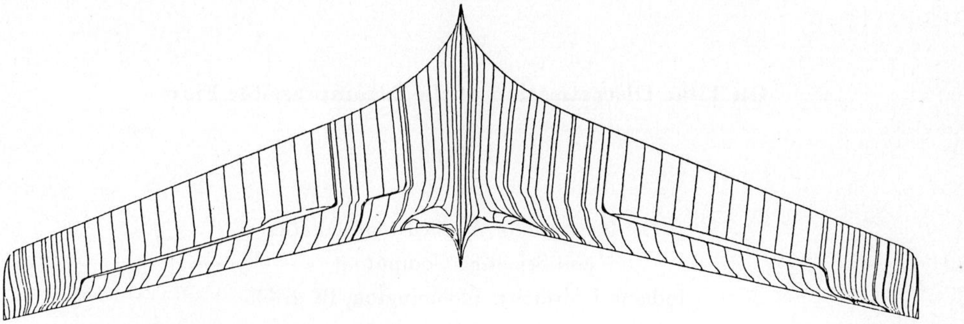

Fig. 4: Skin friction line pattern on upper surface at $\alpha = 0°$ Wind tunnel (right) and free air simulation (left).

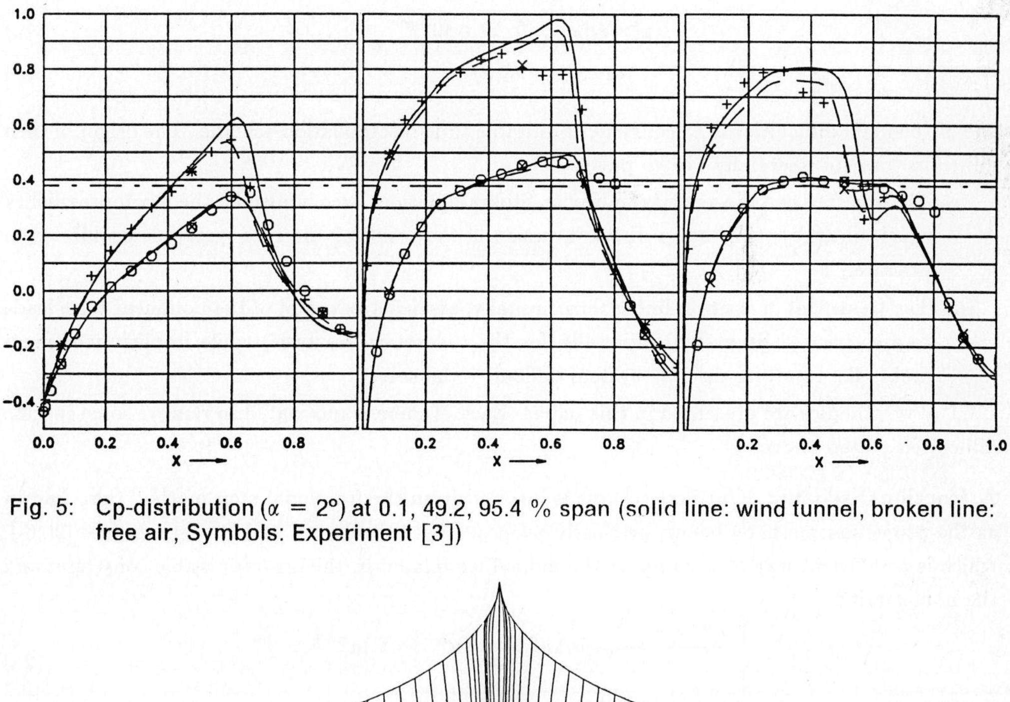

Fig. 5: Cp-distribution ($\alpha = 2°$) at 0.1, 49.2, 95.4 % span (solid line: wind tunnel, broken line: free air, Symbols: Experiment [3])

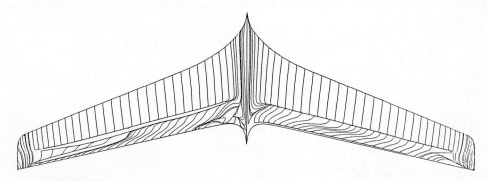

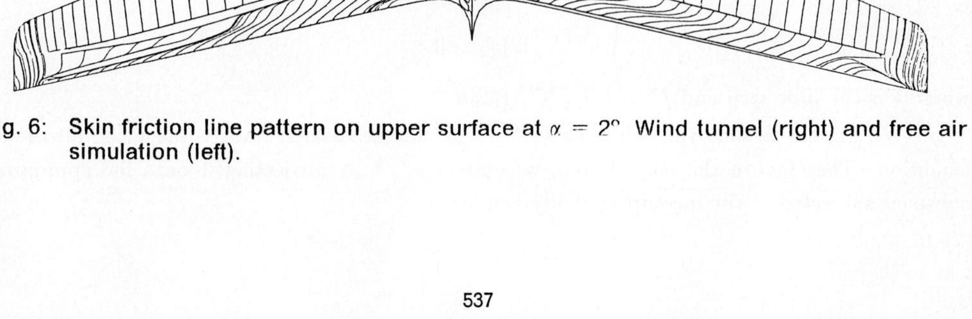

Fig. 6: Skin friction line pattern on upper surface at $\alpha = 2°$ Wind tunnel (right) and free air simulation (left).

On Time Discretization of the Incompressible Flow

JIE SHEN

The Institute for Applied Mathematics
and Scientific Computing
Indiana University, Bloomington, IN 47405

In numerical simulations of incompressible flows represented by the Navier-Stokes equations:

$$\frac{\partial u}{\partial t} - \nu\Delta u + u\nabla u + \nabla p = f(x,t)\ , \quad (x,t) \in Q = \Omega \times [0,T] \tag{1a}$$

$$div u = 0 \ \ in\ Q \tag{1b}$$

$$u(x,t)|_{\partial\Omega} = 0 \ \ \forall t \in [0,T] \tag{1c}$$

$$u(x,0) = u_0(x) \tag{1d}$$

one of the main difficulties is to construct a suitable time discretization scheme. The origin of such difficulty consists essentially of two parts:

(i) The pressure and the velocity in Navier-Stokes equations are coupled by the incompressibility constraint (1b) such that a direct inversion of the resulting discrete system is prohibited in practice.

(ii) The treatment of the non-linear term: usually, explicit treatment of the nonlinear term leads to a restrictive time step, especially for the theoretical analysis, while implicit treatment makes the resulting discrete system difficult to be solved.

Two schemes are presented in this paper. Each of them is motivated to remove some specific difficulties stated above.

A fractional scheme. Our first scheme is inspired from the fractional step method (also known as the projection method) below, originally proposed by A.J. Chorin and R. Temam (see [2],[6]), which is an efficient way to decompose the difficulties related to the incompressible constraint and the nonlinearity.

$$\begin{cases} \dfrac{u^{n+\frac12} - u^n}{k} - \nu\Delta u^{n+\frac12} - (u^{n+\frac12}\cdot\nabla)u^{n+\frac12} = f^n \\ u^{n+\frac12}|_{\partial\Omega} = 0 \end{cases} \tag{2a}$$

$$\begin{cases} \dfrac{u^{n+1} - u^{n+\frac12}}{k} + \nabla p^{n+1} = 0 \\ div u^{n+1} = 0 \\ u^{n+1}\cdot \vec{n}|_{\partial\Omega} = 0 \end{cases} \tag{2b}$$

where k is the time step and $f^n = \frac{1}{k}\int_{nk}^{(n+1)k} f(t)dt$.

We emphasize that the solutions $\{u^{n+\frac12}\}$ do not necessarily satisfy the incompressibility condition. Therefore in the second step, we correct $u^{n+\frac12}$ by projecting it onto the appropriate subspace subjected to the incompressibility condition.

The stability and the convergence results for (2) were established in [6]. But it is well known that (see for instance Remark 7.2 of [7]) this scheme is not appropriate for the approximation of the pressure because we derive from (2b) that: $\frac{\partial p^{n+1}}{\partial \vec{n}}|_{\partial\Omega} = 0$, which is not satisfied by the exact pressure. Numerical experiences suggest that proper Neumann boundary conditions for the pressure should be imposed in order to improve the approximation. A recent work by P.M.Gresho & R.L.Sani [3] confirms this point. Our aim is then to construct a variant of the previous projection method which leads to a suitable approximation for the pressure as well as for the velocity. To this end, we propose here the following scheme:

$$\begin{cases} \dfrac{u^{n+\frac{1}{2}} - u^n}{k} - \dfrac{\nu}{2}\Delta u^{n+\frac{1}{2}} + \dfrac{1}{2}\nabla p^n = \dfrac{1}{2} f^n \\ div u^{n+\frac{1}{2}} = 0 \\ u^{n+\frac{1}{2}}|_{\partial\Omega} = 0 \end{cases} \tag{3a}$$

$$\begin{cases} \dfrac{u^{n+1} - u^{n+\frac{1}{2}}}{k} - \dfrac{\nu}{2}\Delta u^{n+1} + u^{n+\frac{1}{2}} \cdot \nabla u^{n+1} = -\dfrac{1}{2}\nabla p^n + \dfrac{1}{2} f^n \\ u^{n+1}|_{\partial\Omega} = 0 \end{cases} \tag{3b}$$

(3a) is a Stokes like system which can be solved efficiently if we have a fast Stokes solver and (3b) is of the same type as (2a). We derive from (3a) that the pressure approximation p^n satisfies:

$$\frac{\partial p^n}{\partial \vec{n}} = (\nu\Delta u^{n+\frac{1}{2}} + f^n) \cdot \vec{n}|_{\partial\Omega}$$

which corresponds to the boundary condition satisfied by the exact pressure, namely:

$$\frac{\partial p(t_n)}{\partial \vec{n}} = (\nu\Delta u(t_n) + f(n)) \cdot \vec{n}|_{\partial\Omega}$$

We can then expect that this scheme gives a better approximation for the pressure as well as for the velocity.

Let $\mathcal{X} = (H_0^1(\Omega))^d$, $\mathcal{L}^2(\Omega) = (L^2(\Omega))^d$, $\mathcal{H} = \{u \in \mathcal{L}^2(\Omega) : div u = 0, u \cdot \vec{n}|_{\partial\Omega} = 0\}$
$\mathcal{V} = \{u \in \mathcal{X} : div u = 0\}$, $a(u,v) = (\nabla u, \nabla v)$, $\forall\ u, v \in \mathcal{X}$,
and $b(u,v,w) = (u \cdot \nabla v, w) = \sum_{i,j=1}^d \int_\Omega u_i \frac{\partial v_j}{\partial x_i} w_j dx$, $\forall\ u, v, w \in \mathcal{X}$

By the Poincaré inequality, we know that $||\cdot|| \equiv a(\cdot,\cdot)^{\frac{1}{2}}$ is a norm in $\mathcal{X}$. We shall denote the norm of $\mathcal{X}$ (resp. $\mathcal{L}^2(\Omega)$) by $||\cdot||$ (resp. $|\cdot|$).

We claim first a stability result for the scheme (3) which can be proven by an energy method.

Lemma 1. *The scheme (3) is unconditionally stable in the following sense:*

$$|u^{m+1}|^2 + |u^{m+\frac{1}{2}}|^2 + \frac{k\nu}{2}\sum_{n=0}^{m}\{||u^{n+1}||^2 + ||u^{n+\frac{1}{2}}||^2\} + \sum_{n=0}^{m}|u^{n+1} + u^n - 2u^{n+\frac{1}{2}}|^2$$
$$\leq \frac{1}{\nu}||f||^2_{\mathcal{L}^2(0,T,\mathcal{X}')} + |u^0|^2\ ,\ \forall\, 0 \leq m \leq T/k - 1$$

In order to carry out an error analysis, we assume that the solution is sufficiently smooth. Precisely, we assume that:

$$u'(t) \in \mathcal{L}^2(0,T,\mathcal{V}) \ \ and\ \ u''(t) \in \mathcal{L}^2(0,T,\mathcal{V}') \tag{H1}$$

Special cares should be paid when treating the non-linear term. Actually, the term $\sum_{n=0}^{m}|u^{n+1}+u^n-2u^{n+\frac{1}{2}}|^2$ in Lemma 1 plays an essential role for absorbing the splitting error of the non-linear term. We are able to prove the following theorem (see [5] for details):

Theorem 1. *We assume that the solution u of problem (1) satisfies (H1). Then:*

$$k\sum_{n=0}^{m}\{\|u((n+\frac{1}{2})k)-u^{n+\frac{1}{2}}\|^2+\|u((n+1)k)-u^{n+1}\|^2\}$$
$$+|u^m-u(mk)|^2+|u^{m+\frac{1}{2}}-u((m+\frac{1}{2})k)|^2\le ck\ ,\ \forall\,1\le m\le K-1$$

Remark 1.

(i) *We only proved that the scheme (3) is of order $\frac{1}{2}$, but we do not know whether this result is optimum. However, the same proof does not yield a similar result for the scheme (2). One may thus conjecture that the scheme (3) is more accurate than the scheme (2).*

(ii) *The scheme (3) gives full consistent approximation for the pressure only if the velocity field u satisfies the condition: $(u\cdot\nabla)u\cdot\vec{n}|_{\partial\Omega}$. In other cases, the following variant of (3) should give more accurate results:*

$$\begin{cases}\dfrac{u^{n+\frac{1}{2}}-u^n}{k}-\dfrac{\nu}{2}\Delta u^{n+\frac{1}{2}}+\dfrac{1}{2}\nabla p^n=\dfrac{1}{2}(f^n-u^n\cdot\nabla u^n)\\ divu^{n+\frac{1}{2}}=0\\ u^{n+\frac{1}{2}}=0|_{\partial\Omega}\end{cases}\tag{4a}$$

$$\begin{cases}\dfrac{u^{n+1}-u^{n+\frac{1}{2}}}{k}-\dfrac{\nu}{2}\Delta u^{n+1}+\dfrac{1}{2}u^{n+\frac{1}{2}}\cdot\nabla u^{n+1}=-\dfrac{1}{2}\nabla p^n+\dfrac{1}{2}f^n\\ u^{n+1}=0|_{\partial\Omega}\end{cases}\tag{4b}$$

Due to the explicit treatment of the non-linear term in the first time step, we are not expecting to prove unconditional stability of this scheme. But thanks to the implicit process involved in the second step, one can hope that this scheme is more stable than an ordinary explicit one. We should mention that the scheme (4) is of order 1, but we lose in return the unconditional stability.

A modified Crank-Nicolson & Adams-Bashforth scheme. In many cases, one observe that a semi-implicit scheme gives stable results under a time step constraint which is much weaker than what the theoretical results predicted, especially in cases where a smooth solution exists. A natural question one can ask is: can we improve the stability conditions by giving more smoothness assumptions on the solutions?

We will give a positive answer to this question by considering a concrete space discretization, namely, the spectral-Galerkin approximation in two dimensional case (we refer to [1] for a detailed presentation of this method). For other space discretizations, similar results could be obtained by using the same technique.

We assume that u is uniformly bounded in Q, i.e.

(H2) There exists $M>0$ such that $|u|_{L^\infty(Q)}\le M$.

We introduce the truncated function H: $R \to R$ defined as:

$$H(y) = \begin{cases} y \, , & if \ \ |y| \le 2M \\ 2M \, , & if \ \ y \ge 2M \\ -2M \, , & if \ \ y \le -2M \end{cases} \tag{5}$$

Let us set

$$N(u(x)) = \sum_{i=1}^{2} \frac{\partial}{\partial x_i}(H(u_i(x))u(x)) \quad , \quad \forall u \in \mathcal{X}$$

We observe that

$$\begin{cases} |u|_{\mathcal{L}^\infty(Q)} \le 2M \\ u \in \mathcal{V} \end{cases} \quad \text{implies} \ \ N(u) = u \cdot \nabla u$$

We consider now the following modified Crank-Nicolson & Adams-Bashforth scheme where we replace $u \cdot \nabla u$ by $N(u)$:

$$\begin{cases} \text{Find } u_N^n \in \mathcal{V}_N \ \text{ such that:} \\ \frac{1}{k}(u_N^{n+1} - u_N^n, v)_\omega + \frac{\nu}{2} a_\omega(u_N^{n+1} + u_N^n, v) \\ \qquad = -(1.5N(u_N^n) - 0.5N(u_N^{n-1}), v)_\omega + < f^{n+\frac{1}{2}}, v >_\omega \ , \quad \forall v \in \tilde{\mathcal{V}}_N \end{cases} \tag{6}$$

where

ω is either the Chebychev weight function $\Pi_{i=1}^2(1-x_i)^{-\frac{1}{2}}$ or the Legendre weight function 1; $a_\omega(u,v) = (\nabla u, \nabla(v \cdot \omega))$, $(u,v)_\omega = (u, v \cdot \omega)$ and

$\mathcal{V}_N = \{u \in P_N : \ divu = 0, u|_{\partial\Omega} = 0\}$, $\tilde{\mathcal{V}}_N = \{u \in P_N : \ div(u \cdot \omega) = 0, u|_{\partial\Omega} = 0\}$ with

P_N : the set of polynomials of order less or equal to N.

By using integration by parts and the fact $|H(u)| \le 2M$, we find:

$$|(N(u), v\omega)| = |\sum_{i,j=1}^{2}(H(u_i)u_j, \frac{\partial}{\partial x_i}(v_j\omega))| \le cM|u|_\omega ||v||_\omega$$

which is the key point for proving the following lemma.

Lemma 2. *Let $u_0 \in \mathcal{V}_\omega$ and $f \in \mathcal{L}^2(0,T,\mathcal{X}')$. We assume moreover that the solution u of (1) satisfies (H2). Then the scheme (6) is unconditionally stable. More precisely, for $0 \le m \le K$, we have:*

$$|u_N^m|_\omega^2 + k\nu \sum_{n=0}^{m-1} ||u_N^{n+1} + u_N^n||_\omega^2 + \sum_{n=0}^{m-1} |u_N^{n+1} - u_N^n|_\omega^2 \le L_1(f, u_N^1, u_0)$$

We can then prove the following theorem which says that the scheme (6) is of order 2 in time and preserves the spectral accuracy in space.

Theorem 2. *Assuming that the solution u satisfies (H2) and*

$$\begin{cases} u \in C([0,T], \mathcal{H}_\omega^s(\Omega)) \ \ , u' \in C([0,T], \mathcal{H}_\omega^{s-1}(\Omega)) \ \ (s \ge 2) \\ u'' \in \mathcal{L}^2(0,T,\mathcal{H}_{0,\omega}^1(\Omega)) \ \ , u''' \in \mathcal{L}^2(0,T,\mathcal{H}_\omega^{-1}(\Omega)) \end{cases} \tag{H3}$$

Then, we have:

$$|u(mk) - u_N^m|_\omega^2 + k\nu \sum_{n=0}^{m-1} ||u((n+1)k) + u(nk) - u_N^n - u_N^{n+1}||_\omega^2$$
$$\leq L_3(k^4 + |u_N^1 - u(k)|_\omega^2 + N^{2(1-s)}) \ , \ 1 \leq m \leq K$$

By using Theorem 2, we are able to prove the following interesting result (see [4] for details).

Theorem 3. *We suppose that all the conditions of Theorem 2 are satisfied. Then, there exists* $N_0 > 0$ *such that the schemes (6) and the original C.N.-A.B. scheme (with* $N(u)$ *replaced by* $u \cdot \nabla u$ *in (6)) are equivalent as long as* $N > N_0$ *and* $kN^{\frac{1}{2}} \to 0$.

Remark 2. *By giving more smoothness assumptions on the solution u, we succeed in reducing the very restrictive time step constraint for the C.N-A.B. scheme (at least* $kN^4 \leq c$*) to the very weak condition* $kN^{\frac{1}{2}} \to 0$ *which is evidently not a constraint in practice since we should keep the time step k reasonably small to balance the spectral precision of the space discretization.*

Remark 3. *A certain kind of convergence results could be also established for both cases without any additional smoothness assumption such as (H1) and (H3) (see [4],[5] for more details).*

Acknowledgement. *This work was partially supported by AFOSR, contract No. 88-103, and by the Research Fund of Indiana University. A early draft of this work was part of my thesis done at the University of South-Paris under the direction of Professor Roger Temam. I would like to express my deep thanks to him for his continuous guidance and encouragement.*

References.

[1] C.Bernardi, C.Canuto & Y.Maday: Generalized inf-sup condition for Chebychev approximation of the Navier-Stokes equations ; *IAN Report, N.533, Pavia, Italy (1986), submitted to SIAM J. Num. Anal.*

[2] A.J.Chorin: Numerical solution of the Navier-Stokes equations ; *Math. Comp. vol.23, 341-354 (1968)*

[3] P.M.Gresho & R.L.Sani: On pressure boundary conditions for the incompressible Navier-Stokes equations; *UCRL-96741, Lawrence Livermore National Laboratory, (1987)*

[4] J.Shen: On the time discretization of the unsteady Navier-Stokes equations; *To appear*

[5] J.Shen & R. Temam: A new fractional scheme for approximations of incompressible flows; *To appear*

[6] R.Temam: Sur l'approximation de la solution des équations de Navier-Stokes par la méthode des pas fractionnaires II; *Arch. Rat.Mech.Anal. 33, 377-385 (1969)*

[7] R.Temam: *Navier-Stokes equations;* North-Holland, Amsterdam (1979)

A Detailed Analysis of Inviscid Flux Splitting Algorithms for Real Gases with Equilibrium or Finite-Rate Chemistry

Jian-Shun Shuen, Sverdrup Technology, Inc., NASA Lewis Center,Cleveland, Ohio
Meng-Sing Liou, NASA Lewis Research Center, Cleveland, Ohio
Bram van Leer, The University of Michigan, Ann Arbor, Michigan

1. INTRODUCTION

In the past few years several flux-splitting methods by, for example, Steger and Warming[1], van Leer[2], and Roe[3] have been used to solve hyperbolic system of conservation laws, as a means of obtaining accurate weak solutions, for ideal gases. With the renewed interest in high-temperature and chemically reacting flows, these methods have recently been extended to real gases by several researchers[4-8].

Colella and Glaz[4] presented the numerical procedure for obtaining the exact Riemann solution to the real-gas flows. Grossman and Walters[5] extended the formulas of Steger and Warming, van Leer, and Roe. However, the derivations of [5] are not general, introducing unnecessary assumptions, approximations and auxiliary quantities. Vinokur and Liu[6] presented a more careful analysis for the three flux-splitting procedures[1-3] for real gases, while Glaister[7] discussed an extension of Roe splitting. More recently, Liou, van Leer, and Shuen[8] reported a detailed and comprehensive analysis of the real-gas flux-splitting algorithms in conjunction with a second-order TVD scheme. Their methods have yielded impressive results for one-dimensional shock tube and nozzle problems.

Although the analysis of [4-7] have produced useful split-flux formulas for real-gas flows, they all have either assumed chemical equilibrium or adopted a prescribed function for the equation of state(EOS). For flows with finite Damköhler numbers(defined as the ratio of characteristic flow residence time to chemical time), effects of chemical nonequilibrium are important, and finite-rate chemistry has to be included in the analysis.

Our objective of the paper is to show the extension of the known flux-vector and flux-difference splittings to real gases via rigorous mathematical procedures. Formulations of both equilibrium and finite-rate chemistry for real-gas flows are described, with emphasis on derivations of finite-rate chemistry. The split-flux formulas of Steger and Warming, van Leer, and Roe are considered. To eliminate oscillations and to obtain a sharp representation of discontinuities, a second-order upwind-based TVD scheme is adopted.

2. ANALYSIS

Since our split-flux methods for equilibrium flows have already been reported elsewhere[8], we shall concentrate on finite-rate analysis in the following discussions, with only brief mentioning of equilibrium results where it is appropriate. To illustrate how the split-flux formulas are constructed, we first consider the 1D Euler equations,

$$\frac{\partial \mathbf{U}}{\partial t} + \frac{\partial \mathbf{F}(\mathbf{U})}{\partial x} = \mathbf{S} \tag{1}$$

where $\mathbf{U}^t = [\rho, \rho u, \rho E, C_1, C_2, \ldots.C_{N-1}]$, $\mathbf{F}^t = [\rho u, \rho u^2 + p, (\rho E + p)u, C_1 u, C_2 u, \ldots.C_{N-1} u]$, and $\mathbf{S}^t = [0, 0, 0, S_1, S_2, \ldots.S_{N-1}]$, together with $E = e + 0.5u^2$, and $\rho, u, p, e, E, C_i, S_i$ are the density, velocity, pressure, specific internal energy, total energy, and mass concentration and chemical source term for the ith species, respectively. It is noted that since the equation for conservation of mass is already solved, only N-1 species conservation equations need to be considered for a chemical system of N species. In addition, there is an EOS which describes the relationship of the thermodynamic states, and here we assume that the EOS can be written in the general form $p = p(\rho, e, C_1, C_2, ...C_{N-1})$. For the equilibrium gas the EOS can be reduced to $p = p(\rho, e)$, and the speed of sound can then be determined as $a^2 = p_\rho + p_e p/\rho^2$, where $p_\rho = (\frac{\partial p}{\partial \rho})_e$ and $p_e = (\frac{\partial p}{\partial e})_\rho$ are functions of ρ and e. Values of p_ρ and p_e for equilibrium gas are evaluated directly from the EOS.

For chemical nonequilibrium flows, the values of C_i's are now determined by the extent of chemical reactions as well as the transport of fluid, and therefore a priori determination of an EOS is no longer feasible. In this case, pressure p can be expressed as

$$p = R_u T \sum_{i=1}^{N} \frac{C_i}{M_i} \tag{2}$$

where R_u is the universal gas constant and M_i the molecular weight of species i. The speed of sound in the finite-rate approach is now $a^2 = p_\rho + p_e p/\rho^2 + \sum_{i=1}^{N-1} C_i p_{C_i}/\rho$, where $p_e = (\frac{\partial p}{\partial e})_{\rho, C_{i,i=1,N-1}} = \rho R_u/(MC_v)$, $p_\rho = (\frac{\partial p}{\partial \rho})_{e, C_{i,i=1,N-1}} = R_u T/M_N + (R_u/MC_v)\sum_{i=1}^{N-1}(C_i/\rho)(e_i - e_N)$, and $p_{C_i} = R_u T(\frac{1}{M_i} - \frac{1}{M_N}) - (R_u/MC_v)(e_i - e_N)$, where $M = (\frac{1}{\rho}\sum_{i=1}^{N} C_i/M_i)^{-1}$ is the molecular weight of the gas mixture, $C_v = \sum_{i=1}^{N} C_i C_{v_i}/\rho$ is the constant volume specific heat of the gas mixture, and $e_i = \int_{T_r}^{T} C_{v_i} dT + h^o_{f_i}$, C_{v_i}, and $h^o_{f_i}$ are the specific internal energy, the constant volume specific heat, and heat of formation, respectively, of the ith species. Here the temperature T appearing in the formulation is calculated from the flow solutions ρ, e, and C_i's using the following equation

$$\rho e = \sum_{i=1}^{N} C_i \left(\int_{T_r}^{T} C_{v_i} dT + h^o_{f_i} \right) \tag{3}$$

If the EOS is expressed in terms of the flow variables, viz. as $p = p(\rho(\mathbf{U}), e(\mathbf{U}), C_i(\mathbf{U}))$, the Jacobian matrix is readily derived. Now, although the fluxes for real gas flows no longer possess the property of homogeneity, they can be written as a sum of homogeneous and inhomogeneous parts, i.e., $\mathbf{F} = \mathbf{AU} + \mathbf{A}_\rho \mathbf{U}$,where $\mathbf{A} = \frac{\partial \mathbf{F}}{\partial \mathbf{U}}$ has a complete set of eigenvectors. After grouping the eigenvalues of $\mathbf{A}$ according to their signs, the Steger-Warming version of flux-vector splitting for real gases, i.e., the splitting of $\mathbf{F}_h = \mathbf{AU}$, can be accomplished. It is noted that the true speed of sound a is used in the splitting. Since $\mathbf{A}_\rho$ does not have a complete set of eigenvectors[8], an equation that includes $\mathbf{F}_{in} = \mathbf{A}_\rho \mathbf{U}$ alone does not have a hyperbolic character. In consequence, central differencing for the inhomogeneous flux $\mathbf{F}_{in}$ may be appropriate.

When generalizing van Leer's flux-vector splitting, its property of smooth switching is preserved by letting the split fluxes have a common factor $(M \pm 1)^2$. However, the nice property that one eigenvalue vanishes for $| M | \leq 1$, yielding sharp steady shock, is lost for the real gas. On the other hand, the present derivation is far more general than van Leer's original approach, resulting in a family of splitting even for the case of ideal gas.

To construct Roe's flux-difference splitting, one usually defines an average state $\hat{\mathbf{U}}$ such that

$$\Delta \mathbf{F} = \hat{\mathbf{A}} \Delta \mathbf{U}, \quad \hat{\mathbf{A}} = \mathbf{A}(\hat{\mathbf{U}}), \quad \text{and} \quad \hat{\mathbf{U}} = \hat{\mathbf{U}}(\mathbf{U}_L, \mathbf{U}_R), \tag{4}$$

where $\Delta(\cdot) = (\cdot)_R - (\cdot)_L$. For real gases, averages of p_ρ, p_e, and p_{C_i}'s must be introduced in addition to the average state. For the case of finite-rate chemistry, we have chosen the set of average quantities $(\hat{u}, \hat{\rho}, \hat{e}, \hat{H}, \hat{C}_{i,i=1,N-1}, \hat{p}_\rho, \hat{p}_e, \hat{p}_{C_{i,i=1,N-1}})$. Let the Roe-average operator μ be defined as: $\mu(f) = (r_R f_R + r_L f_L)/(r_R + r_L)$, $r = \rho^{\frac{1}{2}}$. Equation (4) is satisfied by $\hat{\rho} = r_L r_R, \hat{f} = \mu(f)$ for $f = u, e, H, C_{i,i=1,N-1}$, and the condition

$$\Delta p = \hat{p}_\rho \Delta\rho + \hat{p}_e \Delta e + \sum_{i=1}^{N-1} \hat{p}_{C_i} \Delta C_i \tag{5}$$

Left to be completed are the definitions of $\hat{p}_\rho, \hat{p}_e$, and $\hat{p}_{C_i}$'s. Here we propose a formula that uses derivative information at the average state $(\hat{\rho}, \hat{e}, \hat{C}_i, ...\hat{C}_{N-1})$, which is guaranteed to lie between states L and R. Introducing $\bar{p}_\rho = p_\rho(\hat{\rho}, \hat{e}, \hat{C}_1, ...\hat{C}_{N-1}), \bar{p}_e = p_e(\hat{\rho}, \hat{e}, \hat{C}_1, ...\hat{C}_{N-1})$, and $\bar{p}_{C_i} = p_{C_i}(\hat{\rho}, \hat{e}, \hat{C}_1, ...\hat{C}_{N-1})$, we divide the residual of Eq. (5), $\delta p = \Delta p - (\bar{p}_\rho \Delta\rho + \bar{p}_e \Delta e + \sum_{i=1}^{N-1} \bar{p}_{C_i} \Delta C_i)$, over the terms in the following manner to complete the definition:

$$\hat{p}_e = \bar{p}_e + 0.5\delta p/\Delta e, \quad \hat{p}_\rho = \bar{p}_\rho + \frac{0.5}{N}\delta p/\Delta\rho, \quad \hat{p}_{C_i} = \bar{p}_{C_i} + \frac{0.5}{N}\delta p/\Delta C_i, \quad i = 1, ...N-1 \tag{6}$$

The definitions given above satisfy Eq. (5) and yield good results at discontinuities, in particular for a contact discontinuity across which the density, internal energy, and species concentrations jump appreciably(see NUMERICAL TEST).

3. NUMERICAL TEST

Extensive tests over a wide range of flow conditions have been conducted to validate the accuracy of the present formulation. Some extreme cases of 1D unsteady shock tube and steady nozzle problems are presented; the performance of these split fluxes for finite-rate chemistry analysis is compared against the equilibrium results.

The Euler equations are integrated using the explicit Lax-Wendroff scheme. However, the chemical source terms appearing in the species equations are treated implicitly to reduce the stiffness caused by the very small chemical time scales as compared to the flow time scales and the time steps required for a cost-effective solution. To obtain a crisp and monotone shock representation, a TVD scheme based on the above split formulas, as described in [8], is employed, along with the super-Bee limiter[9] for steepenning of the contact discontinuity.

For the equilibrium gas cases, the EOS is generated by least-square surface fitting of the calculated results from the program by Gordon and McBride[10] for equilibrium air in the range of 250 k $\leq T \leq$ 12000 k, and 0.1 atm $\leq p \leq$ 100 atm, in which 17 species are included. In the finite-rate approach, in the interest of computational time, only 5 species (O_2, N_2, O, N, NO) and 11 chemical reactions are considered for the dissociating/ recombining air, excluding the effect of ionization. For all the test cases, initial conditions are assumed to be air at equilibrium state.

§3.1 Shock Tube Problem

The initial conditions are those used in [5] and [8]:
for $0 \leq x \leq 0.5$ cm, $p_4 = 100$ atm, $T_4 = 9000$ k,
for 0.5 cm $\leq x \leq 1$ cm, $p_1 = 1$ atm, $T_1 = 300$ k.

Figures 1-3 show the numerical results of the Roe, van Leer, and Steger-Warming splittings for, respectively, $\rho/\rho_4, u/a_4, p/p_4$, and e/e_4, for both the equilibrium and finite-rate cases. The jump across the contact is rather large, about one order of magnitude. This is a difficult case to calculate, as the initial temperatures differ by a factor of 30 and consequently the compositions are completely different. It has been our experience that the TVD scheme can handle large differences in pressure very well, but not as well for large temperature differences. Nevertheless, our numerical results generally show very crisp profiles of shock and contact discontinuity. Among all three splittings, the Roe scheme gives the best results, especially near the contact discontinuity. Because of the finite relaxation time, the finite-rate case shows weaker jump across the contact.

The temperature and species molar fractions obtained using the Roe splitting are shown in Fig.4. The sharp peak in the equilibrium NO molar fraction is a numerical result produced by the smearing of internal energy at the contact discontinuity[8]. This is because the NO molar fraction is a very strong, non-monotone function of temperature(internal energy), and is most stable at some intermediate temperature across the "numerical" contact discontinuity. On the other hand, the oscillations in O and NO molar fractions across the contact in the finite-rate case are the results of numerical inaccuracy caused by the exceedingly large chemical source terms in the species equations. It is noted that the spurious fluctuations in the finite-rate molar fractions can be removed by using smaller time steps in the calculations(CFL number smaller than 10^{-2} is required to removed all the fluctuations for the cases considered in this study. The results shown here are obtained using CFL of 0.6).

§3.2 Steady Nozzle Problem

Calculations using Roe splitting for steady flows in 1D divergent nozzle are given in Figs. 5 and 6. Figure 5 shows the results $\rho/\rho_\infty, u/a_\infty, p/p_\infty$, and e/e_∞ for the nozzle flow with a shock. The nozzle area ratio is 10 and the inflow conditions are T_∞ =6000 k, p_∞ =10 atm. Excellent result is obtained with monotone and sharp resolution across the shock. The finite-rate results agree with those of equilibrium almost exactly due to the high temperature and high pressure involved. To illustrate the nonequilibrium effect, results for air flow, with the same inlet conditions, expanding through a divergent nozzle of area ratio 200 are presented in Fig.6. Initially, the finite-rate results coincide with equilibrium calculations almost exactly. But, because of the very large area ratio of the nozzle, the temperature and pressure rapidly decrease to values at which the flow is essentially chemically frozen. This equilibrium-sudden freezing behavior of the hypersonic nozzle flow, as discussed by Bray[11], is clearly observed in Fig. 6.

4. CONCLUDING REMARKS

Formulations of both equilibrium and finite-rate chemistry for real-gas flows are described. Special care has been exercised to avoid unnecessary assumptions and approximations. The calculated results demonstrate the validity as well as the advantage of the present approach.

5. REFERENCES

1. Steger, J.L. and Warming, R.F., J. Comp. Phys. **40**, 263, 1981.
2. van Leer, B., Lecture Notes in Physics, **170**, 507, 1982.
3. Roe, P.L., J. Comp. Phys. **43**, 357, 1981.
4. Collela, P. and Glaz, H.M., J. Comp. Phys. **59**, 264, 1985.
5. Grossman, B. and Walters, R.W., AIAA paper No. 87-1117-CP, 1987.
6. Vinokur, M. and Liu, Y., AIAA Paper No. 88-0127, 1988.
7. Glaister, P., J. Comp. Phys. **74**, 382, 1988.
8. Liou, M.S., van Leer, B., and Shuen, J.S., J. Comp. Phys. (to appear).
9. Roe, P.L., Ann. Rev. Fluid Mech. **18**, 337,1986.
10. Gordon, S. and McBride, B.J., NASA SP-273, 1976.
11. Bray, K.N.C., Nonequilibrium Flows, edited by P.P. Wegener, Marcel Dekker, Inc., New York, 1970.

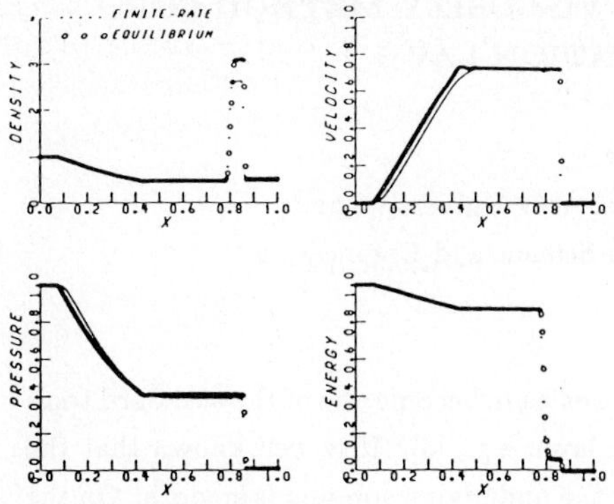

Figure 1. Shock tube problem, Roe flux-difference splitting.

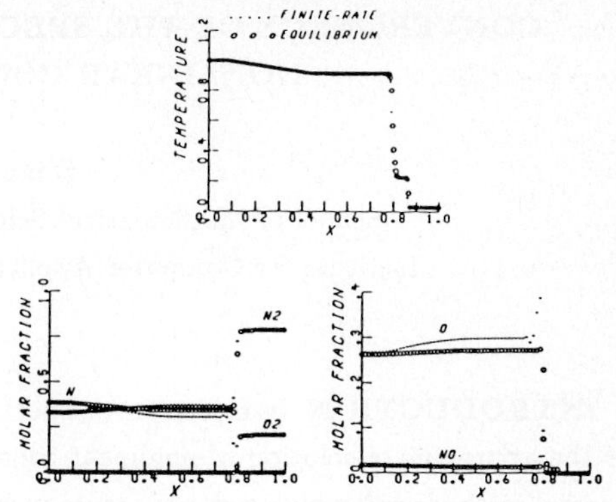

Figure 4. Shock tube problem, temperature and species molar fractions.

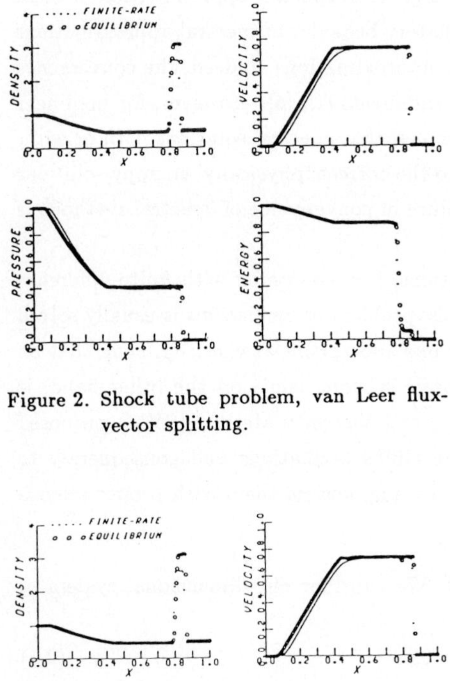

Figure 2. Shock tube problem, van Leer flux-vector splitting.

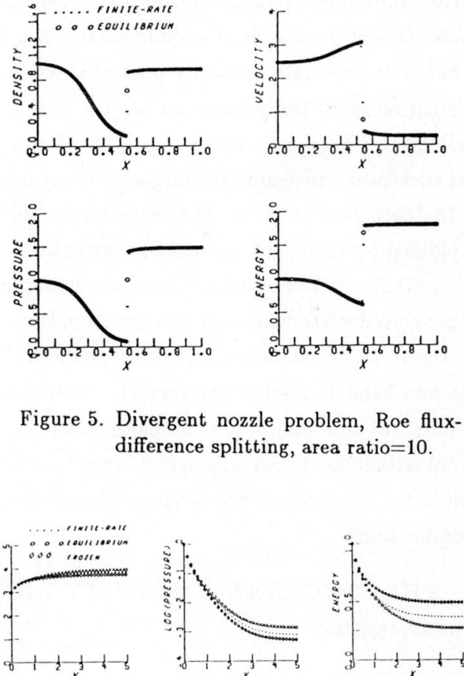

Figure 5. Divergent nozzle problem, Roe flux-difference splitting, area ratio=10.

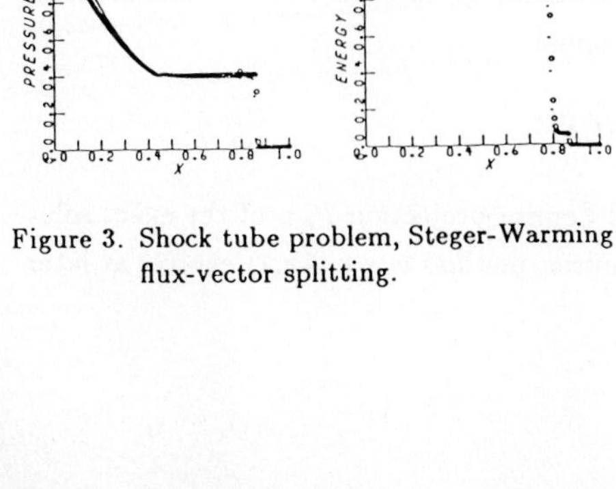

Figure 3. Shock tube problem, Steger-Warming flux-vector splitting.

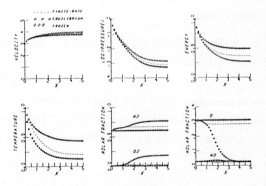

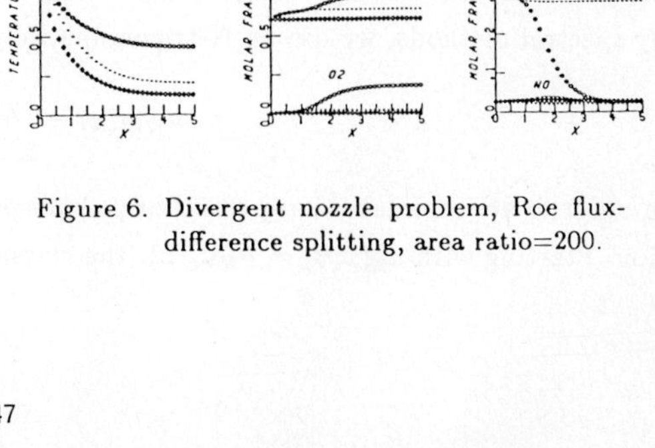

Figure 6. Divergent nozzle problem, Roe flux-difference splitting, area ratio=200.

CONVERGENCE OF THE SPECTRAL VISCOSITY METHOD FOR NONLINEAR CONSERVATION LAWS

Eitan Tadmor
School of Mathematical Sciences, Tel-Aviv University, and
Institute for Computer Applications in Science and Engineering

1. INTRODUCTION. In recent years, spectral methods have become one of the standard tools for the approximate solution of nonlinear conservation laws, e.g., [3]. It is well known that the spectral methods enjoy high order of accuracy whenever the underlying solution is smooth. On the other hand, one of the main disadvantages of using spectral methods for nonlinear conservation laws, lies in the formation of Gibbs phenomena, once spontaneous shock discontinuities appear in the solution; the global nature of spectral methods then pollutes the unstable Gibbs oscillations overall the computational domain and prevent the convergence of spectral approximation in these cases. One of the standard techniques to mask the oscillatory behavior of spectral approximations is based on spectrally accurate post-processing of these approximations. Indeed, the convergence of such recovery techniques can be justified by linear arguments [1], [5]. However, for nonlinear problems we show by a series of prototype counterexamples, that spectral solutions with or without such post-processing techniques, do not converge to the correct 'physically' entropy solutions of the conservation laws. The main reason for this failure of convergence of spectral methods is explained by their lack of entropy dissipation.

A similar situation which involves unstable oscillations, is encountered with finite-difference approximations to nonlinear conservation laws. Here, the problem of oscillations is usually solved by the so called vanishing viscosity method. Namely, one adds artificial viscosity, such that on the one hand it retains the formal accuracy of the basic scheme, while on the other hand, it is sufficient to stabilize the Gibbs oscillations. The Spectral Viscosity Method (SVM) proposed in [6] attempts, in an analogous way, to stabilize the Gibbs oscillations and consequently to guarantee the convergence of spectral approximations, by augmenting them with proper viscous modifications.

2. THE SPECTRAL VISCOSITY METHOD. We consider one-dimensional system of conservation laws

$$\frac{\partial u}{\partial t} + \frac{\partial (f(u))}{\partial x} = 0, \tag{2.1}$$

with prescribed initial data $u_0(x)$. We restrict our attention to the periodic initial-value problem (2.1) which avoids the separate question of boundary treatment. To solve this 2π-periodic problem by spectral methods, we use an N-trigonometric polynomial

$$u_N(x,t) = \sum_{k=-N}^{N} \hat{u}_k(t) e^{ikx},$$

in order to approximate the spectral or pseudospectral Fourier projection $P_N u$ of the exact solution. Starting with $u_N(x,0) = P_N u_o(x)$, the classical Fourier method lets $u_N(x,t)$ evolves at later time according to the approximate model

$$\frac{\partial u_N}{\partial t} + \frac{\partial}{\partial x}[P_N(f(u_N(x,t)))] = 0. \tag{2.2}$$

We have already noted that the convergence to the entropy solution of (2.1), $u_N \underset{N\to\infty}{\longrightarrow} u$ may (and in fact, in some cases must) fail. Instead, we propose to replace (2.2) by appending to it a spectrally accurate vanishing viscosity modification which amounts to

$$\frac{\partial u_N}{\partial t} + \frac{\partial}{\partial x}[P_N f(u_N(x,t))] = \varepsilon \frac{\partial}{\partial x}\left[Q_m(x,t) * \frac{\partial u_N}{\partial x}\right]. \tag{2.3}$$

Here, $Q_m(x,t)$ is a viscosity kernel of the form

$$Q_m(x,t) = \sum_{m \le |k| \le N} \hat{Q}_k(t) e^{ikx}. \tag{2.4}$$

This kind of spectral viscosity can be efficiently implemented in the Fourier rather than in the physical space, i.e.,

$$\varepsilon \frac{\partial}{\partial x}\left[Q_m(x,t) * \frac{\partial u_N}{\partial x}\right] \equiv -\varepsilon \sum_{m \le |k| \le N} k^2 \hat{Q}_k(t) \hat{u}_k(t) e^{ikx}. \tag{2.5}$$

It depends on two free parameters: the viscosity amplitude, $\varepsilon = \varepsilon(N)$, and the effective size of the inviscid spectrum, $m = m(N)$. In [6] it was shown that these two parameters can be chosen so that we have both, i.e., the spectral viscosity retains the spectral accuracy of the overall approximation as $m(N) \uparrow \infty$, and at the same time, it is sufficient to enforce the correct amount of entropy dissipation that is missing otherwise (with $\varepsilon = 0$).

Entropy dissipation is necessary for convergence; the lack of such dissipation in the classical Fourier methods is the main reason for its divergence. On the other hand, under appropriate assumptions, one can use compensated compactness arguments [8] to show that entropy dissipation induced by the SV method is sufficient for convergence.

3. CONVERGENCE. We shall discuss two model problems.

Example 3.1 The scalar case. We consider general nonlinear scalar conservation laws (2.1). The pseudospectral viscosity method collocated at the equidistant points $x_\nu = \frac{2\pi\nu}{N}$, takes the form

$$\frac{\partial}{\partial t} u_N(x_\nu, t) + \frac{\partial}{\partial x}[P_N f(u_N(x_\nu, t))] = -\frac{1}{N} \sum_{|k|=\sqrt{N}}^{N} \frac{(k-\sqrt{N})^2}{N-\sqrt{N}} k^2 \hat{u}_k(t) e^{ikx_\nu}. \tag{4.1}$$

It can be shown, [7], that the spectral viscosity on the right guarantees entropy dissipation and the L^∞-stability of the overall approximation; consequently convergence follows.

Example 3.2 The isenotropic gas dynamics equations

$$\begin{aligned} &\rho_t + (\rho u)_x = 0 \\ &(\rho u)_t + (\rho u^2 + p(\rho))_x = 0 \end{aligned} \tag{4.2}$$

for a polytropic gas $p = Const.\rho^\gamma$, $\gamma < 1$ are approximated in a similar fashion. Under appropriate L^∞-stability assumption (in particular, a uniform bound from the vacuum state), it can be shown [3], [7] that the spectral viscosity solution converges.

References

[1] S. Abarbanel, D. Gottlieb and E. Tadmor, Spectral methods for discontinuous problems, in "Numerical Analysis for Fluid Dynamics II" (K. W. Morton and M. J. Baines, eds.), Oxford University Press, 1986, pp. 129-153.

[2] C. Canuto, M. Y. Hussaini, A. Quarteroni and T. Zang, Spectral Methods with Applications to Fluid Dynamics, Springer-Verlag, 1987.

[3] R. DiPerna, Convergence of approximate solutions to systems of conservation laws, Arch. Rat. Mech. Anal., Vol. 82, pp. 27-70 (1983).

[4] Y. Maday and E. Tadmor, Analysis of the spectral viscosity method for periodic conservation laws, SIAM J. Num. Anal., to appear.

[5] A. Majda, J. McDonough and S. Osher, The Fourier method for nonsmooth initial data, Math. Comp., Vol. 30, pp. 1041-1081 (1978).

[6] E. Tadmor, Convergence of spectral methods for nonlinear conservation laws, SIAM J. Num. Anal., in press.

[7] E. Tadmor, Semi-discrete approximations to nonlinear systems of conservation laws; consistency and L^∞-stability imply convergence, ICASE Report No. 88-41.

[8] L. Tartar, Compensated compactness and applications to partial differential equations, Research Notes in Mathematics 39, Nonlinear Analysis and Mechanics, Heriott-Watt Symposium, Vol. 4 (R. J. Knopps, ed.), Pittman Press, pp. 136-211 (1975).

The Fourier method with spectral vanishing viscosity ... and without spectral viscosity.

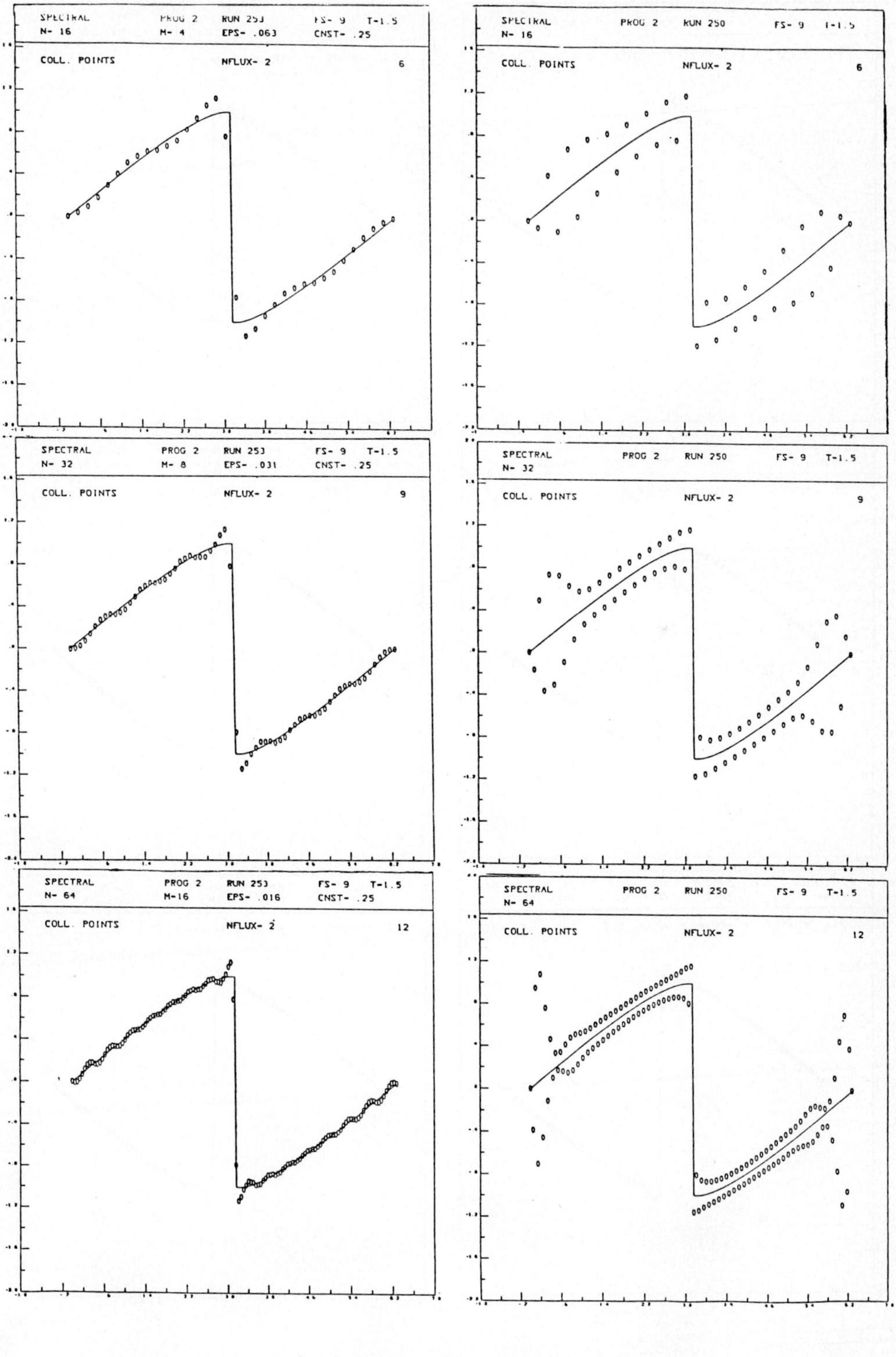

The Fourier method with smooth spectral vanishing viscosity.

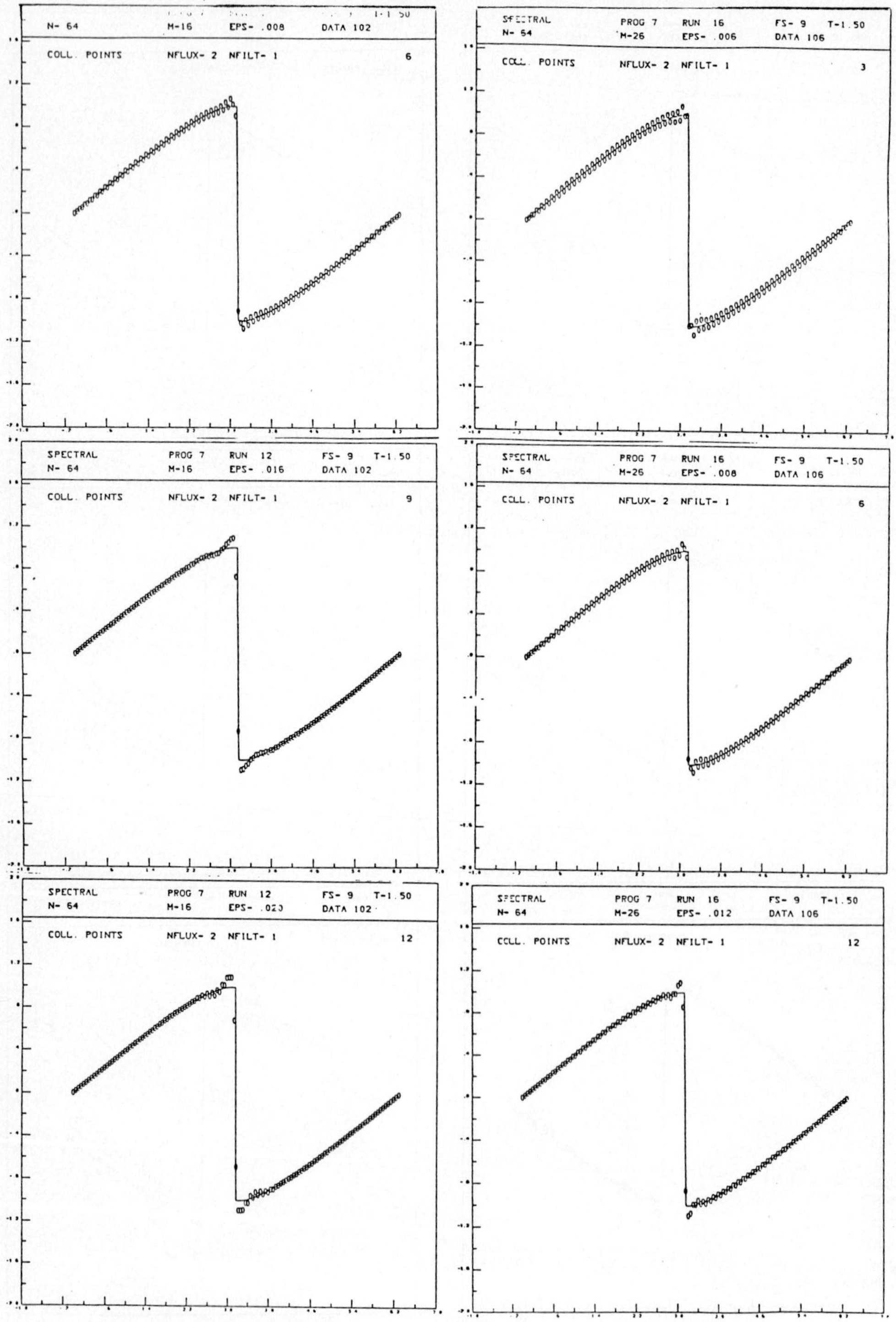

INVISCID AND VISCOUS FLOW SIMULATIONS AROUND THE ONERA-M6 WING BY TVD SCHEMES

Yoko Takakura+, Satoru Ogawa++ and Tomiko Ishiguro++

\+ Scientific Systems Department, Fujitsu Limited, 1-17-25 Shinkamata, Otaku, Tokyo 144, Japan
++ Computational Sciences Division, National Aerospace Laboratory, 7-44-1 Jindaiji-Higashimachi, Chofushi, Tokyo 182, Japan

I. INTRODUCTION

Although the ONERA-M6 wing is a simple and basic configuration, the flow problems around that contain the aerodynamically significant phenomena such as the interaction between shock waves, between shock wave and boundary layer and between two-wall boundary layers and wing tip vortex. In a series of our works[1,2] --- improvements of TVD schemes in general coordinates ---, inviscid flow problems around the ONERA-M6 wing are used as tests. In this paper the simulation indicated by the title are performed, mainly to examine how the TVD schemes capture the above-stated viscous phenomena, at the same time to complement the scheme evaluation for inviscid flow problems. In the viscous flow simulations, the more stable operators to reach steady states would be needed if the turbulent models are used, because the minimum grid size is very small and therefore the time stepping is limited. Here the improvements of the left-hand-side operator and the practicalization of large eddy simulation are also presented.

II. TVD DIFFERENCE SCHEMES

The conservation laws in general coordinate system

$$\frac{\partial \tilde{Q}}{\partial \tau} + \frac{\partial (\tilde{F}^{\ell} - \tilde{F}_v{}^{\ell})}{\partial x^{\ell}} = 0,$$

are numerically solved, where F_v is the viscous term, and ~ denotes the variable multiplied by the volume element $\sqrt{g}$. The TVD numerical fluxes are applied to the inviscid terms and the viscous terms are evaluated by the central-difference approximation.

II-1. TVD NUMERICAL FLUXES

The numerical flux of inviscid terms in the general coordinate system is

$$\bar{F}^{\ell}{}_{i+1/2} \equiv (1/2)[\, \tilde{F}^{\ell}{}_i + \tilde{F}^{\ell}{}_{i+1} + (\sqrt{g}\, R^{\ell}\, \Phi^{\ell})_{i+1/2} \,],$$

and the following two characteristic quantities can be introduced as a substitute of the propagated quantity in the single scaler conservation law:

$$\text{i)} \quad \alpha_{\nu} \equiv (R^{\ell})^{-1}{}_{i+1/2}\, \Delta_{\nu} Q,$$

$$\text{ii)} \quad \alpha_{\nu} \equiv (R^{\ell})^{-1}{}_{\nu}\, \Delta_{\nu} Q, \qquad \nu = i-1/2,\ i+1/2,\ i+3/2.$$

The above treatment is based on the following exact relation [2]:

$$\frac{\partial \tilde{F}^{\ell}}{\partial x^{\ell}} = \frac{\partial \tilde{F}^{\ell}}{\partial Q} \cdot \frac{\partial Q^{\ell}}{\partial x^{\ell}} = \sqrt{g}\, \Lambda^{\ell} \cdot \frac{\partial Q^{\ell}}{\partial x^{\ell}} = \sqrt{g}\, R^{\ell} \Lambda^{\ell} (R^{\ell})^{-1} \cdot \frac{\partial Q^{\ell}}{\partial x^{\ell}}$$

a) Modified Harten-Yee (H-Y) Numerical Flux [3][1]

The elements of Φ^{ℓ} denoted by ϕ^m are

$$\phi^m_{i+1/2} \equiv (1/2)\, \psi(a^{\ell\, m}_{i+1/2})(g^m_i + g^m_{i+1}) - \psi(a^{\ell\, m}_{i+1/2} + \gamma^m_{i+1/2})\, \alpha^m_{i+1/2},$$

with the adjustment quantity for high accuracy $g^m{}_i$ defined by

$$g^m_i = \text{minmod}[\, \alpha^m_{i+1/2}\, ,\ \alpha^m_{i-1/2}].$$

b) Modified Chakravarthy-Osher (C-O) Numerical Flux [4] [1]

$$\Phi'_{i+1/2} = -(\sigma^{+}_{i+1/2} - \sigma^{-}_{i+1/2})$$
$$- \{(1-\phi)/2\}\bar{\sigma}^{-}_{i+3/2} - \{(1+\phi)/2\}\bar{\bar{\sigma}}^{-}_{i+1/2}$$
$$+ \{(1+\phi)/2\}\bar{\sigma}^{+}_{i+1/2} + \{(1-\phi)/2\}\bar{\bar{\sigma}}^{+}_{i-1/2},$$

where

$$\sigma^{\pm}_{i+1/2} = \Lambda^{\pm}_{i+1/2}\ \alpha_{i+1/2},$$
$$\bar{\sigma}^{-}_{i+3/2} = \Lambda^{-}_{i+1/2}\cdot\text{minmod}[\alpha_{i+3/2},\ \beta\alpha_{i+1/2}],$$
$$\bar{\bar{\sigma}}^{-}_{i+1/2} = \Lambda^{-}_{i+1/2}\cdot\text{minmod}[\alpha_{i+1/2},\ \beta\alpha_{i+3/2}],$$
$$\bar{\sigma}^{+}_{i+1/2} = \Lambda^{+}_{i+1/2}\cdot\text{minmod}[\alpha_{i+1/2},\ \beta\alpha_{i-1/2}],$$
$$\bar{\bar{\sigma}}^{+}_{i-1/2} = \Lambda^{+}_{i+1/2}\cdot\text{minmod}[\alpha_{i-1/2},\ \beta\alpha_{i+1/2}].$$

II-2. IMPROVEMENTS OF THE LEFT-HAND-SIDE OPERATOR

To increase the convergence rate, several improvements for the LHS operater are performed:

Operator 1) correct the $\tilde{Q}$-shift operator to the first-order Q-shift operator

Operator 2) modify the operator 1 to the finite-volume-like operator

Operator 3) add the viscous term to the operator 2

The ADI form of these oprarators is written as

$$(x^1\text{-direction})(x^2\text{-direction})(x^3\text{-direction})\,\delta Q = g\ \widehat{\text{RHS}}\ .$$

The one-directional discrete form of each operator is as follows:

Operator 1) $(1/2)h[\tilde{A}_{i+1} - (R\cdot|\tilde{\Lambda}|\cdot R^{-1})_{i+1/2}]\,\delta Q_{i+1}$

$+ [\sqrt{g}_i I + (1/2)h\{(R\cdot|\tilde{\Lambda}|\cdot R^{-1})_{i+1/2} + (R\cdot|\tilde{\Lambda}|\cdot R^{-1})_{i-1/2}\}]\,\delta Q_i$

$+ (1/2)h[-\tilde{A}_{i-1} - (R\cdot|\tilde{\Lambda}|\cdot R^{-1})_{i-1/2}]\,\delta Q_{i-1}$

Operator 2) $(1/2)h\{R\cdot(\tilde{\Lambda}-|\tilde{\Lambda}|)\cdot R^{-1}\}_{i+1/2}\,\delta Q_{i+1}$

$+ [\sqrt{g}_i I + (1/2)h\{\{R\cdot(\tilde{\Lambda}+|\tilde{\Lambda}|)\cdot R^{-1}\}_{i+1/2} - \{R\cdot(\tilde{\Lambda}-|\tilde{\Lambda}|)\cdot R^{-1}\}_{i-1/2}\}]\,\delta Q_i$

$- (1/2)h\{R\cdot(\tilde{\Lambda}+|\tilde{\Lambda}|)\cdot R^{-1}\}_{i-1/2}\,\delta Q_{i-1}$

Operator 3) $(1/2)h[\{R\cdot(\tilde{\Lambda}-|\tilde{\Lambda}|)\cdot R^{-1}\}_{i+1/2}-\tilde{S}_{i+1/2}]\,\delta Q_{i+1}$ +

$[\sqrt{g}_i I+(1/2)h\{\{R\cdot(\tilde{\Lambda}+|\tilde{\Lambda}|)]\cdot R^{-1}\}_{i+1/2}-\{R\cdot(\tilde{\Lambda}-|\tilde{\Lambda}|)\cdot R^{-1}\}_{i-1/2}+(\tilde{S}_{i+1/2}+\tilde{S}_{i-1/2})\}]\,\delta Q_i$

$- (1/2)h[\{R\cdot(\tilde{\Lambda}+|\tilde{\Lambda}|)\cdot R^{-1}\}_{i-1/2}-\tilde{S}_{i-1/2}]\,\delta Q_{i-1}$

where $\delta Q \equiv Q^{n+1}-Q^n$, $h \equiv \Delta t/\Delta x$, and the subscript also indicates the evaluation point of metrix.

III. TURBULENT MODELS

Without turbulent models no steady solution is obtained in viscous flow problem; the numerical separations occur near the wing tip and spread over the wing, which do not observed in physical experiments. In order to perform direct simulation, more than 10^4 grid points would be needed for each direction with the present Reynolds number used, which is impossible in the existing computer performance. The algebraic model [5] is successfully used here for weakly separated flows. However, for strongly separated flows this model does not work well; for example, the solutions of high attack angle case do not converge to steady states. Therefore we have improved the sub grid scale (SGS) viscosity of large eddy simulation (LES) [6] to apply compressible flow problems. Our formulation is

$$\mu_{SGS} = (C\Delta)^2\rho\,|\omega|,\qquad \Delta = \min(\Delta_{x1},\Delta_{x2},\Delta_{x3}),\qquad C \sim 0.5$$

where C is a coefficient, Δ is the length of grid scale and ω is the vorticity.

IV. RESULTS AND DISCUSSION

IV-1. CONDITIONS FOR NUMERICAL EXPERIMENTS

The numerical experiments are carried out for the ONERA-M6 wing [7]. In order to improve the convergence rate, the local time stepping with constant Courant number is adopted. Two types of grid, C-H and C-O, are used. The number of grid points is 191x33x24 or 191x46x24(for two-wall problem) and minimum grid spacing is 10^{-4}. Two typical flow patterns are solved: Case 1, standard test problem $M_\infty=0.8395$, $\alpha=3.06°$, $Re=1.2\times10^7$ and Case 2, problem of higher Mach number and attack angle $M_\infty=0.9298$, $\alpha=6.07°$, $Re=1.2\times10^7$.

IV-2. CHOICE OF CHARACTERISTIC QUANTITIES

Since each ways of i) and ii) in Chap.II-1 yield almost the same numerical solutions with the other, ii) is used for the computational cost, although i) gives closer approximation.

IV-3. COMPARISON OF LEFT-HAND-SIDE OPERATORS

Figure 1 shows the convergence history of each operator in Chap.II-2 with the local time step of constant Courant number 3 in the case of IV-4 b). The Q-shift operators have the better convergency and moreover are capable to take the larger Courant number than the original $\hat{Q}$-shift operators.

IV-4. COMPARISON OF SOLUTIONS IN STANDARD PROBLEM (CASE 1)

The solutions of H-Y and C-O numerical fluxes are compared with experimats.

a) Inviscid Flows The discrepancy of solutions on the C-H and C-O grids exists mainly near the wing tip; the C-O grid yields the solution in better agreement with the experiments. For reference, C_p distributions of the Anderson & Thomas TVD scheme [8] are included in comparison (Fig.2).

b) Viscous Flows As far as the C_p distributions (Fig.3) are concerned, little difference by the grid types is observed. However, when examining the flow patterns (oilflows and streamlines) carefully, the important difference is observed; the C-O grid captures the bubble behind the united shock wave and the wing tip vortex (Fig.4), whereas the C-H grid does not. Almost the same solutions are obtained in both cases of the thin layer and full NS eqs.

c) Two Wall Effect at the Wing-Root In the wind-tunnel experiments a small vertical wall is usually installed at the wing root to protect wing flows from the interaction of wind-tunnel and wing-flow boundary layers. In order to simulate experiments more accurately, a part of the symmetric plane at the wing root is treated as the wall and the full NS eqs. is solved with the two-wall algebraic turbulent model [9]. Then the C_p distributions near the wing root (20% semi-span) and the united point (80% semi-span) tend to agree with experiments better.

IV-5. COMPARISON OF SOLUTIONS IN HIGH ATTACK ANGLE PROBLEM (CASE 2)

Since the algebraic model yields the false separation caused numerically in this case, the LES is performed. In the numerical experiments of case 1, the solutions of LES show the good agreement with experiments. Also in case 2, those show the reasonable agreement for the use of H-Y and C-O numerical fluxes (Fig.5).

V. CONCLUDING REMARKS

Numerical simulations around the ONERA M-6 wing indicate the TVD schemes are applicable to both invisid and viscous flow problems. Especially, the solutions of Euler eqs. in the low attack angle case are excellent. As is widely known, in the case of large separation the NS eqs. must be solved, and at that time, the turbulent models are needed to obtain steady solutions, since without the turbulent models the NS eqs. cause the numerical separations which do not appear in physical experiments. From our experience, however, we cannot but consider that the main effect of turbulent models at present is only to suppress the numerical separation. It would be necessary to seek for the better turbulent model.

In addition this study has been done as a part of the software development in National Aerospace Laboratory.

REFERENCES

[1] Takakura,Y., Ishiguro,T. & Ogawa,S., "On the Recent Difference Schemes for the Three-Dimensional Difference Schemes," AIAA paper 87-1151.

[2] Takakura,Y., Ishiguro,T. & Ogawa,S., "On the TVD Difference Schemes for the Three-Dimensional Euler Equations in General Coordinates," ISCFD-Sydney.

[3] Yee,H.C. & Harten,A., "Implicit TVD schemes for Hyperbolic Conservation Laws in Curvilinear Coordinates," AIAA paper 85-1513.

[4] Chakravarthy,S.R. & Osher,S., "A New Class of High Accuracy TVD Schemes for Hyperbolic Conservation Laws," AIAA paper 85-0363.

[5] Baldwin,B.S. & Lomax,H., "Thin layer Approximation and Algebraic Model for Separated Turbulent Flows," AIAA paper 78-257.

[6] Deardorff,J.W., " a Numerical Study of Three-Dimensional Turbulent Channel Flow at Large Reynolds Numbers," J.Fluid Mech., 41(1970)452-480.

[7] Schmitt,V. & Charpin,F., "Pressure Distributions on the ONERA-M6 WING at Transonic Mach Numbers, " AGARD AR-138-B1, (1979).

[8] Anderson,W.K. & Thomas,J.L., "Multigrid Acceleration of the Flux Split Euler Equations," AIAA paper 86-0274.

[9] Hung,C. & Buning P.G., "Simulation of blunt-fin-induced shock-wave and turbulent boundary-layer interaction," J.Fluid.Mech.154(1985)163-185.

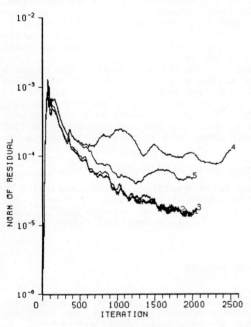

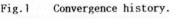

Fig.1 Convergence history.

1~3 Q-shift operator

4 Original ADI

5 Diagonalized original ADI

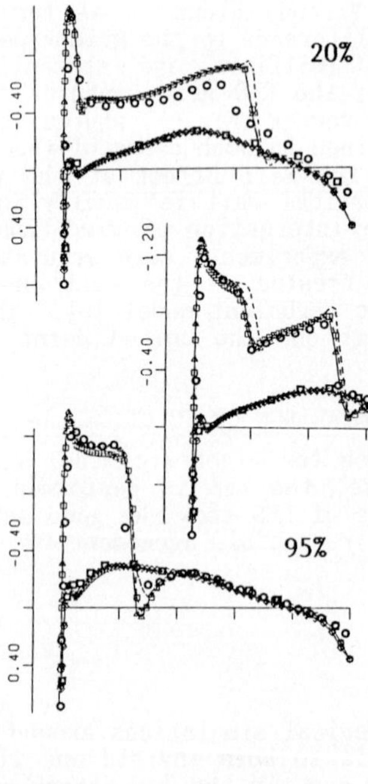

Fig.2

Comparison of schemes for Euler eqs. on C-O grid (Case 1).

△ Harten-Yee, □ Chakravarthy-Osher, — Anderson-Thomas, ○ Experiments

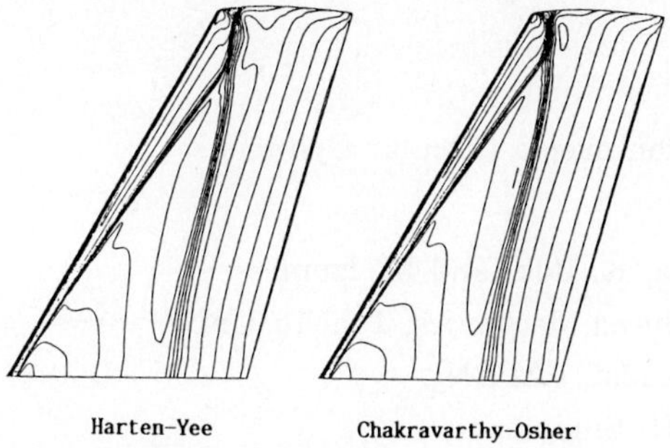

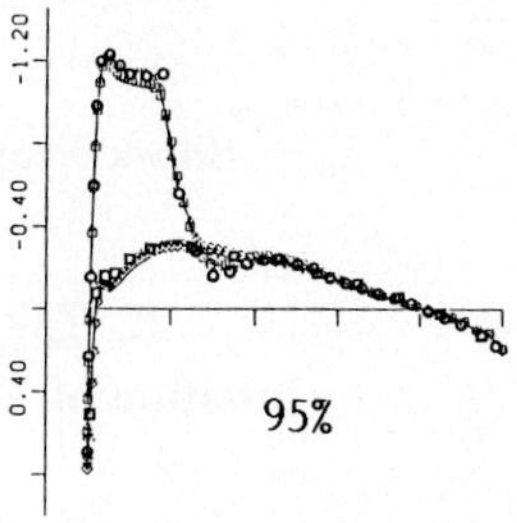

(a) Isobaric contours on upper wing surface.

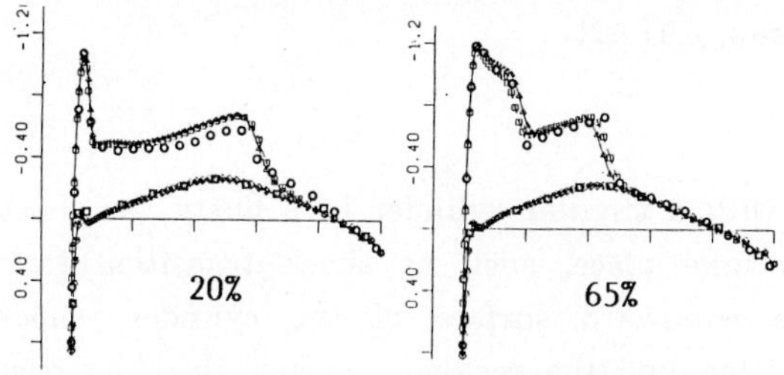

(b) C_p distributions on wing surface.

Fig.3 Comparison of schemes for thin-layer NS eqs. on C-O grid (Case 1).

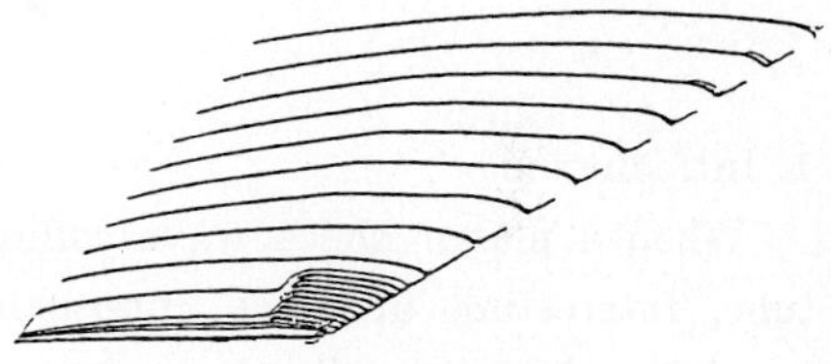

(a) Oilflow on C-H grid.

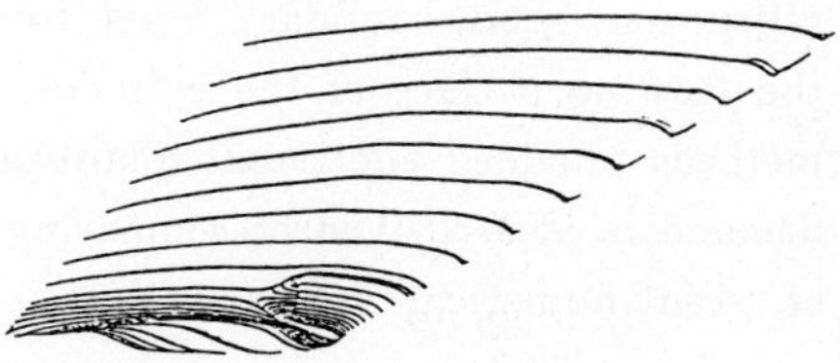

(b) Oilflow on C-O grid.

Fig.4 Comparison of oilflows on C-H and C-O grid (Case 1).

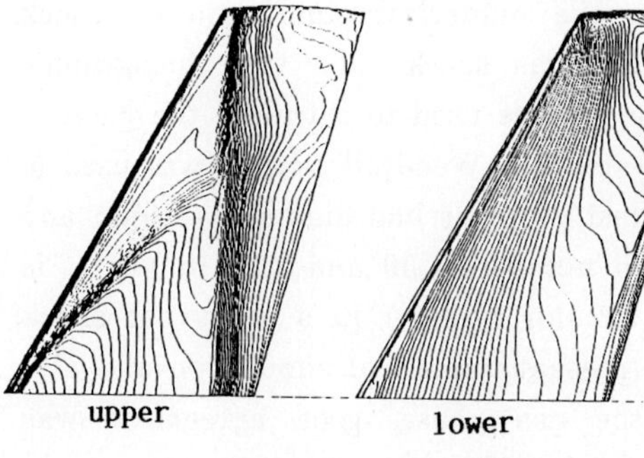

(a) Isobaric contours

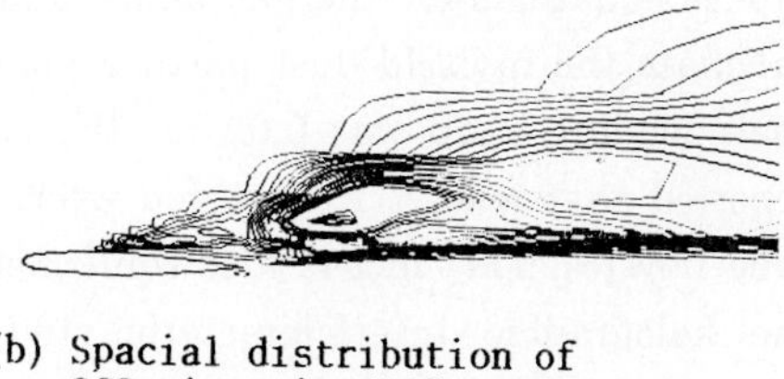

(b) Spacial distribution of SGS viscosity(80% semi-span).

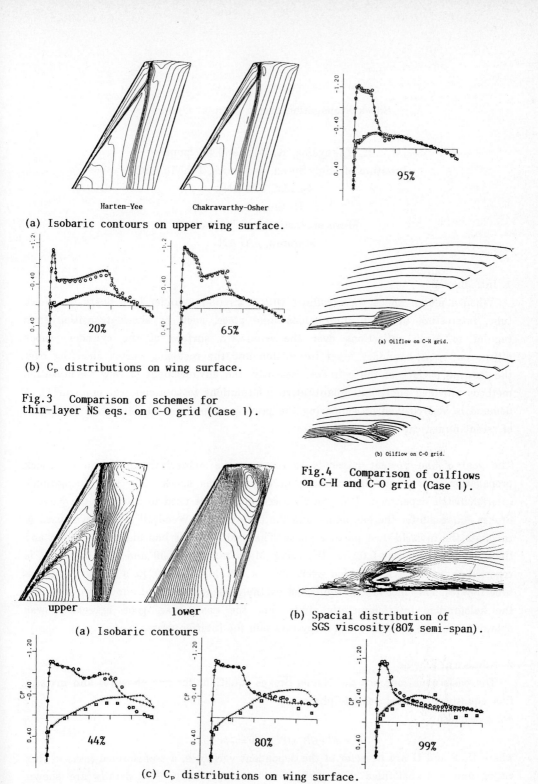

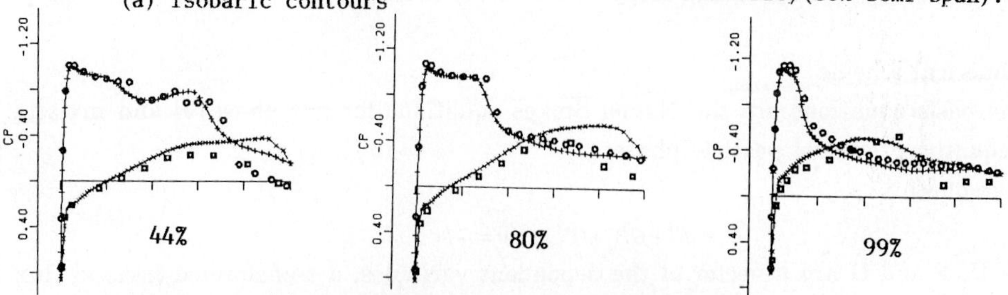

(c) C_p distributions on wing surface.

Fig.5 Solution of LES by H-Y scheme (Case 2)

Shock Propagation over a Circular Cylinder

K. Takayama, K. Itoh, and M. Izumi
Institute of High Speed Mechanics, Tohoku Univ.
Sendai, JAPAN
H. Sugiyama
Muroran Institute of Technology
Muroran, JAPAN

1. Introduction

When a planar shock wave collides with a circular cylinder in a dusty gas shock tube, interesting unsteady interactions take place, such as shock transition from regular to Mach reflections over the windward surface of the cylinder, shock diffraction, shock-boundary layer interaction and the resulting vortex shedding over the leewrad surface of the cylinder. Recently, the advancement of flow visualization methods provided additional quantitative informations to this classical problem [1]. A demand is also revived of comparing the previous experimental data with the result of recent numerical simulations [2].

The goal of the present paper is to compare the numerical simulation on shock propagation over a circular cylinder in a dusty gas shock tube with holographic interfelometric experiments [3]. The TVD scheme [4] was used to calculate the Navier-Stokes equations for the gas phase and the two-step Lax-Wendroff scheme was used to calculate the inviscid dust particle phase. The dust particles had diameter of 5μm and their loading ratio was 0.02 for the shock Mach numbers 1.30 and 2.15 in air. It is reported that a dust free region exists over a circular cylinder in a dusty gas shock tube flow [5]. This fact is also confirmed by the present numerical simulation and also the holographic interfelometoric study. For the gas phase, good agreement was obtained between the numerical isopycnics and the interferograms.

2. Numerical scheme

The basic equations are the Navier-Stokes equations for gas phase [6] and inviscid flow equations for dust particle phase;
For gas phase,

$$U_t+(F^1+G^1)_x+(F^2+G^2)_y=-I \qquad (1)$$

where U, F and G are a vector of the dependent variables, a transformed inviscid flux vector and a transformed viscous flux vector, respectively. The details are shown elsewhere [6]. I denotes the interaction term between gas and particle phase.

For dusty phase,

$$V_t + H_x^{\ 1} + H_y^{\ 2} = I \tag{2}$$

where V, H are a vector of the dependent variables and transformed inviscid flux vector, respectively.

$$V = \begin{vmatrix} \rho_p \\ \rho_p u_p \\ \rho_p v_p \\ e_p \end{vmatrix}, \quad H^1 = u_p V, \quad H^2 = v_p V \tag{3}$$

where ρ_p, u_p, v_p and e_p are particle density, x, y velocity components of the particle and the particle energy. It is assumed that particles are spherical and distrubuted spacially uniformly. The concept of the continuum flow is still valid.

The interaction term I can be given,

$$I = \begin{vmatrix} 0 \\ D_x \\ D_y \\ u_p D_x + v_p D_y + Q \end{vmatrix} \tag{4}$$

where

$$D_y = C^D \rho_p (v - v_p)|v - v_p|, \quad D_x = C^D \rho_p (u - u_p)|u - u_p| \tag{5}$$

C^D is the drag coefficient which is given by a formula [7],

$$C^D = \frac{\pi d}{8\,m}(0.48 + 28Re^{-0.85}) \tag{6}$$

d and m are particle diameter 5μm, weight, Q denotes the heat transfer [7],

$$Q = \rho_p \pi d \mu Cp Nu (T - T_p)/Prm \tag{7}$$

The basic equation Eq. (1) is solved by using the TVD scheme and Eq.(2) is solved by the two step Lax-Wendroff. The boundary conditions are that gas velocity and also particle velocity are zero on the cylinder. As the initial values, the nonequilibrium dusty gas shock wave structure which was obtained by one-dimensional computation is used. The computation was conducted in the polar co-ordenates. The mesh sizes were 241 × 321 for circumferential and radial direction, respectively.

3. Experiment

With these combinations of the parameters, the relaxation lengths were 75 cm for the relaxation lengths were 75 cm for M_s = 1.30 and 40 cm for M_s = 2.15. Experiments were conducted in a 30 mm x 40 mm shock tube having a 10 cm circular cylinder in the test section. Flow visualization was conducted with double exposure holographic interferometry [3]. The dust was fly ash having average diameter of 5μm. The dust loading ratio was adjusted to be 0.02 by a dust circulation system [8]. The loading ratio was monitored by means of the laser light extinction method. The corresponding Reynolds numbers are 6.5×10^4 and 1.6×10^5 for M_s = 1.30 and 2.15, respectively.

4. Results and discussion

Figure 1 shows comparison between a numerical simulation and the experiment. Gas phase isopycnics are compared in Fig.1 (a) and (b). The dust mass concentrations are also compared in Fig.1 (c) and (d). A remarkable agreement was obtained between the gas phase isopycnics. In the numerical simulation, the Reynolds number was adjusted to the experiment. In the experiment the dust free regions are clearly observable as dark regions over the cylinder. It is noted that the contrast of the image indicating the dust free region is different in the non-reconstructed hologram, since even the non-reconstructed hologram already includes the information of the phase difference between double exposures. The shape of the dust free regions agrees very well with the computation. The existence of the dust free regions is attributed to the contrifugal force which would shift the trajectory of the dust particle away from the cylinder surface. This trend is later encouraged significantly by being coupled with the boundary layer separation over the cylinder surfaces.

Figure 2 also shows the case for M_s = 2.15. The dramatic growth of the dust free region can be seen. The results of numerical simulations agree very well with the experiments. When in the dusty gas shock tube flow, the unsteady drag force over the cylinder is evaluated, the enoumous enhancement of the drag force appears.

5. Conclusions

It is concluded that the present TVD scheme applied to the Lax-Wendroff finite difference scheme is of particularly useful to resolve accurately the whole time history of dusty gas shock propagation over a circular cylinder.

Acknowledgements

the authors would like to express their gratitude to Messers O. Onodera, H. Ojima and S. Hayasaka for their help in conducting the experiment and also to Miss K. Sasaya for her typewriting the manuscript.

References

[1] Bryson, E. E. and Gross, R. W. F., J F M, 10 (1961).

[2] Yang, J. Y. et. al., AIAA J., 25 (1987).

[3] Takayama, K., Proc. SPIE, 298 (1983).

[4] Harten, A., J. Comp. Phys. 49 (1983).

[5] Shirozu, M. Master Thesis, Kyushu U niv. Faculty of Engineering (1977).

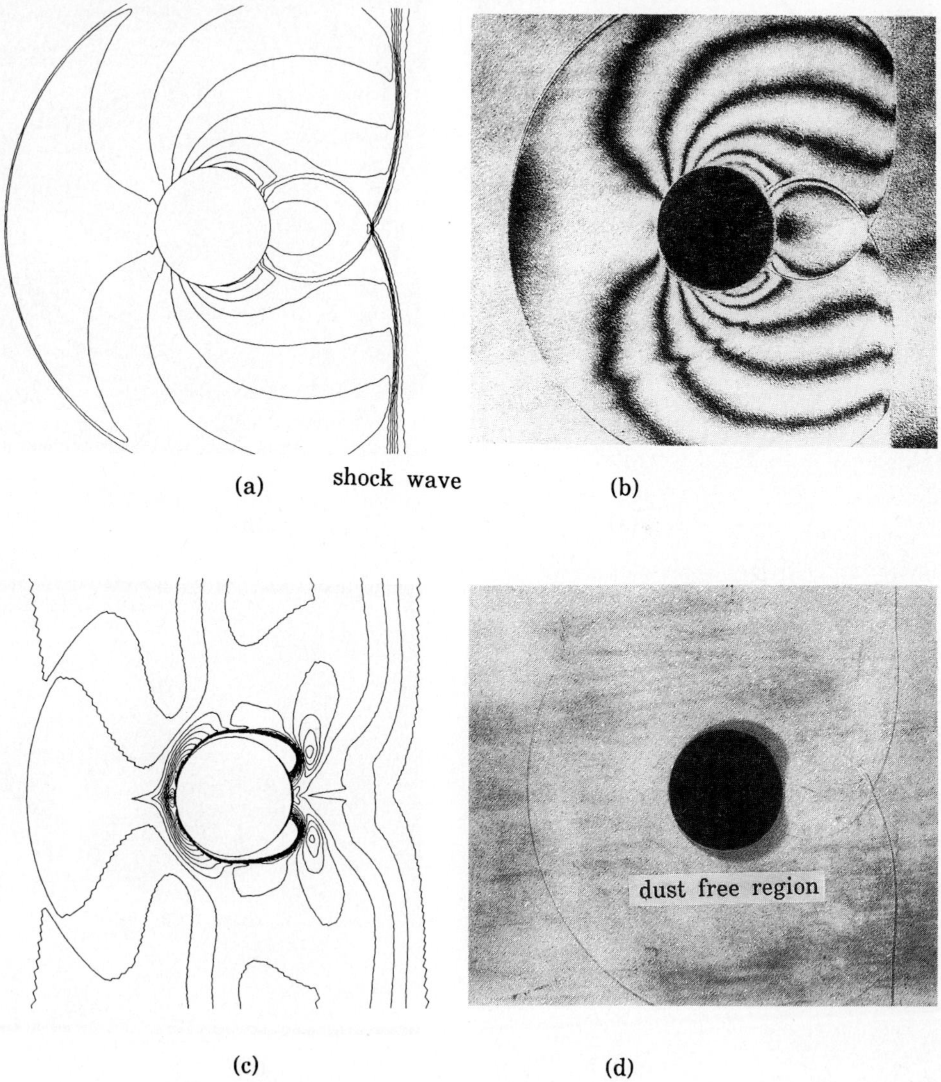

Fig. 1 Comparison of numerical simulation with holographic interferogram, M_S = 1.30, Re = 6.5 × 10^4, d = 5 μm and η = 0.02. (a) gas phase isopycnics, (b) interferogram, (c) dust mass concentration, simulation and (d) non-reconstructed hologram.

[6] Itoh, K. and Takayama, K., Proc. 16th ISTS & W (1988).
[7] Miura, H. and Glass, I. I. UTIAS Report No. 250 (1981).
[8] Sugiyama, H. et al., Trans. JSME, Ser. B, 496 (1987).

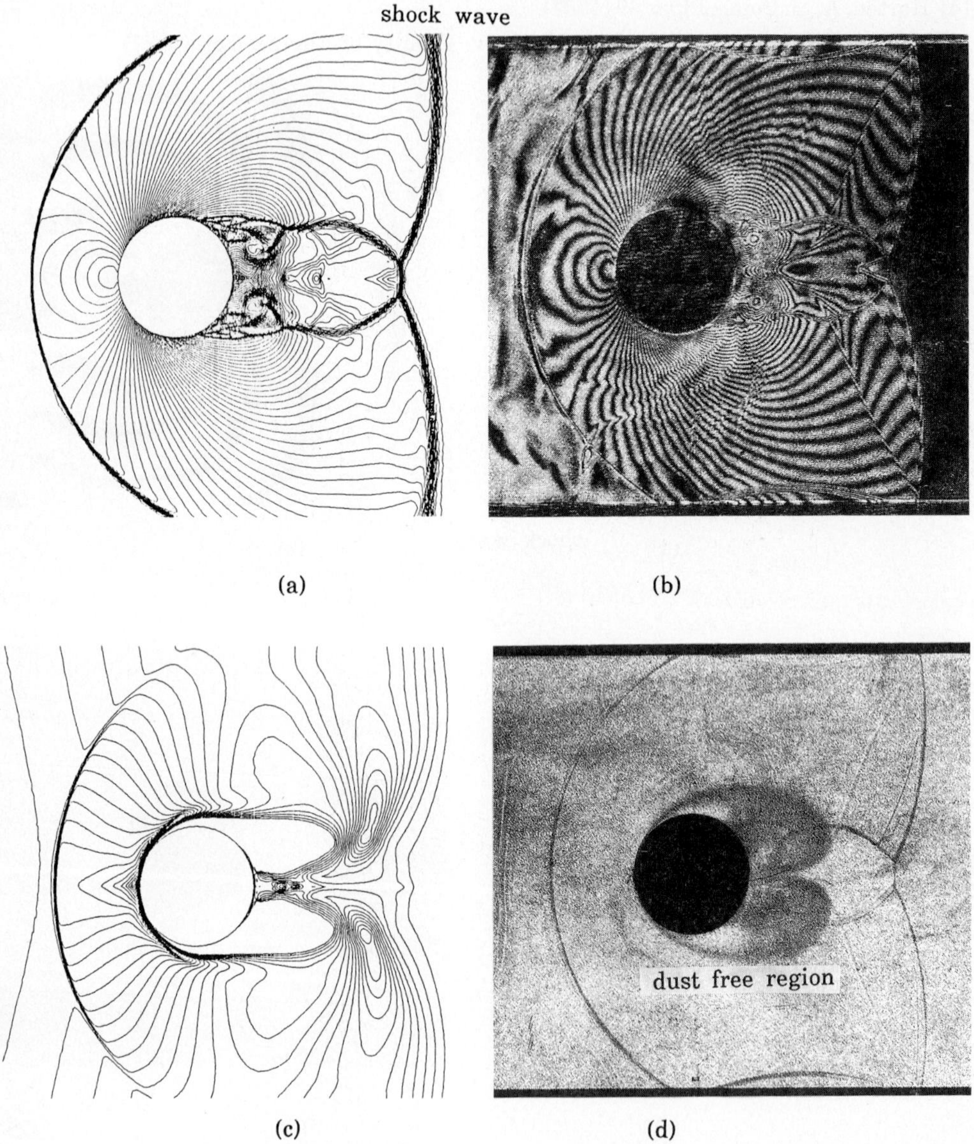

Fig. 2 Comparison of numerical simulation with holographic interferogram, M_S = 2.15, Re = 1.6 × 10^5, d = 5 μm and η = 0.02. (a) gas phase isopycnics, (b) interferogram, (c) dust mass concentration, simulation a nd (d) non-reconstructed hologram.

Three-dimensional Computation of Unsteady Flows around a Square Cylinder

Tetsuro Tamura
ORI, Shimizu Corporation,
2-2-2 Uchisaiwai-cho, Chiyoda-ku, Tokyo 100, Japan
Egon Krause
Aerodynamisches Institut, RWTH Aachen,
Wüllnerstraße zw, 5. u. 7, Aachen, West Germany
Susumu Shirayama and Katsuya Ishii
Institute of Computational Fluid Dynamics,
1-22-3 Haramachi, Meguro-ku, Tokyo 152, Japan
Kunio Kuwahara
The Institute of Space and Astronautical Science,
3-1-1 Yoshinodai, Sagamihara-shi, Kanagawa 229, Japan

INTRODUCTION

To date, the majority of computations have been carried out in the context of two-dimensional flows. Within the given computing resources, a much coarser grid system has to be employed for three-dimensional calculations as compared to two-dimensional flow simulations. It is well documented that, even if a sufficiently large number of grid points are employed, there exist intrinsic disagreements between the two-dimensional computational results and the experimental data. In particular, these discrepancies are conspicuous in the flow details around a bluff body at high Reynolds numbers. Consequently, the important physical quantities of interest, such as the aerodynamic forces acting on the body, may not be predicted accurately if one relies on the two-dimensional flow calculations.

In this paper, we shall analyze unsteady flows past a square cylinder at high Reynolds numbers. We shall acquire the solutions by numerically integrating the two-dimensional and three-dimensional time-dependent incompressible Navier-Stokes equations. We aim to illustrate the quantitative as well as qualitative differences between the two-dimensional and three-dimensional computational results, which are obtained under same conditions for the accuracy of computational discretizations.

In the three-dimensional calculations, the flow is assumed to be periodic in the spanwise directions. We shall bring into focus the character of the three-dimensional flows as the aspect ratio of the cylinder(H/B, H: the span length, B: the breadth of the cylinder) is varied.

PROBLEM FORMULATION

We have employed a direct integration method to obtain numerical solutions to the governing incompressible Navier-Stokes equations. No explicit turbulence models have been incorporated in the present numerical solution procedure. In order to overcome the numerical instability that may arise in high Reynolds number flow calculations, the third-order upwind scheme for the convection terms has been adopted [1].

The governing equations, i.e., the continuity equation and the Navier-Stokes equations, may be expressed as, in a properly non-dimensionalized form,

$$\mathrm{div}\,\boldsymbol{u} = 0, \qquad \cdots\cdots (1)$$

$$\partial \boldsymbol{u}/\partial t + \boldsymbol{u}\cdot \mathrm{grad}\,\boldsymbol{u} = -\mathrm{grad}\,p + 1/Re\,\Delta \boldsymbol{u}, \qquad \cdots\cdots (2)$$

where $\boldsymbol{u}$, p, t and Re denote the velocity vector, pressure, time and the Reynolds number, respectively.

All the spatial derivatives, with the exception of the convection terms, are represented by use of the central finite-differences. The third-order upwind scheme for the convection terms can be written as (3),

$$\left(u\frac{\partial u}{\partial x}\right)_i = u_i\frac{-u_{i+2}+8(u_{i+1}-u_{i-1})+u_{i-2}}{12h} + |u_i|\frac{u_{i+2}-4u_{i+1}+6u_i-4u_{i-1}+u_{i-2}}{4h}, \qquad \cdots\cdots (3)$$

where h is the grid spacing.

Numerical procedures are based on the MAC method[2]. The Poisson equation for pressure is solved by using the SOR scheme. For the temporal integration, the semi-implicit scheme is utilized. This scheme is equivalent to the Euler backward scheme except the nonlinear convection terms. The convection term is linearized as

$$\boldsymbol{u}\cdot \mathrm{grad}\,\boldsymbol{u} = \boldsymbol{u}^{n+1}\cdot \mathrm{grad}\,\boldsymbol{u}^{n+1} \approx \boldsymbol{u}^{n}\cdot \mathrm{grad}\,\boldsymbol{u}^{n+1}. \qquad \cdots\cdots (4)$$

COMPARISONS BETWEEN 2-D and 3-D FLOWS

Figure 1 displays the examples of the grid systems used for 2-D computations and (b) 3-D computations. The square cylinder is placed a little distance forward of the center of the circular computational domain. The approaching flow is taken to be constant U_0 at the far upstream area. The Reynolds number computed is 10^4. The grid points for the two-dimensional flows are (200×100) ~ (400×200), and $200\times100\times$(50~65) for the three-dimensional flows.

First the numerically-computed instantaneous flow patterns around the cylinder are compared with the experimental results obtained by laboratory visualization techniques. Figure 2 presents the 2-D computational results and the laboratory visualization results at $Re=7000$. The experiments were carried out in the water tunnel of the Aerodynamisches Institut by P. Guntermann. The flow was made visuable by injecting fluorescent colors through holes near the four corners in the mid plane. As a whole, it is apparent that these results are in qualitative agreement with each other. However, observing in detail, significant differences in vortex structures appear a little leeward from the cylinder. For computational 2-D flows, we can find a peculiar structure of two-dimensionality, such as vortex-pairings or vortex stretching in the 2-D plane.

Figure 3 depicts the instantaneous streamlines around the square cylinder. In the case of a two-dimensional flow, the formation of the vortices at the side of the cylinder is clearly seen. The separated layer is strongly distorted by the vortex motions in the vicinity of the cylinder surfaces. On the other hand, the results of 3-D calculations exhibit that the separated shear layer extends leeward from the windward corner of the cylinder. Due to the

3-D dissipation and/or the deformation in the spanwise direction, the vortices at the side of the cylinder become weak. There are large discrepancies in the wake of these flow patterns. In the case of three-dimensional flows, it is supposed that the flows gather at one point and change their directions spanwise; that is, some points such as sink or source are found to exist in the wake. Figure 4 illustrates the time variations of the drag coeffcient C_D and the lift coefficient C_L. The results indicate that, for the case of 3-D computations, the time histories of C_D and C_L contain no fine (high-frequency) fluctuations. The averaged value of C_D and the amplitude of C_L for 3-D computations is smaller than that for 2-D case. It is close to the value obtained by the previous experiments. These results are consistent with the flow patterns shown in fig. 3.

THREE-DIMENSIONAL STRUCTURES IN THE NEAR WAKE OF A SQUARE CYLINDER

Figure 5 presents the pressure distribution on the cylinder of $H/B=1.0$ and 1.3 and the pressure contours on the spanwise section in the wake. In the leeward region of the wake, 3-D structures of the flow are shown obviously. There are small and fine vortices behind the cylinder in both cases. There exist no significant differences in these two cases and it is confirmed that the periodic condition does not yeild the undesirable solution caused by the specified periodicity. The surface streamlines are shown in fig. 6. The 3-D effect is discernible at the immediate leeward region of the separation line at the frontal corner of the cylinder. Further downstream, these 3-D structures become finer. Figure 7 exhibits the vorticity contours and the vortex lines around a square cylinder. There exist strong vortices along the separated shear layer. It is seen clearly that the vortices are gradually deformed as their axes are bent in the streamwise direction by the instability of vortex motions.

CONCLUSION

Both 2-D and 3-D time-dependent computations of the flow around a square cylinder have been performed, using the third-order upwind scheme. It is apparent that the flow patterns obtained by the 2-D Navier-Stokes equations have very significant differences from that obtained by the 3-D Navier-Stokes equations, even for the case of 2-D problems. Especially in the near wake, 3-D structures gradually emerge by the vortex instability. These flow characteristics have appreciable effects on the aerodynamic forces, i.e. decreasing the C_D value and the C_L amplitude. Accordingly the 3-D computational result is closer to the experimental value than the 2-D result.

Also, the present computational results illuminate the 3-D flows around a square cylinder by visualizations of various kinds of physical quantities. These provide some fundamental information for the 3-D computation of a 2-D bluff body.

ACKNOWLEDGEMENT

The authors would like to thank Professor Jae Min Hyun of Korea Advanced Institute of Science and Technology, who provided them good and useful suggestions.

REFERENCES

[1]Kawamura, T. and Kuwahara,K., “Computation of High Reynolds Number Flow around a Circular Cylinder with Surface Roughness”, AIAA paper, 84-0340, 1984
[2]Harlow, F. H. and Welch, J.E., “Numerical Calculation of Time-Dependent Viscous Incompressible Flow of Fluid with Free Surface”, Phys. Fluids, Vol.8, 1965

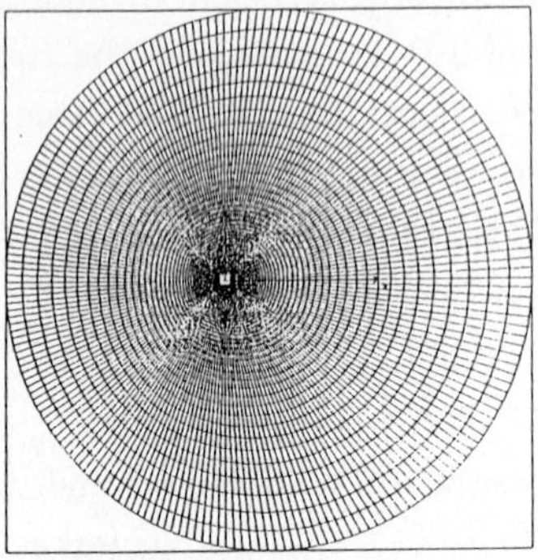

(a) A whole mesh for 2-D computation or a sectional mesh for 3-D computation (200×100 points).

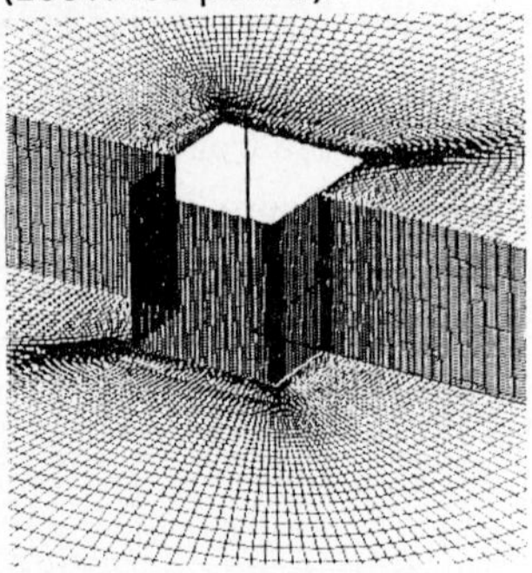

(b) 3-D mesh near the cylinder of $H/B = 1.0$ ($200 \times 100 \times 50$ points).

Fig. 1 Grid systems for a square cylinder.

(a) Experimental visualization.

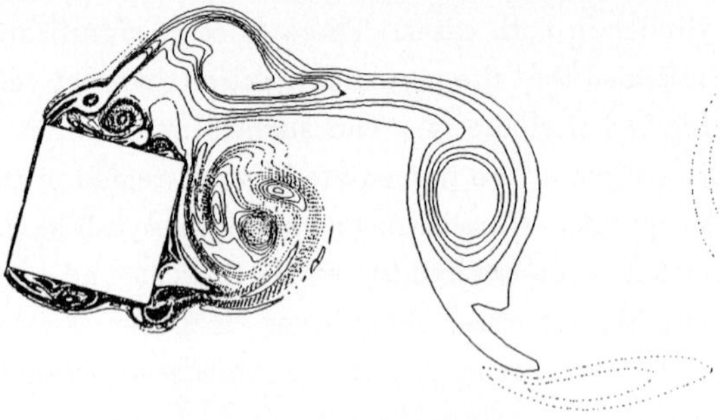

(b) 2-D computational vorticity contours (400×200 grid points).

Fig. 2 Comparison between experimental and computational results of instantaneous flow patterns around a square cylinder with angle of attack (15°) at $Re = 7000$.

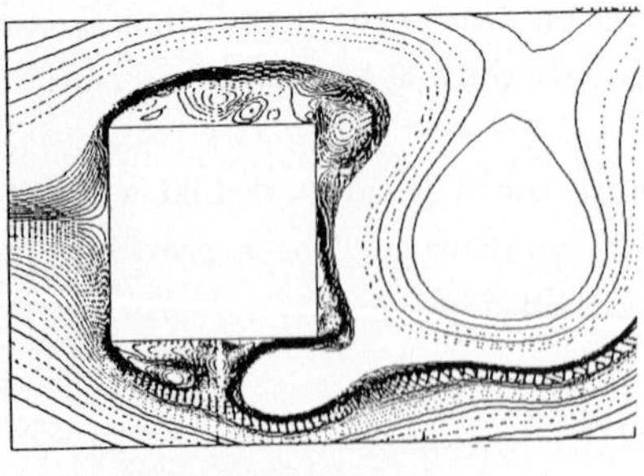

(a) 2-D computational result (200×100 points).

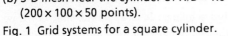

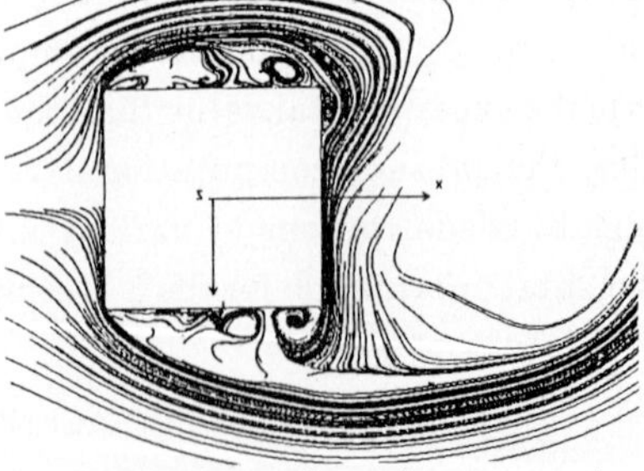

(b) 3-D computational result on a sectional plane ($H/B = 1.0$, $200 \times 100 \times 50$ points).

Fig. 3 Instantaneous streamlines at $Re = 10^4$.

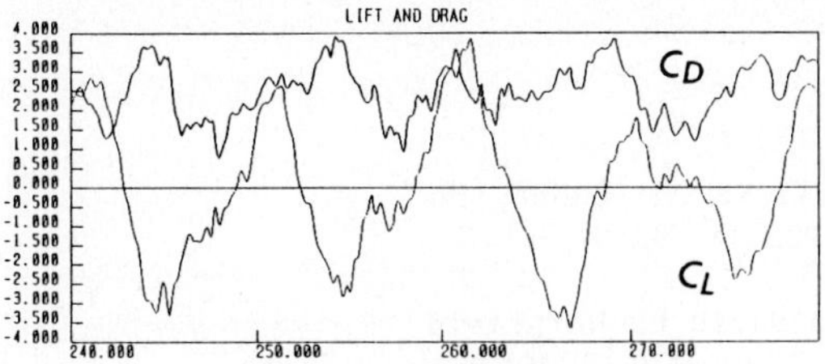

(a) 2-D computational result (200 × 100 points).

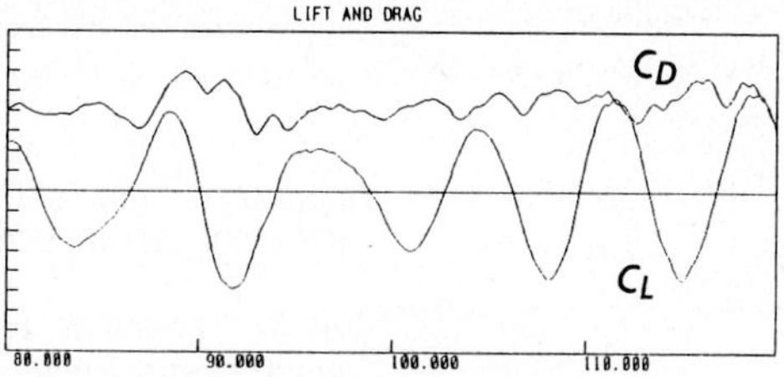

(b) 3-D computational result (*H*/*B* = 1.0, 200 × 100 × 50 points).

Fig. 4 Time histories of the drag coefficient C_D and the lift coefficient C_L at $Re = 10^4$.

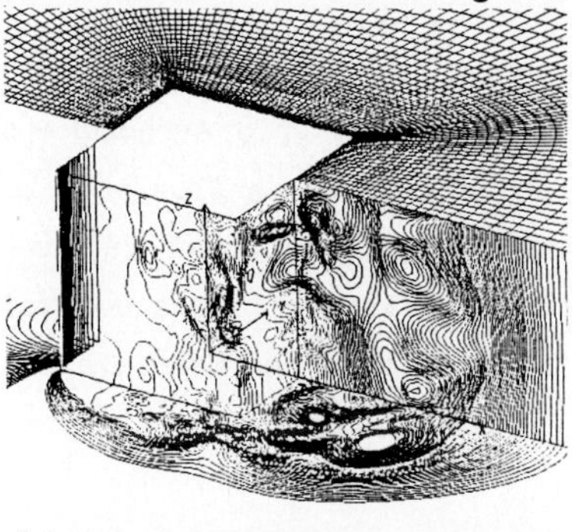

(a) *H*/*B* = 1.0, 200 × 100 × 50 points.

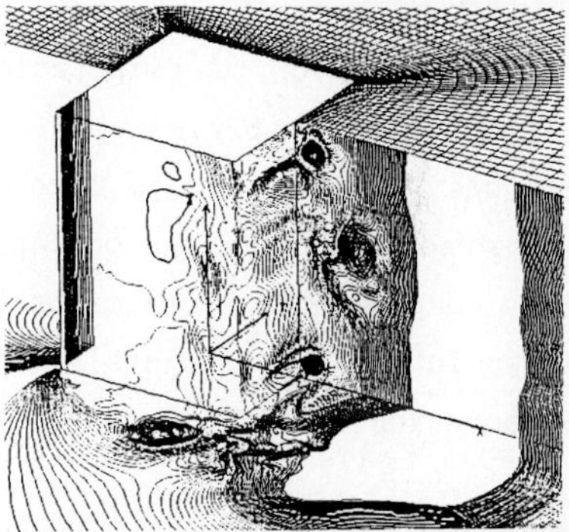

(b) *H*/*B* = 1.3, 200 × 100 × 65 points.

Fig. 5 Instantaneous pressure contours in the near wake of a cylinder at $Re = 10^4$.

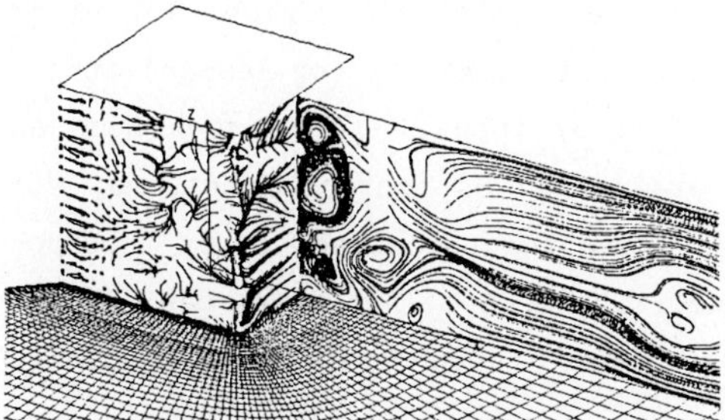

(a) *H*/*B* = 1.0, 200 × 100 × 50 points.

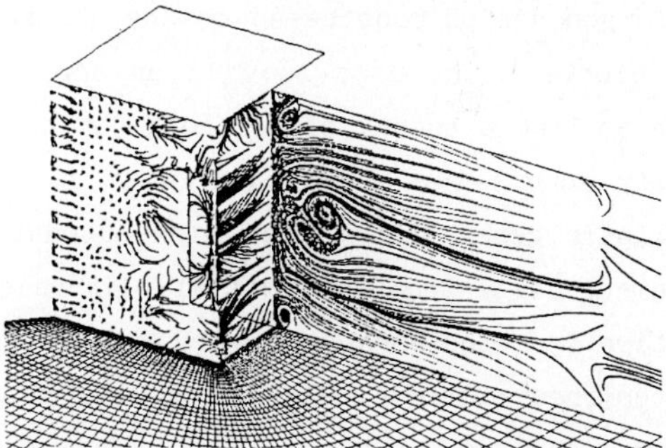

(b) *H*/*B* = 1.3, 200 × 100 × 65 points.

Fig. 6 Instantaneous surface streamlines at $Re = 10^4$.

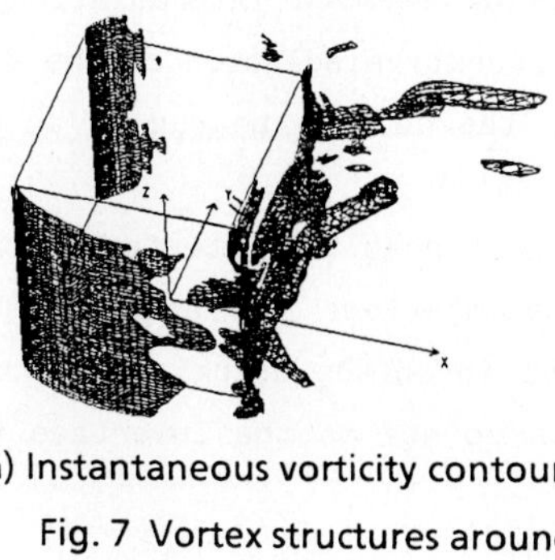

(a) Instantaneous vorticity contours.

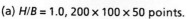

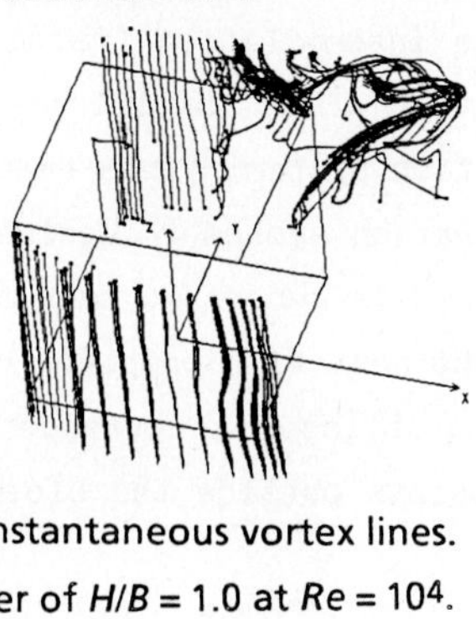

(b) Instantaneous vortex lines.

Fig. 7 Vortex structures around a square cylinder of *H*/*B* = 1.0 at $Re = 10^4$.

TRANSONIC FLOW SOLUTIONS ON GENERAL 3D REGIONS USING COMPOSITE-BLOCK GRIDS

Joe F. Thompson and David L. Whitfield
Department of Aerospace Engineering
Mississippi State University
Mississippi State, MS 39762

The construction of computational fluid dynamics (CFD) codes for complicated regions is greatly simplified by a composite-block grid structure since, with the use of a surrounding layer of points on each block, a flow code is then only required basically to operate on rectangular computational regions. The necessary correspondence of points on the surrounding layers (image points) with interior points (object points) is set up by the grid code and made available to the CFD solution code.

GRID CODE

The present grid code (Refs. 1,2) is a general three-dimensional elliptic grid generation code based on the block structure. This code allows any number of blocks to be used to fill an arbitrary three-dimensional region. Any block can be linked to any other block (or to itself) with complete (or lesser) continuity across the block interfaces as specified by input. This code uses an elliptic generation system with automatic evaluation of control functions, either directly from the initial algebraic grid and then smoothed, or by interpolation from the boundary point distributions. The control functions can also be determined automatically to provide orthogonality at boundaries with specified normal spacing.

The code includes an algebraic three-dimensional generation system based on transfinite interpolation (using either Lagrange or Hermite interpolation) for the generation of an initial grid to start the iterative solution of the elliptic generation system. This feature also allows the code to be run as an algebraic generation system if desired.

The grid is generated such that the grid lines cross the interface from one block to the next with complete continuity. The interface is a branch cut, and the code establishes a correspondence across the interface using a surrounding layer of points outside the blocks. This allows points on the interface to be treated just as all other points, so that there is no loss of continuity. The physical location of the interface is thus totally unspecified, being determined by the code.

An example of the block structure from this code is given in Fig. 1 for a generic body-wing-pylon-store configuration.

The original impetus for using blocked grids was: (1) to simplify the gridding of complex configurations, and (2) permit the solution of large problems requiring many grid points by keeping only the information needed to solve one block in central memory while retaining the information associated with the remaining blocks in secondary memory. Experience exposed a third reason for using blocked grids, having to do with computer cost and/or job turnaround. Supercomputer installations place a high price on the use of central memory, whereas the use of secondary memory is relatively cheap. For example, on a CRAY X-MP 2x4, the cost increases considerably on commercial machines, and the priority decreases considerably on government machines, when more than two million words of central memory are used. By blocking, the use of central memory can be controlled, not only to fit the available memory, but also to control the available budget.

FLOW SOLUTIONS

The three-dimensional Euler and Navier-Stokes equations are transformed to a time-dependent curvilinear coordinate system and are numerically solved using either a flux-vector split or flux-difference split scheme that uses block triangular solution matrices (Refs. 3-7). The numerical solution of the triangular matrices requires a forward and backward pass through the computational domain at each time step. This introduces an inherent back substitution problem that cannot be vectorized along computational coordinates. However, Belk and Janus have succeeded in completely vectorizing the code along diagonal planes using indirect addressing (Refs. 3 and 6).

Both steady and unsteady flow solutions have been obtained using this composite-block grid structure. For steady-state problems there is no approximation at block boundary interfaces. That is, a solution on a blocked grid is the same as the solution on an unblocked (single) grid. For unsteady problems this is not always the case. One cannot block all problems in such a way that information needed for either a forward or backward sweep is available at the appropriate time level. However, even if a problem can be blocked so that needed information is available, it has been found in practice that certain approximations can be made at block boundary interfaces that have negligible influence on unsteady solutions while greatly reducing the number of data transfers to and from secondary memory. The work of Belk (Ref. 6) establishes the methods used for handling the passage of information across block boundary interfaces for both steady and unsteady flow problems.

Numerical results have been obtained using the composite-block grid structure for airfoils, wings, stores, wing-fuselage-pylon-store, multiple stores, vertically and horizontally launched (moving) stores, inlets, nozzles, helicopter rotor, rotor-stator, and single-rotating and counter-rotating propfans. Some of these results are presented in Refs. 3-11. Of particular interest is the counter-rotating propfan propulsion system proposed for transonic flow. The General Electric Unducted Fan (UDF) is shown in Fig. 2. Numerical results for this configuration are presented in Ref. 9. There is little detail experimental data with which to compare results, but an example result is given in Fig. 3, and comparisons with performance coefficients are included in Ref. 9. The computed result of Celestina, Mulac, and Adamczyk (Ref. 12) in Fig. 3 was obtained from their time-averaged Euler method. It would not be expected to agree with experimental data on the forebody because the grid they used did not resolve the forebody region. Contours of constant static pressure downstream of the second blade row as computed by the present unsteady Euler method are shown in Fig. 4.

ADAPTIVE GRID SOLUTIONS

Finally, a dynamically adaptive coupling between the grid and flow codes has been made whereby the grid adjusts to developing gradients in the physical solution (Ref. 13). Here the grid and flow codes are coupled through the control functions in the elliptic grid generation system using the pressure gradient. Results for an ONERA M6 wing at Mach 0.84 and 3.06 degrees angle of attack using 3-block grid structure are shown in Figs. 5 and 6.

ACKNOWLEDGEMENTS

This research was sponsored by the U.S. Air Force Armament Laboratory, Eglin AFB (Grant F08635-84-0228) Dr. Lawrence E. Lijewski, Monitor, and by NASA Lewis Research Center (Grant NAG-3-767) with Dr. Dan Hoyniak as Technical Monitor.

References

1. Thompson, J.F., "A Composite Grid Generation Code for General 3-D Regions", AIAA-87-0275, AIAA 25th Aerospace Sciences Meeting, Reno, 1987, to appear in Journal of Aircraft.

2. Thompson, J.F. and Steger, J.L., "Three Dimensional Grid Generation for Complex Configruations--Recent Progress", AGARD-AG-309, March, 1988.

3. Belk, D.M. and Whitfield, D.L., "3-D Euler Solutions on Blocked Grids Using an Implicit Two-Pass Algorithm", AIAA 87-0450, January, 1987.

4. Belk, D.M. and Whitfield, D.L., "Time-Accurate Euler Equations Solutions on Dynamic Blocked Grids", AIAA 87-1127-CP, June, 1987.

5. Gatlin, B. and Whitfield, D.L., "An Implicit Upwind Finite Volume Method for Solving the Three-Dimensional Thin-Layer Navier-Stokes Equations", AIAA 87-1149-CP, June 1987.

6. Belk, D.M., "Unsteady Three-Dimensional Euler Equations Solutions on Dynamic Blocked Grids", Ph.D. Dissertation, Mississippi State University, August, 1986.

7. Whitfield, D.L., Janus, J.M., and Simpson, L.B., "Implicit Finite Volume High Resolution Wave-Split Scheme for Solving the Unsteady Three-Dimensional Euler and Navier-Stokes Equations on Stationary or Dynamic Grids", Engineering and Industrial Research Station Report MSSU-EIRS-88-2, Mississippi State University, Mississippi State, MS, February 1988.

8. Lijewski, L.E., Thompson, J.F., and Whitfield, D.L., "Computational Fluid Dynamics for Weapon Carriage and Separation", AGARD Fluid Symposium on Store Airframe Aerodynamics, Athens, Greece, October 1985.

9. Whitfield, D.L., Swafford, T.W., Janus, J.M., Mulac, R.A., and Belk, D.M., "Three-Dimensional Unsteady Euler Solutions for Propfans and Counter-Rotating Propfans in Transonic Flow", AIAA 87-1197, June, 1987.

10. Belk, D.M., Janus, J.M., and Whitfield, D.L., "Three-Dimensional Unsteady euler Equations Solution on Dynamic Grids", AIAA Journal, Vol. 25, No. 9, September 1987, pp. 1160-1161.

11. Lijewski, L.E., "Transonic Flow Solutions on a Blunt, Finned Body of Revolution Using the Euler Equations", AIAA Paper No. 86-1082, May 1986.

12. Celestina, M.L., Mulac, R.A., and Adamczyk, J.J., "A Numerical Simulation of the Inviscid Flow Through a Counter-Rotating Propeller", NASA TM-87200, June 1986.

13. Kim, J.K. and Thompson, J.F., "Three Dimensional Adaptive Grid Generation on a Composite Block Grid", AIAA-88-0311, AIAA 26th Aerospace Sciences Meeting, Reno, Nevada, January, 1988.

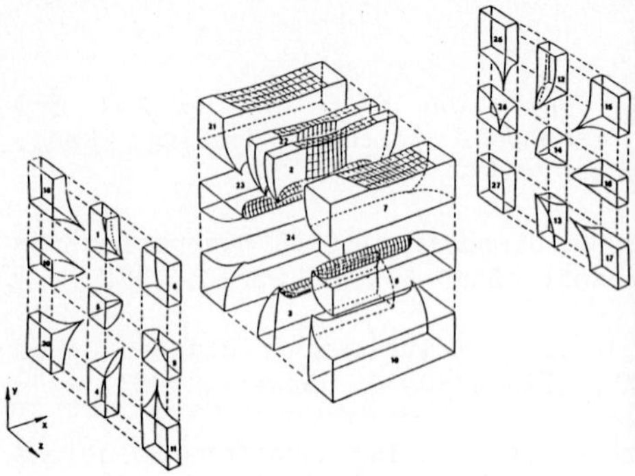

Figure 1 Block Structure:
Wing-Pylon-Store

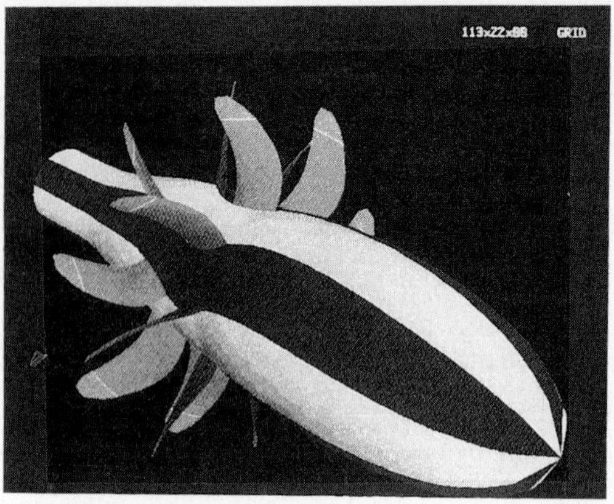

Figure 2 Computational Model of General Electric Unducted Fan (UDF)

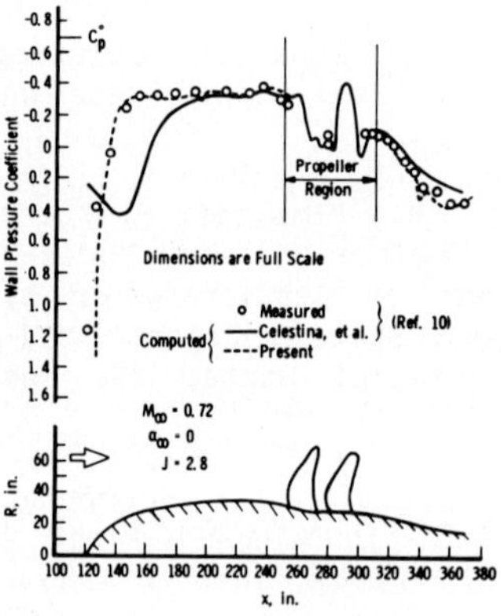

Figure 3 Measured and Computed Surface Pressures for the GE-UDF (M_∞=0.72, α_∞=0, Advance Ratio = 2.8)

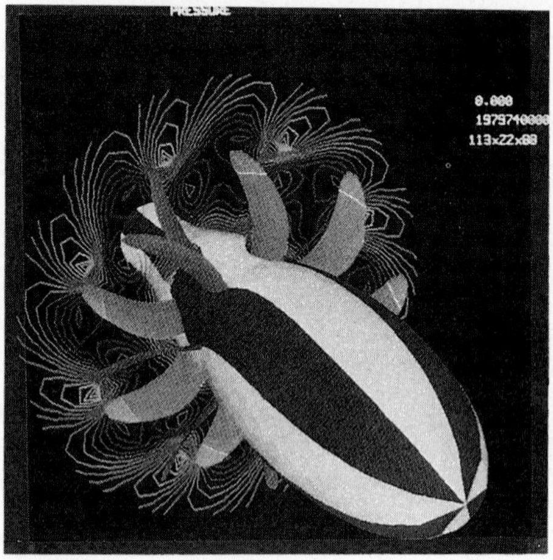

Figure 4 Computed Static Pressure Contours Downstream of Aft Blade Row on UDF

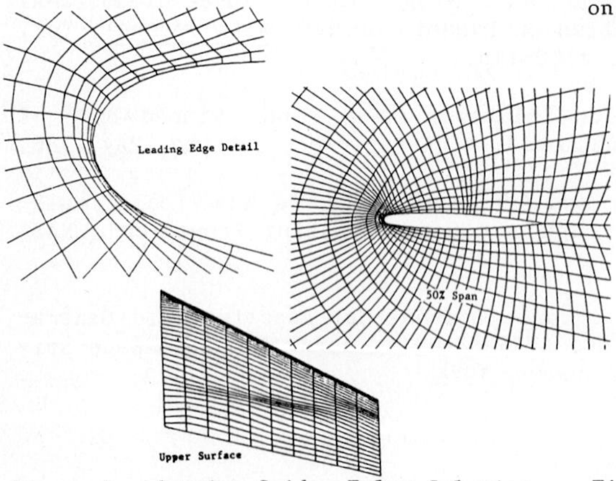

Figure 5 Adaptive Grid: Euler Solution, ONERA M6 Wing, M=0.84, α=3.06

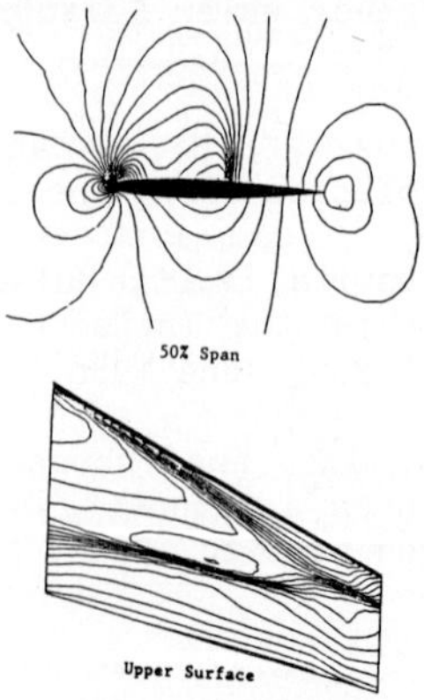

Figure 6 Pressure Contours: Euler Solution, ONERA M6 Wing, M=0.84, α=3.6

STEADY-STATE SOLVING VIA STOKES PRECONDITIONING; RECURSION RELATIONS FOR ELLIPTIC OPERATORS

Laurette S. Tuckerman
Center for Nonlinear Dynamics and Department of Physics
University of Texas, Austin Texas 78712

INTRODUCTION

Recent experiments on Rayleigh-Benard convection in a cylindrical geometry have yielded a rich variety of results concerning wavelength selection, breaking of axisymmetry, and phase turbulence [1,2]. Motivated by these experiments, we have developed a program which numerically simulates the full three-dimensional Boussinesq equations in a cylinder of large aspect ratio.

In order to study the bifurcations undergone by the system, we have brought to bear the full arsenal of tools for studying dynamical systems: time-dependent integration, linear stability analysis, and steady-state continuation. In the past, these tasks have generally been accomplished by using entirely separate programs. However, in [3,4] it was shown that a time-dependent code could easily be adapted to perform linear stability analysis, by iterating the equations linearized about a numerically computed steady state, thus calculating the most unstable (or least stable) eigenvectors via the power method. Here we will show that a time-dependent code can be given the additional capability for direct calculation of stable and unstable steady states via Newton's method.

Our steady-state solver relies on a fast and implicit method for solving the linear Stokes problem, i.e. for inverting the elliptic operators arising from time-stepping of the viscous terms, enforcing incompressibility, and imposing boundary conditions. The need for a fast implicit Stokes solver is not limited to the steady-state code: it is well known that, when integrating the incompressible Navier-Stokes or Boussinesq equations, the linear terms present the most stringent time-stepping criterion, and should therefore be treated implicitly. In the next section, we will present a general result concerning the existence of recursion relations for elliptic operators, and we will demonstrate its application to the solution of Poisson's equation in a cylindrical geometry. The third section will describe the steady-state solver.

INVERSION OF ELLIPTIC OPERATORS VIA RECURSION RELATIONS

We use a pseudospectral method [5,6,7], in which functions are represented by:

$$f(r,\theta,z) = \sum_{k=0}^{K} \sum_{m=0}^{M} \sum_{n=0}^{N} f_{kmn}\, T_k(r)\, e^{im\theta}\, T_n(z) \tag{1}$$

where T_k, T_n are Chebyshev polynomials, and parity restrictions are imposed on $k+m$. Numerical spectral methods, like analytic methods, are difficult to adapt to nonperiodic geometries. Here rigid boundaries at the top, bottom, and sidewalls must be added to the complication of curvilinear coordinates and coordinate singularities. A further challenge arises from the large aspect ratio of the cylinder: the algorithm must be economical in the number K of gridpoints or functions in the radial direction, a direction which is neither periodic (like θ) nor Cartesian (like z).

At the core of our algorithm is a fast solver for elliptic equations such as:

$$\nabla^2 f = g \tag{2}$$

A well known recursion relation [5,7] can be used to reduce the upper triangular matrix corresponding to the (Cartesian) second derivative acting on a Chebyshev series to a tridiagonal diagonally-dominant matrix. Only one spatial direction can be reduced in this way: when two directions are nonperiodic, a Haidvogel-Zang decomposition [7,8] is indicated, in which the (Cartesian-like) z direction is reduced to tridiagonal form and the r direction treated by eigenvector-eigenvalue decomposition, leading to an operation count of $O(3KMN) + O(2K^2MN)$. However, since $K >> N$, what is needed is a recursion relation for the *radial* direction instead.

We have shown in [9] that the standard recursion relation is only one instance of a much larger class of matrices which can be reduced to banded form. The relevant result is:

THEOREM: Let R be an upper triangular matrix of the form:

$$R(k,l) = \begin{cases} \sum_{j=1}^{J} S(k,j)\, T(j,l) & 1 \le k < l \le K \\ R(k,k) & 1 \le k = l \le K \\ 0 & \text{otherwise} \end{cases} \tag{3}$$

with $J < K$. Suppose that the J by J matrices $\hat{S}^k$ defined by $\hat{S}^k(i,j) \equiv S(k+i,j)$ for $1 \le i,\, j \le J$ are invertible for all $k < K-J$ Then there exists an invertible banded matrix B, depending only on S, with J nonzero super-diagonals, such that BR is also banded with J nonzero super-diagonals.

Matrices that differentiate polynomial expansions, or which multiply derivatives by powers of the independent variable (e.g. r) are all of form (3) (see the Appendix of [5]), as are sums of such operators. In these cases S (and T) are monomials, i.e. $S(k,j) = k^{\alpha_j}$, for integers α_j, with $\alpha_i \neq \alpha_j$ if $i \neq j$. Since the monomials k^{α_j} are linearly independent over any range of k, the conditions of the theorem are satisfied. In the tau method, the last rows of the matrix of form (3) are replaced by full rows corresponding to boundary conditions. If only one row is changed (as here) then the bandwidth of BR remains $J+1$ except for the last row. The proofs of these assertions are straightforward [9].

Let us now apply the theorem to the cylindrical Laplacian acting on (1). We define the operators:

$R \equiv r\,\partial_r\, r\,\partial_r$, which is of form (3) and can be reduced by the banded matrix B,

$D_{\theta\theta} \equiv \partial_{\theta\theta}$, which is diagonal with elements $-m^2$, and

$Z \equiv \partial_{zz} = V^{-1} D_{zz} V$, where D_{zz} is diagonal with elements σ_n.

$R, B, D_{\theta\theta}, Z, D_{zz}, V$, and V^{-1} can either be considered to be matrices of order K, M, or N, or extended to order (KMN) via, e.g., $R(k,m,n;\, k',m',n') = R(k,k')\, \delta(m,m')\, \delta(n,n')$.

Equation (2) is then written as:

$$(r^{-2}R + r^{-2}D_{\theta\theta} + Z)\, f = g$$

We first diagonalize in z:

$$(r^{-2}R + r^{-2}D_{\theta\theta} + D_{zz})\, V f = V g$$

and then multiply through by the matrices B and r^2:

$$A\, V f \equiv (BR + B\, D_{\theta\theta} + Br^2\, D_{zz})\, V f = Br^2\, V g \tag{4}$$

For each θ Fourier mode and z eigenfunction, i.e. for each m and n, this can be written as:

$$A_{mn}\,(Vf)_{mn} \equiv (\,BR - Bm^2 + Br^2\,\sigma_n\,)\,(Vf)_{mn} = Br^2\,(Vg)_{mn}$$

All boundary conditions are imposed via the tau method: those in z are built into the matrix Z (i.e. into V, V^{-1}, and D_{zz}) while radial boundary conditions are inserted into the matrices A_{mn}. Since B, BR, and r^2 are all tridiagonal and diagonally dominant, the $K \times K$ matrices A_{mn} are pentadiagonal, with an extra full boundary row, and can be inverted via LU decomposition. The requirements of the algorithm are summarized in the following table.

Operator	Storage	Time for multiplication by operator
V and V^{-1}	$2N^2$	$2KMN^2$
Br^2	$5K$	$5KMN$
A^{-1}	$6KMN$	$6KMN$

For $m \geq 2$, regularity conditions at $r = 0$ must be imposed [6,10]. We do this by setting $f = r^2\,h$ and defining:

$$R' \equiv r^{-1}\partial_r\, r\, \partial_r\, r^2 = r^{-2} R\, r^2$$

which is also of form (3). Diagonalizing (2) in z and multiplying by the same matrix B, we obtain:

$$A'\,V\,h \equiv (\,BR' + B\,D_{\theta\theta} + Br^2 D_{zz}\,)\,V\,h = B\,V\,g$$

which is solved in the same way as (4).

The algorithm for solution of the Stokes problem is built around this Poisson solver. The cylindrical vector Laplacian is decoupled into three scalar elliptic operators [6]. The technique outlined above is used to invert operators such as $I - \varepsilon\nabla^2$, as well as ∇^2. Finally, the velocity field is made divergence-free by the influence matrix method [7,11] with full tau correction, generalized to a cylindrical geometry with two nonperiodic directions [10].

STEADY-STATE SOLVER

A steady-state solver presents several advantages over a time-dependent code. The first is speed: root-finding algorithms such as Newton's method converge quadratically, while time-stepping necessarily follows the dynamics, converging linearly to steady states. The second advantage is more fundamental: a steady-state solver can compute unstable steady states, which a time-dependent code cannot (except for the case in which the instability arises via a symmetry-breaking bifurcation [3,4]). In addition, the steady-state solver may be coupled to standard continuation methods for tracking bifurcations in parameter space, and is exempt from critical slowing down.

To describe our steady-state solver, we represent the differential equation by:

$$\partial_t u = Lu + Nu$$

Here L represents the terms (usually linear) in the equation that are to be integrated by an implicit method, and N represents the terms (usually nonlinear) to be integrated explicitly. We wish to solve:

$$Lu + Nu = 0 \tag{5}$$

Our time-dependent program uses backwards Euler time-stepping for L and forwards Euler time-stepping for N (higher-order time discretizations could also be used):

$$u^{n+1} - u^n = \Delta t\,(Lu^{n+1} + Nu^n)$$

$$u^{n+1} = (I - \Delta t\,L)^{-1}\,(I + \Delta t\,N)\,u^n$$

$$\frac{1}{\Delta t}(u^{n+1} - u^n) = \frac{1}{\Delta t}[\,(I - \Delta t\, L)^{-1}\,(I + \Delta t\, N) - I\,]\, u^n \tag{6}$$

$$\equiv E_{\Delta t}\, u^n$$

The crucial observation is that:

$$E_{\Delta t} = \frac{1}{\Delta t}(I - \Delta t\, L)^{-1}\,[\,(I + \Delta t\, N) - (I - \Delta t\, L)\,]$$

$$= (I - \Delta t\, L)^{-1}\,(N + L)$$

so that, as long as $I - \Delta t\, L$ is invertible,

$$E_{\Delta t}\, u = 0 \tag{7}$$

has the same solutions as (5) for any Δt. The limit $\Delta t \to 0$ yields equation (5), but (7) is numerically more tractable if much larger values of Δt are used, as we shall see below.

We see from (6) that we can act with $E_{\Delta t}$ on a vector by using the time-stepping program. In order to solve (7) by Newton's method, we shall also need the Jacobian $DE_{\Delta t}(U)$ of $E_{\Delta t}$ at U, i.e. the linearization of $E_{\Delta t}$ about U. If L is linear and:

$$Nu = (u \cdot \nabla\,)\, u$$

then:

$$DN(U)\, u = (U \cdot \nabla\,)\, u + (u \cdot \nabla\,)\, U$$

and $DE_{\Delta t}(U)$ is obtained by substituting $DN(U)$ for N in (6). The time-stepping code already computes Nu; with slight modifications it can compute $DN(U)u$ as well. (Indeed, the substitution $N \to DN(U)$ is also precisely that required to perform linear stability analysis [3,4]).

One step of Newton's method consists of solving:

$$DE_{\Delta t}(u^n)\,(u^{n+1} - u^n) = -\,E_{\Delta t}\, u^n \tag{8}$$

for $u^{n+1} - u^n$. The matrix $DE_{\Delta t}(u^n)$ is large and full, necessitating iterative solution of equation (8). However, we can act rapidly with $DE_{\Delta t}(u^n)$ on a vector: the pseudospectral method provides a fast way to multiply by $DN(U)$ in gridspace via Fourier transforms [5,7], while our Poisson/Stokes solver efficiently operates with $(I - \Delta t\, L)^{-1}$. Iterative techniques are able to fully exploit this advantage.

Because $DE_{\Delta t}(u^n)$ is neither symmetric nor definite, iterative solution of (8) is problematic [7,12]. Despite the lack of guarantees, we have experimented with Lanczos and Orthomin methods implemented in the NSPCG (Nonsymmetric Preconditioned Conjugate Gradient) software package [12]. We obtain extremely good results from the biconjugate gradient squared (BCGS) algorithm, a Lanczos-type method.

In the figure below, we compare the time needed to solve (7) for different values of Δt on a Cray X-MP/24. The parameters of the problem are: aspect ratio $\Gamma = 5$, resolution $(K,M,N) = (49,0,15)$, and reduced Rayleigh number $\varepsilon = (Ra - Ra_c)/Ra_c = 1.3$ The initial condition for Newton's method is a steady state at $\varepsilon = 1.2$ (other parameters the same). For details see [13].

Solving (7) to a prescribed level of accuracy usually requires four Newton steps, independent of Δt. However, the number of conjugate gradient iterations (each performing two multiplications by $DE_{\Delta t}(u^n)$) per Newton step may vary greatly. For $\Delta t > 10^{-1}$, about 35 iterations per Newton step are required, leading to a total CPU time of 4 seconds. For $\Delta t < 10^{-2}$, the number of conjugate gradient iterations begins to climb steeply, until, for $\Delta t = 7\times10^{-5}$, as many as 780 iterations are required for one Newton step. For $\Delta t < 7\times10^{-5}$ (and also for $\Delta t > 10^5$) the method breaks down.

Effectively, $(I - \Delta t\, L)^{-1}$ acts as a *preconditioner* for the poorly conditioned matrices $L + DN(u^n)$. For $\Delta t \to 0$, the preconditioning becomes weaker as $(I - \Delta t\, L)^{-1} \to I$, while for $\Delta t \to \infty$, we have $(I - \Delta t\, L)^{-1} \to (-\,\Delta t\, L)^{-1}$.

For purposes of comparison, we note that the time-stepping code, limited by the Courant condition to $\Delta t \leq 2\times10^{-3}$, will converge within the same accuracy to the steady state in 130 CPU seconds, shown as the asterisk in the figure. The combination of quadratic convergence, preconditioning, and biconjugate gradient squared iteration has brought about a drastic reduction in computation time, with almost no additional coding effort.

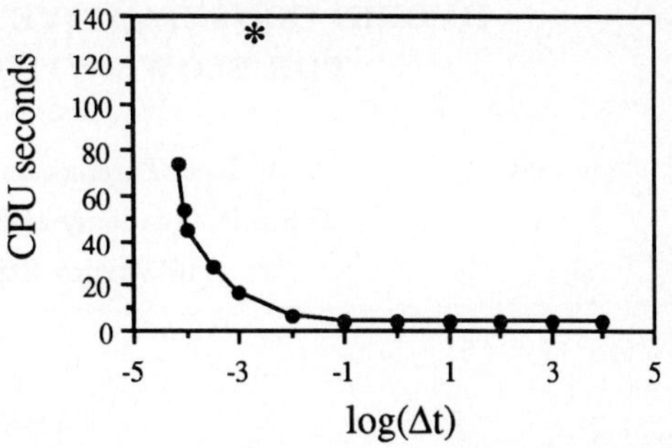

ACKNOWLEDGMENTS

I thank the Center for Numerical Analysis at the University of Texas at Austin for making the NSPCG package available, and especially Tom Oppe for spending many hours patiently explaining its workings to me. Discussions with D. Barkley are gratefully acknowledged. The author was supported by the NSF Mathematical Sciences Postdoctoral Research Fellowship program and the Office of Naval Research Nonlinear Dynamics Program. Computational resources for this work were provided by the University of Texas System Center for High Performance Computing.

REFERENCES

[1] V. Croquette, P. Le Gal, A. Pocheau, and R. Guglielmetti, *Europhys. Lett.* **1** (8) 393 (1986).

[2] V. Steinberg, G. Ahlers, and D.S. Cannell, *Physica Scripta* **T9**, 97 (1985).

[3] L.S. Tuckerman and P.S. Marcus, in *Proceedings of the Ninth Intl. Conf. on Num. Meth. in Fluid Dyn.*, ed. by Soubbaramayer and J.P. Boujot (Springer-Verlag, Berlin, 1985).

[4] P.S. Marcus and L.S. Tuckerman, *J. Fluid Mech.* **185**, 1 (1987) and **185**, 31 (1987).

[5] D. Gottlieb and S.A. Orszag, *Numerical Analysis of Spectral Methods: Theory and Applications* (SIAM Press, Philadelphia, 1977).

[6] A.T. Patera and S.A. Orszag, in *Nonlinear Problems: Present and Future*, ed. by A.R. Bishop, D.K. Campbell, and B. Nicolaenko (North-Holland, Amsterdam, 1982).

[7] C. Canuto, M.Y. Hussaini, A. Quateroni, and T.A. Zang, *Spectral Methods in Fluid Dynamics* (Springer-Verlag, Berlin, 1988).

[8] D.B. Haidvogel and T.A. Zang, *J. Comp. Phys.* **30**, 167 (1979).

[9] L.S. Tuckerman, submitted to *J. Comp. Phys.*

[10] L.S. Tuckerman, *J. Comp. Phys.*, in press.

[11] L. Kleiser and U. Schumann, in *Proc. of the Third GAMM Conf. on Num. Meth. in Fluid Mech.*, ed. by E.H. Hirschel (Vieweg and Sohn, Braunschweig, 1980).

[12] T.C. Oppe, W.D. Joubert, and D.R. Kincaid, Publ. #216 of the Center for Numerical Analysis, University of Texas at Austin.

[13] L.S. Tuckerman and D. Barkley, *Phys. Rev. Lett.*, **61**, 408 (1988).

HYBRID CONSERVATIVE CHARACTERISTIC METHOD FOR FLOWS WITH INTERNAL SHOCKS

L. I. Turchak and V. F. Kamenetsky
U.S.S.R. Academy of Sciences Computing Centre
40 Vavilov Str., 117333 Moscow

ABSTRACT

A conservative hybrid scheme for the inviscid gas equations has been formulated on the basis of the characteristic relations using curvilinear coordinates. A new approximation of the metric coefficients for the moving coordinate system is suggested. The effectiveness of the algorithm is illustrated by solving: 2D and 3D unsteady shock interactions with a moving body, the motion of a body through an explosion field, and the interaction of a freestream with a contact discontinuity.

1. INTRODUCTION

In earlier papers [1,2] the authors solved numerically a number of problems involving supersonic flows with discontinuities past blunt bodies. Numerical simulation was based on a nonconservative shock-fitting grid-characteristic method [3]. In spite of the nonconservative scheme, the results predict the main features of such flows. In some cases, there is good quantitative agreement with experimental data. But for more precise numerical simulation of gasdynamic flows interacting with discontinuities, it is necessary to use conservative algorithms. At the same time, to increase the accuracy of the numerical solution, it is desirable to fit the main discontinuities. Fitting of the main discontinuities is achieved by transformation from the Cartesian coordinate system to a curvilinear one [4,5]. Construction of a conservative algorithm in a curvilinear moving coordinate system creates a number of difficulties connected with the approximation of the metric coefficients. In the present paper we propose a method of approximating the metric coefficients primarily by using the Cartesian coordinates of the corner points of the computational cells. This method has been used to construct a conservative algorithm based on a hybrid scheme [6]; the order of its approximation depends on the smoothness of the solution.

2. NUMERICAL METHOD

Consider the divergent form of the gasdynamic equations equivalent to the conservation laws of mass, momentum, and total energy. In a Cartesian coordinate system, it may be written as:

$$\frac{\partial U}{\partial t} + \frac{\partial F}{\partial x} + \frac{\partial G}{\partial y} + \frac{\partial H}{\partial z} = 0 \tag{1}$$

where

$$U = (\rho, m, n, \ell, E)^T; \; F = (\rho u, mu + P, mv, mw, (E+p)u)^T;$$

$$G = (\rho v, nu, nv + p, nw, (E+p)v); \; H = (\rho w, \ell u, \ell v, \ell w + p, (E+p)w);$$

$$m = \rho u;\ n = \rho v;\ \ell = \rho w;\ E = \rho\ell + 0.5\rho(u^2 + v^2 + w^2).$$

where u, v, w are projections of the vector of the velocity on axes x, y, z; p is the pressure, e is the internal energy.

By transforming from Cartesian coordinates (x, y, z) to curvilinear ones (s, ξ, ϕ), the domain of integration may be made rectangular: $S_1 \leq S \leq S_2$; $\xi_1 \leq \xi \leq \xi_2$; $\phi_1 \leq \phi \leq \phi_2$, where $S_1, S_2, \xi_1, \xi_2, \phi_1, \phi_2$ are constants. In the new coordinate system, (1) has the form

$$\frac{\partial \tilde{U}}{\partial t} + \frac{\partial \tilde{F}}{\partial s} + \frac{\partial \tilde{G}}{\partial \xi} + \frac{\partial \tilde{H}}{\partial \phi} = 0$$

$$\tilde{U} = \frac{U}{J};\ \tilde{F} = \frac{S_t}{J}U + \frac{S_x}{J}F + \frac{S_y}{J}G + \frac{S_z}{J}H;\ \tilde{G} = \frac{\xi_t}{J}U + \frac{\xi_x}{J}F + \frac{\xi_y}{J}G + \tag{2}$$

$$+\frac{\xi_z}{J}H;\ \tilde{H} = \frac{\phi_t}{J}U + \frac{\phi_x}{J}F + \frac{\phi_y}{J}G + \frac{\phi_z}{J}H;\ J = \left|\frac{\partial(S, \xi, \phi)}{\partial(x, y, z)}\right|.$$

In order to conserve the divergence form, the velocity components are kept in Cartesian coordinates.

As mentioned in [5], uniform flow is a solution of Eq. (2) due to the following differential relations valid for a sufficiently smooth coordinate transformation:

$$\frac{\partial}{\partial S}\frac{S_x}{J} + \frac{\partial}{\partial \xi}\frac{\xi_x}{J} + \frac{\partial}{\partial \phi}\frac{\phi}{J} = 0;\ \frac{\partial}{\partial S}\frac{S_y}{J} + \frac{\partial}{\partial \xi}\frac{\xi_x}{J} + \frac{\partial}{\partial \phi}\frac{\phi_y}{J} = 0;$$

$$\frac{\partial}{\partial S}\frac{S_z}{J} + \frac{\partial}{\partial \xi}\frac{\xi_y}{J} + \frac{\partial}{\partial \phi}\frac{\phi_y}{J} = 0;\ \frac{\partial}{\partial t}\frac{1}{J} + \frac{\partial}{\partial S}\frac{S_t}{J} + \frac{\partial}{\partial S}\frac{S_t}{J} + \frac{\partial}{\partial \xi}\frac{\xi_t}{J} + \frac{\partial}{\partial \phi}\frac{\phi_t}{J} = 0 \tag{3}$$

It is therefore natural to claim that a uniform flow will also be a solution of a numerical approximation of system (2).

Let us introduce in the computational domain a uniform grid with steps ΔS, $\Delta\xi$, and $\Delta\phi$. The grid points lie on the intersection of the surfaces $S = S_i$, $\xi = \xi_j$ and $\phi = \phi_\kappa$. The computational cells are bounded by the surfaces $S = S_{i\pm\frac{1}{2}}$, $\xi = \xi_{j\pm\frac{1}{2}}$ and $\phi = \phi_{\kappa\pm\frac{1}{2}}$. Here

$$S_{i\pm\frac{1}{2}} - S_i = \pm\Delta S/2,\ \xi_{j\pm\frac{1}{2}} - \xi_j = \pm\Delta\xi/2,\ \phi_{\kappa\pm\frac{1}{2}} - \phi_\kappa = \pm\Delta\phi/2.$$

The metric coefficients S_t/J, S_x/J, S_y/J and S_zJ are determined at the points $(i\pm\frac{1}{2}, j, \kappa)$. Analogously, ξ_t/J, ξ_x/J, ξ_y/J, ξ_z/J are determined at the points $(i, j\pm\frac{1}{2}, \kappa)$ while ϕ_t/J, ϕ_x/J, ϕ_y/J, ϕ_z/J are determined at the points $(i, j, \kappa\pm\frac{1}{2})$. Finally, we will also consider the corner points of the computational cells $(i\pm\frac{1}{2}, j\pm\frac{1}{2}, \kappa\pm\frac{1}{2})$ which lie on the intersection of the surfaces $S = S_{i\pm\frac{1}{2}}$, $\xi = \xi_{j+\frac{1}{2}}$ and $\phi = \phi_{\kappa\pm\frac{1}{2}}$.

Let us express the metric coefficients not containing time derivatives as

$$\frac{S_x}{J} = z_\phi y_\xi - y_\phi z_\xi \tag{4}$$

Approximating (4) at the point $(i \pm \frac{1}{2}, j, \kappa)$ by central differences, one obtains:

$$\left(\frac{S_x}{J}\right)_{i\pm\frac{1}{2}j\kappa} = \frac{1}{2}\left(\frac{z_{i\pm\frac{1}{2}j+\frac{1}{2}\kappa+\frac{1}{2}} - z_{i\pm\frac{1}{2}j+\frac{1}{2}\kappa-\frac{1}{2}}}{\Delta\psi} + \frac{z_{i\pm\frac{1}{2}j-\frac{1}{2}\kappa+\frac{1}{2}} - z_{i\pm\frac{1}{2}j-\frac{1}{2}\kappa-\frac{1}{2}}}{\Delta\phi}\right)\cdot$$

$$\cdot\frac{1}{2}\left(\frac{y_{i\pm\frac{1}{2}j+\frac{1}{2}\kappa+\frac{1}{2}} - y_{i\pm\frac{1}{2}j-\frac{1}{2}\kappa+\frac{1}{2}}}{\Delta\xi} + \frac{y_{i\pm\frac{1}{2}j+\frac{1}{2}\kappa-\frac{1}{2}} - y_{i\pm\frac{1}{2}j-\frac{1}{2}\kappa-\frac{1}{2}}}{\Delta\xi}\right) -$$

$$-\frac{1}{2}\left(\frac{y_{i\pm\frac{1}{2}j+\frac{1}{2}\kappa+\frac{1}{2}} - y_{i\pm\frac{1}{2}j+\frac{1}{2}\kappa-\frac{1}{2}}}{\Delta\phi} + \frac{y_{i\pm\frac{1}{2}j-\frac{1}{2}\kappa+\frac{1}{2}} - y_{i\pm\frac{1}{2}j-\frac{1}{2}\kappa-\frac{1}{2}}}{\Delta\phi}\right)\cdot$$

$$\cdot\frac{1}{2} = \left(\frac{z_{i\pm\frac{1}{2}j+\frac{1}{2}\kappa+\frac{1}{2}} - z_{i\pm\frac{1}{2}j-\frac{1}{2}\kappa+\frac{1}{2}}}{\Delta\xi} + \frac{z_{i\pm\frac{1}{2}j+\frac{1}{2}\kappa-\frac{1}{2}} - z_{i\pm\frac{1}{2}j-\frac{1}{2}\kappa-\frac{1}{2}}}{\Delta\xi}\right)$$

Using such an approximation, we obtain the finite difference analog of the first three equations (3):

$$\left[\left(\frac{S_x}{J}\right)_{i+\frac{1}{2}j\kappa} - \left(\frac{S_x}{J}\right)_{i-\frac{1}{2}j\kappa}\right]/\Delta S + \left[\left(\frac{\xi_x}{J}\right)_{ij+\frac{1}{2}\kappa} - \left(\frac{\xi_x}{J}\right)_{ij-\frac{1}{2}\kappa}\right]/\Delta\xi +$$

$$+\left[\left(\frac{\phi_x}{J}\right)_{ij\kappa+\frac{1}{2}} - \left(\frac{\phi_x}{J}\right)_{ij\kappa-\frac{1}{2}}\right]/\Delta\phi \equiv 0.$$

The difference approximation of the coefficients S_t/J, ξ_t/J and ϕ_t/J depends on the coordinate transformations used. Here we only mention, that

$$\frac{S_t}{J} = -\left(x_t\frac{S_x}{J} + y_t\frac{S_y}{J} + z_t\frac{S_z}{J}\right); \quad \frac{\xi_t}{J} = -\left(x_t\frac{\xi_x}{J} + y_t\frac{\xi_y}{J} + z_t\frac{\xi_z}{J}\right);$$

$$\frac{\phi_t}{J} = -\left(x_t\frac{\phi_x}{J} + y_t\frac{\phi_y}{J} + z_t\frac{\phi_z}{J}\right).$$

The quantity J^{-1} is determined from the finite difference approximation of the last equation (3), approximating the time derivative by the expression used in the numerical scheme for approximating (2). The numerical scheme proposed here is:

$$\left[\frac{U^{n+1}_{ij\kappa}}{J^{n+1}_{ij\kappa}} - \frac{U^n_{ij\kappa}}{J^n_{ijk}}\right]/\Delta t + \frac{\tilde{F}^n_{i+\frac{1}{2}j\kappa} - \tilde{F}^n_{i-\frac{1}{2}j\kappa}}{\Delta S} + \frac{\tilde{G}^n_{ij+\frac{1}{2}\kappa} - \tilde{G}^n_{ij-\frac{1}{2}\kappa}}{\Delta\xi} +$$

$$+\frac{\tilde{H}^n_{ij\kappa+\frac{1}{2}} - \tilde{H}^n_{ij\kappa-\frac{1}{2}}}{\Delta\phi} - \left[(B_1)^n_{i+\frac{1}{2}j\kappa}(U^n_{i+1j\kappa} - U^n_{ij\kappa}) - (B_1)^n_{i-\frac{1}{2}j\kappa}\cdot\right.$$

$$\left.\cdot\left(U^n_{ij\kappa} - U^n_{i-1j\kappa}\right)\right]/\Delta S - \left[(B_2)^n_{ij+\frac{1}{2}\kappa}\left(U^n_{ij+1\kappa} - U^n_{ij\kappa}\right) - (B_2)^n_{ij-\frac{1}{2}\kappa}\cdot\right.$$

$$\left.\cdot\left(U^n_{ij\kappa} - U^n_{ij-1\kappa}\right)\right]/\Delta\xi - \left[(B_3)^n_{ij\kappa+\frac{1}{2}}\left(U^n_{ij\kappa+1} - U^n_{ij\kappa}\right) - (B_3)^n_{ij\kappa-\frac{1}{2}}\cdot\right.$$

$$\left.\cdot\left(U^n_{ij\kappa} - U^n_{ij\kappa-1}\right)\right]/\Delta\phi = 0$$

$$\tilde{F}^n_{i\pm\frac{1}{2}j\kappa} = \left(\frac{S_t}{J}\right)^n_{i\pm\frac{1}{2}j\kappa}\cdot U^n_{i\pm\frac{1}{2}j\kappa} + \left(\frac{S_x}{J}\right)^n_{i\pm\frac{1}{2}j\kappa}\cdot F^n_{i\pm\frac{1}{2}j\kappa} + \left(\frac{S_y}{J}\right)^n_{i\pm\frac{1}{2}j\kappa} G^n_{i\pm\frac{1}{2}j\kappa} + \left(\frac{S_z}{J}\right)^n_{i\pm\frac{1}{2}j\kappa} H^n_{i\pm\frac{1}{2}j\kappa}$$

where

$$U^n_{i\pm\frac{1}{2}j\kappa},\ F^n_{i\pm\frac{1}{2}j\kappa},\ G^n_{i\pm\frac{1}{2}j\kappa},\ H^n_{i\pm\frac{1}{2}j\kappa}$$

are the symmetrical approximations of vectors U, F, G, H at the points $(i \pm \frac{1}{2}, j, \kappa)$. For example,

$$F^n_{i\pm\frac{1}{2}j\kappa} = 0.5(F^n_{ijk} + F^n_{i+1jk}).$$

Values of $\tilde{G}_{ij\pm\frac{1}{2}\kappa}$ and $\tilde{H}_{ij\kappa\pm\frac{1}{2}}$ are defined similarly. We now define matrices $B_i (i = 1, 2, 3)$. Consider for example matrix B_1. System (2) is hyperbolic and the matrix

$$\frac{\partial \tilde{F}}{\partial U} = \Omega_1^{-1} \| \lambda_{1\ell} \delta_{\ell m} \| \Omega_1,$$

where Ω_1 is the matrix of the left eigenvectors of the matrix $\frac{\partial \tilde{F}}{\partial U}$, $\lambda_{1\ell}$ its eigenvalues, and $\delta_{\ell m}$ the Kronecker delta. Then $B_1 = 0.5\Omega_1^{-1} \| \lambda_{1\ell} g_{1\ell} \delta_{\ell m} \| \Omega_1$, where Ω_1 is the matrix of the left eigenvectors of the matrix $\frac{\partial \tilde{F}}{\partial U}$, $\lambda_{1\ell}$ its eigenvalues, and $\delta_{\ell m}$ the Kronecker delta. Then $B_1 = 0.5\Omega_1^{-1} \| \lambda_{1\ell} g_{1\ell} \delta_{\ell m} \| \Omega_1$. Here $g_{1\ell}$ is the coefficient of hybridness [6]: $g_{1\ell} =$ sign $(\lambda_{1\ell})$ in nonsmooth regions, and

$$g_{1\ell} = \text{sign } (\lambda_{1\ell}) \Delta t (|\lambda_{1\ell}|(\Delta S + |\lambda_{2\ell}|) \Delta \xi + |\lambda_{3\ell}|(\Delta \phi) \cdot J$$

in the region where the solution is smooth; $\lambda_{2\ell}$ and $\lambda_{3\ell}$ are eigenvalues of the matrices $\frac{\partial \tilde{G}}{\partial U}$ and $\frac{\partial \tilde{H}}{\partial U}$. The smoothness or the nonsmoothness of the solution is defined as follows: if

$$(\lambda_{1\ell})_{i+1jn} \cdot (\lambda_{1\ell})_{ij\kappa} > 0$$

and

$$(\lambda_{1\ell})_{i+1j\kappa} > (\lambda_{1\ell})_{ij\kappa}$$

and

$$|(I_{1\ell})_{i+1j\kappa} - (I_{1\ell})_{ij\kappa}| < 0.2/0.3(|I_{1\ell}|_{i+j\kappa} + |I_{1\ell}|_{ij\kappa})$$

the solution is treated as smooth, otherwise it is nonsmooth; $I_{1\ell} = W_{1\ell} U$, where $W_{1\ell}$ is the eigenvector corresponding to the eigenvalue $\lambda_{1\ell}$. Matrices B_2 and B_3 are defined in the same way.

Let us now consider the points which are located on the border of the computational domain. The main difficulties arise when this border is a shock wave or a solid body surface. Along such a border $\xi = \xi_1$, and the ξ derivatives in (2) are written in the following form:

$$\frac{\partial \tilde{G}}{\partial \xi} = U \frac{\partial}{\partial \xi} \frac{\xi t}{J} + F \frac{\partial}{\partial \xi} \frac{\xi_x}{J} + G \frac{\partial}{\partial \xi} \frac{\xi_y}{J} + H \frac{\partial}{\partial \xi} \frac{\xi_z}{J} + A \frac{\partial U}{\partial \xi}$$

where

$$A = \frac{\xi_t}{J} E + \frac{\xi_x}{J} \frac{\partial F}{\partial U} + \frac{\xi_y}{J} \frac{\partial G}{\partial U} + \frac{\xi_z}{J} \frac{\partial H}{\partial U}.$$

We multiply A by the eigenvectors corresponding to the characteristics which intersect points on the surface from the integration domain, and approximate the resulting expressions by one-sided differences. Then, adding the boundary conditions, we obtain the complete system of finite difference equations.

3. RESULTS

The problem of a plane shock wave diffracting over a spherical blunt body moving at supersonic speed was solved first. The solution essentially depends on whether the body and the shock are approaching each other or whether the body overtakes the shock.

In the first case, the incident shock refracts at the bow shock and penetrates the shock layer. Then it reflects from the body surface and moves toward the bow shock. On reaching the front shock, the refracted shock is reflected as a rarefaction wave. The main stages of this interaction are presented in Figs. 1-3 which portray the body, the front and the incident shocks and isobars within the shock layer. The pressure obtained at the stagnation point is compared with the experimental data [7] in Fig. 4.

In the second case, the incident shock penetrates within the shock layer as a rarefaction wave. The subsequent flow is determined by multiple reflection of this wave from the body and the front shock. The present algorithm was also used for the investigation of the interaction of the bow shock with contact discontinuities. Here, when the body penetrates into the higher density gas, there is a phenomenon analogous to the first case discussed above, while in the case of penetration into the lower density gas, it is analogous to the second case. But if the density is sufficiently small, the oncoming flow velocity may fall below the speed of sound. In this case, the front shock leaves the body, and its intensity decreases with time.

Computed results of such variation are presented in Fig. 5. Here curves 1 and 2 show the evolution of the pressure at the stagnation point and behind the front shock on the axis of symmetry; ε is the distance between the front shock and the body measured along the axis of symmetry, and C_x is the resistance coefficient. As may be seen from the illustration, ε does not approach a limit as $t \to \infty$.

Fig. 6 gives numerical results of the 3D shock-interaction problem. Here the angle between the axis of the body movement and the normal to the incident shock is 15°. The time variations of C_x and C_y are given in this Figure.

Finally, we present numerical results for the shock interaction with a non-plane discontinuity, such as that arising from a point explosion. The most interesting case here occurs when the radius of the explosion is comparable to the characteristic length scale of the body. The interaction occurs as follows. The initial stage (Fig. 7) is analogous to that of the first case of the plane shock interaction. When the front shock approaches the central part of the explosion (Fig. 8), it penetrates the region of a sharp decrease in density, thus giving rise to a rarefaction wave. After crossing the central part of the explosion, the front shock penetrates the region of increasing density. A compression wave appears within the shock layer, and transforms later into a shock (Fig. 9). The computed results show that the influence of this shock on the body is comparable to the influence of the refracted spherical shock.

4. CONCLUSION

Using the characteristic properties of the governing gas dynamics equations expressed in conservation form, a conservative numerical algorithm for flows with internal shocks was developed. An approximation of the metrics of the grid transformation in a moving curvilinear coordinate system has been proposed and can be used also for conservative algorithms constructed from other difference schemes. The results obtained for supersonic aerodynamics problems with shock interactions agree with the experimental data. The method is suitable for calculating flows with large spatial gradients; it also allows capturing of internal shocks.

5. REFERENCES

1. Каменецкий В.Ф., Турчак Л.И. Изв. АН СССР, МЖГ, 1984, № 5, с.141-147.
2. Kamenetsky V.F., Turchak L.I. Lec. Not. Phys. 1985, No 218, p.557-56
3. Магомедов К.М., Холодов А.С. ЖВМ и МФ, 1969, 9, № 7, с.373-386.
4. Белоцерковский О.М. ДАН СССР, 1957, 113, № 3, с.509-512
5. Hidman R.G., Katler P., Anderson D. AIAA J, 1981, v.19,No4,p.424-431
6. Семенов А.Ю. ДАН СССР, 1984, 279, № 1, с.34-37.
7. Taylor T.D., Hudgins H.E. AIAA J, 1968, v.6, No 2, p.198-204.

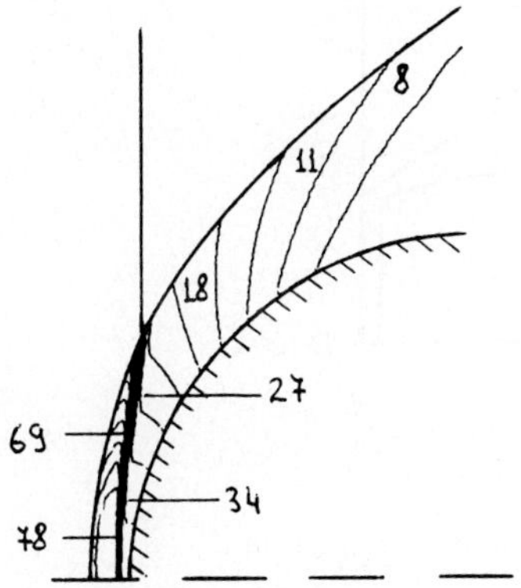

Fig. 1

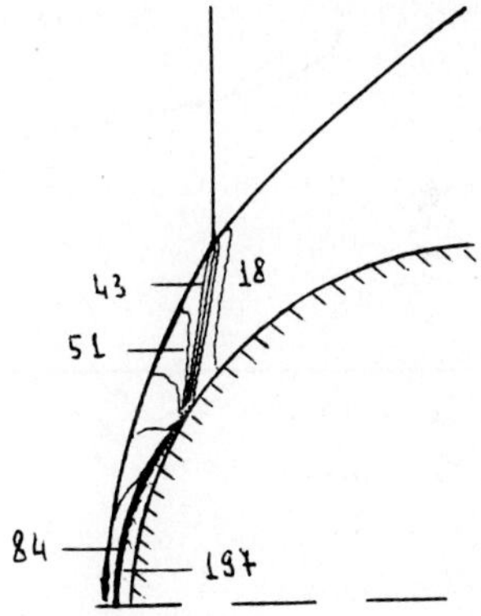

Fig. 2

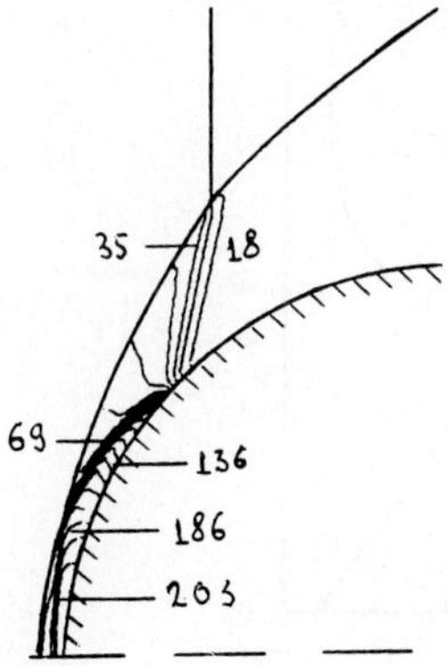

Fig. 3

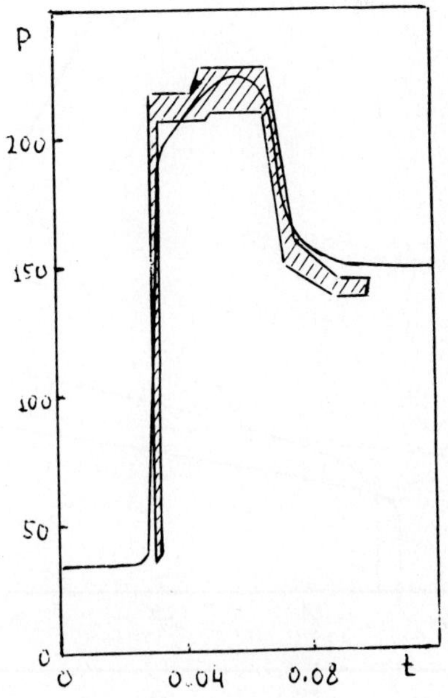

Fig. 4

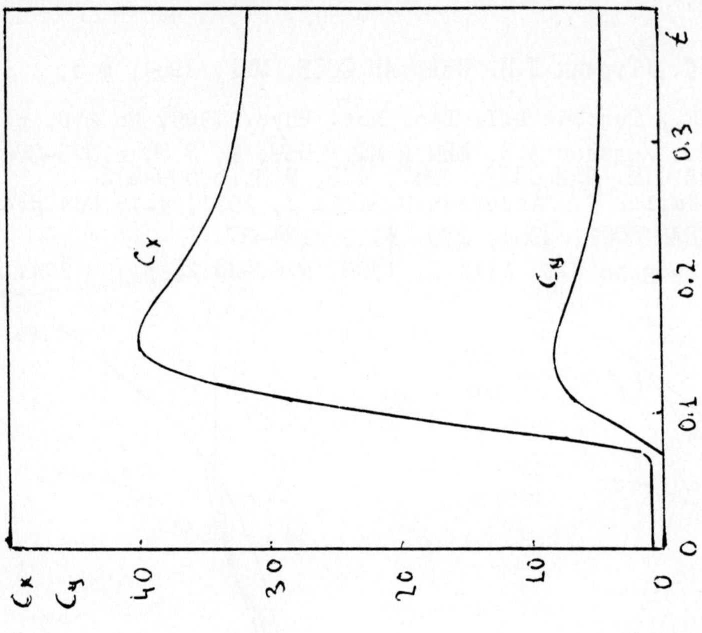
C_x
C_y
40
30
20
10
0
0
0.1
0.2
0.3
t
C_x
C_y

Fig. 6

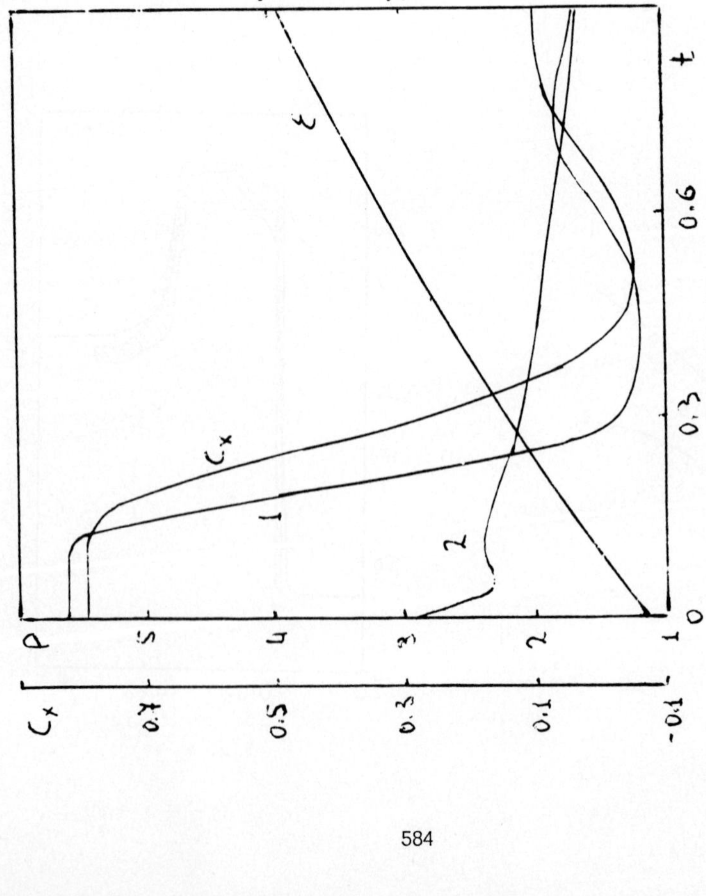
C_x
0.7
0.5
0.3
0.1
-0.1
P
5
4
3
2
1
ε
2.5
2.1
1.7
1.3
0
0.3
0.6
t
C_x
1
2
ε

Fig. 5

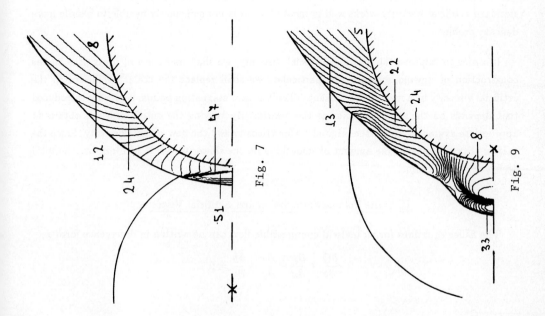

Fig. 7

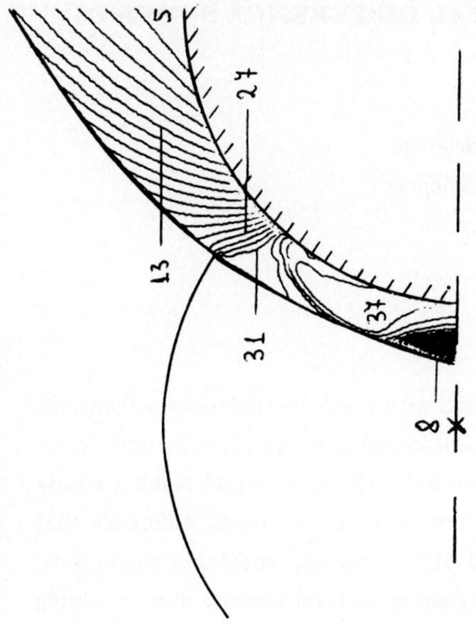

Fig. 8

Fig. 9

IMPROVING THE ACCURACY OF CENTRAL DIFFERENCE SCHEMES

Eli Turkel
School of Mathematical Sciences
Sackler Faculty of Exact Sciences
Tel-Aviv University

I. Introduction

In recent years, central difference schemes have been used with much success to solve transonic flow problems about aerodynamic shapes. These schemes are second order accurate for sufficiently smooth meshes and have an added artificial viscosity to stabilize the scheme and reach a steady state. This artificial viscosity is usually a blend of two terms. One is a fourth difference that stabilizes the even-odd modes that appear with central differences and constants coefficients. Without this viscosity, one cannot reduce the residual beyond some level because of a remaining high frequency mode. The second viscosity term is a nonlinear second difference that limits oscillations in the neighborhood of shocks. A nonlinear shock detector preserves the second order accuracy of the scheme in smooth regions.

An advantage of the artificial viscosity approach is that it allows the user control over the amount of dissipation. Nevertheless, one sometimes finds that there is too much dissipation in the numerical solution. Changing the global constants that appear in the formulas is not sufficient to construct an artificial viscosity that is appropriate in both the shocked and smooth regions of the flow. For some problems, we need to severely limit the viscosity in some smooth regions, e.g., near the trailing edge, while still maintaining stability near the shocks. Hence, although the standard artificial viscosity works well in most cases, it is not sufficiently flexible to handle more delicate problems.

In order to improve the existing artificial viscosity, we shall make use of ideas used in the construction of upwind schemes. In particular, we shall replace the scalar coefficient in the artificial viscosity by a matrix. To prevent difficulties near stagnation points, a cutoff is introduced that depends on the spectral radius of the matrix. By varying the cutoff, one can obtain an appropriate average between the original scalar viscosity and the new matrix viscosity. Since the matrix viscosity reduces the amount of smoothing on the slower waves, it will improve the total accuracy of the scheme.

II. Finite Volume Formulation and Artificial Viscosity

The Euler equations for an inviscid compressible flow can be written in divergence form as

$$\frac{\partial Q}{\partial t} + \frac{\partial f}{\partial x} + \frac{\partial g}{\partial y} + \frac{\partial h}{\partial z} = 0 \qquad (1)$$

where

$$Q = (\rho, \rho u, \rho v, \rho w, E)^t \tag{2a}$$

$$f = (\rho u, \rho u^2 + p, \rho uv, \rho uw, (E+p)u)^t \tag{2b}$$

$$g = (\rho v, \rho uv, \rho v^2 + p, \rho vw, (E+p)v)^t \tag{2c}$$

$$h = (\rho w, \rho uw, \rho vw, \rho w^2 + p, (E+p)w)^t \tag{2d}$$

and for an ideal gas

$$p = (\gamma - 1)[E - \rho(u^2 + v^2 + w^2)/2]. \tag{2e}$$

We can also write (1) in the form

$$\frac{\partial Q}{\partial t} + div(F) = 0. \tag{1b}$$

We integrate (1) over a three dimensional cell and consider $Q_{i,j,k}$ as an approximation to the average of Q over the cell. Hence,

$$\frac{\partial Q_{i,j,k}}{\partial t} + \frac{\int\int\int divF dV}{\int\int\int dV} = 0$$

or using the divergence theorem,

$$\frac{\partial}{\partial t}(VQ)_{ijk} + \int\int \vec{F} \cdot ndS = 0. \tag{3}$$

Hence, the time change of the average Q is governed by the fluxes entering and leaving the cell.

This finite volume approach leads to a pure central difference method for Cartesian grids. Though this scheme is stable for constant coefficient hyperbolic equations it is subject to instabilities that will prevent the convergence to a steady state. To enhance this convergence a fourth difference viscosity is added to the scheme. The fourth difference causes oscillations in the neighborhood of shocks. Hence, a shock detector is constructed and near the shocks the fourth difference is turned off and only the nonlinear second difference is operative. The total artificial viscosity, $\overline{V}$, is the sum of such second and fourth differences in each coordinate direction.

$$\begin{aligned} \overline{V}_{tot} &= \overline{V}^{\xi}_{i+\frac{1}{2},j,k} - \overline{V}^{\xi}_{i-\frac{1}{2},j,k} + \overline{V}^{\eta}_{i,j+\frac{1}{2},k} - \overline{V}^{\eta}_{i,j-\frac{1}{2},k} \\ &\quad + \overline{V}^{\zeta}_{i,j,k+\frac{1}{2}} - \overline{V}^{\zeta}_{i,j,k-\frac{1}{2}} \end{aligned} \tag{4}$$

Hence it is sufficient to describe these terms in the ξ direction. Since we only take differences at neighboring points the artificial viscosity is always in conservation form.

The first difference is defined as

$$D_{i+\frac{1}{2},j,k} = Q_{i+1,j,k} - Q_{i,j,k} \tag{5a}$$

and the second ξ difference is defined as

$$E_{i,j,k} = D_{i+\frac{1}{2},j,k} - D_{i-\frac{1}{2},j,k}. \tag{5b}$$

We then form the second and fourth differences. In particular the fourth difference is formed as a second difference of a second difference with positive weights [3,8]. Hence,

$$\overline{V}^{\xi}_{i+\frac{1}{2},j,k} = \epsilon^{(2)}_{i+\frac{1}{2},j,k} D_{i+\frac{1}{2},j,k} - (\epsilon^{(4)}_{i+1,j,k} E_{i+1,j,k} - \epsilon^{(4)}_{i,j,k} E_{i,j,k}). \tag{6}$$

Let,

$$\nu_{i,j,k} = \left| \frac{p_{i+1,j,k} - 2p_{i,j,k} + p_{i-1,j,k}}{p_{i+1,j,k} + 2p_{i,j,k} + p_{i-1,j,k}} \right|. \tag{7a}$$

$\nu_{i,j,k}$ is used to detect the location of shocks. When $\nu_{i,j,k}$ is large then the fourth difference is reduced. Let,

$$\sigma_{i+\frac{1}{2},j,k} = K^{(2)} \max(\nu_{i-1,j,k}, \nu_{i,j,k}, \nu_{i+1,j,k}, \nu_{i+2,j,k}). \tag{8}$$

We also multiply σ by a function of the Mach number to reduce σ near the surface. Let, $A = \frac{\partial \overline{F}}{\partial Q}, B = \frac{\partial \overline{G}}{\partial Q}, C = \frac{\partial \overline{H}}{\partial Q}$, where $\overline{F}, \overline{G}, \overline{H}$ are the fluxes in the coordinate system (ξ, η, ς). Let λ be a measure of the fluxes. The original code chose λ as

$$\lambda^\xi = \lambda^\eta = \lambda^\varsigma = \rho(A) + \rho(B) + \rho(C) \tag{9a}$$

where ρ is the the spectral radius of the matrix. For problems with a highly stretched mesh it was found [2,3,8] that for increased accuracy one should choose

$$\lambda^\xi = \rho(A), \ \lambda^\eta = \rho(B), \ \lambda^\varsigma = \rho(C). \tag{9b}$$

$K^{(2)}$, $K^{(4)}$ are constants that determine the level of the second and fourth differences. These constants are given as input to the code. Then

$$\epsilon^{(2)}_{i+\frac{1}{2},j,k} = \lambda_{i+\frac{1}{2},j,k} \sigma_{i+\frac{1}{2},j,k} \tag{10a}$$

$$\epsilon^{(4)}_{i,j,k} = \lambda_{i,j,k} \max(0, K^{(4)} - \sigma_{i,j,k}). \tag{10b}$$

In order to imitate the upwind type [7] algorithms we now replace the scalars in (9b) by matrices. Hence,

$$\lambda^\xi = |A|, \ \lambda^\eta = |B|, \ \lambda^\varsigma = |C| \tag{9c}$$

where $|A| = T|\Lambda_\xi|T^{-1}$ when $A = T\Lambda_\xi T^{-1}$ and Λ_ξ is a diagonal matrix with the eigenvalues of A as its entries. This definition of λ can lead to difficulties when an eigenvalue is near zero. Hence, we modified Λ_ξ to be

$$\overline{\Lambda}_\xi = diag(\max(a_i, q\rho(A))) \qquad a_i = \text{e.v. of } A \tag{11}$$

where q is a specified constant. When $q = 1$, then $\Lambda_\xi = \rho(A) \cdot I$ and so (9b) is recovered. When $q = 0$ then no modification of Λ_ξ is done. In general, we found that $q = 0.2$ gives good results.

We point out that the use of (9c) does not allow for a constant enthalpy solution and so enthalpy damping cannot be used [5].

III. Results

We consider the central difference code with Runge-Kutta time stepping [3,5]. As described above we use a matrix valued artificial viscosity which approximates TVD type schemes [6,7,9]. The fourth difference viscosity is still needed to allow the multigrid acceleration to quickly reach a steady state. We consider inviscid flow about a NACA0012 with $M_\infty = 0.8$, $\alpha = 1.25°$. A 192×32 C mesh is used with 128 points on the airfoil. In [1] it is pointed out that the standard code smears the weak shock on the lower surface. In Figure 1, we plot the result for the standard scheme, but without enthalpy damping. In Figure 2, we show the same case but using the matrix viscosity. The convergence rate is slowed down since the fourth order viscosity is not as strong but the shock on the lower surface is sharper. There is an overshoot on the shock on the upper surface. This is due to fact that the cutoff (10b) is not sufficiently sharp. One way to improve this is to replace (7) by

$$\nu_{i,j,k} = \frac{|p_{i+1,j,k} - p_{i,j,k}| - |p_{i,j,k} - p_{i-1,j,k}|}{|p_{i+1,j,k} - p_{i,j,k}| + |p_{i,j,k} - p_{i-1,j,k}| + \varepsilon} \tag{7b}$$

so that $\nu_{i,j,k}$ is one at discontinuities. One can also use the matrix coefficient only for the second difference but use a scalar coefficient for the fourth difference viscosity. This accelerates the convergence to a steady state but smears the shocks. The results presented used a four step Runge-Kutta algorithm with the artificial viscosity frozen after the first stage using a matrix viscosity rather than a scalar viscosity adds about 60% to the total CPU time.

References

[1] Anderson, W., Kyle, Thomas, J. L., van Leer, B., Comparison of Finite Volume Flux Vector Splittings for the Euler Equations, AIAA J., Vol. 24, pp. 1453-1460 (1986).

[2] Caughey, D. A., Turkel, E., Effects of Numerical Dissipation on Finite-Volume Solutions of Compressible Flow Problems, AIAA-88-0621 (1988).

[3] Chima, R. B., Turkel, E., Schaffer, S., Comparison of Three Explicit Multigrid Methods for the Euler and Navier-Stokes Equations, AIAA-87-0602 (1987).

[4] Harten, A., The Artificial Compression Method for Computation of Shocks and Contact Discontinuities III. Self-Adjusting Hybrid Schemes, Math. Comput., Vol. 32, pp. 363-389 (1978).

[5] Jameson, A., Schmidt, W., Turkel, E., Numerical Solutions of the Euler Equations by Finite Volume Methods using Runge-Kutta Time-Stepping Schemes, AIAA-81-1259 (1981).

[6] Osher, S., Tadmor, E., On the Convergence of Difference Approximations to Scalar Conservation Laws, Math. Comput., Vol. 50, pp. 19-51 (1988).

[7] Roe, P. L., Approximate Riemann Solvers Parametric Vectors and Difference Schemes, J. Comput. Phys., Vol. 43, pp. 357-372 (1981).

[8] Swanson, R. C., Turkel, E., Artificial Dissipation and Central Difference Schemes for the Euler and Navier-Stokes Equations, AIAA 8th CFD Conference, AIAA-87-1107-CP (1987).

[9] Tadmor, E., Convenient Total Variation Diminishing Conditions for Nonlinear Differences Schemes, SIAM J. Numer. Anal., to appear.

[10] Turkel, E., Acceleration to a Steady State for the Euler Equations, Numerical Methods for the Euler equations of Fluid Dynamics, pp. 281-311, F. Angrand, et al. (editors), SIAM, Philadelphia (1985).

[11] Turkel, E., van Leer, B., Flux Vector Splitting and Runge-Kutta Methods for the Euler Equations, 9th Inter. Conf. Numer. Methods Fluid Dynamics, Springer-Verlag Lecture Notes Physics, Vol. 208, pp. 566-570 (1984).

[12] van Leer, B., Towards the Ultimate Conservative Difference Scheme, II. Monotonicity and Conservation Combined in a Second-Order Scheme, J. Comput. Phys., Vol. 14, pp. 361-370 (1974).

[13] Yee, H. C., Upwind and Symmetric Shock-Capturing Schemes, NASA TM 89464 (1987).

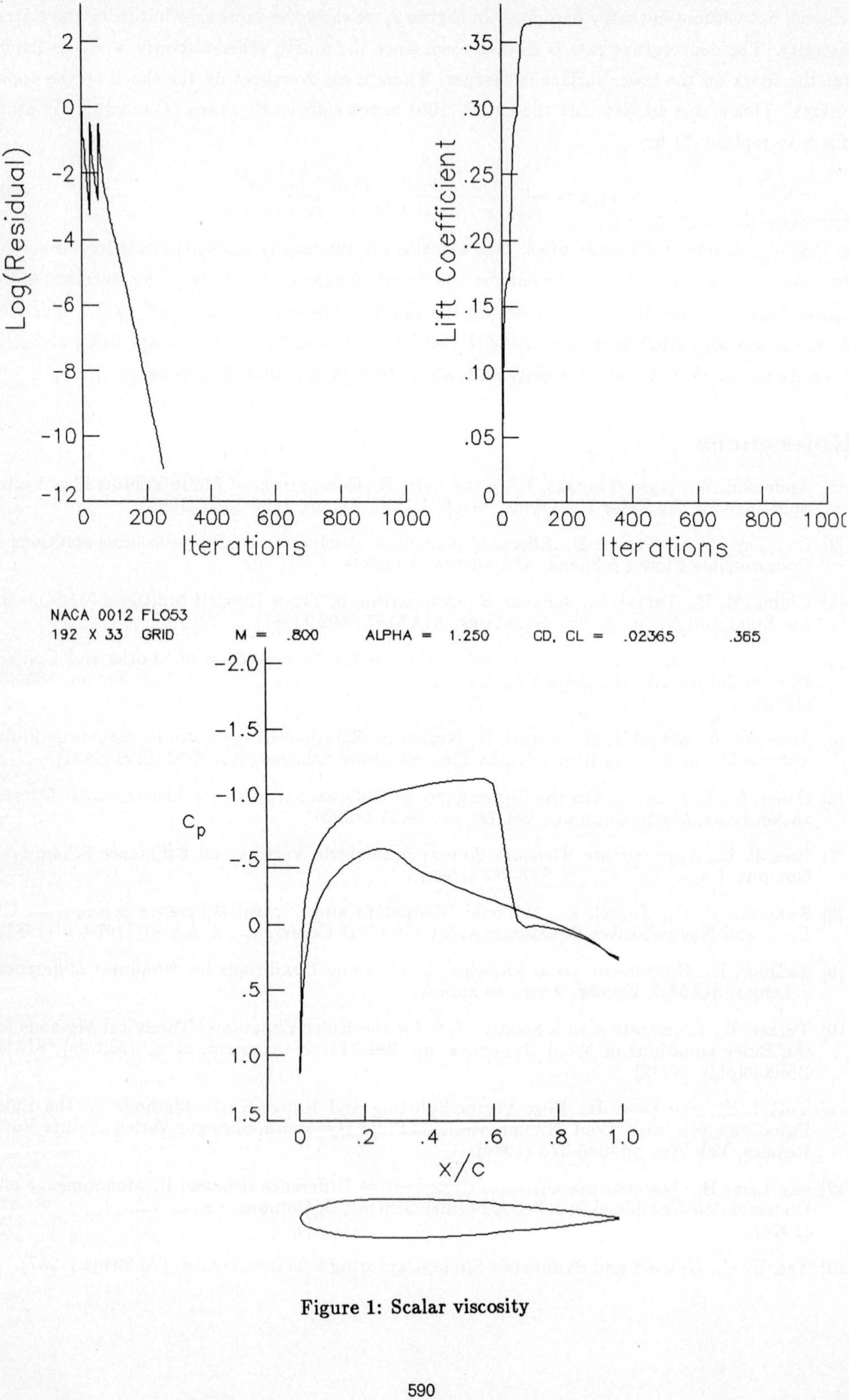

Figure 1: Scalar viscosity

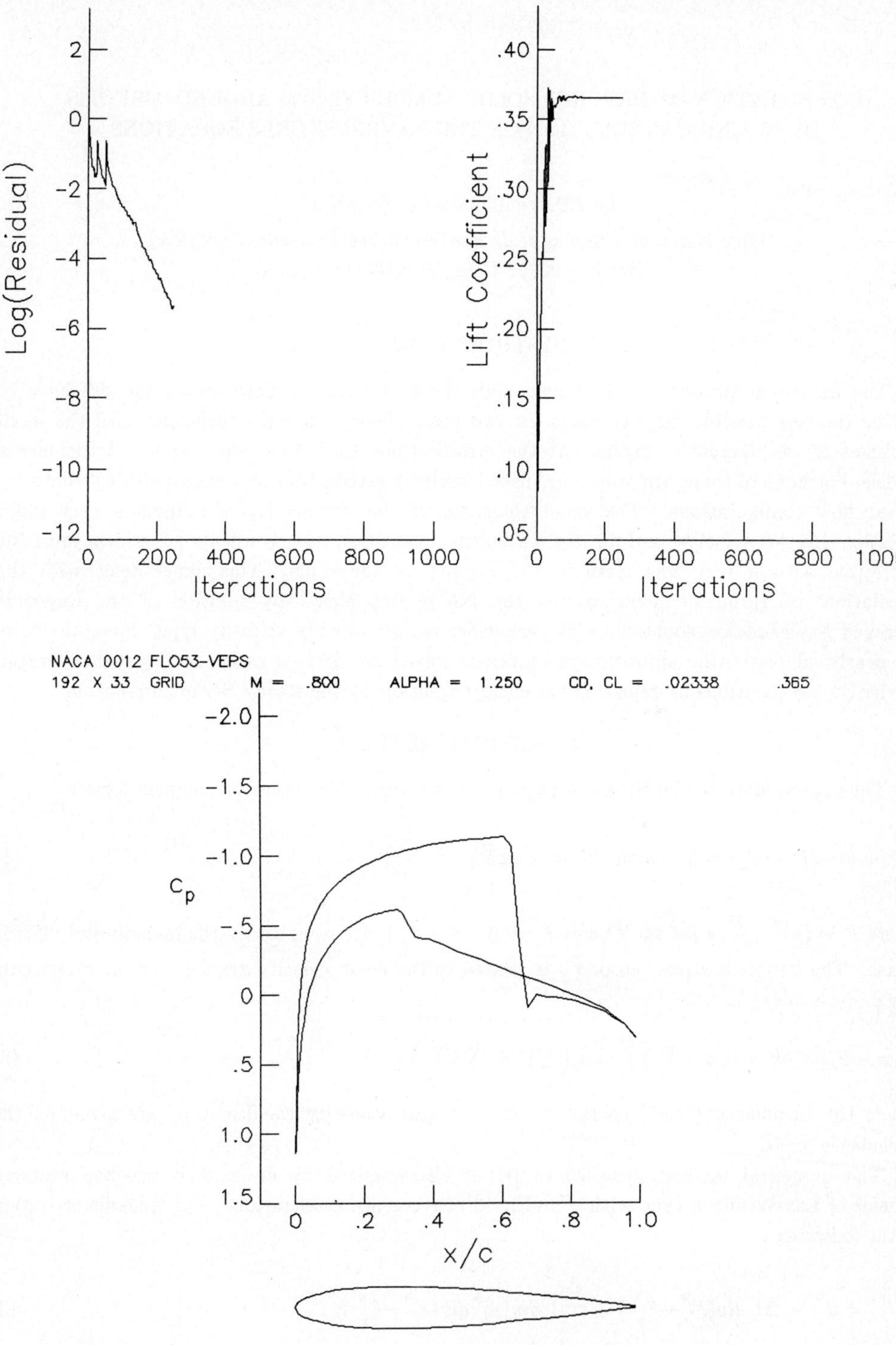

Figure 2: Matrix viscosity

COMPUTATION OF HIGH REYNOLDS NUMBER FLOWS AROUND AIRFOILS BY NUMERICAL SOLUTION OF THE NAVIER-STOKES EQUATIONS

by J.P. Veuillot and L. Cambier

Office National d'Etudes et de Recherches Aérospatiales (ONERA)
BP 72 - 92322 CHATILLON (France)

INTRODUCTION

The numerical simulation of viscous high Reynolds number flows raises the difficulty of finding the best possible representation for two basic phenomena : the turbulence and the small thickness of the dissipative layers. At the present time, there is a wide variety of turbulence models, but none of them has truly established itself by satisfactory agreement with test data for varied flow configurations. The small thickness of the viscous layers induces a very rapid variation in the velocity and in the turbulent quantities, the gradients of which must be discretized with a very fine mesh in the vicinity of the walls. This paper deals with the calculations of turbulent flows around the NACA0012 airfoil by solution of the Reynolds averaged Navier-Stokes equations with turbulence models of eddy viscosity type. Since the flows are nearly adiabatic, the approximate equations solved are derived from the Reynolds averaged Navier-Stokes equations by replacing the energy equation by the steady Bernoulli relation.

NUMERICAL METHOD

The approximate Navier-Stokes equations are written in the following compact form :

$$\frac{\partial u}{\partial t} + div\,(F - F_v) = 0 \quad \text{with} \quad u = (\rho\,,\, \rho\overline{V})\,, \tag{1}$$

where $F = (\rho\overline{V}\,,\, p\overline{\overline{I}} + \rho\overline{V} \otimes \overline{V})$ and $F_v = (0\,,\, \overline{\overline{\tau}} + \overline{\overline{\tau}}_R)$ are respectively the inviscid and viscous fluxes. The Reynolds stress tensor $\overline{\overline{\tau}}_R$ is related to the mean velocity gradient by the Boussinesq law :

$$\overline{\overline{\tau}}_R = -\,2/3\;(\,\rho k + \mu_t\, div\;\overline{V}\,)\,\overline{\overline{I}} + \mu_t(\,\nabla\overline{V} + (\nabla\overline{V})^t\,)\,, \tag{2}$$

where the turbulence kinetic energy k and the eddy viscosity coefficient μ_t are given by the turbulence model.

The numerical method, detailed in [1], is characterized by an explicit one-step centered scheme of Lax-Wendroff type with a multigrid convergence acceleration. The time-discretization is the following :

$$u^{n+1} = u^n - \Delta t\; div(F^n - F_v^n) + \frac{\Delta t^2}{2} div(A^n\, div(F^n - F_v^n))\,, \tag{3}$$

where A denotes the inviscid jacobian matrix ($A = \partial F / \partial u$). Due to an approximation in the time derivative $\partial^2 u / \partial t^2$, the scheme is only first order time accurate in the viscous regions.

The space-discretization is carried out in three stages. i) The viscous terms F_v are calculated at the mesh points using central difference formulas. More precisely, the velocity gradient $\nabla\overline{V}$ at the mesh point M (Fig. 1) is calculated as the mean value of $\nabla\overline{V}$ in the control volume $\omega = (a,b,c,d)$ by the following contour integral :

$$\nabla\overline{V} = \frac{1}{\omega}\int_{\partial\omega} \overline{V} \otimes \overline{n}\, dS .$$

The value of the velocity $\overline{V}$ along each side of ω (for example, ab) is calculated from an arithmetic mean of the velocity at the neighbouring points (for example, B and M).

ii) The first order term $div\ (F - F_v)$ at M is obtained from an arithmetic mean of the values of $div\ (F - F_v)$ at the cell centers a,b,c and d of the four cells surrounding the mesh point M. Each of these values is calculated by a contour integral around the corresponding cell assuming that the fluxes vary linearly along the sides.

iii) The second order term $div\ (A\ div\ (F - F_v))$ is calculated at M by a contour integral around the control volume ω assuming that the quantity $A\ div\ (F - F_v)$ varies linearly along the sides of ω, and with, for example, $A_a = A(\,(u_A + u_B + u_C + u_D)/4)$.

This scheme represents a 21-point difference scheme and reduces to a 9-point scheme for inviscid flows ($F_v = 0$). Artificial dissipative terms consisting of a second order non-linear term and a fourth order linear term are added to improve the stability properties of the scheme and the well-known local time-stepping technique is used.

MULTIGRID ACCELERATION

The multigrid acceleration ([2],[3] and [1]) applied to system (1) consists in integrating on coarse grids the following residual system based on the time derivation of the inviscid equations only :

$$\frac{\partial \delta u}{\partial t} + div\ A(u)\,\delta u = 0 . \tag{4}$$

The residual δu is defined by $\delta u_h = \Delta t\ \partial u/\partial t$ in the fine grid G_h, and initialized in the k-th coarse grid $G_{2^k h}$ by a transfer of the residual obtained in the previous grid $G_{2^{k-1}h}$.

If $T_{2^{k-1}h}^{2^k h}$ denotes a transfer operator from $G_{2^{k-1}h}$ to $G_{2^k h}$ and $I_{2^k h}^{h}$ denotes an interpolation operator from $G_{2^k h}$ to G_h, the solution u in the fine grid is defined after a multigrid cycle with use of p coarse grids by :

$$u_h^{n+1+\alpha} = u_h^{n+1} + \sum_{k=1}^{p} \alpha_k\, I_{2^k h}^{h}\, \delta u_{2^k h}^{n+1/2+\overline{\beta}_k}, \tag{5}$$

with $\delta u_{2^k h}^{n+1/2+\overline{\beta}_k}$ calculated by integration of (4) on the time interval $\beta_k\,\Delta t$ (with $\overline{\beta}_k = \beta_1 + ... + \beta_k$) starting from the initialization :

$$\delta u_{2^k h}^{n+1/2+\overline{\beta}_{k-1}} = T_{2^{k-1}h}^{2^k h}\, \delta u_{2^{k-1}h}^{n+1/2+\overline{\beta}_{k-1}} .$$

In equation (5), $\overline{\alpha} = \alpha_1 + ... + \alpha_p$ denotes the time-advance coefficient of the solution in the fine grid, whereas β_k denotes the time-advance coefficient used for the integration of (4) in the k-th coarse grid. The multigrid cycle allows to advance the solution to the time level $n+1+\overline{\alpha}$ with far less calculations than it is needed with the basic scheme alone to obtain the solution at the same time level. Moreover, the converged solution retains the space accuracy of the Navier-Stokes basic scheme in the fine grid because the quantity δu_h vanishes at convergence.

The influence on the multigrid acceleration efficiency of the number p of coarse grids and of the parameters (α_k , β_k) is detailed in [1] and is illustrated here by figures 2 and 3. In this study the parameters (α_k , β_k) were defined by $\alpha_k = \alpha$ and $\beta_k = 2^{k-1}\beta$, with α and β independent of the coarse grid. Figure 2 shows the effect of p on the convergence (with $\alpha = 1$ and $\beta = 2$), whereas Fig. 3 shows the effect of α and β (with $p = 2$).

RESULTS

The numerical results presented here are related to the computation of the turbulent flow around the NACA0012 airfoil for the following conditions : $M_0 = 0.799$, $\alpha_0 = 2.26\,°$ and $\mathrm{Re}_0 = 9.\ 10^6$, corresponding to a test case proposed for the Viscous Transonic Airfoil Workshop associated with the AIAA 25th Aerospace Sciences Meeting. The calculations have been performed in three C-meshes and with three turbulence models (two algebraic models and a two-equation model).

The first model, proposed by Michel et al. [4], is a one-layer mixing length model in which the mixing length l is given as a universal function of the ratio η/δ (η and δ designating the normal distance to the wall and the boundary layer thickness) by : $l/\delta = 0.085\ tanh(K\eta/0.085\delta)$ with K=0.41. The second model is the very popular two-layer Baldwin-Lomax model [5], modified by introducing the local stress in the Van Driest damping function instead of the wall stress. The third model is the $(k\ , \epsilon)$ model proposed by Jones and Launder [6]. Unlike the original model, the near wall region where the turbulence is not fully developed, is treated with the equilibrium model of Michel et al. with a switch to the $(k\ , \epsilon)$ model when the ratio μ_t/μ is greater than 9.

Table 1 presents the characteristics of the meshes at the leading edge (LE) and at the beginning of the interaction region (IR) (ξ designating the distance along the wall) as well as the lift and drag coefficients obtained with the Michel turbulence model. For this model, Figs. 4-7 represent a partial view of the coarsest mesh and of the finest mesh, and the corresponding Mach number contours. The wall pressure coefficients are plotted in Fig. 8 and compared with experimental data [7]. Using the finest mesh, the comparison of the three turbulence models is illustrated in Fig. 9 which represents the wall pressure and friction coefficients in the interaction region. The three models fail to match the experimental data but it can be noticed that the weakest compression is obtained with the $(k\ , \epsilon)$ model.

REFERENCES

[1] Cambier L., Couaillier V. and Veuillot J-P. - Résolution numérique des équations de Navier-Stokes à l'aide d'une méthode multigrille, La Recherche Aérospatiale No 1988-2.

[2] Ni R.H. - A Multigrid Scheme for Solving the Euler Equations, AIAA J. vol. 20, No 11, nov. 1982.

[3] Couaillier V. - Solution of the Euler Equations : Explicit Scheme Acceleration by a Multigrid Method, ONERA TP No 1985-129.

[4] Michel R., Quémard C. and Durant R. - Application d'un schéma de longueur de mélange à

l'étude des couches limites turbulentes d'équilibre, ONERA NT No 154, 1969.
[5] Baldwin B.S. and Lomax H. - Thin Layer Approximation and Algebraic Model for Separated Turbulent Flows, AIAA paper 78-257, 1978.
[6] Jones W.P. and Launder B.E. - The Prediction of Laminarization with a Two-Equation model of turbulence, Int. J. Heat and Mass Transfer, Vol. 15, pp. 301-314, 1972.
[7] Harris C.D. - Two-Dimensional Aerodynamic Characteristics of the NACA0012 Airfoil in Langley 8-Foot Transonic Pressure Tunnel, NASA TM-81927, 1981.

Mesh	$\Delta\eta_{LE}$	$\Delta\xi_{LE}$	η^+_{IR}	$\Delta\xi/\delta_{IR}$	C_L	C_D
257x65	$5.9\ 10^{-6}$	$1.7\ 10^{-3}$	2.5	1.4	.422	.0399
257x73	$1.6\ 10^{-6}$	$1.7\ 10^{-3}$	1.1	1.5	.422	.0400
513x73	$1.6\ 10^{-6}$	$0.9\ 10^{-3}$	1.1	0.7	.433	.0410

Table 1 - Characteristics of the meshes.

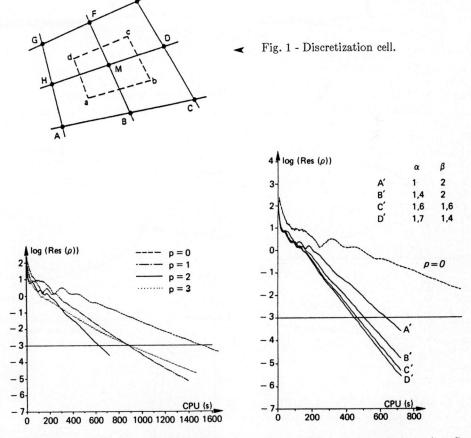

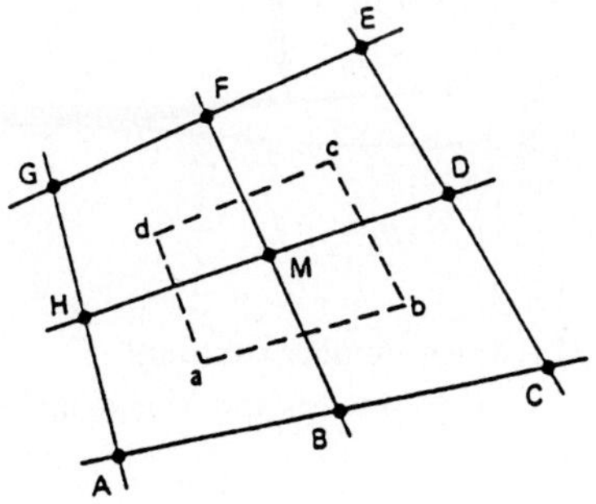

Fig. 1 - Discretization cell.

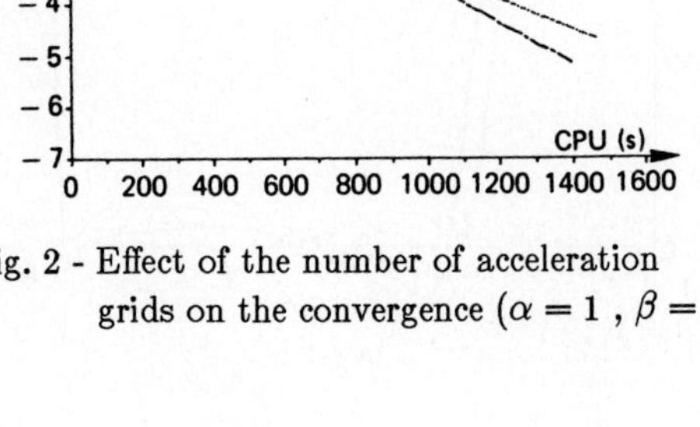

Fig. 2 - Effect of the number of acceleration grids on the convergence ($\alpha = 1$, $\beta = 2$).

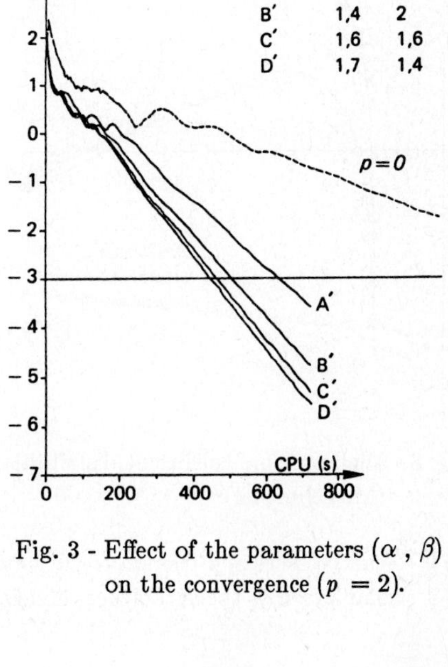

Fig. 3 - Effect of the parameters (α , β) on the convergence ($p = 2$).

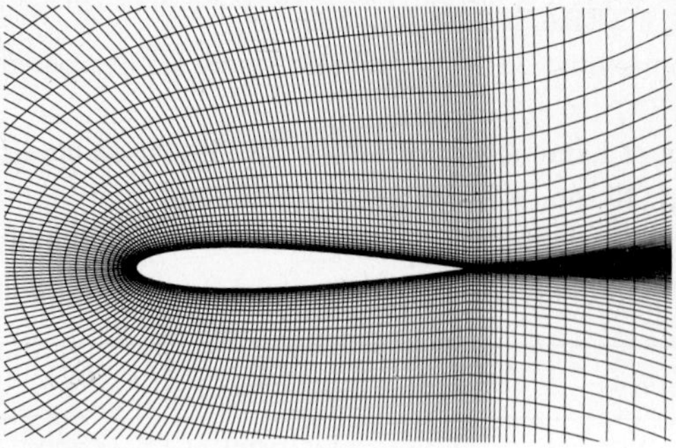

Fig. 4 - Partial view of the 257x65 C-mesh.

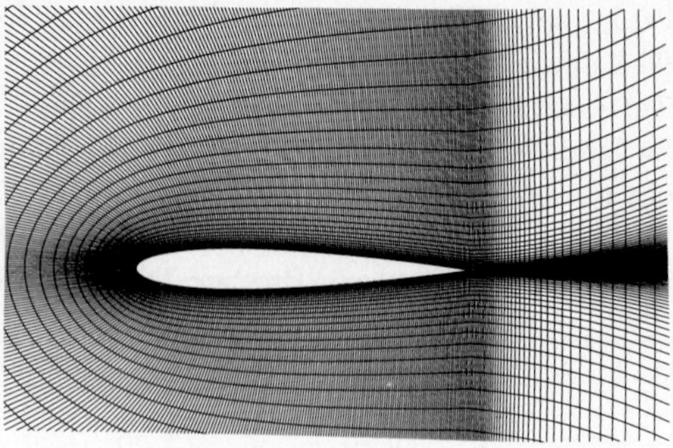

Fig. 5 - Partial view of the 513x73 C-mesh.

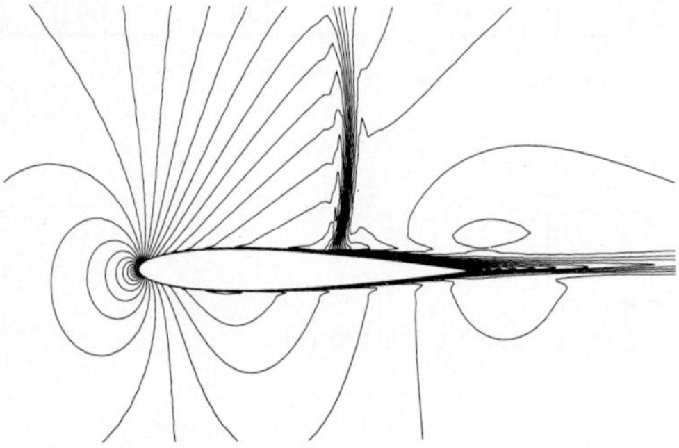

Fig. 6 - Mach number contours
(257x65 mesh and Michel model).

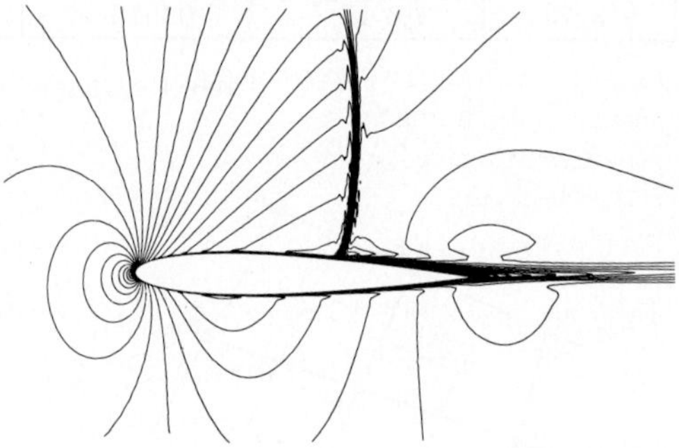

Fig. 7 - Mach number contours
(513x73 mesh and Michel model).

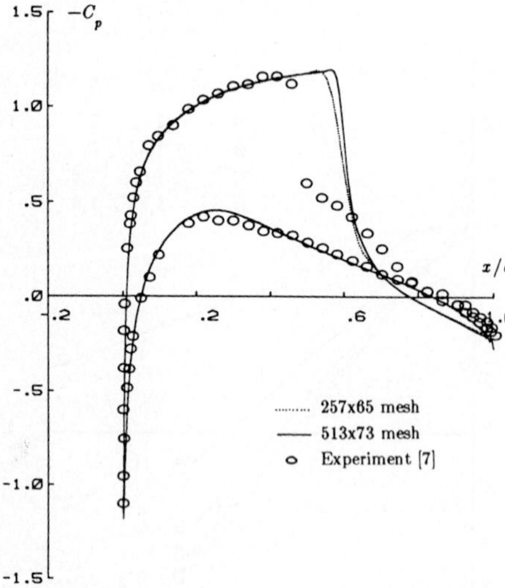

Fig. 8 - Wall pressure coefficient distribution
(Michel model).

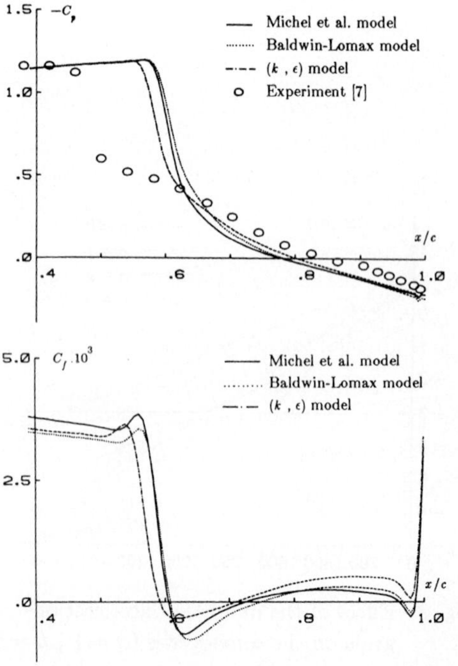

Fig. 9 - Wall pressure and friction coefficients
comparison of the turbulence models.

Diagonal Implicit Multigrid Solution of the Three-Dimensional Euler Equations

Yoram Yadlin and *D. A. Caughey*

Sibley School of Mechanical and Aerospace Engineering
Cornell University
Ithaca, New York 14853

I. Introduction

The effectiveness of the multigrid implementation of an Alternating Direction Implicit algorithm in solving the two-dimensional Euler equations has motivated the attempt to apply it to the solution of the three-dimensional equations. This paper describes the extension of the two-dimensional scheme developed by Caughey [1] to three-dimensional problems, presents results for the calculation of flow past a swept wing, and compares the solution obtained with experimental data.

II. Governing Equations

The Euler equations of inviscid, compressible flow can be written in three space dimensions as:

$$\partial\vec{w}/\partial t + \partial\vec{f_1}/\partial x + \partial\vec{f_2}/\partial y + \partial\vec{f_3}/\partial z = 0\,, \tag{1}$$

where

$$\vec{w} = \{\rho, \rho u_1, \rho u_2, \rho u_3, e\}^T \tag{2}$$

is the vector of conserved dependent variables, and

$$\begin{aligned}
\vec{f_1} &= \{\rho u_1, \rho u_1^2 + p, \rho u_1 u_2, \rho u_1 u_3, u_1(e+p)\}^T\,,\\
\vec{f_2} &= \{\rho u_2, \rho u_2 u_1, \rho u_2^2 + p, \rho u_2 u_3, u_2(e+p)\}^T\,,\\
\vec{f_3} &= \{\rho u_3, \rho u_3 u_1, \rho u_3 u_2, \rho u_3^2 + p, u_3(e+p)\}^T\,,
\end{aligned} \tag{3}$$

are the flux vectors in the x, y, and z coordinate directions, respectively. The pressure p is related to the total energy e by the equation of state:

$$e = p/(\gamma-1) + \rho(u_1^2 + u_2^2 + u_3^2)/2 - \rho H_0\,, \tag{4}$$

where γ is the ratio of specific heats and H_0 is the total enthalpy.

III. Computational Technique

In order to allow for the treatment of arbitrary geometries, the algorithm is implemented within the framework of a finite volume approximation [2]. Under any nonsingular transformation of dependent variables Eqs. 1 can be written as:

$$\partial\vec{W}/\partial t + \partial\vec{F_1}/\partial\xi + \partial\vec{F_2}/\partial\eta + \partial\vec{F_3}/\partial\zeta = 0 \tag{5}$$

where $\vec{W} = h\vec{w}$ is the vector of transformed dependent variables, and

$$\begin{aligned}
\vec{F_1} &= h\{\rho U_1, \rho u_1 U_1 + \xi_x p, \rho u_2 U_1 + \xi_y p, \rho u_3 U_1 + \xi_z p, U_1(e+p)\}^T,\\
\vec{F_2} &= h\{\rho U_2, \rho u_1 U_2 + \eta_x p, \rho u_2 U_2 + \eta_y p, \rho u_3 U_2 + \eta_z p, U_2(e+p)\}^T,\\
\vec{F_3} &= h\{\rho U_3, \rho u_1 U_3 + \zeta_x p, \rho u_2 U_3 + \zeta_y p, \rho u_3 U_3 + \zeta_z p, U_3(e+p)\}^T,
\end{aligned} \tag{6}$$

are the transformed flux vectors. Here, h is the determinant of the Jacobian of the transformation (which corresponds to the cell volume) and U_1, U_2, and U_3 are the contravariant components of the velocity given by

$$\begin{bmatrix} U_1 \\ U_2 \\ U_3 \end{bmatrix} = \begin{bmatrix} \xi_x & \xi_y & \xi_z \\ \eta_x & \eta_y & \eta_z \\ \zeta_x & \zeta_y & \zeta_z \end{bmatrix} \begin{bmatrix} u_1 \\ u_2 \\ u_3 \end{bmatrix}. \tag{7}$$

Artificial dissipation is added as a blend of second and fourth differences of the solution. The fourth differences are necessary to insure convergence to a steady state, while the second differences are necessary to prevent excessive oscillations of the solution in the vicinity of shock waves. Following the two-dimensional implementation of Caughey [1], these dissipation terms are independently scaled to stabilize the one-dimensional problems in each coordinate direction. Following Briley and McDonald [3], and Beam and Warming [4], the time linearized implicit operator is approximated as the product of three one-dimensional factors, resulting in a scheme of the form:

$$\begin{aligned}
&\{I + \theta\Delta t[A^n_{1i,j,k}\delta_\xi - \epsilon^{(2)}_{i,j,k}\delta^2_\xi(1/h) + \epsilon^{(4)}_{i,j,k}\delta^4_\xi(1/h)]\} \\
&\{I + \theta\Delta t[A^n_{2i,j,k}\delta_\eta - \epsilon^{(2)}_{i,j,k}\delta^2_\eta(1/h) + \epsilon^{(4)}_{i,j,k}\delta^4_\eta(1/h)]\} \\
&\{I + \theta\Delta t[A^n_{3i,j,k}\delta_\zeta - \epsilon^{(2)}_{i,j,k}\delta^2_\zeta(1/h) + \epsilon^{(4)}_{i,j,k}\delta^4_\zeta(1/h)]\}\Delta\vec{W}^n_{i,j,k} = \\
&- \Delta t\{\delta_\xi\vec{F}_{1i,j,k} + \delta_\eta\vec{F}_{2i,j,k} + \delta_\zeta\vec{F}_{3i,j,k} \\
&- \epsilon^{(2)}(\delta^2_\xi + \delta^2_\eta + \delta^2_\zeta)\vec{w}_{i,j,k} + \epsilon^{(4)}(\delta^4_\xi + \delta^4_\eta + \delta^4_\zeta)\vec{w}_{i,j,k}\}^n,
\end{aligned} \tag{8}$$

where $A^n_{li,j,k} = \{\partial\vec{F}_l/\partial\vec{W}\}^n_{i,j,k}$ are the Jacobians of the transformed flux vectors with respect to the solution and $\epsilon^{(2)}$and $\epsilon^{(4)}$ are the dissipation coefficients.

For computational efficiency, each factor in Eqs. 8 is diagonalized by a local similarity transformation, yielding a decoupled set of scalar pentadiagonal equations along each line. The diagonalization is done using the modal matrices of the Jacobians A_l $(l = 1,2,3)$. Thus, if Q_l is the modal matrix of A_l, then $Q_l^{-1}A_lQ_l = \Lambda_l$ is a diagonal matrix whose diagonal elements are the eigenvalues of A_l. Applying this transformation and a local linearization at each mesh point, the resulting equations are:

$$\begin{aligned}
&\{I + \theta\Delta t[\Lambda^n_{1i,j,k}\delta_\xi - \epsilon^{(2)}_{i,j,k}\delta^2_\xi(1/h) + \epsilon^{(4)}_{i,j,k}\delta^4_\xi(1/h)]\}Q^n_{1,i,j,k}{}^{-1} \\
Q^n_{2i,j,k}&\{I + \theta\Delta t[\Lambda^n_{2i,j,k}\delta_\eta - \epsilon^{(2)}_{i,j,k}\delta^2_\eta(1/h) + \epsilon^{(4)}_{i,j,k}\delta^4_\eta(1/h)]\}Q^n_{2,i,j,k}{}^{-1} \\
Q^n_{3i,j,k}&\{I + \theta\Delta t[\Lambda^n_{3i,j,k}\delta_\zeta - \epsilon^{(2)}_{i,j,k}\delta^2_\zeta(1/h) + \epsilon^{(4)}_{i,j,k}\delta^4_\zeta(1/h)]\}\Delta\vec{V}^n_{i,j,k} = \\
&- \Delta t Q^n_{1,i,j,k}{}^{-1}\{\delta_\xi\vec{F}_{1i,j,k} + \delta_\eta\vec{F}_{2i,j,k} + \delta_\zeta\vec{F}_{3i,j,k} \\
&- \epsilon^{(2)}(\delta^2_\xi + \delta^2_\eta + \delta^2_\zeta)\vec{w}_{i,j,k} + \epsilon^{(4)}(\delta^4_\xi + \delta^4_\eta + \delta^4_\zeta)\vec{w}_{i,j,k}\}^n,
\end{aligned} \tag{9}$$

where $\Delta\vec{V}^n_{i,j,k} = Q^n_{3,i,j,k}{}^{-1}\Delta\vec{W}^n_{i,j,k}$. The elements of A_l, A_l^{-1} and Λ_l can be expressed explicitly in terms of $\vec{w}$ and the elements of the Jacobian matrix of the coordinate transformation, and are given by Pulliam and Chaussee [5]. The solution of Eqs. 9 is performed sequentially by solving five scalar-pentadiagonal systems along each line in each of the three mesh directions for each time step.

The scheme is incorporated within the multigrid algorithm following the procedure developed by Jameson [6]. The treatment of the explicit boundary conditions in the far field is based on the Riemann invariants of the one-dimensional problem normal to the boundary. On solid surfaces, the pressure is extrapolated from the interior of the field using the normal momentum equation. The implicit boundary conditions are treated in a manner consistent with the characteristic theory, following the two-dimensional implementation of the scheme [1].

IV. Results

The algorithm described above has been applied to the problem of transonic flow past a swept wing mounted on a vertical wall, or symmetry plane. The results presented here have been calculated for the ONERA wing M-6 on a C-grid containing $192 \times 32 \times 32$ mesh cells, in wraparound, normal and spanwise directions, respectively. The grid has been generated by a weak shearing of a square root transformation in each plane of constant z. Figure 1 presents the distribution of cells on the wing surface($x - z$ plane), and Figure 2 presents the M6 wing mounted on the wall;the C-grid is typical to each $x - y$ plane, the far field upstream boundaries are located at 10 and 19 chords in the x and y directions respectively, and at 3.5 semispans in the z direction.

Results have been calculated for a free stream Mach number of 0.839 and 3.06 degrees angle of attack in order to allow comparison with existing wind-tunnel test data [7]. The calculations were performed on an IBM 3090 where 150 time steps required about 7200 seconds of CPU time. Figure 3 presents the convergence histories for the Diagonal ADI scheme on a single grid, and when 5 levels of multigrid are used. The logarithm of the average residual, the drag coefficient, and the number of cells in which the local Mach number is supersonic are plotted as a function of computational work, measured in Work Units. One Work Unit corresponds to the computational labor required for one time step on the fine grid. For both calculations, local time stepping is used at a Courant Number of $C = 16$. The figure illustrates two points: (1) the Diagonal ADI scheme itself is a reasonably efficient time-stepping algorithm; and (2) an appreciable increase in the convergence rate is achieved when multigrid is used. The aerodynamic force coefficients have converged to within plottable accuracy in about 200 time steps for the single-grid case, and in the equivalent of about 30 time steps for the 5-level multigrid case. Figure 4 presents the streamwise pressure distribution at each of the computational stations on the upper and lower surfaces of the wing, and Figure 5 presents contours of p/p_∞ on the upper and lower surfaces of the wing. Figure 6 presents a comparison with wind-tunnel data at two spanwise stations. The calculated results predict very accurately the strength and the location of the shocks. Figure 7 presents contours maps of entropy at different $z - y$ planes, starting at the leading edge and moving downstream. Since entropy is constant along streamlines for a steady, inviscid flow, these plots can be viewed as cuts through stream surfaces, and reveal the generation and evolution of the wing-tip vortex. It is clear that the vortex center moves up and inboard as it flows downstream, as observed in experiments.

V. Acknowledgements

This work has been supported by Grant NAG 2-373 from the NASA Ames Research Center. The computations have been performed at the Cornell National Supercomputer Facility of the Center for Theory and Simulation in Science and Engineering, which is funded, in part, by the National Science Foundation, New York State, and IBM Corporation.

VI. References

[1] Caughey, D. A., "A Diagonal Implicit Multigrid Algorithm for the Euler Equations," **AIAA Paper 87-0354**, 25th Aerospace Sciences Meeting, Reno, Nevada, January 1987.

[2] Jameson, A., Schmidt, W., and Turkel, E., "Numerical Solutions of the Euler Equations by Finite Volume Methods using Runge-Kutta Time-stepping Schemes," **AIAA Paper 81-1259**, Palo Alto, California, June 1981.

[3] Briley, W. R. and McDonald, H., "Solution of the Three-Dimensional Compressible Navier-Stokes Equations by an Implicit Technique," *Proc. of Fourth International Conference on Numerical Methods in Fluid Dynamics,* vol. 35, pp. 105-110, Springer Verlag, New York, 1974.

[4] Beam, R.M. and Warming, R. F., "An Implicit Finite-Difference Algorithm for Hyperbolic Systems in Conservation Law Form," *Journal of Computational Physics,* vol. 22, pp. 87-110, 1976.

[5] Pulliam, T. H. and Chaussee, D. S., "A Diagonal Form of an Implicit Approximate-Factorization Algorithm," *Journal of Computational Physics*, vol. 39, pp. 347-363, 1981.

[6] Jameson, A., "Solution of the Euler Equations by a Multigrid Method," **MAE Report 1613,** Princeton University, June 1983.

[7] **AGARD AR-138.** Experimental Data Base for Computer Program Assessment. 1979.

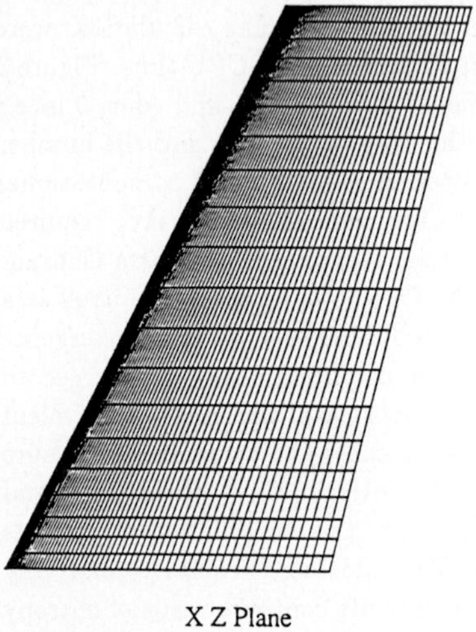

Figure 1: View of Wing.

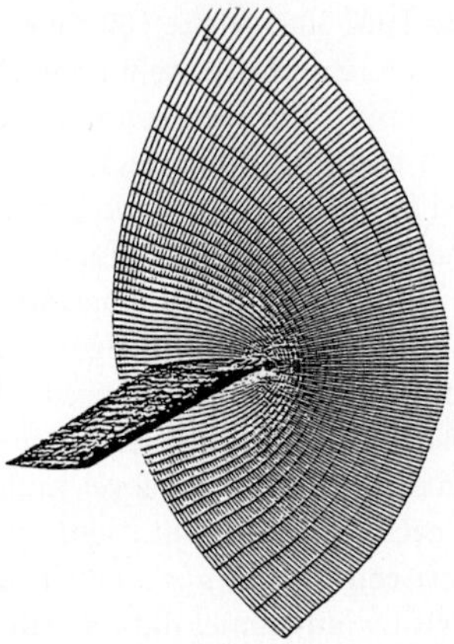

Figure 2: General view of the M6 wing.

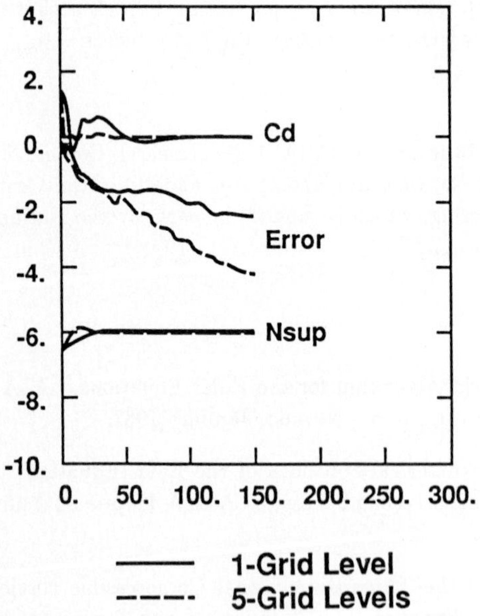

Figure 3: Convergence histories for single-grid and multigrid; Local time-stepping at Courant number 16.

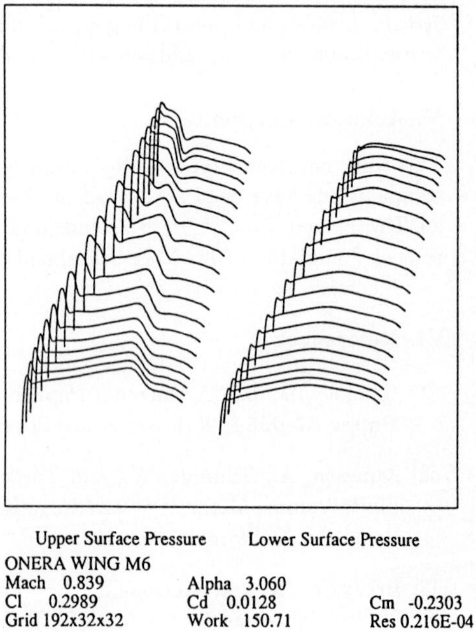

Figure 4: Pressure coefficient distribution in streamwise direction

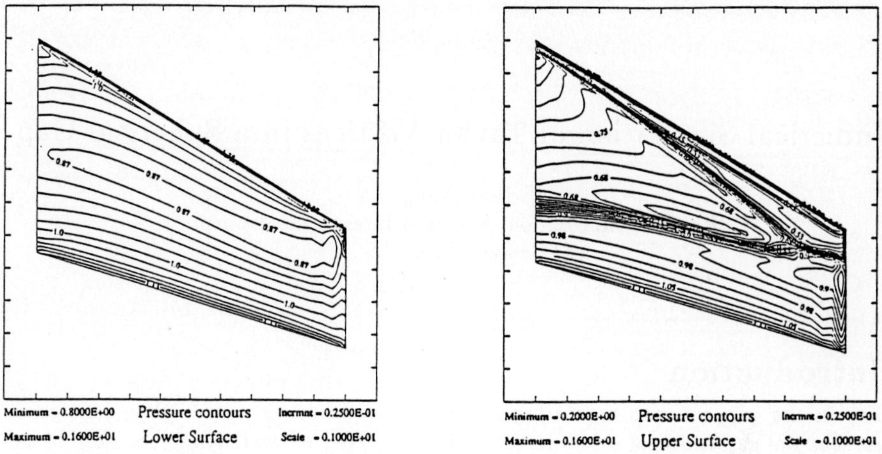

Figure 5: Lines of constant pressure; $M_\infty = 0.839$, $\alpha = 3.06$.

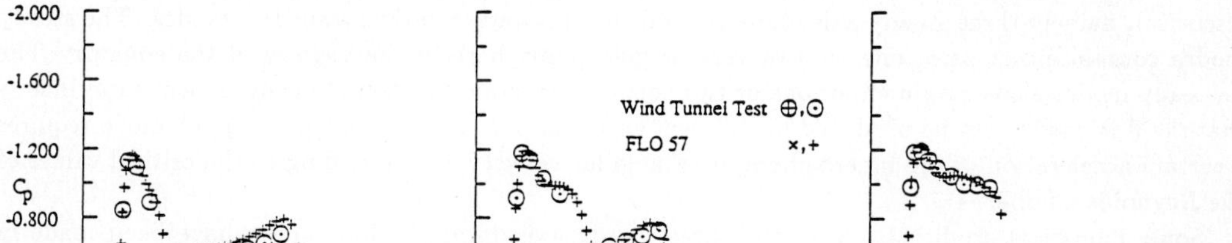

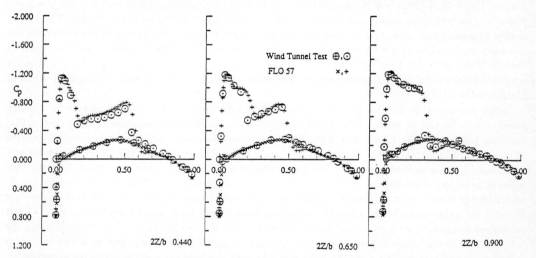

Figure 6: Comparison of measured and computed Pressure coefficients in three chordwise sections of M6 wing. $M_\infty = 0.839$, $\alpha = 3.06$.

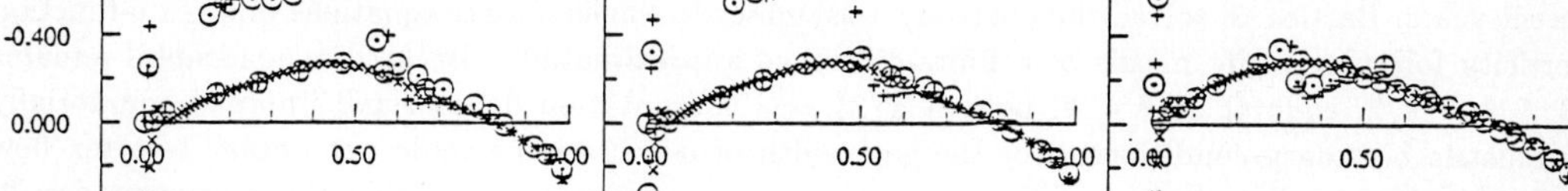

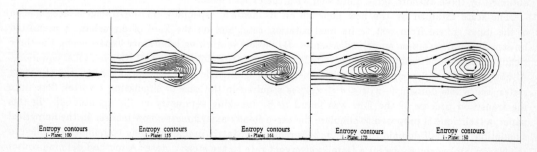

Figure 7: Entropy contours at various spanwise sections (constant percent local chord); leading edge: $i = 97$; trailing edge: $i = 161$; $M_\infty = 0.839$, $\alpha = 3.06$.

Numerical Simulation of Taylor Vortices in a Spherical Gap

R.-J. Yang
Rocketdyne Division, Rockwell International Corp.
Canoga Park, CA

1 Introduction

The flow in the gap between a rotating inner and a stationary outer sphere can generate Taylor vortices. For the gap width of $\sigma = 0.18$ ($\sigma = (R_2 - R_1)/R_1$, R_1 and R_2 are the radii of the inner and outer spheres), Sawatzki and Zierep[1] have observed non-unique flow modes at a supercritical Reynolds number ($Re = \Omega_1 {R_1}^2/\nu$, Ω_1 is the angular velocity of the inner rotating sphere and ν is the kinematic viscosity), namely three steady axisymmetric and two unsteady non-axisymmetric modes. The steady modes contain either zero, one, or two vortices per hemisphere in the vicinity of the equator. The unsteady modes also contain either one or two vortices. Wimmer[2] showed in subsequent experiments that the flow modes can be produced by different accelerations of the inner sphere. Each mode requires a certain acceleration of the inner sphere to an angular velocity corresponding to the critical values of the Reynolds number.

Some numerical studies[3,4,5] on the three steady axisymmetric flow modes have been made in recent years. Bartles [3] solved the unsteady axisymmetric Navier-Stokes equations in stream-function vorticity formulation by means of a finite-difference approximation. He used a quadrantal annulus ($0 \leq \theta \leq \pi/2$, equator at $\theta = 0$, pole at $\pi/2$) as a computation domain and imposed equatorially symmetric boundary conditions. For the gap width of $\sigma = 0.18$, he could not obtain 1-vortex flow without arbitrary disturbances of the symmetry near the equator. The other two steady modes, 0- and 2-vortex flows, were simulated through straightforward calculations. Marcus and Tuckerman [4] used the same formulation while the equations were solved by a pseudo-spectral method. Their numerical computations were carried out in the whole domain ($-\pi/2 \leq \theta \leq \pi/2$). By controlling the accelerations of the inner sphere, they were the first to reproduce 1-vortex flow as a solution of an initial-value problem directly from the basic flow. As with the long development time for experiments, a long computation time was also needed for this mode. Shrauf [5] used the continuation method to investigate how the instability of the spherical Couette flow depends on the gap size. Numerical results obtained by these authors agree quite well with the experiments.

The state attained by the flow depends on its history, especially on the transient acceleration of the inner sphere from rest to its final rotation rate, and on the kind of disturbances permitted by experimental set-up or numerical schemes. By closely mimicking laboratory experiments, the three steady modes can be simulated numerically by solving the full time-dependent Navier-Stokes equations. Although the three flow modes are axisymmetric and symmetric with respect to the equator at steady states, the whole domain ($-\pi/2 \leq \theta \leq \pi/2$) is required in the case of simulating 1-vortex flow since the transition process of the flow was found to be breaking symmetry at the equator [4]. In this paper, a technique is proposed to simulate the three steady axisymmetric flow modes. In the numerical simulation, only half domain ($0 \leq \theta \leq \pi/2$) is required. Matrix preconditioning technique is applied to the Navier-Stokes equations for a faster convergent rate to the steady state. A method defining routes to each flow mode is introduced. The present solutions, in terms of friction torques and vortex sizes, show excellent agreement with the experiments.

2 Governing equations and numerical methods

The three-dimensional, incompressible, Navier-Stokes equations are modified to the following set of governing equations:

$$\frac{1}{\beta}\frac{\partial p}{\partial t} + \frac{\partial u_i}{\partial x_i} = 0 \tag{1}$$

$$\frac{(\alpha+1)u_i}{\beta}\frac{\partial p}{\partial t} + \frac{\partial u_i}{\partial t} + \frac{\partial u_i u_j}{\partial x_j} = -\frac{\partial p}{\partial x_i} + \frac{\partial \tau_{ij}}{\partial x_j} \tag{2}$$

where t is time; x_i are the Cartesian coordinates; u_i are corresponding velocity components; p is the pressure; and τ_{ij} are the stress tensors. The parameters α and β, according to Turkel[6], are introduced to precondition the Navier-Stokes equations and allow for a faster convergence to the steady state. The equations are then transformed to a general coordinate system in which the general coordinates ξ_i are related to the Cartesian coordinate x_i by

$$\xi_i = \xi_i(x, y, z) \tag{3}$$

In the generalized curvilinear coordinates, the governing equations in conservationa-law form are expressed as

$$M\frac{\partial}{\partial t}\hat{D} + \frac{\partial}{\partial \xi_i}(\hat{E}_i - \hat{\Gamma}_i) = 0 \tag{4}$$

The preconditioned matrix M, the pressure and velocity vector $\hat{D}$, the flux vector $\hat{E}$, and the viscous diffusion vector $\hat{\Gamma}$ are described by

$$M = \begin{bmatrix} \beta^{-1} & 0 & 0 & 0 \\ (\alpha+1)u\beta^{-1} & 1 & 0 & 0 \\ (\alpha+1)v\beta^{-1} & 0 & 1 & 0 \\ (\alpha+1)w\beta^{-1} & 0 & 0 & 1 \end{bmatrix}, \quad \hat{D} = \frac{D}{J} = \frac{1}{J}\begin{bmatrix} p \\ u \\ v \\ w \end{bmatrix}, \quad \hat{E}_i = \frac{1}{J}\begin{bmatrix} U_i \\ uU_i + L_{i1}p \\ vU_i + L_{i2}p \\ wU_i + L_{i3}p \end{bmatrix}$$

$$\hat{\Gamma}_i = \frac{\nu}{J}(\nabla\xi_i \cdot \nabla\xi_j)\frac{\partial}{\partial \xi_j}(0, u, v, w)^T$$

where J is the Jacobian of coordinate transformation, and

$$U_i = L_{i1}u + L_{i2}v + L_{i3}w, \quad L_{i1} = (\xi_i)_x,\ L_{i2} = (\xi_i)_y,\ L_{i3} = (\xi_i)_z$$

are the contravariant velocities and the metrics of transformation, respectively.

The governing equations are replaced by an implicit time difference approximation (backward-difference). Nonlinear numerical fluxes at the implicit time level are linearized by Taylor expansion in time about the solution at the known time level. And then spatial difference approximations (central-difference) are introduced. The result is a system of multi-dimensional, linear, coupled difference equations for the dependent variables at the implicit time level. To solve these equations, the approximate-factorization or ADI scheme[7,8] is used. This technique leads to the following system of coupled linear difference equations

$$(M + \Delta t\mathcal{L}_\xi)\Delta D^* = RHS,\ (M + \Delta t\mathcal{L}_\eta)\Delta D^{**} = M\Delta D^*,\ (M + \Delta t\mathcal{L}_\zeta)\Delta D = M\Delta D^{**} \tag{5}$$

where

$$\mathcal{L}_{\xi_i} = J\delta_{\xi_i}(\hat{A}_i - \hat{\Gamma}_i),\ RHS = -\Delta t J\Sigma\delta_{\xi_i}(\hat{E}_i - \hat{\Gamma}_i),\ \hat{A}_i = \partial\hat{E}_i/\partial D,\ \Delta D = D^{n+1} - D^n$$

ΔD^* and ΔD^{**} are intermediate solutions. The system of equation (5) can be written in narrow block-banded matrix structures and be solved by a standard LU decomposition method.

3 Initial and boundary conditions

At a supercritical Reynolds number, the final steady states of the spherical Taylor Couette flow depend on the accelerating history of the inner sphere and initial conditions. The present numerical algorithm uses a preconditioned method and thus transient history has no physical meaning. Disregarding the transient accuracy, the current work is to find routes leading to steady 0-, 1- and 2-vortex flows with a faster convergent rate to the steady states. The computation is conducted in a quadrantal annulus $0 \leq \theta \leq \pi/2$. The initial conditions for all calculations (except some special cases) are given as follows: linear velocity profile across the gap in the circumferential direction, zero velocity components in the meridional plane, and zero static pressure in the entire domain.

The modified Navier-Stokes equations have a symmetric hyperbolic system. The preconditioning matrix is positive definite ($\beta > 0$). The appropriate number of boundary conditions is same as that for the original Navier-Stokes equations. To make the problem well-posed, the following boundary conditions are used: reflection-symmetric condition is applied at equator and pole, no-slip condition is used on inner and outer sphere surfaces, normal pressure gradient on both sphere surfaces is balanced by centrifugal and viscous forces, and a reference pressure is set at a specified point in the computation domain.

4 Computation procedure and results

Spherical Taylor-Couette flow is characterized by two parameters: the Reynolds number,Re, and the gap width, σ. The current computations are performed for $\sigma = 0.18$ and $Re \leq 1500$. Preconditioned parameters $\alpha = 0$ and $\beta = 2$ are used. Numerical convergency is checked by monitoring the friction torques acting on inner and outer sphere surfaces. At a steady state, the friction torques are equal in magnitude. To make sure the steady state solutions are independent of mesh distributions, two grid systems were tested for the case of $Re = 1500$: 111 x 51 and 81 x 41 (J x K, J and K are number of mesh points in azimuthal and radial direction respectively). Both computed solutions were nearly identical to each other. Therefore the 81 x 41 grid system shown in Fig.1 is used throughout these calculations.

In the region near the pole, local flow is similar to the one which occurs in the gap between a rotating disk and a stationary housing. In the region near the equator, local flow is similar to that of cylindrical Taylor-Couette flow. Because of the Eckman pumping effects, centrifugal forces drive fluid along the inner rotating sphere from the pole toward the equator. Near or at the equator plane (depending on flow mode), the fluid is deflected outwards and moved back to the pole in the vicinity of the outer sphere. It should be noted that each flow mode has its own region of existence at steady state. For instance, at $Re = 600$, only 0-vortex flow exists; at $Re = 800$, 1- and 2-vortex flows exist; at $Re = 1500$, the three flow modes exist. The following discussions are based on the assumptions that each flow mode exists at a given Reynolds number in each category. Routes leading to these flow modes are described bellow:

Route to 0-vortex flow

Using initial and boundary conditions depicted in section 3, the current numerical method converges to a 0-vortex flow. It has to be noted that the friction torque for both inner and outer spheres has to reach a constant at steady state. An equilibrium state (both inner and outer friction torques are equal) may be obtained during the process of the calculation. But the friction torque is not a constant during a time interval. Eventually, the friction torque reaches a constant and then another steady flow mode is formed. An example will be given later.

Route to 1-vortex flow

Experimental observations indicated that the vortex size near the equator of 1-vortex flow has same order of the gap width's magnitude. The vortex near the equator is separated from the major portion of the meridional flow through a dividing streamline. Local flow on either side of the streamline reveals symmetric character. For the numerical simulation of the 1-vortex flow, a radial line at one gap width distance away from the equator is chosen as a fictitious inner symmetric boundary. Flow variables are treated as symmetry there. Using conditions described in section 3 and the inner symmetric treatment, a 'rough' 1-vortex flow can be simulated. With the 'rough solution' as the initial condition and taking out the treatment at the specified radial line, the current numerical method converges to a 1-vortex flow.

Routes to 2-vortex flow

There are two rotes which can lead to a steady 2-vortex flow. First, a similar treatment as of 1-vortex flow is applied to the 2-vortex flow simulation. In this case, two radial lines are chosen as fictitious inner symmetric boundaries. The separation distance among the two radial lines and the equator is about one gap width each. Computation with conditions given in section 3 and the inner symmetric treatment at the two radial lines can lead to a 'rough' 2-vortex flow. Taking out the treatment at the two specified radial lines and using the 'rough solution' as the initial condition, the current numerical method converges to a steady 2-vortex flow. This treatment toward the final steady 2-vortex flow is called forced transition. Another route to 2-vortex flow is called natural transition. In this case, a straightforward calculation using initial and boundary conditions given in section 3 can converge to a steady 2-vortex flow. In the process of the natural transition, it is observed that the friction torque for both inner and outer spheres is in equilibrium state but is not a constant with respect to time during a period. Fig.2 shows an example of convergent histories for both forced and natural transition for the case of $Re = 800$. In the case of natural transition, the flow reaches an equilibrium state after 500 time steps (NT=500). But friction torque varies with respect to time until NT=1600. In the case of forced transition, the first 300 time steps were treated as having inner symmetric boundaries at two specified radial lines. After taking out the two fictitious inner boundaries and continuing the calculation, the flow converges to a steady 2-vortex flow. It is obvious that the case of the forced transition led to a much faster convergent rate toward a steady state.

The three steady flow modes behave quite differently near the equator. Figs. 3 and 4 give circumferential skin friction and azimuthal pressure gradient coefficients for the case of $Re = 1500$. Wild differences are clearly shown within $\theta \leq 40^0$. In Figs. 5 and 6, the current results, friction torques and vortex sizes, agree quite well with Marcus' calculations and Wimmer's measurements. From Fig.5, region of existence for the three flow modes is clearly shown. The 0- and 2-vortex flows lie on the same solution branch and the 1-vortex flow lie on a separate branch.

5 Conclusion

A new technique to simulate Taylor vortices in a spherical gap has been developed and tested. The non-unique steady flow modes associated with the Taylor vortices are simulated through special ways. Routes leading to 0-, 1- and 2-vortex flows are designed heuristically. Fictitious symmetric boundaries near the equator are imposed during a time of the calculation. The choice of the location of the fictitious boundaries is determined by either 1- or 2-vortex flow being simulated. The imposition of one or two fictitious symmetric boundaries during the initial calculation generates the state suitable for 1- or 2-vortex flow to exist. After taking out the fictitious boundaries, the flow settles down into its own attractor. By this method, the three steady flow modes can be simulated by using half domain, ie, $0 \leq \theta \leq \pi/2$. The current numerical technique can converge to desired flows very fast.

References

[1] Sawatzki, O. and Zierep, J., Acta Mechanica, Vol. 9, 1970, pp13-35.

[2] Wimmer, M., J. Fluid Mechanics, Vol. 79, 1976, pp317-335.

[3] Bartels, F., J. Fluid Mechanics, Vol. 119, 1982, pp1-25.

[4] Marcus, P. S. and Tuckerman, L. S., J. Fluid Mechanics, Vol. 185, 1987, pp1-65.

[5] Schrauf, G., J. Fluid Mechanics, Vol. 166, 1986, pp287-303.

[6] Turkel, E., J. Computational Physics, Vol. 72, 1987, pp277-297.

[7] Kwak, D., Chang, J. and Shanks, S. P., AIAA Journal, Vol.24, No.3, 1986, pp390-396.

[8] Briley, W. R. and McDonald, H., J. Computational Physics, Vol. 24, 1977, pp372-397.

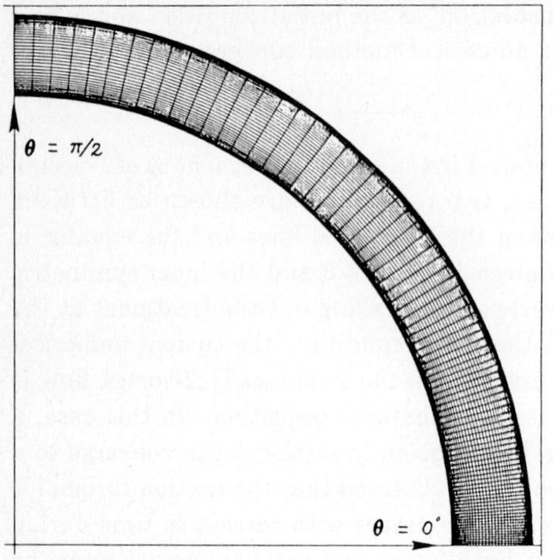

Fig. 1 Computational grid system (81 x 41)

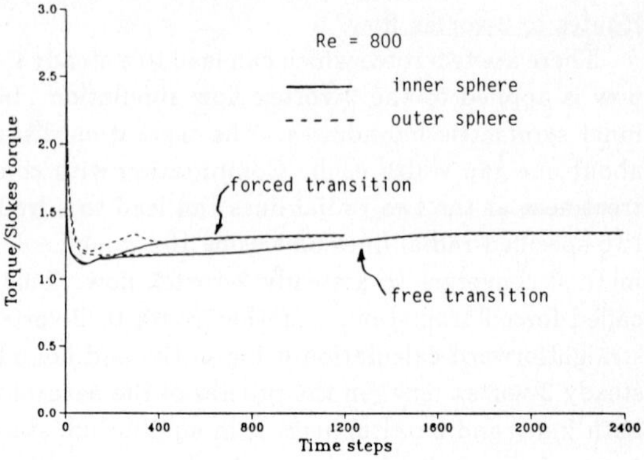

Fig.2 Convergent histories toward 2-vortex flow

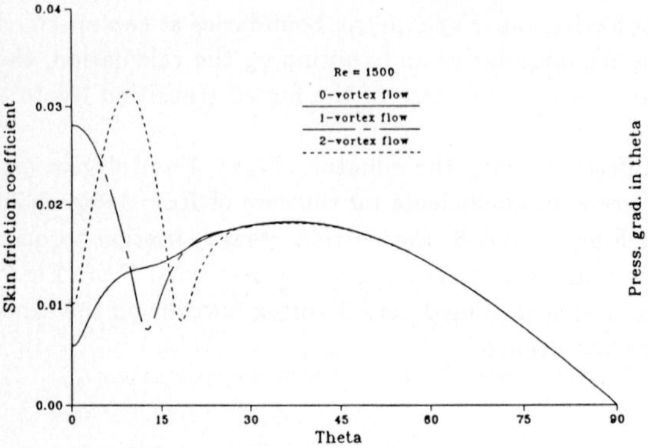

Fig.3 Circumferential skin fricrion coefficient on inner sphere surface

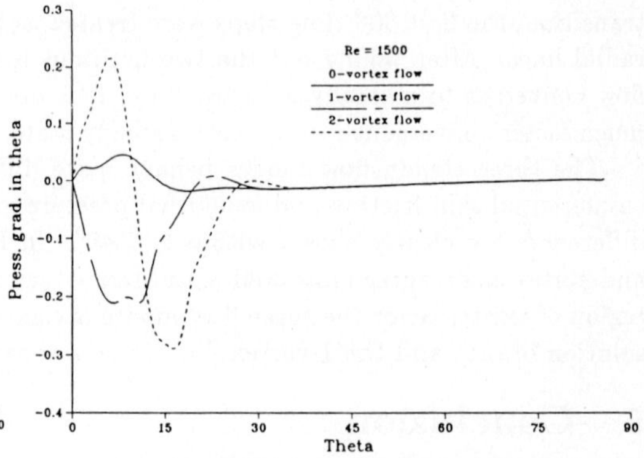

Fig.4 Azimuthal pressure gradient coefficient on inner sphere surface

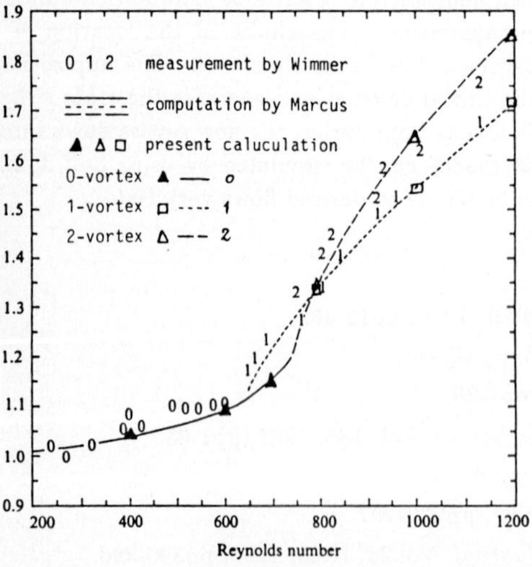

Fig.5 Friction torques

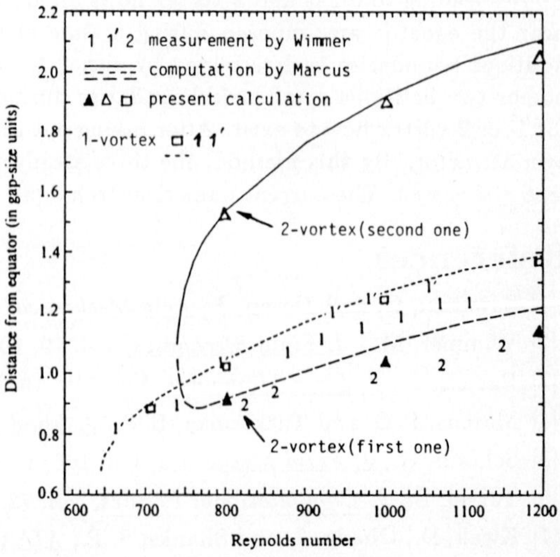

Fig.6 Vortex sizes

Numerical Calculation of Hypersonic Flow by the Spectral Method

Michiru YASUHARA, Yosiaki NAKAMURA and Jian-Ping WANG
Nagoya University, Nagoya 464-01, Japan

1. INTRODUCTION

Recently much attention has been paid to the hypersonic flow problem such as Space Shuttle and Space Plane. We have conducted the experiments on the hypersonic flow by using the shock tunnel with Mach number of 8 at Nagoya University. In particular, the data on the aerodynamic coefficients of the Shuttle-like body and the shape of bow shock have been collected. In parallel with the above experiments, the hypersonic flow has been numerically simulated.

We have so far used the finite difference method and the finite volume method to calculate the hypersonic flow. The Euler equations, the thin layer Navier-Stokes equations, and the PNS equations have been successfully applied to this flow. Those results showed reasonable agreement with the experiment. In the present paper we employed another approach, the spectral method(1).

The spectral method is a promising method to solve the fluid dynamics problem(2,3). In particular, the turbulent flow problems have so far been successfully solved. In addition to this, the Spectral method was applied to other velocity regime flow such as transonic and hypersonic flows. In the fluid dynamics problems, the finite difference method has been extensively used due to its simplicity and easy extension to higher order accuracy. Furthermore, recently the TVD method became popular to obtain the sharp shock wave in the hyperbolic type flow.

However, most researchers in other fields get used to the finite element method due to its generality. Recently the finite element method has been applied to the fluid flow problems. Its strong point is that the grid can be easily generated in the complicated place. In the finite element method the interpolation function is valid within each cell.

On the other hand, in the spectral method the overall region is approximated by one function which is expanded by truncated orthogonal functions. This method has several variations. One of them is referred to as the spectral element method which is similar to the finite element method. Generally, the spectral method is said to be more accurate than any other method when compared on the basis of the same number of grid points.

Here we applied the collocation spectral method to the three-dimensional flow at the hypersonic flow regime. The similar flows were already simulated by several researchers where promising results were obtained. We will also show several results at Mach numbers of 4 and 8 by comparing with other analytical, numerical and experimental results.

Section 2 shows the procedure for applying the spectral method to the hypersonic flow. Section 3 is devoted to calculated results. Finally, the concluding remarks are made in the section 4.

2. Application of the Spectral Method to Hypersonic Flow

We calculated the flow around a sphere and the forward part of a Shuttle-like body by using the spectral method. In the Euler equations the solution vector is (P,u,v,w,S), where P is the logarithm of pressure, and S is entropy.

To remove sharp change in each flow quantity due to the bow shock wave in front of the body, we employed the shock-fitting method. To solve the system of equations, the solution vector was expanded in terms of smooth functions. The Chebyshev polynomials were used in the two directions: from the body to the shock wave, and from the center axis to the downstream boundary. The Fourier expansion was used in the circumferential direction where the half region for sphere and the full region for the Shuttle-like body were considered.

$$Q(X_i, Y_j, Z_k, T^n) = \sum_{G=0}^{L-1} \sum_{F=0}^{N} \sum_{E=0}^{M} Q_{EFG}(T^n)\, Cos\, \frac{\pi E i}{M}\, Cos\, \frac{\pi F j}{N}\, e^{\frac{iG2\pi k}{L}}$$

Once the transformed coefficients for the solution vector Q is calculated, the coefficients for the derivatives of Q such as Qx, Qy, and Qz are easily obtained, using the known relations between them.

Then we extended this procedure to the arbitrary shape of the body. The original equation was transformed to the generalized coordinates.

Furthermore we proceeded to the method where the viscous effect is included. The Navier-Stokes equations were considered and are written as follows:

$$Q_t + E_r + F_\theta + S = 0$$

$$Q = \begin{bmatrix} \rho \\ \rho u \\ \rho v \\ e \end{bmatrix} \qquad E = \begin{bmatrix} \rho u \\ \rho u^2 - \sigma_{rr} \\ \rho v u - \sigma_{r\theta} \\ eu - \kappa \tilde{T}_r \end{bmatrix} \qquad F = \begin{bmatrix} (1/r)\rho v \\ (1/r)\{\rho uv - \sigma_{\theta r}\} \\ (1/r)\{\rho v^2 - \sigma_{\theta\theta}\} \\ (1/r)\{ev - (1/r)\kappa \tilde{T}_\theta\} \end{bmatrix}$$

$$S = \begin{bmatrix} \rho(2u + v\cot\theta)/r \\ \{\rho(2u^2 + uv\cot\theta - v^2) - 2\sigma_{rr} - \sigma_{\theta r}\cot\theta + \sigma_{\theta\theta} + \sigma_{\phi\phi}\}/r \\ \{\rho(3uv + v^2\cot\theta) - 3\sigma_{r\theta} - \sigma_{\theta\theta}\cot\theta + \sigma_{\phi\phi}\cot\theta\}/r \\ \{e(2u + v\cot\theta) - 2\kappa\tilde{T}_r - (1/r)\cot\theta\, \kappa\tilde{T}_\theta\}/r + (1/\gamma)\, p\, \nabla\cdot\mathbf{V} - \Phi \end{bmatrix}$$

As one of the weak points in the spectral method, this method is difficult to apply to the complicated flow. Therefore, we applied the multi-domain method to a hemisphere-cylinder problem.

During these calculations the predictor-corrector method was used to integrate the time dependent differential equations.

3. Results

First the Euler equations were used to solve the flow around the sphere and the forward part of the Shuttle-like body at Mach numbers 4 and 8. The grid points are 8 in the circumferential direction, 9 from the body to the shock, and 9 from the center to the downstream end. Figure 1a shows the pressure contours for M=4 which are compared with Moretti's result(4). Reasonable agreement was ôbtained between them. Our result has more accurate shock stand-off distance than Moretti's. Figure 1b shows the result at M=8. We can see the bow shock comes closer to the body than M=4. The contour shape is similar in both cases.

Figure 2 shows the pressure distributions along the body. At M=4, the result is compared with Moretti's method and the modified Newtonian flow calculation. Except for the low pressure region where the modified Newtonian approximation has poor accuracy, the comparison is generally reasonable.. Figure 2b shows the case for M=8, which also agrees with the other two methods as in M=4.

Then we extended this method to the fully three-dimensional body, the foreward part of the Shuttle-like body. Figure 3 shows the pressure contours for M=4 and M=8. Figure 4 shows the comparison of the bow shock waves with experiments for hemisphere-cylinder and the Shuttle-like body at M=8.

This method was furthermore extended to include the viscous effect, where for simplicity the axisymmetry is assumed. Figure 5a shows the velocity vector distributions. We can see that the boundary layer develops well due to low Reynolds number of Re=1000. Figure 5b shows the result for M=8.

Finally, we applied the multi-domain method to the spectral method to calculate rather complicated body shape, specifically, a hemisphere-cylinder. The result is shown in Fig. 6, where the flow is connected smoothly at the joint between the hemisphere and the cylinder.

4. Concluding Remarks

We first made the three-dimensional spectral method code for the Euler equations at the hypersonic flow regime: M = 4 and 8. Then the generalized coordinates code was developed for three-dimensional body. This code was applied to the forebody of a Shuttle-like body. Furthermore, we calculated the viscous flow assuming axisymmetry. The results solved by the Euler equations show good agreement with experiments for the bow shock wave shape. Finally, we applied the multi-domain method to a hemisphere-cylinder flow problem.

References

1. S.A.Orszag, Spectral Methods for the Problems in Complex Geometries, J. Comput. Phys., 37, 1980, pp. 70-92.
2. M.Y.Hussaini, D.A.Kopriva, M.D.Salas and T.A.Zang, Spectral Methods for the Euler Equations: Part I-Fourier Methods and Shock Capturing, AIAA Journal, 23,

1985, pp. 234-240.

3. M.Y.Hussaini, D.A.Kopriva, M.D.Salas and T.A.Zang, Spectral Methods for the Euler Equations: Part II-Chebyshev Methods and Shock Fitting, AIAA Journal, 23, 1985, pp. 234-240.

4. G.Moretti, Inviscid Blunt Boby Shock Layers (Two Dimensional Symmetric and Axisymmetric Flows), Polytechnic Institute of Brooklin, New York, PIBAL Report, 68-15, 1968.

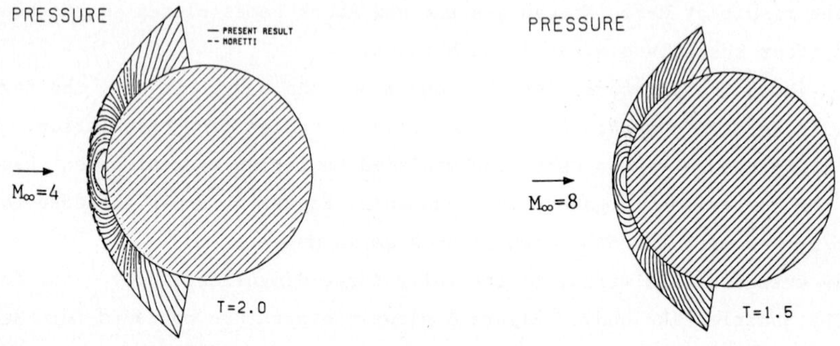

(a) $M_\infty=4$ (b) $M_\infty=8$

Fig. 1 Pressure contours.

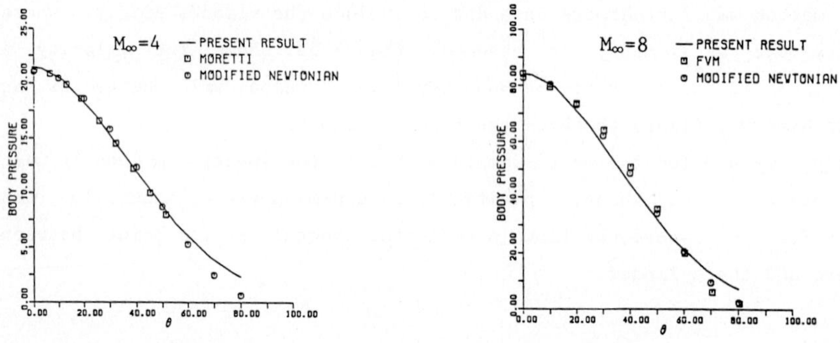

(a) $M_\infty=4$ (b) $M_\infty=8$

Fig. 2 Pressure distributions along the sphere.

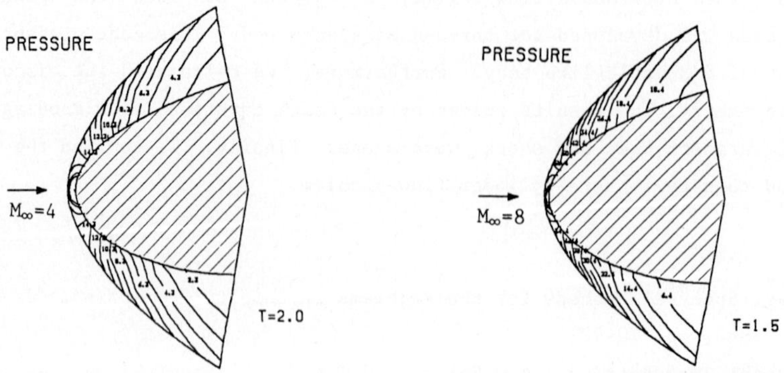

(a) $M_\infty=4$ (b) $M_\infty=8$

Fig. 3 Pressure contours for Shuttle-like body.

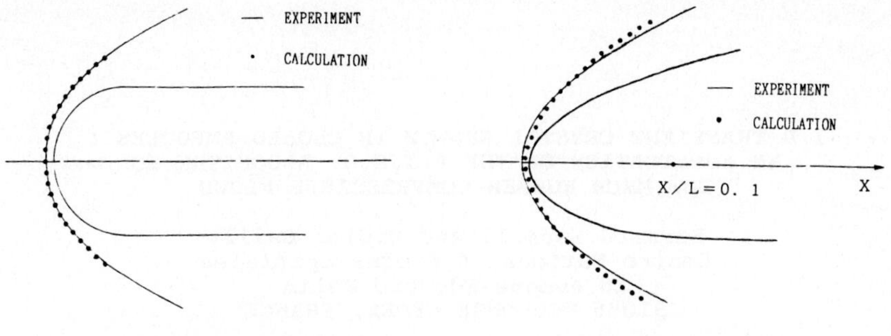

(a) hemisphere-cylinder (b) Shuttle-like body

Fig. 4 Comparison of bow shock with experiment.

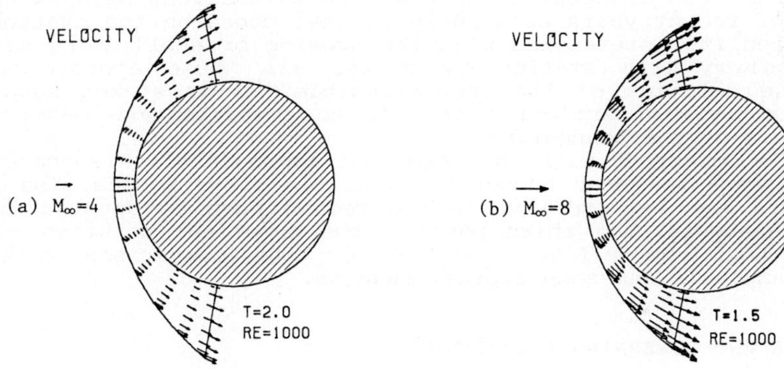

(a) $M_\infty=4$ (b) $M_\infty=8$

Fig. 5 Velocity vectors around sphere for N.S. equations.

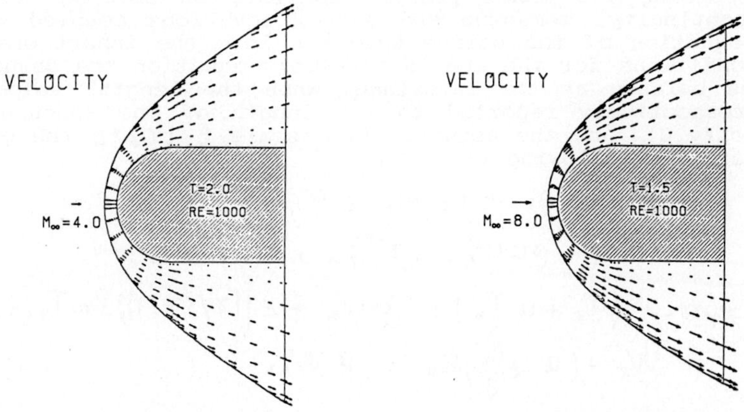

(a) $M_\infty=4$ (b) $M_\infty=8$

Fig. 6 Velocity vectors around hemisphere-cylinder by two-domain method.

1.D TRANSIENT CRYSTAL GROWTH IN CLOSED AMPOULES : AN APPLICATION OF THE P.I.S.O. ALGORITHM TO LOW MACH NUMBER COMPRESSIBLE FLOWS

Bernard Zappoli and Didier Bailly
Centre National d'Etudes Spatiales
18 Avenue Edouard Belin
31055 TOULOUSE CEDEX, FRANCE

1. INTRODUCTION

Crystal growth from a vapour phase has been extensively studied during the course of recent years especially in what concerns the stationnary convection and its interaction with the growing or sublimating crystal surface. Dealing with stationnary states all these approaches are based on the solution of the incompressible Navier stokes equations /1/ with sometimes particular methods to describe non-conservative mass transfers at the boundaries /2/.
This paper is concerned with the transient approach of this phenomenon which leads to a low Mach number compressible flow problem. The first part presents the 1-D model while the second part is devoted to the asymptotic analysis. The third part is the numerical solution of the equations using the P.I.S.O. algorithm, the comparison with the asymptotic analysis and some typical results.

2. THE MODEL AND GOVERNING EQUATIONS

The growing or evaporating crystal is located at x = 0 in a 1-D domain closed at x = 1. When appropriate thermal conditions are set on the crystal end (x = 0) the growth or evaporation of specie A begins together with its diffusion into the other component B of the bulk phase which is innert in regard with the surface reaction
The corresponding 1-D model problem consists in solving for $x \in [0,1]$ the full continuity, momentum and energy equations coupled with the transport equation of the active specie A into the innert one B. The boundary conditions for the specie transport equation are deduced from the species balance at the interface. When the length, velocity and time are respectively reported to the length of the ampoule L, the sound velocity C'_o and the acoustic time scale, $t = t'C'_o/L$ the governing equations are the following :

$$\left\{\begin{aligned} &\rho_t + (\rho u)_x = 0 && (1)\\ &\rho u_t + \rho u u_x = -\gamma^{-1} p_x + 4/3\,\varepsilon\, u_{xx} && (2)\\ &\rho/(\gamma-1)\left(T_t + u T_x\right) = -p u_x + \varepsilon\left[(\gamma/(\gamma-1))\,Pr\,T_{xx} + \tfrac{4}{3}(u_x)^2\right] && (3)\\ &\rho\, W_t + \left(u - \varepsilon\frac{\rho_x}{\rho}\right) W_x = \varepsilon W_{xx} && (4)\end{aligned}\right.$$

$$p = \rho\alpha T \;;\; \alpha = \alpha'/\alpha'_o \;;\; \alpha' = \frac{-M_A + M_B}{M_A M_B} W + \frac{1}{M_B} \qquad (5)$$

$$T = T'/T_o \;;\; \rho = \rho'/\rho'_o$$

M_A and M_B are the molar masses of the components of the mixture and $\varepsilon = P_r t'_A / t'_D$ where $t'_D = L^2/\kappa$ is the heat diffusion time scale. For the sake of simplicity, the Lewis number is supposed equal to 1. The initial conditions correspond to the rest and the thermodynamic equilibrium :

$$\rho = T = 1 \; ; \; W = W_0 \; ; \; u = 0$$

The boundary conditions are the following :

$$x = 0 \quad \begin{cases} T = 1 + f(t) & \text{(heating)} \quad (6) \\ W = \dfrac{W^e(T)}{\rho} & \text{(time dependant equilibrium weight fraction)} \quad (7) \\ u = \varepsilon \dfrac{W_x}{W^e/\rho - 1} & \text{(mass balance)} \quad (8) \end{cases}$$

$$x = 1 \quad \begin{cases} T = 1 & \text{(fixed temperature)} \\ W_x = 0 & \text{(non reacting end)} \\ u = 0 & \text{(mass balance)} \end{cases}$$

3. INITIAL BOUNDARY LAYER SOLUTION

The heating law is considered to give O(1) temperature variations on the $\tau = \varepsilon t$ scale (diffusive) :

$$x = 0 \quad \begin{cases} T = 1 + \varepsilon H'(0) t + O(\varepsilon^2) & (9) \\ W = \dfrac{1 + \varepsilon H'(0) G'(1) t + O(\varepsilon^2)}{\rho} \; ; \; G'(1) = \left.\dfrac{\partial W}{\partial T'}\right|_{T' = T'_0} \end{cases}$$

Matching the time derivative with the diffusive terms in the energy equation and the linear terms in the other one, the following boundary layer normalisation occurs /3/

$$T = 1 + \varepsilon \tilde{T} \; ; \; \rho = 1 + \varepsilon \tilde{\rho} \; ; \; x = \sqrt{\varepsilon}\, z \; ; \; u = \varepsilon^{3/2} \tilde{u} \, , \; W = W_0 + \varepsilon \tilde{W}$$

$$P = 1 + \varepsilon^2 \tilde{P}$$

leading to the following boundary layer equations :

$$\begin{cases} \tilde{\rho}_{0t} + \tilde{u}_{0z} = 0 & (10) \\ \tilde{u}_{0t} = -\gamma^{-1} \tilde{p}_{0z} + 4/3 \, \tilde{u}_{0zz} & (11) \\ \tilde{T}_{0t} = -(\gamma - 1) \tilde{u}_{0z} + \gamma P_r^{-1} \tilde{T}_{0zz} & (12) \\ \tilde{\rho}_0 = -\tilde{T}_0 - \dfrac{A}{A W_0 + B} = -\tilde{T}_0 - \xi \tilde{W}_0 & (13) \\ \tilde{W}_{0t} = \tilde{W}_{0zz} \end{cases}$$

and initial and boundary conditions :

$$t=0 \qquad \tilde{W}_o=\tilde{\rho}_o=\tilde{u}_o=\tilde{p}_o=\tilde{T}_o=0$$

$$z=0 \quad \begin{cases} \tilde{T}_o = H'(0)t \\ \tilde{W}_o = G'(1)H'(0)t - W_o\tilde{\rho}_o(0,t) \\ \tilde{u}_o = \dfrac{\tilde{W}_o}{W_o-1} \end{cases} \qquad (14)$$

$$z \to \infty \begin{cases} \tilde{T}_o \to 0 \\ \tilde{W}_o \to 0 \end{cases}$$

The boundary layer solution for Wo writes :

$$\tilde{W}_o = 4\left[G'(1)H'(0) - W_o\Phi\right] t\, i^2\mathrm{erfc}\left(\frac{z}{2\sqrt{t}}\right) = 4\Psi\, i^2\mathrm{erfc}\left(\frac{z}{2\sqrt{t}}\right) \qquad (15)$$

where it has been supposed a priori that $\tilde{\rho}_o(0,t)=\Phi t$; $\Phi = C^{st}$
from (12), (10) and (13), To is :

$$\tilde{T}_o = C_1 t\, \mathrm{erfc}\left(\frac{z\sqrt{P_r}}{2\sqrt{t}}\right) + C'\tilde{W}_o \quad ; C_1 = 4\left[H'(0) - \xi C_1'\right]; \ C_1' = -(1-P_r^{-1})^{-1}\frac{\gamma-1}{\gamma}\xi$$

from (13) $\tilde{\rho}_o$ is :

$$\tilde{\rho}_o = -\tilde{T}_o - \xi\tilde{W}_o = -C_1 t\, i^2\mathrm{erfc}\left(\frac{z\sqrt{P_r}}{2\sqrt{t}}\right) - (C_1'+\xi)\tilde{W}_o$$

which confirms that $\tilde{\rho}_o(0,t) = \Psi$ t and allows to calculate

$$\Phi = -\frac{1}{4}\ \frac{C_1 + 4\Psi(C_1'+\xi)G'(1)H'(0)}{1 - W_o\Psi(C_1'+\xi)}$$

from (10) and (14) :

$$\tilde{u}_{oz} = -\tilde{\rho}_{ot} \qquad ; \ \tilde{u}_o(0,t) = -\frac{2\xi}{\sqrt{\pi}\,(W_o-1)}\sqrt{t}$$

$$\frac{\tilde{u}_o}{\sqrt{t}} = -\frac{C_1}{2P_r}\left(i\,\mathrm{erfc}\left(\frac{z\sqrt{P_r}}{2\sqrt{t}}\right) - \frac{1}{\sqrt{\pi}}\right) - 2(C_1'+\xi)\Psi\left(i\,\mathrm{erfc}\left(\frac{z\sqrt{P_r}}{2\sqrt{t}}\right) - \frac{1}{\sqrt{\pi}}\right) - \frac{2\xi}{\sqrt{\pi}(W_o-1)} \qquad (16)$$

(11) gives the pressure :

$$\tilde{p}_o = \tilde{p}_w(t) + \gamma\left[C_1\left(\frac{1}{3}-\frac{1}{4P_r}\right)\mathrm{erfc}\left(\frac{z\sqrt{P_r}}{2\sqrt{t}}\right) + \frac{1}{3}(C_1'+\xi)\Psi\,\mathrm{erfc}\left(\frac{z}{2\sqrt{t}}\right) - \frac{2}{\sqrt{\pi}\,t}\left(\frac{C_1}{4\sqrt{P_r}} + (C_1'+\xi)\Psi - \frac{\xi}{W_o-1}\right)\right]$$

where the integration parameter $\tilde{p}_w(t)$ which represents the value of the pressure at the wall could be determined in solving the $O(\varepsilon^2)$ perturbed problem. However the spatial pressure distribution is given and would allow to define the matching conditions with the outer expansion in order to obtain the iniformly valid solution for small time (that is on the acoustic scale). In a same way the governing equations on the τ scale (diffusive scale) could be obtained by a multiple scale expansion procedure. It will be published elswhere very soon ; the boundary layer structure has been given here to compare with the numerical results. It can be emphazised, that taking the limit for $z \to \infty$ (edge of the boundary layer) a non zero velocity is obtained which evidences, here too, a piston effect :

$$\lim_{z\to\infty} u(z,t) = \varepsilon^{3/2} K\sqrt{t} \quad ; \quad K = \frac{C_1}{2P_r\sqrt{\pi}} + \frac{2(C_1'+\xi)\Psi}{\sqrt{\pi}} - \frac{2\xi}{\sqrt{\pi}(W_o-1)}$$

4. THE NUMERICAL SOLUTION

The numerical solution is based on the so called P.I.S.O. algorithm for Pressure Implicit with Splitting of Operators /4/ . The species diffusion equation is first solved by an implicit second order scheme with velocity u^n in order to obtain the boundary conditions at the interface at $x = 0$ from equation (8). Given that boundary value, the P.I.S.O. algorithm is used to compute the velocity, pressure, density and temperature field till all the equations are solved simultaneously and the non linear term in equation (2) is taken into account.
This algorithm is based, for each equation, on one implicit predictor step followed by n explicit corrector steps. The equation for the pressure is an helmholtz equation. The space discretisation is performed on a regular staggered mesh for which the pressure and density are computed on external points. The solution is first obtained for boundary conditions at $x = 0$ corresponding to equation (9), that is for temperature and equilibrium weight fraction variations of order 1 on the τ scale and then for a typical experimental heating law.

4.1. Slow heating

The boundary conditions are given by (9) and the solution is computed for the following values of the parameters.

$$H'(0) = G'(1) = 1 \; ; \; W_0 = 0{,}7 \; ; \; \varepsilon = 4.77\ 10^{-7} \; ; \; Pr = 0.76 \; ; \; \gamma = 1{,}4$$

$$\Delta x = 3{,}33\ 10^{-4} \; ; \; \Delta t = 10^{-2} \; ; \; M_A = M_B .$$

The corrector steps are stopped when $|\Delta u/u|, |\Delta T/T|, |\Delta P/P|$ are smaller than 10^{-2}. For the species transport equation $|\Delta w/w|$ 10^{-8} is considered. The computation is performed on a CYBER 990.
The results for u and t = 0,4 are plotted as a function of z in the boundary layer region on Figure 1 and compared with the value computed by equation (16). The results are in good agreement. The difference in the boundary layer zone (z = O(1)) (essentially a shift of 2 %) comes most probably from a loss of accuracy in the discretisation of the concentration gradient when computing the boundary value for u from equation (14).
The increase in velocity near the crystal is a thermal effect due to the expansion of the fluid layer adjacent to the wall under the thermal boundary layer heating.

4.2. Solution for increasing heating rate

It is supposed that the relative increase in temperature is 10^{-2}, linearly, in 10^{-2} s, which gives $H'(0) = 3.3310^{-4}$.
The other parameters which differ from the previous ones are :
$M_A = 2\ M_B$ and Le = 1,52.

4.2.1. Solution on the acoustic time scale
4.2.1.1. Boundary layer solution

The velocity in the boundary layer is plotted for $0 < t < 1$ with a 0.1 time step. The non linear increase of the velocity at the interface must be emphazised. (fig.2)

4.2.1.2. Outer solution

The velocity and pressure are plotted for $0 < t < 1$ on Figure 3 and 4 as functions of the outer variable x/L. The propagation of the pressure wave at nearly sound velocity is evidenced.

4.2.2. Solution on the diffusion time scale

The velocity and pressure are plotted on Figures 5 and 6 for t = 37.800 as functions of the outer variable x/L. The pressure is now homogeneous due to the acoustic waves fashing back and forth ; the velocity reaches a monotonous shape between u(o) and zero and is hundred times smaller than on Figure 3. When time goes to infinity, the velocity goes to zero as a new equilibrium is reached.

4.3. Concluding remarks

The present results show the efficiency of the P.I.S.O. algorithm to compute low Mach number compressible flow and are in good agreement with the analytical boundary layer analysis results. It would be valuable to apply this numerical code to the 1-D modelling of a real crystal growth experiment or to the effect of g-gitters during a microgravity experiment.

References

/1/ J.P. EXTREMET et Al, Journal of Crystal Growth 82 (1987) 761

/2/ B. ZAPPOLI et Al, in Numerical Methods in Thermal Problems, Vol. 5, Part I (1987) 821

/3/ D.R. KASSOY, S.I.A.M., Vol. 36, n° 3 (1979)

/4/ R.I. ISSA, J. Comp. Phys., Vol. 62, n° 1 (1986).

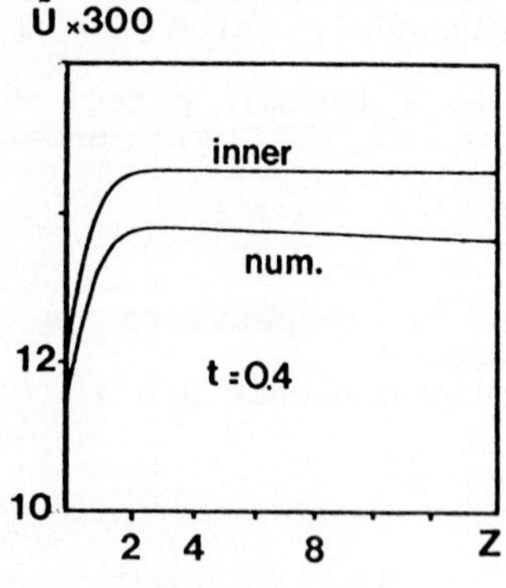

Fig. 1 : Velocity in the initial boundary layer

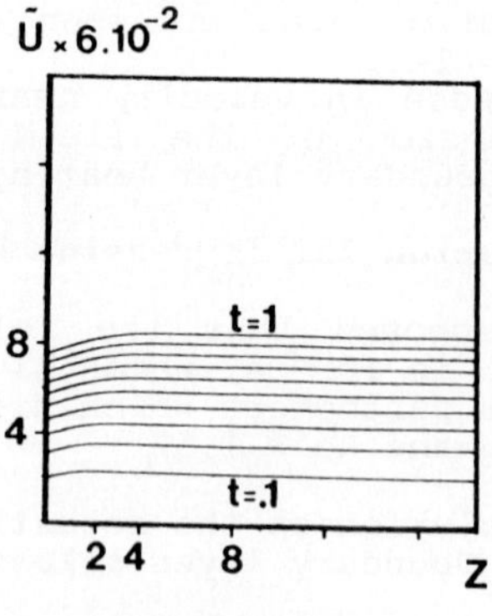

Fig. 2 : Velocity in the boundary layer on the acoustic time scale

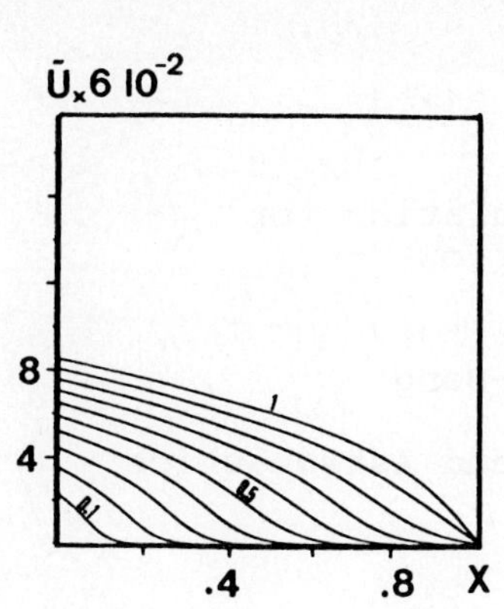

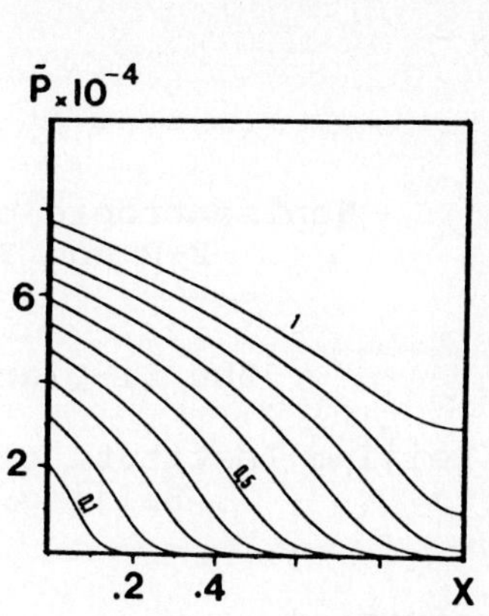

Fig. 3 : Velocity in the core flow on the acoustic time scale

Fig. 4 : Pressure in the core flow on the acoustic time scale

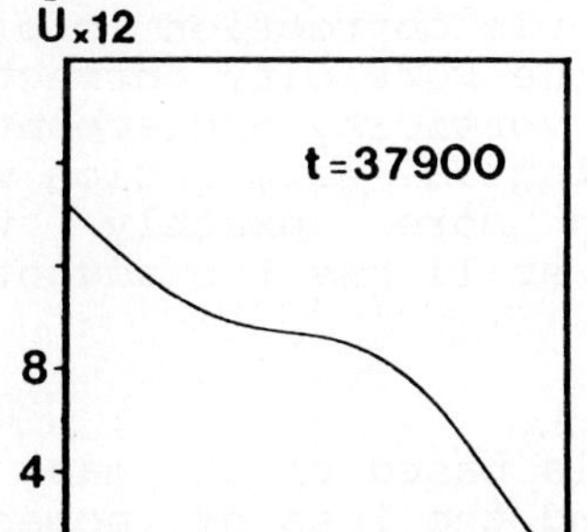

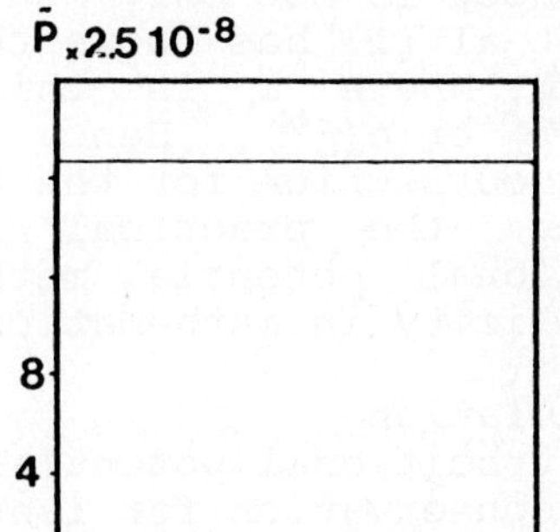

Fig. 5 : Velocity on the diffusion time scale

Fig. 6 : Pressure on the diffusion time scale

Nonisentropic potential calculation for 2-D and 3-D transonic flow

Zhu Zi-Qiang and Bai Xue-Song

Beijing Institute of Aeronautics and Astronautics
Beijing, P.R. China

Introduction

Potential flow method has been widely used in engineering application for its high efficiency, but tranditional potential equation may be invalid in use when there are strong shock waves in the flow field because of the isentropic assumption. The flow after shock is not only nonisentropic, but also rotational. M. Hafez et al [1] has shown that the isentropic correction is $o(\varepsilon^2)$ or $o(\tau^{4/3})$, where τ is thickness, while the vorticity correction is $o(\varepsilon^{7/2})$ or $o(\tau^{7/3})$. Hence neglecting the vorticity and acounting entropy correction for the tranditinal potential formulation will simulate the practical flow situation more exactly than tranditional potential method, while it still has the advantage of simplicity in mathematics aspect.

Formulation

The traditional potential formulation is based on the mass and energy conservation for isentropic flow and the loss of momentum conservation across the shock wave causes the failure of satisfying the Rankine - Hugonoit relation. We introduced a nonisentropic concept to modify the flow field after the shock and at the shock itself, so the third conservation equation is also satisfied.

Shock point operator: Adding to results of several other authors dealing with entropy correction [1-3], a shock point operator was introduced account for entropy correction.

For simplicity, assuming one-dimensional flow, we have

$$A := (\rho u)_x = 0 \tag{1}$$

At shock point i, the Murman operator is

$$D^- A + D \; A = 0 \tag{2}$$

where D^- denotes the backward and D the central differencing of A. Similar to PrOP [3] we have

$$D^- A + DA + B = (\rho u)_{i+1/2} - (\rho u)_{i-3/2} + B = 0 \tag{3}$$

where $B = (-Ks + 1)(\rho u)_{i+1/2}$

$$Ks = \frac{u^2_{i-3/2}}{a^2_*}\left[\frac{1+\frac{\gamma-1}{2}M^2_\infty(1-u^2_{i-3/2})}{1+\frac{\gamma-1}{2}M^2_\infty(1-u^2_{i+1/2})}\right]^{1/(\gamma-1)} \tag{4}$$

One can see that this shock point operator satisfies exactly

$$u_{i-3/2}\cdot u_{i+1/2} = a^2_* \tag{5}$$

which is known as Prandtl's relation.

Meanwhile, we also have

$$Ks\,(\rho u)_{i+1/2} - (\rho u)_{i-3/2} = 0 \quad \text{i.e.}$$

$$(\rho_s u)_{i+1/2} - (\rho_s u)_{i-3/2} = 0 \tag{6}$$

where $\rho_s = \rho\cdot Ks = \left\{K^{-1}\left[1+\frac{\gamma-1}{2}M^2_\infty(1-q^2)\right]\right\}^{1/(\gamma-1)}$

$$K = \frac{1+\frac{\gamma-1}{2}M^2_\infty(1-u^2_{i+1/2})}{1+\frac{\gamma-1}{2}M^2_\infty(1-u^2_{i-3/2})}\left(\frac{a^2_*}{u^2_{i-3/2}}\right)^{(\gamma-1)} \tag{7}$$

in front of the shock Ks = 1. From equations (5) (6), we know this shock point operator satisfies momentum conservation as well as mass conservation together. The comparisons of different shock point operator with the exact solution for the shock strength are given in fig.1. From fig.1 it can be seen that the entropy shock point operator is better than others.

2-D and 3-D case: In 2-D and 3-D case, for the steady uniform incoming flow,the upstream field front of the shock wave is isentropic. After the shock the flow field is not isentropic, but along a streamline the value of entropy is constant. The value at the grid point over the whole flow field after the shock can be obtained by interpolation in the computational space.

The shock strengths on the upper and lower wing surface generally are different, so the K value on the upper and lower wake surface are different. it is necessery to modify the wake condition in the computation. From the condition P1 = P2 along the wake and $K1 \neq k2$, $\rho_1 \neq \rho_2$, we can obtain

$$q^2_2 = c + s\cdot q^2_1$$

where $c = \left(\frac{2}{\gamma-1}\cdot\frac{1}{M^2_\infty} + 1\right)(1 - s)$, $s = (K2 / K1)^{1/\gamma}$

and $\Delta\varphi = \Delta\varphi_T + \int_{x_T}^{x}\psi(M,s)\,dx'$, $\psi(M,s) = \frac{2}{\gamma-1}\cdot\frac{1-s}{M^2_1+\sqrt{s}\,M_1M_2}$

for the isentropic situation $\psi(M, s) = 0$

Results and discussion

2-D case: The computation of 2-D nonisentropic potential flow has been made by using the type of AF2 scheme of the paper [4], which is similar to the Holst's TAIR [5]. In fig.(2a), there are the calculated results for NACA 0012 airfoil at $M_\infty = 0.8$, $\alpha = 1.25$ deg.. It is obvious that the shock position and strength obtained from isentropic potential method are quite different from Euler's equation, while the results of nonisentropic potential method is very close to the Euler's results. Fig.(2b) depicts the comparison of the results with different theories and experiment data for RAE 2822 airfoil at $M_\infty = 0.725$, $\alpha = 2.55$ deg.. It is evident that the results of nonisentropic potential method are closer to the Euler solution and experiment data than the isentropic potential method. From the comparison of mentioned two examples, it can be drawn that the nonisentropic potential method produces results closer to Euler solution, i.e. simulates the real flow better, while its computational efforts are nearly the same as the usual isentropic potential method, the CPU time required is much less than Euler solution.

3-D case: From the comparison of 2-D case it is worthwhile to extend such an attractive formulation to 3-D case. In 3-D calculation the full potential equation was solved by the method [8], which is based on the finite volume method [9]. In fig.(3) there are the calculated results for M6 wing at $M_\infty = 0.92$, $\alpha = 3.07$ deg.. It shows isobar of the nonisentropic potential flow and comparison between nonisentropic, isentropic and experimental pressure distribution at the same spanwise station. As similar to the 2-D case, nonisentropic formulation makes the shock strength smaller, and locates the shock wave upstream to the isentropic ones, the results are closer to the experiment data.

Conclusion

In the present paper a shock point operator is introduced to account for the entropy change across the shock wave. This operator satisfies the conservation relations, which are consistent with shock jump condition. A full potential code with such a non isentropic formulation allows flow computation around airfoil and wing more accurate. Some numerical calculations indicate the present formulation yields results that are closer to the Euler solution and experiment data. The computation time is only a fraction of that required for the Euler code.

Acknowledgments: The authors wish to thank Professor Wu Li-yi for helpful discussion.

Reference

[1] M. Hafez; D. Lovell: "Transonic small disturbance calculations including entropy correction". Symp. on Num. and Phy. Aspects of Aerodynamic flows. Long Beach, 1981

[2] G.H. Klopfer; D. Nixon: "Nonisentropic potential formulation for transonic flows". AIAA J. Vol.22, No.6, 1984

[3] J. Mertens; K.D. Klevenhusen; H.Jacob: "Accurate transonic wave drag prediction using simple physical models". AIAA J. Vol.25, No.6, 1987

[4] Huang Mingke: "A fast algorithm of the finite difference method for computation of the transonic flow past an arbitrary airfoil with the conservative full potential equation". Acta Aerodynamica Sinica. (in chinese) No.2, 1984.

[5] Holst T.L.:"An implicit algorithm for the conservation transonic full potential equation using an arbitrary mesh". AIAA J. Vol.17, No.10, 1979.

[6] Steger J.L.:"Implicit finite difference simulation of flow about two dimensional Geometries". AIAA J. Vol.16, No.7, 1978.

[7] Wang Lei: "The numerical calculation of two dimensional invisid subsonic, transonic and supersonic flow". Ph. D. thesis, BIAA, 1987.

[8] Bai Xue song; Zhu Ziqiang: "Transonic analysis and design using an improved grid". (to be published)

[9] Jameson A.; Caughey D.A.: "A finite volume method for transonic potential flow calculations". AIAA 77-635.

M1 Mach number before the shock
Q   Shock strength
—·— Prop shock point operator
—··— Murman shock point operator
——   Euler solution
——   Entropy shock point operator

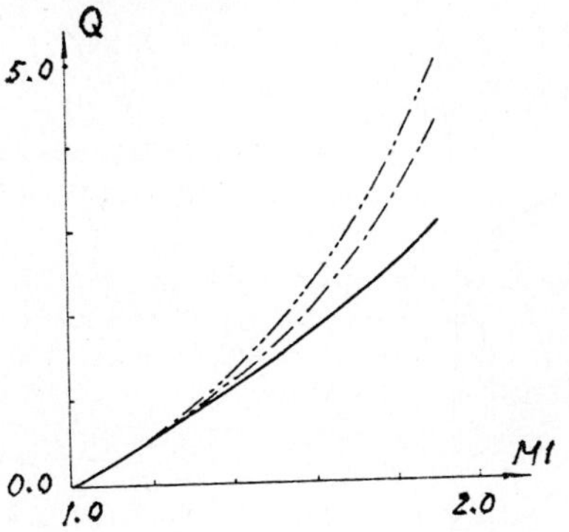

Fig. 1.

Comparison of shock point operators

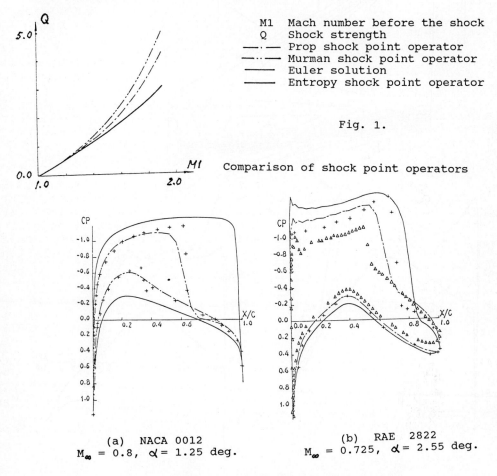

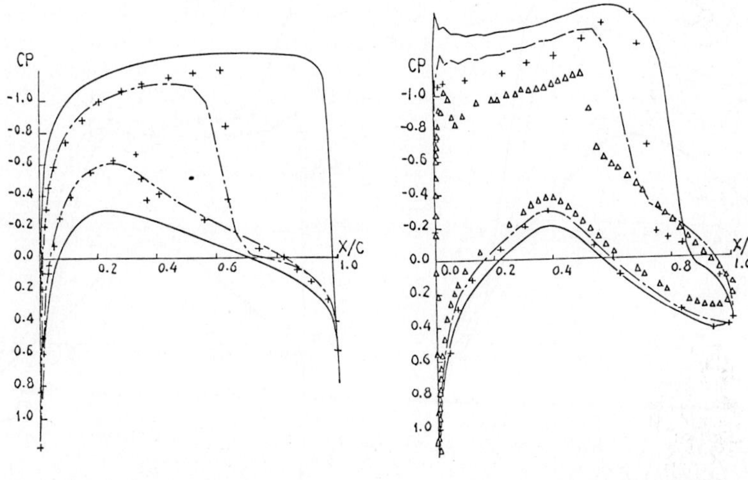

(a) NACA 0012
$M_\infty = 0.8$, $\alpha = 1.25$ deg.

(b) RAE 2822
$M_\infty = 0.725$, $\alpha = 2.55$ deg.

Fig. 2. ——Isentropic ---Nonisentropic +++Euler ▲▲▲Exp.

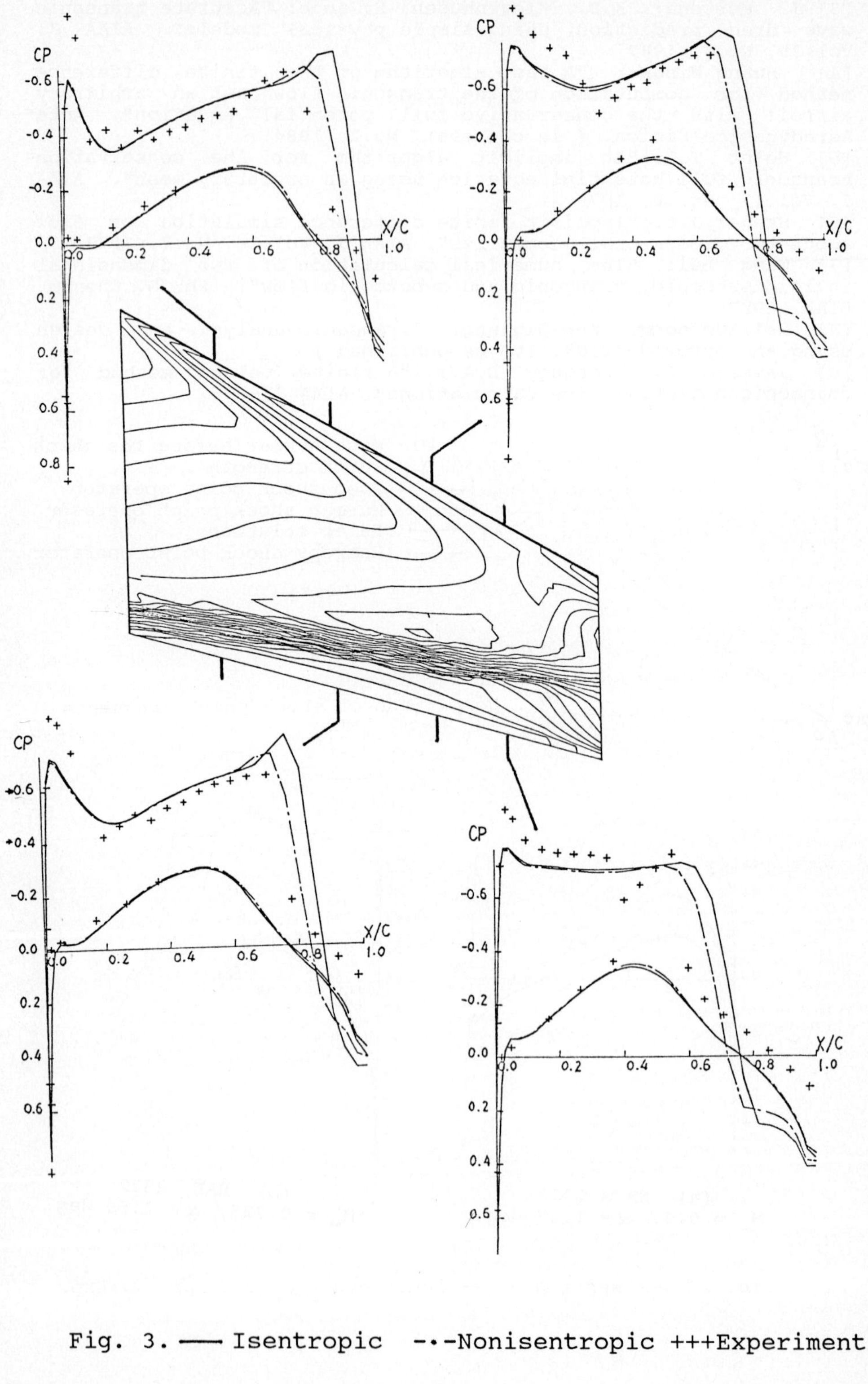

Fig. 3. —— Isentropic -·-Nonisentropic +++Experiment

Lecture Notes in Mathematics

Vol. 1236: Stochastic Partial Differential Equations and Applications. Proceedings, 1985. Edited by G. Da Prato and L. Tubaro. V, 257 pages. 1987.

Vol. 1237: Rational Approximation and its Applications in Mathematics and Physics. Proceedings, 1985. Edited by J. Gilewicz, M. Pindor and W. Siemaszko. XII, 350 pages. 1987.

Vol. 1250: Stochastic Processes – Mathematics and Physics II. Proceedings 1985. Edited by S. Albeverio, Ph. Blanchard and L. Streit. VI, 359 pages. 1987.

Vol. 1251: Differential Geometric Methods in Mathematical Physics. Proceedings, 1985. Edited by P.L. García and A. Pérez-Rendón. VII, 300 pages. 1987.

Vol. 1255: Differential Geometry and Differential Equations. Proceedings, 1985. Edited by C. Gu, M. Berger and R.L. Bryant. XII, 243 pages. 1987.

Vol. 1256: Pseudo-Differential Operators. Proceedings, 1986. Edited by H.O. Cordes, B. Gramsch and H. Widom. X, 479 pages. 1987.

Vol. 1258: J. Weidmann, Spectral Theory of Ordinary Differential Operators. VI, 303 pages. 1987.

Vol. 1260: N.H. Pavel, Nonlinear Evolution Operators and Semigroups. VI, 285 pages. 1987.

Vol. 1263: V.L. Hansen (Ed.), Differential Geometry. Proceedings, 1985. XI, 288 pages. 1987.

Vol. 1265: W. Van Assche, Asymptotics for Orthogonal Polynomials. VI, 201 pages. 1987.

Vol. 1267: J. Lindenstrauss, V.D. Milman (Eds.), Geometrical Aspects of Functional Analysis. Seminar. VII, 212 pages. 1987.

Vol. 1269: M. Shiota, Nash Manifolds. VI, 223 pages. 1987.

Vol. 1270: C. Carasso, P.-A. Raviart, D. Serre (Eds.), Nonlinear Hyperbolic Problems. Proceedings, 1986. XV, 341 pages. 1987.

Vol. 1272: M.S. Livšic, L.L. Waksman, Commuting Nonselfadjoint Operators in Hilbert Space. III, 115 pages. 1987.

Vol. 1273: G.-M. Greuel, G. Trautmann (Eds.), Singularities, Representation of Algebras, and Vector Bundles. Proceedings, 1985. XIV, 383 pages. 1987.

Vol. 1275: C.A. Berenstein (Ed.), Complex Analysis I. Proceedings, 1985–86. XV, 331 pages. 1987.

Vol. 1276: C.A. Berenstein (Ed.), Complex Analysis II. Proceedings, 1985–86. IX, 320 pages. 1987.

Vol. 1277: C.A. Berenstein (Ed.), Complex Analysis III. Proceedings, 1985–86. X, 350 pages. 1987.

Vol. 1283: S. Mardešić, J. Segal (Eds.), Geometric Topology and Shape Theory. Proceedings, 1986. V, 261 pages. 1987.

Vol. 1285: I.W. Knowles, Y. Saitō (Eds.), Differential Equations and Mathematical Physics. Proceedings, 1986. XVI, 499 pages. 1987.

Vol. 1287: E.B. Saff (Ed.), Approximation Theory, Tampa. Proceedings, 1985–1986. V, 228 pages. 1987.

Vol. 1288: Yu. L. Rodin, Generalized Analytic Functions on Riemann Surfaces. V, 128 pages, 1987.

Vol. 1294: M. Queffélec, Substitution Dynamical Systems – Spectral Analysis. XIII, 240 pages. 1987.

Vol. 1299: S. Watanabe, Yu.V. Prokhorov (Eds.), Probability Theory and Mathematical Statistics. Proceedings, 1986. VIII, 589 pages. 1988.

Vol. 1300: G.B. Seligman, Constructions of Lie Algebras and their Modules. VI, 190 pages. 1988.

Vol. 1302: M. Cwikel, J. Peetre, Y. Sagher, H. Wallin (Eds.), Function Spaces and Applications. Proceedings, 1986. VI, 445 pages. 1988.

Vol. 1303: L. Accardi, W. von Waldenfels (Eds.), Quantum Probability and Applications III. Proceedings, 1987. VI, 373 pages. 1988.

Lecture Notes in Physics

Vol. 301: G. Ferenczi, F. Beleznay (Eds.), New Developments in Semiconductor Physics. Proceedings, 1987. VI, 302 pages. 1988.

Vol. 302: F. Gieres, Geometry of Supersymmetric Gauge Theories. VIII, 189 pages. 1988.

Vol. 303: P. Breitenlohner, D. Maison, K. Sibold (Eds.), Renormalization of Quantum Field Theories with Non-linear Field Transformations. Proceedings, 1987. VI, 239 pages. 1988.

Vol. 304: R. Prud'homme, Fluides hétérogènes et réactifs: écoulements et transferts. VIII, 239 pages. 1988.

Vol. 305: K. Nomoto (Ed.), Atmospheric Diagnostics of Stellar Evolution: Chemical Peculiarity, Mass Loss, and Explosion. Proceedings, 1987. XIV, 468 pages. 1988.

Vol. 306: L. Blitz, F.J. Lockman (Eds.), The Outer Galaxy. Proceedings, 1987. IX, 291 pages. 1988.

Vol. 307: H.R. Miller, P.J. Wiita (Eds.), Active Galactic Nuclei. Proceedings, 1987, XI, 438 pages. 1988.

Vol. 308: H. Bacry, Localizability and Space in Quantum Physics. VII, 81 pages. 1988.

Vol. 309: P.E. Wagner, G. Vali (Eds.), Atmospheric Aerosols and Nucleation. Proceedings, 1988. XVIII, 729 pages. 1988.

Vol. 310: W.C. Seitter, H.W. Duerbeck, M. Tacke (Eds.), Large-Scale Structures in the Universe – Observational and Analytical Methods. Proceedings, 1987. II, 335 pages. 1988.

Vol. 311: P.J.M. Bongaarts, R. Martini (Eds.), Complex Differential Geometry and Supermanifolds in Strings and Fields. Proceedings, 1987. V, 252 pages. 1988.

Vol. 312: J.S. Feldman, Th.R. Hurd, L. Rosen, "QED: A Proof of Renormalizability." VII, 176 pages. 1988.

Vol. 313: H.-D. Doebner, T.D. Palev, J.D. Hennig (Eds.), Group Theoretical Methods in Physics. Proceedings, 1987. XI, 599 pages. 1988.

Vol. 314: L. Peliti, A. Vulpiani (Eds.), Measures of Complexity. Proceedings, 1987. VII, 150 pages. 1988.

Vol. 315: R.L. Dickman, R.L. Snell, J.S. Young (Eds.), Molecular Clouds in the Milky Way and External Galaxies. Proceedings, 1987. XVI, 475 pages. 1988.

Vol. 316: W. Kundt (Ed.), Supernova Shells and Their Birth Events. Proceedings, 1988. VIII, 253 pages. 1988.

Vol. 317: C. Signorini, S. Skorka, P. Spolaore, A. Vitturi (Eds.), Heavy Ion Interactions Around the Coulomb Barrier. Proceedings, 1988. X, 329 pages. 1988.

Vol. 319: L. Garrido (Ed.), Far from Equilibrium Phase Transitions. Proceedings, 1988. VIII, 340 pages. 1988.

Vol. 320: D. Coles (Ed.), Perspectives in Fluid Mechanics. Proceedings, 1985. VII, 207 pages. 1988.

Vol. 321: J. Pitowsky, Quantum Probability – Ouantum Logic. IX, 209 pages. 1989.

Vol. 322: M. Schlichenmaier, An Introduction to Riemann Surfaces, Algebraic Curves and Moduli Spaces. XIII, 148 pages. 1989.

Vol. 323: D.L. Dwoyer, M.Y. Hussaini, R.G. Voigt (Eds.), 11th International Conference on Numerical Methods in Fluid Dynamics. XIII, 622 pages. 1989.

oceanography

Diego

bject to